Solvent Extraction
Principles
and Practice

Solvent Extraction Principles and Practice

Second Edition, Revised and Expanded

edited by

Jan Rydberg
Chalmers University of Technology
Göteborg, Sweden

Michael Cox
University of Hertfordshire
Hatfield, Hertfordshire, United Kingdom

Claude Musikas
Commissariat à l'Energie Atomique
Paris, France

Gregory R. Choppin
Florida State University
Tallahassee, Florida, U.S.A.

CRC Press
Taylor & Francis Group
Boca Raton London New York

CRC Press is an imprint of the
Taylor & Francis Group, an **informa** business

Previous edition published as *Principles and Practices of Solvent Extraction* (Rydberg, J. et al., Eds.), Marcel Dekker, 1992.

First published 2004 by Marcel Dekker

Published 2020 by CRC Press
Taylor & Francis Group
6000 Broken Sound Parkway NW, Suite 300
Boca Raton, FL 33487-2742

First issued in paperback 2020

ISBN 13: 978-0-367-57841-1 (pbk)
ISBN 13: 978-0-8247-5063-3 (hbk)

Visit the Taylor & Francis Web site at
http://www.taylorandfrancis.com

and the CRC Press Web site at
http://www.crcpress.com

Library of Congress Cataloging-in-Publication Data
A catalog record for this book is available from the Library of Congress.

Preface

The partition of a solute between two immiscible solvents is a major technique in separation, both in the laboratory and in industry. This technique has been designated *liquid–liquid distribution* by the International Union of Pure and Applied Chemistry (IUPAC), but it is more commonly called *solvent extraction*, the term we use. Solvent extraction has been developed in a broad field of applications: (1) for studying inorganic and organic complex equilibria, either for fundamental understanding of solution equilibria and kinetics or for developing selective separation schemes; (2) for separations in analytical chemistry; (3) for large-scale industrial separation processes in the inorganic, organic, pharmaceutical, and biochemical industries; and (4) for industrial waste treatment. It also serves as a base for other analytical techniques, such as liquid partition chromatography and ion-selective electrodes.

The many uses of solvent extraction make the subject important for university students of chemistry and chemical technology. Some universities offer special courses in solvent extraction, whereas others include it as a minor topic in more comprehensive courses. Pilot-scale experiments on solvent extraction are common in chemical engineering curricula. Because of the breadth of the subject, the treatment in such courses is often scanty, and a satisfactory text is difficult to find in a form suitable for use directly with students.

To meet this demand, we felt it desirable to develop a simple text suitable not only for students but also for chemists in the field. However, no single scientist or engineer can be an expert in all parts of the field. Therefore, it seemed best to develop the book as a joint project among many expert authors, each of whom has decades of experience in research, teaching, or industrial development. The product is a truly international book at a high scientific and technical level. The task of making it homogeneous, consistent, and pedagogical has been undertaken by the editors.

The book is directed to third- to fourth-year undergraduate and postgraduate chemistry and chemical engineering students, as well as to researchers and developers in the chemical industry. The book is also intended for chemical engineers in industry who either have not kept up with modern developments or are considering use of this technique, as well as engineers who already are using this technique but desire to understand it better. Furthermore, the book should be useful to researchers in solvent extraction who wish to learn about its applications in areas other than their own.

The first edition of this book was published more than 10 years ago. Since then four large international conferences have summarized the latest developments in the field of solvent extraction. We have tried to incorporate the latest achievements of a fundamental type in this text, for instance, new types of solvent matrices and industrial applications, without burdening the text with new organic extractants or solvent combinations, which appear almost daily in the specialist literature. As far as possible, we have also rationalized the earlier text, concentrating on or removing outdated information. In doing so we have added new contributors. We therefore hope that this text will be met with the same enthusiasm as the first edition.

After an introduction (Chapter 1), the following five chapters (Chapters 2–6) present the physical principles and formal expressions used in solvent extraction. They are followed by eight chapters (Chapters 7–14) of various industrial applications and two concluding chapters (Chapters 15 and 16) indicating the research frontiers and future developments in technology.

We, the editors, want to express our gratitude to the contributors, who made this book possible through their helpful suggestions and extensive efforts. Prof. Manuel Aguilar, Prof. Eckhart Blass, Prof. José Luis Cortina, Dr. Pier Danesi, Prof. Ingmar Grenthe, Prof. Jan-Olov Liljenzin, Dr. Marian Czerwiński, Prof. Philip Lloyd, Prof. Yizhak Marcus, Prof. Jerzy Narbutt, Dr. Susana Pérez de Ortiz, Dr. Hans Reinhardt, Dr. Gordon Ritcey, Prof. Ana María Sastre, Dr. Wallace Schulz, the late Prof. Tatsuya Sekine, Prof. David Stuckey, Dr. Hans Wanner, and Prof. Ronald Wennersten have worked hard and successfully on their own chapters and have given us much valuable guidance.

Jan Rydberg
Michael Cox
Claude Musikas
Gregory R. Choppin

Contents

Contributors

Manuel Aguilar Department of Chemical Engineering, Universitat Politècnica de Catalunya, Barcelona, Spain

Eckhart F. Blass* Professor Emeritus, Lehrstuhl A für Verfahrenstechnik, Technische Universität Munich, Munich, Germany

Gregory R. Choppin* Department of Chemistry, Florida State University, Tallahassee, Florida, U.S.A.

José Luis Cortina Department of Chemical Engineering, Universitat Politècnica de Catalunya, Barcelona, Spain

Michael Cox* Department of Chemical Sciences, University of Hertfordshire, Hatfield, Hertfordshire, United Kingdom

Marian Czerwiński Zawiercie University of Administration and Management, Zawiercie, Poland

Pier Roberto Danesi Seibersdorf Laboratories, International Atomic Energy Agency, Vienna, Austria

---

*Retired.

Ingmar Grenthe*   Professor Emeritus, Department of Inorganic Chemistry, Royal Institute of Technology, Stockholm, Sweden

Jan-Olov Liljenzin*   Professor Emeritus, Department of Nuclear Chemistry, Chalmers University of Technology, Göteborg, Sweden

Philip J. D. Lloyd   Energy Research Institute, University of Cape Town, Rondebosch, South Africa

Yizhak Marcus*   Professor Emeritus, Department of Inorganic and Analytical Chemistry, The Hebrew University of Jerusalem, Jerusalem, Israel

Claude Musikas*   Département des Procedes de Retraitement, Commissariat à l'Energie Atomique, Paris, France

Jerzy Narbutt   Institute of Nuclear Chemistry and Technology, Warsaw, Poland

Susana Pérez de Ortiz   Department of Chemical Engineering, Imperial College, London, United Kingdom

Hans Reinhardt   MEAB Metallextraktion AB, Göteborg, Sweden

Gordon M. Ritcey   Gordon M. Ritcey & Associates, Inc., Nepean, Ontario, Canada

Jan Rydberg*   Department of Nuclear Chemistry, Chalmers University of Technology, Göteborg, Sweden

Ana María Sastre   Department of Chemical Engineering, Universitat Politècnica de Catalunya, Barcelona, Spain

Wallace W. Schulz   Consultant, Albuquerque, New Mexico, U.S.A.

Tatsuya Sekine[†]   Department of Chemistry, Science University of Tokyo, Tokyo, Japan

David Stuckey   Department of Chemical Engineering, Imperial College, London, United Kingdom

---

*Retired.
[†]Deceased.

Hans Wanner Swiss Federal Nuclear Safety Inspectorate, Villingen-HSK, Switzerland

Ronald Wennersten Department of Chemical Engineering, Royal Institute of Technology, Stockholm, Sweden

Solvent Extraction Principles and Practice

1

Introduction to Solvent Extraction

MICHAEL COX* University of Hertfordshire, Hatfield, Hertfordshire, United Kingdom

JAN RYDBERG* Chalmers University of Technology, Göteborg, Sweden

1.1 WHAT IS SOLVENT EXTRACTION?

The term *solvent extraction*† refers to the distribution of a solute between two immiscible liquid phases in contact with each other, i.e., a two-phase distribution of a solute. It can be described as a technique, resting on a strong scientific foundation. Scientists and engineers are concerned with the extent and dynamics of the distribution of different solutes—organic or inorganic—and its use scientifically and industrially for separation of solute mixtures.

The principle of solvent extraction is illustrated in Fig. 1.1. The vessel (a separatory funnel) contains two layers of liquids, one that is generally water (S_{aq}) and the other generally an organic solvent (S_{org}). In the example shown, the organic solvent is lighter (i.e., has a lower density) than water, but the opposite situation is also possible. The solute A, which initially is dissolved in only one of the two liquids, eventually distributes between the two phases. When this distribution reaches equilibrium, the solute is at concentration $[A]_{aq}$ in the aqueous layer and at concentration $[A]_{org}$ in the organic layer. The *distribution ratio* of the solute

$$D = [A]_{org}/[A]_{aq} \tag{1.1}$$

---

*Retired.

†The International Union of Pure and Applied Chemistry (IUPAC) recommends the use of the term *liquid-liquid distribution*. However, more traditionally the term *solvent extraction* (sometimes abbreviated SX) is used in this book.

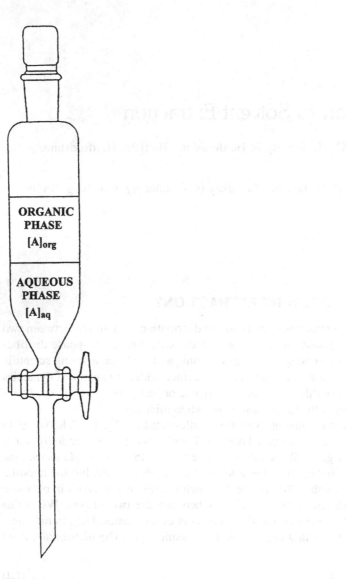

Fig. 1.1 A schematic representation of solvent extraction (liquid-liquid distribution). A solute A is distributed between the upper layer, for example an organic solvent, and the lower layer, an aqueous phase.

is defined as the ratio of "the total analytical concentration of the substance in the organic phase to its total analytical concentration in the aqueous phase, usually measured at equilibrium" [1], irrespective of whether the organic phase is the lighter or heavier one. If a second solute B is present, the distribution ratio for the various solutes are indicated by D_A, D_B, etc. If D_B is different from D_A, A and B can be separated from each other by (single or multistage) solvent extraction. D is also called the *distribution coefficient* or *distribution factor*; we here prefer the expression *distribution ratio*.

For practical purposes, as in industrial applications, it is often more popular to use *the percentage extraction %E* (sometimes named the *extraction factor*), which is given by

$$\%E = 100D/(1+D) \tag{1.2}$$

where D is the distribution ratio of the solute (or desired component). For $D = 1$, the solute is evenly distributed between the two phases. A requirement for practical use of solvent extraction is that a reasonable fraction (percentage) of the desired component is extracted in a single operation (or *stage*).

Solvent extraction is used in numerous chemical industries to produce pure chemical compounds ranging from pharmaceuticals and biomedicals to heavy organics and metals, in analytical chemistry and in environmental waste purification. The scientific explanation of the distribution ratios observed is based on the fundamental physical chemistry of solute–solvent interaction, activity factors of the solutes in the pure phases, aqueous complexation, and complex-adduct interactions. Most university training provides only elementary knowledge about these fields, which is unsatisfactory from a fundamental chemical standpoint, as well as for industrial development and for protection of environmental systems. Solvent extraction uses are important in organic, inorganic, and physical chemistry, and in chemical engineering, theoretical as well as practical; in this book we try to cover most of these important fields.

None of the authors of this book is an expert in all the aspects of solvent extraction, nor do we believe that any of our readers will try to become one. This book is, therefore, written by authors from various disciplines of chemistry and by chemical engineers. The "scientific level" of the text only requires basic chemistry training, but not on a Ph.D. level, though the text may be quite useful for extra reading even at that level. The text is divided in two parts. The first part covers the fundamental chemistry of the solvent extraction process and the second part the techniques for its use in industry with a large number of applications. In this introductory chapter we try to put solvent extraction in its chemical context, historical as well as modern. The last two chapters describe the most recent applications and theoretical developments.

1.2 SOLVENT EXTRACTION IS A FUNDAMENTAL SEPARATION PROCESS

Under normal conditions, matter can appear in three forms of aggregation: solid, liquid, and gas. These forms or *physical states* are consequences of various interactions between the atomic or molecular species. The interactions are governed by *internal chemical properties* (various types of bonding) and *external physical properties* (temperature and pressure). Most small molecules can be transformed between these states (e.g., H_2O into ice, water, and steam) by a moderate change of temperature and/or pressure. Between these physical states— or *phases*—there is a sharp boundary (*phase boundary*), which makes it possible to separate the phases—for example, ice may be removed from water by filtration. The most fundamental of chemical properties is the ability to undergo such phase transformations, the use of which allows the simplest method for isolation of pure compounds from natural materials.

In a gas mixture such as the earth's atmosphere, the ratio of oxygen to nitrogen decreases slightly with atmospheric height because of the greater gravitational attraction of oxygen. However, the gravitational field of the earth is not enough for efficient separation of these gases, which, however, can be separated by ultracentrifugation and by diffusion techniques. In crushed iron ore it is possible to separate the magnetite crystals Fe_3O_4 from the silicate gangue material by physical selection under a microscope or by a magnetic field. In chemical engineering such separation techniques are referred to as *nonequilibrium processes*. Other common nonequilibrium processes are electrolysis, electrophoresis, and filtration.

In contrast to these we have the *equilibrium processes* of sublimation, absorption, dissolution, precipitation, evaporation, and condensation, through which the physical states of solid, liquid, and gas are connected. For example, the common crystallization of salts from sea water involves all three phases. Distillation, which is essential for producing organic solvents, is a two-step evaporation (liquid $\Rightarrow$ gas) condensation (gas $\Rightarrow$ liquid) process.

In Fig. 1.2, phase transformations are put into their context of physical processes used for separation of mixtures of chemical compounds. However, the figure has been drawn asymmetrically in that two liquids (I and II) are indicated. Most people are familiar with several organic liquids, like kerosene, ether, benzene, etc., that are only partially miscible with water. This lack of miscibility allows an equilibrium between two liquids that are separated from each other by a common phase boundary. Thus the conventional physical system of three phases (gas, liquid, and solid, counting all solid phases as one), which ordinarily are available to all chemists, is expanded to four phases when two immiscible liquids are involved. This can be of great advantage, as will be seen when reading this book.

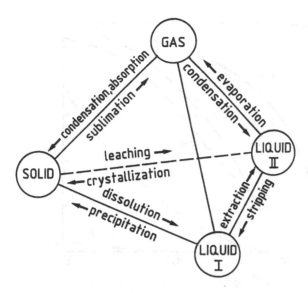

Fig. 1.2 Model of a four-phase system consisting of two liquid phases (e.g., an aqueous and an organic phase) in equilibrium with a gas phase and a solid phase.

Solutes have differing solubilities in different liquids due to variations in the strength of the interaction of solute molecules with those of the solvent. Thus, in a system of two immiscible or only partially miscible solvents, different solutes become unevenly distributed between the two solvent phases, and as noted earlier, this is the basis for the *solvent extraction technique*. In this context, "solvent" almost invariably means "organic solvent." This uneven distribution is illustrated in Fig. 1.3, which shows the extractability into a kerosene solution of the different metals that appear when stainless steel is dissolved in aqueous acid chloride solution. The metals Mo, Zn, and Fe(III) are easily extracted into the organic solvent mixture at low chloride ion concentration, and Cu, Co, Fe(II), and Mn at intermediate concentration, while even at the highest chloride concentration in the system, Ni and Cr are poorly extracted. This is used industrially for separating the metals in super-alloy scrap in order to recover the most valuable ones.

The three main separation processes between solid, gas, and liquid have long been known, while solvent extraction is a relatively new separation technique, as is described in the brief historical review in next two sections. Nevertheless, because all solutes (organic as well as inorganic) can be made more or less soluble in aqueous and organic phases, the number of applications of solvent extraction is almost limitless. Since large-scale industrial solvent extraction is a continuous process (in contrast to laboratory batch processes) and can be

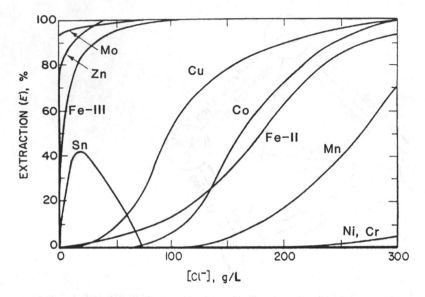

Fig. 1.3 Percentage of extraction of various metals from a solution of dissolved stainless steel scrap, at 40°C. The organic phase is 25% tertiary amine (Alamine 336), 15% dodecanol (Loral C12) and 60% kerosene (Nysolvin 75A). The aqueous phase is a CaCl₂ solution at pH 2.

made more selective than the conventional gas–liquid–solid separation techniques, it offers numerous industrial possibilities to achieve desired separation efficiently and economically.

1.3 EARLY STEPS TOWARD THE USE OF SOLVENT EXTRACTION

Around 500 B.C.E. the Greek philosophers recognized four elements: earth, water, phlogiston (~air), and fire. This view harmonizes with the present concept of three physical states of aggregation (solid, liquid, gas) and heat. Aristotle (~350 B.C.E.) emphasized that these "elements" were not eternal, but could be changed into each other. Five thousand years earlier, scientists already had found that when certain green minerals were heated in a coal fire, metallic copper was obtained. In Aristotle's time it was known how to produce metals such as copper, gold, tin, lead, silver, iron, mercury, and arsenic. Even earlier, by transmuting certain "earths" with fire, ceramics and glasses had been produced. Many of these arts were probably developed by the Egyptians, the first true chemists. The word "*alchemy*" is derived from Arabic and Greek and is supposed to mean "art of transmutations as practiced by the Egyptians." Fruit juices were fermented, oils and fats were squeezed out of vegetables and animal parts, and purified by digestion with earths, bones, etc. Crucibles, retorts, and even distillation equipment seem to have been in use; see Fig. 1.4. We must think of these early alchemists as endlessly mixing, heating, boiling, digesting, cooling, etc., everything they could collect from nature. The purpose of these transmutations varied: for lamps, weapons, pigments, perfumes, and poisons; for cosmetics and medicines to prevent aging and to prolong life (*elixir vitae*); for tanning chemicals, soap, anesthetics, and also for making gold. In fact, they did succeed in producing gold-like metals (e.g., brass). For example, it is known that they heated odorous leaves in alkaline water with fats and oils, so that ointments and perfumes could be enriched in the cooled solidified fat, and that these products were extensively used in the ancient courts, and perhaps even among the general population. If this is considered to be solvent extraction, it is truly one of the oldest chemical techniques. It is also likely that the Egyptians knew how to distill alcohol, long before it is described by the Arab Kautilya and the Greek Aristotle about 300 B.C.E. (See also Ref. [2].)

This experimentation more or less came to a halt during the Greek civilization. The Greeks were philosophers, and not so much experimentalists; Aristotle was a philosopher and a systematizer (systems technician, in modern language), not an experimentalist. The Greeks were followed by the Romans who were administrators, and by the Christians who considered alchemy to be ungodly. Although alchemy was practiced during subsequent centuries, particularly in the Arabian world, it became suspect and was banned by many rulers (though encouraged in secrecy by others). About 500 years ago, alchemy was rather openly revived in Europe, particularly at local courts, and progressed within a few centuries into modern science.

Digestion of various earths (or digested earths) with alcohol produces many organic solvents (ether, acetone, etc.). These solvents could be obtained in pure form through distillation. Such organic solvents could have been produced many thousands years ago, because of the obvious knowledge of distillation [2]. However, 200 years ago only a few pure solvents seem to be known: besides the natural water and oils (and kerosene) only alcohol, ether, and "etheric oils" were acknowledged. It is difficult to trace organic solvents far back in history. The reason may simply be that organic compounds obtained by distillation of mixtures of natural products were found to be rather uninteresting (except for alcohol), because at that time they seemed to have very little practical value, and they certainly could not be used to produce gold. Because solvent extraction requires pure organic solvents of limited aqueous miscibility, it is then understandable why solvent extraction historically is considered (perhaps falsely) to be a newcomer among chemical separation methods.

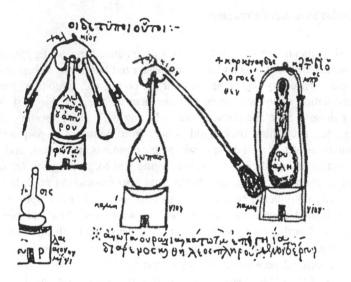

(a)

(b)

Fig. 1.4 (a) Equipment used by alchemists, according to an Alexandrian manuscript (about 300 B.C.E. to A.D. 300). (b) Apparatus for fractionated distillation. Front page of *Philosophi ac Alchimistae Maximi* by Johannes Greininger, Strasbourg, 1531. The original work is ascribed to the great eighth-century Arab alchemist, Abu Musa Jabir.

8

1.4 SOLVENT EXTRACTION IN THE DEVELOPMENT OF MODERN CHEMISTRY

The early nineteenth century saw an enormous development in chemistry. Not only were numerous inorganic and organic compounds synthesized, they were also characterized through skillful analyses. Pure compounds, absolutely essential for the analyst, were identified from measurements of weights and volumes, melting and boiling points, densities, refraction, and crystal structure. It was then found that many inorganic salts, such as chlorides of iron, mercury, and gold, dissolved in and could be recrystallized from alcohols and ethers.

1.4.1 The Distribution Ratio and the Extraction Constant

Inorganic extractions seem to have come into practical use without any great notice. Although Peligot in 1842 reported that uranyl nitrate could be recrystallized from ether, he never mentioned extraction of this salt from aqueous solutions. In textbooks after 1870, however, it is stated briefly that "ether can even withdraw sublimate ($HgCl_2$) from aqueous solution." It was also reported, for example, that cobalt thiocyanate is weakly extracted by ether, better by amyl alcohol, and even better by a mixture of both.

The practical use of solvent extraction for separation and purification of different substances led Berthelot and Jungfleisch to investigate the distribution of a large number of organic and inorganic compounds between ether or carbon disulfide and water. In 1872, they introduced the term *distribution factor* ("coefficient de partage") [Eq. (1.1)] to describe how the distribution of a solute depended on its concentration in the organic and aqueous phases. However, the regularities observed were not general.

About 20 years later, Nernst realized that it was necessary to take into account the different reactions of the solute in each phase, such as dimerization in the organic phase and dissociation in the aqueous phase. Thus the distribution of benzoic acid (HBz) between the organic (benzene) phase and water could be written

Organic (benzene) phase:　$2\,HBz \leftrightharpoons H_2Bz_2$

$$\downarrow\uparrow$$

Aqueous phase:　　　　　$HBz \leftrightharpoons H^+ + Bz^-$

In 1891, Nernst realized that only if the solute has the same molecular weight in the organic phase as in the aqueous phase, the distribution ratio would be independent of the concentration of the solute, or *distribuend*. He proposed the simple relation:

$$K_D = [HBz]_{org}/[HBz]_{aq} \tag{1.3}$$

and demonstrated the validity of this principle for the distribution of a large number of organic and inorganic compounds between organic solvents and water (*the Nernst distribution law*). K_D is referred to as the *distribution constant*. [Note that K_D is sometimes termed *partition coefficient* (or *constant*), and then abbreviated *P*.]

The primary parameter in solvent extraction is *the measured distribution ratio*, where it is up to the writer to define what is being measured, indicating this by an appropriate index. In the Nernst distribution experiment described earlier, the analytically measured concentration of benzoic acid is in the aqueous phase $[Bz]_{aq,tot} = [HBz]_{aq} + [Bz^-]_{aq}$, and in the organic phase $[Bz]_{org,tot} = [HBz]_{org} + [H_2Bz_2]_{org}$. Thus the measured distribution ratio, abbreviated D_{Bz}, becomes

$$D_{Bz} = \frac{[Bz]_{org,tot}}{[Bz]_{aq,tot}} = \frac{[HBz]_{org} + 2[H_2Bz_2]_{org}}{[HBz]_{aq} + [Bz^-]_{aq}} \tag{1.4}$$

From this it follows that the analytically measured distribution ratio is a constant only in systems that contain a single molecular species of the solute (or desired component).

Such considerations were extended to metal complexes in 1902 by Morse, who studied the distribution of divalent mercury between toluene and water at various Hg^{2+} and Cl^- concentrations. By taking complex formation in the aqueous phase into consideration Morse could determine the formation constants of $HgCl^+$ and $HgCl_2$ from distribution measurements, as well as the distribution constant of the neutral complex $HgCl_2$. The overall *extraction reaction* can be written

$$Hg^{2+}(aq) + 2\,Cl^-(aq) \leftrightarrows HgCl_2(org) \tag{1.5a}$$

for which one defines an *extraction constant*, K_{ex}

$$K_{ex} = \frac{[HgCl_2]_{org}}{[Hg^{2+}]_{aq}[Cl^-]^2_{aq}} \tag{1.5b}$$

Although K_{ex} neither describes the intermediate reaction steps (see Chapters 3 and 4) or kinetics of the reaction (Chapter 5), nor explains why or to what extent $HgCl_2$ dissolves in the organic solvent (Chapter 2), the extraction reaction is a useful concept in applied solvent extraction, and K_{ex} values are commonly tabulated in reference works [3–4].

1.4.2 Extractants

During the years 1900 to 1940, solvent extraction was mainly used by the organic chemists for separating organic substances. Since in these systems, the solute (or desired component) often exists in only one single molecular form, such systems are referred to as *nonreactive extraction* systems; here the distribution ratio equals the distribution constant.

However, it was also discovered that many organic substances, mainly weak acids, could complex metals in the aqueous phase to form a complex soluble in organic solvents. A typical reaction can be written

$$M^{z+}(aq) + z \, HA(aq \text{ or } org) \leftrightarrows MA_z(org) + z \, H^+(aq) \tag{1.6}$$

which indicates that the organic acid HA may be taken from the aqueous or the organic phase. This is an example of *reactive extraction*. It became a tool for the analytical chemist, when the extracted metal complex showed a specific color that could be identified spectrometrically. The reagent responsible for forming the extractable complex is termed the *extractant*.

The industrial use of solvent extraction of inorganic compounds grew out of the analytical work. As both areas, analytical as well as industrial, needed both better extractants and an understanding of the reaction steps in the solutions in order to optimize the applications, theoretical interpretations of the molecular reactions in the solutions became a necessity, as will be described in later chapters.

The increased use of computer graphics for modeling molecular structures and chemical reactions has opened a path for the synthesis of tailor-made extractants. Thus the future promises new varieties of extractants with highly selective properties for the desired process.

1.4.3 Industrial Use of Solvent Extraction

A surge in interest in solvent extraction occurred in the decades of the 1940s and 1950s initiated by its application for uranium production and for reprocessing of irradiated nuclear materials in the U.S. Manhattan Project. The first large-scale industrial solvent extraction plant for metals purification was built in 1942 by Mallinckrodt Chemical Co., St. Louis, for the production of ton amounts of uranium by selective extraction of uranyl nitrate by ether from aqueous solutions. The high degree of purity (>99.9%) required for use of uranium in nuclear reactors was achieved. An explosion led to the replacement of the ether by other solvents (dibutylmethanol and methylisobutylketone). At the same time new types of more efficient metal extractants were introduced, e.g., tri-*n*-butylphosphate in 1945 and trioctylamine in 1948. This activity became a great stimulus to the non-nuclear industry, and solvent extraction was introduced as a separation and purification process in a large number of chemical and metallurgical industries in the 1950s and early 1960s. For example, by leaching copper ore with sulphuric acid followed by extraction of this solution with an organic hydroxyaryloxime dissolved in kerosene, several million tons of copper (30% of world production) is now produced annually. This and many other processes are described in later chapters.

For these applications, the technique of solvent extraction had to be further developed and with this a new terminology was also developed. This can be illustrated by considering a process where a desired component in an aqueous solution is extracted with an organic reagent (extractant) dissolved in another organic liquid; note here that the term "organic solvent" is not used because of possible confusion. The term "solvent" could be used for the whole organic phase or for the organic liquid in which the organic extractant is dissolved. Thus the term generally given to the latter is (*organic*) *diluent*.

While in laboratory experiments the extraction vessel may be a test tube, or more conveniently some kind of separation funnel (Fig. 1.1), this is not suited for industrial use. Industry prefers to use continuous processes. The simplest separation unit is then the mixer-settler, or some clever development of the same basic principle, as described in Chapter 9. Figure 1.5 pictures a simple mixer-settler unit, here used for the removal of iron from an acid solution also containing nickel and cobalt (the same systems as in Fig. 1.3). The mixer (or *contactor*) is here simply a vessel with a revolving paddle that produces small droplets of one of the liquid phases in the other. This physical mixture flows into and slowly through the separation vessel, which may be a long tank; through the influence of gravity the two phases separate, so that the upper organic kerosene–octanol–amine phase contains the Fe(III) and the lower aqueous $CaCl_2$ phase contains the Co(II) and Ni(II). Numerous variations of the construction of mixer-settlers (or MS-units, as they are abbreviated) exist (Chapter 9), often several joined together into MS-batteries. Many such will be described later on.

A diagram of a full basic process is given in Fig. 1.6 to illustrate the common terminology. The incoming aqueous solution is called the *feed*. It is

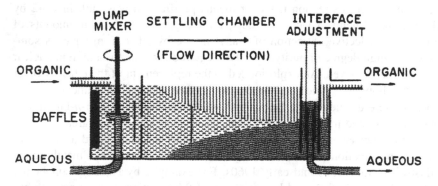

Fig. 1.5 The principle of a mixer-settler unit, e.g., for separation of iron(III) from nickel and cobalt.

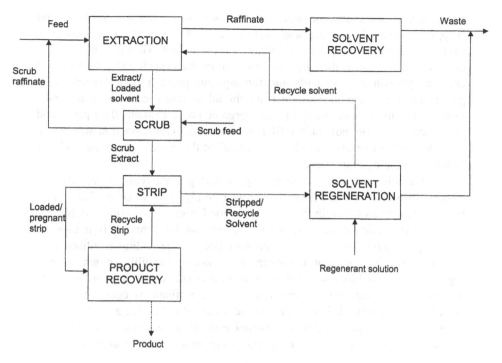

Fig. 1.6 Typical flowsheet of a solvent extraction circuit.

contacted with the (*recycled*) *solvent phase* in a mixer-settler unit. Here we do not indicate the exact type of unit, but only its function (extraction), as commonly is done. After extraction and separation of the phases, the depleted phase becomes the *raffinate* and the enriched solvent phase becomes the *extract* or *loaded (or pregnant) solvent*. The raffinate may undergo a solvent recovery stage to remove any entrained solvent before exiting the process. The extraction process is rarely specific so that other solutes may be co-extracted with the main component. These impurities may be removed with an aqueous *scrub solution* in a scrub stage producing a scrub extract and a scrub raffinate containing the impurities. The latter may return to the feed solution to maintain an overall water balance. The scrubbed extract is now contacted with another aqueous solution to strip or back-extract the desired component. The stripped solvent then may undergo some regeneration process to prepare the solvent phase for recycle. The *loaded (pregnant) strip solution* then is treated to remove the desired product and the strip solution is recycled. One of the important aspects of this flowsheet is that, wherever possible, liquid phases are recovered and recycled. This is important from both an economic and an environmental standpoint.

Figure 1.6 shows a situation where each process—extraction, scrubbing, and stripping—occurs in a single operation or stage. This is generally not efficient because of the finite value of the distribution coefficient. Thus if D has a value of 100 (i.e., E = 99%), then after one extraction, the organic extract phase will contain approximately 99 parts and the aqueous phase 1 part. To achieve a greater extraction, the aqueous raffinate should be contacted with another portion of the solvent after which the new organic phase contains 0.99 parts, and the aqueous raffinate now only 0.01 part. Therefore, two extraction stages will provide 99.99% extraction (with "2 volumes" of the organic phase, but only "1 volume" of the aqueous phase).

Three different ways of connecting such stages are possible: namely, *co-current*, *cross-current*, and *counter-current* (see Fig. 1.7). In co-current extraction, the two phases flow in the same direction between the various contactors. A simple inspection of the diagram will show that with this configuration no advantage is gained over a single contact because, providing equilibrium is reached in the first contactor, the separated flows are in equilibrium when entering the second contactor so no change in relative concentrations will occur. In the second configuration (b), cross-current, the raffinate is contacted with a sample of fresh solvent. This is the classical way of extracting a product in the laboratory when using a separatory funnel and will give an enhanced recovery of the solute. However, on an industrial scale, this is seldom used because it

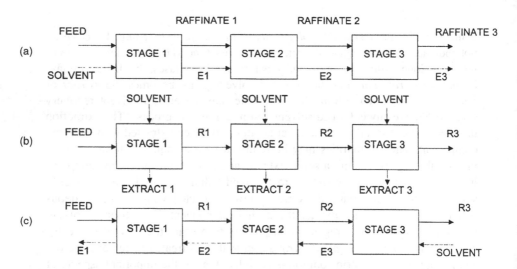

Fig. 1.7 Arrangements of solvent extraction stages: (a) co-current; (b) cross-current; and (c) counter-current.

results in the production of a multitude of product phases containing a reducing concentration of the desired solute. These have to be combined before stripping resulting in a much larger volume of loaded solvent to be treated with consequences for plant size and economics. The third configuration (c), counter-current, is the one generally chosen by industry. The phase volumes remain constant and by feeding the two phases, feed and solvent, at opposite ends of the bank of contactors, the driving force for extraction, i.e., the solute concentration difference between the two phases, is maximized. Chapters in the second part of the book will extend this discussion.

The need to use multiple extraction to achieve efficient extraction required the development of new types of continuously working extractors, especially mixer-settlers and pulsed columns, which were suitable for remotely controlled operations. These new extractors could be built for continuous flow and in multiple stages, allowing very efficient isolation of substances in high yield. A good example is the production of rare earth elements in >99.999% purity in ton amounts by mixer-settler batteries containing hundreds of stages. These topics will be further developed in Chapters 6 and 7.

In the early analytical applications of solvent extraction, optimal extraction or separation conditions were obtained empirically. This was unsatisfactory and general mathematical descriptions were developed by a number of researchers in many countries. This was especially important for large-scale industrial use and is an activity that continues today almost entirely with computers.

1.5 MODERN TRENDS IN APPLICATIONS OF SOLVENT EXTRACTION

What is the future of solvent extraction? We will try to indicate some areas in which we expect important advances in the future. Some are more theoretical, others are applied, but all are related to other areas of chemistry or to other fields of science and technology. The simple answers given here do not require the reading of the subsequent chapters, but do require some good fundamental chemical knowledge.

1.5.1 Solution Theories

A 10% change in the distribution ratio, which can be measured easily and accurately in the distribution range 0.01–100 (the one most commonly used), corresponds to an energy change of only 0.2 kJ/mole. Thus distribution ratio measurements offer a method to investigate low energy reactions in solutions, such as weak solute–solute and solute–solvent interactions, especially in the organic phase. So far, this technique has been only slightly exploited for this purpose. It is particularly noteworthy to find how few thermodynamic studies have been made of the distribution or extraction constants, as enthalpy and entropy values give a good indication of the driving force of the extraction and also may indicate the structure of the molecular species in the organic solvent.

Separation of metals by solvent extraction is usually based on the various complexing properties of the metals (Chapter 3). Separation systems may be chosen on the basis of complexity constants obtained from the literature. However, the literature often shows different values for "same systems" causing considerable concern for process design chemists. There is an obvious need for an objective presentation of the uncertainty in the published equilibrium constants, however conditional they may be.

Although theories of solution (Chapter 2) and formation of extractable complexes (Chapters 3 and 4) now are well advanced, predictions of distribution ratios are mainly done by comparison with known similar systems (Chapter 3). Solvatochromic parameters, solubility parameters, and donor numbers, as discussed in Chapters 2–4, are so far mainly empirical factors. Continuous efforts are made to predict such numbers, often resulting in good values for systems within limited ranges of conditions. It is likely that these efforts will successively encompass greater ranges of conditions for more systems, but much still has to be done. In the future, theory may allow the assignment of exact numbers to the solvatochromic parameters, thus also permitting theoretical predictions of distribution constants for known as well as for hypothetical (not yet synthesized) extractants.

Chemical quantum mechanics (Chapter 16) are now promising to contribute to prediction of distribution ratios, particularly when used in interactive computer modeling of chemical structures and reactions. This will provide better understanding of solute–solvent interactions and the theory of solubility, which is the foundation for prediction of distribution ratios. Progress is likely to speed up as efficient computing programs with large data bases become more easily available on the market at more attainable prices. Such interactive computer research will be able to explain poor extractability due to steric hindrance, hydration, the synergistic effect of specific adducts, etc., and, as a consequence, give clues to better extraction reagents and conditions.

A combination of known extractants can increase the net extraction both in degree and kinetics through synergistic extraction. This subject also is poorly explored, though it is of considerable importance to industrial solvent extraction processes.

1.5.2 Kinetics

All chemical reactions occur with a certain rate. For aqueous systems, in which no redox reactions occur (e.g., simple complex formation), equilibrium is often attained rapidly, which is also true for adduct reactions in the organic phase, all at normal temperatures. However, the transfer of a solute from an aqueous solution to an organic solvent occurs via a phase boundary; such reactions depend on several parameters and may be quite slow. Therefore, the kinetics of solvent extraction, which is of paramount importance to all industrial applications, is largely determined by the interfacial chemistry. To speed up phase transfer, the engineer tries to maximize the interfacial surface by, e.g., violent stirring, producing billions of small droplets. Though the theory is well advanced (see Chapters 5 and 9), the interface is very complicated and its properties are difficult to investigate, particularly at the molecular level. Few applicable techniques are available, and the results are often difficult to interpret. The coalescence of the droplets (to produce two clean phases that are easy to separate) presents another interfacial surface problem, and so-called phase reversals may further complicate the separation. This is particularly important in the development of new types of contactor. Much progress is required here and as such presents a challenge to future scientists and engineers.

The largest research effort in extraction kinetics is likely to be in the development of solvent extraction related techniques, such as various versions of liquid chromatography, liquid membranes, etc. These techniques require a detailed knowledge of the kinetics of the system to predict the degree of separation.

It is a practical fact that most industrial solvent extractions are carried out under nonequilibrium conditions, however close the approach may be; for example, centrifugal contactor-separators (Chapter 9) rarely operate at distribution equilibrium. An interesting possibility is to expand this into extractions further from equilibrium, if the kinetics of the desired and nondesired products are different. Such operations offer a real technlogical challenge.

1.5.3 *Equipment and Processes*

Laboratory solvent extraction studies are often carried out with test tubes, a process that is cheaper but more laborious, time consuming, and usually yields more scattered data, than using semicontinuously stirred baffled beakers or Lewis-type cells. Highly precise distribution data can be obtained efficiently either "mechanically" by continuous flow centrifugal separators, or "physically" by using special phase separating filter devices; some of these methods are described in Chapter 4. Although these techniques are going to be improved, new, more advanced methods are not likely to appear in the near future.

Specific reagents with exotic structures are costly to produce, which requires the experimentalist to use as small amounts as possible of the reagents. This leads to a desire to shift from "milli-experiments" (e.g., mixing 10 mL of each phase) to micro experiments (mixing 50 µL of each phase), which is now taking place, to even smaller sizes (volumes and amounts)—the *nano experiments*. Progress in this field requires a high degree of ingenuity.

Although single-stage laboratory techniques provide the first step toward multistage industrial processes, such process development usually requires small-scale multistage and pilot-plant scale equipment. A large number of excellent designs are available, and we consider further fundamental improvements unlikely.

The industrial application of solvent extraction is a mature technique, and it is now possible to move from laboratory experiments on a new extraction system to full industrial practice with little technological risk. There is a sufficient variety of large-scale equipment available to cope with most problems encountered in application, although much of the equipment remains rather massive. Attempts to miniaturize, for instance, by using centrifugal forces to mix and separate phases, still has to be developed further.

Many industrial processes begin with a leaching step, yielding a slurry that must be clarified before solvent extraction. The solid–liquid separation is a costly step. The solvent extraction of unclarified liquids ("solvent-in-pulp") has been proposed to eliminate solid–liquid separation. The increased revenue and reduced energy cost make this an attractive process, but many problems remain to be solved: loss of metals and extractants to the solid phase, optimization of equipment design, effluent disposal, etc.

An essential step in industrial solvent extraction is the regeneration of the extractant. This can be done in many ways, e.g., by distillation, evaporation, or stripping (back-extraction). While distillation and evaporation do not discriminate between solutes (the diluent is simply removed by heating), stripping, by careful choice of strip solution and conditions, can be made highly selective. Alternatively, all the solutes can be stripped and then subjected to a selective extraction by changing the extractant; examples of both types of process will be found in Chapter 13. The possibilities are many, and it may be worthwhile to explore new paths.

Membrane extraction is a relatively new technique for solvent extraction, in which a solute is transferred from one aqueous phase to another through a membrane holding an extractant dissolved in a diluent. This ingenious scheme has been only slightly explored, though it offers great potential for the future, e.g., for waste water cleaning.

The step from laboratory experiments via pilot plants to industrial scale requires serious consideration of all the points here; practical experience is invaluable in order to avoid mistakes and excess costs, as indicated in Chapter 7.

1.5.4 Organic Chemistry and Biochemistry

There is a continuing demand for selective extractants that can be developed, for example, by tailor making organic molecules with pre-organized metal binding sites. The development of such extractants may be designed on computer based models. Environmental considerations require that the extractants and diluents used are either nontoxic or recovered within the process. The increasing combination of extraction and distillation, as used in biotechnology (Chapter 10), places a whole host of new demands on the diluents employed.

In the extraction of biologically active compounds, care must be taken to avoid the loss of activity that often occurs by contact with organic diluents. Thus a series of systems have been developed specifically with these compounds in mind. The first of these uses mixtures of aqueous solutions containing polymers and inorganic salts that will separate into two phases that are predominately water. A second system uses supercritical conditions in which the original two-phase system is transformed into one phase under special temperature-pressure conditions. Also the active organic compound can be shielded from the organic diluent by encapsulation within the aqueous center of a micelle of surface active compounds. All these systems are currently an active area for research as is discussed in Chapter 15.

1.5.5 Separation of Inorganic Compounds

Separation schemes have been developed for all the elements in the periodic system, both in ionic and molecular forms. Some of these substances are toxic, putting severe restraints on plant operation and the environment. For example, many solvents are bioactive, and increasingly stringent control is needed to prevent their spread in the biosphere. Though the techniques are advanced, demands for improved safety will persist and require new types of equipment or processes.

The production of electricity from nuclear fission energy is accompanied by formation of radioactive waste, of which the larger hazard is the presence of long-lived transuranium isotopes. The problems associated with this waste are still debated, but if the transuranium isotopes could be removed by "exhaustive" reprocessing and transmuted in special nuclear devices, the hazard of the waste would be drastically reduced (Chapter 12). This may require new selective extractants and diluents as well as new process schemes. Research in this field is very active.

The mineral industry, where ores produce pure metals, requires a number of process stages that can be carried out either through pyrometallurgical treatment or by the use of wet chemical separation processes (*hydrometallurgy*). The former produces off gases and slags, which may be environmentally hazardous. The latter introduces chemistry into the mineral industry, and is often met with strong criticism among traditional metallurgists. It is therefore up to the hydrometallurgical chemist to develop schemes for metal production from ores that are economically competitive and environmentally advantageous. Here, the field is rather open for new developments. A specifically interesting trait is the use of "in-situ" leaching of ore bodies, which circumvents conventional mining with its sometimes devastating environmental effects, e.g., in open-pit mining.

In ore processing, the essential minerals are enriched through flotation of the crushed ore, using an aqueous solution containing flotation reagents (we may label them *flotants*), that selectively attach to the mineral surface and—with the aid of air bubbles—lift them up to the surface where they can be skimmed off, while the "gangue" material stays at the bottom. Alternatively, the flotants are made magnetic, allowing the use of magnetic separation as commonly used in the mineral industry. These flotants are chemically very similar to the chelating extractants used in solvent extraction processes. An interesting subject here is collaboration between these two mineral processing fields to produce new and more effective reagents and separation processes.

1.5.6 Environmental Aspects

The future will bring further increase in concern over the environmental impact of chemical operations. The liquid effluents must not only be controlled, they must also be rendered harmless to the environment. This requires removal of the hazardous substances. For many of the dilute waste solutions, solvent extraction has proved to be an effective process. This is even more true for recycling of mixed metals from various industries. Nevertheless, the increasing amounts of wastes from human activities require much more to be done in this field.

In principle, solvent extraction is an environmentally friendly process with no air or water pollution, provided the plant flows are properly designed. It could, therefore, replace many of the present polluting processes. A particular problem, however, is the solvent extraction effluents, which may contain biochemically active substances posing "new" hazards to the environment. These can be handled by various solid sorbents, which then can be incinerated, but the advantage of the solvent extraction process may be lost. There is, therefore, a demand for biodegradable and environmentally benign solvent phases. In the future, additional attention is required to this field.

1.5.7 Spin-off

The principle of solvent extraction—the distribution of chemical species between two immiscible liquid phases—has been applied to many areas of chemistry. A typical one is liquid partition chromatography, where the principle of solvent extraction provides the most efficient separation process available to organic chemistry today; its huge application has become a field (and an industry!) of its own. The design of ion selective electrodes is another application of the solvent extraction principle; it also has become an independent field. Both these applications are only briefly touched upon in the chapter of this book on analytical applications (Chapter 14), as we consider them outside the scope of this work. Nevertheless, fundamental research on solvent extraction will provide further important inpt into these two applications. It is likely that fundamental solvent extraction research will continue to contribute to the development of further selective analytical techniques.

The driving force of the transport of salts, proteins, etc., through the cell membrane from the nucleus to the body fluids, and vice versa, is a complicated biochemical process. As far as is known, this field has not been explored by traditional solution chemists, although a detailed analysis of these transfer processes indicates many similarities with solvent extraction processes (equilibrium as well as kinetics). It is possible that studies of such simpler model systems could contribute to the understanding of the more complicated biochemical processes.

Solvent extraction deals with the transport of chemical substances from one phase into another one, the chemical kinetics of this process, and the final equilibrium distribution of the substances between the two phases. Such transport and distribution processes are the motors that make life in biological systems possible. Fundamental studies of such "solvent extraction" processes contribute to the better understanding of all processes in nature. Here, only the lack of imagination stands in the way of important new scientific discoveries.

1.5.8 Conclusion

Solvent extraction is a mature technique in that extensive experience has led to a good understanding of the fundamental chemical reactions. At the same time, compared to many other chemical separation processes like precipitation, distillation, or pyrometallurgical treatment, the large-scale application of solvent extraction is, nevertheless, a young technique. New reagents are continually being developed, spurred on by computer modeling, and more efficient contacting equipment is coming into use. Considering such factors as demands for higher product purity, less pollution, and the need for recovering substances from more complex matrices and more dilute resources, the efficiency and high selectivity of solvent extraction should make it an increasingly competitive separation process both in research and in industry.

1.6 HOW TO USE THIS BOOK

The first part of the book contains basic solution theory and thermodynamics (Chapter 2); a survey of the effects on the distribution ratio of changes in parameter values, such as concentration of metals, complex formers and other reactants, pH, temperature, etc., (Chapter 3); measurement techniques, data collection, evaluation, and interpretation (Chapter 4); and kinetics (Chapter 5). The ionic strength is an essential factor in all aqueous systems, and how to cope with it is treated in a fundamental way in Chapter 6. These chapters sometimes go into such depth that they may mainly interest only the theoretical chemist or professional engineer; in some chapters we offer some choice by placing the most detailed parts in smaller type. It is not necessary to read these chapters before the applied ones, and some can be read independently, before or after the other more applied chapters.

The second part deals with applications of solvent extraction in industry, and begins with a general chapter (Chapter 7) that involves both equipment, flowsheet development, economic factors, and environmental aspects. Chapter 8 is concerned with fundamental engineering concepts for multistage extraction. Chapter 9 describes contactor design. It is followed by the industrial extraction of organic and biochemical compounds for purification and pharmaceutical uses (Chapter 10), recovery of metals for industrial production (Chapter 11), applications in the nuclear fuel cycle (Chapter 12), and recycling or waste treatment (Chapter 14). Analytical applications are briefly summarized in Chapter 13. The last chapters, Chapters 15 and 16, describe some newer developments in which the principle of solvent extraction has or may come into use, and theoretical developments.

Apart from the first six chapters, each chapter has been written more or less independently but, we hope, consistently.

1.7 SOLVENT EXTRACTION NOMENCLATURE

Because scientists and engineers in chemistry use different symbols, a single set valid for all chapters in this book is not feasible. Moreover, it is desirable that both groups be familiar with the symbols used within the subfields in order to enhance communication. The symbols in this book closely follow IUPAC recommendations (see reference list). In the Appendix of this book, lists are given of the most commonly used symbols, except for those in Chapter 9, which has its own list of symbols.

1.8 LITERATURE ON SOLVENT EXTRACTION

Recommendations of solvent extraction terminology have been published by the International Union of Pure and Applied Chemistry (IUPAC) [1,3–5]. Some excellent monographs on solvent extraction are listed in section 1.8.1; unfortunately, the older ones may only be available in large science libraries. Most of them deal either with a small sector of the field or are very comprehensive. Finally, in section 1.8.2, we have collected edited versions of proceedings of the International Conferences of Solvent Extraction (ISEC), which normally are held every three years.

Because solvent extraction is used extensively as a technique in research, publications with this technique appear in all types of chemical journals. This is also true for its industrial applications that are often described in chemical engineering journals.

1.8.1 General References on Solvent Extraction

1955 L. Alders, *Liquid-liquid Extraction*, Elsevier, New York, 206 p.

1955 E. Hecker, *Verteilungsverfahren in Laboratorium*, Verlag Chemie GmbH, Weinheim, 230 p.

1957 G. H. Morrison and H. Freiser, *Solvent Extraction in Analytical Chemistry*, John Wiley and Sons, New York, 260 p.

1963 R. E. Treybal, *Liquid Extraction*, McGraw-Hill Book Co, New York, 620 p.

1963 A. W. Francis, *Liquid-liquid Equilibriums*, Wiley-Interscience, New York, 288 p.

1964 J. Stary, *The Solvent Extraction of Metal Chelates*, Pergamon Press, New York, 240 p.

1969 Y. Marcus and S. Kertes, *Ion Exchange and Solvent Extraction of Metal Complexes*, Wiley-Interscience, New York, 1037 p.

1970 Y. A. Zolotov, *Extraction of Chelate Compounds* (transl. from Russian Ed. 1968), Humprey, Ann Arbor, 290 p.

1970 A. K. De, S. M. Khopkar and R. A. Chalmers, *Solvent Extraction of Metals*, Van Nostrand Reinhold, New York, 260 p.

1971 C. Hanson (Ed.), Recent *Advances in Liquid-Liquid Extractions*, Pergamon Press, New York, 584 p.

1977 T. Sekine and Y. Hasegawa, *Solvent Extraction Chemistry*, Marcel Dekker, New York, 919 p.

1983 T. C. Lo, M. H. I. Baird and C. Hanson, *Handbook of Solvent Extraction*, Wiley-Interscience, New York, 1980 p.

1984 G. Ritcey and A. W. Ashbrook, *Solvent Extraction. Part I and II*, Elsevier, Amsterdam, 361+737 p.

1992 J. D. Thornton (Ed.), *Science and Practice of Liquid-Liquid Extraction*, Oxford University Press, Oxford 1992, 2 volumes.

1994 J. C. Godfrey and M. J. Slater (Eds.) *Liquid-Liquid Extraction Equipment*, John Wiley and Sons, Chichester 1994, UK.

1.8.2 Proceedings of International Solvent Extraction Conferences (ISEC)

1965 Harwell—*Solvent Extraction Chemistry of Metals* (Eds. H. A. C. McKay, T. V. Healy, I. L. Jenkins and A. Naylor), Macmillan 456 p., London (1965).

1966 Gothenburg—*Solvent Extraction Chemistry* (Eds. D. Dyrssen, J. O. Liljenzin and J. Rydberg), North-Holland 680 p., Amsterdam (1967).

1968 Haifa/Jerusalem—*Solvent Extraction Research* (Eds. A. S. Kertes and Y. Marcus), John Wiley and Sons, 439 p., New York (1969).

1971 The Hague—*Solvent Extraction* (Eds. J. G. Gregory, B. Evans and P. C. Weston), Society of Chemical Industry, 3 vol. 1566 p., London (1971).

1974 Lyon—*Proceedings of ISEC '74*, Society of Chemical Industry, 3 vol. 2899 p. London (1974).

1977 Toronto—*CIM Special Volume 21* Canadian Institute of Mining and Metallurgy, 2 vol. 807 p., Montreal (1979).

1980 Liège—*International Solvent Extraction Conference ISEC'80*, Association d'Ingeneur de l'Université de Liège, 3 vol. 1300 p., Liège (1980).

1983 Denver—*Selected Papers, AIChE Symposium Series*, No 238, Vol. 80, 1984, American Institute of Chemical Engineers, 177 p., New York (1984).

1986 Munich—*Preprints of ISEC '86*, DECHEMA, 3 vol. 2230 p., Frankfurtam Main (1986).

1988 Moscow—*Proceedings of ISEC '88*, 4 vol. 1495 p., Vernadsky Institute of Geochemistry and Analytical Chemistry of the USSR Academy of Sciences, Moscow (1988).

1990 Kyoto—*Solvent Extraction 1990* (Proceedings of ISEC'90) (Ed. T. Sekine) Elsevier, 2 vol. 1923 p., Amsterdam m.m. (1992).

1993 York—*Solvent Extraction in the Process Industries*, Proceedings of ISEC '93,(Eds. D.H.Logsdail and M.J.Slater) Elsevier Applied Science, 3 vol. 1828 p., London & New York (1993).

1996 Melbourne—*Value Adding Through Solvent Extraction*, Proceedings of ISEC '96, (Eds. D. C. Shallcross, R. Paimin, L. M. Prvcic), The University of Melbourne, Parkville, 2 vol. 1684 p., Victoria, Australia (1996).

1999 Barcelona—*Solvent Extraction for the 21st Century* (Proceedings ISEC '99) (Eds. M. Cox, M. Hidalgo, and M. Valiente), Society of Chemical Industry, 2 vols. 1680 p., London (2001).

2002 Cape Town—*Proceedings of the International Solvent Extraction Conference, ISEC '02* (Eds. K.C. Sole, P.M. Cole, J.S. Preston and D.J. Robinson), Chris van Rensburg Publications (Pty) Ltd, 2 vol. p. Johannesburg, South-Africa (2002). Also on CD.

REFERENCES

1. Freiser, H.; and Nancollas, G. H.; *Compendium of Analytical Nomenclature. Definitive Rules 1987.* IUPAC. Blackwell Scientific Publications, Oxford (1987).
2. Blass, E.; Liebl, T.; Häberl, M.; *Solvent Extraction—A Historical Review*, Proc. Int. Solv. Extr. Conf. Melbourne, 1996.

3. Högfeldt, E.; *Stability Constants of Metal-Ion Complexes. Part A: Inorganic Ligands.* IUPAC Chemical Data Series No. 22, Pergamon Press, New York (1982).
4. McNaught, A. D.; and Wilkinson, A.; *IUPAC Compendium of Chemical Terminology*, Second Edition, Blackwell Science (1997).
5. IUPAC, *Quantities, Units and Symbols in Physical Chemistry*, Third Edition, (Ed. Ian Mills), Royal Society of Chemistry, Cambridge 2002.

2
Principles of Solubility and Solutions

YIZHAK MARCUS* The Hebrew University of Jerusalem, Jerusalem, Israel

2.1 INTRODUCTION

Solvent extraction is another name for *liquid–liquid distribution*, that is, the distribution of a solute between two liquids that must not be completely mutually miscible. Therefore, the liquid state of aggregation of matter and the essential forces that keep certain types of liquids from being completely miscible are proper introductory subjects in a study of solvent extraction. Furthermore, the distribution of a solute depends on its preference for one or the other liquid, which is closely related to its solubility in each one of them. Thus, the general subject of solubilities is highly relevant to solvent extraction.

In a solution, the solute particles (molecules, ions) interact with solvent molecules and also, provided the concentration of the solute is sufficiently high, with other solute particles. These interactions play the major role in the distribution of a solute between the two liquid layers in liquid–liquid distribution systems. Consequently, the understanding of the physical chemistry of liquids and solutions is important to master the rich and varied field of solvent extraction.

Solvent extraction commonly takes place with an aqueous solution as one liquid and an organic solvent as the other. Obviously, the extraction process is limited to the liquid range of these substances. Since solvent extraction is generally carried out at ambient pressures, the liquid range extends from about the freezing temperature up to about the normal boiling temperature. If, however, high pressures are applied (as they are in some solvent extraction processes), then the liquid range can extend up to the critical temperature of the substance. Supercritical fluid extraction beyond the critical temperature (such as decaffeination of coffee with supercritical carbon dioxide) is a growing field of applica-

---

*Retired.

tion of solvent extraction. It has the advantages that the properties of the super-critical fluid can be fine-tuned by variation of the pressure, and that this "supercritical solvent" can be readily removed by a drastic diminution of the pressure, but has drawbacks related to the high temperatures and pressures often needed.

Numerous solvents are used in solvent extraction. They can be divided in the context of solvent extraction into different classes as follows (but see Refs. [1] and [2] for some other classification schemes):

Class 1: Liquids capable of forming three-dimensional networks of strong hydrogen bonds, e.g., water, poly- and amino-alcohols, hydroxy-acids, etc.

Class 2: Other liquids that have both active hydrogen atoms and donor atoms (O, N, F; see Chapter. 3), but do not form three-dimensional networks (rather forming chainlike oligomers), e.g., primary alcohols, carboxylic acids, primary and secondary amines, nitro compounds with α-positioned hydrogen atoms, liquified ammonia, etc. They are generally called protic or protogenic substances.

Class 3: Liquids composed of molecules containing donor atoms, but no active hydrogen atoms, e.g., ethers, ketones, aldehydes, esters, tertiary amines, nitro compounds without α-hydrogen, phosphoryl-group containing solvents, etc. (see Table 4.3). They are generally called dipolar aprotic substances.

Class 4: Liquids composed of molecules containing active hydrogen atoms but no donor atoms, e.g., chloroform and some other aliphatic halides.

Class 5: Liquids with no hydrogen-bond forming capability and no donor atoms, e.g., hydrocarbons, carbon disulfide, carbon tetrachloride, supercritical carbon dioxide, etc.

This diversity in solvent properties results in large differences in the distribution ratios of extracted solutes. Some solvents, particularly those of class 3, readily react directly (due to their strong donor properties) with inorganic compounds and extract them without need for any additional extractant, while others (classes 4 and 5) do not dissolve salts without the aid of other extractants. These last are generally used as *diluents* for extractants, required for improving their physical properties, such as density, viscosity, etc., or to bring solid extractants into solution in a liquid phase. The class 1 type of solvents are very soluble in water and are useless for extraction of metal species, although they may find use in separations in biochemical systems (see Chapter 9).

2.1.1 *Properties of Liquids*

If the externally imposed conditions of pressure and temperature permit a substance to be in the liquid state of aggregation, it possesses certain general properties. Contrary to a substance in the solid state, a liquid is fluid; that is, it flows under the influence of forces and is characterized by its fluidity, or the reciprocal, its viscosity. For water, the viscosity is only 0.89 mPa s^{-1} (nearly 1 centipoise) at 25°C [1,2], and water is considered a highly fluid liquid.

Contrary to a fluid in the gaseous state, a liquid has a surface, and is characterized by a surface tension. For water, the surface tension is 72 mN m^{-1} at 25°C [1,2]. Again, contrary to the fluid in the gaseous state, the volume of a liquid does not change appreciably under pressure; it has a low compressibility and shares this property with matter in the solid (crystalline, glassy, or amorphous) state. For water, the compressibility is 0.452 (GPa)$^{-1}$ at 25°C [1,2]. These are macroscopic, or bulk, properties that single out the liquid state from other states of aggregation of matter.

There are also some general microscopic, or molecular, properties that are peculiar to the liquid state. Contrary to the crystalline solid, the particles of a liquid do not possess long-range order. Although over a short range—2 to 4 molecular diameters—there is some order in the liquid, this order dissipates at longer distances. A particle in the liquid is free to diffuse and, in time, may occupy any position in the volume of the liquid, rather than being confined at or near a lattice position, as in the crystalline solid. Contrary to particles in the gaseous state, however, the particles in a liquid are in close proximity to each other (closely packed) and exert strong forces on their neighbors [3].

The close packing of the molecules of a substance in the liquid state results in a density much higher than in the gaseous state and approaching that in the solid state. The density, ρ, is the mass per unit volume, and can be expressed as the ratio of the molar mass M to the molar volume V of a liquid. Table 2.1 lists the values of the properties M and V of representative liquids that are important in the field of solution chemistry and solvent extraction. The densities and molar volumes depend on the temperature, and the latter are given for 25°C. (For a discussion of industrial solvents, see Chapter 12.)

Many liquids used in solvent extraction are polar. Their polarity is manifested by a permanent electric dipole in their molecules, since their atoms have differing electronegativities. Oxygen and nitrogen atoms, for instance, generally confer such dipolarity on a molecule, acting as the negative pole relative to carbon or hydrogen atoms bonded to them. The dipole moment μ characterizes such polar molecules and ranges from about 1.15 D (Debye unit = 3.336 10^{-30} C · m) for diethyl ether or chloroform, to 4.03 D for nitrobenzene, and 5.54 D for hexamethyl phosphoric triamide [1,2]. Water is also a polar liquid, with a moderate dipole moment, $\mu = 1.83$ D. A list of dipole moments of some solvents that are important in solvent extraction is presented in Table 2.1. Substances that do not have a permanent dipole moment (i.e., $\mu = 0$) are called nonpolar. Many hydrocarbons belong to this category, but not all (e.g., toluene has $\mu = 0.31$ D).

When nonpolar liquids are placed in an electric field, only the electrons in their atoms respond to the external electric forces, resulting in some atomic

Table 2.1 Selected Properties of Some[a] Water-Immiscible Solvents

Solvent	M g/mol	V^b mL/mol	μ^c D	ε^b	δ $(J/mL)^{1/2}$
c-Hexane	84.2	108.7	0.	2.02	16.8
n-Hexane	86.2	131.6	0.09	1.88	15.0
n-Octane	114.3	163.5	~0	1.95	15.5
n-Decane	142.3	195.9	~0	1.99	15.8
n-Dodecane	170.4	228.6	~0	2.00	16.0
Decalin (mixed isomers)	138.3	157.4	0.	2.15	18.0
Benzene	78.1	89.9	0.	2.27	18.8
Toluene	92.1	106.9	0.31	2.38	18.8
Ethylbenzene	106.2	123.1	0.37	2.40	18.0
p-Xylene	106.2	123.9	0.	2.27	18.1
Dichloromethane	89.9	64.5	1.14	8.93	20.2
Chloroform	119.4	80.7	1.15	4.89	19.5
Carbon tetrachloride	153.8	97.1	0.	2.24	17.6
1,1-Dichloroethane	99.0	84.7	1.82	10.00	18.3
1,2-Dichloroethane	99.0	79.4	1.83	10.36	20.0
Trichloroethylene	131.4	90.7	0.80	3.42	19.0
Chlorobenzene	112.6	102.2	1.69	5.62	19.8
1,2-Dichlorobenzene	147.0	113.1	2.50	9.93	20.5
Carbon disulfide	76.1	60.6	0.	2.64	20.3
Water[a]	**18.0**	**18.1**	**1.85**	**78.36**	**47.9**[d]
Methanol[a]	32.0	40.7	2.87	32.66	29.3
Ethanol[a]	46.1	58.7	1.66	24.55	26.0
1-Propanol[a]	60.1	75.1	3.09	20.45	24.4
2-Propanol[a]	60.1	76.9	1.66	19.92	23.7
1-Butanol	74.1	92.0	1.75	17.51	23.3
Isoamyl alcohol	88.2	109.2	1.82	15.19	22.1
1-Hexanol	102.2	125.2	1.55	13.30	21.8
1-Octanol	130.2	158.4	1.76	10.34	20.9
2-Ethyl-1-hexanol	130.2	157.1	1.74	4.4	19.4
Diethyl ether	74.1	104.7	1.15	4.20	15.4
Diisopropyl ether	102.2	142.3	1.22	3.88	14.6
Bis(2-chloroethyl) ether	143.0	117.9	2.58	21.20	18.8
Acetone[a]	58.1	74.0	2.69	20.56	22.1
Methyl ethyl ketone	72.1	90.2	2.76	18.11	18.7
Methyl isobutyl ketone	100.2	125.8	2.70	13.11	17.2
Cyclohexanone	98.2	104.2	3.08	15.5	19.7
Acetylacetone	100.1	103.0	2.78	25.7	19.5
Ethyl acetate	88.1	98.5	1.78	6.02	18.2
Butyl acetate	116.2	132.5	1.84	5.01	17.6
Propylene carbonate	102.1	85.2	4.94	64.92	21.8
Nitromethane	61.0	54.0	3.56	35.87	25.7

Table 2.1 Continued

Solvent	M g/mol	V^b mL/mol	μ^c D	ε^b	δ $(J/mL)^{1/2}$
Nitrobenzene	123.1	102.7	4.22	34.78	22.1
Acetonitrile[a]	41.1	52.9	3.92	35.94	24.1
Benzonitrile	103.1	103.1	4.18	25.2	22.7
Quinoline	129.2	118.5	2.18	8.95	22.8
Tributyl phosphate	266.3	273.8	3.07	8.91	15.3

Source: Refs. 1 and 2.
[a]Water miscible.
[b]At 25°C.
[c]Isolated solvent molecules, i.e., in the gaseous phase or dilute solution in an inert solvent.
[d]Behaves in organic-rich aqueous mixtures as if $\delta \approx 30$.

polarization. This produces a relative permittivity (dielectric constant) ε, which is approximately equal to the square of the refractive index (at the sodium D line, n_D), or about 2. Polar molecules, however, further respond to the external electric field by reorienting themselves, which results in a considerably larger relative permittivity. In particular, when cooperative hydrogen bonding takes place (see later discussion), high values of ε are achieved (e.g., 78 for water at 25°C [1,2]). Table 2.1 lists ε values of some liquids that are important in solvent extraction. The attractive forces between charges (such as those on ions) of opposite sign are inversely proportional to the relative permittivity of the liquid medium in which they find themselves. Therefore, the ionic dissociation of electrolytes strongly depends on the relative permittivity of the solvent that is used to dissolve them (see section 2.6).

There are properties of liquids used as solvents, such as the liquid range, viscosity, surface tension, and vapor pressure that, although important from the practical and technical points of view, are not listed in Table 2.1. These can be found in Refs. [1,2]. Tables 12.2, 13.1, and 13.2 list solvents used in extraction with some further information.

2.1.2 Cohesive Forces in Liquids

The ability of even the most inert gases to be liquified at sufficiently low temperatures and high pressures is evidence for the existence of cohesive forces between molecules. Such forces are manifested, in their simplest form, between the atoms of noble gases, such as argon. The motions of the electrons in the argon atom induce a temporary electric dipole in a neighboring argon atom, which in turn strengthens the temporary dipole in the first one. The mutual

interaction of these temporary, mutually induced electric dipoles produces attractive forces, called London forces or *dispersion forces*. These forces are
rather weak, and high pressures and drastic cooling (for argon 4.9 MPa and
−122°C) are required for these forces to be able to overcome the thermal agitation of the atoms that tends to keep them apart. The energy associated with the
dispersion forces decreases with the sixth power of the distance between the
interacting particles, and thus, these forces have a very short range, although
they are present in all liquids. Hydrocarbons are often used as diluents in the
organic phase of a solvent extraction system, and generally are held together as
liquids by these dispersion forces operating between neighboring segments (e.g.,
−CH$_2$−) of different molecules. In fact, in nonpolar substances, the only cohesive forces are the dispersion forces [3].

 In polar liquids, when the electric dipoles are able to arrange themselves
in a "head-to-tail" configuration, that is, when their positive end is on the average more in the vicinity of the negative ends of neighboring molecules (Fig.
2.1a,c), further attractive forces result. The energy associated with these forces

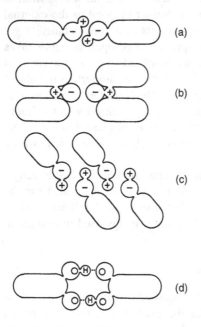

(a)

(b)

(c)

(d)

Fig. 2.1 Dipole and hydrogen bond interactions. A schematic representation of (a)
"head-to-tail" dipole-dipole attractive interactions (e.g., in tri-*n*-octylamine); (b) "head-
to-head" dipole-dipole repulsive interactions caused by steric hindrance (e.g., in dibutyl
sulfoxide); (c) chainlike dipole-dipole interactions (e.g., in 1-octanol); (d) a cyclic, hydrogen-bonded dimer (e.g., in hexanoic acid).

is proportional to the product of the dipole moments of the molecules and is inversely proportional to the sixth power of the mean distance between them (i.e., it has the same range as the dispersion forces). However, the structure of the molecules may be such that they prevent a head-to-tail configuration (see Fig. 2.1b), and the resulting head-to-head configuration causes repulsion between the polar molecules [3].

Some of the liquids that are used in solvent extraction, especially water, interact by means of *hydrogen bonding*. Their molecules have a hydrogen atom attached to a very electronegative element (mainly oxygen and, less effectively, nitrogen), and this hydrogen atom is able to be bound to the electronegative atom (O, N, or F) of a neighboring molecule, forming a hydrogen bridge. Such a bridge requires a rather rigid geometry: a linear configuration of the two electronegative atoms of the neighboring molecules with the hydrogen atom between them. This bond is of considerably greater strength than dispersion and dipole–dipole interactions. If the molecules of a substance can both donate and accept a hydrogen bond, a cyclic dimer may result (see Fig. 2.1d) that is considerably less polar than the monomers of this substance. Occasionally, water again being an example, this hydrogen bonding is a cooperative phenomenon: if such a bond is formed between two molecules, each becomes more likely to participate in further hydrogen bonding with other neighbors [3].

The three kinds of forces described above, collectively known as the cohesive forces that keep the molecules of liquids together, are responsible for various properties of the liquids. In particular, they are responsible for the work that has to be invested to remove molecules from the liquid, that is, to vaporize it. The energy of vaporization of a mole of liquid equals its molar heat of vaporization, $\Delta_V H$, minus the pressure-volume work involved, which can be approximated well by $\mathbf{R}T$, where $\mathbf{R}$ is the gas constant [8.3143 J K^{-1} mol^{-1}] and T is the absolute temperature. The ratio of this quantity to the molar volume of the liquid is its *cohesive energy density*. The square root of the cohesive energy density is called the (Hildebrand) solubility parameter of the liquid, δ:

$$\delta^2 = (\Delta_V H - \mathbf{R}T)/V \tag{2.1}$$

where V is the molar volume. A list of solubility parameters of representative liquids that are important in solvent extraction is given in Table 2.1.

2.2 SOLVENT MISCIBILITY

Solvent extraction takes place through the distribution of a solute or of solutes between two practically immiscible liquids. For a separation to be carried out by solvent extraction, the solute has to transfer from one region of space to another such region, which is physically separated from the first (see Fig. 1.1). In each such region, the solute is dissolved uniformly in a (*homogeneous*) liquid and, at all points, is under the same pressure. Such a region of space is called a *phase*. At equilibrium, at a constant temperature and pressure, a liquid phase is homogeneous and is isotropic: it has the same properties in all directions. In solvent extraction, we are interested in two liquid phases in contact, normally under the same temperature and pressure, which eventually reach *mass transfer equilibrium*. Such an equilibrium is a dynamic process: as many molecules of one solvent enter the other as leave it and return to their original phase as do the molecules (ions) of the solute(s). If each of the two liquids is originally a pure chemical substance and there is (as yet) no solute to distribute, we speak of a *two-component*, two-phase system, but the net effect is that an amount of the substance from one liquid phase is transferred to the other and vice versa. For two practically immiscible liquids this transferred amount is small relative to the total amount of the two phases. There may or may not be a vapor phase present, but this is generally disregarded in the context of solvent extraction.

2.2.1   The Phase Rule

A substance that can be added to a system independently (or removed from it, say by precipitation or vaporization) is called a *component* of such a system. The *phase rule*, summarizing a general behavior of nature, says:

> *The number of degrees of freedom in a system equals the number of components plus two minus the number of phases.*

A two-component, two-phase system, therefore, has two degrees of freedom, that is, two external conditions, such as the temperature and pressure, specify its composition completely. By this we mean that the composition at equilibrium of each of the phases is uniquely determined at a given temperature and pressure. For example, in the diethylether–water system at atmospheric pressure and at 10°C, the ether-rich layer contains 1.164 mass% water (4.6 mol% water) and the water-rich layer contains 9.04 mass% ether (2.4 mol% ether). This holds irrespective of the relative amounts of the liquid phases present, as long as there are two phases. When the temperature is raised to 30°C, the composition of the two phases changes to 1.409 mass% water (5.5 mol% water) and 5.34 mass% ether (1.35 mol% ether) in the two phases, respectively. These compositions are uniquely determined by the (atmospheric) pressure and the temperature, two external conditions (degrees of freedom) for this two-component, two-phase system. [If a third phase, the vapor, is present, then only one degree of freedom remains, and, e.g., the temperature determines the (vapor) pressure attained by the system]. For a two-phase system, each additional component (e.g., a solute) permits a new degree of freedom: here, the concentration of the solute in one of the phases. When this has been fixed, its concentration in the other phase is given by the equilibrium condition and cannot be freely chosen.

For the two-component, two-phase liquid system, the question arises as to how much of each of the pure liquid components dissolves in the other at equilibrium. Indeed, some pairs of liquids are so soluble in each other that they become completely miscible with each other when mixed at any proportions. Such pairs, for example, are water and 1-propanol or benzene and carbon tetrachloride. Other pairs of liquids are practically insoluble in each other, as, for example, water and carbon tetrachloride. Finally, there are pairs of liquids that are completely miscible at certain temperatures, but not at others. For example, water and triethylamine are miscible below 18°C, but not above. Such pairs of liquids are said to have a *critical solution temperature*, T_{cs} For some pairs of liquids, there is a lower T_{cs} (LCST), as in the water–triethylamine pair, but the more common behavior is for pairs of liquids to have an upper T_{cs} (UCST), (Fig. 2.2) and some may even have a closed mutual solubility loop [3]. Such instances are rare in solvent extraction practice, but have been exploited in some systems, where separations have been affected by changes in the temperature.

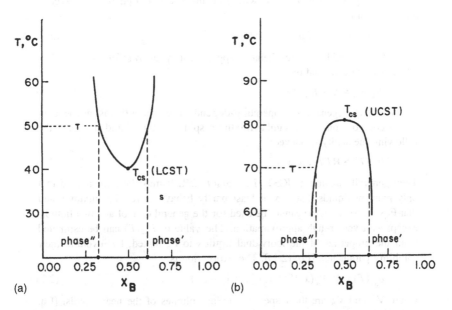

Fig. 2.2 Liquid immiscibility. The guaiacol (A) + glycerol (B) system happens to have a closed miscibility loop. The (phase) coexistence curves are shown on the left-hand side (a) for lower temperatures, at which a lower critical solution temperature (LCST), $T_{cs} = 40°C$, is seen, and on the right-hand side (b) for higher temperatures, where a UCST, $T_{cs} = 82°C$, is seen. The compositions of the A-rich phases″ and the B-rich phases′ are shown at 50°C and 70°C, respectively.

2.2.2 Excess Gibbs Energy of Mixing

The mutual solubility of two liquids A and B depends, in general, on how much the molecules of each liquid tend to attract those of its own kind, relative to their tendency to attract those of the other. This tendency is measured by the excess Gibbs energy of mixing of the two liquids (see section 2.4), $\Delta_M G_{AB}^E$, which is related to the partial vapor pressures p_A and p_B of the two liquids A and B in the mixture. If the composition of the system is given by n_A and n_B moles of the respective components in a given phase, their *mole fractions* in this phase are

$$x_A = n_A / (n_A + n_B) \qquad x_B = 1 - x_A = n_B / (n_A + n_B) \qquad (2.2)$$

The excess Gibbs energy of mixing per mole of mixture is approximately

$$\Delta_M G_{AB}^E = RT[x_A \ln(p_A / x_A p_A^*) + x_B \ln(p_B / x_B p_B^*)] \qquad (2.3)$$

where p_A^* and p_B^* are the vapor pressures of the pure liquids A and B, respectively, at the given temperature T. (The approximation consists of neglecting the interactions in the vapors that is permissible at low vapor pressures.)

The system of two liquids will split into two liquid phases at a given T and composition x if

$$x_A x_B (\partial^2 \Delta_M G_{AB}^E / \partial x^2) < -RT \qquad (2.4)$$

Many liquid mixtures behave approximately as *regular mixtures* [3], which are characterized by

$$\Delta_M G_{AB}^E = x_A x_B b_{AB}(T) \qquad (2.5)$$

$b_{AB}(T)$ being a (generally temperature-dependent) constant that is independent of the composition. Such regular mixtures split into two liquid phases if the following inequality is obeyed:

$$b_{AB}(T) > RT / (2 x_A x_B) \qquad (2.6)$$

They generally exhibit a UCST, $T_{cs} = b_{AB}(T)/2R$. It must be remembered that only pairs of liquids that mix at least partly behave as regular mixtures and that Eqs. (2.5) and (2.6) cannot be used for the general case of any two liquids, except as a very rough approximation. The value of $b_{AB}(T)$ can be estimated from the properties of the individual liquids to be mixed, by invoking their solubility parameters [Eq. (2.1)]. The expression is:

$$b_{AB}(T) = V_A^* V_B^* (x_A V_A^* + x_B V_B^*)^{-1} (\delta_A^2 - \delta_B^2) \qquad (2.7)$$

where V_A^* and V_B^* are the respective molar volumes of the neat liquids. [Eq. (2.7) makes $b_{AB}(T)$ composition dependent and should be used only if the molar volumes are not too disparate, since otherwise the mixture is not a regular one.] Thus, the larger the difference between the solubility parameters of the two liquids, the more likely is $b_{AB}(T)$ to be larger than the right-hand side of Eq. (2.6), and the larger is the tendency of the two-component system to split into two liquid phases [3].

If the mutual miscibility is small, then approximately

$$\ln x_A' \approx -V_B^*(\delta_A - \delta_B)^2 / RT \tag{2.8}$$

where x_A' is the mole fraction of component A in the B-rich phase marked by '. A similar equation holds for x_B'', in the A-rich phase, marked ", with V_A^* replacing V_B^*. Thus, the composition of the two phases can also be estimated by the application of the solubility parameters. These considerations can also be used for an estimate of the distribution of a solute between the immiscible liquids A and B, assuming the conditions for regular mixing to be fulfilled for all the components (see section 2.8.2).

As an example consider the ethylene glycol (A)–benzene (B) system at 25°C, assuming it to be a regular mixture (actually it is not). The relevant quantities are $V_A^* = 55.9$ and $V_B^* = 89.9$ cm^3 mol^{-1}, $\delta_A = 32.4$ and $\delta_B = 16.0$ J$^{1/2}$ cm$^{-3/2}$. Hence for the equimolar mixture $x_A = x_B = 0.5$, according to Eq. (2.7), $b_{AB} = 54.7$ kJ mol^{-1} ≫ $RT/ 2x_Ax_B = 5$ kJ mol^{-1}, so that this system ought to split into two liquid phases, as it does. In the benzene-rich phase (') the ethylene glycol content should be 0.005 wt% according to Eq. (2.8) and the benzene content in the ethylene glycol–rich phase (") should be 0.31 wt%. Most regular mixtures, however, would form a single homogeneous liquid phase, since the condition of Eq. (2.6) is rarely met.

2.2.3 The Mutual Solubility of Solvents

An inspection of Table 2.1 shows that, of the common liquids, water has by far the largest solubility parameter, ~48 J$^{1/2}$ cm$^{-3/2}$ [1,2]. Many organic solvents have appreciably smaller values of δ. Pairs of liquids that have values of δ that are less than 4 units apart have a reasonably large mutual solubility. They may even be completely miscible. Thus, aliphatic hydrocarbons mix with one another or with aromatic hydrocarbons to form a single homogeneous phase, as they also do with most halogenated hydrocarbons. Pairs of liquids having values of δ that are much more than 4 units apart are partially or nearly completely immiscible. This, however, does not apply well when one of the pair is water, unless it is assigned the value $\delta \cong 30$ J$^{1/2}$ cm$^{-3/2}$, which is comparable with that of methanol.

Water mixes completely with the lower alcohols (methanol, ethanol, 1- and 2-propanol, and *t*-butanol), as well as with most polyhydric alcohols and the lower ether alcohols (2-methoxy-ethanol, and so on). Of the ethers, water is miscible with tetrahydrofuran, 1,2-dimethoxyethane, and 1,4-dioxane, but not with the higher members. Among the ketones, water is miscible with acetone only; among the carboxylic acids, water is miscible with those up to and including butanoic acid, but with none of the common esters (an exception being ethylene carbonate). Among nitrogen-containing organic liquid compounds, water is miscible with acetonitrile, but not the higher ones (except dinitriles, such as succinonitrile). Water is miscible with primary, secondary, and tertiary amines, up to a total of about five carbon atoms in all the chains, including pyridine, but not with nitro compounds. Water is also miscible with many difunctional liquid compounds, such as diamines, alcohol amines, amides, and others, and with such compounds as dimethylsulfoxide [1–3].

The higher members of homologous series based on these compounds become increasingly immiscible with water as the chain length increases or as aromatic rings are added. Consequently, these higher members can be used as solvents in extraction systems. Table 2.2 lists the mutual solubilities, in weight percent (wt%), of water and representative organic solvents at 25°C, unless noted otherwise.

2.3   SOLUTE–SOLVENT INTERACTIONS

When a solute particle is introduced into a liquid, it interacts with the solvent particles in its environment. The totality of these interactions is called the *solvation* of the solute in the particular solvent. When the solvent happens to be water, the term used is *hydration*. The solvation process has certain consequences pertaining to the energy, the volume, the fluidity, the electrical conductivity, and the spectroscopic properties of the solute–solvent system. The apparent molar properties of the solute ascribe to the solute itself the entire change in the properties of the system that occur when 1 mol of solute is added to an infinite amount of solution of specified composition. The solvent is treated in the calculation of the apparent molar quantities of the solute as if it had the properties of the pure solvent, present at its nominal amount in the solution. The magnitudes of quantities, such as the apparent molar volume or heat content, do convey some information on the system. However, it must be realized that both the solute and the solvent are affected by the solvation process, and more useful information is gained when the changes occurring in both are taken into account.

2.3.1   Interactions at the Molecular Scale

The solvation process can be envisioned as occurring in several stages, although only the sum of the stage contributions to the overall process is measurable. First, a cavity must be created in the solvent to accommodate the solute. Then the solute is placed in the cavity and permitted to interact with its nearest neighbors, eventually forming coordinate bonds with some of them, forming a new entity, the solvated solute. Finally, this entity may interact further with its surroundings, by orienting solvent molecules, by the formation or disruption of hydrogen bonds, or by other interactions.

If the solute is charged (i.e., if it is an ion), it will orient the dipoles of a polar solvent by its electrical field. Donor–acceptor bonds may be formed if the

Table 2.2 Mutual Solubility of Water and Some Organic Solvents at 25°C

Solvent	Solvent in water wt%	Water in solvent wt%
c-Hexane	0.0055	0.010[a]
n-Hexane	0.00123	0.0111[a]
n-Octane	6.6×10^{-7}	0.0095[a]
n-Decane	5.2×10^{-8}	7.2×10^{-5}
n-Dodecane	3.7×10^{-9}	6.5×10^{-5}
Decalin (mixed isomers)	<0.02	0.0063[a]
Benzene	0,179	0.0635
Toluene	0.0515	0.0334
Ethylbenzene	0.0152	0.043
p-Xylene	0.0156	0.0456
Dichloromethane	1.30	0.198
Chloroform	0.815[a]	0.093
Carbon tetrachloride	0.077	0.0135[b]
1,1-Dichloroethane	5.03[a]	0.096
1,2-Dichloroethane	0.81[a]	0.187
Trichloroethylene	0.137	0.32
Chlorobenzene	0.0488[b]	0.0327
1,2-Dichlorobenzene	0.0156	0.309
Carbon disulfide	0.210[a]	0.0142
1-Butanol	7.45	20.5
Isobutyl alcohol	10	16.9
1-Pentanol	2.19	7.46
Isoamyl alcohol	2.97	9.19
1-Hexanol	0.7061	7.42
1-Octanol	0.0538	
2-Ethyl-1-hexanol	0.07[a]	2.6[a]
Phenol	8.66	28.72
m-Cresol	2.51[c]	
Diethyl ether	6.04	1.468
Diisopropyl ether	1.2[a]	0.57[a]
Bis(2-chloroethyl) ether	1.02[a]	0.1[a]
Methyl ethyl ketone	24.0[a]	10.0[a]
Methyl isobutyl ketone	1.7	1.9
Cyclohexanone	2.3[a]	8.0[a]
Acetylacetone	16.6[a]	4.5[a]
Ethyl acetate	8.08	2.94
Butyl acetate	0.68[a]	1.2[a]
Propylene carbonate	17.5	8.3
Nitromethane	11.1	2.09
Nitrobenzene	0.19[a]	0.24[a]
Benzonitrile	0.2	1[b]
Quinoline	0.609[a]	
Tri-n-butyl phosphate	0.039	4.67

Note. See Table 2.1 for some miscible solvents, marked by [a].
[a] At 20°C.
[b] At 30°C.
[c] At 40°C.

solute and solvent have suitable electron pair donation and acceptance properties. Hydrogen bonds between the solute and neighboring solvent molecules may form if one or the other or both are *protic*, i.e., have hydrogen atoms bonded to electronegative atoms. The solute particle may also undergo changes in the solvation process, its internal structure, if it has one, being affected by the strengthening or weakening of certain bonds, by a redistribution of the partial electrical charges on its atoms, or by the favoring of a certain conformation of a flexible solute molecule.

Spectroscopic measurements may, in certain cases, yield direct information on these interactions. On the other hand, thermodynamic values, obtained by measuring certain bulk properties of the system, require the aid of statistical mechanical methods to be related to specific interactions between the solute and the solvent. However, the thermodynamic aspects of the solute–solvent interactions reflect the preference of the solute for one solvent over another and, thereby, determine distribution of the solute in a solvent extraction system.

2.3.2 *Thermodynamics of the Interactions*

What quantity describes best the totality of these solute–solvent interactions and how can the various contributions to them be estimated? Let the pure solute B be vaporized in an imaginary process to a gas, and let this gas be very dilute, so that it obeys the *ideal gas laws*. In this condition each particle of the solute (molecule or ion) is very remote from any neighbor and has no environment with which to interact. If B is polyatomic, it does have its internal degrees of freedom, such as bond vibrations and rotation of the particle.

If this ideal gaseous particle B is introduced into a liquid A at a given temperature and pressure, all of the solute–solvent interactions become "switched on," the solvation process takes place, and the *Gibbs energy of solvation*, $\Delta_{solv}G_B$, is released. In many cases the process of dissolution of a gaseous solute in a liquid solvent can, indeed, be carried out experimentally—for instance, when propane or carbon dioxide is dissolved in water to give a solution at a given gas pressure.

The Gibbs energy change for the process of dissolution of the gaseous solute B, $\Delta_{soln}G_B$, is the driving force for the material transfer. When equilibrium is reached, $\Delta_{soln}G_B$ becomes zero (since, at equilibrium, no more net transfer occurs). The following equation then holds:

$$\Delta_{soln}G_{B\,eq} = 0 = \Delta_{soln}G_B^{\circ} + RT \ln \left[c_B(l) / c_B(g) \right]_{eq} \qquad (2.9)$$

where $\Delta_{soln}G_B^{\circ}$ is the *standard molar Gibbs energy of solvation* of the solute B, defined by Eq. (2.9) and c_B is its molar concentration (moles per unit volume) in the designated phases, (l) for liquid and (g) for gaseous. Equilibrium is generally assumed for the processes discussed elsewhere in this chapter, and is only emphasized here for Eq. (2.9) by the subscript eq for $\Delta_{soln}G_B$ and the ratio of concentrations, $c_B(l)/c_B(g)$. The concentration in the gas phase is, of course, given by the pressure according to the ideal gas law $c_B(g) = n_B/V = P/RT$. The ratio of the concentrations is known as the *Bunsen coefficient*, $K_{B(B,A)}$, for the solubility of the gaseous solute B in liquid A. At low pressures and concentrations $K_{B(B,A)}$ depends only on the temperature and is independent of the pressure and the concentration.

Whether obtained from an actual experimentally feasible process or from a thought process, $\Delta_{soln}G_B^{\circ}$, which is obtained from Eq. (2.9) by re-arrangement, pertains to the solvation of the solute and expresses the totality of the solute–solvent interactions. It is a *thermodynamic function of state*, and so are its derivatives with respect to the temperature (the standard molar entropy of solvation) or pressure. This means that it is immaterial how the process is carried out, and only the initial state (the ideal gaseous solute B and the pure liquid solvent) and the final state (the dilute solution of B in the liquid) must be specified.

Because it is a function of state, $\Delta_{soln}G_B^{\circ}$ may be considered to be made up additively of the contributions from the various stages in which the transfer of the solute particle from the gaseous state into the liquid solvent has been envisaged by the foregoing to take place.

The first stage is the creation of the cavity in the liquid to accommodate the solute. Obviously, work must be done against the cohesive forces of the liquid that hold its molecules together. This work should be proportional to the required size of the cavity, and increase as the volume of the solute increases. An expression for this work would be

$$\Delta_{cav}G = A_{cav}V_B\delta_A^2 \tag{2.10}$$

where A_{cav} is a proportionality coefficient, V_B is the molar volume of the solute B, and δ_A^2 is the cohesive energy density of the solvent A. Thus, in a series of solvents for a given solute, the positive contribution of cavity formation to $\Delta_{soln}G_B^{\circ}$ increases with the squares of the solubility parameters of the solvents. For a series of solutes and a given solvent, it increases with the molar volumes of the solutes.

It is more difficult to estimate the contribution from the dispersion forces to the solute–solvent interactions. Their energy increases with the product of the polarizabilities of the partners, but decreases strongly with the distance between them (being proportional to the inverse sixth power of the distance). The polarizability is related to the molar volume, hence, to the third power of the linear dimension of the solute or the solvent. Hence, the product of the polarizabilities depends on the sixth power of the distance between the centers of the interacting molecules. Consequently, these tendencies balance each other. For large molecules (e.g., metal chelates, liquid hydrocarbons), it is better to consider the interactions between adjacent segments of neighboring molecules, which are at a constant mean distance from one another in the liquid solution. The contribution from the dispersion forces (negative, because they are attractive) is proportional to the surface areas of the interacting molecules, or to the number N of segments present, and depends on their chemical natures. If A represents again the liquid solvent, then [3]

$$\Delta_{disp} G = \sum A_{iBA} N_{iB} N_{iA} \tag{2.11}$$

where A_{iBA} is the (negative) interaction Gibbs energy of a pair of segments of kind i, and the summation extends over all the different kinds of segments. A methylene group, a halogen atom, a $-CH=CH-$ group of an aromatic ring, or some other functional group, generally serves as a segment.

There may be additional, specific interactions between the solute and the solvent. Hydrogen bonds may be formed between them, particularly in protic solvents, i.e., solvents that contain hydrogen atoms bonded to oxygen (more rarely, nitrogen) atoms, such as water, alcohols, carboxylic acids, or acidic phosphoric esters. Hydrogen bonds are formed with solute anions, with hydrated ions in general (having an outer surface of water molecules), and with neutral solutes that have a very basic atom with a lone pair of electrons that can accept a hydrogen bond. Also, donor–acceptor bonds can be formed if the solvent has a very basic atom in its structure donating a pair of unshared electrons, if it is suitably exposed, and the solute is a cation or some acidic neutral molecule accepting this pair.

A generalized equation for $\Delta_{solv} G_B^\circ$ is [4]

$$\Delta_{solv} G_B^\circ = A_0 + A_\pi \pi^* + A_\alpha \alpha + A_\beta \beta + A_\delta \delta^2 \tag{2.12}$$

which describes the value of $\Delta_{solv} G_B^\circ$ for a given solute (characterized by A_0, A_π, A_α, A_β, and A_δ) with a series of solvents. The solvents are characterized by their Taft–Kamlet *solvatochromic parameters*: π^* for polarity–polarizability, α for hydrogen bond donation acidity, and β for hydrogen bond acceptance basicity. Values of these solvatochromic parameters have been tabulated for many solvents. (Table 2.3 gives the values for selected solvents.) As before, δ is the solubility parameter. The first two terms on the right-hand side of Eq. (2.12) express $\Delta_{disp} G$, the next two the hydrogen-bonding interactions, and the last one $\Delta_{cav} G$.

As an example, consider phenol as the solute and water and toluene as two solvents. The parameters for phenol are $A_\pi = 5.7$, $A_\alpha = -12.9$, $A_\beta = -18.3$, and $A_\delta = 0.0091$, whereas A_0 is unspecified, but a negative quantity. With the solvent parameters from Tables 2.1 and 2.3, the standard Gibbs energy of solvation of phenol in water becomes $A_0 + 3.39$, and in toluene $A_0 + 4.11$ kJ mol^{-1}. It is seen that $\Delta_{solv} G_B^\circ$ is lower in water than in toluene, so that the transfer of phenol from water to toluene entails an increase in $\Delta_{solv} G_B^\circ$. The consequence of this is that phenol prefers water over toluene, since work would be required to make this transfer. It should be remembered that the standard Gibbs energies of solvation refer to the state of infinite dilution of the solute (solute–solute

Table 2.3 Solvatochromic Parameters for Some Solvents and (Monomeric) Solutes (in Parentheses)

Substance	π^*	α	β	Substance	π^*	α	β
c-Hexane	0	0	0	2-Butanone	0.60	0.06	0.48
n-Hexane	−0.11	0	0	Cyclohexanone	0.68	0	0.53
Benzene	0.55	0	0.10	Acetophenone	0.81	0.04	0.49
Toluene	0.49	0	0.11	Acetic acid	0.64	1.12	0.45
p-Xylene	0.45	0	0.12	Hexanoic acid	0.52	1.22	0.45
Dichloromethane	0.82	0.13	0.10	Benzoic acid	0.80	(0.87)	(0.40)
Chloroform	0.58	0.20	0.10	Ethyl acetate	0.45	0	0.45
Carbon tetrachloride	0.21	0	0.10	Butyl acetate	0.46	0	0.45
1,2-Dichloroethane	0.73	0	0.10	Propylene carbonate	0.83	0	0.40
Chlorobenzene	0.68	0	0.07	Formamide	0.97	0.71	0.48
Water	1.09	1.17	0.47	Dimethylformamide	0.88	0	0.69
Water monomer	(0.39)	(0.32)	(0.15)	Tetramethylurea	0.79	0	0.80
Methanol	0.60	0.98	0.66	Hexamethyl phosphoramide	0.87	0	1.00
Methanol monomer	(0.39)	(0.38)	(0.41)	Acetonitrile	0.66	0.19	0.40
Ethanol	0.54	0.86	0.75	Benzonitrile	0.88	0	0.37
Ethanol monomer	(0.39)	(0.36)	(0.47)	Nitromethane	0.75	0.22	0.06
1-Propanol	0.52	0.84	0.90	Nitrobenzene	0.86	0	0.30
2-Propanol	0.48	0.76	0.84	Triethylamine	0.09	00.71	
1-Hexanol	0.40	0.80	0.84	Tri-n-butylamine	0.06	0	0.62
Trifluoroethanol	0.73	1.51	0	Dimethyl sulfoxide	1.00	0	0.76
Trifluoroethanol monomer		(0.59)	(0.10)	Diphenyl sulfoxide		0	0.70
Phenol	0.72	1.65	0.30	Sulfolane	0.90	0	0.39
Diethyl ether	0.24	0	0.47	Triethyl phosphate	0.69	0	0.77
Diisopropyl ether	0.19	0	0.49	Tributyl phosphate	0.63	0	0.80
Tetrahydrofuran	0.55	0	0.55	Triethyl phosphine oxide		0	1.05
Dioxane	0.49	0	0.37	Triphenyl phosphine oxide		0	0.94
Anisole	0.70	0	0.32	Pyridine	0.87	0	0.64
1,2-Dimethoxyethane	0.53	0	0.41	Quinoline	0.93	0	0.64
Acetone	0.62	0.08	0.48				

Source: Ref. 2.

interactions being, therefore, absent), and that a reaction such as ionic dissociation of the phenol in water is ignored in this example.

Extensive tables of solute parameters are beyond the scope of this book. Equations (2.10) to (2.12) are meant to show the nature of the dependencies of the additive terms on various quantities. They enable the prediction of tendencies of solute–solvent interactions for a given solute with a series of solvents or for a series of solutes with a given solvent.

2.3.3 *Solvation of Electrolytes*

If the solute B is an electrolyte that dissociates into ions in the solvent in question, the solvation of each ion separately may be considered in the hypothetical transfer process, although experimentally only quantities pertaining to the entire electrolyte can be measured [5]. The ions of a crystalline salt are brought from their standard solid state into the ideal gas state by the investment of the lattice Gibbs energy (approximately the lattice energy for a salt dissociating into one cation and one anion). The transfer of an ion from the ideal gas to the liquid is accompanied by reorientation of the dipoles of the solvent molecules. The contribution to $\Delta_{solv}G_B^\circ$ of this reorientation is given by the *Born equation*:

$$\Delta_{el}G = -A_{el}z^2r^{-1}(1-1/\varepsilon) \qquad (2.13)$$

where $A_{el} = N_Ae^2/8\pi\varepsilon_0 = 69.5 \text{ kJ mol}^{-1}$ (with r in nm), z is the charge on the ion in proton charge units, r is a distance related to the radius of the ion in nm (Table 2.4), and ε is the relative permittivity (dielectric constant) of the solvent. The Born equation should be applied only from a distance r from the center of the ion that is larger than the radius of the bare ion by Δr, which depends on the sizes of the ion B and the solvent molecules, A. Closer to the ion, the phenomenon of dielectric saturation sets in (Fig. 2.3), and the solvent cannot reorient itself freely because the binding to the ion becomes too strong. The contribution from this *inner shell of solvation* to $\Delta_{solv}G_B^\circ$ must be estimated by other means (see, e.g., Ref. [5]), but for polar solvents can be taken to be only slightly dependent on the nature of the solvent (except that it may depend on the thickness of this inner shell). In any event, Eq. (2.13) shows that the contribution of the reorientation of the solvent by the ion depends on the square of the charge on the ion and, reciprocally, on both the size of the ion and the relative permittivity of the solvent.

Terms for the electrostatic interactions [Eq. (2.13)] for the region outside the first solvation shell, and an appropriate one for the inner region, must be added to Eq. (2.12) for each ion of an electrolyte B, for the evaluation of $\Delta_{solv}G_B^\circ$. Since cations do not accept hydrogen bonds and anions do not donate them, except when protonated, like HSO_4^-, the term in α of the solvent becomes unimportant for cations and that in β of the solvent for anions.

Table 2.4 Crystal Ionic Radii and Standard Molar Gibbs Free Energies of Hydration of Ions

Ion	r (nm)	$-\Delta_{hyd}G°$ (kJ/mol)	Ion	r (nm)	$-\Delta_{hyd}G°$ (kJ/mol)	Ion	r (nm)	$-\Delta_{hyd}G°$ (kJ/mol)
H^+		1056				F^-	0.133	472
Li^+	0.069	481	Al^{3+}	0.053	4531	Cl^-	0.181	347
Na^+	0.102	375	Sc^{3+}	0.075	3801	Br^-	0.196	321
K^+	0.138	304	Y^{3+}	0.090	3457	I^-	0.220	283
Rb^+	0.149	281	La^{3+}	0.105	3155	OH^-	0.133	439
Cs^+	0.170	258	Ce^{3+}	0.101	3209	SH^-	0.207	303
NH_4^+	0.148	292	Nd^{3+}	0.099	3287	CN^-	0.191	305
Me_4N^+	0.280	412	Gd^{3+}	0.094	3385	SCN^-	0.213	287
Ag^+	0.115	440	Ho^{3+}	0.090	3480	N_3^-	0.195	287
Ph_4As^+	0.425	−32	Lu^{3+}	0.086	3522	BF_4^-	0.230	200
Mg^{2+}	0.072	1838	Pu^{3+}	0.101	3245	ClO_3^-	0.200	287
Ca^{2+}	0.100	1515	Am^{3+}	0.100	3297	BrO_3^-	0.191	340
Sr^{2+}	0.113	1386	Cr^{3+}	0.062	4010	IO_3^-	0.181	408
Ba^{2+}	0.136	1258	Fe^{3+}	0.065	4271	ClO_4^-	0.240	214
Mn^{2+}	0.083	1770	Ga^{3+}	0.062	4521	MnO_4^-	0.240	245
Fe^{2+}	0.078	1848	In^{3+}	0.079	3989	NO_2^-	0.192	339
Co^{2+}	0.075	1922	Tl^{3+}	0.088	3976	NO_3^-	0.179	306
Ni^{2+}	0.069	1992	Bi^{3+}	0.102	3486	$CH_2CO_2^-$	0.232	373
Cu^{2+}	0.073	2016	Ce^{4+}	0.080	6129	BPh_4^-	0.421	−42
Zn^{2+}	0.075	1963	Th^{4+}	0.100	5823	CO_3^{2-}	0.178	1479
Cd^{2+}	0.095	1763	U^{4+}	0.097	6368	$C_2O_4^{2-}$	0.21	1200
Hg^{2+}	0.102	1766	Zr^{4+}	0.072	6799	SO_3^{2-}	0.20	1303
Pb^{2+}	0.118	1434	Hf^{4+}	0.071	6975	SO_4^{2-}	0.230	1090
UO_2^{2+}	0.28	1229				CrO_4^{2-}	0.240	958
						PO_4^{3-}	0.238	2773

Source: Ref. 6.

These considerations are important, since solvent extraction often involves aqueous ions that must be transferred from an aqueous solution to some organic solvent. (The latter may contain complexing reagents that interact with the ions.) This transfer can be envisaged as taking place by steps in which the first one involves the freeing of an ion from its hydration before it is permitted to react with the solvent or reagent into which it is transferred. This freeing requires the investment of work that is the negative of the standard Gibbs energy of hydration of the ion [5,6]. The specification of this quantity requires splitting the value of an appropriate electrolyte into individual values for the cation and the anion. This cannot be accomplished by purely thermodynamic means, and an extrathermodynamic assumption or convention is needed. Several such assump-

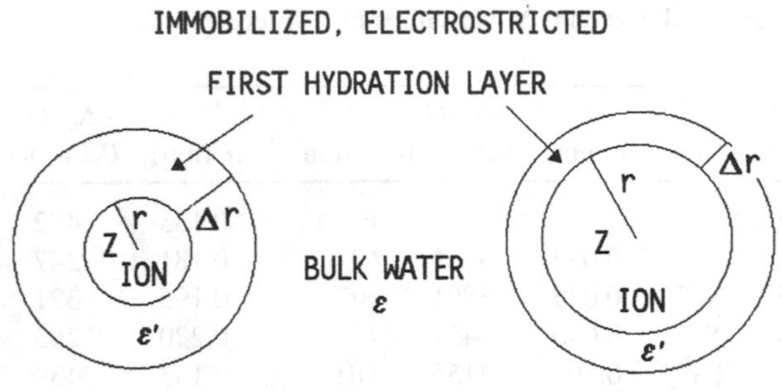

IMMOBILIZED, ELECTROSTRICTED

FIRST HYDRATION LAYER

Fig. 2.3 A schematic representation of the hydration layer near a small ion (left) and a large ion (right), showing the region where the water is dielectrically saturated (with a low relative permittivity ε'), hence electrostricted (squeezed) and immobilized. The thickness of this layer, Δr, depends reciprocally on the size of the ion.

tions that lead to consistent results have been proposed. One of them involves a *reference electrolyte*, of which the cation and anion are large, spherical, univalent, and similar in all respects except the sign of the charge. Such a reference electrolyte is tetraphenylarsonium tetraphenylborate (TATB); the standard molar Gibbs energy of hydration of each of its ions is $\Delta_{hyd}G° = -38 \pm 6$ kJ mol^{-1}. On this basis, values of $\Delta_{hyd}G°$ of other ions can be obtained, consistent with certain other extrathermodynamic assumptions, and are listed in Table 2.4 [6].

The differences in the solvation abilities of ions by various solvents are seen, in principle, when the corresponding values of $\Delta_{solv}G°$ of the ions are compared. However, such differences are brought out better by a consideration of the standard molar Gibbs energies of transfer, $\Delta_t G°$ of the ions from a reference solvent into the solvents in question (see further section 2.6.1). In view of the extensive information shown in Table 2.4, it is natural that water is selected as the *reference solvent*. The TATB reference electrolyte is again employed to split experimental values of $\Delta_t G°$ of electrolytes into the values for individual ions. Tables of such values have been published [5–7], but are outside the scope of this text. The notion of the standard molar Gibbs energy of transfer is not limited to electrolytes or ions and can be applied to other kinds of solutes as well. This is further discussed in connection with solubilities in section 2.7.

2.4 THERMODYNAMICS OF SOLUTIONS

Thermodynamics is the branch of science dealing with the energetics of substances and processes. It describes the tendency of processes to take place spontaneously, the effects of external conditions, and the effects of the composition of mixtures on such processes. Thermodynamics is generally capable of correlating a variety of data pertaining to widely changing conditions by relatively simple formulae. One approach to such a correlation involves the definition of a hypothetical ideal system and the subsequent consideration of deviations of real systems from the ideal one. In many cases, indeed, such deviations are relatively small and can be ignored in a first approximation. Such examples are, for instance, a gas under low pressure or a dilute solution of a solute in some solvent. In many other instances (unfortunately in many that pertain to practical solvent extraction), such an approximation is far from being valid, and quite incorrect estimates of properties of the real systems can result from ignoring the deviations from the ideal.

2.4.1 Ideal Mixtures and Solutions

2.4.1.1 One Liquid Phase

Consider two liquid substances that are rather similar, such as benzene and toluene or water and ethylene glycol. When n_A moles of the one are mixed with n_B moles of the other, the composition of the liquid mixture is given by specification of the mole fraction of one of them [e.g., x_A, according to Eq. (2.2)]. The energy or heat of the mutual interactions between the molecules of the components is similar to that of their self interactions, because of the similarity of the two liquids, and the molecules of A and B are distributed completely randomly in the mixture. In such mixtures, the *entropy of mixing*, which is a measure of the change in the molecular disorder of the system caused by the process of mixing the specified quantities of A and B, attains its maximal value:

$$\Delta_M S_{AB} = -R[x_A \ln x_A + x_B \ln x_B] \qquad (2.14)$$

per mole of mixture. This is the molar entropy of mixing expected for an *ideal mixture*. The molar heat of mixing of such a mixture, $\Delta_M H_{AB}$, is zero, since no net change in the energies of interaction takes place on mixing. Therefore, the *molar Gibbs energy of mixing*, in the process that produces an ideal mixture, is:

$$\Delta_M G_{AB} = \Delta_M H_{AB} - T\Delta_M S_{AB} = RT[x_A \ln x_A + x_B \ln x_B] \qquad (2.15)$$

Since the mole fractions are quantities that are smaller than unity, their logarithms are negative, $\Delta_M S_{AB}$ is positive, and $\Delta_M G_{AB}$ is negative, the process of mixing being a spontaneous process under the usual conditions. Equations (2.14) and (2.15) suffice to define an ideal (liquid) mixture.

The solute and the solvent are not distinguished normally in such ideal mixtures, which are sometimes called *symmetric ideal mixtures*. There are, however, situations where such a distinction between the solute and the solvent is reasonable, as when one component, say, B, is a gas, a liquid, or a solid of limited solubility in the liquid component A, or if only mixtures very dilute in B are considered ($x_B \ll 0.5$). Such cases represent *ideal dilute solutions*.

The molar Gibbs energy G of an ideal mixture, whether symmetric or dilute, consists of the molar Gibbs energies of the pure components A and B, weighted by their mole fractions, plus the molar Gibbs energy of mixing, $\Delta_M G_{AB}$.

$$G = x_A G_A^* + x_B G_B^* + \Delta_M G_{AB} \tag{2.16}$$

The *chemical potential*, μ, of a component of the mixture is the partial derivative of the Gibbs energy of the mixture with respect to the number of moles of this component present, the number of moles of all the other components being held constant, as are also the temperature and the pressure. For the component A in a mixture containing also B, C, . . . , the chemical potential is $\mu_A = (\partial G/\partial n_A)_{P,T,n_B,n_C\ldots}$, whether the mixture is ideal or not. In the ideal mixture the chemical potential of A is thus obtained from Eqs. (2.15) and (2.16) on carrying out the partial differentiation, yielding:

$$\mu_A = \mu_A^\circ + RT \ln x_A \tag{2.17}$$

As the mole fraction of A in the mixture increases toward unity, the second term in Eq. (2.17) tends toward zero, and the chemical potential of A tends toward the *standard chemical potential*, $\mu_A^\circ = G_A^*$, the molar Gibbs energy (or chemical potential) of A in the realizable *standard state* of pure A, in the sense that pure liquid A is a known chemical substance. A similar equation holds for the component B.

Consider a dilute ideal solution of the solute B (which could be gaseous, liquid, or solid at the temperature in question) in the solvent A. Suppose that more concentrated solutions do not behave ideally and, in particular, the state of pure liquid B cannot be attained by going to more and more concentrated solutions (e.g., by removing A by volatilization). It is possible to define a standard chemical potential pertaining to a hypothetical standard state of the ideal infinitely dilute solution as the limit:

$$\mu_B^\infty = \lim_{x_B \to 0} (\mu_B - RT \ln x_B) \tag{2.18}$$

Although μ_B tends to $-\infty$ as x_B tends to 0 (and $\ln x_B$ also tends to $-\infty$), the difference on the right-hand side of Eq. (2.18) tends to the finite quantity μ_B^∞, the standard chemical potential of B. At infinite dilution (practically, at high dilution) of B in the solvent A, particles (molecules, ions) of B have in their surroundings only molecules of A, but not other particles of B, with which to interact. Their surroundings are thus a constant environment of A, independent of the actual concentration of B or of the eventual presence of other solutes, C, D, all at high dilution. The standard chemical potential of the solute in an ideal dilute solution thus describes the solute–solvent interactions exclusively.

A solution in which the solvent A obeys the relationship of Eq. (2.17) and the solute B obeys the expression

$$\mu_B = \mu_B^\infty + RT \ln x_B \tag{2.19}$$

over a certain (low) concentration range of B in A is said to obey *Raoult's law* for the solvent and *Henry's law* for the solute in this concentration range (Fig. 2.4).

The vapor pressure of the solvent in such a case is given by

$$p_A = x_A p_A^* \tag{2.20}$$

where p_A^* is the vapor pressure of pure liquid A at the given temperature. Conversely, when Eq. (2.20) is followed by the vapor pressure of the solvent, Raoult's law is said to be valid and Eq. (2.17) is obeyed. The vapor pressure of the solute is also proportional to its mole fraction, if Henry's law is obeyed. However, the proportionality constant is not its vapor pressure in the pure liquid state, which may not be attainable at the given temperature. Instead, the expression

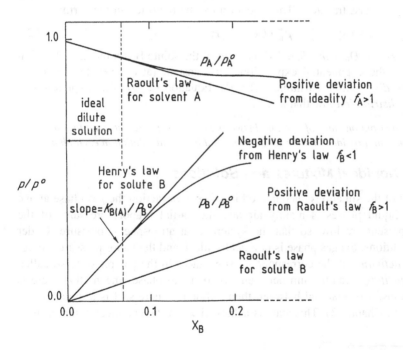

Fig. 2.4 The vapor pressure diagram of a dilute solution of the solute B in the solvent A. The region of ideal dilute solutions, where Raoult's and Henry's laws are obeyed by the solvent and solute, respectively, is indicated. Deviations from the ideal at higher concentrations of the solute are shown. (From Ref. 3.)

$$p_B = K_{B(A)} x_B \tag{2.21}$$

is followed, where $K_{B(A)}$ is called the *Henry's law constant*, and depends on the chemical natures of both B and A, as the molecular description of the foregoing solutions requires. These considerations apply also when there are several solutes present at low concentrations, so that Raoult's law may apply to the solvent and Henry's law to each of the solutes.

2.4.1.2 Two Liquid Phases

Consider now two practically immiscible solvents that form two phases, designated by $'$ and $''$. Let the solute B form a dilute ideal solution in each, so that Eq. (2.19) applies in each phase. When these two liquid phases are brought into contact, the concentrations (mole fractions) of the solute adjust by mass transfer between the phases until equilibrium is established and the chemical potential of the solute is the same in the two phases:

$$\mu_{B''} = \mu_{B''}^{\infty} + RT \ln x_{B''} = \mu_{B'} = \mu_{B'}^{\infty} + RT \ln x_{B'} \tag{2.22}$$

(It is the *difference in the chemical potentials* of the solute that is the driving force for the mass transfer.) This equation can be rewritten in the form:

$$x_{B'} / x_{B''} = \exp[(\mu_{B''}^{\infty} - \mu_{B'}^{\infty}) / RT] = P_B \tag{2.23}$$

where $x_{B'} / x_{B''} = D_B$ is *the distribution ratio* of the solute B (on the mole fraction scale) and the exponential expression is a constant (at a given temperature), called the *distribution constant* of B,* P_B. Equation (2.23) is an expression of *Nernst's distribution law* that states:

> The distribution ratio of a solute between two liquid phases at equilibrium is a constant, provided that the solute forms a dilute ideal solution in each phase.

2.4.2 *Nonideal Mixtures and Solutions*

For most of the situations encountered in solvent extraction the gas phase above the two liquid phases is mainly air and the partial (vapor) pressures of the liquids present are low, so that the system is at atmospheric pressure. Under such conditions, the gas phase is practically ideal, and the vapor pressures represent the *activities* of the corresponding substances in the gas phase (also called their *fugacities*). Equilibrium between two or more phases means that there is no net transfer of material between them, although there still is a dynamic exchange (cf. Chapter 3). This state is achieved when the chemical potential μ as

---

*There is no internationally recommended symbol for the distribution constant, but the symbol P (for *p*artition constant) is common usage in this context. Coordination (complex) chemists (as in Chapter 3) prefer the symbol K_D (for the German "*K*onstante" for "constant"); sometimes other symbols are used. The important point is that the symbol should be properly defined in the text.

well as the activity a of each substance are the same in all the phases. It is the inequality of the activities of a substance in two phases that causes some of the substance to transfer from the one (where it is higher) to the other phase, until equality is achieved. The activity of a pure liquid or solid substance is defined as unity. In any mixture, whether ideal or not, the activity of a component A, a_A, is related to its chemical potential by [3]:

$$\mu_A = \mu_A^\circ + RT \ln a_A \tag{2.24}$$

Therefore, if component A is the solvent in a solution, and it obeys Raoult's law when the solution is dilute in all solutes, then $a_A = x_A$ under these conditions. Otherwise, the solution is nonideal, and the deviation from the ideal is described by means of the *activity coefficient*

$$f_A = a_A / x_A \tag{2.25}$$

where a_A is given by Eq. (2.24) or can also be expressed by the ratio of the vapor pressures, p_A/p_A^* (neglecting interactions in the vapor phase, otherwise the fugacities have to be employed). As the solution is made more dilute and approaches the dilute ideal solution and the pure solvent, the activity coefficient f_A approaches unity:

$$\lim_{x_A \to 1} f_A = 1 \tag{2.26}$$

Equation (2.24) can now be rewritten as

$$\mu_A = \mu_A^* + RT \ln x_A + \mu_A^E = \mu_A^* + RT \ln x_A + RT \ln f_A \tag{2.27}$$

where μ_A^E is the *excess chemical potential* of A in the solution and is equal to $RT \ln f_A$, expressing the deviation of the solution from ideality.

As already mentioned, in solutions that are not symmetric mixtures it is conventional to treat the solute B (or solutes B, C, ...) differently from the treatment accorded to the solvent A. The activity coefficient is still defined as the ratio a_B/x_B, but the limit at which it equals unity is taken as the infinite dilute solution of B in A:

$$\lim_{x_B \to 0} f_B = 1 \tag{2.28}$$

so that

$$\mu_B = \mu_B^\infty + RT \ln x_B + \mu_B^E = \mu_B^\infty + RT \ln x_B + RT \ln f_B \tag{2.29}$$

The activity coefficients f_A and f_B may be smaller or larger than unity; hence, the excess chemical potentials μ_A^E and μ_B^E may be negative or positive. Depending on the sign of μ_A^E, the solution is said to exhibit positive or negative deviations from Raoult's law. Similarly, positive or negative deviations from Henry's law can be noted (see Fig. 2.4).

The considerations that lead to Eq. (2.22) also apply to the case of non-ideal solutions, except that the terms $RT \ln f_{B'}$ and $RT \ln f_{B''}$ must be added to the left- and right-hand sides. When rearranged into the form of Eq. (2.23), the result is

$$D_B = x_{B'} / x_{B''} = P_B(f_{B''} / f_{B'}) \tag{2.30}$$

Therefore, the distribution ratio of B remains constant only if the ratio of the activity coefficients is independent of the total concentration of B in the system, which holds approximately in dilute solutions. Thus, although solutions of metal chelates in water or nonpolar organic solvents may be quite nonideal, Nernst's law may still be practically obeyed for them if their concentrations are very low ($x_{chelate} < 10^{-4}$). Deviations from Nernst's law (constant D_B) will in general take place in moderately concentrated solutions, which are of particular importance for industrial solvent extraction (see Chapter 12).

When no distinction between the solvent and the solute in a liquid mixture is made, then nonideal mixtures can still be described by means of an expression similar to Eq. (2.16), but with the addition of a term that is the *excess molar Gibbs energy of mixing*, ΔG_{AB}^E (see also section 2.2). Thus the Gibbs energy per mole of mixture is:

$$G = x_B G_A' + x_B G_B' + RT[x_A \ln x_A + x_B \ln x_B] + \Delta G_{AB}^E \tag{2.31}$$

where $\Delta G_{AB}^E = x_A \mu_A^E + x_B \mu_B^E = RT [x_A \ln f_A + x_B \ln f_B]$ is a function of the composition. It is this quantity that has been used in Eq. (2.5) to define the variable $b_{AB}(T)$, which describes the tendency of the mixture to separate into two liquid phases. The excess Gibbs energy is positive ($>2RT$) for mixtures that tend to separate, but is negative when the components have a strong mutual attraction of their molecules. For some mixtures, ΔG_{AB}^E is positive over a part of the composition range and negative over another part.

When a partial derivative of ΔG_{AB}^E relative to the number of moles of a component, say, A, is taken, the result is RT times the (natural) logarithm of the activity coefficient of this component, $RT \ln f_A$. When the partial derivative relative to the temperature is taken, the negative of the excess entropy of mixing, $-\Delta S_{AB}^E$, results. For a regular solution (see section 2.2) this quantity is zero, so that the (excess) heat (enthalpy) of mixing is $\Delta H_{AB}^E = \Delta G_{AB}^E$. The qualifier "excess" for the enthalpy of mixing is put in parentheses, since ideal mixtures have no enthalpies of mixing, so that if heat is evolved or absorbed when two liquids are mixed, the mixture is not ideal and the observed heat (enthalpy) of mixing is always an excess quantity. The enthalpy of mixing, ΔH_{AB}^E, may or may not be temperature dependent itself, and may be positive, negative, or may change sign within the composition range. A negative ΔH_{AB}^E is, again, indicative of strong mutual interactions of the molecules of the components.

An example of such behavior is met with in aqueous ethanol (EtOH) at, say, 75°C. At $x_{EtOH} < 0.24$, the heat of mixing is negative (heat is evolved on mixing the components), but at higher ethanol contents it is positive (heat is absorbed and the mixture cools). In the equimolar mixture $\Delta H_{AB}^E = 220$ but $\Delta G_{AB}^E = 960$ J mol^{-1}, due to a negative entropy of mixing, but since

$2x_{\text{EtOH}}x_{\text{water}}RT = 1446$ J mol^{-1} for this mixture, a single liquid phase is stable. However, for aqueous 1-butanol (BuOH) at 25°C, $\Delta G_{\text{AB}}^{\text{E}} \geq 2x_{\text{BuOH}}x_{\text{water}}RT = 1220$ J mol^{-1} for the near equimolar mixture, although $\Delta H_{\text{AB}}^{\text{E}} = 500$ J mol^{-1} only, so the mixture is on the verge of splitting into two liquid phases (it splits when $x_{\text{BuOH}} = 0.485$).

2.4.3 Scales of Concentration

The composition of a mixture need not be given in terms of the mole fractions of its components. Other scales of concentration are frequently used, in particular, when one of the components, say, A, can be designated as the solvent and the other (or others), B, (C, . . .) as the solute (or solutes). When the solute is an electrolyte capable of dissociation into ions (but not only for such cases), the *molal scale* is often employed. Here, the composition is stated in terms of the number of moles of the solute, m, per unit mass (1 kg) of the solvent. The symbol m is used to represent the molal scale (e.g., 5 m = 5 mol solute/1 kg solvent). The conversion between the molal and the *rational scale* (i.e., the mole fraction scale, which is related to ratios of numbers of moles [see Eq. (2.2)] proceeds according to Eqs. (2.32a) or (2.32b) (cf. Fig. 2.4):

$$m_{\text{B}} = x_{\text{B}} / M_{\text{A}} (1 - x_{\text{B}}) \approx x_{\text{B}} / M_{\text{A}} \tag{2.32a}$$

$$x_{\text{B}} = m_{\text{B}} / (m_{\text{B}} + 1 / M_{\text{A}}) \approx m_{\text{B}} M_{\text{A}} \tag{2.32b}$$

where M_{A} is the molar mass of the solvent A, expressed in kilograms per mole (kg mol^{-1}). For aqueous solutions, for instance, $M_{\text{A}} = 0.018$ kg mol^{-1} and $1/M_{\text{A}} = 55.5$ mol kg^{-1}. The approximate equalities on the right-hand sides of Eqs. (2.32a) and (2.32b) pertain to very dilute solutions. Activity coefficients on the molal scale are designated by γ, so that Eq. (2.29) becomes

$$\mu_{\text{B}} = \mu_{\text{B(m)}}^{\infty} + RT \ln m_{\text{B}} + RT \ln \gamma_{\text{B}} \tag{2.33}$$

where, again, $\mu_{\text{B(m)}}^{\infty} = \lim_{(m_{\text{B}} \to 0)} (\mu_{\text{B}} - RT \ln m_{\text{B}})$ is the standard chemical potential of B on the molal scale, and $\lim_{(m_{\text{B}} \to 0)} \gamma_{\text{B}} = 1$.

Another widely used concentration scale is the *molar scale*, that describes the number of moles of the solute, c, per unit volume of the solution [i.e., per 1 dm^3 = 1 L (liter)]. The symbol M = mol/L is used to represent this scale [e.g., 0.5 M = 0.5 mol solute/1 liter (L) of solution]. Conversion between the molar and molal scales is made according to Eqs. (2.34a) and (2.34b) by means of the density of the solution, ρ, in kg L^{-1}(= g mL^{-1} = g cm^{-3}; see Fig. 2.5):

$$c_{\text{B}} = m_{\text{B}} \rho / (1 + m_{\text{B}} M_{\text{B}}) \approx m_{\text{B}} \rho_{\text{A}} / (1 + m_{\text{B}} M_{\text{B}}) \tag{2.34a}$$

$$m_{\text{B}} = c_{\text{B}} / (\rho - c_{\text{B}} M_{\text{B}}) \approx c_{\text{B}} / (\rho_{\text{A}} - c_{\text{B}} M_{\text{B}}) \tag{2.34b}$$

where M_{B} is the molar mass of the solute (in kg mol^{-1}). The approximate equalities on the right-hand sides of Eqs. (2.34a) and (2.34b) pertain to very dilute

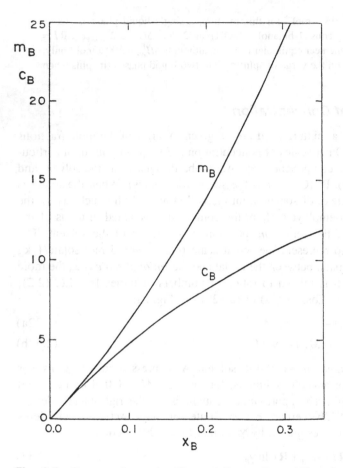

Fig. 2.5 Concentration scales. The molality m_B [in mol/(kg solvent)] and molarity c_B [in mol (L solution)$^{-1}$] of solute B, which has a molar mass M_B 0.100 kg mol^{-1} and a molar volume $V_B = 0.050$ L mol^{-1}, in the solvent A, which has $M_A = 0.018$ kg mol^{-1} and $V_A = 0.018$ L mol^{-1} (water), are shown as a function of the mole fraction x_B of the solute. Note that the molarity tends toward a maximal value ($1/V_B$), whereas the molality tends toward infinity as x_B increases toward unity.

solutions. For aqueous solutions at 25°C, $\rho_A = 0.997$ kg L^{-1}, and a generally negligible error is made if ρ_A is replaced by 1 for very dilute aqueous solutions. The activity coefficient on the molar scale is designated by y, so that Eq. (2.29) becomes

$$\mu_B = \mu_{B(c)}^{\infty} + RT \ln c_B + RT \ln y_B \qquad (2.35)$$

where, as before, $\mu_{B(c)}^{\infty} = \lim_{(c_B \to 0)} (\mu_B - RT \ln c_B)$ is the standard chemical potential of B on the molar scale, $\lim_{(c_B \to 0)} y_B = 1$.

As an example for such concentration scale conversions, consider 20 wt% tri-n-butyl phosphate (TBP, B) in toluene (A). The mole fraction of the TBP is $x_B = (20/M_B)/[(20/M_B) + (80/M_A)] = 0.080$, its molality is $m_B = (0.200/M_B)/0.800 = 0.94$ m, and its molarity is $c_B = 0.94$ $\rho_A/(1 + 0.94\,M_B) = 0.65$ M (M_B is taken in kg mol^{-1}).

The quantity μ_B is independent of the concentration scale used, being a true property of the solution, but the three standard chemical potentials $\mu_B^{\infty}{}_{(x)}$, $\mu_B^{\infty}{}_{(m)}$, and $\mu_B^{\infty}{}_{(c)}$ are not equal. Consequently, differences between the standard chemical potentials of a solute in the two liquid phases employed in solvent extraction also depend on the concentration scale used. Thus, P_B defined in Eq. (2.23) is specific for the rational concentration scale, and does not equal corresponding quantities pertaining to the other scales. Therefore, Eq. (2.30) might be rewritten with a subscript (x), to designate the rational scale, [i.e., with $D_{B(x)}$ and $P_{B(x)}$]. Similar expressions would then be

$$D_{B(m)} = m_{B'} / m_{B^*} = P_{B(m)}(\gamma B''/\gamma B') \tag{2.36a}$$

for the molal scale, and

$$D_{B(c)} = c_{B'} / c_{B^*} = P_{B(c)}(yB''/yB') \tag{2.36b}$$

for the molar scale. The subscripts are often omitted, if the context unequivocally defines the concentration scale employed.

2.5 AQUEOUS ELECTROLYTE SOLUTIONS

In many practical solvent extraction systems, one of the two liquids between which the solute distributes is an aqueous solution that contains one or more electrolytes. The distributing solute itself may be an electrolyte. An *electrolyte* is a substance that is capable of *ionic dissociation*, and does dissociate at least partly to ions in solution. These ions are likely to be solvated by the solvent (or, in water, to be hydrated) [5]. In addition to ion–solvent interactions, the ions will also interact with one another: repulsively, if of the same charge sign, attractively, if of the opposite sign. However, ion–ion interactions may be negligible if the solution is extremely dilute. The electrolyte $C_{\nu_+}A_{\nu_-}$, is made up of ν_+ positive ions, or *cations*, C^{z+}, and ν_- negative ions, or *anions*, A^{z-}, or of altogether $\nu = \nu_+ + \nu_-$ ions, and is designated in the following by the symbol CA. A general principle that is always obeyed in electrolyte solutions is that of the electroneutrality of the solution as a whole. This means that the number of cations weighted by their charge numbers z_+ (the number of protonic charges each has) equals that of the anions similarly weighted with z_-. Per unit volume of the solution this can be written as:

$$\sum c_+ z_+ + \sum c_- z_- = 0 \qquad (2.37)$$

(Note that z_- is a negative number, so that $v_+ z_+ + v_- z_- = 0$.) The electroneutrality principle is equivalent to stating that it is impossible to produce a solution
that contains, for example, only cations or an excess of positive charge (i.e.,
that ions cannot be considered as independent components of solutions). Only
entire electrolytes are components that can be added to a solution.

It is expedient to consider the mean ionic properties of the electrolytes.
An electrolyte that dissociates into v ions has a mean molality $m_\pm = (v_+ m_+ + v_- m_-)/v$ and a mean molarity $c_\pm = (v_+ c_+ + v_- c_-)/v$.

A useful concept that is used when the activities of electrolytes are calculated is that of the *ionic strength* of the solution. This is defined (on the molar
scale) as:

$$I = \frac{1}{2}\sum c_i z_i^2 \qquad (2.38)$$

where the summation extends over all the cations and anions present in the
solution. Since the charges appear to the second power in the expression for the
ionic strength, all the terms are positive. An electrolyte that is completely dissociated to univalent ions is designated as a 1:1 electrolyte, examples being aqueous HCl and $LiNO_3$. If only a single 1:1 electrolyte is present at a molar concentration c_{CA}, then $I = c_{CA}$. According to this notation, $MgCl_2$, and $(NH_4)_2SO_4$ are
2:1 and 1:2 electrolytes, respectively, for which $I = 3\,c_{CA}$. For 1:1, 2:2, ... (i.e.,
symmetric) electrolytes, $m_\pm = m_{CA}$, and $I = c_{CA}$, $4\,c_{CA}$, ... The relationship between m_{CA} and c_{CA} is given by Eq. (2.34.)

2.5.1 Electrolyte Activity*

The chemical potential of an electrolyte may formally (but not experimentally)
be considered as being made up from contributions of the individual kinds of
ions:

$$\mu_{CA} = v\mu_\pm = v_+\mu_+ + v_-\mu_- \qquad (2.39)$$

Mean ionic activities, molalities (or molarities), and activity coefficients, designated with subscript $\pm$, are used in equations corresponding to Eqs. (2.33) or
(2.35) for electrolyte solutions:

$$\mu_{CA} = v\mu_\pm^\infty + vRT \ln m_\pm + vRT \ln \gamma_\pm \qquad (2.40)$$

Electrolyte solutions are typical nonsymmetrical solutions, in that the solvent is treated according to Eq. (2.27) and the solute according to Eq. (2.40).

---

*See also Chapter 6.

The mole fraction concentration scale is generally used for the solvent water, designated by subscript w and having molar mass M_w.

The activity of water is related formally to the molality of the electrolyte by means of the *osmotic coefficient*, φ, of the solution:

$$\ln a_w = \ln(p_w / p_w^*) = -v\,m_{\pm}M_w\varphi \tag{2.41}$$

As the solution becomes more dilute in the electrolyte (or, in all electrolytes present, where $v\,m_{\pm}$ is replaced by $\Sigma\,v_i\,m_{\pm i}$), $\ln a_w$ approaches zero (a_w approaches unity), and φ approaches unity.

A definite relationship exists between $\gamma_{\pm}$ and φ (the *Gibbs–Duhem relationship*), which may be expressed by the excess Gibbs energy of the solution of an electrolyte:

$$\Delta G^E = v\,m_{\pm}RT[1 - \varphi + \ln\gamma_{\pm}] \tag{2.42}$$

This expression is analogous to Eq. (2.3), in that $(1 - \varphi)$ expresses the contribution of the solvent and $\ln\gamma_{\pm}$ that of the electrolyte to the excess Gibbs energy of the solution. The calculation of the mean ionic activity coefficient of an electrolyte in solution is required for its activity and the effects of the latter in solvent extraction systems to be estimated. The osmotic coefficient or the activity of the water is also an important quantity related to the ability of the solution to dissolve other electrolytes and nonelectrolytes.

Electrostatic and statistical mechanics theories were used by Debye and Hückel to deduce an expression for the mean ionic activity (and osmotic) coefficient of a dilute electrolyte solution. Empirical extensions have subsequently been applied to the Debye–Hückel approximation so that the expression remains approximately valid up to molal concentrations of 0.5 m (actually, to ionic strengths of about 0.5 mol L^{-1}). The expression that is often used for a solution of a single aqueous 1:1, 2:1, or 1:2 electrolyte is

$$\log\gamma_{\pm} = -A|z_+z_-|I^{1/2}/[1+1.5\,I^{1/2}]+bI \tag{2.43}$$

A is a constant that depends on the temperature, and at 25°C $A = 0.509$, and b has the value 0.2. (Note that although I is given on the molar scale, the $\gamma_{\pm}$ obtained from the semi-empirical Eq. (2.43) is on the molal scale.) However, Eq. (2.43) does not seem to be valid for 2:2, 3:1, or higher electrolytes (i.e., those with highly charged ions) at practical concentrations. A much more complicated expression results from the Debye–Hückel theory for the osmotic coefficient of the solution, but in the range of validity of Eq. (2.43), φ differs from unity by no more that about -5%. Other modifications of the theory have been suggested to extend its validity beyond the $I = 0.5$ mol L^{-1} of Eq. (2.43), but their consideration [3,8] is outside the scope of this book. If I for the solution is >0.5 mol L^{-1}, and for higher-charge-type electrolytes (2:2, 3:1, etc.), experimental values of $\gamma_{\pm}$ and φ can be obtained from suitable compilations [8].

2.5.2 Mixtures Containing Electrolytes

It is important to be able to estimate the values of $\gamma_\pm$ for aqueous electrolytes in their mixtures, say B and C. An expression that has been found to be valid under wide conditions is *Harned's rule*:

$$\log \gamma_{\pm B} = \log \gamma_{\pm B}^0 (I_m) - \alpha_{B(C)} I_{mC} \qquad (2.44a)$$

and

$$\log \gamma_{\pm C} = \log \gamma_{\pm C}^0 (I_m) - \alpha_{C(B)} I_{mB} \qquad (2.44b)$$

where in Eq. (2.44a) the activity coefficient of electrolyte B in its mixture with C is $\gamma_{\pm B}$, the latter electrolyte contributing I_{mC} to the total ionic strength on the molal scale, $I_m = \frac{1}{2}\Sigma\, m_i z_i^2$, and $\gamma_{\pm B}^0(I_m)$ is the activity coefficient of B when present alone at molal ionic strength I_m. A similar expression, Eq. (2.44b), holds for $\log \gamma_{\pm C}$ of electrolyte C in the mixture.

> Consider, for example an aqueous solution that is 1 m in HCl (electrolyte B) and 2 m in LiCl (electrolyte C), at a total ionic strength on the molal scale of $I_m = 3$ m. The activity coefficient of HCl in 3 m HCl is 1.316 [its logarithm is $\log \gamma_{\pm HCl}^°(I_m) = 0.119$) and $\alpha_{HCl(LiCl)} = 0.004$ at this total molality. Hence, log $\gamma_{\pm HCl} = 0.119 - 2 \times 0.004 = 0.111$ or $\gamma_{\pm HCl} = 1.291$ in this mixture. Similarly, $\gamma_{\pm LiCl}^°(I_m) = 1.174$ (log $\gamma_{\pm LiCl}^°(I_m) = 0.070$) for 3 m LiCl and $\alpha_{LiCl(HCl)} = -0.013$, so that log $\gamma_{\pm LiCl} = 0.070 - 1 \times (-0.013) = 0.083$ or $\gamma_{\pm LiCl} = 1.211$ in this mixture. Thus, the activity coefficient of HCl is lower and that of LiCl is higher in the mixture than in the single electrolyte solutions at the same total molality. If the HCl is present at trace concentration in 3 m LiCl, its activity coefficient would be somewhat reduced to $\gamma_{\pm HCl,trace} = 1.279$, whereas that of trace LiCl in 3 m HCl would be somewhat increased to $\gamma_{\pm LiCl,trace} = 1.285$.
>
> The values of $\alpha_{B(C)}$ and $-\alpha_{C(B)}$ are interrelated. For very simple systems they are equal, but generally they are not and must be obtained from the literature [9]. They also depend on the total ionic strength, I, but not on the composition relative to the two electrolytes. It must also be noted that Eq. (2.44) is a semi-empirical expression, and many systems do not obey it, requiring added terms in I_{mC}^2 or I_{mB}^2.

When one electrolyte, say, B, is present at very low concentrations in the presence of the other, say, C, then $I_{mC} \approx I_m$, and if this ionic strength is fixed, then according to Eq. (2.44) the activity coefficient $\gamma_{\pm B}$ of the trace electrolyte B becomes independent of its own concentration in the mixture. That is, log $\gamma_{\pm B} = \log \gamma_{\pm B}^°(I_m) - \alpha_{B(C)} I_m \neq f(m_B)$. The same is true if there are several electrolytes present, all at concentrations much lower than that of C, which is fixed.

This is the basis of the *ionic medium method*, where an electrolyte C, such as sodium perchlorate, is kept in the solution at a fixed and high concentration (e.g., 3 mol L^{-1}). This practice permits the concentrations of reactive electrolytes, that is, those that provide ions that participate in reactions (contrary to the "inert" ions of the medium electrolyte) to be varied below certain limits at will, without changes in their activity coefficients. These are determined entirely by the natures of the electrolytes and the fixed concentration of C. Being constant quantities under these conditions, the activity coefficients can be incorporated into *conditional equilibrium constants* for the reactions where the ions participate. (It becomes also immaterial for the use of the constant ionic medium whether the ionic strength is specified in terms of a constant molality or molarity, although the numerical values of the equilibrium constants depend on this specification.)

The nonideality of electrolyte solutions, caused ultimately by the electrical fields of the ions present, extends also to any nonelectrolyte that may be present in the aqueous solution. The nonelectrolyte may be a co-solvent that may be added to affect the properties of the solution (e.g., lower the relative permittivity, ε, or increase the solubility of other nonelectrolytes). For example, ethanol may be added to the aqueous solution to increase the solubility of 8-hydroxyquinoline in it. The nonelectrolyte considered may also be a reagent that does not dissociate into ions, or one where the dissociation is suppressed by the presence of hydrogen ions at a sufficient concentration (low pH; cf. Chapter 3), such as the chelating agent 8-hydroxyquinoline.

In any event, the activity coefficient of the nonelectrolyte, designated by subscript N, generally follows the *Setchenov equation*:

$$\log y_N = \log y_N^\circ + k_{N,CA} c_{CA} \tag{2.45}$$

where y_N° is the activity coefficient of the nonelectrolyte on the molar scale in the absence of the electrolyte CA, and $k_{N,CA}$ is a proportionality constant that depends on the natures of N and the electrolyte CA, but is independent of their concentrations. The larger the molar volume and the lower the polarity of the nonelectrolyte N, generally the larger $k_{N,CA}$ is. The better hydrated the ions of the electrolyte CA are, the larger, again, $k_{N,CA}$ is. Since the values of $k_{N,CA}$ are generally positive, the activity coefficient of the nonelectrolyte in the solution increases in the presence of the electrolyte, and for a given activity of the former (e.g., determined by equilibrium with excess insoluble nonelectrolyte), its concentration must decrease. The nonelectrolyte is then said to be salted-out by the electrolyte. As an illustration, consider the salting-out of benzene by CsCl, LiCl, and $BaCl_2$: the corresponding values of $k_{N,CA}$ are 0.088, 0.141, and 0.334. These values are ordered according to the strength of the hydration of these salts (see Table 2.4). A poorly hydrated salt, such as tetramethylammonium bromide even salts-in benzene, with $k_{N,CA} = -0.24$. Equation (2.45) may be obeyed up to electrolyte concentrations of several moles per liter. The contributions of several electrolytes are, within limits, additive.

2.6 ORGANIC SOLUTIONS

Solutions in organic solvents or in mixed aqueous–organic solvents, on the whole, behave not very differently from purely aqueous solutions. In particular, if the relative permittivity of the solvent (see Table 2.1 for values) is roughly $\varepsilon > 40$, electrolytes are, more or less, completely dissociated into ions. On the other hand, if $\varepsilon < 10$, only slight or practically no ionic dissociation takes place. Furthermore, solutions in organic or mixed aqueous–organic solvents lack the three-dimensional cooperative hydrogen bond network that characterizes aqueous solutions. Many "anomalous" properties of aqueous solutions that depend on this structured nature of water become normal properties in organic solvents.

2.6.1 Electrolyte Transfer

The presence of an organic solvent in an aqueous mixture affects the activity coefficient of an electrolyte relative to its value in water, even when the electrolyte is present at such low concentrations that, in the absence of the co-solvent, effectively $y_\pm$ would have been unity. Accordingly, in mixed aqueous–organic solvents there is a *primary medium effect* on the activity coefficient, reflecting the interactions of the ions with their mixed-solvent surroundings, compared with their interactions with a purely aqueous environment. If the content of water in the mixture is continuously reduced until the pure organic component is reached, this effect is expected to increase and reach a value characteristic of the pure solvent. This quantity can be expressed in terms of a standard molar *Gibbs energy of transfer* of the electrolyte CA from water, aq, to a new medium, org (which may also be a nonaqueous solvent that is immiscible with water): $\Delta_t G^\circ(CA,aq \rightarrow org)$ (see section 2.3); here, we use the symbols aq and org to indicate the pure liquids. Thus, when an infinitely dilute solution of the electrolyte CA in the new medium is compared with a corresponding solution in water, we have

$$\Delta_t G^\circ(CA,aq \rightarrow org) = \mu_{CA}^\infty(org) - \mu_{CA}^\infty(aq) \tag{2.46}$$

Note that the μ_{CA}^∞ are the standard molar Gibbs energies of solvation of the (combined ions of the) electrolyte [5]. Alternatively, a transfer activity coefficient can be defined as

$$\ln \gamma_t^0(CA,aq \rightarrow org) = \Delta_t G^\circ(CA,aq \rightarrow org)/RT \tag{2.47}$$

The superscript zero designates the standard state of infinite dilution, where only solute–solvent but no solute–solute interactions take place. As mentioned in section 2.3, tables of $\Delta_t G^\circ(ion,aq \rightarrow org)$ [5–7] may be used to obtain the values for ions, which can then be combined to give the value of $\Delta_t G^\circ(CA,aq \rightarrow org)$, for the desired electrolyte. For example, for the solvent org = nitrobenzene (PhNO$_2$), $\Delta_t G^\circ(Na^+,H_2O \rightarrow PhNO_2) = 36$ kJ mol^{-1}, $\Delta_t G^\circ(Cs^+, H_2O \rightarrow PhNO_2) = 18$ kJ mol^{-1}, and $\Delta_t G^\circ(I_3^-, H_2O \rightarrow PhNO_2) = -23$ kJ mol^{-1}. Therefore, $\Delta_t G^\circ = 13$ kJ mol^{-1} for NaI$_3$ and -5 kJ mol^{-1} for CsI$_3$ transferring from water to nitrobenzene, signifying the preference of the former salt for water (positive $\Delta_t G^\circ$ and of the latter salt for nitrobenzene (negative $\Delta_t G^\circ$).

When nonnegligible concentrations of the electrolyte are present in the organic solvent, ion–ion interactions superimpose on the ion–solvent ones, or *the secondary medium effect*. Although an equation similar to Eq. (2.43) may be used for determining the activity coefficient in the new medium, it is necessary to employ the appropriate value of A in this equation that depends on the relative permittivity of the medium: $A(\text{org}) = A(\text{aq})(\varepsilon_{aq}/\varepsilon_{org})^{3/2}$. Unless very water-rich mixed solvents are used, different numerical values of the parameters in the denominator and the second term on the right-hand side of Eq. (2.43) have to be employed.

2.6.2 Ion Pair Formation

A further complication that sets in when organic or mixed aqueous–organic solvents are used, which is aggravated when the relative permittivity of the medium, ε, falls below 40, is *ion pairing*. This phenomenon does occur in purely aqueous solutions, mainly with higher-valence-type electrolytes: 2:2 and higher, and with 2:1 or 1:2 electrolytes only at high concentrations. Ion pairs may also form in aqueous solutions of some 1:1 electrolytes, provided the ions are poorly hydrated and can approach each other to within <0.35 nm. Such ion pairs are of major importance in solvents that are relatively poor in water or that are nonaqueous.

Ion pairing is due to electrostatic forces between ions of opposite charges in a medium of moderate to low relative permittivities. It should be distinguished from complex formation between metal cations and anionic ligands, in which coordinative bonds (donation of an electron pair) takes place. One distinguishing feature is that, contrary to complex formation, the association is nondirectional in space. The association of a cation and an anion to form an ion pair can, however, be represented as an equilibrium reaction by analogy to complex formation with an equilibrium constant K_{ass} [3,5]. If α is the fraction of the electrolyte that is dissociating into ions and therefore $(1 - \alpha)$ is the fraction that is associated, then

$$K_{ass} = (1-\alpha)y_N / c_{CA}\alpha^2 y_{\pm}^2 \qquad (2.48)$$

where y_N is the activity coefficient of the undissociated part [treated according to Eq. (2.45) or set, with good approximation, equal to unity] and $y_{\pm}$ is the mean ionic activity coefficient of the dissociated part of the electrolyte. This quantity can be calculated from the extended Debye–Hückel Eq. (2.43), provided that the value of A appropriate for the medium having the relative permittivity ε is used and that the ionic strength I takes into account the incomplete dissociation: αc replaces c in Eq. (2.38).

According to *Bjerrum's theory*, the association constant is proportional to $a \cdot b$ times a certain function of b, $Q(b)$. The quantity a is the ionic distance parameter (i.e., the distance between the centers of the cation and the anion). The quantity b is proportional to the absolute value of the product of the charges of the cation and the anion and inversely proportional to the absolute temperature, T, the relative permittivity (dielectric constant) of the solvent, ε, and to the distance a. At 25°C

$$\log b = \log|z_+z_-| + 1.746 - \log \varepsilon - \log a(\text{in nm}) \tag{2.49}$$

The function $Q(b) = {}_2\int^b t^{-4}e^t\,dt$ (t is an auxiliary variable) mentioned earlier can be approximated by e^b/b^4 at high values of b, or be read from a figure [3]. The value of K_{ass} at 25°C is given by

$$\log K_{ass}(\text{in L mol}^{-1}) = 3\log|z_+z_-| + 6.120 - 3\log \varepsilon + \log Q(b) \tag{2.50}$$

Finally, values of α can be obtained from K_{ass} by means of Eq. (2.48), provided that $y_\pm$ is calculated iteratively [use Eqs. (2.34), (2.38), and (2.43)].

Solvent-separated ion pairs, in which the first solvation shells of both ions remain intact on pairing may be distinguished from solvent-shared ion pairs, where only one solvent molecule separates the cation and the anion, and contact ion pairs, where no solvent separates them (Fig. 2.6). The parameter a reflects the minimum distance by which the oppositely charged ions can approach each other. This equals the sum of the radii of the bare cation and anion plus 2, 1, and 0 diameters of the solvent, respectively, for the three categories of ion pairs. Since a appears in Eq. (2.49), and hence, also in $Q(b)$, it affects the value of the equilibrium constant, K_{ass}. The other important variable that affects K_{ass} is the product $T\varepsilon$ and, at a given temperature, the value of the relative permittivity, ε. The lower it is, the larger b is and, hence, also K_{ass}.

When the relative permittivity of the organic solvent or solvent mixture is $\varepsilon < 10$, then ionic dissociation can generally be entirely neglected, and potential electrolytes behave as if they were nonelectrolytes. This is most clearly demonstrated experimentally by the negligible electrical conductivity of the solution, which is about as small as that of the pure organic solvent. The interactions between solute and solvent in such solutions have been discussed in section 2.3, and the concern here is with solute–solute interactions only. These take place mainly by dipole–dipole interactions, hydrogen bonding, or adduct formation.

2.6.3   Other Kinds of Interactions

Dipole-dipole interactions between polar solute particles in organic solvents may be either repulsive ("head-head" interactions) or attractive ("head-tail" interactions). The former are rare and are due to steric hindrance around one of the poles of the dipole, which causes the ends of neighboring molecules that carry partial charges of the same sign to approach it (see Fig. 2.1b). The attractive dipole interactions lead to solute aggregation, either to dimers, where the "head" of each of the two partners is near the "tail" of the other (see Fig. 2.1a), or to larger, chainlike or cyclic aggregates (oligomers), for which each member of the chain or cycle interacts with a neighbor at both ends (see Fig. 2.1c). The

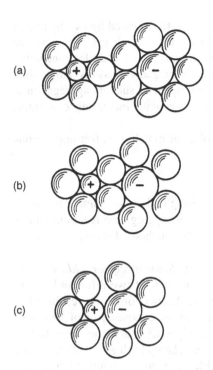

Fig. 2.6 Ion pairs. A two-dimensional representation of (a) a solvent-separated pair of ions, each still retaining its intact shell of solvating solvent molecules; (b) a solvent-sharing ion pair, which has lost some of the solvent between the partners, so that one layer of solvent shared between them separates them; (c) a contact ion pair, the cation and anion being contiguous.

dipole moment, μ, is proportional to the spatial separation of the (partial) charges that cause the particle to be dipolar and is generally considered to "reside" midway between these charges. The energy of dipole–dipole interactions depends on the product of the squares of the dipole moments, μ, of the partners (i.e., on μ^2, if there is a single kind of interacting species) and inversely on the sixth power of the distance between the centers of the dipoles. Such an energy is not very large, ranging between 1 and 10 kJ mol^{-1} (i.e., 0.4 to 4 times the mean thermal energy, $\mathbf{R}T$, that tends to disperse the aggregates). Tertiary amine salts in hydrocarbon solvents are typical examples of such aggregated solutes.

Hydrogen bonding occurring between solute particles also leads to aggregation, when the molecule of the solute contains both a hydrogen atom bonded to an electronegative atom and another electronegative atom that can accept a hydrogen bond. Typical of such solutes are carboxylic acids and acidic phosphate esters. Cyclic dimers are formed (see Fig. 2.1d), bonded by two hydrogen bridges, the energy of each being about 20 kJ mol^{-1}. Such cyclic dimers are quite stable. If the numbers of ionizable hydrogen atoms and electronegative accepting sites do not match, then noncyclic aggregates, possibly larger than dimers, are formed by hydrogen bonding. One aspect of such an aggregation is a drastic increase in the viscosity of the solution with increasing concentrations.

The mean aggregation number of a solute at molality m, forming various oligomers (i-mers, $i = 1, 2, 3, \ldots$), is

$$\tilde{n} = m / \sum_i m_i = \sum_i i m_i / \sum_i m_i \tag{2.51}$$

the molality of each oligomeric species being weighted by the number of monomers it contains. The equilibrium constants for the aggregation reactions can be obtained from the dependence of $\tilde{n}$ on m by the methods discussed elsewhere in this book (see Chapter 3).

The third kind of solute–solute interaction, *donor–acceptor adduct formation*, tends to form 1:1 adducts between the molecules of two different kinds of solute, rather than to lead to self-aggregation. Adduct formation results when one partner has a donor atom (i.e., an atom with an exposed pair of unbonded electrons), such as the nitrogen atom in trioctylamine, and the other has an acceptor atom (i.e., one with an orbital that can take up such a pair of electrons), such as antimony in antimony pentachloride or even the hydroxyl hydrogen atom in octanol. Although adduct formation formally can be said to take place in dipole–dipole interactions between two different kinds of molecules, this is rare, since the strong dipoles required naturally participate also in self-aggregation. But if one kind of molecule is a donor only and the other is an acceptor only, as occurs when octanol (the oxygen atom of which has two exposed nonbonded pairs of electrons) in the foregoing example is replaced by antimony pentachloride, then self-aggregation is precluded, and only mutual adduct formation can take place (see following).

Hydrogen bond formation between dissimilar molecules is an example of adduct formation, since the hydrogen atom that is bonded to an electronegative atom, such as oxygen or nitrogen, is a typical acceptor atom. The ability of molecules to donate a hydrogen bond is measured by their Taft–Kamlet *solvatochromic parameter*, α, (or α_m for the monomer of self-associating solutes) (see Table 2.3). This is also a measure of their acidity (in the Lewis sense, see later, or the Brønsted sense, if protic). Acetic acid, for instance, has $\alpha = 1.12$, compared with 0.61 for phenol. However, this parameter is not necessarily correlated with the acid dissociation constant in aqueous solutions. The ability of molecules to accept a hydrogen bond is measured by the Taft–Kamlet *solvatochromic parameter*, β, (or β_m for the monomer of self-associating solutes) (see Table 2.3). This, too, is a measure of their basicity (in the Lewis sense), also measured by the Gutmann donor number DN (discussed later). Thus, pyridine has $\beta = 0.64$, compared with 0.40 for acetonitrile, but again, this measure is not directly correlated to the base dissociation constant (protonation constant) in aqueous solutions. The tendency for a hydrogen bond to be formed between two dissimilar solute molecules A and B increases with the sum of the products of their donating and accepting parameters: $\alpha_A \beta_B + \alpha_B \beta_A$.

Adduct formation does not require formation of a hydrogen bond, and other acceptor atoms (or molecules) are known (e.g., transition metal cations in general, $SbCl_5$, I_2, and so on). They are collectively called *Lewis acids*, and they react with electron pair donors, that are collectively called *Lewis bases*.

The enthalpy change during donor–acceptor adduct formation has been re-lated by Drago to the sum of two terms: (1) the product of the electrostatic proper-ties of the acid and the base, E_A and E_B; and (2) the product of their tendency toward covalent bonding, C_A and C_B [10]. For the particular case, where the ac-ceptor is specified to be $SbCl_5$ (and the inert solvent is 1,2-dichloroethane), the negative of this enthalpy change (in kcal mol^{-1}, 1 cal = 4.184 J) is the Gutmann donor number, DN [2,11]. These concepts are further discussed in Chapter 3.

2.7 SOLUBILITIES IN BINARY SYSTEMS

Binary systems are systems that involve two components, that is, two substances that can be added individually. (Ions are not components; they have to be added in combinations as electroneutral electrolytes; see section 2.5.) The phase rule (see section 2.2) states that at a given temperature and pressure, when only a single liquid phase is present, there will be one degree of freedom, and we can choose the composition of this phase, made up from the two components, at will. This does not preclude, however, the phenomenon of *saturation*, where beyond a certain amount of solute in the liquid mixture a new phase appears. When two phases are present at the given temperature and pressure in the binary system, there are no longer any degrees of freedom, and the compositions of the phases are fixed. The new phase may be a solid or a second liquid. Generally, the solid phase is the pure solid solute, but in rare circumstances, it is a solid solution of the two components in each other; occasionally, it is a pure solid solvate of the original solute by the solvent. When a second liquid phase separates out, it is generally a saturated solution itself, rather than a pure liquid. We then have a solvent-rich dilute solution of the solute and a solute-rich concentrated solution of it. As an example, consider a solution of phenol in water at 25°C. At this tempera-ture, pure phenol is a solid (its melting point is 40.9°C), but when equilibrated with water at 25°C, the saturated aqueous layer contains 8.66% by mass of phenol, and the phenol-rich layer contains 28.72% by mass of water. However, on a mole fraction basis, both layers appear to be water-rich phases, the one having x_{water} = 0.982 and the other (the "phenol" phase) having x_{water} = 0.678.

The solubility of a solute in a solvent is given by the composition of the saturated solution at a given temperature and pressure. The solubility may be expressed on any of the concentration scales: the molar (mol L^{-1}; i.e., per liter of the saturated solution), the molal (mol kg^{-1}; i.e., per kilogram of the pure solvent), the mole fraction, the mass fraction (wt%; see Table 2.2), or the volume fraction scales. Only if the solute is a gas is the pressure of any significance. For liquid and solid solutes, however, the temperature is the only variable that ordinarily needs to be specified.

2.7.1 Gas Solubility

The solubility $s_{B(A)}$ of a gaseous solute B in a liquid A is given by its Bunsen coefficient $K_B(B) = [c_{B(A)}/c_B(g)]_{eq}$ [see Eq. (2.9)]. For a binary system, ignoring the vapor pressure of the solvent, the partial pressure p_B of the solute gas equals the total pressure p. For sufficiently low pressures (< 1 MPa) the gas behaves as an *ideal gas*, obeying the ideal gas law $pV = n_BRT$; hence, $c_B(g) = n_B/V = p/RT$ is proportional to its pressure. According to Eq. (2.9) the solubility of the gas in the liquid phase (A) is

$$s_{B(A)} = c_{B(A),eq} = p_B(RT)^{-1}\exp(-\Delta_{solv}G^{\circ}{}_B / RT) \tag{2.52}$$

The solubility of the gas is thus proportional to its pressure. Gas solubilities are generally very low, except when the gas interacts chemically with the solvent, e.g., ammonia in water. Combination of Eqs. (2.32b) and (2.34b) with this fact leads to the mole fraction of the saturated gas solution: $x_{B(A),eq} = (M_A/\rho_A)s_{B(A)} = V_As_{B(A)}$, where M_A, ρ_A, and V_A are the molar mass, the density, and the molar volume of the liquid A. For a given value of the solute–solvent interaction, measured by $\Delta_{solv}G^{\circ}{}_B$, the solubility of a gaseous solute, measured by the number of moles of solute accommodated by a given number of moles of solvent, is higher the larger the molar volume of the solvent.

2.7.2 Solubility of Nonelectrolytes

Solvent extraction rarely involves gases, so that other cases should now be considered. Most liquid organic solutes are completely miscible with, or at least highly soluble in, most organic solvents. The case of a liquid solute that forms a solute-rich liquid phase that contains an appreciable concentration of the solvent is related to the mutual solubility of two solvents, and has been discussed in section 2.2. This leaves solid solutes that are in equilibrium with their saturated solution. It is expedient to discuss organic, nonelectrolytic solutes separately from salts or other ionic solutes.

Solid organic solutes often have a limited solubility, not only in water, but also in organic solvents. Their solubility can be handled by assuming that work must first be invested to convert the solute into a supercooled liquid at the temperature of interest (say, 25°C, $T = 298.15$ K). As a good approximation, the product of the heat capacity change on fusion, $\Delta_F C_p$, and the difference between the temperature of fusion, T^F, and T is a measure of this work (the Gibbs energy of fusion at T). This virtual supercooled liquid would have the same solubility as an ordinary liquid of similar structure.

The particular case of the solubilities of organic solutes in water can be dealt with by rather simple equations, based on a general equation for solvent-dependent properties, applied to solubilities, distribution ratios, rate constants, chromatographic retention indices, spectroscopic quantities, or heats of association [4] [see Eq. (2.12) for an example of its application]. For the molar solubilities of (liquid) aliphatic solutes B in water at 25°C the equation

$$\log s_{B(water)} = \log[c_B(g) / K_B(B)] = 0.69 - 3.44(V_B / 100) +$$
$$0.43\,\pi_B^* + 5.15\beta_{mB} \qquad (2.53)$$

has been found to hold well for 115 solutes (the first equality holding for gaseous solutes). Here, V_B is the molar volume of the solute (in cm^3 mol^{-1}), π_B^* is its polarity–polarizability index, and β_B is its hydrogen bond acceptance ability (the subscript m denoting the β value for the monomer, if the solute B is capable of self-association). Tables 2.1 and 2.3 present values of these parameters for some representative substances. For example, the solubility of butyl acetate in water is 0.0437 mol L^{-1} (i.e., log $s_B = -1.36$); Eq. (2.53) gives $0.69 - 3.44 \times 1.325 + 0.43 \times 0.46 + 5.15 \times 0.45 = -1.35$ or 0.0444 mol L^{-1} for the solubility. The first equality on the right-hand side of the equation holds for volatile solutes, for which $c_B(g) = p_B/2450$, p_B being the saturated vapor pressure of the neat solute in kPa. The qualitative meaning of Eq. (2.53) is that the solubility decreases with increasing solute molar volumes (the coefficient of V_B is negative) but increases somewhat with increasing polarity-polarizability (π^*) and in particular with basicity (ability to accept a hydrogen bond, β_m).

For the solubility of aromatic solutes in water at 25°C, a different equation holds:

$$\log s_B = 0.57 - 5.58(V_{B,vdW} / 100) + 3.85\beta_B - 0.0100(T^F - 298.15) \quad (2.54)$$

has been found to hold for 70 solid and liquid solutes. The last term in Eq. (2.54) is to be used only for solutes that are solid at 298.15 K. Note that Eq. (2.54) involves the van der Waals volume of the solute, $V_{B,vdW}$, rather than its molar volume, V_B, since the former is more readily estimated than the latter for solid solutes. Note also that the term in π_B^* is missing from Eq. (2.54), which is fortunate, since it is difficult to estimate such values for solids. Again, using the β_B parameter from Table 2.3 and the van der Waals volume of p-xylene (67.1 cm^3 mol^{-1}), Eq. (2.54) predicts log $s_B = 0.57 - 5.58 \times 0.671 + 3.85 \times 0.12 = -2.71$, or a solubility of 0.00194 mol L^{-1}, compared with the experimental solubility of 0.00186 mol L^{-1}. For the solid solute p-dinitrobenzene the parameters are $V_{B,vdW} = 77.1$ cm^3 mol^{-1}, $\beta_B = 0.55$, $T^F = 455$ K; hence, the calculated value of log $s_B = -3.34$, compared with the experimental value -3.33.

When polychlorinated biphenyls and polycyclic aromatic hydrocarbons are included with other aromatic solutes, the slightly modified equation

$$\log s_B = 0.27 - 5.29(V_{B,vdW} / 100) + 3.93\beta_B - 0.0099(T^F - 298.15) \quad (2.55)$$

holds for 139 aromatic solutes. The differences noted between Eq. (2.54) [or Eq. (2.55)] for aromatic solutes and Eq. (2.53) for aliphatic ones are significant, but have not as yet been explained. However, the qualitative trends are the same.

2.7.3 Solubility of Salts

As is well known, water is the best medium for the solubility of salts. This is due to several factors. The high relative permittivity permits complete ionic dissociation when the ionic charges are not too high. The resulting ions are effectively solvated, the cations due to the fairly high electron pair donation ability of water molecules, and the anions due to the very high hydrogen bond donation ability of water molecules. Furthermore, due to the small size of these molecules a relatively large number of these can thus solvate the ions. As a consequence, the attractive electrostatic forces between the ions in the crystal lattice of a salt are compensated by the solvation forces of the dissociated ions, and this permits reasonable solubilities. Nonaqueous solvents generally have many of these properties of water to a lesser degree (have lower relative permittivities and anion solvating abilities and are larger, hence fewer, around the ions), and, therefore, commonly show lower salt solubilities.

The solubility of an ionic solute, s_{CA}, may be expressed in terms of its *solubility product*, K_{sp}. The equilibrium between a pure solid salt, $C_{v+}A_{v-}$ and its saturated solution in a solvent where it is completely dissociated to ions (generally having $\varepsilon > 40$; see section 2.6) is governed by its standard molar Gibbs energy of dissolution

$$-\Delta_{soln} G° = \mathbf{R}T \ln K_{sp} = \mathbf{R}T \ln (a_C^{v+} a_A^{v-}) \tag{2.56}$$

For a sparingly soluble electrolyte CA, for which the ionic strength produced in the saturated solution is very low (e.g., $<10^{-4}$), the approximations $a_C \approx v_+ s_{CA}$ and $a_A \approx v_- s_{CA}$ can be made. With the notation $v = v_+ + v_-$, this approximation leads to

$$s_{CA} = [K_{sp} /(v_+^{v^+} v_-^{v^-})]^{1/v} \tag{2.57a}$$

or

$$K_{sp} = (v_+^{v^+} v_-^{v^-}) s_{CA}^v \tag{2.57b}$$

that relate the solubility of the salt, s_{CA}, to its solubility product, K_{sp}.

Equation (2.57) presumes not only that the solution is very dilute, but also that the solute ions do not undergo side reactions, such as hydrolysis or self-complexation in the saturated solution. A striking case for which, in spite of the low solubility, these conditions are not obeyed is mercury(I) chloride (calomel, Hg_2Cl_2) in water. Its solubility product at 25°C is $K_{sp} = 1.43 \times 10^{-18}$ mol^3 L^{-3}; its solubility is $s_{CA} = 8.4 \times 10^{-6}$ mol L^{-1}, which is not equal to $(K_{sp}/4)^{1/3}$. This is due to the hydrolysis of the Hg_2^{2+} ion and its disproportionation to form dissolved

Hg^0 and Hg(II) (this is strongly complexed by the Cl^- ions present). The precaution of taking into account any side reactions must be followed when it is desired to deduce the actual molar solubilities from solubility products; in particular, in higher-charge-type salts.

Solubility products can be derived indirectly from standard electrode potentials and other thermochemical data, and directly from tabulated standard Gibbs energies of formation, $\Delta_f G°$, of the ions in aqueous solution [12]. Thus, the use of

$$\Delta_{soln} G° = v_+ \Delta_f G°(C^{z+}, aq) + v_- \Delta_f G°(A^{z-}, aq) - \Delta_f G°(C_{v+} A_{v-}) \qquad (2.58)$$

and Eq. (2.56) gives the corresponding K_{sp} value in aqueous solutions.

When solubility products in nonaqueous solvents are desired, tables [5–7] of the Gibbs energies of transfer of the ions from water to the desired solvent, org, must be consulted. For any ion

$$\Delta_f G°(ion, org) = \Delta_f G°(ion, aq) + \Delta_t G°(ion, aq \rightarrow org) \qquad (2.59)$$

In other words, values of $\Delta_t G°(ion,aq \rightarrow org)$, weighted by the stoichiometric coefficients for both cation and anion, must be added to $\Delta_{soln} G°(aq)$ from Eq. (2.58) to obtain $\Delta_{soln} G°(org)$, which in turn gives K_{sp} and s_{CA} in the solvent org from Eqs. (2.56) and (2.58) [12].

The lower the relative permittivity of the solvent org, the greater is the deviation of the activity coefficient of the electrolyte from unity at a given low ionic strength, and the smaller must K_{sp} be for the approximation [Eq. (2.57)] to remain valid. On the other hand, the values of $\Delta_t G°(ion,aq \rightarrow org)$ are positive for the halide and many other anions, and are positive or negative, but smaller absolutely than the former, for univalent cations. This means that for 1:1 electrolytes, the value of $\Delta_{soln} G°(org)$ is more positive than that of $\Delta_{soln} G°(aq)$ and K_{sp} in the organic solvent org is smaller than that in water. This trend is generally enhanced with decreasing values of ε of the solvents. Most alkali metal halides are much less soluble in common organic solvents than in water, although there are exceptions (some lithium salts and some iodides).

Divalent or higher-valent cations and, in particular, transition metal cations, are likely to be covalently solvated by solvents that are strong electron pair donors (have large solvatochromic β values). This solvation often persists in crystals, so that the salt that is in equilibrium with the saturated solution in such solvents may not be the anhydrous salt (nor the salt hydrate). Equation (2.56) omits any consideration of the solvent of crystallization and pertains to the solventless (anhydrous) salt. For a salt hydrated by n water molecules in the crystal, the activity of water raised to the nth power must multiply the right-hand side of Eq. (2.56) for it to remain valid. A similar consideration applies for salts crystallizing with other kinds of solvent molecules, the activity of the solvent in the saturated solution replacing that of water. Such situations must be avoided when Eq. (2.59) is to be applied for the calculation of solubilities. For instance, the solubility of lead iodide, PbI_2, in water is only 1.65×10^{-3} mol L^{-1}, but those of the solvates $PbI_2 \cdot DMF$ and $PbI_2 \cdot 2DMSO$ are 0.71 and 0.99 mol L^{-1} in N,N-dimethylformamide (DMF) and dimethylsulfoxide (DMSO), respectively.

2.7.4 Heats of Solution

When a solute is dissolved in a solvent, heat may either be evolved (as with sulfuric acid in water, where strong heating is observed) or absorbed (as with ammonium nitrate in water, where strong cooling is observed). Thus the dissolution may be either exothermic or endothermic, and the *standard molar enthalpy of solution*, $\Delta_{soln}H°$, is then either negative or positive. The sign of $\Delta_{soln}H°$ for a solid solute depends on the balance between its *lattice enthalpy* (or the cohesive enthalpy of a liquid solute) and its enthalpy of solvation or that of the ions produced by it.

Since, for a nonionic solute B, we have $-RT \ln s_B = \Delta_{soln}G°$, it follows that the temperature coefficient of its solubility is

$$ds_B / dT = s_B(\Delta_{soln}H_B° / RT^2) \tag{2.60}$$

For an ionic solute dissociating into v ions, the temperature coefficient is $1/v$ times the right-hand side of Eq. (2.60). Again, it is assumed that the solubility is sufficiently low for the mean ionic activity coefficient to be effectively equal to unity and independent of the temperature. When this premise is not met, then corrections for the heat of dilution from the value of the solubility to infinite dilution must be added to $\Delta_{soln}H°_B$ in Eq. (2.60).

The qualitative trend predicted by this equation is that, when the heat of solution is negative (the dissolution is exothermic, i.e., heat is evolved, the enthalpy of solvation is more negative than the lattice enthalpy is positive), the solubility diminishes with increasing temperatures. The opposite trend is observed for endothermic dissolution. An analogue of Eq. (2.58), with H replacing G, and the same tables [12] can be used to obtain the required standard enthalpies of solution of ionic solutes. No general analogues to Eqs. (2.53)–(2.55) are known as yet.

2.8 SOLUBILITIES IN TERNARY SYSTEMS

2.8.1 Ternary Phase Diagrams

Ternary systems of interest to solvent extraction are generally two-phase systems, but occasionally ternary systems that consist of a single phase must be considered. According to the phase rule (see section 2.2), a single-phase three-component system at a given temperature and pressure has two degrees of freedom, so that the mole fractions of two of the components, say, x_A and x_B, can be selected at will, that of the third being given by difference: $x_C = 1 - x_A - x_B$.

It is customary to depict such a system at a given temperature on a triangular phase diagram, where a composition is indicated by a point in an equilateral triangle (Fig. 2.7a). The sums of the distances of such a point from the sides of the triangle are independent of the position of the point (and equal to the height of the triangle). These distances represent the corresponding mole fractions. An apex of the triangle represents a pure component; a side represents a binary mixture of the components at its two ends. The addition (removal) of a component from a given composition (point in the triangle) is represented by movement along the straight line connecting the point with the apex for this component toward this apex (or away from it). The two degrees of freedom available in this system can be taken as freedom to choose a composition from the entire two-dimensional area of the triangle.

Suppose now that, as the temperature is changed, the system at certain compositions splits into two (liquid) phases. There are then composition ranges where the system remains a single homogeneous (liquid) phase, but also another region where two phases coexist. A line on the triangular diagram encloses such a region of nominal compositions of the system. When a composition within this region is chosen, two phases result, the compositions of which are located on the enclosing line. The one remaining degree of freedom can be taken as the freedom to choose a point on that line, but the composition of the other phase in equilibrium with it, also on the enclosing line, is now fixed. A straight *tie-line* connects the points on the enclosing line that correspond to phases in mutual equilibrium. The amounts of the two phases at equilibrium for a nominal composition that is situated on any tie-line (i.e., within the two-phase region) is at the inverse ratio of the distances of the point for this composition from the ends of the tie-line (the *lever rule;* see Fig. 2.7b).

Suppose now that the temperature is changed further. The enclosed two-phase region may grow and change shape. It may be bounded on one or two sides (rarely all three sides) by the sides of the triangle. If this is the case, the corresponding binary mixtures show miscibility gaps. As long as only two phases coexist, there is still one degree of freedom. If, however, a further phase forms, for example, by one component crystallizing (on cooling below its freezing temperature), then the system becomes invariant, and its composition is fixed uniquely by the fact that there are now three phases in equilibrium at the temperature of interest.

In ternary systems of relevance to solvent extraction these may be two (partly) immiscible solvents and one (solid) solute, or two (solid) solutes and a single solvent. The latter system may constitute a partial system of quaternary or higher mixtures that involve two liquid phases, which are solvent extraction systems. In principle, however, a system of two solid solutes and a liquid solvent could split into two liquid phases, one rich in the one solute, the second in the other. In general, when one solute crystallizes out from a ternary (two solutes

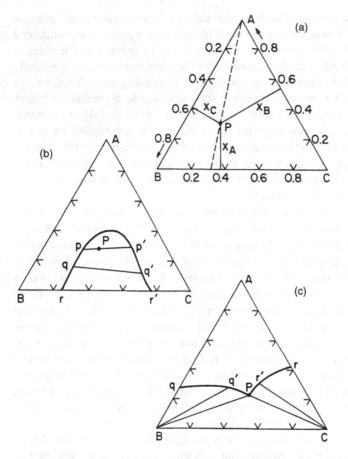

Fig. 2.7 Ternary phase diagrams with components A, B, and C. (a) A homogeneous system, the composition at point P being given by the lengths of the lines marked with Xs; the dashed line shows the changes occurring when solvent A is added (upward) or removed (downward). (b) A heterogeneous system, splitting into two liquid phases, with compositions marked with r, q, p, p′, q′, and r′, along the phase boundary; r and r′ denote the BC miscibility gap; at nominal composition P there are Pp′ moles of a phase of composition p for every pP moles of one of composition p′, at the two ends of the tie-line. (c) A heterogeneous system with two solid solutes, B and C, and solvent A; r and q denote the solubilities in the binaries AC and AB; at composition r′ the solution contains both B and C and is saturated with solid C; at the invariant point P the solution is saturated with both B and C.

plus solvent) system, the composition changes along a saturation line, the saturated solution becoming poorer in this solute. The other solute has its own saturation line. The point at which the two saturation lines meet is, again, an invariant point of the system, since now there are two solids in equilibrium with a saturated solution (i.e., three phases in all). The composition of the solution that is saturated relative to the two solid solutes is thus given uniquely by the fixed temperature (see Fig. 2.7c).

General solvent extraction practice involves only systems that are unsaturated relative to the solute(s). In such a ternary system, there would be two almost immiscible liquid phases (one that is generally aqueous) and a solute at a relatively low concentration that is distributed between them. The single degree of freedom available in such instances (at a given temperature) can be construed as the free choice of the concentration of the solute in one of the phases, provided it is below the saturation value (i.e., its solubility in that phase). Its concentration in the other phase is fixed by the equilibrium condition. The question arises of whether or not its distribution between the two liquid phases can be predicted.

Given a nonionic solute that has a relatively low solubility in each of the two liquids, and given equations that permit estimates of its solubility in each liquid to be made, the distribution ratio would be approximately the ratio of these solubilities. The approximation arises from several sources. One is that, in the ternary (solvent extraction) system, the two liquid phases are not the pure liquid solvents where the solubilities have been measured or estimated, but rather, their mutually saturated solutions. The lower the mutual solubility of the two solvents, the better can the approximation be made. Even at low concentrations, however, the solute may not obey Henry's law in one or both of the solvents (i.e., not form a dilute ideal solution with it). It may, for instance, dimerize or form a regular solution with an appreciable value of $b(T)$ (see section 2.2). Such complications become negligible at very low concentrations, but not necessarily in the saturated solutions.

2.8.2 The Fragment Hydrophobicity Model*

Several models have been suggested for the estimation of the distribution ratios of nonionic solutes between water and (practically) immiscible organic solvents. One model takes 1-octanol to represent, in general, lipophilic ("fat-liking") media, which hydrophobic ("water-fearing") solutes would prefer over water. Such media may be oils, biological lipid membranes, and, somewhat less suitably, hydrocarbon solvents.

---

*See also Hansen solubility parameters, section 4.13.3.

The logarithm of the *distribution constant*, log P, (i.e., of the distribution ratios D extrapolated to infinite dilution) in such 1-octanol/water systems was shown by Hansch and Leo to be additive relative to the contributions of a number N_i of certain fragments of the solute molecules and a number N_j of certain structure factors. Each hydrocarbon fragment contributes some hydrophobicity (its tendency to avoid water) or lipophilicity (tendency to prefer nonpolar organic solvents) to the solute, and adds a positive quantity f_i to log P; polar fragments provide negative f_i values, and certain structural features, F_j, may change the hydrophobicity (e.g., by the shielding of fragments at branching points from water).

$$\log P(\text{1-octanol/water}) = \sum N_i f_i + \sum N_j F_j \qquad (2.61)$$

Table 2.5 lists the additive values for representative fragments f and structural factors F. As an example, the infinite dilution distribution ratio of C_2H_5 $C(O)OC_2H_5$, ethyl propionate, between 1-octanol and water is obtained as follows. The carboxylate group contributes -1.49 to log P, each ethyl group contributes $0.89 + 0.66 - 0.12 = 1.43$ (for the methyl and methylene groups and

Table 2.5 Selected Fragment Factors f and Structure Factors F for the Calculation of log P(1-octanol/water)

Fragment	f	Fragment	f
>C<	0.20	—N<	−2.18
>CH—	0.43	—NH—	−2.15
—CH$_2$—	0.66	—NH$_2$	−1.54
—CH$_3$	0.89	—OH	−1.64
—H	0.23	—SH	−0.23
—F	−0.38	—C(O)H	−1.10
—Cl	0.06	—C(O)OH	−1.11
—Br	0.20	—C(O)N(H)—	−2.71
—I	0.59	—C(O)NH$_2$	−2.18
—NO$_2$	−1.16		
—O—	−1.82		
—S(O)—	−3.01	**Function**	F
>SO$_2$	−2.67		
—OP(O)O$_2$<	−2.29	double bond	−0.55
—S—	−0.79	triple bond	−1.42
—CN	1.27	n-Membered chain	−0.12($n-1$)
—C(O)N<	−3.04	Member of ring	−0.09 n
—C(O)—	−1.90	Multiple (n) halogens:	
—C(O)O—	−1.49	on same carbon	0.17 n
—CH— (aromatic)	0.355	on adjacent carbons	0.28($n-1$)
—C< (aromatic ring fusion point)	0.225	Fragmented attached to aromatic ring	−1.0

the chain structural feature, respectively), and altogether the calculated log P is $-1.49 + 2 \times 1.43 - 0.12 = 1.25$ or the distribution ratio is 17.8. The experimental value of log P is 1.21 and of the distribution ratio 16.2, and it is seen to be estimated with sufficient accuracy.

The entries in Table 2.5 pertain to the distribution between water and 1-octanol; for any other organic solvent other solute f and F values should be used. However, 1-octanol is a fair representative of organic solvents in general, and the effect of using some other solvent can be dealt with by the use of a "solvent effect" correction, which is much smaller than the solute dependency given by the log P values. An approximation to such solvent effect corrections may be made by using the solubility parameters, accounting in effect for the differing amounts of work done against the cohesive energies of the solvents. Thus, for the distributions of a solute B (also an electrolyte CA) between water and two solvents A_1 and A_2 (where 1 might be 1-octanol):

$$\log (D_{B(A_1)} / D_{B(A_2)}) = V_B(\delta_{A1}^2 - \delta_{A2}^2)/RT \qquad (2.62)$$

If $D_{B(A1)}$ can be equated with P calculated from the entries in Table 2.5, then $D_{B(A2)}$ in any other solvent A_2 can be estimated from Eq. (2.62). Equation (2.62) is actually a combination of four expressions of the form of Eq. (2.8) (see section 2.2.2), two for water and solvent A_1 and two for water and solvent A_2, presuming them to be immiscible pairs of liquids. It employs concentrations on the mole fraction scale, and assumes that the systems behave as regular solutions (which they hardly do). This eliminates the use of the solubility parameter δ of water, which is a troublesome quantity (see Table 2.1). Solvent A_1 need not, of course, be 1-octanol for Eq. (2.62) to be employed, and it suggests the general trends encountered if different solvents are used in solvent extraction.

As an illustration of the application of Eq. (2.62), consider the distribution of acetylacetone, having a molar volume of 103 cm^3 mol^{-1}, between various solvents and water, compared with the distribution between 1-octanol and water, for which log $P = 1.04$. The values of log D calculated from Eq. (2.62) at infinite dilution of the acetylacetone are 1.29, 0.80, 0.69, 0.51, and 0.14 for 1-pentanol, chloroform, benzene, carbon tetrachloride, and n-hexane, respectively. The experimental values of log D are 1.13, 0.77, 0.76, 0.50, and −0.05 for these solvents. The order is the same, and even the numerical values are not very different, considering that solute–solvent interactions other than the work required for cavity formation have not been taken into account.

2.8.3 The Distribution of Nonelectrolytes

A general expression for the distribution of nonelectrolytes takes into account all the differences in the solute–solvent interactions in the aqueous and the organic phases, provided the solutions are sufficiently dilute for solute–solute interactions to be negligible. In such cases, the energetics of the relevant interactions consist of the work required for the creation of a cavity to accommodate the solute, and the Gibbs energy for the dispersion force interactions and for the donation and acceptance of hydrogen bonds. For a given solute distributing between water (W) and a series of water-immiscible solvents:

$$\log D = A_V V_{\text{solute}} \Delta \delta^2 + A_\pi \pi^*_{\text{solute}} \Delta \pi^* + A_\alpha \alpha_{\text{solute}} \Delta \beta + A_\beta \beta_{\text{solute}} \Delta \alpha \qquad (2.63)$$

where the difference factors are $\Delta \delta^2 = \delta_{\text{solvent}}^2 - \delta_W^2$, $\Delta \pi^* = \pi^*_{\text{solvent}} - \pi^*_W$, $\Delta \alpha = \alpha_{\text{solvent}} - \alpha_W$, and $\Delta \beta = \beta_{\text{solvent}} - \beta_W$, i.e., the differences between the relevant properties of the organic solvent and those of water. The first term on the right-hand side is related to Eqs. (2.10) and (2.62) and the next three terms to Eq. (2.12). Interestingly, it was established [13] that the coefficients A_V, A_π, A_α, and A_β are independent of the solutes and solvents employed—i.e., are "universal"—provided that the water content of the organic solvent at equilibrium is approximately $x_W < 0.13$. Such water-saturated solvents can be considered to be still "dry" and to have properties the same as those of the neat solvents. The values of the coefficients are $A_V = 2.14$ (for δ^2 in J cm^{-3} and $V_{\text{solute}} = 0.01 V_X$, where V_X is the additive McGowan–Abraham intrinsic volume [2]), $A_\pi \approx 0$, $A_\alpha = -7.67$, and $A_\beta = -4.62$. The substantial absence of a term with the polarity/polarizability values π^*_{solute} and π^*_{solvent} is explained by the fact that the cavity formation term in V_{solute} and $\Delta \delta^2$ effectively takes care also of the polar interactions. For "wet" solvents, i.e., those with $x_W > 0.13$ when saturated with water, such as aliphatic alcohols from butanol to octanol and tri-n-butyl phosphate, Eq. (2.63) is still valid with practically the same A_V, A_π, and A_β, but with $A_\alpha = 1.04$, provided that δ^2, α, and β are taken as those measured (or calculated) for the water-saturated solvents.

> The applicability of Eq. (2.63) was tested for some 180 individual solutes (with up to 10 carbon atoms) and 25 "dry" solvents by stepwise multivariable linear regression and for 28 solutes and 9 dry solvents by target factor analysis; essentially the same conclusions and universal coefficients were obtained by both methods. As an example of the application of Eq. (2.63), the distribution of succinic acid between water and chloroform (a "dry" solvent with $x_W = 0.0048$) and tri-n-butyl phosphate (a "wet" solvent with $x_W = 0.497$) may be cited [13]. For the former solvent $\log D = -1.92$ (experimental) vs. -1.98 (calculated) and for the latter $\log D = 0.76$ (experimental) vs. 0.88 (calculated). Obviously, succinic acid with two carboxylic groups that strongly donate hydrogen bonds (assigned $\alpha = 1.12$ as for acetic acid) prefers the basic (in the Lewis basicity, hydrogen-bond-accepting sense) tri-n-butyl phosphate ($\beta = 0.82$, measured for the "wet" solvent) over water ($\beta = 0.47$) and naturally also chloroform ($\beta = 0.10$).

2.8.4 The Distribution of Electrolytes

The distribution ratios of electrolytic solutes, CA, that are practically completely dissociated into ions in both water and the solvent org, can be approximated by

$$D_{CA} = (1/v) \exp[\Delta_t G°(CA,aq \to org)/RT] \qquad (2.64)$$

This expression arises from Eq. (2.40), noting that the chemical potential μ_{CA} must be the same in the phases aq and org for equilibrium to be maintained, and that $\Delta_t G°(CA,aq \to org) = \mu^\infty_{CA,org} - \mu^\infty_{CA,aq}$ by definition. The approximation is that the activity coefficients of CA are taken to be equal (near unity) at the low concentrations where Eq. (2.64) is valid.

This equation would be exact if the two phases org and aq were completely immiscible.

The standard molar Gibbs energy of transfer of CA is the sum $v_+\Delta_t G°(C)$ + $v_-\Delta_t G°(A)$, where the charges of the cation C^{z+} and anion A^{z-} and the designation of the direction of transfer, (aq $\to$ org), have been omitted. The values for the cation and anion may be obtained from tables [5–7], which generally deal with solvents org that are miscible with water and not with those used in solvent extraction. However, $\Delta_t G°(C)$ depends primarily on the β solvatochromic parameter of the solvent and $\Delta_t G°(A)$ on its α parameter, and these can be estimated from family relationships also for the latter kind of solvents.

The general qualitative rules are that ions prefer water if they are multiply charged and small, but otherwise they may prefer solvents that have large β values (donor properties for cations) or α values (acceptor properties for anions). Consider, for example, the small cation Li^+ and multiply charged cation Cu^{2+}: these cations prefer water ($\beta = 0.47$) over such solvents as methyl ethyl ketone ($\beta = 0.48$), propylene carbonate ($\beta = 0.40$), benzonitrile ($\beta = 0.41$), nitrobenzene ($\beta = 0.30$), and 1,2-dichloroethane ($\beta = 0.00$), and also 1-butanol (see later), but prefer the more highly basic (although water-miscible) solvents dimethylsulfoxide ($\beta = 0.76$), N,N-dimethylformamide ($\beta = 0.69$), and hexamethyl phosphoric triamide ($\beta = 1.06$) over water. [Note that 1-butanol ($\beta = 0.88$) is exceptional, being capable of hydrogen bond donation.] However, with propylene carbonate, the preference is reversed for the large univalent cations Cs^+ and tetramethylammonium, $(CH_3)_4N^+$. For the tetraalkylammonium ions with n-propyl or longer alkyl chains, the preference for the organic solvents is quite general. For the anions, water is preferred over most solvents, an exception being 2,2,2-trifluoroethanol, with $\alpha = 1.51$, larger than that of water (1.17). However, very large anions may show a preference for the organic solvents, as for ClO_4^- with propylene carbonate and hexamethyl phosphoric triamide and for I_3^- with most organic solvents. A numerical example is given at the end of section 2.6.1, where the transfer of Na^+, Cs^+, and I_3^- from water to nitrobenzene is discussed. The negative value of $\Delta_t G°$ for CsI_3 is related to the ability of this salt to be extracted into nitrobenzene from water, contrary to NaI_3, which cannot (having a positive $\Delta_t G°$).

A common situation is that the electrolyte is completely dissociated in the aqueous phase and incompletely, or hardly at all, in the organic phase of a ternary solvent extraction system (cf. Chapter 3), since solvents that are practically immiscible with water tend to have low values for their relative permittivities ε. At low solute concentrations, at which nearly ideal mixing is to be expected for the completely dissociated ions in the aqueous phase and the undissociated electrolyte in the organic phase (i.e., the activity coefficients in each phase are approximately unity), the distribution constant is given by

$$\log P_{CA} = \log c_{CA}(\text{org}) - \nu \log c_{CA}(\text{aq}) \tag{2.65}$$

The *distribution ratio* for the electrolyte, D_{CA}, however, is given by

$$\log D_{CA} = \log[c_{CA}(\text{org})/c_{CA}(\text{aq})] = \log P_{CA} - (\nu-1)\log c_{CA}(\text{aq}) \tag{2.66}$$

which is seen to be a concentration-dependent quantity. Even if the constant $\log P_{CA}$ can be predicted, $\log D_{CA}$ cannot, unless some arbitrary choice is made for the aqueous concentration.

As examples, take the extraction of nitric acid by tri-n-butyl phosphate or of hydrochloric acid by diisopropyl ether. For the former, at low acid concentrations, the acid is completely dissociated in the aqueous phase, but is associated in the organic phase. Therefore, D_{HNO_3} increases with the nitric acid concentration according to Eq. (2.65), up to such concentrations at which the dissociation of this acid in the aqueous phases becomes significantly lower than complete, and then D_{HNO_3} flattens out. For hydrochloric acid, the situation is the opposite: it is dissociated in the ether phase up to about 3 M aqueous acid, so that D_{HCl} is constant, but at higher concentrations it associates in the organic phase, and D_{HCl} starts to increase, again according to Eq. (2.65), when the nonidealities in both phases are taken into account.

2.9 CONCLUSION

This chapter provides the groundwork of solution chemistry that is relevant to solvent extraction. Some of the concepts are rather elementary, but are necessary for the comprehension of the rather complicated relationships encountered when the solubilities of organic solutes or electrolytes in water or in nonaqueous solvents are considered. They are also relevant in the context of complex and adduct formation in aqueous solutions, dealt with in Chapter 3 and of the distribution of solutes of diverse kinds between aqueous and immiscible organic phases dealt with in Chapter 4.

For this purpose it is necessary to become acquainted, at least in a cursory fashion, with the physical and chemical properties of liquids (section 2.1.1) and the forces operating between their molecules (section 2.1.2). Since solvent extraction depends on the existence of two immiscible liquid phases, solvent miscibility, i.e., their mutual solubility, is an important issue (section 2.2). The solvation of a solute that is introduced into a solvent, i.e., its interactions with the solvent, is described (section 2.3.1) and the thermodynamics thereof are elaborated (section 2.3.2), the case of electrolytes that dissociate into ions being given special attention (section 2.3.3).

Solution chemistry depends strongly on thermodynamic relationships that have to be mastered in order to make full use of most other knowledge concerning solvent extraction. Therefore, a comprehensive section (2.4) is devoted to this subject, dealing with ideal (section 2.4.1) as well as nonideal (section 2.4.2) mixtures and solutions. Then again, as solvent extraction in, e.g., hydrometallurgy, deals with electrolytes and ions in aqueous solutions, the relevant thermodynamics of single electrolytes (section 2.5.1) and their mixtures (section 2.5.2) has to be understood. In the organic phase of solvent extraction systems, complications, such as ion pairing, set in that are also described (section 2.6.2).

The amount of a solute that can be introduced into a solvent depends on its solubility, be it a gas (section 2.7.1), a solid nonelectrolyte (section 2.7.2), or an electrolyte (section 2.7.3). Ternary systems, which are the basic form of solvent extraction systems (a solute and two immiscible solvents), have their own characteristic solubility relationships (section 2.8.1).

In conclusion, therefore, it should be perceived that this groundwork of solution chemistry ultimately leads to the ability to predict at least semi-quantitatively the solubility and two-phase distribution in terms of some simple properties of the solute and the solvents involved. For this purpose, Eqs. (2.12), (2.53), (2.61), and (2.63) and the entries in Tables 2.3 and 2.5 should be particularly useful. In the case of inorganic ions, the entries in Table 2.4 are a rough guide to the relative extractabilities of the ions from aqueous solutions. The more negative the values of $\Delta_{hyd}G°$ of the ions, the more difficult their removal from water becomes, unless complexation (see Chapter 3) compensates for the ion hydration.

Although theories of solution (this chapter) and formation of extractable complexes (see Chapters 3 and 4) now are well advanced, predictions of distribution ratios are mainly done by comparison with known similar systems. Solvatochromic parameters, solubility parameters, and donor numbers, as discussed in Chapters 2–4, are so far mainly empirical factors. Continuous efforts are made to predict such numbers, often resulting in good values for systems within limited ranges of conditions. It is likely that these efforts will successively encompass greater ranges of conditions for more systems, but much still has to be done.

Because a 10% change in the distribution ratio, which can be measured easily and accurately in the distribution range 0.01–100, corresponds to an energy change of only 0.2 kJ/mole, distribution ratio measurements offer a method to investigate low energy reactions in solutions, such as weak solute–solute and solute–solvent interactions in the organic phase. So far, this technique has been only slightly exploited for this purpose. It is particularly noteworthy to find how few thermodynamic studies have been made of the distribution or extraction constants, as enthalpy and entropy values give a good indication of the driving force of the extraction and indicate the structure of the molecular species in the organic solvent.

REFERENCES

1. Riddick, J. A.; Bunger, W. B.; Sasano, T. K. *Organic Solvents*, Fourth Edition, John Wiley and Sons, New York, 1986.
2. Marcus, Y. *The Properties of Solvents*, John Wiley and Sons, Chichester, 1998.
3. Marcus, Y. *Introduction to Liquid State Chemistry*, John Wiley and Sons, Chichester, 1977; 204–207.
4. Kamlet, J. M.; Abboud, J.-L.; Abraham, M. H.; Taft, R. W. *J. Org. Chem.*, 1983, 48: 2877.
5. Marcus, Y. *Ion Solvation*, John Wiley and Sons, Chichester, 1985.
6. Marcus, Y. *Ion Properties*, M. Dekker, New York, 1997.
7. Marcus, Y. *Pure Appl. Chem.*, 1983, **55**:977.
8. Robinson, R. A.; Stokes, R. H. *Electrolyte Solutions*, Second Revised Ed., Butterworth, London, 1970.
9. Harned, H. S.; Robinson, R. A. *Multicomponent Electrolyte Solutions*, Pergamon Press, Oxford, 1968.
10. Drago, R. S. *Struct. Bonding*, 1973; 15:73.
11. Marcus, Y. *J. Solution Chem.*, 1984; 13:599.
12. Wagman, D. D.; et al., *NBS Tables of Chemical Thermodynamic Properties*, *J. Phys. Chem. Ref. Data*, 1982, Vol. 11, Suppl. 2.
13. Marcus, Y. *J. Phys. Chem.*, 1991;95:8886; Marcus, Y. *Solvent Extract. Ion Exch.*, 1992, 10: 527; Migron, Y.; Marcus, Y.; Dodu, M. *Chemomet. Intell. Lab. Syst.*, 1994, 22: 191.

3
Complexation of Metal Ions

GREGORY R. CHOPPIN* Florida State University, Tallahassee, Florida, U.S.A.

3.1 METAL ION COMPLEXATION

Chapter 2 discussed the various forms of interaction between solute and the solvent molecules (see section 2.3), which leads to a certain solubility of the solute in the solvent phase. It was also described how the ratio of the solubility of the solute between two immiscible solvents could be used to estimate distribution ratios (or constants) for the solute in the particular system (see section 2.4). It was also pointed out that in the case of aqueous solute electrolytes, specific consideration had to be applied to the activity of the solute in the aqueous phase, a consideration that also was extended to solutes in organic solvents.

However, in many solvent extraction systems, one of the two liquids between which the solute distributes is an aqueous solution that contains one or more electrolytes, consisting of positive and negative ions that may interact with each other to form complexes with properties quite different from the ions from which they are formed. Such complexation is important to the relative extractive properties of different metals and can provide a sufficient difference in extractability to allow separation of the metals.

In this chapter, the factors underlying the strength of metal–ligand interaction are reviewed. An understanding of how these factors work for different metals and different ligands in the aqueous and in the organic phases can be of major value in choosing new extraction systems of greater promise for possible improvement in the separation of metals. The discussion in this chapter builds on the principles of solubility of Chapter 2 and provides additional theoretical

---

*Retired.

background for the material in the chapters that deal with the distribution equilibria, kinetics, and practices of solvent extraction systems.

3.1.1 Stability Constants

The extent of metal ion complexation for any *metal–ligand* system is defined by the equilibrium constant that is termed the *stability constant* (or *formation constant*) metal–ligand interaction. The term "metal" in metal–ligand systems simply refers to the *central metal ion* (or *cation*), and the term "ligand" (from Latin *ligare*, to bind) to the *negative ion* (*anion*) or to a neutral electron donor molecule, which binds to the central atom. Since most ligands bind to the metal ion in a regular sequence, equilibria are established for the formation of 1:1, 1:2, 1:3, etc., metal-to-ligand ratios. In some systems, *polynuclear* (i.e., polymetal) complexes such as 2:1, 2:2, 3:2, etc., form. In this chapter, the primary concern is with the more common *mononuclear* complexes in which a single metal atom has one or more ligands bonded to it. These relatively simple complexes can serve to illustrate the principles and correlations of metal ion complexation that apply also to the polynuclear species. Defining M as the metal and L as the ligand (with charges omitted to keep the equations simpler), the successive complexation reactions can be written as:

$$M + L = ML \tag{3.1a}$$
$$ML + L = ML_2 \tag{3.1b}$$

Or, generally,

$$ML_{n-i} + L = ML_n \tag{3.1c}$$

The equilibrium constants for these stepwise reactions are expressed generally as K_n:

$$K_n = [ML_n]/[ML_{n-1}][L] \tag{3.2a}$$

Sometimes a complexation reaction is considered to occur between M and an acidic ligand HL

$$M + HL = ML + H \tag{3.1d}$$

in which case the *protonated stepwise formation constant* is

$$*K_n = [ML][H]/[M][HL] \tag{3.2b}$$

It is common to use the *overall stability constant*, β_{pqr}, where p = number of metal, q = number of hydrogens (for protonated species), and r = number of ligands. The value of q is negative for hydroxo species. For the reaction to form ML,

$$M + L = ML; \beta_{101} = [ML]/[M][L] = K_1 \tag{3.3}$$

However, the formation of multiligand complexes is more complex: e.g.;

$$M + 3L + ML_3; \qquad \beta_{103} = [ML_3]/[M][L]^3 = K_1 K_2 K_3. \tag{3.4}$$

This can be generalized as:

$$M + nL = ML_n; \qquad \beta_{10n} = [ML_n]/[M][L]^n = \prod_{i=1}^{n} K_i \tag{3.5}$$

The β_{pqr} is sometimes referred to as the *complexity constant*.

Some ligands retain an ionizable proton. For example, depending on the pH of the solution, metals may complex with HSO_4^-, SO_4^{2-} or both. In the formation of $MHSO_4$, the stability constant may be written as:

$$M + HSO_4 = MHSO_4; \qquad K_1 = [MHSO_4]/[M][HSO_4] \tag{3.6a}$$

or

$$M + H + SO_4 = MHSO_4; \qquad \beta_{111} = [MHSO_4]/[M][H][SO_4] \tag{3.6b}$$

where

$$\beta_{111} = K_{111}/K_{a2}$$

and K_{a2} is the dissociation acid constant for HSO_4^-. Alternately, some complexes are hydrolyzed and have one or more hydroxo ligands. For such complexes, usually in neutral or basic solutions, the reaction may be represented as:

$$M + H_2O + L = M(OH)L + H \tag{3.7a}$$

and

$$\beta_{1-11} = [M(OH)L]/[M][H]^{-1}[L] \tag{3.7b}$$

or

$$*\beta_{111} = [M(OH)L][H]/[M][L] \tag{3.7c}$$

In general, we use β_n to represent β_{10n} in this text.

3.1.2 Use of Stability Constants

If the concentration $[L]$ of "free" (uncomplexed, unprotonated, etc.) ligand at equilibrium can be measured by spectrometry, a selective ion electrode, or some other method, the stability constant can provide direct insight into the relative concentrations of the species. This is achieved by rewriting Eq. (3.5) in the form:

$$[ML_n]/[M] = \beta_n \cdot [L]^n \tag{3.7d}$$

For example, if $\beta_1 = 10^2$ and $[L] = 10^{-2}$ M, there are equal concentrations of M and ML in the solution since $\beta_1 \cdot [L] = 10^2 \, 10^{-2} = 1$. This relationship is independent of the amount of metal in the solution and the equality $[M] = [ML] = 1$ is valid whether the total metal concentration is 10^{-9} M or is $1M$ *as long as the free ligand concentration* $[L] = 0.01M$. For example, if the ligand is present only as L and ML, and if the total metal concentration, $[M] + [ML]$, is 10^{-9} M, the total ligand concentration is $[L] + ML = 0.01 + 0.5 \times 10^{-9} = 0.01$ M. However, if the total metal is 1.00 M and $[M] = [ML]$, the total ligand is $0.01 + 0.50 = 0.51$ M.

Consider a system in which $\beta_1 = 10^2$, $\beta_2 = 10^3$ and $\beta_3 = 2 \times 10^3$. The concentrations of each complex can be calculated relative to that of the free (uncomplexed) metal for any free ligand concentration. If $[L] = 0.1$ M,

$$[ML]/[M] = \beta_1 \cdot [L] = 10^2 \times 10^{-1} = 10$$

$$[ML_2]/[M] = \beta_2 \cdot [L]^2 = 10^3 \times 10^{-2} = 10$$

$$[ML_3]/[M] = \beta_3 \cdot [L]^3 = 2 \times 10^3 \times 10^{-3} = 2$$

Thus, abbreviating the total metal concentration $\Sigma[ML_n] = [M_T]$, and defining the mole fraction of each metal species (i.e., of M, ML, ML_2, and ML_3) as the ratio of the concentration of that species to the total metal concentration by $K_{ML,i} = [ML_i]/[M_T]$, then

$$K_M = 1/(1+10+10+2) = 0.43$$

$$K_{ML} = 10/(1+10+10+2) = 0.43$$

$$K_{ML} = 10/(1+10+10+2) = 0.43$$

$$K_{ML} = 2/(1+10+10+2) = 0.087$$

Such calculations can be done for a series of free ligand concentrations to generate a family of *formation curves* of concentration or mole fraction of metal complex species as a function of the concentration of the free ligand. Such curves are shown in Figs. 3.1 and 3.2. These calculations are particularly useful for trace level values of metal as they require only knowledge of the free ligand concentrations and the β_n. Values of stability constants can be found in Refs. [1,2].

3.1.3 Stability Constants and Thermodynamics

The examples in section 3.1.2 of calculations using stability constants involve concentrations of M, L, and ML_n. Rigorously, a stability constant, as any thermodynamic equilibrium constant should be defined in terms of standard state conditions (see section 2.4). When the system has the properties of the standard state conditions, the concentrations of the different species are equal to their activities. However, the standard state conditions relate to the ideal states described in Chapter 2, which can almost never be realized experimentally for solutions of electrolytes, particularly with water as the solvent. For any conditions other than those of the standard state, the activities and concentrations are related by the activity coefficients as described in Chapter 2, and especially

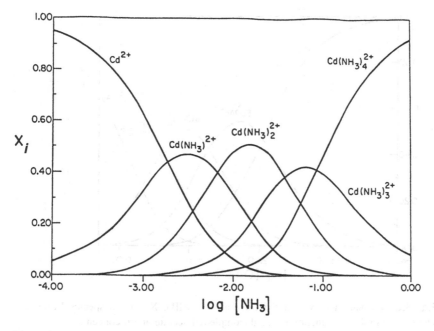

Fig. 3.1 Formation curves of the fraction of Cd(II), X_i, in the species $Cd(NH_3)_i^{2+}$, where $i = 0$ to 4, as a function of the uncomplexed concentration of NH_3.

as discussed in Chapter 6. Thus, for the thermodynamic constant, K_1°, for the reaction

$$M + L = ML$$

one obtains

$$K_1^\circ = \frac{a_{ML}}{a_M a_L} = \frac{[ML]y_{ML}}{[M]y_M [L]y_L} = \frac{[ML]y_{ML}}{[M][L]y_M y_L} = \beta_1 \cdot y_{ML} / y_M y_L = \beta_1^\circ \quad (3.8)$$

Sometimes the activities are replaced by braces, e.g., $a_{ML} = \{ML\}$; here we prefer the former formality. The estimation of y_i values by equations such as Eqs. (2.38) and (2.43) is unreliable for ionic strengths above about 0.5 M for univalent cations and anions and at even lower ionic strengths for polyvalent species. Consequently, values of β_i° are rarely calculated, except for very exact purposes (see Chapter 6). Instead, measurements of equilibrium concentrations of the species involved in the reaction are used in a medium of fixed ionic strength where the ionic strength, I, is defined as $I = \frac{1}{2} \Sigma c_i z_i^2$ M [see Eq. (2.38)]. A solution of fully ionized $CaCl_2$ of 0.5 M concentration has an ionic strength $I = \frac{1}{2} \langle 0.5 \times 2^2 + 2 \times 0.5 \times (-1)^2 \rangle$ or 1.5 M. Stability constants are reported, then, as measured in solutions of 0.1 M, 0.5 M, 2.0 M, etc., ionic strength. In practice,

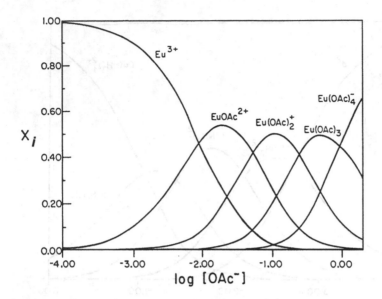

Fig. 3.2 Formation curves of the fraction of Eu(III), X_i, in the species $Eu(OAc)_i^{3-i}$, where $i = 0$ to 4, as a function of the uncomplexed acetate anion concentration.

the concentrations of the complexing metal and ligand may be so low (e.g., 0.01 M) that if an inert (i.e., noncomplexing) electrolyte is added at higher concentration (e.g., 1.00 M), it can be assumed that the ionic strength remains constant during the reaction. However, such a practice, referred to as a *constant ionic medium*, is not always possible if the complexation is so weak that relatively high concentrations of metal and ligand are required.

Consider the complexation of a trivalent metal by a dinegative anion. Assume $M(ClO_4)_3$ is 0.01 M and Na_2SO_4 is 0.05 M. To obtain an ionic strength of 1.00 M, if no complexation occurs, requires a concentration of $NaClO_4$ calculated to be:

$$[NaClO_4] = 1.00 - 1/2 \langle 0.01 \times (3)^2 + 3 \times 0.01(-1)^2 + 2 \times 0.05 \times (1)^2$$

$$+ 0.05 \times (2)^2 \rangle = 1.00 - 0.21 = 0.79 \text{ M}$$

If all the M^{3+} is complexed,

$$[MSO_4^+] = 0.01 \text{ M},$$

$$[SO_4^{2-}] = 0.05 - 0.1 = 0.4 \text{ M}$$

$$[Na^+] = 2 \times 0.05 = 0.10 \text{ M},$$

$$[ClO_4^-] = 3 \times 0.01 = 0.3 \text{ M and}$$

$$[NaClO_4] = 0.79$$

The ionic strength would be:

$$I = 1/2\langle 0.01 \times (1)^2 + 0.04 \times (-2)^2 + (0.10 + 0.79)(1)^2$$
$$+ \ (0.03 + 0.79)(-1)^2 \rangle = 0.94 \text{ M}$$

Therefore, a 6% error in the ionic strength arises from not adding additional $NaClO_4$ during the measurement of the stability constant. Such a change in the ionic strength might result in significant changes in the y_i values of Eq. (3.8). If the ionic strength is maintained constant during the measurement, Eq. (2.43) indicates that the values of y_i should remain constant. Thus, Eq. (3.8) can be written for constant ionic strength as:

$$\beta_n^\circ = \beta_n \cdot y_{ML} / y_M y_L^n = \beta_n \cdot C \qquad (3.9)$$

Since C has a particular value for each value of ionic strength, I, the β_n values can only be used for solutions of that value of I. In fact, the problem can be even more complicated. Equation (2.43) implies that y_i is determined by the value of I and is not dependent on the electrolyte used to obtain the ionic strength. For example, β values have been measured in solutions with KNO_3 as the added inert electrolyte while other experimenters have used $NaClO_4$, $LiCl$, $NaCl$, etc. In $I \le 0.1$ M solutions, the differences in β values from these different solutions are small but when the ionic strength increases above 0.1 M, significant differences may arise. Therefore, both I, the ionic strength, and the salt that is used to set the ionic strength should be specified for β values. For further details see Chapter 6 and Ref. [3].

3.2 FACTORS IN STABILITY CONSTANTS

Many factors play a role in establishing the value of the stability constants of a particular metal–ligand system. J. Bjerrum [4] considered such factors and their effect on the successive stability constants. The ratio of two stepwise constants is defined as:

$$\log K_n / K_{n+1} = T_n / T_{n+1} \qquad (3.10)$$

Bjerrum divided this $T_{n,n+1}$ value into two terms: $S_{n,n+1}$, which accounts for statistical effects and $L_{n,n+1}$, which accounts for all effects attributable to the nature of the ligand, including electrostatic effects.

3.2.1 Statistical Effect

When simply hydrated, the number of water molecules bonded to the metal (central) ion correspond to a value N, the *coordination number*, which is also termed the *hydration number*. In complexation, ligands displace the hydrate waters, although not necessarily on a 1:1 basis. Charge, steric, and other effects may cause the maximum number of ligands to be less than N. For example, in aqueous solutions Co^{2+} is an octahedral hexahydrate, $Co(H_2O)_6^{2+}$, but complexes with chloride to form the tetrahedral species $CoCl_4^{2-}$. Trivalent actinides have large hydration numbers—usually 8 or 9—but may form complexes with $N <$ 8 with bulky anionic ligands.

In his treatment of *the statistical effect*, Bjerrum assumed that the coordination number was the same for hydration and complexation. If a ligand with a single binding site approaches the metal, its probability of interacting is proportional to the number of available bonding sites on the metal. On the other hand, the probability of dissociating a ligand is proportional to the number of ligands present in the complex. Bjerrum concluded that the stepwise stability constant, K_n, should be proportional to the ratio of the available free sites at the complex to the number of occupied sites in the complex. For, example, for the complex ML, the probability of ML formation is N and that of ML dissociation is 1. Thus

$$K_1 \propto N/1$$

Similarly, for the ML_2 formation by $ML + L$, only $N - 1$ sites are available on the M since one site is already occupied, but the probability of dissociation of L from ML_2 is proportional to 2:

$$K_2 \propto (N-1)/2$$

In general, for $ML_{n-1} + L = ML_n$

$$K_n \propto (N-n+1)/n \tag{3.11}$$

From these statistical arguments the ratio of any two successive constants would be expressed by:

$$\frac{K_{n+1}}{K_n} = \frac{N-n}{N-n+1} \cdot \frac{n}{n+1} \tag{3.12}$$

For a metal cation of $N = 6$ that complexes with a monodentate ligand (one with a single binding site), the statistical effect gives the following relationship for K_1 through K_4 and for β_1 through β_4

$$K_2 = \frac{5}{12}K_1; \qquad\qquad \beta_2 = \frac{5}{12}(\beta_1)^2$$

$$K_3 = \frac{8}{15}K_2 = \frac{2}{9}K_1; \qquad \beta_3 = \frac{5}{54}(\beta_1)^3$$

$$K_4 = \frac{9}{16}K_3 = \frac{1}{8}K_1; \qquad \beta_4 = \frac{5}{432}(\beta_1)^4$$

Thus, for a hypothetical system in which $\beta_1 = 100$, we can calculate $K_2 \cong 42$, and $\beta_2 = 4200$, while $K_3 = 22$ and $\beta_3 = 9.2 \times 10^4$, etc. This can be compared to the experimental values reported for cadmium complexation by chloride,* i.e., $\beta_1 = 30$, $\beta_2 = 150$, $\beta_3 = 100$. The disagreement shows that factors other than the statistical effect serve to further reduce K_2, K_3, etc., in the $Cd^{2+} + Cl^-$ complexation system. However, the statistical effect calculations can provide an upper limit on the K_{n+1} / K_n ratio as the ligand effects would act to further lower K_{n+1} in all but a few unusual situations.

For bidentate chelate ligands, if there are a maximum of 6 coordination sites, only 3 ligands can be accommodated. In this system, $K_2 = 1/3\ K_1$ and $K_3 = 1/3\ K_1$ [or $K_3 = 1/9\ (K_1)^2$].

3.2.2 Electrostatic Effect

In section 2.3.3, the *Born equation* was introduced in discussing the electrostatic interactions when an ion is transferred from a vacuum to a solvent of dielectric constant ε. The Born equation can be used also to estimate the electrostatic interaction between an anion and a cation in a solution. The equation in this case has the form:

$$\Delta G_{el} = -A_{el} z_+ z_- / \varepsilon r \qquad (3.13)$$

where A_{el} is the constant (section 2.3.3), z_+ and z_- are the cationic and anionic charges, ε is the dielectric constant, and r is the distance between the charge centers (i.e., the sum of the radii of the cation and anionic bonding group). Although Eq. (3.13) has the proper form to calculate ΔG_{el}, in practice, empirical values of ε and/or r are obtained by putting an experimental value of ΔG_{el} into Eq. (3.13). These empirical parameters can be used, in turn, to calculate ΔG_{el} for other similar systems.

Equation (3.13) indicates that for complexation systems in which the bonding is strongly electrostatic (minimal covalent contribution), for the same ligand (so the structural effects are constant) the stability constant should be related to the charge of the cation divided by r. Since the anionic radius is constant, log β_n for a series of metal–anion complexes (e.g., lanthanide fluorides) could be expected to correlate with z_n/r_n, where r_n is the cationic radius. Figure 3.3 shows the correlation of log β_1 for a number of metals complexed with the fluoride anion and the z_n/r_n of the metals. The good correlation confirms the strongly electrostatic bonding in the 1:1 complexes of these metals with the very electronegative F^- anion. In Fig. 3.3, and many of the successive figures, no "points" are given as the spread, i.e., statistical or methodological uncertainty is fairly large in many systems.

Another useful correlation can be derived from Eq. (3.13). The proton

---

*The literature may contain several differing equilibrium constants for the same reaction [1,2] as measured in different laboratories or by different techniques.

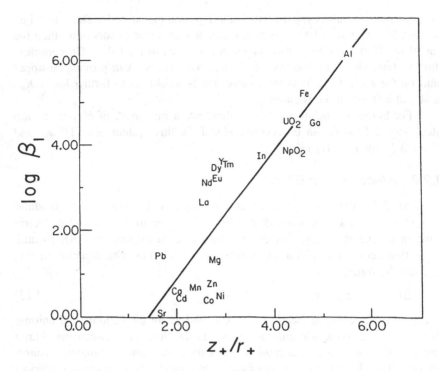

Fig. 3.3 Correlation of log β_1 of fluoride complexation vs. z_+/r_+ of the metal.

associates with ligands by electrostatic bonding, therefore, Eq. (3.13) should be applicable to proton association as well as to metal complexation when the latter is strongly electrostatic. If this is correct, log β_1 for metal complexation should correlate with log K_a for proton dissociation (or K_a^{-1} for proton association, i.e., pK_a). This assumes no structural effects in the metal complex so such a correlation can be expected only with series of structurally similar ligands, and particularly for the first stability constant. Figure 3.4 shows the excellent correlation found for log β_1 for samarium (III) with pK_a of a series of monocarboxylate ligands.

This correlation also exists for ligands with more than one binding site and for successive stability constants. In Fig. 3.5 the log K_1 and log K_2 values for UO_2^{2+} complexation correlate well with the sum $pK_{a1} + pK_{a2}$ for a series of salicylate type ligands. Salicylic acid has a hydroxy and a carboxylic acid group ortho to each other on a benzene ring. The protons of both groups can be dissociated and the UO_2^{2+} binds to both groups in the resulting complex. The related ligands in Fig. 3.5 have other species such as Cl, NO_2, SO_3, etc., bonded also to the benzene ring but these do not interact with the UO_2^{2+}. Figure 3.6 shows the structure of the UO_2 complex with the 3,5-dibromosalicylate ligand.

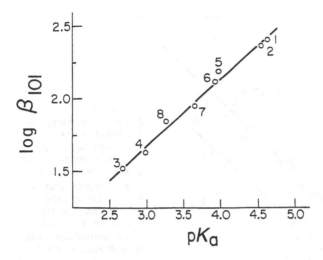

Fig. 3.4 Relationship between the stability constant, β_{101}, for formation of SmL^{2+} and the acid constant, pK_a, of HL: (1) propionic acid; (2) acetic acid; (3) iodoacetic acid; (4) chloroacetic acid; (5) benzoic acid; (6) 4-fluorobenzoic acid; (7) 3-fluorobenzoic acid; and (8) 3-nitrobenzoic acid.

3.2.3 Geometric Effects

Among the many other factors that are grouped into the $T_{n,n+1}$ factor in section 3.4, the geometry of the complex plays a significant role in many complexes. The importance of the relative sizes of the cations and anions in determining the geometrical pattern in ionic crystals has been long recognized and the same steric constraints can be expected to be present in complexes. In Table 3.1, the coordination number and the geometric pattern are listed for various ratios of the radius of the metal cation to that of the anionic ligand. Co^{2+} and H_2O have a radius ratio of about 0.5 and, as predicted, octahedral $Co(H_2O)_6^{2+}$ is the hydrated species. Co^{2+} and Cl^- have a radius ratio close to 0.3 and, again as predicted $CoCl_4^-$ is the complex which forms in tetrahedral solutions of high chloride concentration. Figure 3.7 shows common stereochemistries of the elements [5].

For electrostatic interaction, the radius ratio can be used as a guide to the possible geometry of the complex. However, when the metal–ligand bond has a significant *covalent contribution*, the geometry is fixed by the necessity to have the bonding orbitals of the metal and the donor atom of the ligand to overlap. Generally, this involves *hybrid bond orbitals* of the metal. The sp^3 and dsp^2 hybrid orbitals result in coordination number 4 with tetrahedral (sp^3) and square planar (dsp^2) geometry. For coordination number 6 the d^2sp^3 hybrid

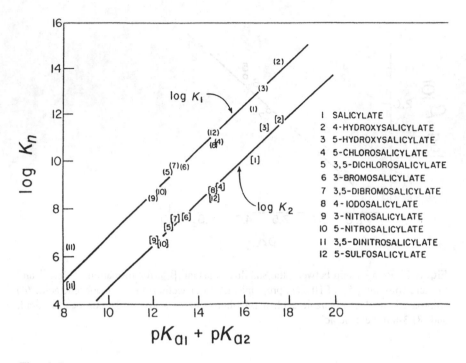

Fig. 3.5 Correlation of log K_n for formation of UO_2L_n ($n = 1$ or 2) with ΣpK_a of a series of salicylates.

orbitals produce an octahedral configuration. Thus $NiCl_4^{2-}$ is tetrahedral while $Ni(CN)_4^{2-}$ is square planar and $Ni(NH_3)_6^{2+}$ is octahedral. Both $NiCl_6^{4-}$ and $Ni(CN)_6^{4-}$ would have too much electrostatic repulsion to form the octahedral complexes. The reason why the cyanide forms a planar complex is because CN^- tends to form bonds that are more covalent than those formed by Cl^-, and that the dsp^2

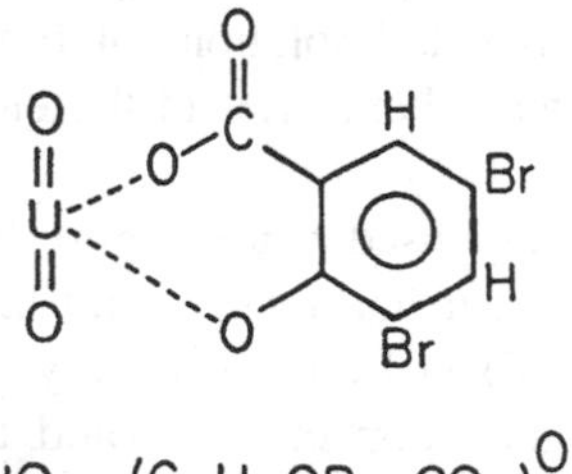

$$UO_2 \ (C_6H_2OBr_2CO_2)^0$$

Fig. 3.6 Structure of the 1:1 complex uranyl dibromosalicylate.

Table 3.1 Range of Radius Ratio Values for Various Cationic Coordination Numbers

Coordination number	Geometric pattern	Radius ratio (r_+/r_-)
2	Linear	≤0.15
3	Triangular	0.15 – 0.22
4	Tetrahedral	0.22 – 0.41
4	Planar	0.41 – 0.73
6	Octahedral	0.41 – 0.73
8	Cubic	>0.73

configuration (square planar) is associated with complexes with more covalent nature than the tetrahedral complexes of sp^3 configuration. However, neutral ligands such as H_2O and NH_3 can form octahedral structures $Ni(H_2O)_6^{2+}$ and $Ni(NH_3)_6^{2+}$. Figure 3.8 shows examples of these geometric structures.

In summary, the geometry of transition metal complexes is determined by the necessity (1) to group the ligands about the metal to minimize electrostatic repulsions and (2) to allow overlap of the metal and ligand orbitals. The first

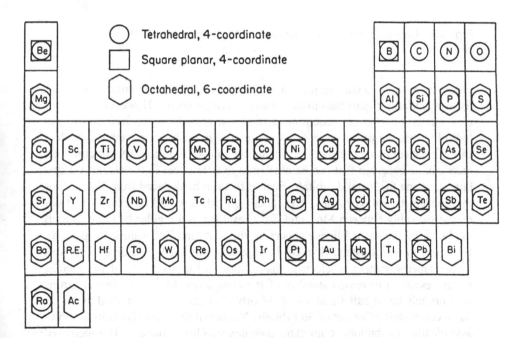

Fig. 3.7 Common stereochemistries of the elements. (After Ref. 4.)

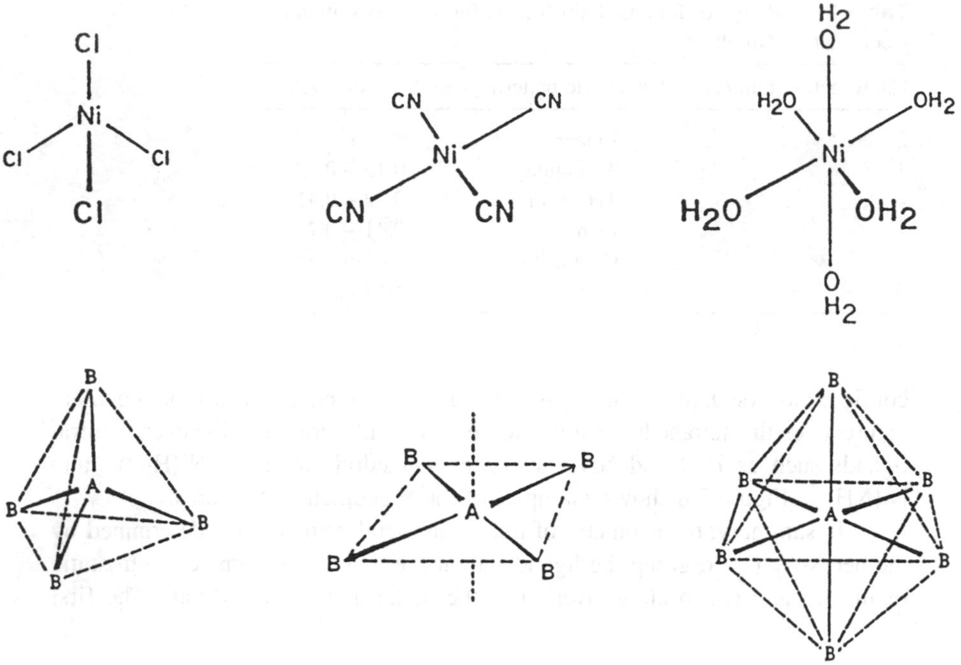

Fig. 3.8 Geometric structures of tetrahedral $NiCl_4^{2-}$, planar $Ni(CN)_4^{2-}$ and octahedral $Ni(H_2O)_6^{2+}$.

requirement favors a tetrahedral configuration for coordination number 4, as the ligands are farther apart than in the square planar geometry. However, if overlap of orbitals is a stronger requirement, and a d orbital can be included in the hybridization, the dsp^2 square planar geometry gives the more stable complex. In the octahedral complexes of coordination number 6, secondary structural effects can be observed that can be attributed to differences in the electron distribution among the metal d orbitals. Such effect can be deduced from Fig. 3.9 for the divalent transition metals. The metal ion radius would be expected to decrease smoothly between Mn^{2+} and Zn^{2+}, resulting in a regular increase in $\log\beta$ as shown by the dashed lines connecting the Mn^{2+} and Zn^{2+} values for $\beta_{103}(en)$. However, the electrostatic field of the ligand anions (or ligand dipoles) causes an asymmetric pattern of electron distribution in the d orbitals of the metals, which results in increased stability of the complexes. Mn^{2+} with five electrons in d orbitals has a half-filled set of 3d orbitals while Zn^{2+} with 10 d electrons has a completely filled set of 3d orbitals. Neither of these configurations allows asymmetric distribution, so no extra complex stability is present. The increased

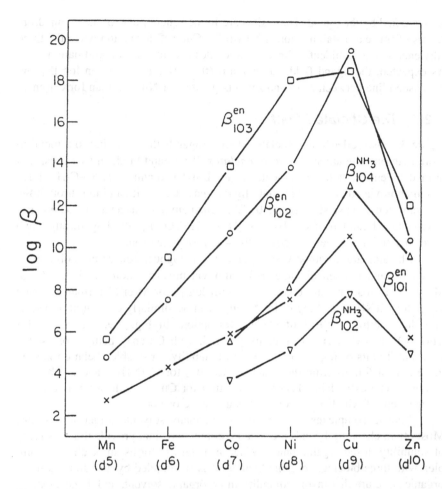

Fig. 3.9 Variation of log β_{101} for complexation by ammonia (NH_3) and ethylenediamine (en) of some first-row divalent transition metal. The electron occupations of the 3d orbitals are listed for each metal.

complex stability between Mn and Zn due to the asymmetric electron distribution is known as *ligand field stabilization* and is a maximum for metals with d^3 and d^8 configurations.

Notice in Fig. 3.9 that the expected maximum at the d^8 configuration (Ni^{2+}) is not present inasmuch as the $d^9(Cu^{2+})$ complexes are more stable. This extra stability is due to the *Jahn–Teller effect* by which d^9 systems have exceptional stability when only two or four of the six coordinate sites of the metal are occupied by the ligand. These sites are in the square plane of the octahedron. A loss of this extra stabilization is shown for $Cu(en)_3^{2+}$ formation when the three ethylenediamine (a bidentate ligand) molecules occupy all six coordination sites. As expected, the ligand field stabilization (the difference between log β_{103} and the dashed line between Mn^{2+} and Zn^{2+}) is greater for $Ni(en)_3^{2+}$ than for $Cu(en)_3^{2+}$.

3.2.4 The Chelate Effect

Figure 3.9 also reflects the "chelate effect." Ligands that can bind to a metal in more than one site are said to form *chelates*. If a ligand binds in two sites, it is termed bidentate, if three, terdentate, etc. Ethylenediamine, $H_2N-C_2H_4-NH_2$, (often abbreviated en) is a bidentate ligand inasmuch as it can bind through the two nitrogen atoms; see Fig. 3.10. Ethylenediaminetetraacetate (EDTA) is a hexadentate ligand $[(OOC \cdot CH_2)_2 \, NC_2H_4N(CH_2 \cdot COO)_2]^{4-}$ binding through two nitrogens and an oxygen of each of the 4 carboxylate groups.

Chelates are commonly stronger than analogous, nonchelate complexes. Ammonia and ethylenediamine both bind via nitrogen atoms and log β_2 for $M(NH_3)_2^{2+}$ formation can be compared with log β_1 for $M(en)^{2+}$ formation while log β_4 for $M(NH_3)_4^{2+}$ and log β_2 for $M(en)_2^{2+}$ can be similarly compared, to ascertain the stabilizing effect of chelate formation. In Fig. 3.9, the curves for $Cu(NH_3)_2^{2+}$ and $Cu(en)^{2+}$ are roughly parallel, with $Cu(en)^{2+}$ being more stable by about 2 units of log β. The increase in stability for double chelation is seen to be almost 6 units from the values of log β_{104} for $Cu(NH_3)_4^{2+}$ and log β_{102} for $Cu(en)_2^{2+}$. The extra Jahn–Teller stabilization for Cu^{2+} complexation is roughly independent of whether the complex is a chelate or not.

Chelate complexes are very useful in many solvent extraction systems. Most often, chelating ligands are organic compounds that provide the possibility of solubility in the organic phase. When a metal ion forms an octahedral complex with three bidentate organic ligands, it is surrounded by an outer "skin" of organic structure that favors solubility in an organic solvent, and, hence, extraction from the aqueous phase.

Fig. 3.10 Structures of $Cd(NH_3)_4^{2+}$ and $Cd(en)_2^{2+}$.

3.3 MODELS OF COMPLEX FORMATION

3.3.1 Early Models

A. Werner had clarified many of the structural aspects of complexes, including chelates, in the early part of this century. A further major advance in developing a theory of how and why metal ions form complexes was made by N. V. Sidgwick in 1927. He noted that the number of ligands that bond to a metal ion could be explained by assuming the metal ion accepted an electron pair from each ligand until the metal ion completed a stable electronic configuration. He called the metal the "acceptor" and the ligand the "donor." A few years earlier, G. N. Lewis had introduced a generalized acid-base theory in which "the basic substance furnishes a pair of electrons for chemical bond; the acid substance accepts such a pair" [6]. Combining these views, the acceptor (metal) becomes an "acid" and the donor (ligand) is a "base." In these models, complexation was one class of the general acid-base reactions.

In the 1950s it was recognized that metal cations could be divided into two general classes. The first, or "class a," cations behaved like H^+ and their stability constants in aqueous solution had the order:

F $\gg$ Cl>Br>I

O $\gg$ S>Se>Te

The second, or "class b," cations had stability constants that followed the reverse sequence: i.e.,

I $\gg$ Br > Cl > F

Te ~ Se ~ S $\gg$ O

Complexes of "class a" metals are more ionic, while those of the "class b" metals are more covalent. Generally, the metals that form tetrahedral complexes by using sp^3 hybrid orbitals are "class a" types. Those forming square planar complexes by using dsp^2 hybrid orbitals are normally "class b" types.

3.3.2 Hard/Soft Acids and Bases (HSAB)

A popular model for predicting whether a metal and a ligand are likely to react strongly or not is the *hard-soft acid-base principle* [7]. In general, hard acids are cations that are in the "class a" group and the log of their stability constants shows a correlation with the pK_a of the ligand base of the complex. Soft acids are usually in "class b" and their log β values correlate with the redox potential, $E°$, or the ionization potential, IP, of the ligand base. Ligands that are hard bases tend to have higher pK_a values, while soft bases have large $E°$ or IP values. The important HSAB principle states that *hard acids react strongly with hard bases and soft acids react strongly with soft bases*. This is only a general principle, as some hard acid/hard base pairs do not interact strongly and neither do some of the soft acid/soft base pairs. The exceptions to the general rule have factors other than inherent acidity and/or basicity, which are more important in the interaction. Nevertheless, the HSAB principle has proven a very useful model for a large variety of complexation reactions, partially because of its simplicity.

Commonly, soft acids have low-lying acceptor levels while soft bases have high-lying donor orbitals. As a consequence, the bond in the complex results from sharing the electron pair. Softness is thus associated with a tendency to covalency and soft species generally have large polarizabilities. In hard (metal) acids, the acceptor levels are high and in hard bases, the donor levels are low. The large energy difference is so large that the cation acid cannot share the electron pair of the base. The lack of such covalent sharing results in a strongly electrostatic bond.

The principal features of these species can be listed as follows:

1. Hard species: difficult to oxidize (bases) or reduce (acids); low polarizabilities; small radii; higher oxidation states (acids); high pK_a (bases); more positive (acids) or more negative (bases) electronegativities; high charge densities at acceptor (acid) or donor (base) sites.
2. Soft species: easy to oxidize (bases) or reduce (acids); high polarizability; large radii; small differences in electronegativities between the acceptor and donor atoms; low charge densities at acceptor and donor sites; often have low-lying empty orbitals (bases); often have a number of d electrons (acids).

Following these guidelines, cations of the same metal would be softer for lower oxidation states, harder for higher ones. In fact, Cu^+ and Tl^+ are soft while Cu^{2+} is borderline and Tl^{3+} is hard. Even though Cs^+ has a large radius and low charge density, the low ionization potential is sufficient to give it hard acid characteristics. This illustrates that the properties listed may not be possessed by all hard (or soft) species, but can serve as guides as to what characteristics can be used to predict the acid-base nature of species.

3.3.3 Qualitative Use of Acid-Base Model

Using these properties, a number of species have been placed in the hard, soft, or borderline categories in Table 3.2. This table can be used to predict, at least qualitatively, the strength of complexation as measured by the stability constants. For example, Pu^{4+} is a hard acid, F^-, a hard base, and I^-, a soft base. This leads to the prediction that $\log \beta_1(PuF^{3+})$ would be larger than $\log \beta_1(PuI^{3+})$; the experimental $\log \beta_1$ values are 6.8 and <1.0, respectively. By contrast, since Cd^{2+} is a soft acid, $\log \beta_1(CdF^+)$ could be expected to be less than $\log \beta_1(CdI^+)$; the respective values are 0.46 and 1.89. However, many metals of interest such as

Table 3.2 Partial List of Hard and Soft Acids and Bases

A. Acids:		
Hard		
	+1 ions	H⁺, Li to Cs
	+2 ions	Mg to Ba, Fe(II), Co, Mn
	+3 ions	Fe(III), Cr(III), Ga, In, Sc, all Ln (III), all actinides (III)
	+4 ions	Ti, Zr, Hf, all actinides (IV)
	−yl ions	Cr (VI), VO, MoO₃, AnO₂, Mn(VII)
Borderline	+2 ions	Fe, Co, Ni, Cu, Zn, Sn, Pb
	+3 ions	Sb, Bi, Rh, Ir, Ru, Os, R3C⁺, C₆H₅⁺
Soft		BH₃
	+1 ions	Cu, Ag, Au, Hg, CH₃Hg, I
	+2 ions	Cd, Hg, Pd, Pt
B. Bases:		
Hard		H₂O, ROH, NH₃, RNH₂, N₂H₄, R₂O, R₃PO, (RO)₃PO,
	−1 ions	OH, RO, RCO₂, NO₃, ClO₄, F, Cl,
	−2 ions	O, R(CO₂)₂, CO₃, SO₄
	−3 ions	PO₄
Borderline		C₆H₅NH₂, C₅H₅N
	−1 ions	N₃, NO₂, Br
	−2 ions	SO₃,
Soft		C₂H₄, C₆H₆, CO, R₃P(RO)₃P, R₃As, R₂S
	−1 ions	H, CN, SCN, RS, I.
	−2 ions	S₂O₃

Fe^{2+}, Co^{2+}, Ni^{2+}, Cu^{2+}, and Zn^{2+} are borderline, and, therefore, their complexing trends are less easily predicted. For F^-, we find the order of $\log \beta_1$ to be

$$Cu > Zn \sim Fe > Mn > Ni > Co$$

whereas for thiocyanate, SCN^-, it is

$$Cu \gg Ni > Co > Fe > Zn > Mn$$

For borderline metals and/or borderline ligands, the order is determined by other factors such as dehydration (removal of hydrate water), steric effects, etc.

Table 3.3 shows some stability constants for hard and soft types of metal ligand complexes. The constants are experimental values [1,2] and, as such, contain some uncertainties. Nevertheless they support well the HSAB principle.

The HSAB principle is useful in solvent extraction, as it provides a guide to choosing ligands that react strongly (high $\log \beta$) to give extractable complexes. For example, the actinide elements would be expected to, and do, complex strongly with the β-diketonates, $R_3C-CO-CH_2-CO-CR_3$ (where R = H

Table 3.3 Stability Constants for Hard and Soft Types of Metal Complexes and Halide Ions

Ligand[a]	Hard type ion: In^{3+}				Soft type ion: Hg^{2+}			
	$\log K_1$	$\log K_2$	$\log K_3$	$\log K_4$	$\log K_1$	$\log K_2$	$\log K_3$	$\log K_4$
F^- (hardest)	3.70	2.56	2.34	1.10	1.03			
Cl^- ↓	2.20	1.36			6.74	6.48	0.85	1.00
Br^- ↓	1.93	0.67			9.05	8.28	2.41	1.26
I^- (softest)	1.00	1.26			12.87	10.95	3.78	2.23

[a]For In^{3+} the ionic medium is 1M $NaClO_4 + F^-$, Cl^-, and Br^-, and 2M $NaClO_4 + I^-$. For Hg^{2+} the ionic medium is 0.5M $Na(X,ClO_4)$.

or an organic group), as the bonding is through the oxygens (hard base sites) of the enolate isomer anion. The formation of a chelate structure by the metal–enolate complex results in a relatively strong complex, and if the R groups on the β-diketonates are hydrophobic, the complex would be expected to have good solubility in organic solvents. In fact, many β-diketonate complexes show excellent solvent extraction properties.

The solvent extraction of the β-diketonate system can be used to illustrate the importance of the high coordination numbers of the actinides and lanthanides (see section 3.2) in which hydrophobic adducts are involved. The charge on a trivalent actinide or lanthanide cation is satisfied by three β-diketonate ligands (each is −1). However, the three β-diketonate ligands would occupy only six coordination sites of the metal while the coordination number of these metal cations can be 8 or 9. Thus, the neutral ML_3 species can coordinate other ligand bases. The alkyl phosphates, $(RO)_3PO$, are neutral hard bases that react with the ML_3 species to form $ML_3 \cdot S_n$ ($n = 1$ to 3). Tributyl phosphate is widely used as such a *neutral adduct-forming ligand* of actinides and provides enhanced extraction as the $ML_3 \cdot S_n$ species is more hydrophobic than the hydrated ML_3 and, thus, more soluble in the organic phase. However, the role of coordination sphere saturation by hydrophobic adducts is not limited to synergic systems. In the Purex process for processing irradiated nuclear fuel, uranium and plutonium are extracted from nitric acid solution into kerosene as $UO_2(NO_3)_2 \cdot (TBP)_2$ and $Pu(NO_3)_4 \cdot (TBP)_2$. In these compounds the nitrate can be bidentate and so fills four or fewer of the coordinate sites for the uranium and eight or fewer for the plutonium. Uranium (U) in UO_2^{2+} usually has a maximum coordination number of 6, while that of Pu^{4+} can be 9 or even 10. The addition of the two TBP adduct molecules in each case causes the compound to be soluble and extractable in the kerosene solvent.

3.3.4 Quantitative Models of Acid-Base Reaction

The major disadvantage of the HSAB principle is its qualitative nature. Several models of acid-base reactions have been developed on a quantitative basis and have application to solvent extraction. Once such model uses *donor numbers* [8], which were proposed to correlate the effect of an adduct on an acidic solute with the basicity of the adduct (i.e., its ability to donate an electron pair to the acidic solute). The reference scale of donor numbers of the adduct bases is based on the enthalpy of reaction, ΔH, of the donor (designated as B) with $SbCl_5$ when they are dissolved in 1,2-dichloroethane solvent. The donor numbers, designated DN, are a measure of the strength of the $B-SbCl_5$ bond. It is further assumed that the order of DN values for the $SbCl_5$ interaction remains constant for the interaction of the donor bases with all other solute acids. Thus, for any donor base B and any acceptor acid A, the enthalpy of reaction to form B:A is:

$$\Delta H_{B:A} = a\, DN_{B:SbCl5} + b \tag{3.14}$$

From Eq. (3.14), it is possible to calculate $-\Delta H_{B:A}$ for a base B and an acid A if $-\Delta H_{B:A}$ has been measured for two other donors. For example, the interaction of $UO_2(HFA)_2$ (HFA = hexafluoroacetone) with pyridine and with benzonitrile has been measured in $CHCl_3$ solvent. The equilibrium constants were determined to be log $K_{B:A}$ 0.88 in pyridine and -2.79 in benzonitrile, assuming that log $K_{B:A} \propto -\Delta H_{B:A}$. With the use of the donor number for these ligands in Table 3.4, the values of a and b are calculated to be 0.04 and -4.85, respectively. In turn, with these values and the donor number that is given for acetonitrile (AcN), 59.0 in Table 3.4, the value of log K for the reaction

Table 3.4 Donor Numbers (kJ·mol^{1-})

Donor	Donor Number	Donor	Donor Number
Benzene	0.42	Diphenyl phosphonic chloride	93.7
Nitromethane	11.3	Tributylphosphate	99.2
Benzonitrile	49.8	Dimethoxyethane	100.4
Acetonitrile	59.0	Dimethylformamide	111.3
Tetramethylene sulfone	61.9	Dimethylacetamide	116.3
Dioxane	61.9	Trimethylsulfoxide	124.7
Ethylene carbonate	63.6	Pyridine	138.5
n-Butylnitrile	69.5	Hexamethyl phosphoramide	162.4
Acetone	71.1	Ethylenediamine	230.2
Water	75.3	t-Butylamine	240.6
Diethylether	80.4	Ammonia	246.9
Tetrahydrofuran	83.7	Triethylamine	255.3

$$UO_2(HFA)_2 + AcN = UO_2(HFA)_2 \cdot AcN$$

in $CHCl_3$ is estimated to be 3.3. Experimentally, it was measured to be 3.5.

The donor number model has been used for correlating a number of systems. The success of this approach, to a large extent, is based on its application to hard systems. It is unreasonable to expect that a single order of reaction strengths for donors would be applicable to all acidic solutes. Also, the assumption of constant entropy, which is required if $\log \beta_{B:A} \propto -\Delta H_{B:A}$, would not be universally valid. In general, it can be expected to have more limited value for soft systems than for hard ones. Nevertheless, for oxygen and nitrogen donors and, particularly, in nitrogen and oxygen solvents, it has proven to be very useful.

3.4 THERMODYNAMICS OF COMPLEXATION

The *free energy of complexation* reaction, ΔG_n, is defined by (cf. section 2.4):

$$\Delta G_n = \Delta G_n^\circ - RT \ln [ML_n]/[M][L]^n \tag{3.15a}$$

where

$$\Delta G_n^\circ = -RT \ln \beta_n \tag{3.15b}$$

and β_n is the stability constant for the complexation. The deviation from equilibrium conditions (defined by β_n) determines the direction of reaction that goes spontaneously in the direction of $\Delta G_n^\circ < O$. The free energy term is the difference between the enthalpy of reaction, ΔH_n, and the entropy changes for the reaction, ΔS_n; that is

$$\Delta G_n^\circ = \Delta H_n^\circ - T\Delta S_n^\circ \tag{3.16}$$

The free energy is calculated from the stability constant, which can be determined by a number of experimental methods that measure some quantity sensitive to a change in concentration of one of the reactants. Measurement of pH, spectroscopic absorption, redox potential, and distribution coefficient in a solvent extraction system are all common techniques.

The enthalpy of complexation can be measured directly by reacting the metal and ligand in a calorimeter. It can also be determined indirectly by measuring $\log \beta_n$ at different temperatures and applying the equation

$$d \ln \beta_n / dT = -\Delta H_n^\circ / RT \tag{3.17}$$

The temperature variation method is used often in solvent extraction studies. It can give reliable values of the enthalpy if care is taken to verify that the change in temperature is not causing new reactions, which perturb the system. Also, a reasonable temperature range (e.g., $\Delta T \sim 50°C$) should be used so that it is possible to ascertain the linearity of the $\ln \beta_n$ vs. $1/T$ plot upon which Eq. (3.17) is based.

3.4.1 Enthalpy–Entropy Compensation

The thermodynamics of complexation between hard cations and hard (O, N donor) ligands often are characterized by positive values of both the enthalpy and entropy changes. A positive ΔH value indicates that the products are more stable than the reactants, i.e., destabilizes the reaction, while a positive entropy favors it. If $T\Delta S_n^\circ > \Delta H_n^\circ$, ΔG_n° will be negative and thus $\log\beta_n$ positive, i.e., the reaction occurs spontaneously. Such reactions are termed "entropy driven" since the favorable entropy overcomes the unfavorable enthalpy.

Complexation results in a decrease in the hydration of the ions which increases the randomness of the system and provides a positive entropy change ($\Delta S_{\text{dehyd}} > 0$). The dehydration also causes an endothermic enthalpy ($\Delta H_{\text{dehyd}} > 0$) as a result of the disruption of the ion–water and water–water bonding of the hydrated species. The interaction between the cation and the ligand has a negative enthalpy contribution ($\Delta H_{\text{reaction}} < 0$) due to formation of the cation–anion bonds. This bonding combines the cation and the anion to result in a decrease in the randomness of the system and gives a negative entropy contribution ($\Delta S_{\text{reaction}} < 0$). The observed overall changes reflect the sum of the contributions of dehydration and cation–ligand combination. If the experimental values of ΔH and ΔS are positive, the implication is that the dehydration is more significant in these terms than the combination step and vice versa. The reaction steps can be written (with charges omitted and neglecting intermediates):

1. Dehydration: ΔH_h, ΔS_h

$$M(H_2O)_p + L(H_2O)_q = M(H_2O)_m + L(H_2O)_s + (p+q-m-s)H_2O \quad (3.18a)$$

2. Combination: ΔH_c, ΔS_c

$$M(H_2O)_m + L(H_2O)s = ML(H_2O)_{m+s} \quad (3.18b)$$

3. Net reaction: ΔH_r, ΔS_r

$$M(H_2O)_p + L(H_2O)_q = ML(H_2O)_{m+s} + (p+q-m-s)H_2O \quad (3.18c)$$

and, therefore:

$$\Delta H_r = \Delta H_h + \Delta H_c = \Delta H^\circ$$
$$\Delta S_r = \Delta S_h + \Delta S_c = \Delta S^\circ \quad (3.19)$$

It has been shown that for many of the hard–hard complexation systems, there is a linear correlation between the experimental ΔH° and ΔS° values. Figure 3.11 shows such a correlation for actinide complexes. These ΔH and ΔS data involve metal ions in the +3, +4, and +6 oxidation states and a variety of both inorganic and organic ligands. Such a correlation of ΔH° and ΔS° has been termed *the compensation effect*. To illustrate this effect, reconsider the

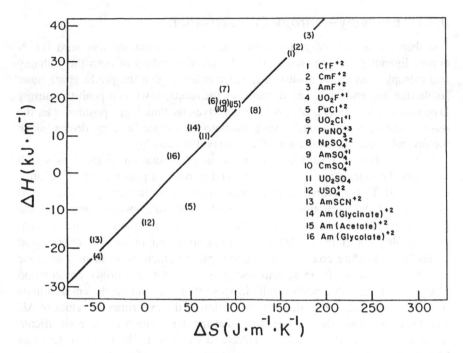

Fig. 3.11 Correlation of ΔH and ΔS of formation of a series of 1:1 complexes at different ionic strengths.

step reactions above for complexation. The net reaction has the thermodynamic relations:

$$\Delta G_r = \Delta G_c + \Delta G_h = (\Delta H_c + \Delta H_h) - T(\Delta S_c + \Delta S_h) \qquad (3.20)$$

The compensation effect assumes $\Delta G_h \sim 0$; the positive values of ΔH_r ($=\Delta H_c + \Delta H_h$) and ΔS_r ($=\Delta S_c + \Delta S_h$) mean $|\Delta H_h| > |\Delta H_c|$ and $|\Delta S_h| > |\Delta S_c|$. So Eq. (3.20) can be rewritten:

$$\Delta G_r \approx \Delta G_c \approx \Delta H_r - T \cdot \Delta S_r \approx \Delta H_c + T \cdot \Delta S_c \qquad (3.21)$$

The complexation, interpreted in this fashion, implies that:

1. The free energy change of the total complexation reaction, $\Delta G°$, is related principally to step 2 [Eq. (3.18b)], the combination subreaction;
2. The enthalpy and entropy changes of the total complexation reaction, $\Delta H°$ and $\Delta S°$, reflect, primarily, step 1 [Eq. (3.18a)], the dehydration subreaction.

This discussion is important to solvent extraction systems, as it provides further insight into the aqueous phase complexation. It also has significance for the organic phase reactions. In organic solvents, solvation generally is weaker than for aqueous solutions. As a consequence, the desolvation analogous to step 1 would result in small values of $\Delta H_{(solv)}$ and $\Delta S_{(solv)}$ (note that we term the aqueous phase solvation *hydration*). Therefore, $\Delta H_c = \Delta H_{solv}$ and $\Delta S_c = \Delta S_{solv}$, which means that $\Delta H°$ is more often negative, while $\Delta S°$ may be positive or negative but relatively small.

3.4.2 Inner Versus Outer Sphere Complexation

Complexation reactions are assumed to proceed by a mechanism that involves initial formation of a species in which the cation and the ligand (anion) are separated by one or more intervening molecules of water. The expulsion of this water leads to the formation of the "inner sphere" complex, in which the anion and cation are in direct contact. Some ligands cannot displace the water and complexation terminates with the formation of the "outer sphere" species, in which the cation and anion are separated by a molecule of water. Metal cations have been found to form stable inner and outer sphere complexes and for some ligands both forms of complexes may be present simultaneously.

Often, it is difficult to distinguish definitely between inner sphere and outer sphere complexes in the same system. Based on the preceding discussion of the thermodynamic parameters, ΔH and ΔS values can be used, with cation, to obtain insight into the outer vs. inner sphere nature of metal complexes. For inner sphere complexation, the hydration sphere is disrupted more extensively and the net entropy and enthalpy changes are usually positive. In outer sphere complexes, the dehydration sphere is less disrupted. The net enthalpy and entropy changes are negative due to the complexation with its decrease in randomness without a compensatory disruption of the hydration spheres.

These considerations lead, for example, to the assignment of a predominantly outer sphere character to Cl^-, Br^-, I^-, ClO_3^-, NO_3^-, sulfonate, and trichloroacetate complexes and an inner sphere character to F^-, IO_3^-, SO_4^{2-}, and acetate complexes of trivalent actinides and lanthanides. The variation in $\Delta H°$ and $\Delta S°$ of complexation of related ligands indicates that those whose pK_a values are <2 form predominantly outer sphere complexes, while those for whom $pK_a > 2$ form predominantly inner sphere complexes with the trivalent lanthanides and actinides. As the pK_a increases above 2, increasing predominance of inner sphere complexation is expected for these metals.

3.4.3 Thermodynamics of Chelation

In the preceding section, the positive entropy change observed in many complexation reactions has been related to the release of a larger number of water molecules than the number of bound ligands. As a result, the total degrees of freedom of the system are increased by complexation and results in a positive

Table 3.5 Thermodynamic Parameters of Reaction of Cadmium(II)-Ammonia
Complex with Ethylenediamine

Product	log K	$\Delta H°$(kJ·mol^{-1})	$\Delta S°$ (J·m·K^{-1})
Cd(en)$^{+2}$	0.9	+0.4	5.4
Cd(en)$_2^{+2}$	2.2	−3.4	+15.0

$$K = \frac{[Cd(en)_n^{+2}][NH_3]^{2n}}{[Cd(NH_3)_{2n}^{+2}[en]^n}$$

value of the entropy change. In many systems, a similar explanation can be given for the enhanced stability of chelates. Consider the reaction of Cd(II) with ammonia and with ethylenediamine to form $Cd(NH_3)_4^{2+}$ and $Cd(en)_2^{2+}$, whose structures are given in Fig. 3.10. Table 3.5 gives the values of log K, $\Delta H°$, and $T\Delta S°$ for the reactions

$$Cd(NH_3)_2^{2+} + en = Cd(en)^{2+} + 2NH_3 \tag{3.22}$$

$$Cd(NH_3)_4^{2+} + 2en = Cd(en)_2^{2+} + 4NH_3 \tag{3.23}$$

The enthalpy value of Eq. (3.23) is very small as might be expected if two Cd−N bonds in $Cd(NH_3)_2^{2+}$ are replaced by two Cd−N bonds in $Cd(en)^{2+}$. The favorable equilibrium constants for reactions [Eqs. (3.22) and (3.23)] are due to the positive entropy change. Note that in reaction, Eq. (3.23), two reactant molecules form three product molecules so chelation increases the net disorder (i.e., increase the degrees of freedom) of the system, which contributes a positive $\Delta S°$ change. In reaction Eq. (3.23), the ΔH is more negative but, again, it is the large, positive entropy that causes the chelation to be so favored.

As the size of the chelating ligand increases, a maximum in stability is normally obtained for 5 or 6 membered rings. For lanthanide complexes, oxalate forms a 5-membered ring and is more stable than the malonate complexes with 6-membered rings. In turn, the latter are more stable than the 7-membered chelate rings formed by succinate anions.

3.5 SUMMARY

The important role of thermodynamics in complex formation, ionic medium effects, hydration, solvation, Lewis acid-base interactions, and chelation has been presented in this chapter. Knowledge of these factors are of great value in understanding solvent extraction and designing new and better extraction systems.

REFERENCES

1. Martell, A. E.; Smith, R. M.; *Critical Stability Constants*, Vol. 1–5, Plenum Press, New York, (1974–1982).
2. Högfeldt, E. *Stability Constants of Metal-Ion Complexes, Part A: Inorganic ligands.* IUPAC Chemical Data Series No. 22, Pergamon Press, New York (1982).
3. Robinson, R. A.; Stokes, R. H.; *Electrolyte Solutions*, Second Revised Edition, Butterworth, London (1970).
4. Bjerrum, J.; *Metal Amine Formation in Aqueous Solution*, P. Haase and Son, Copenhagen (1941).
5. Moeller, T.; *Inorganic Chemistry*, John Wiley and Sons, New York (1952).
6. Lewis, G. N.; *Valence and the Structures of Atoms and Molecules*, The Chemical Catalog Co., New York (1923).
7. Pearson, R. G.; Ed. *Hard and Soft Bases*, Dowden, Huchinson and Ross, East Stroudsburg, PA, (1973).
8. Gutmann, V.; *The Donor-Acceptor Approach to Molecular Interactions*, Plenum Press, New York (1980).

4

Solvent Extraction Equilibria

JAN RYDBERG*   Chalmers University of Technology, Göteborg, Sweden

GREGORY R. CHOPPIN*   Florida State University, Tallahassee, Florida, U.S.A.

CLAUDE MUSIKAS*   Commissariat à l'Energie Atomique, Paris, France

TATSUYA SEKINE[†]   Science University of Tokyo, Tokyo, Japan

4.1 INTRODUCTION

The ability of a solute (inorganic or organic) to distribute itself between an aqueous solution and an immiscible organic solvent has long been applied to separation and purification of solutes either by extraction into the organic phase, leaving undesirable substances in the aqueous phase; or by extraction of the undesirable substances into the organic phase, leaving the desirable solute in the aqueous phase. The properties of the organic solvent, described in Chapter 2, require that the dissolved species be electrically neutral. Species that prefer the organic phase (e.g., most organic compounds) are said to be *lipophilic* ("liking fat") or *hydrophobic* ("disliking water"), while the species that prefer water (e.g., electrolytes) are said to be *hydrophilic* ("liking water"), or *lipophobic* ("disliking fat"). Because of this, a hydrophilic inorganic solute must be rendered hydrophobic and lipophilic in order to enter the organic phase.

Optimization of separation processes to produce the purest possible product at the highest yield and lowest possible cost, and under the most favorable environmental conditions, requires detailed knowledge about the solute reactions in the aqueous and the organic phases. In Chapter 2 we described physical factors that govern the solubility of a solute in a solvent phase; and in Chapter 3, we presented the interactions in water between metal cations and anions by

This chapter is a revised and expanded synthesis of Chapters 4 (by Rydberg and Sekine) and 6 (by Allard, Choppin, Musikas, and Rydberg) of the first edition of this book (1992).
*Retired.
†Deceased.

which neutral metal complexes are formed. This chapter discusses the equations that explain the extraction data for inorganic as well as organic complexes in a quantitative manner; i.e., the measured solute distribution ratio, D_{solute}, to the concentration of the reactants in the two phases. It presents chemical modeling of solvent extraction processes, particularly for metal complexes, as well as a description of how such models can be tested and used to obtain equilibrium constants.

The subject of this chapter is broad and it is possible to discuss only the simpler—though fundamental—aspects, using examples that are representative. The goal is to provide the reader with the necessary insight to engage in solvent extraction research and process development with good hope of success.

4.1.1 The Distribution Law

The *distribution law*, derived in 1898 by W. Nernst, relates to the distribution of a solute in the organic and in the aqueous phases. For the equilibrium reaction

$$A \ (aq) \rightleftarrows A \ (org) \tag{4.1a}$$

the Nernst distribution law is written

$$K_{D,A} = \frac{\text{Concentration of Species A in organic phase}}{\text{Concentration of Species A in aqueous phase}} = \frac{[A]_{org}}{[A]_{aq}} \tag{4.1b}$$

where brackets refer to concentrations; Eq. (4.1) is the same as Eqs. (1.2) and (2.23). $K_{D,A}$ is the *distribution constant* (sometimes designated by P, e.g., in Chapter 2; see also Appendix C) of the solute A (sometimes referred to as the *distribuend*). Strictly, this equation is valid only with pure solvents. In practice, the solvents are always saturated with molecules of the other phase; e.g., water in the organic phase. Further, the solute A may be differently solvated in the two solvents. Nevertheless, Eq. (4.1) may be considered valid, if the mutual solubilities of the solvents (see Table 2.2) are small, say <1%, and the activity factors of the system are constant. If the solute is strongly solvated, or at high concentration (mole fraction >0.1), or if the ionic strength of the aqueous phase is large (>0.1 M) or changes, Eq. (4.1) must be corrected for deviations from ideality according to

$$K_{D,A}^0 = \frac{y_{A,org}[A]_{org}}{y_{A,aq}[A]_{aq}} = \frac{y_{A,org}}{y_{A,org}} K_{D,A} \tag{4.2}$$

where y's are activity coefficients [see Eq. (2.25)]. For aqueous electrolytes, the activity factors vary with the ionic strength of the solution (see sections 2.5, and 3.1.3, and Chapter 6). This has led to the use of the constant ionic medium method (see Chapter 3); i.e., the ionic strength of the aqueous phase is kept constant during an experiment by use of a more or less inert "bulk" medium like $NaClO_4$. Under such conditions the activity factor ratio of Eq. (4.2) is assumed to be constant, and K_D is used as in Eq. (4.1) as conditions are varied at a constant ionic strength value. In the following derivations, we assume that the activity factors for the solute in the aqueous and organic solvents are constant. Effects due to variations of activity factors in the aqueous phase are treated in Chapter 6, but no such simple treatment is available for species in the organic phase (see Chapter 2).

The assumption that the activity factor ratio is constant has been found to be valid over large solute concentration ranges for some solutes even at high total ionic strengths. For example, the distribution of radioactively labeled $GaCl_3$ between diethyl ether and 6M HCl was found to be constant ($K_{D,Ga} \approx 18$) at all Ga concentrations between 10^{-3} and 10^{-12} M [1].

In the following relations, tables, and figures, the temperature of the systems is always assumed to be 25°C, if not specified (temperature effects are discussed in Chapters 3 and 6, and section 4.13.6). We use *org* to define species in the organic phase, and no symbol for species in the aqueous phase (see Appendix C).

4.1.2 The Distribution Ratio

The IUPAC definition of the *distribution ratio*, *D*, is given in the introduction to Chapter 1 and in Appendix C. For a metal species M it can be written

$$D_M = \frac{\text{Concentration of all species containing M in organic phase}}{\text{Concentration of of all species containing M in aqueous phase}} = \frac{[M]_{t,org}}{[M]_{t,aq}} \tag{4.3}$$

When M is present in various differently complexed forms in the aqueous phase and in the organic phase, $[M]_t$ refers to the sum of the concentrations of all *M* species in a given phase (the subscript *t* indicates total M). It is important to distinguish between the distribution constant, K_D, which is valid only for a single specified species (e.g., MA_2), and the distribution ratio, D_M, which may involve sums of species of the kind indicated by the index, and thus is not constant.

4.1.3 Extraction Diagrams

Solvent extraction results are presented typically in the form of diagrams. This is schematically illustrated in Fig. 4.1a for three hypothetical substances, A, B, and C. The distribution ratio is investigated as a function of the concentration of some reactant Z, which may be pH, concentration of extractant in the organic phase (e.g., an organic acid HA, $[HA]_{org}$), the extractant anion concentration in the aqueous phase (e.g., $[Cl^-]$), salt concentration in the aqueous phase, etc. The

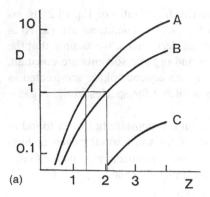

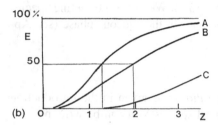

Fig. 4.1 Liquid-liquid distribution plots. (a) The distribution ratios D for three different substances A, B, and C, plotted against the variable Z of the aqueous phase. Z may represent pH, concentration of extractant in organic phase ($[HA]_{org}$), free ligand ion concentration in the aqueous phase ($[A^-]$), aqueous salt concentration, etc. (b) Same systems showing percentage extraction $\%E$ as a function of Z. D and Z are usually plotted on logarithmic scale.

range of D is best measured from about 0.1–10, though ranges from about 10^{-5}–10^4 can be measured with special techniques (see section 4.15).

In many practical situations, a plot like Fig. 4.1a is less informative than one of percentage extraction, $\%E$, where:

$$\%E = 100D/(1+D) \tag{4.4}$$

Such a plot is shown in Fig. 4.1b for the same system as in Fig. 4.1a. Percentage extraction curves are particularly useful for designing separation schemes. A series of such curves has already been presented in Fig. 1.3.

A convenient way to characterize the S-shaped curves in Figs. 1.3 or 4.1b, where the extraction depends on the variable Z, is to use the log Z value of 50% extraction, e.g., $\log[Cl^-]_{50}$. The pH_{50}-value indicates $-\log[H^+]$ for 50% extraction. This is shown in Fig. 4.1 for distribuends A and B.

Very efficient separations are often needed in industry, and a single extraction stage may be insufficient. The desired purity, yield, etc. can be achieved by multiple extractions, as discussed in Chapter 7 (see also section 1.2). In the design of separation processes using multistage extractions, other extraction diagrams are preferred. Only single stage extraction is discussed in this chapter, while multistage extraction is discussed in the second part (Chapters 7–14) of this book.

4.2 THERMODYNAMICS OF EXTRACTION SYSTEMS

Extraction from aqueous solutions into organic solvents can be achieved through different chemical reactions. Some may seem very complicated, but usually occur through a number of rather simple steps; we assume this in making a model of the system. The subdivision of an extraction reaction into its simpler steps is useful for understanding how the distribution ratio varies as a function of the type and concentration of the reagents. Often these models allow equilibrium constants to be measured.

As solute, we consider both nonelectrolytes (abbreviated as A or B, organic or inorganic), and electrolytes (e.g., as metal-organic complexes, metal ions rendered soluble in organic solvents through reactions with organic anions A^- and with adduct formers B). The system of equations shown later is only valid as long as no species are formed other than those given by the equations, all concentrations refer to the *free concentrations* (i.e., uncomplexed), and activity factors and temperatures are constant. Further, we assume that equilibrium has been established. It may be noted that the use of equilibrium reactions mean that the reactions take place in the aqueous phase, the organic phase or at the interface, as is illustrated in the next examples, but do not show any intermediates formed; this information can be obtained by kinetic studies, as described in Chapter 5, or by "fingerprinting" techniques such as molecular spectroscopy.

Before a detailed analysis of the chemical reactions that govern the distribution of different solutes in solvent extraction systems, some representative practical examples are presented to illustrate important subprocesses assumed to be essential steps in the overall extraction processes.

4.2.1 Case I: Extraction of Uranyl Nitrate by Adduct Formation

This is a purification process used in the production of uranium. The overall reaction is given by

$$UO_2^{2+} + 2\,HNO_3 + 2\,TBP(org) \rightarrow UO_2(NO_3)_2(TBP)_2(org) \qquad (4.5)$$

where TBP stands for tributylphosphate. The organic solvent is commonly kerosene. In Table 4.1 this extraction process is described in four steps. In Table

Table 4.1 Schematic Representation of the Hypothetical Steps in U(VI) Extraction by TBP and Their Associate ΔG_i^o of Reaction

	First step	Second step	Third step	Fourth step
Organic phase (TBP + diluent)		TBP $\downarrow$		$UO_2(TBP)_2(NO_3)_2$ $\uparrow$
Aqueous solution ($HNO_3 + UO_2^{2+} +$ H_2O)	$2\,HNO_3 + UO_2^{2+} \rightarrow$ $UO_2(NO_3)_2 + 2H^+$	$\downarrow$ TBP	$UO_2(NO_3)_2 + 2TBP$ $\rightarrow UO_2(TBP)_2(NO_3)_2$	$\uparrow$ $UO_2(TBP)_2(NO_3)_2$
Start $\rightarrow$ Final $\Delta G_{ex}^o < 0$	$\Delta G_1^o > 0$	$\Delta G_2^o > 0$	$\Delta G_3^o \neq 0$	$\Delta G_4^o \ll 0$

4.1, the sign of the free energy change, ΔG^0, in each step is given by qualitatively known chemical affinities (see Chapter 2). The reaction path is chosen beginning with the complexation of U(VI) by NO_3^- in the aqueous phase to form the uncharged $UO_2(NO_3)_2$ complex (Step 1). Although it is known that the free uranyl ion is surrounded by water of hydration, forming $UO_2(H_2O)_6^{2+}$, and the nitrate complex formed has the stoichiometry $UO_2(H_2O)_6(NO_3)_2$, water of hydration is not listed in Eq. (4.5) or Table 4.1, which is common practice, in order to simplify formula writing. However, in aqueous reactions, water of hydration can play a significant role. As the reactive oxygen (bold) of tributylphosphate, **O**$P(OC_4H_9)_3$, is more basic than the reactive oxygen of water, TBP, which slightly dissolves in water (Step 2), replaces water in the $UO_2(H_2O)_6(NO_3)_2$ complex to form the *adduct complex* $UO_2(TBP)_2(NO_3)_2$. This reaction is assumed to take place in the aqueous phase (Step 3). *Adduct formation* is one of the most commonly used reactions in solvent extraction of inorganic as well as organic compounds. (Note: the term *adduct* is often used both for the donor molecule and for its product with the solute.) The next process is the extraction of the complex (Step 4). Even if the solubility of the adduct former TBP in the aqueous phase is quite small (i.e., D_{TBP} very large), it is common to assume that the replacement of hydrate water by the adduct former takes place in the aqueous phase, as shown in the third step of Table 4.1; further, the solubility of the adduct $UO_2(TBP)_2(NO_3)_2$ must be much larger in the organic than in the aqueous phase (i.e., $D_{UO_2(TBP)_2(NO_3)_2} \gg 1$), to make the process useful. Other intermediate reaction paths may be contemplated, but this is of little significance as ΔG_{ex}^0 depends only on the starting and final states of the system. The use of such a thermodynamic representation depends on the knowledge of the ΔG_i^0 values as they are necessary for valid calculations of the process.

The relation between ΔG_{ex}^0 and K_{ex} is given by

$$\Delta G_{ex}^o = \Sigma \Delta G_i^o = -RT \ln K_{ex} \tag{4.6}$$

Omitting water of hydration, the equilibrium constant for the net extraction process in Eq. (4.5) is K_{ex}, where

$$K_{ex} = \frac{\left[UO_2(NO_3)_2(TBP)_2\right]_{org}}{\left[UO_2^{2+}\right]\left[HNO_3\right]^2\left[TBP\right]_{org}^2} \tag{4.7}$$

The *extraction constant*, K_{ex}, can be expressed as the product of several equilibrium constants for other assumed equilibria in the net reaction:

$$K_{ex} = \Pi K_i = \beta_{2,NO_3} K_{DR}^{-2} \beta_{2,TBP} K_{DC} \tag{4.8}$$

where β_{2,NO_3} is the *complex formation constant* of $UO_2(NO_3)_2$, and $\beta_{2,TBP}$ the formation constant of the extractable $UO_2(NO_3)_2(TBP)_2$ complex from from $UO_2(NO_3)_2$ and TBP. K_{DR} and K_{DC} are the *distribution constants of the uncharged species*, the reagent and the extractable complex, respectively.

K_{ex} determines the efficiency of an extraction process. It depends on the "internal chemical parameters" of the system, i.e., the chemical reactions and the concentration of reactants of both phases. The latter determine the numerical value of the *distribution factor for the solute*, which for our example is

$$D_U = \frac{[U]_{tot,org}}{[U]_{tot,aq}} = \frac{\left[UO_2(NO_3)_2(TBP)_2\right]_{org}}{\left[UO_2^{2+}\right] + \left[UO_2(NO_3)_n^{2-n}\right]} \tag{4.9a}$$

In the aqueous phase we have included the $UO_2(NO_3)_n^{2-n}$ complexes but excluded the $UO_2(NO_3)_2(TBP)_2$ complex, because the concentration of the last complex in the aqueous phase is negligible compared to the other two. In dilute solutions, the nitrate complex can be negleted compared to the free UO_2^{2+} concentration. In the latter case the U distribution equals

$$D_U = K_{ex}\left[HNO_3\right]^2\left[TBP\right]_{org}^2 \tag{4.9b}$$

Of the reaction steps, only the first three have values of $\Delta G^0 > 0$; however, the large negative value of the fourth step makes the overall reaction ΔG_{ex}^0 negative, thus favoring the extraction of the complex. The first step can be measured by the determination of the dinitrato complex in the aqueous phase. The second is related to the distribution constant $K_{D,TBP}$ in the solvent system. Also, the formation constant of the aqueous $UO_2(NO_3)_2(TBP)_2$ can be measured (for example by NMR on ^{31}P of TBP in the aqueous phase). Thus, ΔG_4^0 can be derived.

4.2.2 Case II: Synergistic Extraction of Uranyl Ions by Chelation and Adduct Formation

Solvent extraction is a powerful technique in research on metal complexes. Consider a metal complexed by a chelate compound (see Chapter 3), where the chelate is a weak organic acid. For example, the uranyl ion can be neutralized

by two TTA⁻ (Appendix D:5e) anions to form the neutral $UO_2(TTA)_2(H_2O)_2$
complex. This complex is extractable into organic solvents, but only at high
concentrations of the TTA anion.

A large adduct formation constant increases the hydrophobicity of the
metal complex and thus the distribution ratio of the metal. This is commonly
referred to as a *synergistic effect*. Figure 4.2 illustrates the extraction of the
$UO_2(TTA)_2$ complex from 0.01 M HNO_3 into cyclohexane. Because the linear
O−U−O group is believed to have five to seven coordination sites, where only

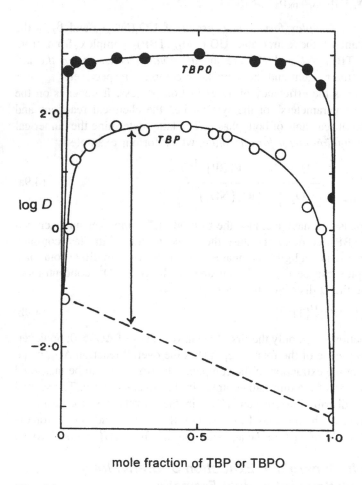

mole fraction of TBP or TBPO

Fig. 4.2 Synergistic extraction: Distribution of U(VI) between 0.01 M HNO_3 and mixtures of thenoyltrifluoroacetone (TTA) and tributylphosphate (TBP), or tributylphos-
phineoxide (TBPO), at constant total molarity ($[TTA]_{org}$ plus $[TBP]_{org}$ or $[TBPO]_{org} =$
0.02 M) in cyclohexane. (From Ref. 2.)

four are occupied in this complex, the uranyl group is coordinativaly unsaturated. At the left vertical axes of Fig. 4.2, the free coordination sites are occupied by water and/or NO_3^-, only; and the U(VI) complex is poorly extracted, log D_U about -1. When TBP or TBPO (tributylphosphine oxide*) [both indicated by B] are added while [HTTA] + [B] is kept constant, the D_U value increases to about 60 for TBP and to about 1000 for TBPO. At the peak value, the complex is assumed to be $UO_2(TTA)_2B_{1 \text{ or } 2}$. The decrease of D_U at even higher [B] is due to the corresponding decrease in [TTA$^-$], so that at the right vertical axes of Fig. 4.2 no U(VI)—TTA complex is formed. For this particular case, at much higher nitrate concentrations, the U(VI) is complexed by NO_3^- and is extracted as an adduct complex of the composition $UO_2(NO_3)_2 B_{1-2}$, as discussed earlier for Case I.

The primary cause for synergism in solvent extraction is an increase in hydrophobic character of the extracted metal complex upon addition of the adduct former. Three mechanisms have been proposed to explain the synergism for metal + cheland† + adduct former. In the first suggested mechanism, the chelate rings do not coordinately saturate the metal ion, which retains residual waters in the remaining coordination sites and these waters are replaced by other adduct-forming molecules. The second involves an opening of one or more of the chelate rings and occupation by the adduct formers of the vacated metal coordination sites. The third mechanism involves an expansion of the coordination sphere of the metal ion upon addition of adduct formers so no replacement of waters is necessary to accommodate the adduct former. As pointed out before, it is not possible from the extraction constants to choose between these alternative mechanisms, but enthalpy and entropy data of the reactions can be used to provide more definitive arguments.

The HTTA + TBP system can serve to illustrate the main points of thermodynamics of synergism. The *overall extraction reaction* is written as:

$$M^{n+} + n \text{ HTTA(org)} + p \text{ TBP(org)} \leftrightarrows M(TTA)_n(TBP)_p(\text{org}) + p \text{ H}^+ \quad (4.10a)$$

We assume that the first step in the extraction equation is complexation in the aqueous phase

$$M^{z+} + n \text{ TTA}^- \leftrightarrows M(TTA)_n^{z-n} (\text{aq}) + n \text{ H}^+ \quad (4.10b)$$

leading to the formation of the uncharged complex $M(TTA)_n$, which immediately dissolves in the organic phase due to its high hydrophobicity/lipophilicity

$$M(TTA)_n (\text{aq}) \leftrightarrows M(TTA)_n (\text{org}) \quad (4.10c)$$

---

*TBPO = $(C_4H_9)_3PO$, see Appendix D, example 16, at the end of this book.
†*Cheland* or *chelator* is the chelating ligand.

The adduct formation reaction in the organic phase (the "synergistic reaction") is obtained by subtracting Eqs. (4.10b) and (4.10c) from Eq. (4.10a):

$$M(TTA)_n(org) + pTBP(org) \leftrightarrows M(TTA)_n(TBP)_p(org) \qquad (4.10d)$$

Thermodynamic data for the extraction reactions of Eqs. (4.10a) and (4.10c) allow calculation of the corresponding values for the synergistic reaction of Eq. (4.10d). Measurements of the reaction

$$UO_2(TTA)_2(org) + TBP(org) \leftrightarrows UO_2(TTA)_2 \cdot TBP(org) \qquad (4.11)$$

at different temperatures gives log $K = 5.10$, $\Delta H^0 = -9.3$ kJ $\cdot$ mol^{-1}, $T\Delta S^\circ = 20.0$ kJ $\cdot$ mol^{-1}.

In another experiment, it was found for Th(TTA)$_4$

$$Th(TTA)_4(org) + TBP(org) \leftrightarrows Th(TTA)_4 \cdot TBP(org) \qquad (4.12)$$

the corresponding values: log $K = 4.94$, $\Delta H^\circ = -14.4$ kJ.mol^{-1}, $T\Delta S^\circ = 13.7$ kJ. mol^{-1}.

Both UO$_2$(TTA)$_2$ and Th(TTA)$_4$ have two molecules of hydrate water when extracted in benzene, and these are released when TBP is added in reactions Eqs. (4.11) and (4.12). The release of water means that two reactant molecules (e.g., UO$_2$(TTA)$_2 \cdot$ 2H$_2$O and TBP) formed three product molecules (e.g., UO$_2$(TTA)$_2 \cdot$ TBP and 2H$_2$O). Therefore, ΔS is positive. Since TBP is more basic than H$_2$O, it forms stronger adduct bonds, and, as a consequence, the enthalpy is exothermic. Hence, both the enthalpy and entropy changes favor the reaction, resulting in large values of log K.

4.2.3 Case III: Maintaining Metal Coordination Number

A guiding principle for the solvent extraction chemist is to produce an uncharged species that has its maximum coordination number satisfied by lipophilic substances (reactants). For trivalent lanthanides and actinides (Ln and An, respectively), the thermodynamic data suggest a model in which addition of one molecule of TBP displaces more than one hydrate molecule:

$$An(TTA)_3(H_2O)_3 \xrightarrow{\text{TBP}} An(TTA)_3(TBP)(H_2O)_{1-2}$$
$$\xrightarrow{\text{TBP}} An(TTA)_3(TBP)_{1-3} \qquad (4.13)$$

This scheme of steps reflects the ability of some metals, like the trivalent actinides and lanthanides, to vary their coordination number; since the trivalent Ln and An may go from 9 to 8 and, finally, back to 9. The last step reflects the operation of the third mechanism proposed for synergism.

Th(TTA)$_4$ can be dissolved in dry benzene without hydrate water. The values of the reaction of Eq. (4.12) in the system are: log $K = 5.46$, $\Delta H^\circ = -39.2$

kJ · mol^{-1}, $T\Delta S° = -8.0$ kJ · mol^{-1} · K^{-1}. The negative entropy is understandable as the net degrees of freedom are decreased (two reactant molecules combine to form one product molecule). However, the $\Delta H°$ value is much more negative.

These equations do not provide complete definition of the reactions that may be of significance in particular solvent extraction systems. For example, HTTA can exist as a keto, an enol, and a keto-hydrate species. The metal combines with the enol form, which usually is the dominant one in organic solvents (e.g., $K = [\text{HTTA}]_{enol}/[\text{HTTA}]_{keto} = \sim6$ in wet benzene). The kinetics of the keto → enol reaction are not fast although it seems to be catalyzed by the presence of a reagent such as TBP or TOPO. Such reagents react with the enol form in drier solvents but cannot compete with water in wetter ones. HTTA · TBP and TBP · H$_2$O species also are present in these synergistic systems. However, if extraction into only one solvent (e.g., benzene) is considered, these effects are constant and need not be considered in a simple analysis.

In section 4.13.3 we return briefly to the thermodynamics of solvent extraction.

4.3 OVERVIEW OF EXTRACTION PROCESSES

Many organic substances as well as metal complexes are less extracted from aqueous solutions into organic solvents than expected from simple considerations such as the amount of organic matter in the solute or their solubility in organic solvents. Such substances are *hydrated* (see Chapter 3). More basic donor molecules can replace such water, forming adducts. For the most common oxygen-containing adduct molecules, the efficiency of the replacement depends on the *charge density*, also referred to as *basicity*, of the oxygen atoms. The sequence in which these donor groups are able to replace each other is

$$\text{RCHO} < \text{R}_2\text{CO} < \text{R}_2\text{O} < \text{ROH} < \text{H}_2\text{O} \approx (\text{RO})_3\text{PO}'$$
$$< \text{R}''\text{R}'\text{NCOR} \approx (\text{RO})_2\text{RPO} < \text{R}_3\text{PO}$$

where R stands for organic substituend. In Chapter 3 the basicity was presented in form of *donor number*. The larger the difference between the donor number of water and the adduct former, the larger the adduct formation constant. Often the donor property has to be rather strong, which is the case for many phosphoryl compounds (like TBP, TBPO, TOPO, etc.), because the concentration of H$_2$O in the aqueous phase is very large (often >50 M), even though H$_2$O is only a moderately strong donor.

Table 4.2 gives a survey of the most common extraction processes. In general, Type I extraction refers to the distribution of nonelectrolytes, without (A) or with adduct former (B). Type II refers to extraction of (mainly organic) acids, Type III to the extraction of metal complexes, and Type IV to the special (but common) use of solvent extraction for evaluation of formation constants

Table 4.2   Symbolic Survey of Fundamental Liquid-Liquid Distribution Processes[a]

Type	Description	Scheme
Type I-A	*Nonelectrolyte extraction*[b] Solute A extracted into organic phase (solvent) (Equilibrium governed by the Nernst distribution law) Solute is the nonelectrolyte A in water	A $\updownarrow$ A
Type I-B	*Nonelectrolyte adduct formation and extraction*[c] Adduct AB in organic phase (plus eventually B) Solute A and adduct former (or extractant) B	$B \quad AB$ $(\updownarrow) \; \updownarrow$ $A + B \leftrightarrows AB$
Type II-A	*Extraction of nonadduct organic acids* Acid and dimer (and possible polymers) in organic phase Acid dissociation in aqueous phase	$HA \leftrightarrows HA, H_2A_2 + \ldots$ $\updownarrow$ $HA \leftrightarrows H^+ + A^-$
Type II-B	*Extraction of acid as adduct* Acid adduct (and acid and adduct former) in organic phase Acid dissociation in aqueous phase	$HAB \; (\leftrightarrows) \; B \; (+) \; HA$ $\updownarrow \qquad (\updownarrow) \quad (\updownarrow)$ $HAB \; \leftrightarrows \; B \; + HA \leftrightarrows H^+ + A^-$
Type III-B[d]	*Extraction of saturated metal complex* Neutral, coordinatively saturated metal complex in organic phase Metal ion M^{z+} is complexed by $z\,A^-$ ligands	MA_z $\updownarrow$ $M^{z+} + zA^- \leftrightarrows MA_z$

Type III-C *Adduct extraction of unsaturated metal complex*

Coordinatively saturated metal complex in organic phase (and B)

Formation of saturated metal complex trough adduct former B[e]

$$M^{z+} + zA^- + bB \leftrightarrows MA_zB_b$$

Type III-D *Liquid anion exchange extractions*

Organic phase with anion exchanger and metal complex

Metal with complexing anions L^- and organic amine

$$M^{z+} + nL^- + pRNH^+L^- \leftrightarrows (RNH^+)_p ML_n^{-p}$$

Type III-E *Extraction of ion pairs, and other unusual complexes*

Ion pair $C_1^+A_2^-$ (and counter species) in organic phase

Aqueous cation C_1^+ and anion A_2^- associated into ion pair $C_1^+A_2^-$

$$C_1^+A_1^- + C_2^+A_2^- \leftrightarrows C_1^+A_2^- + C_2^+A_1^-$$

Type IV *Hydrophilic complex formation and solvent extraction*

Coordinatively saturated metal complex in organic phase

Formation of extractable and nonextractable complexes

$$M^{z+} + zA^- + nX^- \leftrightarrows MA_z + MX_n^{z-n}$$

[a]The *organic phase* (*solvent, diluent*) is assumed to be "inert" (shaded area). The *aqueous phase* (nonshaded area) is unspecified, but may contain various *salting agents*, not considered here.

[b]A *nonelectrolyte solute* is denoted A, an *electrolyte solute* is assumed to be the cation M^{z+} and anion A^-, L^-, or X^-.

[c]The *extractant* (or *reactant*) is denoted A^- (from *acid* HA), or *ligand* L^-, and by B (for *adduct*).

[d]Type III-A (denoted Class A in first edition of this book) is closely related to and covered by Type I-A.

[e]If B is undissociated HA, the self-adduct $MA_z(HA)_x$ may be formed.

for hydrophilic complexes. An arrow within parentheses suggests a reaction of secondary importance. Our three examples are all of Type III-C, but contain also elements of Type I-A (the distribution of TBP) and II-A (the distribution of the weak acid HTTA), though the presence of undissociated acid (HTTA) or the acid adduct (TBP-HTTA) is not discussed. In evaluations of experiments, all molecular species present and all equilibria must be taken into account, as demonstrated subsequently for a number of cases.

Solutes containing metals can further be classified according to the type of ligand; N refers to the maximum coordination number of the metal relative to the ligand:

Class A: Type MX_N. (Note: We generally assume that the ligand is monovalent.) A small number of almost purely covalent inorganic compounds that are extracted by nonsolvating organic solvents. As these complexes are nonelectrolytes and almost as inert as the solutes of Type I-A, they are treated jointly in section 4.4.

Class B: Type MA_z. Neutral coordinatively saturated complexes formed between the metal ion and a lipophilic organic acid. This class contains the large group of metal-organic chelate compounds. For monbasic acids forming bifunctional chelates, $z = N/2$. They belong to the extraction Type III-B, treated in section 4.8.

Class C: Type MA_zB_b or ML_zB_b. These Type III-C complexes are discussed in section 4.2. They are neutral complexes formed between the metal ion and ligands A^- or L^-, where the neutral complex MA_z or ML_z is coordinatively unsaturated ($N > z$ or $2z$) and acts as an acceptor for uncharged organic compounds (adducts B) containing lipophilic donor groups. If the system does not contain any donor molecules B, the water of hydration may be replaced by undissociated HA (assuming the ligand A^- to be a dissociation product of HA), at least at high HA concentrations; the $MA_z(HA)_x$ complexes are refered to as *self-adducts*. Both types of complexes are discussed in section 4.9.

Class D: Ion pairs, consisting of the metal bound in an anionic complex (e.g., ML_n^{z-n}, where $n > z$) and one or more large organic (usually monovalent) cations (symbolized by RNH^+); the extracted complex is written $(RNH)_{n-z}ML_n$. These complexes are treated in section 4.10.

Class E: Metal complexes that do not fit into these categories; e.g., other types of ion pairs and chlatrate compounds (see section 4.11).

All metal ions in water are hydrated, and at higher pH most of them also hydrolyze. It can be difficult to distinguish between the hydrolyzed and the complexed species, as well as their self-adducts. For such systems, plots of D_M against $[A^-]$ at various pH and total concentrations of [HA] show three types of curves: (a) for the simple chelate MA_n, (b) for the self-adduct $MA_n(HA)_b$, and

(c) for the (mixed) hydroxide $MA_n(OH)_p$ see Fig. 4.3. It should be noted that the mixed $MA_n(OH)_p$ complexes include the $MA_n + M(OH)_p$ complexes. As mixed complexes are more difficult to determine, they are less often described. However, it is important to realize that if metal hydroxy complexes are formed and not corrected for, the result of the investigation can be misleading. A test of the system according to Fig. 4.3 rapidly establishes the type of metal complexation.

Because metals differ in size, charge, and electronic structure, no two metals behave exactly the same in the same solvent extraction system, not even for the same class of solutes. Nevertheless, there are systematic trends in the formation and extraction of these complexes, as described in Chapter 3. Here, the emphasis is on models that give a quantitative description of the extraction within each type or class.

In the subsequent discussion, the following simplifications are made:

1. *The systems behave "ideally,"* i.e., the activity factors are assumed to be unity, unless specifically discussed;
2. *The metal extracted is in trace concentration*: $[M]_t \ll [Extractant]_t$, as this simplifies the equations;
3. *The reactants are at very low concentrations in both phases.*

These are great simplifications in comparison with the industrial solvent extraction systems described in later chapters. Nevertheless, the same basic reactions occur also in the industrial systems, although activity factors must be introduced or other adjustments made to fit the data, and the calculation of free

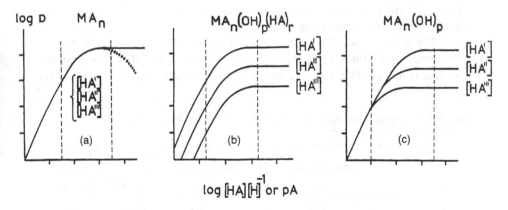

Fig. 4.3 Extraction curves for various types of metal chelate complexes, when log D_M is plotted against free ligand ion concentration, $pA = -\log[A^-]$, or against $[HA][H^+]^{-1}$. From such plots, the general type of metal chelate complex may be identified: (a) type MA_n, (see also Fig. 4.10); (b) type $MA_n(OH)_p(HA)_r$, (see also Figs. 4.14 and 4.30); (c) type $MA_n(OH)_p$, (see also Fig. 4.19). (From Refs. 3a and 3b.)

ligand concentrations are more complex. Some of these simplifications are not used in later chapters.

4.4 EXTRACTION OF INERT MOLECULES (TYPE I-A)

Here, and in later sections, we begin with same kind of rectangular figure to indicate the type of extraction: to the right we indicate the distribution of the solutes in a two-phase system (the organic phase is shaded); the system is also briefly described by the text to the left, and—of course—in detail in the main text.

Solute A extracted into organic phase (solvent)	A
(Nernst distribution law for regular mixtures and solvents:)	$\downarrow\uparrow$
The non-electrolyte solute A in water	A

If the solute A does not undergo any reaction in the two solvents, except for the solubility caused by the "solvation" due to the nonspecific cohesive forces in the liquids, the distribution of the solute follows the Nernst distribution law, and the equilibrium reaction can be described either by a distribution constant $K_{D,A}$, or an (equilibrium) extraction constant K_{ex}:

$$A(aq) \leftrightarrows A(org); \quad K_{D,A} = K_{ex} = [A]_{org}/[A]_{aq} \qquad (4.14)$$

K_{ex} always refers to a two-phase system. The *measured distribution ratio* for the solute A, D_A, equals $K_{D,A}$, and is a constant independent of the concentration of A in the system. Only "external" conditions influence the $K_{D,A}$ value. In "external" conditions we include the organic solvent, in addition to physical conditions like temperature and pressure.

   The noble gases and the halogens belong to the same type of stable molecular compounds: RuO_4, OsO_4, $GeCl_4$, $AsCl_3$, $SbCl_3$, and $HgCl_2$. The simplest example is the distribution of the inert gases, as given in Table 4.3. The larger

Table 4.3 Distribution Ratios of Some Gases Between Organic Solvents and 0.01 M NaClO$_4$ at 25°C

Solvent	Permittivity ε	Solute			
		Xenon	Radon	Bromine	Iodine
Hexane	1.91	41	80	14.5	36
Carbontetrachloride	2.24	35	59	28	86
Chloroform	4.90	35	56	37	122
Benzene (π-bonds)	2.57	27	55	87	350
Nitrobenzene	34.8	14	21	41	178

Source: Ref. 4.

Rn is extracted more easily than the smaller Xe, because the work to produce a cavity in the water structure is larger for the larger molecule. The energy to produce a cavity in nonpolar solvents is much less, because of the weaker interactions between neighboring solvent molecules. Energy is released when the solute leaves the aqueous phase, allowing the cavity to be filled by the hydrogen-bonded water structure. Thus the distribution constant increases with increasing inertness of the solvent, which is measured by the dielectric constant (or relative permittivity). The halogens Br_2 and I_2 show an opposite order due to some low reactivity of halogens with organic solvents. Very inert solvents with low permittivity, such as the pure hydrocarbons, extract inert compounds better than solvents of higher permittivity; conversely, liquids of higher permittivity are better solvents for less inert compounds. Molar volumes should be used for accurate comparisons; such data are found in Table 2.1 and in Ref. [6].

In benzene, the distribution constant depends on specific interactions between the solute and the benzene pi-electrons. Table 4.4 shows the importance of the volume effect for the mercury halide benzene system (Cl<Br<I).

Undissociated fatty acids (HA) behave like inert molecules. Figure 4.4 shows the distribution ($D_{HA} = K_{D,HA}$) between benzene and 0.1 M $NaClO_4$ of fatty acids of different alkyl chain lengths (C_n, $n = 1$ to 5); the distribution constant for an acid with chain length n is given by the expression log $K_{D,HA} = -2.6 + 0.6n$. Similar correlations between $K_{D,HA}$ and molecular size or chain length are observed also for other reagents (e.g., normal alcohols).

For organic solutes, not only the size but also the structure is of importance. Table 4.5 gives distribution constants for substituted oxines. When the substitution increases the size of the molecule, the distribution constant increases. The variations within Table 4.5 and position of the substitution in the oxine molecule reflect structural effects. Table 4.6 shows distribution constants for β-diketones. The increasing K_D with molecular size for the series acetylacetone, benzoylacetone, and dibenzoylmethane, reflects the decreasing solubility in the water phase, mainly governed by the increased energy necessary to overcome the solute-solvent interactions for the larger extractant molecules in water. Thenoyltrifluoroactetone has a greater hydrophilic character than the other β-diketones due to its O− and F− atoms, which interact with the water molecules,

Table 4.4 Distribution Constants for Mercury Halides Between Benzene and 0.5 M $NaClO_4$

Solute	Log K_D	Solute	Log K_D
$HgCl_2$	−0.96	HgClBr	−0.42
$HgBr_2$	0.15	HgICl	0.28
HgI_2	1.79	HgIBr	0.79

Source: Ref. 5.

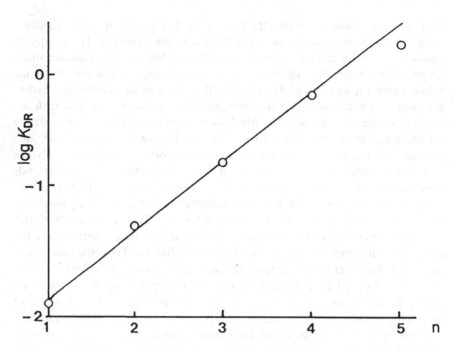

Fig. 4.4 Distribution constants $K_{D,HA}$ of fatty acids as a function of the number n of carbon atoms in the alkyl chain (C_1 is acetic acid) in the system 0.1 M NaClO$_4$/benzene. (From Ref. 7.)

Table 4.5 Dissociation, K_a, and Distribution, $K_{D,HA}$, Constants for Substituted Oxines[a]

Reagent	pK_a	log $K_{D,HA}$
Oxine	9.7	2.7
2-Methyloxine	10.0	3.4
5-Methyloxine	9.9	3.3
5-Acetyloxine	7.8	2.8
4,7-Dichlorooxine	7.4	3.9
7,7-Diiodooxine	8.0	4.2
5-Chloro-7-iodooxine	7.9	3.9

[a]Aqueous phase 0.1 M NaClO$_4$; organic phase chloroform at 25°C.
Source: Refs. 8a, b.

Table 4.6 Dissociation K_a and Distribution Constants $K_{D,HA}$ for β-Diketones; Aqueous Phase 0.1 M NaClO$_4$[a]

		log K_{DR}		
Reagent (solute)	pK_a	C$_6$H$_6$	CHCl$_3$	CCl$_4$
Acetylacetone, AA; CH$_3$ R CH$_3$	8.76	0.76	1.36	0.51
Benzoylacetone, BZA; CH$_3$ R C$_6$H$_5$	8.74	3.15	3.60	2.81
Dibenzoylmethane, DBM; C$_6$H$_5$ R C$_6$H$_5$	9.35	5.34	5.40	4.51
Thenoyltrifluoroacetone, TTA; R'R CF$_3$	6.3	1.61	1.84	1.54

R is $-C\ CH_2\ C-$ R' is thenoyl, H$_3$C$_3$SC$-$
$\quad\quad \| \quad\quad \|$
$\quad\quad O \quad\quad O$

[a]Organic phases 0.1 M in solute; 25°C.
Source: Ref. 4.

leading to a reduction in the distribution constant. Thus, either the size effect related to the water structure or the presence of hydrophilic groups in the solute determines the general level of its distribution constant.

4.5 EXTRACTION OF ADDUCT-FORMING NONELECTROLYTES (TYPE I-B)

Adduct *AB* in organic phase (plus evt. *B*)	*B*	*AB*
	(↓↑)	↓↑
Solute A and adduct former (extractant) B	$A + B \leftrightarrows AB$	

The extraction of a solute A may be improved by its reaction with another solute ("extraction reagent", or *extractant*), B, forming an *adduct compound*, AB. This occurs through chemical interaction between A and B.

$$B \leftrightarrows B(org) \qquad K_{D,B} = [B]_{org} / [B] \qquad\qquad (4.15a)$$

$$A + B \leftrightarrows AB \qquad K_{ad} = [AB]/[A][B] \qquad\qquad (4.15b)$$

where K_{ad} is the *adduct formation constant* (in the aqueous phase)

$$AB \leftrightarrows AB(org) \qquad K_{D,AB} = [AB]_{org}/[AB] \qquad\qquad (4.15c)$$

and $K_{D,AB}$ the adduct distribution constant. The extraction constant for the overall reaction is

$$A(aq) + B(org) \leftrightharpoons AB(org) \quad K_{ex} = [AB]_{org}/[A]_{aq}[B]_{org}$$

$$= K_{D,AB} K_{ad} K_{D,B} \quad (4.15d)$$

and also

$$D_A = K_{ex}[B]_{org} \quad (4.15e)$$

For the extraction reaction it may suffice to write the reaction of Eq. (4.15d), though it consists of a number of more or less hypothetical steps. As mentioned, equilibrium studies of this system cannot define the individual steps, but supplementary studies by other techniques may reveal the valid ones. Equation (4.15) indicates that the reaction takes place at the boundary (*interface*) between the aqueous and organic phases. However, it is common to assume that a small amount of B dissolves in the aqueous phase, and the reaction takes place in the steps

$$A(aq) + B(aq) \rightarrow AB(aq) \rightarrow AB(org)$$

These equations allow definition of a distribution constant for the species AB, $K_{D,AB}$ [see Eq. (4.15c)]. Distribution constants can also be defined for each of the species A, B and AB ($K_{D,A}$, etc.) but this is of little interest as the concentration of these species is related through K_{ex}. A large K_{ex} for the system indicates that large distribution ratios D_A can be obtained in practice. As shown in Eq. (4.15), the concentration of B influences the distribution ratio D_A.

Consider first the extraction of hexafluoroacetylacetone (HFA) by TOPO by Example 1, and, second, the extraction of nitric acid by TBP (Example 7). The principles of volume and water-structure effects, discussed for the solute A in section 4.4, are also important in the distribution of the adducts.

Example 1: Extraction of hexafluoroacetylacetone (HFA) by trioctylphospine oxide (TOPO).

Abbreviating HFA (comp. structure 5e, Appendix D) by HA, and TOPO by B, we can write the relevant reactions

$$HA(aq) \leftrightharpoons HA(org) \qquad\qquad K_D = [HA]_{org}/[HA] = D_0 \quad (4.16a)$$

$$HA(org) + B(org) \leftrightharpoons HAB(org) \quad K_{ad1} = [HAB]_{org}/[HA]_{org}[B]_{org} \quad (4.16b)$$

$$HA(org) + 2B(org) \leftrightharpoons HAB_2(org) \quad K_{ad2} = [HAB_2]_{org}/[HA]_{org}[B]_{org}^2 \quad (4.16c)$$

assuming that 2 adducts are formed, HAB and HAB$_2$, the latter containing 2 TOPO molecules. Equation (4.16a) denotes the distribution of "uncomplexed HA" by D_0. Combining these equations yields

$$D \cdot D_0^{-1} = 1 + K_{ad1}[B] + K_{ad2}[B]^2 \quad (4.16d)$$

Figure 4.5 shows the relative distribution, $\log D \cdot D_0^{-1}$, of hexafluoroacetylacetone as a function of the concentration of the adduct former TOPO. HFA

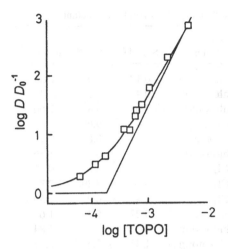

Fig. 4.5 Relative increase, D/D_o in extraction of hexafluoroacetylacetone (HFA) into hexane from 0.1M NaClO$_4$ at pH = 2, at different concentrations of the adduct trioctylphosphine oxide (TOPO) in the organic phase. The fitted curve is $D/D_o = 1 + 10^{4.22}$ [TOPO]$_{org}$ + $10^{7.51}$[TOPO]$^2_{org}$. (From Ref. 9.)

is a moderately weak acid, while TOPO associates strongly with hydrogen-bond donors in nonpolar solvents like hexane. The constants were determined to log $K_{ad1} = 4.22$ and log $K_{ad2} = 7.51$. Thus even at moderately low TOPO concentrations, the "dimer adduct" dominates.

4.6 EXTRACTION OF NONADDUCT ORGANIC ACIDS (TYPE II-A)

Acid and dimer (and possibly polymers) in organic phase	$HA \leftrightarrows \frac{1}{2} H_2A_2 + \cdots$ $\downarrow\uparrow$
Acid dissociation in aqueous phase (and protonation)	$(H^+AH) \leftrightarrows HA \leftrightarrows H^+ + A^-$

Tables 4.5–4.7 and Fig. 4.6 list organic acids commonly used as metal extractants. When the acids are not protonated, dissociated, polymerized, hydrated, nor form adducts, the distribution ratio of the acid HA is constant in a given solvent extraction system:

$$HA(aq) \leftrightarrows HA(org) \qquad K_{D,HA} = [HA]_{org}/[HA] = D_{HA} \qquad (4.17)$$

This is shown by the horizontal trends in Fig. 4.6, for which Eq. (4.17) is valid; i.e., the distribution constant $K_{D,HA}$ equals the measured distribution ratio. When

Table 4.7 Physical Properties of Some Commonly Used Acidic Extractants[a]

Acid	Organic diluent[b]	$S_{org}(M)$[c]	$-\log K_a$	$\log K_{DR}$
Salicylic acid	Chloroform	0.17	2.9	0.5
Cupferron	Chloroform	0.4	4.2	2.3
8-Hydroxyquinoline (oxine) (OQ)	Chloroform	2.63	9.8[d]	2.66
D:o	CCl_4	—	9.66	2.18
Acetylacetone (AA)	Benzene	∞	8.85	0.78
D:o	Chloroform	—	—	1.36
D:o	CCl_4	—	8.67	0.51
Benzoylacetone (BA)	CCl_4	—	8.39	2.81
Benzoyltrifluoroacetone (BTFA)	CCl_4	—	6.03	2.39
Thenoyltrifluoroacetone (TTA)	Benzene	5.27	6.3	1.6
D:o	Chloroform	—	—	1.84
1-Nitroso-2-naphthol	Chloroform	1.35	7.6	2.97
Di(2-ethylhexyl)phosphoric acid[f]	*n*-Octane	∞	1.4	3.44
Mono(2-ethylhexyl)phosphoric acid	*n*-Octane	∞	1.3[e]	—
Dinonyl naphthalene sulfonic acid				

[a]The aqueous phase is mostly 0.1 M $NaClO_4$ at 25°C.
[b]The choice of organic diluent only affects the distribution constant, not the acid dissociation constant.
[c]S_{org} is solubility in M in organic solvent.
[d]Log $K_{aH} = 5.00$.
[e]K_{a1}.
[f]Dimerization constant: log $K = 4.47$; see also section 4.6.3.

HA is used for the extraction of a metal, $K_{D,HA}$ is abbreviated K_{DR}, for the *d*istribution constant (of the unmodified) *r*eagent (or extractant).

Figure 4.7a shows the effect of aqueous salt concentrations on the D_{HA} value of acetylacetone at constant total HA concentration and pH. The salt has two effects: (1) it ties up H_2O molecules in the aqueous phase (forming hydrated ions) so that less free water is available for solvation of HA; and (2) it breaks down the hydrogen bond structure of the water, making it easier for HA to dissolve in the aqueous phase. Figure 4.7 shows that the former effect dominates for NH_4Cl while for $NaClO_4$ the latter dominates. We describe the increase of the distribution ratio with increasing aqueous salt concentration as a *salting-out* effect, and the reverse as a *salting-in* effect.

Figure 4.7b shows D_{HA} for the extraction of acetylacetone into $CHCl_3$ and C_6H_6 for two constant aqueous $NaClO_4$ concentrations at pH 3, but with varying concentrations of HA. Acetylacetone is infinitely soluble in both $CHCl_3$ and C_6H_6; at $[HA]_{org} = 9$ M, about 90% of the organic phase is acetylacetone (M_w 100), so the figure depicts a case for a changing organic phase. Figure 4.7b also

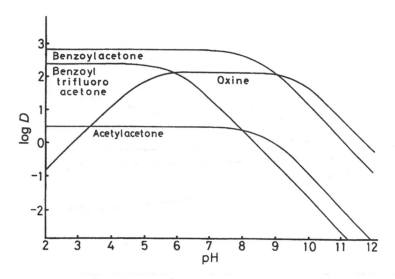

Fig. 4.6 Distribution ratios calculated by Eq. (4.22) for acetylacetone (HAA); benzoylacetone (HBA); bezoyltrifluoroacetone (HBTFA); and oxine (8-hydroxyquinoline, HOQ), in the system 0.1 M $NaClO_4$ /CCl_4, using the following constants. (From Refs. 8a, b.)

	HAA	HBA	HBTFA	HOQ
$\log K_D$	0.51	2.81	2.39	2.18
$\log K_a$	8.67	8.39	6.03	9.66
$\log K_{aH}$	—	—	—	5.00

indicates different interactions between the acetylacetone and the two solvents. It is assumed that the polar $CHCl_3$ interacts with HA, making it more soluble in the organic phase; it is also understandable why the distribution of HA decreases with decreasing concentration (mole fraction) of $CHCl_3$. C_6H_6 and aromatic solvents do not behave as do most aliphatic solvents: in some cases the aromatics seem to be inert or even antagonistic to the extracted organic species, while in other cases their pi-electrons interact in a favorable way with the solute. For acetylacetone, the interaction seems to be very weak. The salting-in effect is shown both in Figs. 4.7a. and 4.7b.

4.6.1 Dissociation

Acids dissociate in the aqueous phase with a dissociation constant K_a

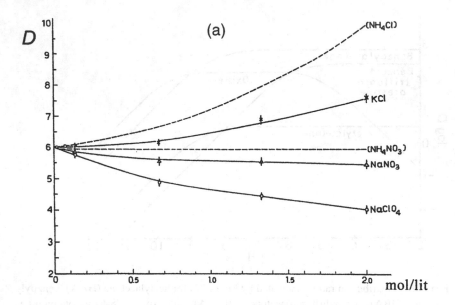

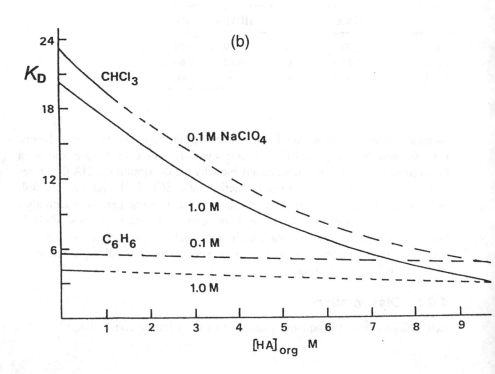

$$HA \leftrightharpoons H^+ + A^- \qquad K_a = [H^+][A^-]/[HA] \tag{4.18}$$

The distribution ratio incorporates the K_a for extraction of acids, HA, as:

$$D_A = [HA]_{org}([HA]+[A^-])^{-1} = K_{D,HA}(1+K_a \ [H^+]^{-1}) \tag{4.19}$$

Index A indicates that the distribution ratio refers to the concentration of all species of A in the organic and in the aqueous phase. In Fig. 4.6 the distribution of the β-diketones is constant in the higher hydrogen ion concentration range (lower pH) where they are undissociated. In the higher pH region, D_A becomes inversely proportional to the hydrogen ion concentration due to increase in the concentration of the dissociated form of the acid A^-, in agreement with Eq. (4.19.)

The *free ligand concentration*, $[A^-]$, is an important parameter in the formation of metal complexes (see Chapter 3 and section 4.8). In a solvent extraction system with the volumes V and V_{org} of the aqueous and organic phases, respectively, $[A^-]$ is calculated from the material balance:

$$\log [A^-] = \log K_a - \log[H^+] + \log \ (m_{HA,t} \ / \ V_{org}) - \log \ F \tag{4.20a}$$

where

$$F = K_{D,HA} + V \ V_{org}^{-1} + K_a \ [H^+] \ V \ V_{org}^{-1} \tag{4.20b}$$

$m_{HA,t}$ is the total amount (in moles) of HA (reagent) added to the system. Often $m_{HA,t} \ V_{org}^{-1}$ is abbreviated $[HA]_{org}^o$, indicating the original concentration of HA in the organic phase at the beginning of the experiment (when $[HA]_{aq} = 0$). When $V_{org} = V$ and pH $\ll$ pK_a, $F = 1 + K_{D,HA}$. From Eq. (4.20) it can be deduced that $[A^-]$ increases with increasing pH, but tends to become constant as the pH value approaches that of the pK_a value. In the equations relating to the extraction of metal complexes, HA is often identical with reagent R; the indexes may be changed accordingly, thus e.g., $K_{D,HA} \equiv K_{DR}$. (Note: Various authors use slightly different nomenclature; here we follow reference Appendix C.)

4.6.2 Protonation

At low pH, some organic acids accept an extra proton to form the H_2A^+ complex. This leads to a decrease in the D_A value at pH < 6, as shown in Fig. 4.6:

Fig. 4.7 Distribution ratio D_{HA} of undissociated acetylacetone. (a) Distribution between benzene and aqueous phase containing different inorganic salts; 25°C. (b) Distribution between CHCl$_3$ (upper curves) or C$_6$H$_6$ (lower curves) and aqueous phase 0.1 and 1.0 M in NaClO$_4$ as a function of $[HA]_{org}$. The uncertainty at the lowest D values is ±1 for CHCl$_3$ and ±0.2 for C$_6$H$_6$. (From Ref. 10.)

$$HA + H^+ \leftrightharpoons H_2A^+ \quad K_{aH} = [HA][H^+]/[H_2A^+] \tag{4.21}$$

Because H_2A^+ is ionic, it is not extracted into the organic phase, and thus the distribution ratio becomes

$$D_A = [HA]_{org}([H_2A^+]+[HA]+[A^-])^{-1}$$
$$= K_{D,HA}(K_{aH}^{-1}[H^+]+1+K_a[H^+]^{-1})^{-1} \tag{4.22}$$

as illustrated in Fig. 4.6 for oxine.

4.6.3 Dimerization

Figure 2.1 illustrates a number of orientations by which two linear acids may form a dimer. The partial neutralization of the hydrophilic groups leads to increased solubility of the acid in the organic solvent, but is not observed in the aqueous phase. The dimerization can be written as:

$$2HA(org) \leftrightharpoons H_2A_2(org) \quad K_{di} = [H_2A_2]_{org}/[HA]_{org}^2 \tag{4.23}$$

The distribution ratio for the extraction of the acid becomes:

$$D_A = ([HA]_{org} + 2[H_2A_2]_{org})([HA]+[A^-])^{-1}$$
$$= K_{D,HA}(1+2K_{di}K_{D,HA}[HA]_{aq}(1+K_a[H]^{-1})^{-1} \tag{4.24}$$

The last term can be expressed in several different ways. Because the distribution ratio D_A reflects the analytical concentration of A in the organic phase, the dimer concentration is given as $2[H_2A_2]$, although it is a single species (one molecule). Figure 4.8 illustrates how the dimerization leads to an increase of acid distribution ratio with increasing aqueous acid concentration. For propionic acid $\log K_a = -4.87$, $\log K_{D,HA} = -1.90$ and $\log K_{di} = 3.14$ in the system. The extraction increases as the size of the acid increases. A dimeric acid may form monobasic complexes with metal ions, as is illustrated by the formulas in Appendix D:14 b–d, for the $M(H(DEHP)_2)_3$ and $UO_2((DEHP)_2)_2$ complexes. The situation may be rather complex. For example, at very low concentrations in inert solvents, dialkylphosphates $(RO)_2POOH$ act as a monbasic acid, but at concentrations >0.05 M they polymerize, while still acting as monobasic acids (i.e., like a cation exchanger). The degree of dimerization/polymerization depends on the polarity of the solvent [11b].

4.6.4 Hydration

In solvent extraction, the organic phase is always saturated with water, and the organic extractant may become hydrated. In the extraction of benzoic acid, HBz (Appendix D:2), it was found that the organic phase contained four different

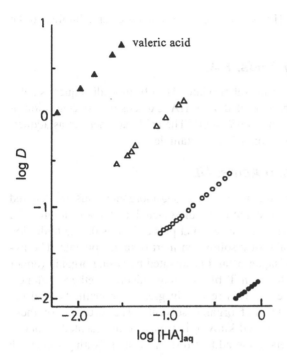

Fig. 4.8 Distribution ratios (from bottom to top) of acetic (C_2 ●), propionic (C_3 O), butyric (C_4 Δ), and valeric (C_5 ▲) acids (carbon chain length C_n) between carbon tetrachloride and water as a function of the acid concentration in the aqueous phase, $[HA]_{aq}$. (From Ref. 11a.)

species: the monomer HBz, the monomer hydrate HBz · H_2O, the dimer H_2Bz_2, and the dimer hydrate $H_2Bz_2(H_2O)_2$. Only by considering all these species is it possible to explain the extraction of some metal complexes with this extractant.

4.7 EXTRACTION OF ACIDS AS ADDUCTS (TYPE II-B)

Acid adduct (and acid and adduct former) in organic phase	HAB ($\leftrightarrows$	B	$+ HA$)
	↓↑	(↓↑)	(↓↑)
Acid HA dissociating and forming adduct in aqueous phase	HAB $\leftrightarrows$	B	$+ HA \leftrightarrows H^+ + A^-$

The solubility of organic acids in water is due to the hydrophilic oxo- and hydroxo-groups of the acid that form hydrogen bonds with water molecules. If the hydrogen ion of the acid is solvated by a donor organic base, B, in the organic phase, the adduct B–HA is likely to have much greater solubility in the organic phase.

4.7.1 Weak (Organic) Acids, HA

In extraction of weak organic acids (abbreviated HA) from acidic aqueous solutions, the concentration of undissociated acid, HA, exceeds the concentration of its dissociated anions, A^-, as long as $pK_a > pH$. The acid may then act as adduct-forming nonelectrolyte; see section 4.5 and Example 1.

4.7.2 Strong (Inorganic) Acids, HL

To avoid confusion with weak organic acids, strong inorganic acids are denoted by HL. Most strong acids are completely dissociated and both cations and anions are hydrated in aqueous solutions even at pHs as low as 0. The hydration makes them lipophobic and almost insoluble in inert organic solvents. The hydrogen ion is a Lewis acid (Chapter 3) and is solvated by strong organic (donor or base) molecules, such as those in Table 4.8 (e.g., alcohols, ethers, ketones, esters, amines, phosphoryls, etc.). This results in greater lipophilicity, and the acid becomes more soluble in inert organic solvents. The structure of these "solvated hydrogen salts" is not well known, but may be represented symbolically by $HB_b^+L^-$, where B refers to the adduct former or the solvating solvent; b may have a value of 1–4.

The order of extractability changes with aqueous acidity, but in general follows the order $HClO_4 \approx HNO_3 > HI > HBr > HCl > H_2SO_4$ (see Table 4.8). Since the hydration energies of the acids follow the opposite order, dehydration is an essential step in the solvent extraction process. This order of acids has a practical significance: acids higher can be replaced by the acids lower in the sequence; e.g., HF and HNO_3 are extracted from acidic stainless steel pickling waste solutions into kerosene by addition of H_2SO_4 (see Chapter 14 of this book).

The extraction of most acids is accompanied by extraction of water. In the extraction of HNO_3 by TBP into kerosene, many different species have been identified, several of which involve hydration. The ratio of acid:adduct is not very predictable. For example, $HClO_4$ apparently is extracted into kerosene with 1–2 molecules of TBP, HCl into ethylether with one molecule of ethylether, etc. Also, the extracted acid may dimerize in the organic solvent, etc. Example 2 illustrates the complexity of the extraction of HNO_3 by TBP into kerosene.

Assuming that B is almost insoluble in the aqueous phase, the equilibrium reaction can be written in two ways:
1. The *interface extraction model* assumes that HA reacts with B at the interface. Thus

$$HA(aq) + bB(org) \leftrightarrows HAB_b \ (org) \qquad\qquad (4.25a)$$

Table 4.8 Basicity (Electron Pair-Donating Tendency) of Some Ions and Molecules (R is an alkyl or aryl group)

Basicity of some common anions
relative to the (hard type) actinide cations
$ClO_4^- < I^- < Br^- < Cl^- < NO_3^- < SCN^- < acetate^- < F^-$

Basicity of some organic molecules

Amine compounds		$R_3N^a < R_2NH^b < RNH_2^c < NH_3$
Arsine compounds		R_3As
Phosphine compounds		R_3P
Oxo-compounds	Phosphoryls	$(RO)_3PO^d < R'(RO)_2PO^e < R_2'(RO)PO^f < R_3'PO^g$
	Arsenyls	R_3AsO^h
	Carbonyls	$RCHO < R_2CO\ (\le R_2O < ROH < H_2O)^i$
	Sulfuryls	$(RO)_2SO_2^j < R_2SO_2^k < (RO)_2SO^l < R_2SO^m$
	Nitrosyls	$RNO_2^n < RNO^o$

Substitutions causing *basicity decrease of oxo compounds*
$(CH_3)_2CH— < CH_3(CH_2)_n— < CH_3— < CH_3O— < ClCH_2—$

[a-c]Tertiary, secondary, and primary amines.
[d]tri-R phosphate.
[e]di-R-R' phosphonate.
[f]R-di-R' phosphinate.
[g]tri-R' phosphine oxide.
[h]arsine oxide.
[i]ether and hydroxo compounds.
[j]sulfates.
[k]sulfones.
[l]sulfites.
[m]sulfoxides.
[n]nitro compounds.
[o]nitroso compounds.
Source: Ref. 12.

Since $[HA]_{aq}$ and $[B]_{org}$ are easily measurable quantities, it is common to define the extraction constant K_{ex} for this model:

$$K_{ex} = [HAB_b]_{org}\ [HA]^{-1}\ [B]_{org}^{-b} \tag{4.25b}$$

2. The *organic phase reaction model* assumes all reactions take place in the organic phase. Thus one assumes

$$HA(org) + bB(org) \leftrightarrows HAB_b(org) \tag{4.26a}$$

The equilibrium constant for this reaction is

$$K_{ad,bB} = [HAB_b]_{org}[HA]_{org}^{-1}\ [B]_{org}^{-b} \tag{4.26b}$$

where $K_{ad,bB}$ is the (organic phase) *adduct formation constant*. The distribution ratio of the acid in this system becomes

$$D_A = ([HA]_{org} + [HAB_b]_{org})([HA] + [A^-])^{-1}$$

$$= K_{D,HA}(1 + \Sigma K_{ad,bB}[B]^b_{org})/(1 + K_a[H^+]) \qquad (4.26c)$$

Equation (4.25b) becomes identical to Eq. (4.26b) if K_{ex} is replaced by $K_{D,HA}$ $K_{ad,bB}$. Equilibrium measurements do not allow a decision between the two reaction paths.

Example 2: Extraction of nitric acid by pure TBP.

Many metals can be extracted from nitrate solutions by TBP. In those systems it is important to account for the HNO_3-TBP interactions. The next set of equations were derived by [13] and are believed to be valid for the extraction of HNO_3 at various nitrate concentrations into 30% TBP in kerosene. Abbreviating HNO_3 as HL, and TBP as B, and including hydration for all species without specification, one derives

1. The formation of an acid monoadduct:

$$H^+ + L^- + B(org) \leftrightarrows HLB(org) \qquad (4.27a)$$

For simplicity, we write the adduct HLB, instead of HB^+L^-. The extraction constant is

$$K_{ex1} = [HLB]_{org}[H]^{-1}[L]^{-1}[B]^{-1}_{org} \qquad (4.27b)$$

2. The formation of a diacid monoadduct:

$$2H^+ + 2L^- + B(org) \leftrightarrows (HL)_2B(org) \qquad (4.28a)$$

$$K_{ex2} = [(HL)_2B]_{org}[H]^{-2}[L]^{-2}[B]^{-1}_{org} \qquad (4.28b)$$

3. Ion pair association:

$$H^+ + L^- \leftrightarrows H^+L^- \qquad (4.29a)$$

$$K_{ass} = [H^+][L^-]/[H^+L^-] \qquad (4.29b)$$

This reaction only occurs under strong acid conditions, and the equilibrium constant may be <1.

4. The distribution of nitric acid is then given by

$$D_{NO3} = ([HLB]_{org} + 2[(HL)_2B]_{org} + \ldots)([L^-] + [H^+L^-])^{-1} \qquad (4.30)$$

5. Furthermore, the dimerization of TBP in the organic phase must be taken into account:

$$2B \leftrightarrows B_2 \qquad (4.31a)$$

$$K_{di} = [B_2]_{org}/[B]^2_{org} \qquad (4.31b)$$

yielding the total concentration of TBP in the organic phase

$$[TBP]_{org,t} = [B]_{org} + 2K_{di}[B]^2_{org} \qquad (4.32)$$

6. The distribution ratio in terms of only [H$^+$] and monomeric [B]$_{org}$ can then be expressed by

$$D_{NO3} = K_{ass}^{-1} \, [B]_{org} \, (K_{ex1} + 2 \, K_{ex2} \, [H^+]^2)$$ (4.33)

In this equation it is assumed [H$^+$]=[L$^-$] (electroneutrality in the aqueous phase). Equation (4.33) has been tested, and the results agreed with >2300 experiments under varying conditions, see Fig. 4.9. The example illustrates the rather complicated situation that may occur even in such "simple" systems as the extraction of HNO$_3$ by TBP.

4.8 EXTRACTION OF COORDINATIVELY SATURATED METAL CHELATE TYPE COMPLEXES (TYPE III-B)

Neutral, coordinatively saturated metal complex in organic phase	MA_z $\updownarrow$
Metal ion M^{z+} is complexed by $z\ A^-$ ligands to form neutral MA_z.	$M^{z+} + zA^- \leftrightarrows MA_z$

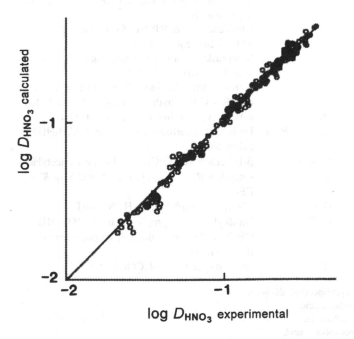

Fig. 4.9 Test of the equations in Example 2 for extraction of 0.01–0.5 M nitric acid with 30% TBP in kerosene at temperatures 20–60°C. (From Ref. 13.)

Section 3.2 describes an important class of organic ligands that are able to complex a metal ion through two or more binding sites of "basic" atoms, like O, N, or S, to form *metal chelates*. Table 4.9 presents the various types, their number of acidic groups, the chelate ring size, and the coordinating atoms. Neutral chelate compounds are illustrated in Appendix D: 5h, 14b and 14d. In Appendix D:5h the ring size is 6 for the complex between Cu^{2+} and each of the two acetylacetonate anions, Aa^-, while for the dimeric HDEHP ligand in the figures in Appendix D:14b and 14c the ring size becomes 7 for the M^{3+} and UO_2^{2+} complexes (see the structures in Appendix D). As discussed in Chapter 3,

Table 4.9 Some Organic Compounds Functioning as Polydentate Anions in Metal Extraction

Chelate ring size	Acidic groups	Coordinating atoms	Compound group and examples
4	1	O, O	Carboxylic acid, RCOOH; e.g., perfluorobutyric (C_3F_7COOH), salicylic $C_6H_4(OH)COOH$, cinnamic ($C6H5(CH)_2COOH$) acids
4	1	O, O[a]	Di(alkyl or aryl)phosphoric and phosphinic acids, RR'PO(OH); e.g., HDEHP[b]; corresponding thioacids[a]
4	1	S, S	Dithiocarbamate, RR'NC(S)SH, xanthate, ROC(S)SH; e.g., NaDDC[c]
4	2	O, O[a]	Mono(alkyl or aryl)phosphoric and phosphinic acid, RPO(OH)$_2$; e.g., H_2MEHP[d]
5	1	O, O	Nitrosohydroxylamine, RN(NO)OH; e.g., cupferron ($R = C_6H_5$); hydroxamic acid, RC(O)NHOH
5	1	O, N	8-Hydroxyquinoline (oxine), C_9NH_6OH
5 or 6	1	S, N, or N, N	Diphenylthiocarbazone (dithizone), $C_6H_5NHNC(SH)NNC_6H_5$
6	1	O, O	β-Diketone, RC(O)CHC(OH)R'; e.g., acetylacetone ($R = R' = CH_3$), HTTA[e] ($R = C_4SH_3$, $R' = CF_3$)
6	1	O, O	1-Nitroso-2-naphthol, $C_{10}H_6(NO)OH$
6	2	O, O	Di(alkyl or aryl)pyrophosphate, RP(O)(OH) OP(O)(OH)R'; e.g., dioctylpyrophosphate ($R = R' = C_8H_{17}O$)
>5 or 2 × 4	2	O, O	Dicarboxylic acids, R(COOH)2

[a]O, S, or S, S for the corresponding thioacids.
[b]Di(2-ethylhexyl)phosphoric acid.
[c]Sodium diethyldithiocarbamate.
[d]Mono(2-ethylhexyl)phosphoric acid.
[e]Thenoyltrifluoroacetone.
Source: Ref. 12.

chelation provides extra stability to the metal complex. The formation and extraction of metal chelates are discussed extensively also in references [14–16].

In Chapter 3 we described how an uncharged metal complex MA_z is formed from a metal ion M^{z+} (*central atom*) through a stepwise reaction with the anion A^- (*ligand*) of a monobasic organic acid, HA, defining a *stepwise formation constant* k_n, and an *overall formation constant* β_n, where

$$\beta_n = [MA_n^{z-n}]/[M^{z+}][A^-]^n \qquad (4.34)$$

The MA_z complex is lipophilic and dissolves in organic solvents and the *distribution constant* K_{DC} is defined (index C for complex):

$$K_{DC} = [MA_z]_{org}/[MA_z] \qquad (4.35)$$

Taking all metal species in the aqueous phase into account, the distribution of the metal can be written (omitting the index aq for water)

$$D_M = \frac{[MA_z]_{org}}{\Sigma\,[MA_n^{z-n}]} = \frac{K_{DC}\beta_z[A^-]^z}{\Sigma\,\beta_z[A^-]^z} \qquad (4.36)$$

The distribution ratio depends only on the free ligand concentration, which may be calculated by Eq. (4.20). Most coordinatively saturated neutral metal complexes behave just like stable organic solutes, because their outer molecular structure is almost entirely of the hydrocarbon type, and can therefore be extracted by all solvent classes 2–5 of Chapter 2. The rules for the size of the distribution constants of these coordinatively saturated neutral metal complexes are then in principle the same as for the inert organic solutes of section 4.4. However, such complexes may still be amphilic due to the presence of electronegative donor oxygen atoms (of the chelating ligand) in the chelate molecule. In aqueous solution such complexes then behave like polyethers rather than hydrocarbons. Narbutt [17] has studied such outer-sphere hydrated complexes and shown that the dehydration in the transfer of the complex from water to the organic solvent determines the distribution constant of the complex. This is further elaborated in Chapter 16.

Example 3: Extraction of Cu(II) by acetylacetone.

Simple β-diketones, like acetylacetone (Appendix D: 5d) can coordinate in two ways to a metal atom, either in the uncharged keto form (through two keto oxygens), or in dissociated anionic enol form (through the same oxygens) as shown in Appendix D: 5c, 5h. It acts as an acid only in the enolic form. Figure 4.10 shows the extraction of Cu(II) from 1 M NaClO$_4$ into benzene at various concentrations of the extractant acetylacetone (HA) [18]. Acetylacetone reacts with Cu(II) in aqueous solutions to form the complexes CuA$^+$ and CuA$_2$. Because acetylacetone binds through two oxygens, the neutral complex CuA$_2$ contains two six-membered chelate rings; thus four coordination positions are taken up, forming a planar complex (Appendix D: 5f). This complex is usually considered to be coordinatively saturated, but two additional very

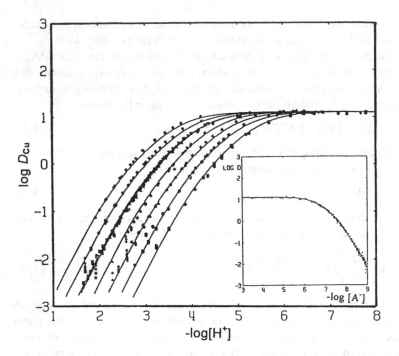

Fig. 4.10 Extraction of Cu(II) from 1 M NaClO₄ into benzene as a function of pH (large figure) and of free acetylacetonate ion concentration (insert) at seven different total concentrations of acetylacetone ([HA]$_{aq}$ 0.05–0.0009 M). (From Ref. 18.)

weak bonds can be formed perpendicular to the plane; we can neglect them here.

The distribution of copper, D_{Cu}, between the organic phase and water is then described by

$$D_{Cu} = \frac{[CuA_2]_{org}}{[Cu^{2+}] + [CuA^+] + [CuA_2]}$$

(4.37)

One then derives

$$D_{Cu} = \frac{K_{DC} \, \beta_2 \, [A^-]^2}{1 + \beta_1 \, [A^-] + \beta_2 [A^-]^2} = \frac{K_{DC} \beta_2 [A^-]^2}{\Sigma \, \beta_n [A^-]^n}$$

(4.38)

where K_{DC} refers to the distribution constant of the uncharged complex CuA₂.

In Eq. (4.37), log D is a function of [A⁻], *the free ligand concentration*, only, and some constants. In Fig. 4.10, log D_{cu} is plotted vs. log [H⁺] (= –pH). Through Eqs. (4.17) and (4.18) it can be shown that log D_{Cu} is a function of pH only at constant [HA]$_{org}$ (or [HA]$_{aq}$), while at constant pH the log D_{Cu} depends only on [HA]$_{org}$ (or [HA]$_{aq}$).

In the insert of Fig. 4.10, log D is plotted as a function of log $[A^-]$, where $[A^-]$ has been calculated from pH, $[HA]_{org}^0$, phase volumes, and K_a and K_{DR} (same as $K_{D,HA}$) for HA by means of Eq. (4.20); it is found that all curves coincide into one at high pH (high $[A^-]$), as expected from Eq. (4.38). The distribution curve approaches two asymptotes, one with a slope of 2 and one horizontal (zero slope). From Eq. (4.38) it follows that, at the lowest $[A^-]$ concentration (lowest pH), the concentration of CuA^+ and CuA_2 in the aqueous phase becomes very small; Eq. (4.38) is then reduced to

$$\lim_{[A^-] \to 0} D_{Cu} = [CuA_2]_{org} / [Cu^{2+}] = K_{DC} \, \beta_2 [A^-]^2 \tag{4.39}$$

At the highest A^- concentrations a horizontal asymptote is approached:

$$\lim_{[A^-] \to \infty} D_{Cu} = [CuA_2]_{org} / [CuA_2] = K_{DC} \tag{4.40}$$

The horizontal asymptote equals the distribution constant of CuA_2, i.e., K_{DC}. From the curvature between the two asymptotes, the stability constants β_1 and β_2 can be calculated.

This example indicates that in solvent extraction of metal complexes with acidic ligands, it can be more advantageous to plot log D vs. log$[A^-]$, rather than against pH, which is the more common (and easy) technique.

In order to calculate D_M from Eq. (4.36), several equilibrium constants as well as the concentration of free A^- are needed. Though many reference works report stability constants [19, 20] and distribution constants [4, 21], for practical purposes it is simpler to use the extraction constant K_{ex} for the reaction

$$M^{z+}(aq) + zHA(org) \leftrightarrows MA_z(org) + zH^+ (aq) \tag{4.41a}$$

in which case the MA_n^{z-n} complexes in the aqueous phase are neglected. The relevant extraction equations are

$$K_{ex} = [MA_z]_{org} \, [H^+]^z \, [M^{z+}]^{-1} \, [HA]_{org}^{-z} \tag{4.41b}$$

and

$$D_M = K_{ex}[HA]_{org}^z \, [H^+]^{-z} \tag{4.41c}$$

Thus only one constant, K_{ex}, is needed to predict the metal extraction for given concentrations $[H^+]$ and $[HA]_{org}$. Tables of K_{ex} values are found in the literature (see references given).

Equation (4.41) is valid only when the complexes MA_n^{z-n} can be neglected in the aqueous phase. Comparing Eqs. (4.37b) and (4.41c), it is seen that no horizontal asymptote is obtained even at high concentrations of A^-, or HA and H. Thus, for very large distribution constant of the uncharged complex (i.e., $\geqslant 1000$) a straight line with slope $-z$ is experimentally observed, as in the case for the Cu(II)-thenoyltrifluoroacetone (HTTA) system (Appendix D: 5g).

Example 4: Extraction of Cu(II) by Thenoyltrifluoroacetone.

Figure 4.11a shows the distribution of Cu(II) between three organic solvents and water in the presence of isopropyltropolone (HITP) or thenoyltrifluoroacetone (HTTA) as a function of pH [22]. The straight line of slope -2 in the pH-plot fits Eq. (4.41c); thus $z = 2$. It indicates that the aqueous phase does not contain any significant concentrations of the complexes CuA^+ or CuA_2, yet CuA_2 must be formed in considerable concentrations, otherwise there would be no extraction of Cu(II). The line also corresponds to the asymptote Eq. (4.39), or (i.e., $[A^-] \propto [H^+]^{-1}$). Thus the conclusion is that the aqueous phase is completely dominated by Cu^{2+}, while the organic phase contains only CuA_2. This leads to the copper distribution ratio

$$D_{Cu} = [CuA_2]_{org}/[Cu^{2+}] \tag{4.42}$$

which is valid for the reaction

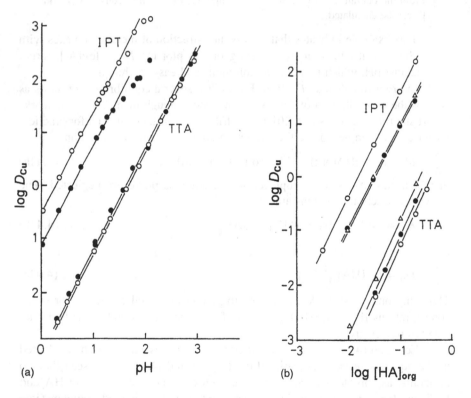

Fig. 4.11 Distribution of Cu(II) between hexone (Δ), carbon tetrachloride ($\bullet$), or chloroform ($\bigcirc$) and 0.1 M NaClO$_4$ in the presence of isopropyltropolone (IPT) or thenoyltrifluoroacetone (TTA); (a) as a function of $-\log[H^+]$ at constant $[TTA]_{org} = 0.1$ M; (b) as a function of $[TTA]_{org}$ at constant $[H^+] = 0.1$M. (From Ref. 22.)

$$Cu^{2+} + 2HA(org) \leftrightarrows CuA_2(org) + 2 H^+ \qquad (4.43a)$$

The extraction constant is

$$K_{ex} = [CuA_2]_{org} [H^+]^2 /[Cu^{2+}] [HA]_{org}^2 \qquad (4.43b)$$

and combining with Eq. (4.42)

$$D_{Cu} = K_{ex} [HA]_{org}^2 [H^+]^{-2} \qquad (4.44)$$

From Eq. (4.44) a plot of log D against $-\log[H^+]$ should yield a straight line with slope +2, as in Fig. 4.11a, and a plot against log $[HA]_{org}$ as in Fig. 4.11b should also yield a straight line with slope +2. Continued in Example 14.

This example illustrates a case of considerable analytical importance, especially for the determination of complex formation constants for hydrophilic complexes, as discussed in section 4.12, when the equilibrium constants for the stepwise metal-organic complexes are of secondary interest. K_{ex} values are tabulated in several reference works. K_{ex} is a conditional constant and only valid provided no other species are formed besides the extracted one.

The distribution constant of the neutral complex MA_z, K_{DC}, has been referred to several times. In favorable cases, when both the organic and the aqueous phases are dominated by the same uncharged complex over a larger concentration region, K_{DC} can be directly measured, as is the case for most of the data in Table 4.10 [22–23b]. Otherwise K_{DC} can be estimated or calculated from K_{ex} data combined with β_n, K_a, and K_{DR} [see Eqs. (4.8) and (4.46)].

4.9 EXTRACTION OF METAL COMPLEXES AS ADDUCTS (TYPE III-C)

Coordinatively saturated metal adduct complex in organic phase (and B)	B ($\downarrow\uparrow$)	MA_zB_b $\downarrow\uparrow$
Formation neutral complex cordinatively saturated by adduct former B	$M^{z+} + zA^- + bB \leftrightarrows MA_zB_b$	

If the neutral metal complex is coordinatively unsaturated, it forms MA_z $(H_2O)_x$ in the aqueous phase, where $2z + x$ (A being bidentate) equals the maximum coordination number. In the absence of solvating organic solvents, this complex has a very low distribution constant. Obviously, if water of hydration can be replaced by organic molecules B, the result is a more lipophilic adduct complex MA_zB_b; many adduct formers are listed in Appendix D and several tables. Depending on the ligand, several types of such adducts exist: (i) type MA_zB_b, where A and B are different organic structures; (ii) type MX_zB_b, where MX_z is a neutral inorganic compound (salt); and (iii) type $MA_z(HA)_b$, where A and HA are the basic and neutral variant of the same molecule (so-called *self-adducts*).

Table 4.10 Distribution Constants for Acetylacetone (HA) and Some Metal Acetylacetonates Between Various Organic Solvents and 1 M $NaClO_4$ at 25°C

Organic solvent[a]	ε Solvent	$\log K_{DR}$ HA	$\log K_{DC}$ ZnA$_2$	CuA$_2$	NpA$_4$
n-Hexane (1)	1.88	−0.022	−1.57	—	0.5
Cyclohexane (3)	2.02	0.013	−1.16	−0.04	0.8
Carbon tetrachloride (4)	2.24	0.52	−0.39	0.85	2.7
Mesitylene[b] (6)	2.28	0.44	—	0.43	—
Xylene (7)	2.27	0.57	−0.47	0.80	—
Toluene (8)	2.38	0.66	−0.37	0.85	—
Benzene (9)	2.28	0.77	−0.21	1.04	3.3
Dibutylether (2)	3.06	—	−1.05	—	—
Methylisobutylketone (5)	13.1	0.77	−0.15	0.61	—
Chloroform (10)	4.9	1.38	0.83	2.54	—
Benzonitrile (11)	25.2		0.21	—	—

[a]Numbers in parentheses refer to Figs. 4.23 and 4.26.
[b]1,3,5-trimethylbenzene.
Source: Refs. 22–23.

4.9.1 Metal-Organic Complexes with Organic Adduct Formers, Type MA$_z$B$_b$

The extraction of the metal complex adduct can be written

$$M^{z+}(aq) + zHA(org) + bB(org) \leftrightarrows MA_zB_b(org) + zH^+(aq) \qquad (4.45a)$$

The extraction constant is defined by

$$K_{ex} = \frac{[MA_zB_b]_{org}[H^+]^z}{[M^{z+}][HA]_{org}^z[B]_{org}^b} \qquad (4.45b)$$

or

$$K_{ex} = D_M[H^+]^z[HA]_{org}^{-z}[B]_{org}^{-b} \qquad (4.45c)$$

Thus the distribution of the metal, D_M, is shown to depend on the concentrations of H^+ and HA_{org} to the power z of the charge of the metal ion, and on the concentration of B to the power of b (i.e., number of adduct formers in the extracted complex).

It can be shown that

$$K_{ex} = K_a^z \beta_z K_{DR}^{-z} K_{DC} K_{ad,bB} \qquad (4.46)$$

where (omitting ionic charges) $K_a = [H][A]/[HA]$ is the acid dissociation constant [see Eq. (4.18)], $K_{DR} = [HA]_{org}/[HA]$ is the distribution constant for the undissociated acid HA [see Eq. (4.17)], and

$$K_{ad,bB} = [MA_zB_b]_{org}/[MA_z]_{org}^b \tag{4.47}$$

is the formation constant for the adduct MA_zB_b in the organic phase [see Eqs. (4.15), (4.16), and (4.26)]. The five parameters K_a, β_n, K_{DR}, K_{DC}, and $K_{ad,bB}$ are in principle unrelated, even though it may not always be possible to change one without affecting the others, as each molecular species may take part in several equilibria. Without considering the independent parameters, it is often difficult to understand why K_{ex} varies in the fashion observed, and it may be impossible to predict improvements of the system. A good example is the extraction of Zn(II) by β-diketones and TBP:

> Example 5: The extraction of Zn(II) by β-diketones and phosphoryl adduct formers.
>
> Figure 4.12 illustrates the extraction of Zn(II) from 1 M $NaClO_4$ into carbon tetrachloride by β-diketones (HA) in the presence of the adduct formers

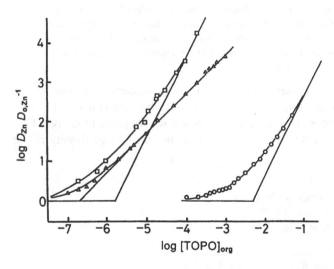

Fig. 4.12 Enhancement of Zn(II) extraction, $D\ D_o^{-1}$, from 1 M $NaClO_4$ into carbon tetrachloride containing the complexing extractants acetylacetone ($\bigcirc$), trifluoroacetone ($\triangle$), or hexafluoroacetone ($\square$) as a function of the concentration of the adduct former trioctyl phosphine oxide (B). The curves are fitted with Eq. (4.50) using the constants $\log K_{ad1} = 3.07$ (AA), 6.70 (TFA), 7.0 (TFA), and $K_{ad2} = 4.66$ (AA), nil (TFA), 11.6 (HFA). (From Ref. 24.)

TBP and TOPO (B) [12]. The extracted neutral complex is ZnA_2B_b. The distribution ratio becomes

$$D_{Zn} = \frac{[ZnA_2]_{org} + [ZnA_2B]_{org} + [ZnA_2B_2]_{org} + \cdots}{[Zn] + [ZnA] + [ZnA_2] + \cdots} \quad (4.48a)$$

To analyze these systems, the overall extraction reaction must be broken into its partial reactions, or by introducing Eq. (4.47), to obtain

$$D_{Zn} = \frac{K_{DC}\beta_2[A]^2 (1 + K_{ad,1}[B]_{org} + K_{ad,2}[B]_{org}^2 + \cdots)}{\Sigma \beta_n [A]^n} \quad (4.48b)$$

where the adduct formation constant is defined by

$$K_{ad,b} = [ZnA_2B_b]_{org} [ZnA_2]_{org}^{-1} [B]_{org}^{-b} \quad (4.49)$$

In the absence of any adduct former, D_{Zn} is given as a function of the free ligand concentration by Eq. (4.36), i.e.,0 the parentheses in Eq. (4.48b) equals 1; denoting this D_{Zn}-value as D_o, and introducing it into Eq. (4.48) gives

$$D'_{Zn} = D_o (1 + K_{ad,1}[B]_{org} + K_{ad,2}[B]_{org}^2 + \cdots) \quad (4.50)$$

D_{Zn}' (instead of D'_{Zn}) indicates that this expression is valid only at constant $[A^-]$, or, better, constant $[H^+]$ and $[HA]_{org}$ [see Eqs. (4.36) and (4.41c)]. In Fig. 4.12, log $D'_{Zn} D_0^{-1}$ is plotted as a function of log $[B]_{org}$. The distribution ratio proceeds from almost zero, when almost no adduct is formed, towards a limiting slope of 2, indicating that the extracted complex has added two molecules of B to form ZnA_2B_2. From the curvature and slope the $K_{ad,b}$-values were determined (see section 4.10). The calculation of the equilibrium constants is further discussed under Example 13.

Tables 4.11–4.13 presents adduct formation constants according to Eq. (4.47). For the alkaline earths TTA complexes in carbon tetrachloride in Table 4.11, the TBP molecules bond perpendicular to the square plane of the two TTA rings, producing an octahedral complex. The higher the charge density of the

Table 4.11 Adduct Formation Constants for the Reaction $M(TTA)_2 + bTBP \rightleftarrows M(TTA)_2(TBP)_b$ in CCl_4 Showing Effect of Charge Density

M^{z+} (r pm)	log K_{ad1}	log K_{ad2}
Ca^{2+} (100)	4.11	8.22
Sr^{2+} (118)	3.76	7.52
Ba^{2+} (135)	2.62	5.84

Source: Ref. 4.

Table 4.12 Adduct Formation Constants for the Reaction $EuA_3(org)$ + $bTBP(org)$ ⇄ $EuA_3(TBP)_b(org)$, Eq. (4.26a), Where org = $CHCl_3$, and *HA* Substituted Acetylacetone Ligands

Ligand (A)	Log K_a	Log K_{ad1}	Log K_{ad2}
Acetylacetone, AA	8.76	1.90	—
Benzoylacetone, BZA	8.74	1.60	—
Trifluoroacetone, TFA	—	3.32	4.64
Benzoyltrifluoroacetone, BTA	—	3.64	5.28
2-Furoyltrifluoroacetone, FTA	—	3.50	5.00
Thenoyltrifluoroacetone, TTA	6.3	3.34	5.28

Source: Ref. 4.

central atom, the stronger is the adduct complex with TBP. Note: *Charge density* refers to the electrostatic interaction between ions of opposite charge according to the Born equation [see (2.13), and (3.13)] (based on the Coulomb interaction). It is mostly given as the ratio between the ionic charge, z_+ (or z_+z_-) and the ionic radius r_+ (or r_+r_-). Table 4.12 compares the adduct formation of the europium β-diketone complexes with TBP in chloroform. In $Eu(TTA)_3$ the TTAs only occupy six of the eight coordination positions available; the two empty positions have been shown to be occupied by water. Though the tendency is not strong, the stronger the acid (i.e., the larger its electronegativity), the larger is the adduct formation constant. Table 4.13 compares the adduct formation tendency of the $Eu(TTA)_3$ complex with various adductants. The basicity of the donor oxygen atom increases in the order as shown in Table 4.13, as does the adduct formation constant; this is in agreement with the order of basicity in Table 4.8 and the donor numbers of Table 3.3; see also section 4.2.

Table 4.13 Constants for Formation of $Eu(TTA)_3B_b$ Adducts According to Eq. (4.26a)

Adduct forming ligand (B)	Chloroform		Carbon tetrachloride	
	log K_{ad1}	log K_{ad2}	log K_{ad1}	log K_{ad2}
TTA(self-adduct)	0.56	—	>0.5	—
Hexone	1.16	1.52	1.71	2.34
Quinoline	3.29	—	3.48	5.16
TBP	3.63	5.40	5.36	8.96
TOPO	5.40	7.60	7.49	12.26

Source: Ref. 4.

   The difference between the two solvent systems is likely a result of $CHCl_3$ solvating the Eu complex to some extent, while CCl_4 is inert. This has two effects: the K_{DC} value increases due to the solvation by $CHCl_3$ (not shown in the table), while the adduct formation constant K_{ad} decreases as the solvation hinders the adduct formation; the more inert solvent CCl_4 causes an opposite effect, a lower K_{DC} and a larger K_{ad}.

4.9.2   Metal Inorganic Complexes with Organic Adduct Formers, Type MX_zB_b

The extraction of mineral salts is generally less complicated than the extraction of mineral acids. Metal salts with monodentate univalent anions like Cl^-, ClO_4^-, SCN^-, and NO_3^- are strongly hydrated in the aqueous phase and have quite small, if any, solubility in inert solvents. In order to extract these acids, they must either form an adduct with a strongly basic extractant like TBP or TOPO, or be in solvating solvents such as ethers, ketones, alcohols, or esters. Examples of extracted metal salt adducts are: $Br_3(EtO)_b$, $PaCl_3(MIBK)$, $UO_2(NO_3)_2(TBP)_2$, $Co(ClO_4)_2(octanol)_b$, etc. (b is uncertain). Tables of the extraction of a large number of metal salts by solvating solvents or commercial adduct formers dissolved in kerosene are given (see Ref. [5]).

   It has been shown [25] that monomeric metal hydroxides can be extracted by strong donor molecules; e.g., in the form of the adducts $Ln(OH)_3(TOPO)_b$ into $CHCl_3$, where b is 2–3. Under favorable conditions, the D_{Ln} value may exceed 1, though the fraction of hydroxide is quite low.

   When the solvent is a good solvater, the determination of the solvation number b is difficult, unless the dependence of the extractant concentration on the solvent can be obtained. Solvation numbers can be obtained in mixtures of a solvating extractant and an inert diluent like hexane. Further, in these systems the extraction of the metal commonly requires high concentrations of salt or acid in the aqueous phase, so the activity coefficients of the solutes must be taken into account.

   Example 6: Extraction of Zn(II) thiocyanate complexes by TOPO.
      Figure 4.13 shows the extraction of Zn(II) from aqueous thiocyanate (L^-) solutions into 0.001 M TOPO (B) in hexane; the aqueous phase is 1.0 M $Na(SCN^-, ClO_4^-)$ at a pH around 5 [26]. Zn(II) is known to form a number of weak $Zn(SCN)_n^{n-2}$ complexes in the aqueous phase. The uncharged one is assumed to accept TOPO to form the adducts $Zn(SCN)_2(TOPO)_b$, where $b = 1$ or 2.
      Assume that the reaction between the neutral complex and the solvating molecule takes place at the interface (to assume the reaction to take place in the organic phase would be unrealistic, as the zinc thiocyanate is insoluble in hexane); thus the etxraction reaction is

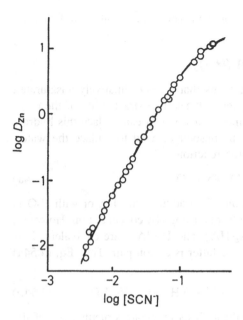

Fig. 4.13 Distribution ratio of Zn(II) when extracted from 1 M Na(SCN⁻,ClO₄⁻) into 0.001 M TOPO in hexane, as a function of aqueous SCN⁻ concentration. The following equilibrium constants were obtained with Eq. (4.53): β_1 3.7, β_2 21, β_3 15, K_{ex} 2.5 10^7 for $b = 2$. (From Ref. 26.)

$$ZnL_2 + b\,B(org) \leftrightarrows ZnL_2B_b(org) \qquad (4.51)$$

for which we may define an equilibrium constant $K_{ex,bB}$.

Because more than one solvated species may be extracted, the distribution ratio becomes

$$D_{Zn} = \frac{[ZnL_2B]_{org} + [ZnL_2B_2]_{org} + \cdots}{[Zn] + [ZnL] + [ZnL_2] + [ZnL_3] + \cdots} \qquad (4.52)$$

Inserting the partial equilibrium constants [see Eq. (4.46)],

$$D_{Zn} = \frac{\beta_2[L]_2 \; \Sigma \, K_{ex,bB} \; [B]^b_{org}}{1 + \Sigma \, \beta_n[L]^n} \qquad (4.53)$$

where both summations are taken from 1. The solvation number b can be determined from the dependence of D on $[B]_{org}$ while $[L]$ is kept constant. From the slope of the line in Fig. 4.13 at low SCN⁻ concentrations, it follows that $b = 2$; thus only one adduct complex is identified: $Zn(SCN)_2(TOPO)_2$. The authors were able to calculate the formation constants β_n from the deviation of the curve from the straight the line at constant $[B]_{org}$, assuming b constant. With these equilibrium constants, the line through the points was calculated with Eq. (4.53).

4.9.3   Self-Adducts, Type $MA_z(HA)_b$

Hydration only occurs in neutral complexes that are coordinatively unsaturated by the organic ligand. The hydrate water reduces the extractability of the complex. In the absence of strong donor molecules, which can replace this hydrate water, there is still a chance for the undissociated acid to replace the water, leading to a *self-adduct* according to the reaction

$$MA_z(H_2O)_w + bHA \leftrightarrows MA_z(HA)_b + wH_2O \qquad (4.54a)$$

This competition between the formation of an adduct with HA or with H_2O is observed as an increased extraction with increasing HA concentration. However, stoichiometrically the complexes $MA_Z(HA)_b$ and H_bMA_{Z+b} are equivalent. Formally the former is a self-adduct and the latter is an ion pair. Thus Eq. (4.54a) could be written

$$MA_z(H_2O)_w + bHA \leftrightarrows M(H_2O)_{w-2}A_{z+b}^{-b} + b\ H^+ + w - 2b\ H_2O \qquad (4.54b)$$

assuming the HA can replace 2 H_2O through its bidentate structure. All of the hydrate water can be displaced by the organic ligand, with formation of negatively charged chelate complexes.

Chemical equilibrium experiments, e.g., distribution ratio measurements, cannot distinguish between these two types of complexes; however, they may be identified by fingerprinting techniques like NMR, IR, or x-ray structure determinations. Existence of similar adducts like MA_zB_b support the existence of self-adducts. The case of promethium(III) acetylacetone is an interesting illustration of this problem.

Example 7: Extraction of Pm(III) by acetylacetone.

Figure 4.14 shows the distribution ratio log D_{Pm} for promethium in the system acetylacetone (HA), benzene/1 M $NaClO_4$ as a function of $-\log[A^-] = pA$ at various starting concentrations, $[HA]_{org}^o$, of HA in the organic phase. The coordination number of Pm(III) with respect to oxygen is reported to be 8 or 9. In the aqueous phase all stepwise complexes up to PmA_4^- can therefore be expected to be formed. The vacant coordination sites may be filled with water (forming a hydrate) or undissociated acetylacetone (forming a self-adduct). The last assumption is suppported by the fact that a large number of adducts of type LnA_3B_b are known.

Omitting water of hydration, the distribution ratio becomes

$$D_{Pm} = \frac{[PmA_3]_{org} + [PmA_3HA]_{org} + [PmA_3(HA)_2]_{org} + \cdots}{[Pm] + [PmA] + [PmA_2] + \cdots + [PmA_4]} \qquad (4.55a)$$

which is abbreviated, using the earlier relations, to

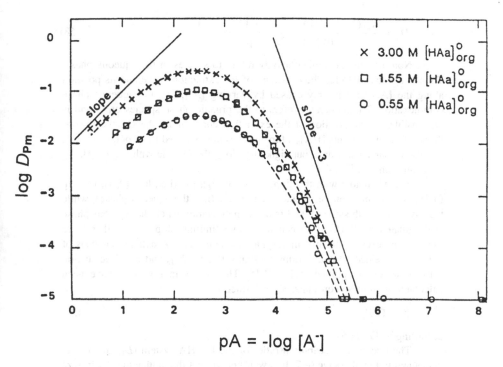

Fig. 4.14 Extraction of Pm(III) by acetylacetone (HAa) from 1 M NaClO₄ into benzene at three different original concentrations of HAa in the organic phase. pA = −log[Aa⁻] is calculated according to Eq. (4.20). The analysis of the system yielded the constants log β_1 5.35, log β_2 9.20, log β_3 13.22, log β_4 14.06, K_{ad1} 7, K_{ad2} 3, and K_{DC} 0.008, shown for *Pm* in Fig. 4.15. (From Refs. 27a,b.)

$$D_{Pm} = \frac{K_{DC}\beta_3[A]^3\,(1 + K_{ad1}[HA]_{org} + K_{ad2}[HA]_{org}^2 + \cdots)}{\Sigma\beta_3[A]^n} \qquad (4.55b)$$

where K_{DC} refers to the distribution of uncharged PmA₃ between the benzene and the aqueous phase

$$K_{DC} = [PmA_3]_{org}/[PmA_3] \qquad (4.56)$$

$K_{ad,b}$ is the equilibrium constant for the self-adduct formation in the organic phase, i.e.,

$$PmA_3B(org) + b\,HA(org) \leftrightarrows PmA_3(HA)_b\,(org) \qquad (4.57a)$$

$$K_{ad,b} = [PmA_3(HA)_b]_{org}[PmA_3]_{org}^{-1}\,[HA]_{org}^{-b} \qquad (4.57b)$$

For pedagogic reasons we rewrite Eq. (4.55b)

$$D_{Pm} = \frac{K_{DC}(1 + K_{ad,1}[HA]_{org} + K_{ad,2}[HA]_{org}^2 + \cdots)}{(\beta_3 [A]^3)^{-1} \Sigma \beta_n [A]^n} \tag{4.58}$$

Note that the denominator only refers to species in the aqueous phase. Thus at constant $[HA]_{org}$, the curvature of the extraction curve and its position along the $[A^-]$ axes is only caused by varying $[A^-]$, i.e., the aqueous phase complexation. The numerator refers to the organic phase species only, and is responsible for the position of the extraction curve along the D-axis; thus at three different constants $[HA]_{org}$, three curves are expected to be obtained, with exactly the same curvatures, but higher up along the D-scale with higher $[HA]_{org}$ concentration; see Fig. 4.14.

The asymptote with the slope of -3 at high pA (i.e., low $[A^-]$), fits Eq. (4.58), when Pm^{3+} dominates the denominator (i.e., the aqueous phase), while the asymptote with slope of $+1$ fits the same equation when the aqueous phase is dominated by PmA_4^-. Between these two limiting slopes, the other three PmA_n complexes are formed in varying concentrations. A detailed analysis of the curves yielded all equilibrium constants K_{ex}, K_n, $K_{ad,b}$ and K_{DC} (see section 4.14.3), which are plotted in Fig. 4.15. The curves in Fig. 4.14 have been calculated with these constants. K_n is defined by

$$\beta_n = \Pi \, K_n \tag{4.59}$$

according to Eq. (3.5).

The maximum distribution ratio for the Pm-HA system (D_{Pm} about 0.1) is reached in the pH range 6–7. It is well known that the lanthanides hydrolyze in this pH region, but it can be shown that the concentration of hydrolyzed species is <1% of the concentration of PmA_n species for the conditions of Fig. 4.14, and can thus be neglected.

Self-adducts are rather common, and have been identified for complexes of Ca, Sr, Ba, Ni, Co, Zn, Cd, Sc, Ln, and U(VI) with acetylacetone, thenoyltrifluoro-acetone, tropolone, and oxine. Table 4.14 lists some self-adduct constants. The fact that the constants vary with the organic solvent indicates that the self-adduct

Fig. 4.15 The system La(III) acetylacetone (HA) – 1M $NaClO_4$/benzene at 25°C as a function of lanthanide atomic number Z. (a) The distribution ratio D_{Ln} (stars, right axis) at $[A^-] = 10^{-3}$ and $[HA]_{org} = 0.1$ M, and extraction constants K_{ex} (crosses, left axis) for the reaction $Ln^{3+} + 4HA(org) \leftrightarrows LnA_3HA(org) + 3H^+$. (b) The formation constants, K_n, for formation of LnA_n^{3-n} lanthanide acetylacetonate complexes (a break at $_{64}Gd$ is indicated); circles $n = 1$; crosses $n = 2$; triangles $n = 3$; squares $n = 4$. (c) The self-adduct formation constants, K_{ad}, for the reaction of $LnA_3(org) + HA(org) \leftrightarrows LnA_3HA(org)$ for org = benzene. (A second adduct, $LnA_3(HA)_2$, also seems to form for the lightest Ln ions.) (d) The distribution constant K_{DC} for hydrated lanthanum triacetylacetonates, $LnA_3(H_2O)_{2-3}$, between benzene and 1M $NaClO_4$. (From Ref. 28.)

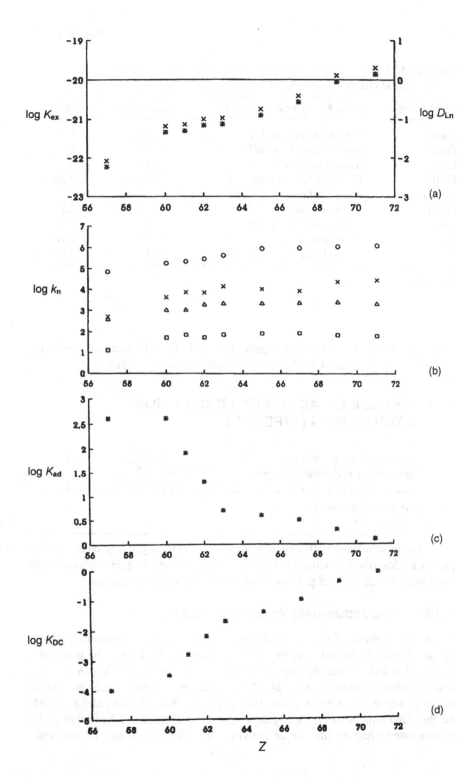

Table 4.14 Constants for the Formation of Some Metal Self-Adducts
According to $MA_z(\text{org}) + b\, HA(\text{org}) \leftrightarrows\ \rightarrow MA_z(HA)_b$

Metal ion M_{z+}	Ligand and adduct former HA	Organic solvent	Log K_{ad}
Zn(II)	Isopropyltropolone, IPT	$CHCl_3$	1.72
Zn(II)	Isopropyltropolone, IPT	CCl_4	1.9
La(III)	Acetylacetone, AA	C_6H_6	2.6
Eu(III)	Thenoyltrifluoroacetone, TTA	$CHCl_3$	0.56
Eu(III)	Thenoyltrifluoroacetone, TTA	CCl_4	0.5
Eu(III)	Isopropyltropolone, IPT	$CHCl_3$	2.1
Eu(III)	Isopropyltropolone, IPT	CCl_4	, 2.0
U(VI)	Acetylacetone, AA	$CHCl_3$	1.16
U(VI)	Thenoyltrifluoroacetone, TTA	C_6H_6	0.50

Source: Ref. 4.

reaction probably occurs in the organic phase. Table 4.14 contains both very
inert (e.g., CCl_4), polar ($CHCl_3$), and pi-bonding solvents (C_6H_6).

4.10   METAL EXTRACTION BY LIQUID ANION EXCHANGERS (TYPE III-D)

Organic phase with anion exchanger and metal complex	RNH^+L^- $\downarrow\uparrow$	$(RNH^+)_pML_n^{-p}$ $\downarrow\uparrow$
Aqueous phase: metal with complexing anions L^- and amine	$M^{z+} + nL^- + pRNH^+L^- \leftrightarrows (RNH^+)_pML_n^{-p}$	

Metals that react with inorganic ligands to form negatively charged complexes as described in Chapter 3 can be extracted into organic solvents with
large organic cations in a process referred to as *liquid anion exchange*.

4.10.1   Two Industrially Important Cases

Extraction of metals from chloride solutions into kerosene containing trioctyl
amine (TOA), as illustrated by Fig. 1.3, has found application in the production
of pure Co and Ni metals from their ores, as discussed in Chapter 11. The
aqueous phase contains the Co and Ni in addition to other metals in chloride
solution, where the chloride concentration goes to 300 g/L (i.e., about 8.5 M).
At this high aqueous ionic strength, a rigorous mathematical treatment of the
system would require the use of activities in order to explain the extraction

diagram; however, here it suffices to use concentrations. An interesting feature in Fig. 1.3 is the differences between metals in various oxidation state, notably Fe(II) and Fe(III), the latter being much more easily extracted than the former because Fe(III) forms much stronger complexes with Cl^- ions. The requirement for extraction of metals by an aminelike TOA, or by its amine salt, $TOAH^+Cl^-$, is the formation of negatively charged metal chloride complexes in the aqueous phase. The complex formation from M^{z+} to MCl_n^{z-n}, where $z - n \leq -1$, proceeds in the stepwise manner described in Chapter 3. For example, Fe(III) forms the complex $FeCl_4^-$, while cobalt forms $CoCl_4^{2-}$. The extraction reaction for Co can be written

$$CoCl_4^{2-} + 2\ RNH^+Cl^-(org) \leftrightharpoons (RNH)_2CoCl_4(org) + 2\ Cl^- \qquad (4.60a)$$

Amine extraction is used also in another important industrial process, the extraction of uranium from sulphuric acid leached ores, which uses trilauryl amine (TLA). In that case, the extraction reaction is

$$UO_2(SO_4)_4^{2-} + 2\ (RNH^+)_2SO_4^{2-}(org)$$
$$\leftrightharpoons (RNH)_4UO_2(SO_4)_3(org) + SO_4^{2-} \qquad (4.60b)$$

These processes are referred to as liquid anion exchange, because the aqueous anionic metal complex replaces the anions Cl^- or SO_4^{2-} of the large organic amine complex. The amine complex is an ion pair, or salt, RNH^+L^-, where R is a large hydrocarbon group. The amine salt is highly soluble in the organic (e.g., kerosene) phase, and almost insoluble in the aqueous phase. The reaction is presumed to occur either at the interface or in the aqueous phase with a low concentration of dissolved RNH^+L^-. Before discussing these reactions in terms of molecular species, it is necessary to consider more specific aspects of liquid anion exchangers.

4.10.2 Properties of Liquid Anion Exchangers

As mentioned earlier, hydrogen ions can be solvated by strong donor molecules like TBP, sometimes leading to the extraction of a solvated hydrogen salt, $HB_b^+L^-$; in Example 2, TBP extracts nitric acid, HNO_3, in the form of the adduct complex $TBP \cdot HNO_3$, which could be considered as the complex $HTBP^+NO_3^-$ ion pair. The amines form stronger adducts with hydrogen ions; in fact, they are so strong that they remain protonated while exchanging the anion. A classical reaction is the formation of the ammonium ion NH_4^+ when NH_3 is dissolved in water.

The organic amines (the *amine base* RN) have a nitrogen atom N attached to a large organic molecule R usually containing >7 aliphatic or aromatic carbon atoms. They are highly soluble in organic solvents (diluents) and almost insoluble in water. In contact with an aqueous phase containing HL, the amine base

RN reacts with the acid HL to form RNH^+L^-, but extracts with an excess amount of acid HL (over the 1:1 HL:RN ratio) into the organic solvent, and also with additional water. The practical concentration of amine in the organic solvent is usually less than 20%; at higher concentrations the amine salt solutions become rather viscous. The amine salts dissociate in highly polar solvents, while in more inert diluents they easily polymerize to form micelles, and at higher concentrations a third phase. For example, in xylene $(TLA \cdot HBr)_n$ aggregates with $n = 2, 3,$ and 30 have been identified, and in other systems aggregation numbers above 100 have been reported. $TLA \cdot HNO_3$ is mainly trimeric in m-xylene at concentrations 0.002–0.2 M, but larger aggregates are formed at higher amine concentrations. These aggregates seem to behave like monofunctional species, each extracting only one anionic metal complex. The aggregation can be reduced, and the third phase formation avoided, by using aromatic diluents and/or by adding a *modifier*, usually another strong Lewis base (e.g., octanol or TBP). Such additions often lead to considerable reduction in the K_{ex} value.

Four types of organic amines exist, as shown in Table 4.8: primary amines RNH_3^+, secondary $R_2NH_2^+$, tertiary R_3NH^+, and quaternary R_4N^+(Appendix D). The hydrocarbon chain R is usually of length C_8–C_{12}, commonly a straight aliphatic chain, but branched chains and aromatic parts also occur. In general the amines extract metal complexes in the order tertiary > secondary > primary. Only long-chain tertiary and—to a smaller extent—quarternary amines are used in industrial extraction, because of their suitable physical properties; trioctylamine (TOA, 8 carbons per chain) and trilauryl amine (TLA, 12 carbons per chain) are the most frequently used. For simplicity we abbreviate all amines by RN, and their salts by RNH^+L^-.

The tertiary and quaternary amine bases are viscous liquids at room temperature and infinitely soluble in nonpolar solvents, but only slightly soluble in water. The solubility of the ion-pair RNH^+L^- in organic solvents depends on the chain length and on the counterion, L^-: the solubility of $TLA \cdot HCl$ in wet benzene is 0.7 M, in cyclohexane 0.08 M, in $CHCl_3$ 1.2 M, and in CCl_4 0.7 M. Nitrate and perchlorate salts are less soluble, as are lower molecular weight amines.

The formation of the ion pair salt can be written

$$RN(org) + H^+ + L^- \leftrightarrows RNH^+L^-(org) \quad K_{ex,am} \tag{4.61}$$

Table 4.15 gives the equilibrium constants $K_{ex,am}$ (for *ex*traction, *am*ine) for this reaction with trioctylamine in various solvents. Although the ion pairs are only slightly soluble in water, they can exchange the anion L^- with other anions, X^-, in the aqueous phase. (Note that we use L^- to indicate any anion, while X^- is used for (an alternative) inorganic anion.)

$$RNH^+L^-(org) + X^- \leftrightarrows NH^+X^-(org) + L^- \quad K_{ex,ch} \tag{4.62}$$

Table 4.15 Equilibrium Constants $K_{ex,am}$ for Formation
of Trioctyl Amine Salts [Eq. (4.61)]

Anion	Solvent	Log $K_{ex,am}$
F^-	Toluene	3.0
Cl^-	Toluene	5.9
Cl^-	Carbontetrachloride	4.0
Cl^-	Benzene	4.1
Br^-	Toluene	8.0
NO_3^-	Toluene	6.6
NO_3^-	Carbontetrachloride	5.0
SO_4^{2-} [a]	Carbontetrachloride	6.7
SO_4^{2-} [a]	Benzene	8.3

[a]The equilibrium constant refers to the formation of $(R_3NH)_2SO_4$.
Source: Ref. 5.

The equilibrium constant ($K_{ex,ch}$ for *ex*traction, ex*ch*ange) for this reaction increases in the order $ClO_4^- < NO_3^- < Cl^- < HSO_4^- < F^-$. From Chapter 3 it follows that the formation constants for the metal complex ML_n^{z-n} usually increases in the same order. Therefore, in order to extract a metal perchlorate complex, very high ClO_4^- concentrations are required; the perchlorate complex is easily replaced by anions higher in the sequence.

All negatively charged metal complexes can be extracted by liquid anion exchangers, independent of the nature of the metal and the complexing ligand. Liquid anion exchange has extensive industrial application, and many examples are given in later chapters. Lists of extraction values are found in other works [e.g., 5, 21].

4.10.3 Equations for Liquid Anion Exchange Extractions

The discussion on the properties of the amine salts serves to explain why, in metal amine extractions, it is difficult to obtain simple mathematical relations that agree well with the experimental data. While there is no difficulty, in principle, in obtaining reasonably good values for the formation of the negatively charged metal complexes in the aqueous phase, there is a major problem in defining the organic phase species, which may consist of free amine [RN], monomeric [RNHL], and polymeric $[RNHL]_n$ amine salt, and several extracted metal complexes $[(RNH^+)_p(ML_n^{p-})]$. A contributing difficulty in practice is the need to use high ligand concentrations, $[L^-]$, in the aqueous phase in order to obtain the negatively charged complexes (see Fig. 1.3).

In general the metal M^{z+} reactions with a monobasic anion L^- can be written

$$M^{z+} + n\,L^- \leftrightarrows ML_n^{z-n} \qquad \beta_n = [ML_n^{z-n}]/[M^{z+}]\,[L^-]^n \qquad (3.5, 4.34)$$

When $z - n = p$ is negative, a negatively charged metal complex has been formed, which can be extracted

$$ML_n^{-p} + p\,RNH^+L^-(org) \leftrightarrows (RNH^+)_p^{p+}\,ML_n^{-p}(org) + p\,L^- \qquad (4.63a)$$

We define the extraction constant by

$$K_{ex} = \frac{[(RNH)_p^{+p}ML_n^{-p}]_{org}\,[L^-]^p}{[ML_n^{-p}]\,[RNHL]_{org}^p} \qquad (4.63b)$$

A priori, it must be assumed that the aqueous phase contains all the stepwise complexes ML_n^{z-n}. Thus the distribution ratio is

$$D_M = \frac{[(RNH)_p^{+p}ML_n^{-p}]_{org}}{\Sigma\,[ML_n^{z-n}]} = K_{ex}\,\frac{\beta_p\,[L^-]^p\,[RNHL]_{org}^p}{1 + \Sigma\beta_p\,[L^-]^p} \qquad (4.64)$$

The distribution of M depends on both the free amine salt in the organic phase and the concentration of free L^- in the aqueous phase until all metal in the aqueous phase is bound in the ML_n^{-p} complex. At constant amine concentration, Eq. (4.64) indicates that a plot of D_M vs. $[L^-]$ would have a linear slope p if the denominator of Eq. (4.64) is $\ll 1$; i.e., the metal species in the aqueous phase are dominated by the uncomplexed metal ion M^{z+}. At higher $[L^-]$ concentrations, where the ML_n^{-p} complex begins to dominate in the aqueous phase, the D_M value becomes equal to $K_{ex}\,[RNHL]_{org}^p$. Equations (4.64) and (4.4) show that S-shaped curves result for metals with large K_{ex} values. In a plot of D_M vs. $[RNHL]_{org}$ a straight line of slope p is obtained only at constant $[L^-]$. From such measurements both p, K_{ex} and β_p can be evaluated. The following example illustrates this.

Example 8: Extraction of Trivalent Actinides by TLA.

    In an investigation of the extraction of trivalent actinides, An(III) from 0.01 M nitric acid solutions of various $LiNO_3$ concentrations into o-xylene containing the tertiary amine salt trilaurylmethylammonium nitrate, TLMA HNO_3, Van Ooyen [29] found that the amine was monomeric only at very low concentrations (≤ 0.1M in the organic phase) but at higher concentration formed both dimers and trimers.

    Using trace concentrations of Ce(III) and An(III) a log-log plot of D_M against the nitrate ion activity, $m\gamma_\pm = [LiNO_3]^{1/2}$, had a slope of approximately 3, Fig. 4.16b. From Eq. (4.64) this slope corresponds to the p-value of 3 when the aqueous phase is dominated by the free metal ion, which is not an unreasonable assumption at low nitrate concentrations.

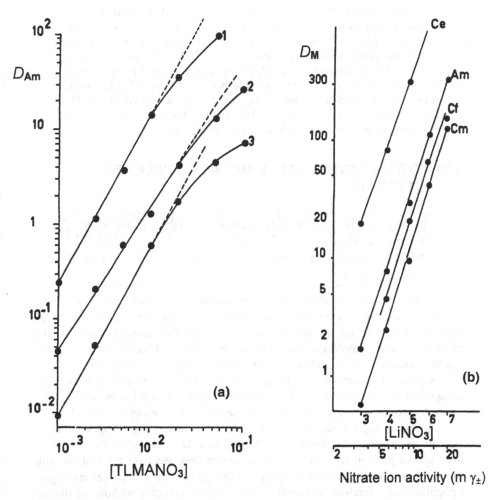

Fig. 4.16 Distribution ratio of M^{3+} ions between the trilaurylmethyl ammonium nitrate (TLMA) in *o*-xylene and aqueous phases of varying $LiNO_3$ concentrations. (a) As a function of $TLMANO_3$ concentration at 1–7 M, 2–5 M, 3–3 M $LiNO_3$. (b) Extraction of Eu(III) and tree actinide(III) ions at 0.1 M $TLMANO_3$ in *o*-xylene and varying aqueous salt concentrations. (From Ref. 29.)

In plots of log D_{Am} against [TLMA · NO$_3$]org at different nitrate concentrations, the curves in Fig. 4.16a had straight slopes of 1.5–1.8 at low concentrations of TLMA NO$_3$, but bending at higher concentrations, was explained by the formation of polymeric amine species. If Eq. (4.64) is valid, these slopes correspond to the number of TLMA HNO$_3$ groups attached to the extracted Am species. Thus Van Ooyen described his complex as $\{(TLMA\ NO_3)_n\}_2$ Am(NO$_3$)$_3$, for which $n = 1$ only at the very low amine concentrations. Thus for $n = 1$ the complex could as well be written (TLMA)$_2$Am(NO$_3$)$_5$. The Ce and other An complexes would have similar configurations.

4.11 OTHER EXTRACTABLE METAL COMPLEXES (TYPE III-E)

Ion pair (and possibly "counter species") in organic phase	$(C^+X^-$ and $Y^+A^-) C^+A^-$ $\downarrow\uparrow$
Aqueous cation C^+ and anion A^- associated into ion pair C^+A^-	$C^+ + A^- \leftrightarrows C^+A^-$

There are a few types of complexes that do not fit well into the previous classifications: monovalent metals that form extractable complexes with large organic monobasic anions, and, conversely, monovalent inorganic anions that form extractable complexes with large organic cations. Though the amine-type liquid anion exchangers could fit into the latter group, it is simpler to treat them as a separate homogenous group (class III-D). The large monovalent ion pair, tetraphenylarsonium tetraphenylborate (Ph$_4$As$^+$ Ph$_4$B$^-$), which has been suggested as a reference in solvent extraction (see Chapter 2), also belongs to this class.

Of some importrance is the extraction of alkali ions by tetraphenylborate and by crown ethers (Appendix D: 21), of fluoride by tetraphenylantimonium ions, and of perrhenate by tetraphenylarsonium ions. Because most of the volume of these complexes is taken up by organic groups, the complexes are highly lipophilic and, therefore, extractable into organic solvents without additional solvation. Though these systems have limited applications, the crown ethers offer some interesting extraction systems. The ethers form cage compounds ("clathrates"); i.e., the metal cation fits into a cage formed by the crown structure. Because the cage can be designed to fit almost any ion of a certain size, rather selective extractions are possible with this system, as described in Chapter 15.

Example 9: Extraction of K$^+$ by tetraphenylborate.
Consider the extraction of K$^+$ from aqueous solution into the organic solvent nitrobenzene by addition of NaPh$_4$B. Abbreviating Ph$_4$B$^-$ by A$^-$, the extracted complex is the ion pair K$^+$A$^-$. In an inert solvent this is a stable ion pair, but in a highly polar solvent like nitrobenzene, the ion pair may dissociate. The organic phase may thus contain both solvated K$^+$ and K$^+$A$^-$ species, while

the only potassium species in the aqueous phase is free K^+.
The extraction equilibria may be written

$$K^+ + A^- \leftrightharpoons K^+ A^-(org) \quad \text{Equil. const. } K_{ex} \tag{4.65}$$

while for the reaction in the organic phase

$$K^+ A^-(org) \leftrightharpoons K^+(org) + A^-(org) \quad \text{Equil. const. } K_{ass} \tag{4.66}$$

The distribution ratio of K^+ becomes

$$D_K = ([K^+ A^-]_{org} + [K^+]_{org}) [K^+]^{-1} = K_{ex}[A^-](1 + K_{ass}/[A^-]_{org}) [K]^{-1} \tag{4.67}$$

When A^- is added into the aqueous phase as a Na^+ salt in large excess to K^+, the dissociation in the organic phase becomes negligible and Eq. (4.67) is reduced to

$$D_K = K_{ex}[A^-] \tag{4.68}$$

Figure 4.17 shows the distribution ratio of K^+ when a large excess of Na^+A^- is added to the system. Although the extracted complex should be com-

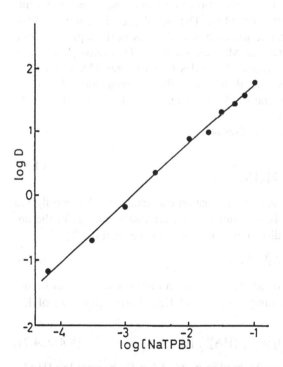

Fig. 4.17 Distribution ratio of potassium(I) between nitrobenzene and water as a function of initial aqueous tetraphenylborate concentration in 0.1M NaClO₄. (From Refs. 30a,b.)

pletely dissociated in the nitrobenzene, the ionic concentration of the organic phase is large enough to suppress dissociation, and the distribution ratio thus becomes proportional to the concentration of the extractant in the aqueous phase.

4.12 STUDIES OF HYDROPHILIC COMPLEXATION BY SOLVENT EXTRACTION (TYPE IV)

Coordinatively saturated metal complex in organic phase	MA_z $\updownarrow$
Formation of extractable and nonextractable complexes	$M^{z+} + zA^- + nX^- \leftrightarrows MA_z + MX_n^{z-n}$

Solvent extraction has become a common technique for the determination of formation constants, β_n, of aqueous hydrophilic metal complexes of type MX_n, particularly in the case when the metal is only available in trace concentrations, as the distribution can easily be measured with radioactive techniques (see also section 4.15). The method requires the formation of an extractable complex of the metal ion, which, in the simplest and most commonly used case, is an uncharged lipophilic complex of type MA_z. The metal-organic complex MA_z serves as a probe for the concentration of metal M^{z+} ions in the aqueous phase through its equilibrium with the free M^{z+}, section 4.8.2. This same principle is used in the design of metal selective electrodes (see Chapter 15). Extractants typically used for this purpose are β-diketones like acetylacetone (HAA) or thenoyltrifluoroacteone (TTA), and weak large organic acids like dinonyl naphtalene sulphonic acid (DNNA).

The pertinent distribution equation is

$$D_M = \frac{[MA_z]_{org}}{[M]\{1 + \Sigma\beta_n[A]^n + \Sigma\beta_p[X]^p\}} \tag{4.69}$$

(summed from 1 to n or p) where the formation constants for MA_n are β_n and those for the MX_p complexes β_p and summations are taken from 1. In the absence of MX_p complexes, the distribution ratio of M is defined by D_M^o:

$$D_M^o = [MA_z]_{org} /[M] (1 + \Sigma\beta_n[A]^n) \tag{4.70}$$

If the conditions are chosen so that the aqueous complexes of HA can be neglected [see Eq. (4.41) and Example 4], D_M^o of Eq. (4.70) equals D_M of Eq. (4.41c); i.e.,

$$D_M^o = D_M \text{ [of Eq. (4.41c)]} = K_{ex} [HA]_{org}^z [H^+]^{-z} \tag{4.41c, 4.71}$$

for which linear correlations are obtained in plots of log D_M against log $[HA]_{org}$ at constant pH, or against pH at constant $[HA]_{org}$. Dividing Eq. (4.69) by Eq. (4.70) gives

$$D^\circ/D = 1 + \beta_1[X] + \beta_2[X]^2 + \beta_3[X]^3 + \cdots \tag{4.72}$$

Thus a plot of D°/D against $[X]$ provides a polynomial from which the β_ps can be calculated (see section 4.14).

Example 10: Determination of formation constants for Be(II) oxalate complexes.

Figure 4.18 shows the extraction of beryllium with HTTA into hexone as a function of the concentration of oxalate ions, Ox^{2-}, in the aqueous phase. Here the conditions have been chosen so that the aqueous complexes of TTA can be neglected as has been described earlier, and thus Eq. (4.71) is valid, and the relation $D_M^\circ = K_{ex} [HA]_{org}^2 [H]^{-2}$ so that

$$\log D / D^\circ = \log D \, K_{ex}^{-1} [HA]_{org}^{-2} [H]^2$$
$$= -\log (1 + 10^{3.55}[Ox^{2-}] + 10^{5.40}[Ox^{2-}])^2 \tag{4.73}$$

The parameter values on the right-hand side of the equation, are the β_n values, and were determined by a curve fitting technique (section 4.14). Introducing these values in Eq. (4.71) gives the solid line through the points in Fig. 4.18. The advantage of this "strange" ordinate is that the curvature only reflects the oxalate complexation.

Hydrolysis of some metals begins at very low pH (e.g., Zr at pH ~ 0.4, Th(IV) and U(VI) at pH 1–2, the lanthanides at pH 3–5, etc.). However, if complex

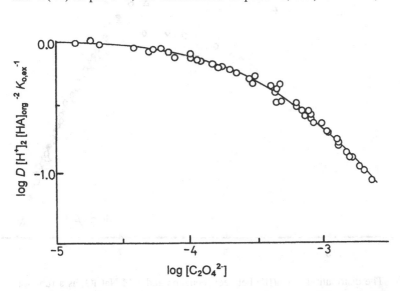

Fig. 4.18 Decrease in distribution ratio of Be(II) as a function of oxalate ion (Ox^{2-}) concentration due to formation of aqueous $BeOx_n^{2-2n}$ complexes. The extraction system is 0.03 M TTA in methylisobutylketone and 1.0 M Na(0.5 Ox^{2-}, ClO_4^-). See Eq. (4.72) for ordinate function. (From Ref. 31.)

formation strongly dominates over hydrolysis, the latter can be neglected. In the overview in section 4.3, it was shown in Fig. 4.3 that the general shape of the extraction curves indicates the type of complexes formed. Thus, the extraction of gallium(III) with acetylacetone, Fig. 4.19, indicates a behavior according to Fig. 4.3c. in the plot of log D vs. free ligand concentration, [A⁻].

Example 11: Formation of hydrolyzed Ga(III) acetylacetonates.

Figure 4.19 shows the extraction of Ga(III) by acetylacetone into benzene at various concentrations of total acetylacetone, $[HA]_t$, and constant ionic strength, using the AKUFVE technique (see section 4.15.3). By comparing with Fig. 4.3, one may guess that hydroxy complexes are formed (diagram c). If the general complex is designated $Ga_mA_n(OH)_p$, a first investigation showed

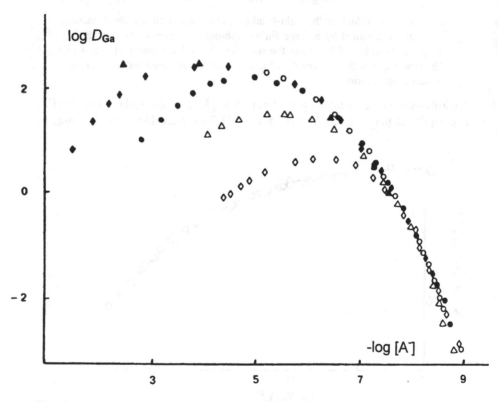

Fig. 4.19 The distribution of Ga(III) between benzene and 1 M NaClO₄ as a function of pA = −log[A⁻] at 6 different acetylacetone (HA) concentrations (0.06–0.0006 M).The different curves are due to different extent of hydrolysis in the aqueous phase. The concentration of HA decreases from the upper left corner, where $[HA]_{aq}$ is 0.06 M (▲), toward the right lower corner, where it is 0.0006 M (◇). (From Ref. 32.)

that no polynuclear complexes were formed, i.e., $m = 1$, as the extraction curves were independent of the Ga(III) concentration. This was also expected as the total concentration of Ga(III) ranged from 10^{-4} to 10^{-6} M. Next, using a computer program, the formation constants for the following complexes were determined: GaA^{2+}, GaA_2^+, GaA_3, $Ga(OH)_3$ and $Ga(OH)_4^-$ [32]. It was concluded that no mixed complexes with both A^- and OH^- were formed. The organic phase contained only GaA_3 (which was hydrated).

In using this technique accurately, it is necessary to consider possible sources of errors. A rather extensive summary of this technique applied to hydrophilic actinide complexation by practically all inorganic ligands, as well as a considerable number of weak organic acids, has recently been presented [33a,33b].

4.13 EFFECTS AND CORRELATIONS

Because of the many parameters involved in solvent extraction, chemical as well as physical, it is a difficult task to draw reliable conclusions about the reactions that are responsible for the observed distribution values. In the introductory part of this chapter, section 4.2, the extraction reaction described by K_{ex} was shown to be a product of a number of parameters related to the different steps involved in the formation of the extracted complex. Sections 4.4–4.11 have described these various subreactions, which are described by Eq. (4.46):

$$K_{ex} = K_a^z \; \beta_z \; K_{DR}^{-z} \; K_{DC} \; K_{ad,bB} \qquad \text{(4.46, 4.74)}$$

Examples of how K_{ex} data may be interpreted in more fundamental terms follow. In the lack of information on the independent parameters of Eq. (4.74), tenable conclusions can be drawn about their size and effects from known correlations between the parameters. This section explores the effect of such correlations.

4.13.1 Effects of the Dissociation Constant of Acidic Ligands and Formation Constants of the Corresponding Metal Complexes

4.13.1.1 The Ligand (Acid) Dissociation Constant K_a

According to Eq. (4.74) K_{ex} increases to the z power of the dissociation constant, K_a, for the extractant acid HA. Thus it appears desirable to use stronger acids as extractants, i.e., those with large K_a (or small pK_a). Thus factors that affect the dissociaton constant of an acid are of importance for extraction behavior.

The acidity of organic acids increases (i.e., pK_a decreases) by electrophilic substitution in the carbon chain. For example, replacing an H atom in the CH_3 group of acetic acid (pK_a 4.75) by I, Br, Cl, and F atoms yields monohalide acetic acids with pK_a of 3.0, 2.9, 2.8, and 2.7, respectively. The mono-, di-, and trichloroactic acids have pK_a 2.8, 1.48, and about 1.1, respectively. Other electrophilic groups which can increase acidity are HO−, NC−, NO$_2$−, etc. Conversely, nucleophilic substitution by aliphatic or aromatic groups usually has little effect on the pK_a, though it may affect the distribution constant K_{DR} (see section 4.13.2): e.g., the addition of a CH$_3$− or C$_6$H$_5$− group in acetic acid changes the pK_a to 4.9 (propionic acid) and 4.3 (phenyl acetic acid). The further the substitution is from the carboxylic group, the less the effect: e.g., while pK_a for benzoic acid is 4.19, a Cl in orthoposition yields pK_a 2.92 and in the parapos- ition, 3.98.

These general rules hold rather well for the acidic organic extractants and can be used to extrapolate from related compounds to new ones as well as to develop new extractants. The effect of various substituends on pK_a is extensively discussed in textbooks on organic chemistry [e.g., 14, 34].

4.13.1.2 The Complex Formation (or Stability) Constant β$_n$

K_{ex} also increases with increasing formation constant of the uncharged metal complex, β$_z$. Thermodynamic factors and geometrical aspects that influence the size of β$_n$ are discussed in Chapter 3. Some further observations follow.

To achieve the highest possible D value, the concentration of the extracted uncharged complex (MA$_z$, MA$_z$B$_b$ or B$_b$ML$_{z+b}$) must be maximized. A large value for the formation constant of the neutral complex favors this goal. In Chapter 3 it was pointed out that for hard acids the complex formation constant increases with the charge density of the metal ion, provided there is no steric hindrance. This is seen for the metal fluoride complexes in Fig. 3.4 (no steric hindrance). For the lanthanide acetylacetonates in Fig. 4.15b, the increase in β$_n$ from $_{57}$La to $_{64}$Gd is due to a reduction in the ionic radius and, accordingly, to the increased charge density. It should be noted that the diminishing lanthanide size leads to a successive diminution of the coordination number (from 9 to 8 for H$_2$O) for La → Gd; for Gd → Lu the coordination number is probably con- stant, as is inferred from Fig. 4.15c; see also [35].

4.13.1.3 Correlations Between K_a and β$_n$

Because the H$^+$ ion acts as a hard metal ion, one should expect a close correla- tion between the formation constants for HA and for MA$_z$. Since K_a is defined as the *dissociation* constant of HA, while β$_z$ is the *formation* constant of MA$_z$, a correlation is therefore expected between pK_a (= −logK_a) and logβ$_z$. The corre- lations in Figs. 3.5 and 3.6 show that large logβ$_n$ values are usually observed for organic acids with large logK_a. Since K_{ex} is directly proportional to both β$_z$ and K_a^z [see Eq. (4.74)], and a large β$_z$ value is likely to be accompanied by a small K_a value (large pK_a); in Eq. (4.74) the two parameters counteract each other. Thus in order to obtain high free ligand concentrations, which favors the formation of the MA$_z$ complex, [H$^+$] must be low, i.e., pH must be high. However, at high pH the metal may be hydrolyzed. A careful balance must be struck between β_z, K_a, and pH to achieve an optimum maximum concentration of the uncharged complex.

4.13.2 Effects of Molecular Volume and Water Structure Upon K_D

Chapters 2 and 3 factors are discussed which affect the distribution of uncharged species between the organic and aqueous phases, the most important being (1) the interaction of the species (complex or extractant) with the solvents (solvation); and (2) energy released or required when an uncharged species leaves its original phase (cavity closure) and that required when it enters the new phase (cavity formation). Sections 4.4 and 4.5 discuss the distribution of inert species, including undissociated acid HA, K_{DR}. From Eq. (4.74) it is seen that K_{ex} is negatively influenced by a large distribution constant for the extractant reagent, but favored by a large K_{DC} for the extraction of the metal complex.

4.13.2.1 Metal Complexes, K_{DC}

There are rather limited data on systematics for the distribution constants K_{DC} of metal complexes. Table 4.10 gives distribution constants for Zn(II), Cu(II), and Np(IV) acetylacetonates in various solvents. The zinc complex has an octahedral coordination: four oxygen atoms from the ligands and two oxygen atoms from water molecules can occupy six positions in an almost spherical geometry around the zinc atom. The copper complex has a square planar configuration with four oxygen atoms from the acetylacetones in the plane; a fifth (and possibly sixth) position perpendicular to the plane and at a longer distance from the copper atom can be occupied by an oxygen from a water molecule. The zinc complex has a larger degree of hydration in the aqueous phase and consequently a lower distribution constant. The NpA$_4$ structure is likely a square antiprism with no hydrate water. It has a larger molecular volume than the two other complexes, and its larger distribution constant is in agreement with this, reflecting the release of more water structuring energy when leaving the aqueous phase.

Two further observations can be made from Table 4.10: (1) the distribution constant seems to increase with increasing polarity of the solvent; (2) when the reagent acetylacetone has a large K_{DR}, the metal complexes also show a large K_{DC} value.

The lanthanide ions, Ln^{3+}, are known to contract with increasing atomic number (Z), from La^{3+} with a hydrated radius of 103 pm to Lu^{3+} of 86 pm (lanthanide contraction). Thus one expects that the neutral LnA$_3$ complex becomes smaller with increasing atomic number, and consequently that β_3 should increase and K_{DC} decrease with increasing Z. Figure 4.15d shows that measured

K_{DC} values for the lanthanide acetylacetonates are the reverse from that expected for the size effect. The explanation is likely that the neutral complex with the formula LnA_3 is coordinatively unsaturated, which means that a hydrated complex exists in the aqueous phase (possibly also in the organic phase). The more coordinatively unsaturated lanthanide complexes (of the larger ions) can accommodate more water and thus are more hydrophilic. The result is a K_{DC} several orders of magnitude lower for the lightest, La, than for the heaviest, Lu.

The tetravalent metal acetylacetonates [36] have all coordination numbers (CN) 8 or 9 in the neutral complexes. For Th (radius 109 pm at CN 9) there is often one molecule of water in ThA_4, whereas for the corresponding complexes of U(IV) and Np(IV) (radius 100 and 98 pm, respectively, CN = 8), there seems to be none. The measured distribution constants (log K_{DC}) are for Th 2.55 and for U and Np 3.52 and 3.45, respectively, in 1 M $NaClO_4$/benzene. This agrees with the greater hydrophilic nature of the $ThA_4 \cdot H_2O$ complex.

4.13.2.2 Correlations Between K_{DR} and K_{DC}

Qualitatively, the distribution constants of the complexing reagent and the corresponding neutral metal complex are related due to their similar outer surface toward the solvents. Thus it is reasonable that there should be good correlation between K_{DR} and K_{DC} for homologous series of reagents or complexes. This is indeed the case, as is seen when $K_{D,R}$ is plotted vs. $K_{D,C}$, in Fig. 4.20 for a complex with a variation of the solvent, and in Fig. 4.21 for complexes of related reagents with the same solvent. This correlation can be related to the solubility parameter concept discussed later.

4.13.3 The Solubility Parameter

In Chapter 2 we learned that the cohesive forces that keep molecules together in a solution can be described by

$$\delta^2 = (\Delta H_V - \mathbf{R}T)/V \qquad\qquad (2.1;\ 4.75a)$$

where δ^2 is the *cohesive energy density*, and δ, *the solubility parameter*, V, the molar volume and ΔH_V, the molar heat of vaporization. This holds for regular solutions [37] that show ideal entropy effects in mixing solute and solvent, and no interactions occur besides the cohesive forces between the solute and solvent molecules. [Note: Regular solutions exhibit heat changes when mixed, while for ideal solutions, the heat of mixing is zero (see section 2.4.1).] There is no change of state in association or in orientation. Thus the work required to produce a cavity, ΔG_{cav}, by the solute in the solvent is given by

$$\Delta G_{cav} = A_{cav}\ V_B\ \delta^2 \qquad\qquad (2.10;\ 4.75b)$$

where A_{cav} is a proportionality factor and V_B is the molar volume of the solute B.

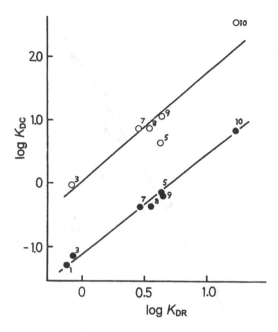

Fig. 4.20 Distribution constants K_{DC} for uncharged metal complexes MA_z, vs. distribution constants K_{DR} for the corresponding undissociated reagent, the acid ligand HA, for various organic solvents and 1 M NaClO$_4$: open circles Cu(II), solid circles Zn(II). Numbers refer to solvents listed in Table 4.10. (From Ref. 36.)

Pioneering work on the application of this theory for correlating and predicting distribution ratios was done in the 1960s [38a–40c]. Several reviews on the use of this theory for two-phase distribution processes are also available [41,42]. Recently this theory has been refined by the use of *Hansen solubility parameters* [6,43,44], according to which

$$\delta_{total}^2 = \delta_{dispersion}^2 + \delta_{polar}^2 + \delta_{hydrogen}^2 \qquad (4.75c)$$

where interactions between the solute and the solvent are described by contributions from the various types of cohesive forces. In general, the dispersion (or London) forces dominate. In Fig. 4.22 the measured distribution ratios of an americium complex between an aqueous nitrate solution and various organic solvent combinations are compared with distribution ratios calculated from tabulated Hansen parameter values [45].

Few solubility parameters are available for the metal-organic complexes discussed in this chapter. Another approach is then necessary. The distribution constant for the reagent (extractant), **R**, can be expressed as:

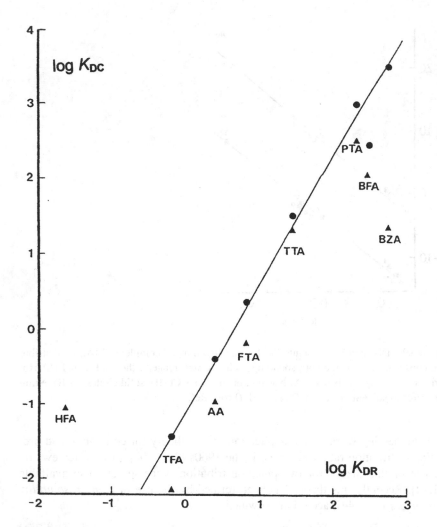

Fig. 4.21 Distribution constants, K_{DC}, for uncharged metal complexes MA_z vs. distribution constants K_{DR} for the corresponding undissociated acid ligand HA; solid circles Zn(II), solid triangles Co(III). Variation with ligand composition: HFA hexafluoroacetylacetone, TFA trifluoroacetylacetone, AA acetylacetone, FTA 2-furoyltrifluoroacetone, TTA 2-thenoyltrifluoroacetone, PTA pivaloyltrifluoroacetone, BFA benzoyltrifluoroacetone, BZA benzoylacetone. (From Ref. 36.)

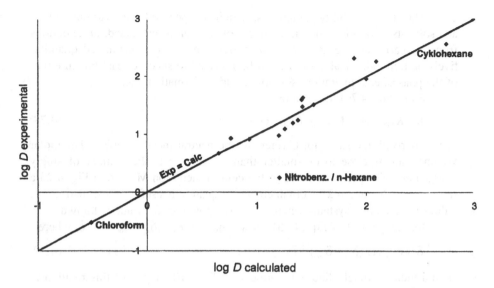

Fig. 4.22 Comparison of measured and calculated distribution ratios D_{Am} of americium(III)-terpyridine-decanoic acid complexes between 0.05 M HNO_3 and various organic solvent combinations. The calculated values are obtained with the Hansen partial solubility parameters. (From Ref. 45.)

$$\mathbf{R}T \ln K_{DR} = V_R[(\delta_{aq} - \delta_R)^2 - (\delta_{org} - \delta_R)^2] + \mathbf{R}T\ V_R(1/V_{org} - 1/V_{aq}) \quad (4.76)$$

where K_{DR} is given in terms of mole fractions. The solubility parameter of water has to be empirically estimated. The thermodynamic value is 48, but, since the system may be sensitive to the choice of δ_{aq}, it is better to select a "reference δ_{aq}" in each set of experiments, assuming constant values for the aqueous systems in the two different two-phase systems. Molar volumes and solubility parameter values are listed in Table 2.1 and Refs. [6,44].

The overall enthalpy and entropy changes for the distribution reaction (i.e., transfer of the metal complex from the aqueous to the organic phase) can be obtained from the temperature dependence of K_{DR} according to

$$\mathbf{R}\,T \ln K_{DR} = -\Delta H + T\Delta S \qquad (4.77)$$

An enthalpy term corresponding to $-V_R[(\delta_{aq} - \delta_R)^2 - (\delta_{org} - \delta_{aq})^2]$ and an entropy term corresponding to $\mathbf{R}\,V_R(1/V_{org} - 1/V_{aq})$ can be obtained formally from Eq. (4.76). If the distribution constant for a standard system is used as a reference, changes in enthalpy and entropy relative to this standard system can be assessed by replacing δ_{aq} and V_{aq} with the corresponding data for the organic diluent in the reference system, δ_{org}^{ref} and V_{org}^{ref}.

Thus from solubility parameters, which are specific for the various solutes and solvents, and molar volumes, values for K_{DR} can be estimated, or deviations from regularity can be assessed. These deviations can be estimated quantitatively and, in individual systems, can be ascribed to specific reactions in either of the phases, e.g., hydration, solvation, adduct formation, etc.

From Eq. (4.76) the relation

$$\log K_{DC} = (V_C / V_R) \log K_{DR} + \text{const.} \tag{4.78}$$

can be derived; the subscript C refers to the neutral metal complex. The molar volume ratio is close to or smaller than z, where z is the number of singly charged anionic species attached to the central metal ion M^{z+}. From Fig. 4.23 a ratio of ca. 1.5 (for $z = 2$) is obtained. This equation is useful for estimating K_{DR} values in extraction systems where the corresponding K_{DC} value is known.

Rearrangement of Eq. (4.75) shows that $\log K_{DR}/(\delta_{aq} - \delta_{org})$ vs. δ, where

$$\delta = \delta_{org} \, RT(\delta_{aq} - \delta_{org})^{-1} \, (V_{org}^{-1} - V_{aq}^{-1}) \tag{4.79}$$

should yield a straight line with slope $V_R / RT \ln 10$. A plot of this relation in Fig. 4.23 demonstrates a satisfactory correlation between distribution constants and solvent parameters, indicating the usefulness of the solubility parameter concept in predicting K_{DR} as well as K_{DC} values.

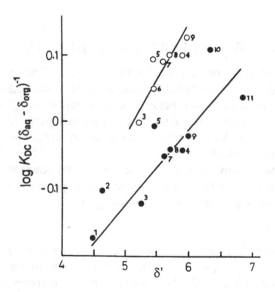

Fig. 4.23 Application of the regular solution theory for correlation of distribution constants for ZnA_2 and CuA_2 with solvent properties (solubility parameters); the numbers refer to the solvents listed in Table 4.10. (From Ref. 22.)

4.13.4 Effects of the Solvent (Diluent) Composition

The composition of the organic phase solvent (diluent) influences the distribution of both the neutral metal organic complex MA_z and the complexing reagent HA, through similar interactions ranging from that of cavity formation for very inert diluents like hexane, through dipole-dipole interactions, pi electron interaction, and hydrogen bonding for the more reactive solvents. The expected degree of interaction can be judged from values of permittivity and solubility parameters. Previous tables have illustrated some of these aspects: Table 4.1 on the distribution of inert solutes (hole formation), Table 4.7 on distribution of organic acids (polarity), and especially Table 4.10 on distribution of metal acetylacetonates (solvent permittivity), discussed in section 4.13.2.1. These cases involve the aqueous phase, so some phenomena may be attributed to the interaction with water molecules.

Figure 4.24 shows a decrease in K_D for acetylacetone with increasing number of methyl groups in the substituted benzene solvent. Since acetylacetone is more soluble in aromatics than in aliphatics, as illustrated by the K_{DR} sequence:

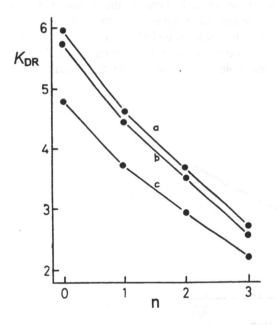

Fig. 4.24 Distribution constants K_{DR} of acetylacetone between methyl-substituted benzenes at different aqueous ionic strengths: (a) 0.001 M, (b) 0.1 M, and (c) 1.0 M $NaClO_4$ at 25°C. The number of methyl groups, n, changes the aromatic character at $n = 0$ to more aliphatic at $n = 3$. (From Ref. 46.)

hexane 0.95, cyclohexane 1.03, ethylbenzene 3.31, and benzene 5.93 (aqueous phase 1 M NaClO$_4$), the decrease observed in Fig. 4.24 may simply be an effect of the increasing aliphatic character of the solvents.

4.13.5 Effects of the Aqueous Medium

The effect of the water activity of the aqueous phase, discussed in Chapters 2, 3, and 6, is determined by the total concentration and nature of the salts. Generally, the distribution constant for a neutral metal complex would increase with increasing ionic strength, as illustrated in Fig. 4.25. This *salting-out effect* is often ascribed to a reduction in free water available for hydration. On the other hand, the salt also breaks down the water structure, which could reduce the energy to form a hole in the phase.

4.13.6 Effects of the Temperature

The introduction of a neutral complex into a solution phase involves a number of processes that can be associated with large changes in enthalpy (solvation processes) and in entropy (solvent orientation and restructuring), leading to considerable temperature effects; see Chapter 2. In general, the distribution of a neutral metal complex increases with increasing temperature for complexes with significant hydrophobic character. The metal acetylacetonate systems in Fig. 4.26 illustrate this effect. In these particular systems, the free energy of distribution between the organic and aqueous phases is dominated by the enthalpy term.

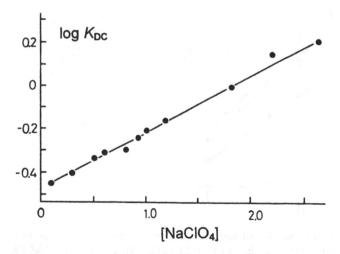

Fig. 4.25   The distribution constant K_{DC} for zinc diacetylacetonate, ZnA$_2$, between benzene and various concentrations of NaClO$_4$ in the aqueous phase; 25°C. (From Ref. 36.)

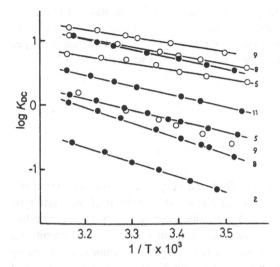

Fig. 4.26 Distribution constant K_{DC} for metal diacetylacetonates, MA_2, between organic solvents and 1 M $NaClO_4$ as a function of temperature: Cu(II) (open circles) and Zn(II) (solid circles) with various solvents (numbers refer to the solvents given in Table 4.10). (From Ref. 36.)

From measurements of the temperature dependency of the equilibrium constant, thermodynamic parameters may be deduced (section 3.4). Very few enthalpy and entropy constants have been derived for the distribution reaction $MA_z(aq) \leftrightarrows MA_z(org)$ of neutral complexes; such investigations give information about hydration and organic phase solvation.

4.13.7 Further Structural Considerations

In Chapters 2 and 3 several physicochemical factors of importance to solvent extraction have been described, and many of their effects have been illustrated in this chapter. Here, we summarize some observed regularities. The effect of various structures on the bond strengths in metal organic complexes has been extensively treated in other publications [47–49].

4.13.7.1 Chelate Ring Size

Formation of metal-organic chelate complexes results in stronger complexation (i.e., larger β_n values) compared to interaction with monodentate ligands (Chapter 3). The common types of bidentate ligands are presented in Table 4.9; the chemistry of these complexes has been extensively discussed in the literature [14,47]. Chapter 3 presents the most important factors in the formation of such complexes: (1) the type of binding atom; (2) the chelate ring size (or "bite");

and (3) the number of chelate rings formed (mono- or polydentate). The following diagram illustrates the general structure of UO_2^{2+} and Am^{3+} diamide chelate complexes: two oxygen binding atoms; ring size is $5 + n$, 2–3 chelate rings per complex; the $-O-M-O-$ part is commonly referred to as the "claw."

```
   |            |
   N   CₙH₂ₙ   N
  / \   / \   / \
  CH         CH
  O           O
   \         /
        M
```

The extractants TBOA, TBMA, and TBSA are very similar, but their structural differences (see formulas in Table 4.16) allow the formation of only one type of metal chelate complex: 5-, 6-, and 7-membered rings, respectively. Similarily, the reactants DMDOMA and DMDOSA form only 6- and 7-membered chelates. Table 4.16 shows that extraction (i.e., largest K_{ex} value) is favored by 6-membered rings. This is not unexpected as the K_{ex} values in this case reflect the stability constants β_n acc. to Eq. (4.74).

The electron shells of the M^{3+} elements with unfilled or partially filled 3d, 4d, and 5f orbitals contract as these shells are being filled with electrons, increasing the charge density (z/r) of the cation, leading to increasing stability constants for the MA_3 complexes with increasing atomic number. For bidentate ligands the f-electron (lanthanide) MA_3 complexes are coordinatively unsaturated, i.e., only six of eight available coordination sites are filled. Therefore, the two remaining sites are occupied by H_2O or HA or some other donor molecules B, leading to the self-adduct MA_3HA^{-1} or adduct MA_3B_{1-2}. It has been shown that the adduct formation constants for the reaction $MA_3 + B \rightarrow MA_3B$ decrease

Table 4.16 Consultants for Extraction of Actinide Amides from Nitrate Solutions into an Organic Solvent Showing the Effect of Chelate Ring Size

Metal ion	Extractant[a]	K_{ex}	Ring size	Ref.
UO_2^{2+}	TBOA	0.005	5	[50]
UO_2^{2+}	TBMA	10.6	6	[50]
UO_2^{2+}	TBSA	9.3	7	[50,51]
Am^{3+}	DMDOMA	0.47	6	[52]
Am^{3+}	DMDOSA	<0.0002	7	[52]

[a]TBOA $(C_4H_9)_2NCO_2$; TBMA $[(C_4H_9)_2NCO]_2CH_2$; TBSA $[(C_4H_9)_2NCO]_2(CH_2)_2$; DMDOMA $(CH_3C_8H_{17}NCO)_2CH_2$; DMDOSA $(CH_3C_8H_{17}NCO)_2(CH_2)_2$.

in the same order [27a–28], see Fig. 4.15; i.e., the stronger the complex, the weaker the adduct.

The increase, by adduct formation, in the coordination number of the central atom is possible not only for weak but also for strong chelates, provided their ligand bites are relatively short. To make room in the inner sphere of the metal ion for another adduct-forming ligand, three chelating ligands of small bite angle can easily be shifted away without significant energy-consuming distortion of the chelate rings [53a,b]. The electronic structure of the central atom is also of key importance for synergism, as illustrated by the easy adduct formation of metal ions with unfilled or partially filled d or f orbitals, contrary to, e.g., p-block elements. Inner- and outer-sphere complexation is further discussed in the next two paragraphs and in Chapter 16.

4.13.7.2 Donor Ligand Effects

Table 4.17 shows extraction constants for some metal ions with three alkyl phosphates substituted by 0, 1, and 2 sulfur atoms, but with almost identical aliphatic branchings. Section 3.3 discusses hard and soft acids and bases (HSAB theory). According to this theory, hard acids form strong complexes with hard bases, while weak acids form strong complexes with weak bases. In Table 4.17, the metals are ordered in increasing hardness, the sub II.b group being rather soft (Table 3.2). Presumably the K_{ex} values reflect this pattern, as they are proportional to K_a^z and β_z. In Table 4.17, the acidity of the acids increases (i.e., K_a increases) in order $R \cdot PSSH < R \cdot POSH < R \cdot POOH$ (consult Tables 4.8 and 4.9), as sulfur is less basic than oxygen. In general, the K_{ex} increases for for the dialkyl phosporic acid (hard donor ligand) with increasing metal charge density within each group as predicted in Chapter 3. For the soft metals, K_{ex} also increases with increasing softness of the ligand, while the opposite effect is seen for the hard Ln-metal ions. The divalent subgroup II.b metals prefer to bind to sulfur rather than to oxygen because they have a rather soft acceptor character, while the hard metals III.b prefer to bind to the hard O-atom of the ligand.

Table 4.17 Comparison of the Extraction Constants K_{ex} for the IIb-Subgroup Divalent Ions, and IIIb Lanthanide Ions, with Sulfur or Oxygen Dialkyl Phosphoric Acids[a]

Metal ion	Hg^{2+}	Cd^{2+}	Zn^{2+}	$_{57}La^{3+}$	$_{63}Eu^{3+}$	$_{71}Lu^{3+}$
Charge density Z^2/r	3.6	4.1	5.4	7.83	8.74	9.68
log K_{ex} for $R \cdot POOH$	−2.20	−1.80	−1.20	−2.52	−0.44	2.9
log K_{ex} for $(C_4H_9O)_2 \cdot POSH$	5.40	3.70	0.70	−4.78	−4.23	0.34
log K_{ex} for $(C_4H_9O)_2 \cdot PSSH$	4.40	3.49	2.40		−8.28	

[a]Values estimated from Refs. 5, 20, 54, 55, and 56. R is $(C_4H_9CH(C_2H_5)CH_2O)_2$.

4.13.7.3 Inner and Outer Sphere Coordination

The extractabilities of metal-organic complexes depend on whether inner or outer sphere complexes are formed. Case I, section 4.2.1, the extraction of uranyl nitrate by TBP, is a good example. The free uranyl ion is surrounded by water of hydration, forming $UO_2(H_2O)_6^{2+}$, which from nitric acid solutions can be crystallized out as the salt $UO_2(H_2O)_6^{2+}(NO_3)_2^{2-}$, though it commonly is written $UO_2(NO_3)_2(H_2O)_6$. Thus, in solution as well as in the solid salt, the UO_2^{2+} is surrounded by 6 H_2O in an *inner coordination sphere*. In the solid nitrate salt, the distance d_{U-O}(nitrate) between the closest oxygen atoms of the nitrate anions, $(O)_2NO$, and the U-atom is longer than the corresponding distance, d_{U-O}(water), to the water molecules, OH_2, i.e., d_{U-O}(nitrate) $> d_{U-O}$(water); thus the nitrate anions are in an *outer coordination sphere*.

When the adduct former TBP is added, the $OP(OC_4H_9)_3$ groups are closer to the U-atoms than the OH_2's, d_{U-O}(TBP) $< d_{U-O}$(water), two waters are expelled, although the OP- bond only occupies one coordination site. The charge distribution around the UO_2-group allows the nitrate group to enter the inner coordination sphere of the uranyl complex. The resulting configuration is shown in Fig. 4.27. Thus, the extraction reaction can more correctly be written

$$UO_2(H_2O)_6^{2+} + 2\ NO_3^- + 2\ TBP(org)$$
$$\leftrightarrows UO_2(TBP)_2(NO_3)_2(org) + 6\ H_2O \qquad (4.5b)$$

In many systems it is difficult to distinguish between inner and outer sphere complexation. In that case, knowledge about the thermodynamic parameters ΔH and ΔS may be of help. For inner sphere complexes, the hydration is disrupted more extensively and the net enthalpy and entropy changes are positive. In Eq. (4.5b) it is noted that the number of entities increases from five to seven, an entropy increase. This entropy effect (i.e., $-T\Delta S$) is obviously larger than the enthalphy effect, as the overall energy change ΔG is negative. In outer sphere complexes, the dehydration is less disrupted, and the net enthalpies and entropies are negative owing to the complexation with its decrease in randomness without a compensary disruption of the hydration spheres.

Migration experiments have shown that the hydrated cations not only carry with them the water in the inner coordination sphere, but also one or more shells of additional water molecules, for typical total values of 10–15. When the metal ion leaves the aqueous phase in the solvent extraction step, this ordered coordinated water returns to the bulk water structure, contributing an additional factor to consider in evaluating the thermodynamics of extraction.

4.13.7.4 Steric Hindrance

The closer the reactive ligand atoms get to the central metal ion, the stronger the bond, and, consequently, the greater the formation constant. This can be

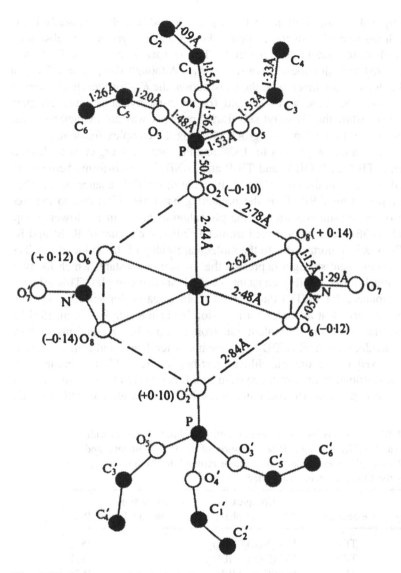

Fig. 4.27 Structure of the $UO_2(NO_3)_2 \cdot 2(C_2H_5O)_3PO$. The uranyl oxygens are situated perpendicular to the plane shown around the central atom. (From Ref. 12.)

studied by making substitutions in the organic ligand, which structurally interferes with the formation of the complex. For example, Dyrssen [8] substituted
oxines (8-hydroxyquinolinols), Appendix D.6 in various positions, Table 4.5,
and measured their distribution constants, K_{DC}. Although the pK_a and $K_{D,HA}$ in
Table 4.5 did not vary much with the substitution, the K_{DC} for the ThA$_4$ complex
with unsubstituted oxine was 425, and for the 5-methyl substituted complex
around 1000, while the 2-methyl substituted complex was not extractable. This
can be attributed to the 2-methyl group blocking the complex formation.

Table 4.18 lists K_{ex} values for UO$_2^{2+}$ and Pu^{4+} with four organophosphorous
extractants, TBP and DOBA, and TiBP and DOiBA, the *iso*-forms being more
branched. The data in the upper half of the table shows little change in K_{ex} when
TBP is replaced by TiBP. Thus the branching has little effect due to the free
rotation of the substituents around the phosphorous atom. In the lower group,
the amide DOBA, and its branched isomer DOiBA, the nature of R, R′ and R″
in R″R′NCOR is important due to the molecular rigidity of the amide group. The
branched substituent strongly depresses the extraction constant, which likely is
due to a much lower stability constant for the branched complex. This effect is
more pronounced for Pu(IV) than for UO$_2^{2+}$. In a search for selective systems
for separating trivalent actinides form fission lanthanides, Spjuth et al. [59] investigated the extraction of trivalent ions from aqueous HNO$_3$ solutions containing 2-bromodecanoic acid (HBDA) by means of terphenyl and triazine derivatives dissolved in the organic diluent *tert*-butylbenzene (TBB). Figure 4.28
shows the distribution ratio of the system as a function HBDA in TBB with and
without the terpyridine (formula in Fig. 4.29). This is an illustration of the

Table 4.18 Constants for Extraction of Actinide Nitrate Adducts with
Phosphoric Acid Trialkylesters (Free Rotation of the *P* Substituents) and
with *N,N*-Dialkylamides (Restricted Rotation Around the Amide Moiety)
Showing the Effect of Steric Hindrance

Metal ion	Extractant	Complex in organic phase[a]	Extraction constant K_{ex}	Ref.
UO$_2^{2+}$	TBP	UO$_2$(NO$_3$)$_2$(TBP)$_2$	28.3	[57]
UO$_2^{2+}$	TiBP	UO$_2$(NO$_3$)$_2$(TiBP)$_2$	26	[57]
Pu^{4+}	TBP	Pu(NO$_3$)$_4$ (TBP)$_2$	11.7	[57]
Pu^{4+}	TiBP	Pu(NO$_3$)$_4$ (TiBP)$_2$	8.9	[57]
UO$_2^{2+}$	DOBA	UO$_2$(NO$_3$)$_2$(DOBA)$_2$	5.75	[58]
UO$_2^{2+}$	DOiBA	UO$_2$(NO$_3$)$_2$(DOiBA)$_2$	0.55	[58]
Pu^{4+}	DOBA	Pu(NO$_3$)$_4$ (DOBA)$_2$	0.235	[58]
Pu^{4+}	DOiBA	Pu(NO$_3$)$_4$ (DOiBA)$_2$	0.003	[58]

[a]TBP (C$_4$H$_9$)$_3$PO; TiBP [(CH$_3$)$_2$CHCH$_2$O]$_3$PO; DOBA [C$_4$H$_9$CH(C$_2$H$_5$)CH$_2$]$_2$NCOC$_3$H$_7$;
DOiBA [C$_4$H$_9$CH(C$_2$H$_5$)CH$_2$]$_2$NCOCH(CH$_3$)$_2$.

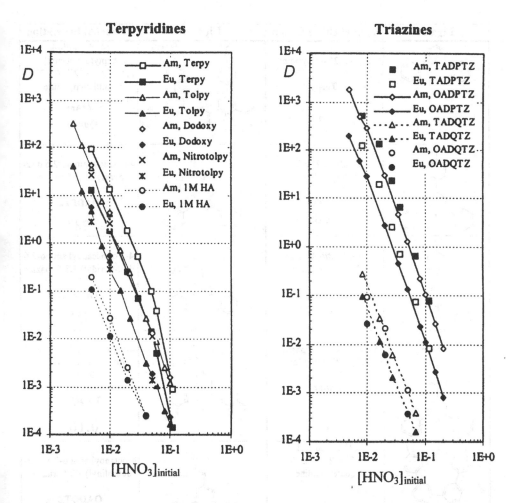

Fig. 4.28 Extraction of Eu(III) and Am(III) from K,HNO₃ with 0.02 M oligopyridine or triazine and 1 M 2-bromodecanoic acid in *tert*-butylbenzene. (From Ref. 59.)

necessity to achieve synergistic effects by using adduct formers for extraction of metal ions that otherwise are too strongly hydrated to be extractable. The structures of the terphenyls (adduct formers) tested are listed in Fig. 4.29 with the distribution ratios achieved for Am(III), D_{Am}. For example, in the system with DODOXY, D_{Am} is 4.2, while with TADPTZ it is 359. The triazine TAD-PTZ is more branched (the terphenyl DODOXY more linear) and has a slightly larger molar volume, but the difference may also be attributed to the greater basicity (i.e., smaller K_a) compared to terphenyl. However, the 100 times larger

Ligand	Name/Abbreviation	Ligand	Name/Abbreviation
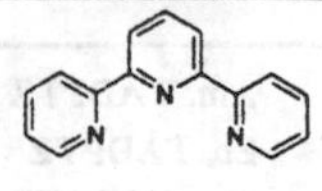	2,2'-6',2''-terpyridine **Terpy** 13	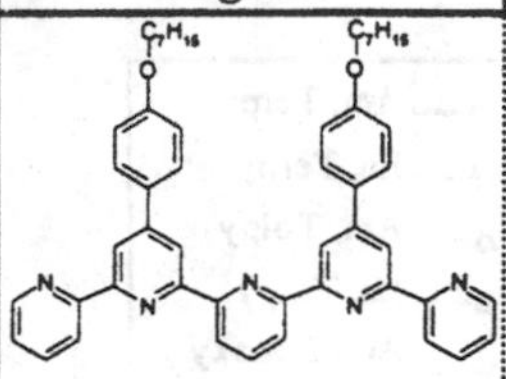	(4',4''')-di-(4-heptyloxyphenyl)-2,2':6'2'':6'2''':6'''2''''-quinquepyridine **D_{Am}** **Quinque**
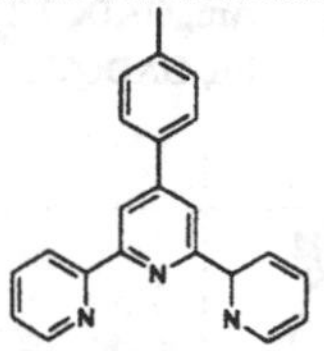	4'-tolyl-2,2':6'2''-terpyridine **Tolpy** 2.8	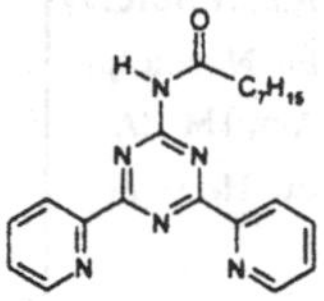	4-octanoyl amino-2,6 di(2-pyridyl) 1,3,5 triazine **OADPTZ** 287
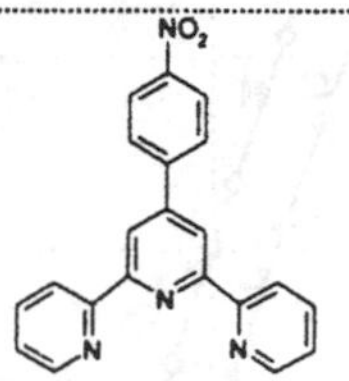	4'-(4-nitrophenyl)-2,2':6'2''-terpyridine **Nitrotolpy** 2.6	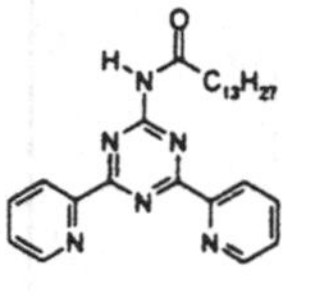	4-tetradecanoyl amino-2,6 di(2-pyridyl) 1,3,5 triazine **TADPTZ** 359
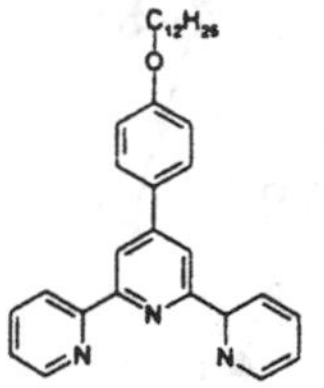	4'-(4-dodecyloxyphenyl)-2,2':6'2''-terpyridine **Dodoxy** 4.2	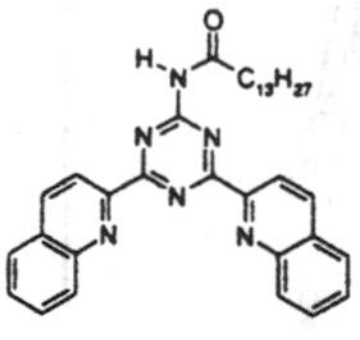	4-tetradecanoyl amino-2,6 di(2-quinolinyl) 1,3,5 triazine **TADQTZ** 0.16
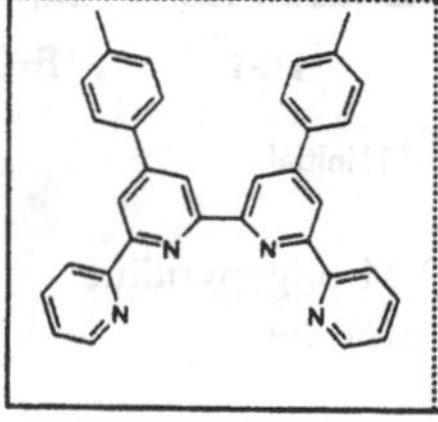	(4',4'')-ditolyl-2,2':6'2'':6'2'''-quaterpyridine **Quater** 490	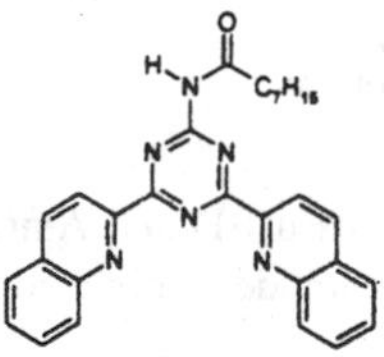	4-octanoyl amino-2,6 di(2-quinolinyl) 1,3,5 triazine **OADQTZ** 0.09

Fig. 4.29 Structure of various oligopyridine and triazine adducts, and distribution ratios (inserts) for Am(III) complexes with these oligopyridines (0.02 M) and 2-bromodecanoic acid (0.00025 M) in *tert*-butylbenzene and 0.01 M HNO_3. (From Ref. 59.)

extraction by the triazine cannot be explained by the HSAB theory; various explanations are suggested [59]. Steric hindrance can be predicted by modeling the molecular structures and chemical reactions. Advanced programs require mainframe computers, but useful programs are available for desktop computers [49].

4.14 CALCULATION OF EQUILIBRIUM CONSTANTS

Earlier sections presented chemical models for the extraction of acids and metals into organic solvents, and show that these models, expressed mathematically, agree with experimental data at trace metal concentrations and at constant activity coefficients. These models provide a rationale for understanding the chemical principles of solvent extraction.

From plots of the distribution ratio against the variables of the system— [M], pH, $[HA]_{org}$, [B], etc.—an indication of the species involved in the solvent extraction process can be obtained from a comparison with the extraction curves presented in this chapter; see Fig. 4.3. Sometimes this may not be sufficient, and some additional methods are required for identifying the species in solvent extraction. These and a summary of various methods for calculating equilibrium constants from the experimental data, using graphical as well as numerical techniques is discussed in the following sections. Calculation of equilibrium constants from solvent extraction is described in several monographs [60–64].

4.14.1 Identification of Species

An initial step in data analysis is to develop an equation that represents the experimental data reasonably. Although previous sections dealt with this issue, the approach assumed that certain species are formed. Two alternatives to this procedure are discussed here, both yielding the approximate stoichiometry of the complexes formed in the system. The most elementary is referred to as *Job's method*, while the *ligand number method*, developed by J. Bjerrum, is slightly more advanced.

4.14.1.1 Job's Method

When a metal M with a ligand A forms an extractable complex of the chemical form M_mA_a, the extraction is at maximum when the molar ratio of these two in the system is m:a. Similarily, if the extracted species is MA_aB_b, a plot of the mole ratios of A and B yields a maximum at the ratio a:b. This is illustrated in Fig. 4.2 for the extraction of U(VI) in the TTA-TBP (or TBPO) system. The curves are almost symmetrical around the mole fractions 0.5:0.5, indicating that the extracted species has a 1:1 ratio of TTA to TBP. Because the complex must contain 2 TTA (for electroneutrality), there should also be 2 TBP (or TBPO) molecules in the complex, i.e., MA_2B_2.

This method is useful when only one species is extracted, but it has little value for the study of solvent extraction systems that contain several complex species.

4.14.1.2 Ligand Number Method

This method [65–67] is useful for identifying the average composition of the metal species in the system. Consider Eq. (4.36a) for the extraction of MA_z, and assume—for the moment—that only one species, MA_n, exists in the aqueous phase. Taking the derivative of the logarithm of Eq. (4.36) yields

$$d \log D / d \log [A] = z - n \tag{4.80}$$

In 1941, J. Bjerrum [65] developed the useful concept of *average ligand number*, $\bar{n}$, defined as the mean number of ligands per central atom:

$$\bar{n} = \Sigma \, n \, [MA_n]/[M]_t \tag{4.81}$$

It can be shown [66a–67] that $\bar{n}$ equals n in Eq. (4.79), which can be rewritten

$$\bar{n} = z - d \log D / d \log [A] \tag{4.82}$$

For example, in Fig. 4.10 (Example 3), the slope of the plot of log D_{Cu} vs. log [A$^-$] can be used to conclude what species dominate the system at a given [A$^-$] value. The relation in Eq. (4.82) indicates an asymptote of slope 2, so the aqueous phase is dominated by uncomplexed Cu^{2+} ($n = 0$), while for slope 0 the neutral complex CuA_2 dominates the system ($n = 2$). Equation (4.82) shows that any tangent slope of the curve (i.e., $d \log D / d \log [A]$) yields the difference $z - \bar{n}$ in these simple systems.

> Example 12: Extraction of Th(IV) by acetylacetone.
> Figure 4.30 shows a smoothed curve of measurements of the distribution of Th(IV) from 0.1 M $NaClO_4$ into chloroform containing the extractant acetylacetone (HA) [66a,b]. Taking the derivative of this curve according to Eq. (4.79) the average ligand number is derived as shown in the lower insert. Th(IV) is successively complexed by A$^-$ forming ThA^{3+}, ThA_2^{2+}, ThA_3^+ and uncharged ThA_4, which is extracted. At pA > 8.5 the $\bar{n}$-value is zero, i.e., Th is uncomplexed, while at pA < 3.5 the average ligand number of 4 is reached, i.e., Th is fully complexed as ThA_4. See also Example 15.

The average ligand number can be used to obtain approximate equilibrium constants, as described by [65], assuming that at half integer $\bar{n}$-values the two adjacent complexes dominate: e.g., at $\bar{n} = 0.5$ the species M^{z+} and MA^{z-1} dominate, while at $\bar{n} = 1.5$ MA^{z-1} and MA_2^{z-2} dominate, etc.. The following expression for the stepwise formation constant is approximately valid at

$$\bar{n} = n - 0.5, \; \log K_n \approx -\log [A]_{\bar{n}} \tag{4.83}$$

This method of obtaining an estimate of the formation constants was done as a first step in the Th(IV)-acetylacetone system in Fig 4.30, where in the lower

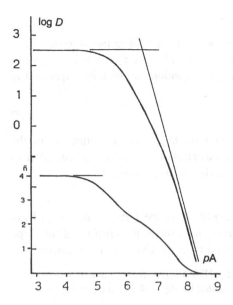

Fig. 4.30 Upper curve: the distribution of Th(IV) between benzene and 0.1 M NaClO₄ as a function of aqueous acetylacetonate ion concentration; $pA = -\log[A^-]$; the asymptotes have slopes 0 and −4. Lower curve: The average number of ligands per central atom, $\bar{n}$, in same system, as obtained from a derivation of the log $D(pA)$ curve. Using the ligand number method, the following equilibrium constants were estimated (with values from graphical slope analysis within parenthesis): $\log K_1$ 8.0 (7.85), $\log K_2$ 7.6 (7.7), $\log K_3$ 6.4 (6.3), and $\log K_4$ 5.1 (5.0). Log K_{D4} 2.50 is obtained from the horizontal asymptote. (From Refs. 66a,b.)

figure, $\bar{n}$ is plotted against $-\log [A^-]$, yielding the preliminary log K_n values given in Fig. 4.3. In a similar manner, the adduct formation constants can be determined:

Example 13: The extraction of Zn(II) by β-diketones and phosphoryl adduct formers (cont. of Example 5).

Consider the formation of adducts of the type MA_zB_b, as described in Example 5 (Fig. 4.12). The derivative of Eq. (4.48b) with respect to [B] at constant [A] yields

$$d \log D / d \log [B] = \bar{b} \tag{4.84}$$

where b is the average number of adduct forming molecules in the molecule at given [B] value. In Fig. 4.12 the asymptote has a slope of 2, indicating that a maximum of two molecules of TBP (or TOPO) bind to the neutral metal complex. From the lower slopes of the curve, the average number, b, can be estimated according to Eq. (4.83).

4.14.2 Graphic Slope Analysis

For a rational application of slope analysis it is important to measure the variation of the distribution ratio D with one component x at a time while the others, C, are kept constant. The solvent extraction equation can then be expressed in the form of a simple polynomial of type

$$y = a_0 + a_1 x + a_2 x^2 + \cdots \qquad (4.85)$$

where y is a function of the distribution ratio $D_{const.C}$ and x a function of the variable (e.g., pH, the free ligand ion concentration $[A^-]$, the concentration of free extractant HA or adduct former B in the organic phase, etc.).

4.14.2.1 Linear Plots

When the distribution equation can be expressed in the form $y = a_0 + a_1 x$, from a plot of y vs. x the intercept on the y axis yields the a_0 parameter and the slope the a_1 parameter. This treatment is referred to as the *limiting value method*.

> Example 14: The extraction of Cu(II) by HTTA (cont. from Example 4).
>   Example 4 (Fig. 4.11a), is represented by Eq. (4.44), which in logarithmic form is written
>
> $$\log D_{Cu} = \log K_{ex} + 2 \log [HA]_{org} + 2 \, pH \qquad (4.86)$$
>
> For constant $[HA]_{org}$ (Fig. 4.11a), a plot of $\log D_{Cu}$ vs. pH yields a line of slope 2, which intercepts pH $= 0$ at $\log D_{Cu} = \log K_{ex} + 2\log [HA]_{org}$. Another plot of $\log D_{Cu}$ against $\log [HA]_{org}$ at constant pH yields a line that intercepts $[HA]_{org} = 0$ at $\log D_{Cu} = \log K_{ex} + 2pH$ (Fig. 4.11b). In either case, the HTTA system yields $\log K_{ex} = -1.25$ for the CHCl$_3$, and -1.08 for the CCl$_4$ system. The corresponding values for the IPT system are 1.60 and 0.95, respectively.

> Example 15: Extraction of Th(IV) by acetylacetone (cont. from Example 12).
>   In Example 12, it was concluded from Fig. 4.30 that the aqueous phase contained all ThA$_n$-complexes with $0 < n < 4$, and the organic phase only the uncharged ThA$_4$ complex. We can therefore write the distribution of Th(IV) between chloroform and water
>
> $$D_{Th} = [ThA_4]_{org} / \Sigma \, [ThA_n] = K_{DC} \, \beta_4 \, [A]^4 / \Sigma \, \beta_n [A]^n \qquad (4.87)$$
>
> This is rearranged to yield
>
> $$D_{Th}^{-1} = F_0 = (1 + a_1 x^{-1} + a_2 x^{-2} + a_3^{x-3} + a_4 x^{-4}) / K_{DC} \qquad (4.88a)$$
>
> where K_{DC} is the distribution constant of ThA$_4$, x equals $[A]$, and $a_{4-n} = \beta_n / \beta_4$. In a plot of F_0 against x^{-1}, the intercept becomes K_{DC}^{-1} and the slope is a_1 / K_{DC}. In the next step the function F_1 is calculated
>
> $$F_1 = (K_{DC} \, D^{-1} - 1)x = a_1 + a_2 x^{-1} + a_3 x^{-2} + \cdots \qquad (4.88b)$$

A plot of F_1 vs. x^{-1} yields a_1 at the intercept and a_2 as the slope. In third step, $F_2 = (F_1 - a_1) x = a_2 + a_3 x^{-1} + a_4 x^{-2}$ is calculated, yielding a_2 and a_3, etc. Using this technique all four a_n values are obtained, from which one can deduce the β_n values [actually the K_n values in Eq. (3.5)] and the K_{DC} value.

Dyrssen and Sillén [68] pointed out that distribution ratios obtained by conventional batchwise techniques are often too scattered to allow the determination of as many parameters as used in Examples 15 and 16. They suggested a simplified graphic treatment of the data, based on the assumption that there is a constant ratio between successive stability constants, i.e., $K_n/K_{n+1} = 10^{2b}$, and that all distribution curves can be normalized so that $N^{-1} \log \beta_N = a$, where N is the number of ligands A^- in the extracted complex. Thus, the distribution curve $\log D_M$ vs. $\log[A^-]$ is described by the two parameters a and b, and the distribution constant of the complex, K_{DC}. The principle can be useful for estimations when there is insufficient reliable experimental data.

4.14.2.2 Nonlinear Plots

It is not possible to obtain simple linear relations between D and the variables when both the aqueous or organic phase contain several metal species. Instead a double polynomial such as

$$D = Y = \frac{b_0 + b_1 + b_2 y^2 + b_3 y^3 + \cdots}{a_0 + a_1 x + a_2 x^2 + a_3 x^3 + \cdots} = \frac{\Sigma\, b_i y^i}{\Sigma\, a_n x^n} \tag{4.89}$$

is obtained for which there is no simple solution.

Example 16: Extraction of U(VI) by acetylacetone.
  Fig. 4.31 shows the extraction of U(VI) by acetylacetone (*HA*) from nitric acid solutions, D_U, as a function of $pA = -\log[A^-]$ at six different total concentrations of HA. A comparison with Fig. 4.3 indicates that we can expect complexes of the type $MA_n(OH)_p(HA)_r$, which formally is equivalent to $H_{r+p}MA_{n+r}$. The extraction reaction is of the type Eq. (4.86), as one expects a series of self-adduct complexes $MA_n(HA)_r$ in the organic phase, and one or two series of stepwise complexes MA_n and $M(OH)_p$ in the aqueous phase. Thus the extraction reaction is of the type Eq. (4.89) with a ratio of several polynomials. Even in this case, a graphic extrapolation technique was useful by determining a set of intermediary constants for each constant $[HA]_{org}$ from $\log D_U$ vs. pA. Plotting these intermediate constants vs. $[HA]_{org}$, a new set of constants were obtained, from which both the stepwise formation constants $\beta_{n,p}$ and the adduct formation constants $K_{ad,1}$ and $K_{ad,2}$ for the reaction $UO_2A_2(H_2O)_{2-3} + nHA \leftrightarrows UO_2A_2(HA)_{1-2}$ in the organic phase. A more detailed analysis also allowed the determination of formation constants for $UO_2(OH)_p$ ($p = 1$ and 2).
  There may be other explanations to the extraction results. However, with the model used and the constants calculated, $\log D_U$ could be correctly predicted over the whole system range $[U]_{tot}$ 0.001–0.3 M, pH 2–7, and $[HA]_{org}$ 0.01–1.0 M. See also Example 17.

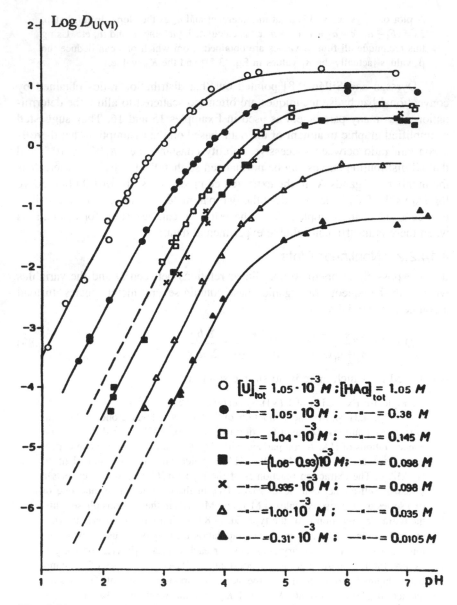

Fig. 4.31 Extraction of U(VI) by acetylacetone (HA) from 0.1 M NaClO₄ into chloroform as a function of pH at different total concentrations of HA. (From Ref. 69.)

4.14.3 Numerical Methods

The most complicated equations presented in this chapter are of the types shown by Eq. (4.89), where a series of complexes are formed in the organic phase and, at the same time, one or a series of complexes in the aqueous phase. More complicated equations would be obtained if mixed complexes and/or polynuclear complexes are present, and if varying activity factors are introduced, but these cases are not discussed here. Thus the solvent extraction equations can be expressed by Eq. (4.89), where x and y are independent variables and a_n and b_n are unknown independent parameters. The polynomial in the denominator refers to the formation of aqueous phase complexes; in the case that the metal forms several series of complexes with different ligands A^-, L^-, etc., the denominator contains several polynomials: $c_0 + c_1 z + \cdots$, etc. The polynomial in the numerator always refers to the formation of organic phase complexes.

There is no exact numerical solution to Eq. (4.89) when both x and $y > 0$. In practice, therefore, one variable must be kept constant (or zero), while the value of the other changes. This was described for Example 16. Thus, if x is kept constant, the double polynomial is reduced to a simple one

$$y = C_x \, \Sigma \beta_i y^i \quad (C_x^{-1} = a_0 + a_1 x + a_2 x^2 + \cdots) \tag{4.90}$$

where C_x is a constant at constant x. The problem is therefore reduced to the determination of the parameters of a simple polynomial, to which many numerical methods have been applied with varying degrees of success. The main requirement is that all computed parameters should be positive numbers (or zero).

In order to solve an algebraic system of n parameters, only n equations are needed (a minimum with no error estimates). When the solvent extraction reaction can be described by Eq. (4.90), there are as many equations as there are experimental points. Commonly, in solvent extraction 10–50 points are needed to cover the whole concentration range of interest, while the number of unknown parameters in simple cases is <5. In evaluating the parameters, it is important to use the complete suite of experimental data, as that gives greater significance to the a_n or b_n values.

There are several ways to solve a large number of polynomials like Eq. (4.90) for the a_n parameters, e.g., by minimizing the sum of the residuals, the least square method (LSQ) being the standard procedure. It is an objective regression analysis method, which yields the same results for a given chemical system, independent of method of investigation, provided there are no errors in the technique or fundamental chemical assumptions. Because the method may be applied in somewhat different ways, it may sometimes give slightly different results. Therefore, in judging a set of equilibrium constants, it is useful to learn how an author has applied the minimization technique.

In Eq. (4.91) x is usually the free ligand concentration $[A^-]$, or concentration $[B]$ of the adduct former, while y is a simple function of the distribution ratio D. The least square method requires that S (the *weighted squared residuals*) in expression Eq. (4.91) is minimized

$$S = \sum_{i=1}^{L} w_i \left(\sum_{n=0}^{N} (a_n x_i^n) - y_i \right)^2 \tag{4.91}$$

where $N + 1$ is the number of parameters, and L the number of experimentals points; $N + L - 1$ is referred to as the number of degrees of freedom of the system; each point has a value x_i / y_i. Because experiments are carried out over a large range of D, [A^-] and/or [B] values, the points carry different algebraic weight (e.g., the value 1000 obscures a value of 0.001). Therefore, in order to use the LSQ technique properly, *each point must be correctly weighted*, w_i. This can be done in several ways, the most common being to weight it by y_i, or by σ_i^{-2}, or by a percentage value of y_i ; σ_i is the standard deviation in the measurement of y_i. The difference $a_n x_i^n - y_i$ (the *residual*) is not zero, because the difference is to be taken between a measured value, y_i (meas.), and the corresponding calculated value, y_i (calc.), by the $a_n x_i^n$ function [i.e., y_i(calc.) − y_i(meas.)], using the actual a_n values at the time of the operation.

The principle of the LSQ technique is to compute the set of positive a_n values that give the smallest sum of the residuals; Eq. (4.91) reaches the S_{min} value. If the residuals equal zero (which rarely occurs in practice), S_{min} would be zero, and there would be a perfect fit between the experimental points and the calculated curve.

There are several mathematically different ways to conduct the minimization of S [see Refs. 70–75]. Many programs yield errors of internal consistency (i.e., the standard deviations in the calculated parameters are due to the deviations of the measured points from the calculated function), and do not consider external errors (i.e., the uncertainty of the measured points). The latter can be accommodated by weighting the points by this uncertainty. The overall reliability of the operation can be checked by the χ^2 (*chi square*) *test* [71], i.e., $S_{min}/(L + N − 1)$ should be in the range 0.5–1.5 for a reasonable consistency between the measured points and the calculated parameters.

Example 17: Extraction of Pm(III) by acetylacetone.

In Example 7, it was concluded that a number of self-adducts PmA$_3$ (HA)$_b$ were formed in the organic phase (0 < b < 2) in addition to the PmA$_n$ complexes in the aqueous phase (0 < n < 4); some extraction curves are given in Fig. 4.14. Equation (4.55) is of the same form as Eq. (4.89); rewriting Eq. (4.55b)

$$Y = D_{Pm} \; \Sigma \beta_n [A]^n / \beta_3 [A]^3 = K_{DC} (1 + K_{ad1} [HA]_{org} + \cdots) \tag{4.92}$$

which can be subdivided into

$$Y_{[A]} = C_i \; K_{DC} (1 + K_{ad1} [HA]_{org} + \cdots) \tag{4.93a}$$

and

$$Y_{[HA]org} = C_2 \, D_{Pm} \, \Sigma\beta_n[A]^n / \beta_3[A]^3 \tag{4.93b}$$

$Y_{[HA]org}$ contains the measured D_{Pm}, the [A] values, and the β_n values. It can thus in principle be treated as an Example 16 to yield the β_n values and the C_2 values (one for each $[HA]_{org}$), remembering that $a_0 = 1$. Because $Y_{[A]} / C_1 = Y_{[HA]org}/C_2 = Y'$, a second plot of Y' against $[HA]_{org}$ yields the K_{D3} and K_{adb} values. The analysis of the system yielded $K_{DC} = 0.008$, $K_{ad.1} = 7$ and $K_{ad.2} = 3$. In these calculations, a SIMPLEX program was used [75].

In comparisons of equilibrium constants collected from the literature (e.g., Fig. 4.22 or [47]), or correlations of data for a large number of systems (e.g., Figs. 4.20–4.23), it is desirable to present both the statistical uncertainty of each "point," which is often given by the standard deviation (one or several σ's) of the point, and the general reliability (statistical significance) of the whole correlation [76], for which the chi-square test offers a deeper insight into the reliability of the experimental results [77]. More advanced statistical tests for systems of our kind have been described by Ekberg [78].

4.15 EXPERIMENTAL DETERMINATION OF DISTRIBUTION RATIOS

The studies described above have the purpose of identifying the reacting species in a solvent extraction process and developing a quantitative model for their interactions. The fundamental parameter measured is the distribution ratio, from which extraction curves are derived. Solvent extraction work can still be carried out with simple batchwise (or point-by-point) technique, but continuous on-line measurements give faster and more accurate results.

4.15.1 Stirred Cell Semicontinuous Techniques

Each point-by-point experiment requires a complete set of mixing, separation, sampling, and analysis. This usually leads to scattered results, though it may not be critical, if the D values cover a limited range from 0.1–10. However, the more the D values deviate from 1, the more accurate must be the measurements; also the number of points required for a reliable extraction curve usually increases. To reduce the uncertainty and labor involved with the batch technique, the stirred cell technique has become popular.

Figure 4.32 shows a typical thermostated stirred cell. The liquid volumes are commonly 50 + 50 mL. The stirrer may consist of a single paddle at the interface, or a double paddle, one in the center of each phase. The cell may contain baffles, etc., to improve the mixing. Stirring may be violent, completely destroying the interface and producing very small ($\ll 1$ mm) droplets, or slow

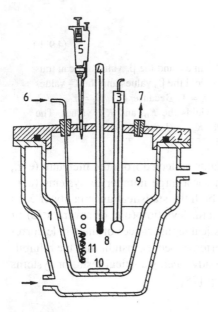

Fig. 4.32 A thermostated double jacket (1) cell for solvent extraction studies (heavier phase 8, lighter phase 9) under nonoxidizing conditions, using a hydrogen gas inlet (6) to a Pd-black catalyst (11), pH glass electrode (3), magnetic stirrer (10), connections for additions (5). Alternative constructions contain rotating paddles and fixed pipings connected to the two phases for frequent sample withdrawals.

in order not to destroy the interface. The stirring rate is optimized to the time for reaching equilibrium and complete phase separation. The experiments are either carried out with intermittent violent stirring, in which case samples are withdrawn after each complete phase separation, or with mild stirring during which it is possible to continously withraw samples. Equal volumes are sampled each time, commonly <1 mL. The simple stirred cell has been improved by introducing phase discriminating membranes in the sampling outlet [79]. This is particulary advantageous for kinetic experiments and is further described in Chapter 5. The sampling of the stirred cell can be automated, so that at regular intervals pH and temperature are recorded, and samples withdrawn for automatic analysis of concentration of interesting species in more or less standard fashion. It is also possible to use ion-sensitive probes in one of the phases instead of sampling.

4.15.2 *Centrifugal Extraction-Separation Systems*

A different approach to solvent extraction experimentation, referred to as the AKUFVE principle, was developed in the 1960s [80a–d]. The AKUFVE is illustrated in Fig. 4.33. Efficient mixing of the two phases and their additions is achieved in the mixing vessel and at the inflow into the phase separator, which consists of a *continuous flow centrifuge*, which in a special separation chamber and at very high rotational velocity (5,000–50,000 rotations per minute) separates the mixture into two very pure phases, containing <0.01% of entrainment of one phase in the other phase. The H-centrifuge may be made of Pd-stabilized

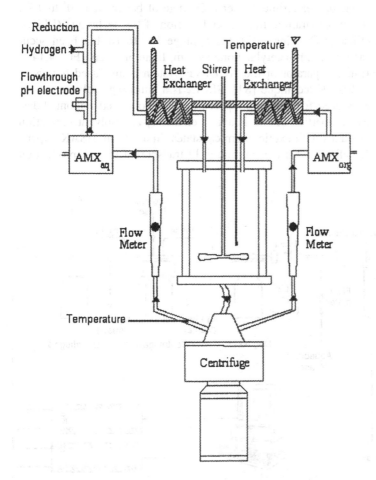

Fig. 4.33 The AKUFVE solvent extraction apparatus: Efficient mixing is achieved in the separate mixing vessel, from which the mixture flows down into the continuous liquid flow centrifugal separator (the H-centrifuge, hold-up time <1 s). (From Refs. 83a,b.) The outflow from the centrifuge consists of two pure phases, which pass on-line detectors, AMXs, for on-line detectors or continuous sampling. (From Refs. 80a–80d, 81.)

Ti, or PEEK (polyether ether ketone) to allow measurements under very corrosive conditions. The separated phases pass AMX gadgets for on-line detection (radiometric, spectrophotometric, etc.) or phase sampling for external measurements (atomic absorption, spectrometric, etc.), depending on the system studied. The aqueous phase is also provided with cells for pH measurement, redox control (e.g., by reduction cells using platinum black and hydrogen, metal ion determination, etc.) and temperature control (thermocouples).

The AKUFVE technique allows a large number of points (50–100) to be determined in a one-day experiment over a D-range of better than 10^3 to 10^{-3}, not counting time of preparation. In a special version of this technique (LISOL for LIquid Scintillation On Line) [82], the D range 10^{-5} to 10^4 has been accurately covered, as for the Pm-acetylacetone system, Example 7 and Fig. 4.14.

The centrifugal separator of the AKUFVE system is also used for phase separation in the SISAK technique [84]. SISAK is a multistage solvent extraction system that is used for studies of properties of short-lived radionuclides, e.g., the chemical properties of the heaviest elements, and solvent extraction behavior of compounds with exotic chemical states. In a typical SISAK experiment, Fig. 4.34, radionuclides are continuously transported from a production

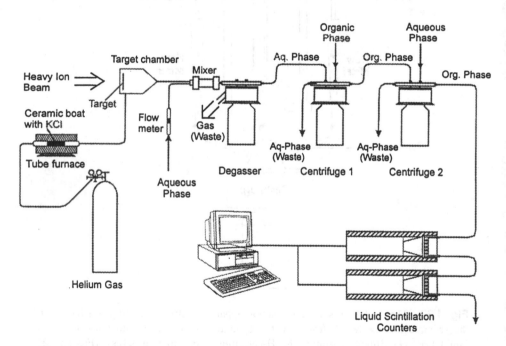

Fig. 4.34 A typical SISAK setup used for studies of α-decaying nuclides, e.g., transactinides. (From Ref. 84.)

site in an accelerator or a research reactor to the SISAK equipment via a gas-jet transport system. The nuclides are dissolved in an aqueous phase that is fed into a centrifuge battery comprising 1–4 solvent extraction steps. The product solution leaving the last step is pumped, e.g., to a nuclear radiation detection system. The transport time from the target site to the detection system depends on the centrifuge size, number of centrifuge steps, and flow rate. For a one-step chemistry, i.e., dissolution step and a single centrifuge extraction, and maximum flow rates, the overall transport time is around 2.5 s. This fast transport has allowed detailed γ-spectroscopic studies of radionuclides with half-lives around 1 s. Recently the SISAK equipment was succesfully applied to studies of the heaviest elements, and solvent extraction data were obtained for element 104 Rf [85]. Fast chemical separation systems have been developed for quite a large number of elements [86].

Centrifugal extractors have been designed for a number of industrial uses (see succeeding chapters). In some cases they have been scaled down to laboratory size but mainly been used for developing industrial multistage processes.

4.15.3 Liquid Partition Chromatography

In liquid partition chromatography a solute distributes itself between a mobile liquid phase and an immobile solvent attached to a solid matrix (in the earliest experiments a claylike *kieselguhr*). It is outside the scope of this chapter to discuss this technique, which, however, is briefly described in Chapter 15.

4.16 CONCLUSION AND FINAL COMMENTS

The solvent extraction process is usually described by a single net reaction, defined by the extraction constant K_{ex}. Variations in K_{ex} caused by modifications of the solvent system, such as changes in the temperature or aqueous ionic strength, or by replacing one solvent by another, or making substitutions in the extractant molecule, may be explained by careful consideration of the parameters of the system. However, such studies are difficult and not always sufficient for predicting new systems. A better foundation for understanding the extraction process is to consider the steps in the process contributing to the net extraction reaction, particularily when these steps are governed by regularities. Knowledge of these regularities helps in interpreting systems as well as in predicting new ones. Extension of these distribution data to thermodynamic constants is likely to give benefits in increased chemical knowledge of the behavior of solutes in different solvent systems. Advanced quantum chemistry calculations and computer modeling of extraction processes can help us in designing new, selective solvent extraction systems, as well as in interpreting extraction phenomena (see Chapter 16).

REFERENCES

1. Grahame, D. C.; Seaborg, G. T. *J. Am. Chem. Soc.* **1938**, *60*:2524.
2. Irving, H. p. 91, *ISEC'66, Solvent Extraction Chemistry* Dyrssen, D.; Liljenzin, J. O.; Rydberg, J. Eds.; North Holland, Amsterdam (1967).
3a. Rydberg, J. *Arkiv Kemi* 1954, *8*:101.
3b. Rydberg, J. *Rec. Trav. Chim.* **1956**, *75*:737.
4. Sekine, T.; Hasegawa, Y. *Solvent Extraction Chemistry*, Marcel Dekker, New York, 1977.
5. Marcus, Y.; Kertes, S. *Ion Exchange and Solvent Extraction of Metals*, Wiley-Interscience, Chichester, 1969.
6. Barton, A. F. M. *CRC Handbook of Solubility Parameters and Other Cohesion Parameters*, CRC Press, 1991.
7. Kojima, I.; Yoshida, M.; Tanaka, M. *J. Inorg. Nucl. Chem.* **1970**, *32*:987.
8a. Dyrssen, D. *Svensk Kem. Tidskr.* **1956**, *68*:212.
8b. Dyrssen, D. *Svensk Kem. Tidskr.* **1952**, *64*:213.
9. Sekine, T. et al. *Bull. Chem. Soc. Jap.* **1983**, *56*:700.
10. Rydberg, J. *Svensk Kem. Tidskr.* **1950**, *62*:179.
11a. Sekine, T.; Isayama, M.; Yamaguchi, S.; Moriya, H. *Bull. Chem. Soc. Jap.* **1967**, *40*:27.
11b. Dyrssen, D. Private communication. See also *Acta Chem. Scand.* **1960**, *14*:1091.
12. Ahrland, S.; Liljenzin, J. O.; Rydberg, J. *The Chemistry of the Actinides.* In *Comprehensive Inorganic Chemistry*, Pergamon Press, London, 1973.
13. Jassim, T. M.; Fridemo, L.; Liljenzin, J. O. in T. M. Jassim, Co-Extraction of Pertechnetate with some Metal Nitrates in TBP-Nitric Acid Systems. *Diss. Chalmers Techn. Univ.*, Gothenburg (1986).
14. Martell, A. E.; Calvin, M. *Chemistry of the Metal Chelate Compounds*, Prentice-Hall, Englewood Cliffs, N.J., 1952.
15. Stary, J. *The Solvent Extraction of Metal Chelates*, Pergamon Press, New York, 1964.
16. Zolotov, Yu. *Extraction of Chelate Compounds*, Humprey Sci. Publ., Ann Arbor, London, 1970.
17a. Narbutt, J. *J. Inorg. Nucl. Chem.* **1981**, *43*:3343.
17b. Narbutt, J. *J. Phys. Chem.* **1991**, *95*:3432.
17c. Narbutt, J.; Fuks, L. *Radiochimca Acta* **1997**, *78*:27.
18. Liljenzin, J. O.; Stary, J.; Rydberg, J. *ISEC'68, Solvent Extraction Research*, Kertes, A. S.; Marcus, Y., Eds., J. Wiley and Sons, New York, 1969.
19. Sillén, L. G.; Martell, A.; Högfeldt, E., Eds., *Stability Constants of Metal-Ion Complexes*. The Chemical Society Spec. Publ. No. 25, Burlington House, London, 1971.
20. Martell, A. E.; Smith, R. M. *Critical Stability Constants*, Vol. 1–5, Plenum Press, New York 1974–1982.
21. Marcus, Y.; Kertes, A. S.; Yanir, E. *Equilibrium Constants of Liquid-liquid Distribution Reactions*, Part 1, Butterworths, London, 1974.
22. Sekine, T.; Dyrssen, D. *J. Inorg. Nucl. Chem.* **1964**, *26*:1727.
23a. Allard, B.; Johnsson, S.; Rydberg, J. *ISEC'74, Proc. Int. Solvent Extraction Conf.*, Lyon, 1974.

23b. Allard, B.; Johnsson, S.; Narbutt, J; Lundquist, R. *ISEC '77, CIM Special Volume 21*, 1977.

24. Sekine, T.; Ihara, N. *Bull. Chem. Soc. Jap.* **1971**, *44*:2942.

25. Cecconie, T.; Freiser, H. *Anal. Chem.* **1990**, *62*:622.

26. Moryia, H.; Sekine, T. *Bull. Chem. Soc. Jap.* **1971**, *44*:3347.

27a. Rydberg, J.; Albinsson, Y. *J. Solv. Extr. Ion Exch.* **1989**, *7*:577.

27b. Albinsson, Y.; Rydberg, J. *Radiochim. Acta* **1989**, *48*:49.

27c. Albinsson, Y. *Acta Chem. Scand.* **1989**, *43*:919.

28. Albinsson, Y. *Development of the AKUFVE-LISOL Technique. Solvent Extraction Studies of Lanthanide Acetylacetonates.* Diss. Chalmers Univ. Techn., Gothenburg (1988).

29. Van Ooyen, J.; Dyrssen, D.; Liljenzin, J. O.; Rydberg, J. (Eds.) *Solvent Extraction Chemistry (ISEC 1966)*, North-Holland Publ. Co., Amsterdam, 1967.

30a. Sekine. T.; Dyrssen, D. *Anal. Chim. Acta* **1969**, *45*:433.

30b. Sekine, T.; Fukushima, T.; Hasegawa, Y. *Bull. Chem. Soc. Japan* **1970**, *43*:2638.

31. Sekine, T.; Sakairi, M. *Bull. Chem. Soc. Jap.* **1967**, *40*:261.

32. Liljenzin, J. O.; Vadasdi, K.; Rydberg, J. *Coordination Chemistry in Solution* Högfeldt, E. Swedish Nat. Res. Council; Trans.Roy. Inst.Techn., 280, Stockholm, 1972.

33a. Rydberg, J. *Rev. Inorg. Chem.* **1999**, *19*:245.

33b. Rydberg, J. *Min. Pro. Ext. Met. Rev.* **2000**, *21*:167.

34. Carey, F. A.; Sundberg, R. J. *Advanced Organic Chemistry*, Third Edition, Plenum Press, New York, 1990.

35. Fuks, L.; Majdan, M. *Min. Pro. Ext. Met. Rev.* **2000**, *21*:25.

36. Allard, B. *The Coordination of Tetravalent Actnide Chelate Complexes with β-Diketones*, Diss. Chalmers Univ. Techn. Gothenburg, 1975.

37. Hildebrand, J. H.; Scott, R. C. *Solubilities of Nonelectrolytes*, Dover Publishers, 1964.

38a. Siekierski, S. *J. Inorg. Nucl. Chem.* **1962**, *24*:205.

38b. Siekierski, S.; Olszer, R. *J. Inorg. Nucl. Chem.* **1963**, *24*:1351.

39. Buchowski, H. *Nature* **1962**, *194*:674.

40a. Wakahayashi, T.; Oki, S.; Omori, T.; Suzuki, N. *J. Inorg. Nucl. Chem.* **1964**, *26*: 2255.

40b. Omori, T.; Wakahayashi, T.; Oki, S.; Suzuki, N. *J. Inorg. Nucl. Chem.* **1964**, *26*: 2265.

40c. Oki, S.; Omori, T.; Wakahayashi, T.; Suzuki, N. *J. Inorg. Nucl. Chem.* **1965**, *27*: 1141.

41. Irving, H. M. N. H. *Ion Exchange and Solvent Extraction, Vol. 6*, (Marinsky, J.; Marcus, Y. Eds.) Marcel Dekker, New York, 1973.

42. Siekierski, S. *J Radioanal. Chem.* **1976**, *73*:335.

43. Hansen, C.; Skaaup, K. *Dansk Kemi* **1967**, *48*:81.

44. Hansen, C. *Hansen Solubility Parameters: A Users Handbook,* CRC Press, 2000.

45. Nilsson, M. *Influence of Organic Phase Composition on the Extraction of Trivalent Actinides*, Dipl. work, Dept. Nucl. Chem., Chalmers Univ. of Techn., Gothenburg, 2000.

46. Johansson, H.; Rydberg, J. *Acta Chem. Scand.* **1969**, *23*:2797.

47. Hancock, R. D.; Martell, A. *Chem. Rev.*, **1989**, *89*:1875–1913.

48. Hancock, R. D. *Prog.Inorg.Chem.* **1989**, *36*:187–290.
49. CambridgeSoft Corp., 100 Cambridge Park Drive, Cambridge, Mass. (1999).
50. Siddall, T. H., III; Good, M. L. *J. Inorg. Nucl. Chem.* **1967**, *29*:149.
51. Charbonnel, M. C.; Musikas, C. *Solv. Extr. Ion Exch.* **1989**, *7*:1007.
52. Musikas, C.; Hubert, H. *Solv. Extr. Ion Exch.* **1987**, *5*:877.
53a. Narbutt, J.; Krejzler, J. *Inorg. Chim. Acta* **1999**, *286*:175.
53b. Narbutt, J.; Czerwinski, M.; Krejzler, J. *Eur. J. Inorg. Chem.* **2001**, 3187 pp.
54. Handley, T. H.; Dean, J. A. *Anal. Chem.* **1962**, *35*:1312.
55. Patee, D.; Musikas, C.; Faure, A.; Chachaty, C. *J. Less Common Metals* **1986**, *122*: 295.
56. Handley, T. H. *Anal. Chem.* **1963**, *35*:991.
57. Schultz, W. W.; Navratil, J. (Eds.), *Science and Technology of TBP*, Vol. 1, CRC Press, Boca Raton, (1984).
58. Condamines, N. Commissariat l'Energie Atomique, Report R5519, 1990.
59. Spjuth, L. *Solvent Extraction Studies with Substituted Malonamides and Oligopyridines*, Diss. Chalmers Univ. Techn., Gothenburg, 1999.
60. Rossotti, F. J. C.; Rossotti, H. *Determination of Stability Constants*, McGraw Hill, New York, 1961.
61. Froneus, S. *Technique of Inorganic Chemistry* (Jonassen, H. B.; Weissberger, A., Eds.) Wiley-Interscience, Chichester, 1963.
62. Beck, M. T. *Chemistry of Complex Equilibria*, Van Nostrand Reinhold, New York, 1970.
63. Hartley, F. R.; Burgess, C.; Alcock, R. *Solution Equilibria*, Ellis Hoorwood, John Wiley and Sons, New York, 1980.
64. Meloun, M.; Havel, J.; Högfeldt, E. *Computation of Solution Equilibria*, John Wiley Sons, New York, 1988.
65. Bjerrum, J. *Metal Amine Formmation in Aqueous Solutions*, Diss. Copenhagen, 1941.
66a. Rydberg, J. *Acta Chem. Scand.* **1950**, *4*:1503.
66b. Rydberg, J. *Arkiv Kemi* **1953**, *5*:413.
67. Irving, H.; Williams, R. J. P.; Rossotti, F. J. C. *J. Chem. Soc.* **1955**, *1955*:1906.
68. Dyrssen, D.; Sillén, L. G. *Acta Chem. Scand.* **1953**, *7*:663.
69. Rydberg, J. *Arkiv Kemi* **1954**, *8*:113.
70. Deming, W. E. *Statistical Adjustment of Data*, John Wiley and Sons, NewYork, 1948.
71. Fisher, R. A. *Statistical Methods for Research Workers*, Hafner Publishing, New York, 1958.
72a. Sullivan, J. C.; Rydberg, J.; Miller, W. F. *Acta Chem. Scand.* **1959**, *13*:2023.
72b. Rydberg, J. *Acta Chem. Scand.* **1960**, *14*:57.
72c. Rydberg, J. *Acta Chem. Scand.* **1961**, *15*:1723.
73a. Sillen, L. G. *Acta Chem. Scand.* **1962**, *16*:159.
73b. Liem, D. H. *Acta Chem. Scand.* **1971**, *25*:1521.
73c. Liem, D. H. *Acta Chem. Scand.* **1979**, *A33*:481.
74. Dixon, L. C. W. *Nonlinear Optimization*, English University Press, 1972.
75. Albinsson, Y.; Mahmood, A.; Majdan, M.; Rydberg, J. *Radiochim. Acta* **1989**, *48*:49.

76. Meinrath, G.; Ekberg, C.; Landgren, A.; Liljenzin, J. O. *Talanta*, **2000**, *51*:231.
77. Rydberg, J.; Sullivan, J. *Acta Chem. Scand.* **1959**, *13*:2057.
78. Ekberg, Ch. *Uncertainties in Actinide Solubility Calculations Illustrated Using the Th−OH−PO₄ System*, Diss. Chalmers tekniska högskola, Göteborg, 1999.
79. Watari, H.; Cunningham, L.; Freiser, H. *Anal. Chem.* **1982**, *54*:2390.
80a. Reinhardt, H.; Rydberg, J. *Solvent Extraction Chem.* (*ISEC*) 1966, North-Holland Publ. Co. Amsterdam, 612 pp.
80b. Reinhardt, H.; Rydberg, J. *Chem. Ind.* **1970**, 488 pp.
80c. Rydberg, J., et al. *Acta Chem. Scand.* **1969**, *23*:647.
80d. Rydberg, J., et al., *Acta Chem.*, **1969**, 2773, 2781, 2797 pp.
81. MEAB Metal Extraction Co Ltd, V. Frölunda, Sweden.
82. Albinsson, Y.; Ohlsson, L. E.; Persson, H.; Rydberg, J. *Appl. Radiat. Isot.* **1988**, *39*:113.
83a. Rydberg, J.; Reinhardt, H. *Centrifuge for complete phase separation of two liquids*, U.S.Patent 3,442,445, 1966.
83b. Reinhardt, H. *En kontinuerlig separator för två-fas vätskeblandningar*, Diss. Chalmers Univ. Techn., Gothenburg, 1973.
84. Persson, H.; Skarnemark, G.; Skålberg, M.; Alstad, J.; Liljenzin, J. O.; Bauer, G.; Heberberger, F.; Kaffrell, N.; Rogowski, J.; Trautmann, N. *Radiochim. Acta* **1989**, *48*:177.
85. Skarnemark, G., et al., *J. Nucl. Radiochem. Sci.* **2002**, *3*:1.
86. Skarnemark, G. *J. Radioanl. Nucl. Chem.* **2000**, *243*(1):219.

5
Solvent Extraction Kinetics

PIER ROBERTO DANESI International Atomic Energy Agency,
Vienna, Austria

5.1 GENERAL PRINCIPLES

The kinetics of solvent extraction is a function of both the various chemical
reactions occurring in the system and the rates of diffusion of the various species
that control the chemistry of the extraction process.

5.1.1 Rate-Controlling Role
of the Chemical Reactions

The dependence of the kinetics on the chemical reactions is easily understood
by considering that the final products of any extraction process are usually in a
chemical state different from the initial unreacted species. This is true even for
the simple partition of neutral molecules between two immiscible liquid phases,
where the chemical change is in the solvation environment of the extracted
species. More drastic chemical changes take place in the extraction of a metal
cation from an aqueous solution by a chelating extractant dissolved in an organic
diluent. In the extraction some of the solvation water molecules can be removed
from the metal ion, and a new coordination compound, soluble in the organic
phase, is formed with the chelating group of the extractant. In addition, the
extracting reagent can undergo an acid dissociation reaction and, together with
the extractant-metal complex, can undergo changes in aggregation in the organic
phase. Consequently, whenever at least one of the chemical steps of the overall
reaction mechanism is slow enough, compared with the diffusion rate, the kinetics of extraction would depend on the rate of the slow chemical reactions.

For solvent extraction systems, we have two additional complications. First, the chemical reactions can take place, at least in principle, in two bulk phases, since we are dealing with two immiscible liquid layers. Second, the chemical reactions can occur in the two-dimensional region called the liquid–liquid interface, that separates the two immiscible liquids, or in a thin volume region very close to it. When interfacial chemical reactions are important, the situation is analogous to that describing the kinetics of the chemical reactions encountered in heterogeneous catalysis and in some electrode processes. Although a relatively large number of sophisticated techniques are available for studying chemical reactions at solid–fluid interfaces, very few tools have been developed to investigate chemical changes occurring at liquid–liquid interfaces. Our knowledge of such interfacial reactions, therefore, is still limited and is based largely on indirect evidence and speculations.

5.1.2 Rate-Controlling Role of the Diffusional Processes

To understand the dependence of the extraction kinetics on the rate of diffusion of the various chemical species that participate in the extraction reaction, we first have to distinguish between diffusion in the bulk phases and diffusion through the thin layers adjacent to the interface. In most solvent extraction processes of practical interest, both the aqueous and the organic phase are efficiently stirred. It follows that transport of material from the bulk of the phases up to a region very close to the interface can be considered instantaneous and that diffusion in the bulk of the phases can be neglected. Nevertheless, diffusional processes can still have an appreciable influence on the solvent extraction kinetics. For example, even when the two phases are vigorously stirred, it is possible to describe interfacial diffusion by assuming the existence of two stagnant thin layers of finite thickness located on the aqueous and organic side of the interface. This model of the interface, often referred to as the two-film theory [1,2], is extremely useful for describing extraction kinetics that are controlled by diffusion occurring in proximity to the interface. The two-film model is used throughout our treatment. Other models and theories of higher complexity exist for describing diffusional transport in proximity to the interface. These theories (penetration, surface renewal, boundary layer) are described in detail in more advanced books [3,4].

The two films are schematically described in Fig. 5.1, in which the presence of an interfacially absorbed layer of extractant molecules is also shown. δ_o and δ_w represent the thickness of the organic and aqueous films, respectively. In these layers the liquid phases are considered completely stagnant (i.e., no movement of the fluids takes place in spite of the mechanical energy that is dissipated in the two-phase system to provoke mixing of the aqueous and organic phases).

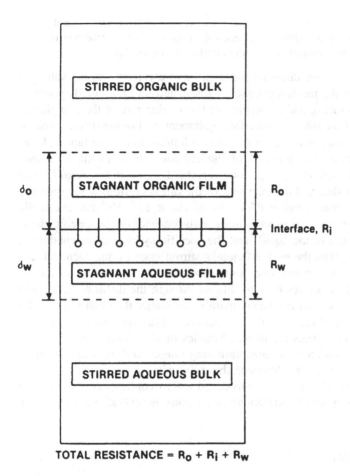

TOTAL RESISTANCE = R_O + R_i + R_W

Fig. 5.1 Interfacial diffusion films. δ_o and δ_w are the thickness of the organic and aqueous films, respectively. The presence of an adsorbed layer of extractant molecules at the interface is also shown.

These two thin liquid films, which are also called diffusion films, diffusion layers, or Nernst films, have thicknesses that range between 10^{-2} and 10^{-4} cm (in this chapter centimeter-gram-second (CGS) units are used, since most published data on diffusion and extraction kinetics are reported in these units; comparison with literature values is, therefore, straightforward).

The description of the diffusion films as completely stagnant layers, having definite and well-identified thicknesses, represents only a practical approximation useful for a simple mathematical description of interfacial diffusion. A more realistic physical description should consider that, starting at a given distance from the liquid interface, the renewal of the organic and the aqueous fluids becomes progressively less as the interface is approached.

The thickness of the diffusion films is, up to a certain limiting value, an inverse function of the mechanical energy dissipated into the liquid system to stir it. This means that δ_o and δ_w depend on the stirring rate of the two phases, on the geometry of the solvent extraction equipment, and of the stirrers used, as well as on the viscosity and density of the two liquids. For given liquid phases and a fixed apparatus, the thickness of the diffusion films initially decreases with increase in the rate of mixing. However, the decrease does not go beyond a certain limiting value, and the thickness of the diffusion films never seems to go down to zero, with limiting thickness of about 10^{-3}–10^{-4} cm eventually reached with all systems. The exact limiting value depends on the specific physicochemical properties of the liquids and the specific hydrodynamic conditions. Even when dealing with the most efficiently stirred systems, the chemical species that have to be transported to or away from the interface for the extraction reaction to take place always have to diffuse through the diffusion films. The time required for this often can be comparable or longer than that required for the actual chemical changes in the extraction. As a consequence, diffusion through these films controls the overall kinetics of extraction. Experience has shown that film diffusion is the predominating rate-controlling factor in many practical extraction processes. Although these diffusion processes occur through very thin films, the films must be considered macroscopic, since the diffusion distances involved exceed the molecular dimensions by several orders of magnitude.

5.2 DIFFUSION

Diffusion is that irreversible process by which matter spontaneously moves from a region of higher concentration to one of lower concentration, leading to equalization of concentrations within a single phase.

The laws of diffusion correlate the rate of flow of the diffusing substance with the concentration gradient responsible for the flow. In general, in a multicomponent system, the process is described by as many diffusion equations as the number of chemical species in the system. Moreover, the diffusion equations are intercorrelated. Nevertheless, when dealing with solutions containing extractable species and extracting reagents at concentrations much lower than those of the molecules of the solvents (dilute solutions), diffusion is sufficiently well described by considering only those species for which the concentrations appreciably change during the extraction reaction.

The *diffusion flow*, *J*, of a chemical species is defined as the amount of matter of this species passing perpendicularly through a reference surface of unit area during a unit time. The dimensions of J are those of mass per square centimeter per second (mass $cm^{-2} s^{-1}$). When the concentrations of the diffusing species (c) are expressed in moles per cubic centimeter (mol cm^{-3}), the correlation between the flux across a unit reference area located perpendicularly to the linear coordinate x (along which diffusion occurs) and the concentration gradient, $\partial c / \partial x$, will be given by the Fick's first law of diffusion, that is,

$$J = -D \partial c / \partial x \qquad (5.1)$$

D is the diffusion coefficient of the species under consideration. Its dimensions are given in $cm^2 s^{-1}$. Equation (5.1) implies that D is independent of concentration; it is a constant, characteristic of each diffusing species, in a given medium, at constant temperature. This is only approximately true, but holds sufficiently well for most diffusing species of interest in solvent extraction systems. The values of D for the majority of the extractable species fall in the range $10^{-5}–10^{-6}$ $cm^2 s^{-1}$ in both water and organic solutions. When dealing with viscous phases or bulky organic extractants, or with complexes that may undergo polymerization reactions, the D values can drop to 10^{-7} or 10^{-8} $cm^2 s^{-1}$.

Equation (5.1) is extremely useful to evaluate the flux whenever the concentration gradient can be considered constant with time (i.e., when we can assume a steady state). When a steady state cannot be assumed, the concentration change with time must also be considered. The non–steady-state diffusion is expressed by Fick's second law of diffusion:

$$\partial c / \partial t = D \partial^2 c / \partial x^2 \qquad (5.2)$$

Diffusion is a complex phenomenon. A complete physical description involves conceptual and mathematical difficulties associated with the need to involve theories of molecular interactions and to solve complicated differential equations [3–6]. Here and in sections 5.8 and 5.9, we present only a simplified picture of the diffusional processes, which is valid for limiting conditions. The objective is to make the reader aware of the importance of this phenomenon in connection with solvent extraction kinetics.

For steady-state diffusion occurring across flat and thin diffusion films, only one dimension can be considered and Eq. (5.1) is greatly simplified. Moreover, by replacing differentials with finite increments and assuming a linear concentration profile within the film of thickness δ, Eq. (5.1) becomes

$$J = -D \, dc / dx = -D(c_2 - c_1) / \delta = (c_2 - c_1) / \Delta \qquad (5.3)$$

where

$$\Delta = \delta / D \qquad (5.4)$$

is a diffusional parameter dependent on the thickness of the diffusion film and the value of the diffusion constant of the diffusing species. The units of Δ are cm^{-1} s.

Some simple considerations about the variation of concentrations with time that can be expected in diffusion-controlled processes, can be made by considering the imaginary one-liquid-phase system shown in Fig. 5.2. The imaginary system consists of two hypothetical well-stirred aqueous reservoirs of volumes V_1 and V_2, containing different concentrations of the same solute, in contact through a nonstirred aqueous liquid film across which the solute moves only under the influence of molecular diffusion. Since c_1 and c_2 are allowed to vary with time, diffusion through the film of thickness δ tends to equalize c_1 and c_2. This means that J, c_1 and c_2 are a function of time. If a linear concentration gradient is always assumed and the amount of diffusing matter inside the film is negligible relative to that in the reservoirs, the flux equation J can be integrated to yield a simple correlation between c_1, c_2 and time (t). By considering that the correlation between the flux and the concentration variations within the reservoirs is

$$-V_1\, Q^{-1} dc_1 / dt + V_2\, Q^{-1}\, dc_2 / dt = J \tag{5.5}$$

where Q is the area of the diffusion film (equal on both sides), and, assuming, for simplicity, that $V_1 = V_2 = V$, Eqs. (5.3) and (5.5) can be combined to give

$$VQ\Delta dc_1 / dt = c_1 - c_2 \tag{5.6}$$

For the initial conditions at $t = 0$, $c_2 = 0$, and $c_1 = c_1^0$, the integration of Eq. (5.6) is straightforward. Separation of the variables leads to

$$Qdt / V\Delta = dc_1(c_1^0 - 2c_2) \tag{5.7}$$

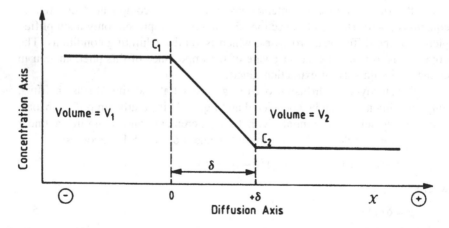

Fig. 5.2 Concentration profile of a solute diffusing across an unstirred liquid film of thickness δ in contact with two well-stirred reservoirs. The same liquid phase is assumed throughout this imaginary model system.

Upon integration and insertion of Eq. (5.4), one obtains

$$\ln\{(2c_1 - c_1^0)/c_1^0\} = -2QDt/V\delta \tag{5.8}$$

Two conclusions follow from Eq. (5.8):

1. For $t = \infty$, $(2c_1 - c_1^0) = 0$, that is

$$c_1(\text{at } t = \infty) = c_2(\text{at } t = \infty) = c_1^0/2. \tag{5.9}$$

This means that at equilibrium the concentrations in the two reservoirs are equal.
2. The condition of equal concentrations is approached with a rate that is directly proportional to the contact area (Q) and the diffusion coefficient (D), and inversely proportional to the volume of the reservoir (V) and the thickness of the diffusion layer (δ).

Although these conclusions are intuitive and trivial and are based on several crude approximations introduced into an imaginary model system consisting of only one liquid phase, the simple mathematical treatment developed helps the reader better appreciate the fundamental role that film diffusion plays in controlling the rate of solvent extraction. When two different liquid phases (aqueous and organic) are in contact, two liquid films can be assumed to exist on the two opposite sides of the interface, and diffusion through both must be considered. The presence of two phases at which species can be unevenly distributed, and with interfacial chemical reactions of varying degree of complexity, will in fact appear in the extraction rate equations only as modifications of this simple diffusion equation.

5.3 RATE LAWS AND MECHANISMS

5.3.1 Rate Laws and Reaction Mechanisms

For a general treatment of the various rate laws, the reader is referred to several specialized books existing on the subject [7,8]. However, care must be exercised in extrapolating the information obtained from the study of reactions in solutions to solvent extraction systems. Solvent extraction deals with two solutions in contact and in the presence of the interface, which has to be crossed by a solute during the extraction, a fact that requires special consideration.

In this section, we briefly summarize the fundamental concepts that are useful for better understanding those specific solvent extraction rates and mechanisms, which are presented in some detail in subsequent sections.

The immediate result of a kinetic study is a rate law. For a general reaction with the stoichiometric equation

$$xX + yY = zZ + wW \tag{5.10}$$

the rate is expressed by

$$\frac{-d[X]}{dt}, \frac{-d[Y]}{dt}, \frac{+d[Z]}{dt} \quad \text{or} \quad \frac{+d[W]}{dt} \tag{5.11}$$

where t indicates time and the square brackets concentrations. The rate of the reaction is given as function of the composition of the system:

$$rate = -d[X]/dt = f\{[X], [Y], [Z], [W]\} \tag{5.12}$$

The first goal of any kinetic study is to devise experiments that establish the algebraic form of the rate law and to evaluate the rate constants. Rate laws of the form

$$-d[X]/dt = k[X]^x[Y]^y[Z]^z[W]^w \tag{5.13}$$

are relatively easy to treat mathematically when x, y, z, and w are small integers.

Rate laws can be derived by measuring concentration variations as function of time or the initial rates as function of the initial concentrations. Unfortunately, there is no general method for finding the rate law and the reaction order from measurements of concentration vs. time or other types of measurements. Usually, a trial-and-error procedure is used, based upon intelligent guesses.

Once a rate law has been defined and the rate constants evaluated, the next step is to correlate it with the most likely mechanism of the extraction reaction.

Since the experimental kinetic data refer to a reaction rate and how this is affected by variables, such as concentration, temperature, nature of the solvent, presence of other solutes, structural variations of the reactants, and so forth, the assignment of a mechanism is always only indirectly derived from primary data. Therefore, it is not surprising that more than one mechanism has often been proposed to explain the same rate law and that reaction mechanisms, which were once consistent with all experimental information available on a system, have later on been considered erroneous and have been disregarded, or drastically modified, as long as new experimental evidence was accumulated. In general, the stoichiometry of the reaction, even when this is a simple one, cannot be directly related with its mechanism, and when the reaction occurs through a series of elementary steps, the possibility that the experimental rate law may be interpreted in terms of alternative mechanism increases. Therefore, to resolve ambiguities as much as possible, one must use all the physicochemical information available on the system. Particularly useful here is information on the structural relations between the reactants, the intermediate, and the reaction products.

Table 5.1 presents the mathematical description of selected very simple rate laws.

Irreversible first-order reactions [Eq. (5.14) in Table 5.1] depend only on

Table 5.1 Common Rate Expressions

	Irreversible first-order reaction	
Reaction	X + other reactants (OR) → products (P)	(5.14a)
Rate expression	$-d[X]/dt = k[X]$	(5.14b)
Solution	$[X] = [X]_0 e^{-kt}$	(5.14c)
Conditions	OR = const.; reverse rate: zero; $[X] = [X]_0$ at $t = 0$	
	Reversible first-order reactions	
Reaction	$X + OR \underset{k_2}{\overset{k_1}{\rightleftharpoons}} Y + OP$ (other products)	(5.15a)
Rate expression	$-d[X]/dt = k_1[X] - k_2[Y]$	(5.15b)
Solution	$[X] = [X]_0 (k_1 + k_2)^{-1} (k_2 + k_1 e^{-kt})$	(5.15c)
Conditions	$k = k_1 + k_2$; OR and OP = const.; at time $t = 0$, $[Y] = 0$ and $[X] = [X]_0$	(5.15d)
	Series first-order reactions	
Reaction	$X \overset{k_1}{\rightarrow} Y \overset{k_2}{\rightarrow} Z$	(5.16a)
Rate expression	$d[Y]/dt = k_1[X]_0 e^{-k_1 t} - k_2[Y]$	(5.16b)
Solution	$[Y] = [X]_0 k_1 (k_2 - k_1)^{-1} (e^{-k_1 t} - e^{-k_2 t})$	(5.16c)
Conditions	no Y or Z initially present; $[X] = [X]_0$ at $t = 0$	

the concentration of a reactant X, and the rate of the reverse reaction, *rate (reverse)*, is always equal to zero. Simple distribution processes between immiscible liquid phases of noncharged components characterized by very large values of the partition constant can be formally treated as first-order irreversible reactions.

Reversible first-order reactions are given by Eq. (5.15) of Table 5.1. Equation (5.15c) can be rewritten in a more elegant form by considering that at equilibrium ($t = \infty$) the net rate is equal to zero, that is

$$k_1[X]_{eq} = k_2[Y]_{eq} \qquad (5.17)$$

where subscript eq indicates equilibrium concentrations. The equilibrium constant of the reaction is

$$K = k_1 / k_2 = [Y]_{eq} / [X]_{eq} \qquad (5.18)$$

Equation (5.15c) can then be reformulated as

$$\ln\{([X]-[X]_{eq})([X]_0 - [X]_{eq})^{-1}\} = -(k_1 + k_2)t \qquad (5.19)$$

Equation (5.19) shows that the rate law is still first-order, provided the quantity ([X] − [X]$_{eq}$) is used instead of [X]. A plot of ln([X] − [X]$_{eq}$) vs. t will then be a straight line of slope -($k_1 + k_2$). The individual rate constants of the reaction can still be evaluated from the slope of such a plot, providing the equilibrium constant [Eq. (5.18)] is also available. Many distribution processes between immiscible liquid phases of noncharged species, as well as extraction of metal ions performed at very low metal concentrations, can be treated as first-order reversible reactions when the value of the equilibrium (partition) constant is not very high.

Equations (5.16) of Table 5.1 refer to *series first-order reactions*. Of interest for the solvent extraction kinetics is a special case arising when the concentration of the intermediate, [Y], may be considered essentially constant (i.e., $d[Y]/dt = 0$). This approximation, called the stationary state or steady-state approximation, is particularly good when the intermediate is very reactive and present at very small concentrations. This situation is often met when the intermediate [Y] is an interfacially adsorbed species. One then obtains

$$[Y] = [X]_0 e^{-k_1 t} k_1 / k_2 \qquad (5.20)$$

and

$$[Z] = [X]_0 \{1 - (1 + k_1 / k_2) e^{-k_1 t}\} \qquad (5.21)$$

$k_2 \gg k_1$ means a very reactive intermediate present at very low concentration; $t \gg 1/k_2$ indicates that the induction period of the reaction has been passed, that is, the initial phase of the reaction, when the rate of formation of Z is slow, is immeasurably short in comparison with the time response of the measuring technique.

5.4 SOME KINETICS AND MECHANISMS WITH COMPLEX IONS

5.4.1 *Introduction*

In the solvent extraction of metal ions, the composition of the coordination sphere often changes, either because of the formation of complexes between the metal ions and a complexing reagent, preferentially soluble in an organic phase, or because of the replacement of a ligand in the aqueous phase metal complex with another more lipophilic one in the organic phase. Complexes that rather readily exchange ligands or water molecules with other ligands are termed labile. Inert complexes do not exhibit ligand exchange or do so very slowly. The terms *labile* and *inert* refer to the velocity of the reactions and should not be confused with the terms *stable* and *unstable*, which refer to the thermodynamic tendency of species to exist under equilibrium conditions. Although, in many cases, thermodynamic stability parallels the inertness of a complex, there are many examples of complex ions that are thermodynamically stable, but kinetically labile, or the reverse.

Kinetic studies of ligand exchange reactions seek to elucidate the mechanisms of such reactions. In the following section, information on the mechanisms that control solvent exchange and complex formation is reported. Although the information has been derived from studies of one-phase liquid systems, much of it can be safely extrapolated to solvent extraction systems whenever interfacial film diffusion processes can be neglected as the rate-determining processes.

5.4.2 Mechanisms

Solvent exchange and complex formation are special cases of *nucleophilic substitution* reactions.

Generally, when a coordination bond is broken in a metal ion complex, the electron pair responsible for the bond accompanies the leaving group (commonly, the organic ligand). It follows that reactions involving solvent exchange and complex formation have as reactive centers electron-deficient groups. Therefore, they can all be considered as nucleophilic substitutions (in symbols S_N).

The basic classification of nucleophilic substitutions is founded on the consideration that when a new metal complex is formed through the breaking of a coordination bond with the first ligand (or water) and the formation of a new coordination bond with the second ligand, the rupture and formation of the two bonds can occur to a greater or lesser extent in a synchronous manner. When the rupture and the formation of the bonds occur in a synchronous way, the mechanism is called *substitution nucleophilic bimolecular* (in symbols S_N2). On the other extreme, when the rupture of the first bond precedes the formation of the new one, the mechanism is called *substitution nucleophilic unimolecular* (in symbols S_N1). Mechanisms S_N2 and S_N1 are only limiting cases, and an entire range of intermediate situations exists.

Inorganic chemists prefer a slightly more detailed classification and subdivide the range of possible mechanisms into the following four groups.

1. *D mechanism.* This is the *limiting dissociative mechanism*, and a transient inter-mediate of reduced coordination number is formed. The intermediate persists long enough to discriminate between potential nucleophiles in its vicinity. Here we are dealing with an S_N1 limiting process, since the dissociation of the metal-ligand bond fully anticipates the formation of the new bond. The vacancy in the coordination shell that occurs as a result of the dissociation is then taken by the new ligand.

2. *I_d mechanism.* This is the *dissociative interchange mechanism* and is similar to the previous one in the sense that dissociation is still the major rate-controlling factor. Therefore, we are still dealing with an S_N1 process. Nevertheless, differently from the D mechanism, no experimental proof exists that an intermediate of lower coordination is formed. The mechanism involves a fast outer sphere association between the initial complex and the

incoming nucleophile (the new ligand), and when the complex metal ion and the incoming ligand have opposite charges, this outer sphere associate is an ion pair. As a result of the formation of this new outer sphere associate, the new group is now suitably placed to enter the primary coordination sphere of the metal ion as soon as the outgoing group has left.

3. I_a mechanism. This is the *associative interchange mechanism* and is similar to the I_d mechanism, in the sense that here, also, no proof exists that an intermediate of different coordination is formed. However, differently from I_d, we have significant interaction between the incoming group and the metal ion in the transition state. In this instance, the process is partially of the S_N2 type.

4. A mechanism. This is the pure *associative mechanism*, and here we are dealing with a process entirely of the S_N2 type. The rate-determining step of the mechanism is the association between the complex metal ion and the entering ligand, leading to the formation of an intermediate of increased coordination number that can be experimentally identified. This mechanism is often operative in ligand-displacement reactions occurring with planar tetracoordinated complexes.

In practice, there is a continuous gradation of mechanisms from D, through I_d and I_a to A, depending on the extent of interaction between the metal cation and the incoming group in the transition intermediate. Moreover, the diagnosis of the mechanism is not a straightforward and unambiguous process. A variety of methods have to be used, depending on circumstances, since an approach based only on rate-law consideration can easily lead to false conclusions.

5.4.3 Water Exchange and Complex Formation from Aquoions

The rate at which solvent molecules are exchanged between the primary solvation shell of a cation and the bulk solvent is of primary importance in the kinetics of complex formation from aquocations. In both water exchange and complex formation, a solvent molecule in the solvated cation is replaced with a new molecule (another water molecule or a ligand). Therefore, strong correlations exist between the kinetics and mechanisms of the two types of reactions.

The basic mechanism for the formation of complexes from aquocations [9] is based on the following observations:

For a given metal ion the rates and the activation parameters for complex formation are similar to those for water exchange, with the complex formation rate constants usually about a factor of 10 lower than those for water exchange,

For a given metal ion the rates show little or no dependence on the identity of the ligand.

This means that, at least as a first approximation, the complex formation mechanism can be described by the rapid equilibrium formation of an ion pair (for oppositely charged species) or of an outer sphere association complex between the aquometal ion and the ligand, followed by a rate-determining interchange step (D or I_d mechanism). Here, the ligand bonds to the complex cation only after a water molecule has been released from the primary coordination sphere. For a six-coordinated single-charged cation, the mechanism can be expressed as

$$M(H_2O)_6^+ + L^- \rightarrow (L^-, M(H_2O)_6^+) \rightarrow kM(H_2O)_5 + H_2O$$

$$\text{fast} \quad \text{outer sphere} \quad \text{slow}$$
$$\text{associate} \tag{5.22}$$

The rate law corresponding to such a mechanism is:

$$\text{rate} = \{K_{ass}k[M^+][L^-]\} / \{1 + K_{ass}[L^-]\} \tag{5.23}$$

In Eqs. (5.22–5.23), coordination water molecules are omitted and K_{ass} indicates the ion pair (or outer sphere association) constant. In general K_{ass} is a small number (≤ 10) and Eq. (5.23) simplifies to

$$\text{rate} = K_{ass}k[M^+][L^-] \tag{5.24}$$

Rate Eq. (5.23) indicates that the process is second-order, even if the rate-controlling step is unimolecular. Furthermore, since the constant K_{ass} differs only slightly from species to species, depending mostly on the charge density, only a small dependence on the nature of the incoming ligand can be expected.

Once the value of K_{ass} is estimated, either from theoretical considerations or from separate experimental measurements, the rate constants of the water exchange rate-controlling step can be calculated. The values of k are in good agreement with the independently evaluated rate constants of the water exchange process:

$$M(H_2O)_6^+ + H_2O^* \rightarrow M(H_2O)^*(H_2O)_5 + H_2O \tag{5.25}$$

In Eq. (5.25) H_2O^* represents a water molecule initially present outside the coordination sphere of the metal ion, which, as a result of the exchange, has entered the first coordination sphere. It follows that the degree of kinetic reactivity of aquometal ions with complexing agents parallels their kinetic lability toward water exchange. Moreover, since the water exchange rate constants of most metal ions are known, predictions on the rate of complex formation of aquometal ions can be made.

Rates of water exchange at metal ions vary tremendously with the nature of the cation. The range is from almost 10^{10} s^{-1}, for Cu^{2+} or Cr^{2+}, down to less than 10^{-7} s^{-1} for Rh^{3+}. The values for many metal aquocations are shown in Fig.

5.3 on a logarithmic scale. They have been obtained by nuclear magnetic resonance (NMR) and sound absorption, for the rapidly exchanging ions, and by isotopic labeling for the more inert ones.

Generally, in solvent extraction, for the simplest case of an organic complexing reagent (ligand) reacting with an aquometal ion, when the water exchange rate constant is $<10^2$ s^{-1}, the reaction can be considered slow enough that the complex formation rate may compete with the rate of diffusion through the interfacial films in controlling the overall extraction kinetics. In exceptional cases, dealing with very efficiently stirred phases (in which the thickness of the diffusion films may even be reduced to about 10^{-4} cm), complexation reactions with rate constants as high as 10^6 s^{-1} can be rate limiting in solvent extraction. In all other cases, the ions can be considered as reacting instantaneously relative to the rate of film diffusion. The rate of diffusion also is the predominating rate-controlling process when the aquocations are extracted into the organic phase with their entire coordination sphere of water molecules. These considerations do not hold when other slow chemical processes, such as hydrolysis reactions of the metal cation, polymerizations in the organic phase of the extractant or the metal complex, adsorption-desorption processes at the interphase, keto-enol

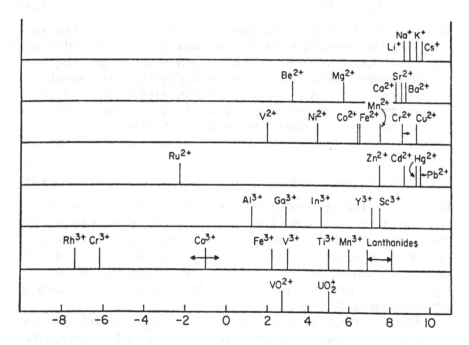

Fig. 5.3 Logarithm of water exchange constants (s^{-1}) for aquometal ions at 25°C. (From Ref. 26.)

tautomerizations, or others, control the chemistry of the solvent extraction system. In these cases, it is possible that the slow chemical control predominates over the faster diffusional process. In such systems, solvent extraction rate data can still provide information on the rate and mechanism of the chemical processes.

5.4.4 Ligand Displacement Reactions

For ligand displacement reactions occurring in octahedral complexes, very few generalizations can be made, since the reaction mechanisms tend to be specific to each chemical system. This is unfortunate because many solvent extraction processes are indeed ligand displacement reactions of octahedral complexes in which L is the extractant [see Eq. (5.22)]. However, it has been experimentally observed, at least for aqueous-phase reactions, that the variation of the rates with the identity of the ligand correlates well with the variation in the thermodynamic stability of the complex. Therefore, whenever the complex is not extracted unchanged into the organic phase, thermodynamically very stable complexes can be expected to react slowly with the extractant. An example is the extraction of trivalent lanthanide or actinide cations from aqueous solutions by diethylhexyl phosphoric acid (HDEHP). If the aqueous medium contains only weakly complexing ligands (Cl^- or NO_3^-), the extraction kinetics are very fast. On the other hand, in the presence of polyaminocarboxylic acids, such as EDTA that are powerful complexing agents, the extraction reaction proceeds only slowly.

For planar tetracoordinated complexes, the mechanistic problem is more straightforward. A large amount of experimental evidence on ligand substitution reactions for complexes of transition metals with electronic configuration d^8, such as Pt(II), Pd(II), Au(III), Ni(II), and Rh(I), has been accumulated. The transition state is a five-coordinated species having the structure of a trigonal bipyramid on which the entering group lays on the trigonal plane. Ligand-substitution reactions with planar tetracoordinated complexes are often slow in comparison with the rate of diffusion through the interfacial diffusional films. It follows that the kinetics of solvent extraction of planar tetracoordinate complexes in a well-stirred system are often entirely controlled by the rate of the chemical process. Studies on the rate of extraction can then be directly used to obtain information on the rate of the biphasic ligand-substitution reactions, neglecting the complications from diffusional contributions.

5.5 THE LIQUID–LIQUID INTERFACE

As mentioned in section 5.1, when the two immiscible aqueous and organic phases are in contact and transfer of chemical species occurs from one phase to the other, the diffusional transport in proximity of the interface can be described by assuming the presence on either side of the interface of two stagnant films (the diffusional films). Although these films are thin when compared with the bulk phases, their thickness is still macroscopic (about 10^{-3} or 10^{-4} cm) in comparison with the molecular dimensions and the range of action of the molecular and ionic forces. Thus, these films have a finite volume, and the concentrations of chemical species within them must always be considered as volume concentrations. Moreover, the physicochemical properties of the liquids within these films, such as density, viscosity, dielectric constant, charge distribution, are the same as those of the bulk phases. However, the situation changes when the distance from the interface approaches the order of magnitude of the molecular dimensions and the range of action of the molecular and ionic forces. In this region, generally defined as the interface, the physicochemical properties differ from those of the bulk phases and, therefore, any chemical change occurring herein is likely to be affected by the different environment. The difference will be particularly enhanced by the presence on the interface of adsorbed layers of polar or ionizable extractant molecules. The extractant, because of its simultaneous hydrophobic-hydrophilic (low water affinity–high water affinity) nature, necessary to maintain a high solubility in low dielectric constant diluents, and selective complexing power relative to water-soluble species, tends to orient itself with the polar (or ionizable) groups facing the aqueous side of the interface. The rest of the molecule, having a prevalent hydrophobic character, will be directed instead toward the organic phase.

Unfortunately, little direct information is available on the physicochemical properties of the interface, since real interfacial properties (dielectric constant, viscosity, density, charge distribution) are difficult to measure, and the interpretation of the limited results so far available on systems relevant to solvent extraction are open to discussion. Interfacial tension measurements are, in this respect, an exception and can be easily performed by several standard physicochemical techniques. Specialized treatises on surface chemistry provide an exhaustive description of the interfacial phenomena [10,11]. *The interfacial tension*, γ, is defined as that force per unit length that is required to increase the contact surface of two immiscible liquids by 1 cm^2. Its units, in the CGS system, are dyne per centimeter (dyne cm^{-1}). Adsorption of extractant molecules at the interface lowers the interfacial tension and makes it easier to disperse one phase into the other.

Interfacial tension studies are particularly important because they can provide useful information on the interfacial concentration of the extractant. The simultaneous hydrophobic-hydrophilic nature of extracting reagents has the resulting effect of maximizing the reagent affinity for the interfacial zone, at which both the hydrophobic and hydrophilic parts of the molecules can minimize their free energy of solution. Moreover, as previously mentioned, a preferential orientation of the extractant groups takes place at the interface. Consequently, most solvent extraction reagents are interfacially adsorbed and produce a lowering of the organic-aqueous interfacial tension. The extent of this adsorption is a function of the chemical nature and structure of the hydrophilic and hydrophobic groups, of the extractant bulk concentration, and of the physico-chemical properties of the diluent. The extractant interfacial concentration can be extremely high, depending on the way molecules can pack themselves at the interface, and even moderately strong surfactants can form an interface fully covered with extractant molecules when their bulk concentration is as low as 10^{-3} M or less. It follows that, when an extractant is a strong surfactant and exhibits low solubility in the water phase, the interfacial zone is a region where a high probability exists that the reaction between an aqueous soluble species and an organic soluble extracting reagent can take place. This is why a strong surface activity of an extractant can sometimes lend support to extraction mechanisms involving interfacial chemical reactions.

Interfacial concentrations can be evaluated from interfacial tension measurements by utilizing the Gibbs equation

$$d \prod / d \ln c = n_i kT \tag{5.26}$$

In Eq. (5.26), $\prod$ is the interfacial pressure of the aqueous-organic system, equal to $(\gamma_0 - \gamma)$ [i.e., to the difference between the interfacial tensions without the extractant (γ_0) and the extractant at concentration c (γ)], c is the bulk organic concentration of the extractant, and n_i is the number of adsorbed molecules of the extractant at the interface. The shape of a typical $\prod$ vs. $\ln c$ curve is shown in Fig. 5.4; n_i can be evaluated from the value of the slopes of the curve at each c. However, great care must be exercised when evaluating interfacial concentrations from the slopes of the curves because Eq. (5.26) is only an ideal law, and many systems do not conform to this ideal behavior, even when the solutions are very dilute. Here, the proportionality constant between $d\prod/d \ln c$ and n_i is different from kT. Nevertheless, Eq. (5.26) can still be used to derive information on the bulk organic concentration necessary to achieve an interface completely saturated with extractant molecules (i.e., a constant interfacial concentration). According to Eq. (5.26), the occurrence of a constant interfacial concentration is indicated by a constant slope in a $\prod$ vs. $\ln c$ plot. Therefore, the value of c at which the plot $\prod$ vs. $\ln c$ becomes rectilinear can be taken as the bulk concentration of the extractant required to fully saturate the interface.

Many extractants reach a constant interfacial concentration at bulk organic concentrations far below the practical concentrations that are generally used to perform extraction kinetic studies. This means that when writing a rate law for an extraction mechanism that is based on interfacial chemical reactions, the interfacial concentrations can often be incorporated into the apparent rate constants. This leads to simplifications in the rate laws and to ambiguities in their interpretation, which are discussed in later sections.

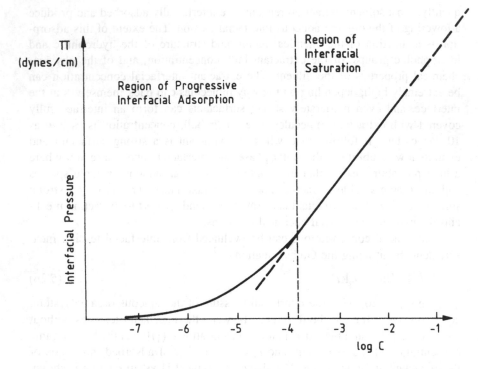

Fig. 5.4 Typical interfacial pressure (Π) vs. logarithm of bulk organic concentration (log c) plotted for an extractant exhibiting surface-active properties.

It is interesting to observe that alkylammonium salts, alkylarylsulfonic acids, hydroxyoximes, alkylphosphoric acids, and alkylhydroxamic acids, as well as neutral extractants such as crown ethers and tributyl phosphate, all form water-organic interfaces saturated with extractant molecules when their bulk organic concentration is larger than 10^{-3} M.

The liquid–liquid interface is not a sharply defined surface, but rather an ill-defined region the boundaries of which extend above and below the layer of the interfacially adsorbed extractant molecules. Nevertheless, it is often convenient to assume a mathematical dividing surface located where the physico-chemical properties of the two-phase system experience the sharpest discontinuity. This imaginary surface of zero thickness, which we refer to as the *interface*, is useful for defining the concentrations of interfacially adsorbed species, which can be expressed in moles per square centimeter (mol cm^{-2}). On the other hand, when in later sections we refer to interfacial concentrations, we mean volume concentrations in that region which is immediately adjacent to this imaginary dividing surface (i.e., at the extreme limit of the diffusional films).

The physical depth of the microscopic interfacial region can be estimated to correspond to the distance over which interfacial molecular and ionic forces exert their influence. Although molecules or ions experience no net forces in the interior of the bulk phases, these forces become unbalanced as the ions or molecules move toward the interface. On the aqueous side of the interfaces, where monolayers of charged molecules or polar groups can be present, these forces can be felt over several nanometers. On the organic side of the interface, where van der Waals forces are mainly operative, the interfacial region generally extends for tens of nanometers. The van der Waals forces decrease with the seventh power of the intermolecular distances, so molecules experience essentially symmetric forces, once they are a few molecular diameters away from the interface.

Although an experimentally verified molecular model of the interface is not yet available, a qualitative description of the interfacial zone can be attempted on the basis of speculations that extrapolate to the liquid–liquid interface results that are predicted from existing and verified physicochemical theories. Considering that extracting reagents can be present at the interface with their polar or ionizable groups facing the water side, a more structured water can be expected in the interfacial zone. The polar heads of the extractant tend to polarize the water molecules close to the interface, producing a more structured and viscous interfacial water, as depicted in Fig. 5.5. This interfacial water, having the possible depth of several molecular layers, is the result of the water penetration of the hydrophilic groups of the extractant, which hydrogen bond with the water molecules close to the interface. As a consequence, interfacial water should represent a more viscous environment to diffusing species than ordinary bulk water. The aqueous side of the interface can then be envisioned, at the microscopic level, as a very particular medium that still retains the chemical properties of bulk water, but for which the physical properties are much more like those of a strongly structured solvent, such as glycerol [12]. This interfacial zone should be characterized by lower fluidity, lower dielectric constant, higher stiffness, and be more compact than bulk water. The extent of these effects is a function of the surface-active properties of the extractant for a given organic diluent–aqueous electrolyte solution system. They can slow ligand exchange reactions, with octahedral aquoions occurring at the interface, since the more viscous interfacial environment slows the associative component of the overall interchange mechanism that controls this type of reactions. The associative step is, by contrast, very fast and, therefore, not rate controlling for most ligand-substitution reactions of octahedral aquoions occurring in bulk water.

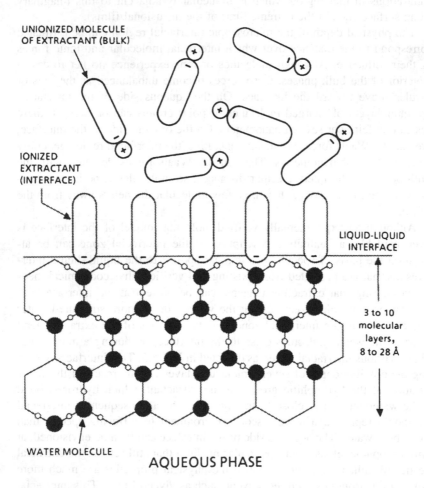

Fig. 5.5 Simplified picture of the interfacial structuring effect caused by an ionizable extractant adsorbed at the organic-water interface.

5.6 RATE-CONTROLLING EXTRACTION REGIMES

5.6.1 Definition of the Extraction Regimes

Section 5.1 describes how, in a stirred system, solvent extraction kinetics can be controlled only by slow chemical reactions or only by diffusion through the interfacial films. An intermediate situation can also occur whereby both the rates of chemical reactions and of interfacial film diffusion simultaneously control the rate of extraction.

When one or more of the chemical reactions is sufficiently slow in comparison with the rate of diffusion to and away from the interface of the various species taking part in an extraction reaction, such that diffusion can be considered instantaneous, the solvent extraction kinetics occur in a *kinetic regime*. In this case, the extraction rate can be entirely described in terms of chemical reactions. This situation may occur either when the system is very efficiently stirred and when one or more of the chemical reactions proceeds slowly, or when the chemical reactions are moderately fast, but the diffusion coefficients of the transported species are very high and the thickness of the two diffusion films is close to zero. In practice the latter situation never occurs, as diffusion coefficients in liquids generally do not exceed 10^{-4} cm^2 s^{-1}, and the depth of the diffusion films apparently is never less than 10^{-4} cm.

If, however, all chemical reactions in the biphasic system are very fast in comparison with the diffusional processes, the solvent extraction kinetics are said to occur in a *diffusional regime*. In this situation, the rate of extraction can be described simply in terms of interfacial film diffusion. Chemical engineers define this as "mass transfer with instantaneous chemical reaction." A diffusional regime may control the extraction rate also in the presence of relatively slow chemical reactions whenever the degree of stirring of the two phases is so poor that thick diffusional films can be assumed on the two sides of the interface. When both chemical reactions and film diffusion processes occur at rates that are comparable, the solvent extraction kinetics are said to take place in a *mixed diffusional-kinetic regime*, which in engineering, is often referred to as "mass transfer with slow chemical reactions." This is the most complicated case, since the rate of extraction must be described in terms of both diffusional processes and chemical reactions, and a complete mathematical description can be obtained only by simultaneously solving the differential equations of diffusion and those of chemical kinetics.

The unambiguous identification of the extraction rate regime (diffusional, kinetic, or mixed) is difficult from both the experimental and theoretical viewpoints [12,13]. Experimental difficulties exist because a large set of different experimental information, obtained in self-consistent conditions and over a very broad range of several chemical and physical variables, is needed. Unless simplifying assumptions can be used, frequently the differential equations have no analytical solutions, and boundary conditions have to be determined by specific experiments.

5.6.2 Identification of the Extraction Regimes

The experimental identification of the regime that controls the extraction kinetics is, in general, a problem that cannot be solved by reference to only one set of measurements. On the contrary, in some systems, a definite situation cannot be obtained even when the rate of extraction is studied as a function of both hydrodynamic parameters (viscosity and density of the liquids, geometry of the apparatus, geometry of stirrers, or stirring rate) and concentrations of the chemical species involved in the extraction reaction. This difficulty is because sometimes extraction rates may show the same dependence on hydrodynamic and concentration parameters, even though the processes responsible for the rate are quite different. For a correct hypothesis on the type of regime that controls the extraction kinetics, it is necessary to supplement kinetic investigations with other information concerning the biphasic system. This may be the interfacial tension, the solubility of the extractant in the aqueous phase, the composition of the species in solution, and so forth.

The criteria that are most often used to distinguish between a diffusional regime and a kinetic regime are

1. Comparison of the heat transfer and the mass transfer coefficients
2. Evaluation of the activation energy of extraction
3. The reference substance method
4. Dependency of the rate of extraction on the rate of stirring of the two phases.

Criteria 1–3 are difficult to apply and have been used only occasionally to evaluate the extraction regime.

1. According to this criterion, when, in the same apparatus, the same dependence of the heat transfer coefficient and the mass transfer coefficient on the stirring rate of the phases is observed, the conclusion can be reached that the extraction occurs in a diffusional regime.
2. The rationale behind the use of this method is that, when the rate is controlled by a chemical reaction, the activation energy is generally higher than that expected for a diffusion-controlled process (diffusion coefficients vary only slightly with temperature). However, this criterion is not always very meaningful, since many chemical reactions occurring in solvent extraction processes exhibit activation energies of only a few kilocalories per mole, i.e., have the same order of magnitude as those of diffusional processes.
3. The reference substance method is based on the addition to the solution, containing the species for which the transfer rate is going to be investigated, of another inert component for which the rate of extraction is known to be controlled only by diffusion. By following the simultaneous transfer of the species of interest and of the reference component as function of the hydrodynamic conditions in the extraction apparatus, a diffusional regime will be indicated by a similar functional dependence, whereas a kinetic regime is indicated by a sharply different one.
4. This criterion is the simplest, as proved by its widespread use. A typical

curve of rate of extraction vs. stirring rate is shown in Fig. 5.6. Such curves are generally obtained both when constant interfacial-area-stirred cells (Lewis cells) and vigorously mixed flasks are used (see section 5.10 for a description of the various techniques). In general, a process occurring under the influence of diffusional contributions is characterized by an increase of the rate of extraction, as long as the stirring rate of the two phases is increased (see Fig. 5.6, zone A). When, on the other hand, the rate of extraction is independent of the stirring rate (see Fig. 5.6, zone B), it is sometimes possible to assume that the extraction process occurs in a kinetic regime.

The rationale behind this criterion is that an increase in stirring rate produces a decrease in thickness of the diffusion films. Moreover, since at low stirring rates the relationship between stirring rate and the inverse thickness of the diffusion films (at least in a constant interfacial-area-stirred cell) is approximately linear, the first part of such a plot will usually approximate a straight line. In any case, the rate of extraction will increase with the rate of stirring, as long as a process is totally or partially diffusion controlled. When, eventually, the thickness of the diffusion films is reduced to zero, only chemical reactions can be rate controlling, and the rate of extraction becomes independent of the stirring rate. Unfortunately, this kind of reasoning can lead to erro-

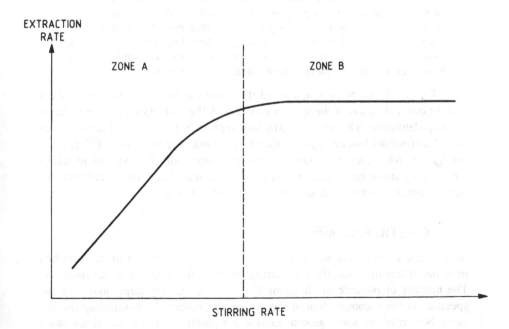

Fig. 5.6 Typical extraction rate vs. stirring rate for constant interfacial–area-stirred cell.

neous conclusions with both constant interfacial-area-stirred cells and vigorously stirred flasks.

In the case of constant interfacial–area-stirred cells, although zone A is certainly an indication that the process is controlled by diffusional processes, the opposite is not true for zone B. In fact, in spite of the increased stirring rate, it may happen that the thickness of the diffusion films never decreases below a sufficiently low value to make diffusion so fast that it can be completely neglected relative to the rate of the chemical reactions. This effect, sometimes called "slip effect," depends on the specific hydrodynamic conditions of the apparatus in which the extraction takes place and simulates a kinetic regime.

As far as vigorously stirred vessels are concerned, strong ambiguities are also faced when this criterion is applied. Unless the overall interfacial area is known and taken into consideration, an increase in the rate of extraction can take place with the increase in stirring rate when the system is in a kinetic regime determined by interfacial chemical reactions. The increase observed in zone A indicates here an increase in the number of droplets of the dispersed phase (proportional to the overall interfacial area) and not a decrease in the thickness of the diffusion film. Moreover, the plateau region of zone B does not necessarily prove that the extraction occurs in a kinetic regime. Although at high stirring rates the number of droplets of the dispersed phase eventually becomes constant, since the rate of drop formation equals the rate of drop coalescence, lack of internal circulation and poor mixing can occur inside the droplets of the dispersed phase. This is particularly true with small droplets and in the presence of extractants that are strong surfactants. Therefore, here also, a plateau region may simulate a diffusion-controlled regime. It is then apparent that criterion 4 can also lead to ambiguous or misleading conclusions.

Finally, it has to be emphasized that both the hydrodynamic parameters and the concentrations of the species involved in the extraction reaction simultaneously determine whether the extraction regime is of kinetic, diffusional, or mixed diffusional-kinetic type. It, therefore, is not surprising that different investigators, who studied the same chemical solvent extraction system in different hydrodynamic and concentration conditions, may have interpreted their results in terms of completely different extraction regimes.

5.7 KINETIC REGIME

In a kinetic regime system, the kinetics of solvent extraction can be described in terms of chemical reactions occurring in the bulk phases or at the interface. The number of possible mechanisms is, in principle, very large, and only the specific chemical composition of the system determines the controlling mechanism. Nevertheless, some generalizations are possible on considerations based on the nature of the reagents involved in the extraction reaction. Since, in many metal extractions, ligand substitution reactions take place, rate laws similar to those for complexation reactions in solution may be expected. During most extraction processes, coordinated water molecules or ligands are substituted in part or wholly by molecules of a more organophilic ligand (the extractant) or of the organic diluent. When the extractant shows little solubility in the aqueous phase and is a strong surfactant, the ligand substitution reaction may take place at the interface. The highest concentration of extractant molecules is at the interface, so the probability of reaction is higher than in the bulk phases.

In this section, we describe three simple cases of rates and mechanisms that have been found suitable for the interpretation of extraction kinetic processes in kinetic regimes. These simple cases deal with the extraction reaction of a monovalent metal cation M^+ (solvation water molecules are omitted in the notation) with a weakly acidic solvent extraction reagent, BH. The overall extraction reaction is

$$M^+ + BH(org) \rightarrow MB(org) + H^+ \tag{5.27}$$

with equilibrium extraction constant, K_{ex} equal to

$$K_{ex} = [\overline{MB}]_{eq}[H^+]_{eq} / [M^+]_{eq}[\overline{BH}]_{eq} \tag{5.28}$$

Ideal behavior of all solute species will be assumed.

Case 1: *The rate-determining step of the extraction reaction is the aqueous phase complex formation between the metal ion and the anion of the extracting reagent.* Even if at very low concentration, BH will always be present in the aqueous phase because of its solubility in water. The rate-determining step of the extraction reaction is as follows:

$$M^+ + B^- \underset{k_{-1}}{\overset{k_1}{\longleftrightarrow}} MB \quad (slow) \tag{5.29}$$

with a reaction rate expression:

$$rate = -d[M^+] / dt = k_1[M^+][B^-] - k_{-1}[MB] \tag{5.30}$$

To derive an expression for the rate in terms of easily measurable quantities (initial concentrations), the following (instantaneously established) equilibria have to be included (cf. Chap. 4):

1. $BH \leftrightarrow \overline{BH};$ $K_{DB} = [\overline{BH}]/[BH]$ (5.31)

2. $BH \leftrightarrow B^- + H^+;$ $K_a = [B^-][H^+]/[BH]$ (5.32)

3. $MB \leftrightarrow MB(org);$ $K_{DM} = [\overline{MB}]/[MB]$ (5.33)

By substituting Eqs. (5.31–5.33) into Eq. (5.30), we obtain the rate expression

$$\text{rate} = k_1 K_a / K_{DB}^{-1} [M^+][\overline{BH}][H^+]^{-1} - k_{-1} K_{DM}^{-1} [\overline{MB}] \tag{5.34}$$

In the right-hand side of Eq. (5.34), the first term represents the forward rate of extraction and the second term the reverse rate of extraction.

The values of the apparent rate constants of the extraction reaction,

$$k_1 K_a K_{DB}^{-1} \quad \text{and} \quad k_{-1} K_{DM}^{-1} \tag{5.35}$$

permit the evaluation of the rate constants of the aqueous complex formation reaction only if K_{DB}, K_{DM}, and K_a are known. Here, comparisons can be made with literature values of the rate constants for reaction in aqueous solutions between the same metal ion and aqueous soluble ligands containing the same complexing groups as the extracting reagent. Agreement between the values supports, but does not prove, the validity of the proposed mechanism.

The apparent values of the rate constants of the solvent extraction reaction are usually evaluated by measuring the rate of extraction of M^+ as function of $\overline{[BH]}$ (at $[H^+]$ constant), of $[H^+]$ (at $\overline{[BH]}$ constant), and of $\overline{[MB]}$ (at $[H^+]$ and $\overline{[BH]}$ constant). The experimental conditions are usually chosen in such a way that the reaction can be assumed pseudo-first-order for $[M^+]$. The apparent rate constants are evaluated from the slope of the straight lines obtained by plotting

$$\ln\,([M^+]-[M^+]_{eq})/([M^+]_0 -[M^+]_{eq}) \text{ vs. } t \tag{5.36}$$

Subscript 0 indicates initial concentrations in the aqueous phase.

It must be observed that when this mechanism holds, the rate of the extraction reaction is independent of the interfacial area, Q, and the volume of the phases, V. The expected logarithmic dependency of the forward rate of extraction on the specific interfacial area ($S = Q/V$), the organic concentration of the extracting reagent and the aqueous acidity, is shown in Fig. 5.7, case 1.

Case 2: *The rate-determining step of the extraction reaction is the interfacial formation of the complex between the metal ion and the interfacially adsorbed extracting reagent.* Here, the rate-determining step of the extraction reaction can be written as

$$M^+ + B^-(ad) \underset{k_{-2}}{\overset{k_2}{\longleftrightarrow}} MB(ad) \qquad \text{(slow)} \tag{5.37}$$

where (ad) indicates species adsorbed at the liquid–liquid interface. The rate of the reaction that follows is

$$\text{rate} = -d[M^+]/dt = k_2 S_w [M^+][B^-]_{ad} - k_2 S_o [MB]_{ad} \tag{5.38}$$

where S_w and S_o represent the specific interfacial areas [i.e., the ratios between the interfacial area Q (cm^2) and the volumes of the aqueous (V_w) and organic (V_o) phase, respectively]. Throughout the following treatments we will always assume, for simplicity, that $V_w = V_o = V$, and

KINETIC REGIMES

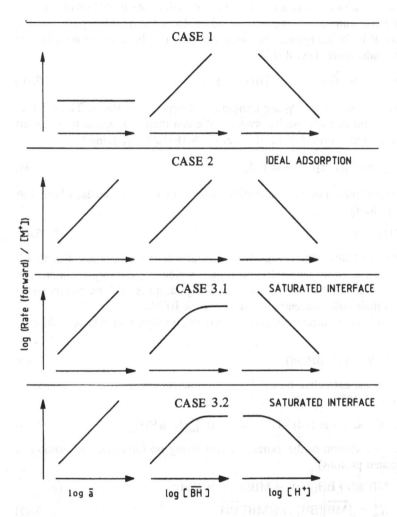

Fig. 5.7 Logarithm of the forward extraction rate (normalized to the extractable metal concentration) vs. log interfacial area (*a*), log $\overline{[BH]}$, and log $[H^+]$ for some exemplary kinetic regimes.

$$Q/V = S = S_W = S_o \tag{5.39}$$

To derive a rate of extraction in terms of easily measurable quantities, we express the concentration of the reagent adsorbed at the liquid–liquid interface as function of its bulk organic concentration. This can be done by utilizing the Langmuir's adsorption law: that is,

$$[BH]_{ad} = (\alpha_2[\overline{BH}]/\gamma_2)/(1+[\overline{BH}]/\gamma_2) \tag{5.40}$$

In Eq. (5.40), α_2 and γ_2 are Langmuir adsorption constants. Their values are characteristic of each specific system. We can then distinguish two regions of adsorption. One defined as ideal, where $1 \gg [\overline{BH}]/\gamma_2$, yielding

$$[BH]_{ad} = (\alpha_2[\overline{BH}]/\gamma_2) = \alpha[\overline{BH}] \tag{5.41}$$

and one of complete interface saturation with extractant molecules, where $1 \gg [BH]/\gamma_2$, yielding

$$[BH]_{ad} = \alpha_2 \tag{5.42}$$

With extractants exhibiting strong surface active properties, Eq. (5.42) holds true in the entire concentration range, which is generally of interest in practical studies, and the interface becomes fully saturated with extractant molecules when their bulk concentration is as low as 10^{-3} M.

The following, instantaneously established, equilibria then have to be considered:

1. $BH(org) \leftrightarrow BH(ad)$ \hfill (5.43)

adsorption of the extractant molecules at the interface described by Langmuir's law [see Eq. (5.40)]

2. $BH(ad) \leftrightarrow B^-(ad) + H^+; \quad K_a^* = [B^-]_{ad}[H^+][BH]_{ad}^{-1}$ \hfill (5.44)

interfacial dissociation of the extractant (involving no interfacial adsorption of the dissociated protons)

3. $MB(ad) + BH(org) \leftrightarrow MB(org) + BH(ad);$

$$K_{ex}^0 = ([\overline{MB}][BH]_{ad})/([MB][\overline{BH}]) \tag{5.45}$$

fast replacement at the interface of the interfacially adsorbed metal complex with bulk molecules of the extractant. Equations (5.40), (5.44), (5.45) can be inserted into Eq. (5.38) to obtain:

$$\text{rate} = S\, k_2\alpha_2\, K_a^*[M^+][H^+]^{-1} - S\, k_{-2}\, K_{ex}^{0\;-1}\, \alpha_2[\overline{MB}]^{-1}[\overline{BH}] \tag{5.46}$$

which holds for a fully saturated interface, and

$$\text{rate} = S\, k_2\, \alpha\, K_a^*[M^+][\overline{BH}][H^+]^{-1} - S\, k_{-2}\, K_{ex}^{0\;-1}\, \alpha[\overline{MB}] \tag{5.47}$$

which holds for ideal adsorption.

The equilibrium constant of the extraction reaction [see Eq. (5.28)] is equal to

$$K_{ex} = k_2 \, K_a^* K_{ex}^0 \, k_{-2}^{-1} \tag{5.48}$$

for both ideal adsorption and for a saturated interface. The expression for K_{ex} is derived from either Eq. (5.46) or Eq. (5.47), since at equilibrium, rate (forward) = rate (reverse).

Equations (5.34) and (5.47) have the same dependence on the concentration variables. The only difference is that for the interfacial reaction the apparent rate constants are directly proportional to the ratio $S = Q/V$. This dependency can be experimentally verified by measuring the apparent rate of extraction as a function of the interfacial area and the volume of the phases. A plot of the apparent rate constant of the forward rate of extraction vs. S must yield a straight line through the origin of the axes when case 2 holds.

Equation (5.46) shows a further difference relative to Eq. (5.34). Equation (5.46) indicates a zero reaction order relative to $\overline{[BH]}$ in the forward extraction rate, reflecting the complete saturation of the interface with the extracting reagent.

The determination of the true rate constants, k_1 and k_{-2}, is generally more difficult than for case 1, since information on α, α_2, K_a^* and K_{ex}^0 can be difficult to obtain.

Case 3: *There are two interfacial rate-determining steps, consisting of: (1) formation of an interfacial complex between the interfacially adsorbed molecules of the extractant and the metal ion; and (2) transfer of the interfacial complex from the interface to the bulk organic phase and simultaneous replacement of the interfacial vacancy with bulk organic molecules of the extractant.* For this mechanism, we distinguish two possibilities. The first (case 3.1) describes the reaction with the dissociated anion of the extracting reagent, $B^-(ad)$. The second (case 3.2) describes the reation with the undissociated extractant, $BH(ad)$.

Case 3.1: The first mechanism is represented by the following equations:

$$M^+ + B^-(ad) \underset{k_{-3}}{\overset{k_3}{\leftrightarrow}} MB(ad) \quad \text{(slow)} \tag{5.49}$$

$$MB(ad) + BH(org) \underset{k_{-4}}{\overset{k_4}{\leftrightarrow}} BH(ad) + MB(org) \quad \text{(slow)} \tag{5.50}$$

The rate equations holding for the two slow steps are:

$$\text{rate } 1 = S \, k_3 \, [M^+][B^-]_{ad} - S \, k_{-3} \, [MB]_{ad} \tag{5.51}$$

$$\text{rate } 2 = S\, k_4\, [\text{MB}]_{\text{ad}} [\overline{\text{BH}}] - S\, k_{-4}\, [\text{BH}]_{\text{ad}} [\overline{\text{MB}}] \tag{5.52}$$

When the metal concentration is sufficiently low and the concentration of the interfacially adsorbed metal complex is consequently low, for the stationary condition at $[\text{MB}]_{\text{ad}}$

$$\text{rate } 1 = \text{rate } 2 \tag{5.53}$$

Equation (5.53) can then be solved for [MB], after having introduced the substitution:

$$[\text{B}^-]_{\text{ad}} = K_a^* \, [\text{BH}]_{\text{ad}} \, [\text{H}^+]^{-1} \tag{5.54}$$

and considering that, for a fully saturated interface $[\text{BH}]_{\text{ad}} = \alpha_2$. The expression for $[\text{MB}]_{\text{ad}}$ found in this way is

$$[\text{MB}]_{\text{ad}} = \{k_3\, K_a^* \, \alpha_2 \, [\text{M}^+][\text{H}^+]^{-1} + k_{-4}\, \alpha_2 [\overline{\text{MB}}]\} / \{k_{-3} + k_4 [\overline{\text{BH}}]\} \tag{5.55}$$

Substituting the value of $[\text{MB}]_{\text{ad}}$ into either Eq. (5.51) or Eq. (5.52) gives

$$\text{rate} = \{S\, k_3 k_4 K_a^* \alpha_2 [\overline{\text{BH}}][\text{M}^+][\text{H}^+]^{-1}\}\{k_{-3} + k_4 [\overline{\text{BH}}]\}^{-1}$$
$$- \{S\, k_{-3} k_{-4} \alpha_2 [\overline{\text{MB}}]\}\{k_{-3} + k_4 [\overline{\text{BH}}]\}^{-1} \tag{5.56}$$

An expression for the equilibrium constant of the extraction reaction can be obtained from Eq. (5.56) by considering that, at equilibrium, one can set rate (forward) = rate (reverse). Thus

$$K_{\text{ex}} = K_3\, k_4\, K_a^* \, k_{-3}^{-1}\, k_{-4}^{-1} \tag{5.57}$$

Comparison of Eqs. (5.34), (5.47), and (5.56) shows that if $[\overline{\text{BH}}]$ is sufficiently small to allow introduction of the approximation $k_4\, [\overline{\text{BH}}] \ll k_{-3}$ into the denominator of Eq. (5.56) (i.e., for a low and restricted concentration range of the extractant), the three rate equations have exactly the same dependence on $[\text{M}^+]$, $[\overline{\text{BH}}]$, and $[\text{H}^+]$, although the extraction mechanisms are characterized by different rate-determining steps:

1. A homogeneous reaction in the aqueous phase with the anion of the extractant [see Eq. (5.34)].
2. An interfacial reaction with the anion of the extractant, which is ideally adsorbed at the interface [see Eq. (5.47)].
3. Two sequential interfacial reactions: the first one being the reaction with the anion of the extractant that saturates the interface; the second one being the slow desorption of the interfacial complex from the interface [see Eq. (5.56)].

The difference among these mechanisms can be seen only by measuring the dependence of the rate of extraction on the specific interfacial area, S, and by using the broadest possible concentration range of the reactants. If a dependence on S exists, case 1 can be excluded. If in addition the extraction rate is initially first order and then becomes zero order when $\overline{[BH]}$ increases, case 2 can also be ruled out.

Case 3.2: When the interfacially reactive species are the undissociated molecules of the extractant adsorbed at the interface [i.e., the first rate-determining step of the two-step mechanism is the reaction between the metal ion M^+ and $HB(ad)$], the following equations will hold:

$$M^+ + BH(ad) \leftrightarrow MB(ad) + H^+ \quad \text{(slow)} \tag{5.58}$$

$$MB(ad) + BH(org) \leftrightarrow BH(ad) + MB(org) \quad \text{(slow)} \tag{5.59}$$

The rate equations for the slow steps [see Eqs. (5.58) and (5.59)] are

$$\text{rate } 1 = k_3^* \, S[M^+][BH]_{ad} - k_{-3}^* \, S[MB]_{ad}[H^+] \tag{5.60}$$

$$\text{rate } 2 = k_4^* \, S[MB]_{ad}\overline{[BH]} - k_{-4}^* \, S[BH]_{ad}\overline{[MB]} \tag{5.61}$$

By assuming that the stationary condition, (rate 1 = rate 2), can be applied and by introducing the same substitutions as those used to derive Eq. (5.56), we obtain (for a fully saturated interface):

$$[MB]_{ad} = \{\alpha_2 k_3^* \, [M^+] - \alpha_2 k_{-4} \, \overline{[MB]}\} \, \{k_{-3}^* \, [H^+] + k_4 \, \overline{[BH]}\}^{-1} \tag{5.62}$$

The insertion of Eq. (5.62) into Eq. (5.61) gives

rate = rate (forward) − rate (reverse) =

$$= \{S \, k_3^* k_4 \alpha_2 \overline{[BH]}[M^+]\} \, \{k_{-3}^*[H^+] + k_4\overline{[BH]}\}^{-1}$$
$$- \{S \, k_{-3}^* k_{-4} \alpha_2 \overline{[MB]}[H^+]\} \, \{k_{-3}^* + k_4\overline{[BH]}\}^{-1} \tag{5.63}$$

Equation (5.63) has the same dependence on $[M^+]$, $\overline{[BH]}$, and $[H^+]$ as Eqs. (5.34) and (5.47) only when the simplification $k_4\overline{[BH]} \ll k_{-3}\,[H^+]$ can be introduced in the denominator.

Figure 5.7 shows the logarithmic dependencies expected between the normalized rates of forward extraction, rate (forward)/$[M^+]$, and S, $\overline{[BH]}$ and $[H^+]$, for selected cases.

Although the three described cases represent rates and mechanisms that have often been used in describing solvent extraction systems that were claimed to occur in the absence of diffusional contributions, we have to emphasize that many other mechanisms and rates are possible [12]. However, the three foregoing cases are useful in showing the type of ambiguities often faced when trying to assign a reaction mechanism on the basis of an experimentally determined solvent extraction rate law. These ambiguities are only partly reduced by studying the extraction rate in a very broad concentration range of the reagents and by varying S. Much additional information on the species present and on the physicochemical properties of the system may be necessary to identify and support a mechanism. For example, an appreciable water solubility and weak adsorption at the interface of the extracting reagent generally favor cases like 1. Alternatively, highly water-insoluble and a strongly interfacially adsorbed chelating extractant increase the possibility that cases 2 and 3 may describe the mechanism of extraction.

The simplified chemical system (M^+, $\overline{BH}$, B^-) used here to derive the equations of cases 1–3, although useful to shown similarities, differences, and ambiguities that can be encountered when dealing with the rates and mechanisms of solvent extraction processes, only approximately describes the very complex and diverse chemical systems that have been reported in the literature. With minor changes and modifications, it is not difficult to derive other equations suitable for describing many other more complex situations. For example, with the help of case 1, the equations describing the kinetic of extraction of divalent metal ions, such as Zn^{2+}, Ni^{2+} and Co^{2+}, by dithizone and its derivatives can be derived. Minor modifications also have to be introduced in cases 2 and 3 to obtain the equations that describe the extraction of metal ions, such as Cu^{2+} and Fe^{3+}, by hydroxyoximes; Fe^{3+} by hydroxamic acids; and trivalent lanthanide and actinide cations by alkylphosphoric acids. Other chemical systems in which neutral or negatively charged metal species are extracted by neutral extractants or by alkylammonium salts have been described by mechanisms involving aqueous-phase complexation or interfacial reactions (one-step or two-step mechanisms) as rate-determining steps. Examples are the extraction of $UO_2(NO_3)_2$ by tributyl phosphate (TBP), or $FeCl_4^-$ and $Pu(NO_3)_4$ by trialkylammonium salts. The equations reported in cases 1–3 can be easily modified to adapt them to the specific chemical features of the reagents involved [13]. In most instances, these modifications involve writing the correct charge for the extracted species (including zero and negative charges), eliminating the acid dissociation step of the extractant whenever this is appropriate, including additional equilibrium equations, and additional simple rate-determining steps. However, examples of completely different mechanisms and rates, which in no way can be related to cases 1–3, are abundant in the literature.

5.8 DIFFUSIONAL REGIME

Let us now consider the systems in a diffusional regime. Given the simplified picture of the interface described by the two-film theory (see Fig. 5.1), we can now visualize that any species going from one phase to the other, crossing the liquid–liquid interface, will encounter a total resistance R, which is the sum of three separate resistances. Two of these resistances are of a diffusional nature and depend on the fact that diffusion through the stagnant interfacial films can be a slow process. These resistances are indicated in Fig. 5.1 as R_w and R_o. R_w

includes the diffusional contributions on the water side of the interface, assumed to occur entirely through the stagnant film of thickness, δ_w. R_o includes the contributions on the organic side, where the thickness of the stagnant film is δ_o. The resistance associated with the crossing of the interface is indicated as R_i. As a consequence, the total transfer resistance, R, is

$$R = R_w + R_i + R_o \tag{5.64}$$

This equation holds at the steady state. In a diffusional regime and in absence of rigid interfacial films, R_i is generally negligible relative to R_w and R_o.

Consider two simple cases of extraction processes in which kinetics are controlled by interfacial film diffusion (the solutions are always considered stirred). The two cases are treated with the simplifying assumptions introduced in section 2 (i.e., steady-state and linear concentration gradients throughout the diffusional films).

Case 4: *The interfacial partition between the two phases of unchanged species is fast. The rate is controlled by the diffusion to and away from the interface of the partitioning species.* In the absence of an interfacial resistance, the partition equilibrium of A between the aqueous and organic phase, occurring at the interface, can be always considered as an instantaneous process. Here, A is any species, neutral or charged, organic or inorganic. This instantaneous partition process (interfacial equilibrium) is characterized by a value of the partition coefficient equal to that measured when the two phases are at equilibrium.

The following equations describe the extraction process:

$$A \leftrightarrow A(org) \tag{5.65}$$

with an extraction equilibrium constant K_{DA}, reached at $t = \infty$;

$$A_i \leftrightarrow A_i(org) \quad \text{(fast)} \tag{5.66}$$

interfacial equilibrium, holding at any time (subscript i indicates species in contact with the interface); thus

$$K_{DA} = [A]_{eq} / [A]_{eq} = [A]_i / [A]_i \tag{5.67}$$

The symbol $[\ ]_i$ indicates concentrations at the extreme limit of the diffusional film (i.e., volume concentrations in the region in direct contact or very close to the liquid-liquid interface). Because of the fast nature of the distribution reaction, local equilibrium always holds at the interface.

The diffusional process can be treated by applying the first Fick's diffusion law to the two diffusional processes occurring on both sides of inter face, under the simplifying assumptions of steady-state and linear concentration gradients shown in Fig. 5.8. The diffusional fluxes through the aqueous, J_w and organic, J_o, diffusion films will then be

$$J_w = -D_A \partial[A]/\partial x = -D_A([A]_i - [A])/\delta_w \tag{5.68}$$

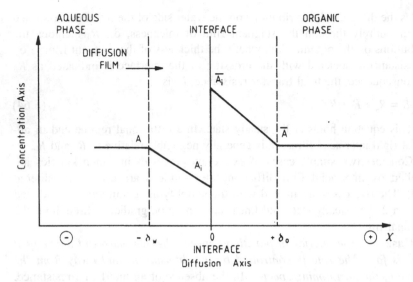

Fig. 5.8 Linear concentration profiles for the partition of A between an aqueous and an organic phase.

$$J_o = -D_{\overline{A}} \, \partial \, [\overline{A}]/\partial x = -D_A([\overline{A}]-[\overline{A}]_i)/\delta_o \tag{5.69}$$

D_A and $D_{\overline{A}}$ are the diffusion coefficients of the distributing species in the aqueous and organic layers, respectively. By considering that, at the steady state, $J_w = J_o = J$ and by setting

$$\Delta_w = \delta_w/D_A, \quad \Delta_o = \delta_o/D_{\overline{A}} \tag{5.70}$$

Equations. (5.68) and (5.69) can be solved for the interfacial concentrations to yield

$$[A]_i = [A]-J\Delta_w \tag{5.71}$$

$$[\overline{A}]_i = J\Delta_o +[\overline{A}] \tag{5.72}$$

By inserting Eqs. (5.71) and (5.72) into Eq. (5.67), solving for J, and considering the correlation between flux and rate [see Eq. (5.5)], we obtain

$$\text{rate} = -d[A]/dt = \text{rate (forward)} - \text{rate (reverse)}=$$

$$=S \, K_{DA}[A]/(K_{DA}\Delta w+\Delta_o) - S[\overline{A}]/(K_{DA}\Delta_w+\Delta_o) \tag{5.73}$$

The diffusion-controlled, extraction kinetics of A, therefore, can be described as a pseudo-first-order rate process with apparent rate constants

$$k_1 = K_{DA}[A]/(K_{DA}\Delta_w +\Delta_o) \text{ and } k_{-1} = 1/(K_{DA}\Delta_w +\Delta_o) \tag{5.74}$$

The rate equation is indistinguishable from that of an extraction process occurring in a kinetic regime, which is controlled by a slow, interfacial partition reaction:

$$A \underset{k_{-1}}{\overset{k_1}{\leftrightarrow}} A(org) \quad \text{(slow)} \tag{5.75}$$

The diffusion-controlled process [see Eqs. (5.66), (5.68), and (5.69)] can be experimentally differentiated from the process occurring in the kinetic regime [see Eq. (5.75)] only by measuring the variations of k_1 and k_{-1} with δ_w and δ_o. Otherwise, the identical rate laws will not permit one to distinguish between the two mechanisms.

Equation (5.73) can be integrated to obtain

$$\ln \left(([A]_o - [A]_{eq}) / ([A] - [A]_{eq}) \right) = S(K_{DA} + 1)t /$$

$$(K_{DA}\Delta_w + \Delta_o) = S(k_1 + k_{-1})t \tag{5.76}$$

Equations (5.74) and (5.76) show that when K_{DA} is very high (i.e., the partition of A is all in favor of the organic phase):

$$k_1 \sim D_A / \delta_w \quad \text{and} \quad k_{-1} \sim 0$$

and the extraction rate is controlled only by the aqueous-phase diffusional resistance. On the other hand, when K_{DA} is very small, it is

$$k_1 \sim 0 \quad \text{and} \quad k_{-1} \sim D_{\bar{A}} / \delta_o$$

and the rate is only controlled by the organic-phase diffusional resistance.

Case 5: *A fast reaction between the metal cation and the undissociated extracting reagent occurs at the interface (or in proximity of the interface). The rate is controlled by the diffusion to and away from the interface of the species taking part in the reaction.*

The overall extraction reaction (extraction stoichiometry) is represented by the equation:

$$M^+ + BH(org) = MB(org) + H^+ \tag{5.77}$$

with equilibrium extraction constant K_{ex} chemical reaction occurring at or near the interface is considered to be always at equilibrium, although bulk phases reach this condition only at the end of the extraction process. The condition of interfacial (local) equilibrium is expressed by

$$M_i^+ + BH_i(org) \leftrightarrow MB_i(org) + H_i^+ \quad \text{(fast)} \tag{5.78}$$

with interface equilibrium constant K_i, which alternatively can be written as the sum of the two equilibria.

$$M_i^+ + B_i^-(org) \leftrightarrow MB_i(org), \quad K_1 = \overline{[MB]_i} / [M^+]_i [B^-]_i \tag{5.79}$$

$$\text{BH}_i(\text{org}) \leftrightarrow \text{H}_i^+ + \text{B}_i^-(\text{org}), \qquad K_2 = [\text{B}^-]_i [\text{H}^+]_i / [\overline{\text{BH}}]_i \tag{5.80}$$

with

$$K_{\text{ex}} = K_1 K_2 = ([\overline{\text{MB}}]_{\text{eq}} [\text{H}^+]_{\text{eq}}) / ([\overline{\text{BH}}]_{\text{eq}} [\text{M}^+]_{\text{eq}})$$

$$= ([\overline{\text{MB}}]_i [\text{H}^+]_i) / ([\overline{\text{BH}}]_i [\text{M}^+]_i) = K_i \tag{5.81}$$

Following the previous line of reasoning, the rate of extraction can be obtained by considering the diffusional fluxes through the diffusion films of M^+, H^+, MB, and BH, expressed through Fick's law, and introducing the chemical reaction [see Eq. (5.78)] as boundary condition to the differential equations describing diffusion through the aqueous and organic diffusion films. Under the usual simplifying assumptions of steady-state and linear concentration gradients and referring to Fig. 5.9, where the concentration gradients of the four species M^+, H^+, BH, and MB are schematically indicated, the rate of extraction of M^+ can be derived. Equations (5.82a–d) describe the diffusional fluxes of M^+, H^+, BH, MB through the stagnant films:

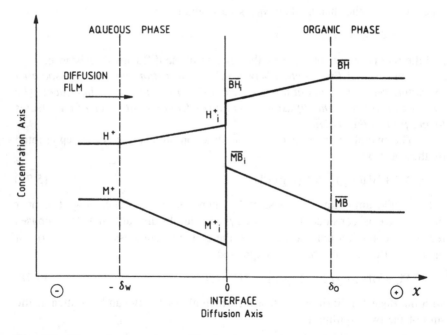

Fig. 5.9 Linear concentration profiles for the extraction of M^+ by BH(org) in presence of an interfacial chemical reaction.

$$J_{M^+} = -D_{M^+} \partial [M^+]/\partial x = -D_{A^+}([M^+]_i - [M^+])/\delta_w \qquad (5.82a)$$

$$J_{H^+} = -D_{H^+} \partial [H^+]/\partial x = -D_{H^+}([H^+]_i - [H^+])/\delta_w \qquad (5.82b)$$

$$J_{MB} = -D_{MB} \partial [\overline{MB}]/\partial x = -D_{AB}([\overline{MB}] - [\overline{MB}]_i)/\delta_o \qquad (5.82c)$$

$$J_{BH} = -D_{BH} \partial [\overline{BH}]/\partial x = -D_{BH}([\overline{BH}] - [\overline{BH}]_i)/\delta_o \qquad (5.82d)$$

Furthermore, the four fluxes are not independent and, at the steady state, are related by the condition:

$$J_{M^+} = J_{\overline{MB}} = -J_{H^+} = -J_{\overline{BH}} \qquad (5.83)$$

indicating that, although M^+ and MB diffuse in the positive direction of the x axis, H^+ and BH diffuse in the opposite one. Moreover,

$$-d[M^+]/dt \; V/Q = J = J_{M^+} \qquad (5.84)$$

By solving Eqs. (5.82a–d) for the interfacial concentrations [ ]$_i$ after taking into account the correlation between the fluxes [see Eqs. (5.83) and (5.84)] and introducing the simplification that the diffusion coefficients depend only on the nature of the phase, that is,

$$\Delta_w = \delta_w/D_{M^+} = \delta_w/D_{H^+} \qquad (5.85a)$$

$$\Delta_o = \delta_o/D_{\overline{MB}} = \delta_o/D_{\overline{BH}} \qquad (5.85b)$$

we obtain

$$[M^+]_i = [M^+] - J\Delta_w \qquad (5.86a)$$

$$[H^+]_i = [II^+] + J\Delta_w \qquad (5.86b)$$

$$[\overline{MB}]_i = [\overline{MB}] + J\Delta_o \qquad (5.86c)$$

$$[\overline{BH}]_i = [\overline{BH}] - J\Delta_o \qquad (5.86d)$$

Equations (5.86a–d) can then be inserted into Eq. (5.81) to yield

$$K_{ex} = [\overline{MB}] + J\Delta_o)([H^+] + J\Delta_w)/([M^+] - J\Delta_w)(\overline{BH}] - J\Delta_o) \qquad (5.87)$$

Equation (5.87) can be solved for J to derive an expression for J (and the extraction rate) as function of the four concentration variables in the bulk phases, the equilibrium constant of the extraction reaction, and the two diffusional parameters Δ_w and Δ_o.

A simple analytical solution for J (and the extraction rate) can be derived from Eq. (5.87) by neglecting the term in J^2 (very low fluxes) and isolating J:

$$\text{rate} = \text{rate (forward)} - \text{rate (reverse)} = -d[M^+]/dt = JS =$$

$$= \{S \; K_{ex}[M^+][\overline{BH}]\} \{\Delta_w([\overline{MB}] + K_{ex}[\overline{BH}]) + \Delta_o([H^+] + K_{ex}[M^+])\}^{-1} -$$

$$- \{S \; [\overline{MB}][H^+]\} \{\Delta_w([\overline{MB}] + K_{ex}[\overline{BH}]) + \Delta_o([H^+] + K_{ex}[M^+])\}^{-1} \qquad (5.88)$$

In the limiting condition of very low metal ion concentrations, the terms containing [M$^+$] and [MB] can be neglected in the denominator of Eq. (5.88). This means that [BH] and [H$^+$], in practice, are constant with time, and their initial and equilibrium concentrations are essentially the same. The rate equation then becomes

$$\text{rate} = \{S\, K_{ex}[\text{M}^+][\overline{\text{BH}}]_{eq}\}\,\{\Delta_w K_{ex}[\overline{\text{BH}}]_{eq} + \Delta_o[\text{H}^+]_{eq}\}^{-1} -$$
$$- \{S\,[\overline{\text{MB}}][\text{H}^+]_{eq}\}\,\{\Delta_w K_{ex}[\overline{\text{BH}}]_{eq} + \Delta_o[\text{H}^+]_{eq}\}^{-1} \qquad (5.89)$$

By dividing the numerators and denominators in Eq. (5.89) by [H$^+$]$_{eq}$ and setting

$$K_{ex}^* = K_{ex}[\overline{\text{BH}}]_{eq}/[\text{H}^+]_{eq} \qquad (5.90)$$

where K$_{ex}^*$ is the conditional equilibrium constant of equilibrium [see Eq. (5.77)] at [H$^+$] and [BH] constant, Eq. (5.89) becomes identical with Eq. (5.73) when K_{DA} is replaced by K_{ex}. As expected, the rate of extraction of species present at high dilution in a system of fixed composition can be formally treated as the simple distribution of an unchanged species between two immiscible phases.

The comparison of Eq. (5.89) with Eq. (5.63) also indicates that the rate laws obtained for a system in a kinetic regime controlled by slow two-step interfacial chemical reactions [see Eqs. (5.58) and (5.59)] and for a system in diffusional regime controlled by slow film diffusion processes coupled to an instantaneous reaction occurring at the interface (or in the proximity of the interface) [see Eqs. (5.82a–d) and (5.78)], have the same functional dependence on the concentration variables when low fluxes and low metal concentrations are involved. Alternatively, this result can be formulated by stating that, because of diffusional contributions to the extraction rate, a fast extraction reaction occurring with a simple stoichiometry can mimic a two-step interfacial chemical reaction occurring in absence of diffusional contributions [14,15]. This type of ambiguity, in the past, has led solvent extraction chemists (including this author) to sometimes erroneously derive extraction mechanisms invoking a series of two slow interfacial chemical reactions as rate determining steps in systems for which the two phases were insufficiently stirred. However, a distinction between the two cases can occasionally be made if it is possible to conduct kinetic experiments in which the values of δ_w and δ_o are varied. When the rate is independent of δ_w and δ_o, Eq. (5.63) rather than Eq. (5.89) of this section describes the rate of extraction. Nevertheless, since the film thicknesses never seem to go to zero, the ambiguity may not be resolved.

Film diffusion usually is slower than the rate of many ligand-substitution reactions. Therefore, when rate laws, such as Eqs. (5.63) or (5.89), are found for the extraction kinetics of metal species, preference should, in general, be given to an interpretation in which the determining step of the extraction reaction is film diffusion.

When the fast reactions occurring in the system have stoichiometries different from the simple one shown by Eq. (5.78), analytical solutions of the diffusion equations are difficult to obtain. Nevertheless, numerical solutions can be obtained by iterative routines, and the results are conceptually similar to those described. The additional complications introduced by non–steady-state diffusion and nonlinear concentration gradients can be similarly handled.

5.9 MIXED DIFFUSIONAL–KINETIC REGIME

A mixed diffusional–kinetic regime occurs when one or more of the chemical steps of the reaction mechanism proceed with a velocity that is comparable with that of the diffusional processes through the interfacial films. The other two regimes are thus limiting cases of this more general one. Up to a certain limit, the diffusional processes in an isothermal system can be accelerated by stirring, which reduces the thickness of the diffusion films, and the rate of chemical reactions can be increased by increasing the concentrations of the reactants. The range of experimental conditions that determine the nature of the extraction regime is specific for each chemical system and depends on the hydrodynamic conditions of the apparatus used for the extraction.

The analytical description of a mixed regime is rather complicated, as all difficulties and ambiguities associated with both the kinetic and diffusional regimes may be present. To fully describe the extraction kinetics, it is necessary to solve simultaneously the equations of diffusion and of chemical kinetics. Furthermore, the slow chemical reactions can be of two types: homogeneous, wherein the chemical changes occur in the entire volume of a phase or in a thin layer, generally not too far from the interface; and heterogeneous, wherein the chemical changes occur at the interface. The description of a diffusion regime coupled to a homogeneous chemical reaction is particularly complicated, since the differential equations describing the rate of formation of new chemical species have to be solved simultaneously with the system of differential equations describing the diffusional processes. On the other hand, when the slow chemical reactions are heterogeneous (i.e., interfacial), the rates of the chemical reactions appear in the boundary conditions of the differential diffusion equations.

A mathematically simple case, that occurs frequently in solvent extraction systems, in which the extracting reagent exhibits very low water solubility and is strongly adsorbed at the liquid interface, is illustrated. Even here, the interpretation of experimental extraction kinetic data occurring in a mixed extraction regime usually requires detailed information on the boundary conditions of the diffusion equations (i.e., on the rate at which the chemical species appear and disappear at the interface).

Thus, the chemical kinetics at the interface must be known from separate experiments carried out in a pure kinetic regime. In fact, although the problem has some similarities with that of a diffusional regime associated with a fast interfacial chemical reaction, for which the equilibrium law was set up as an interfacial boundary condition to the diffusion of the species, here, the further complication exists that the interfacial chemical rate laws cannot be a priori known and have to be themselves derived from kinetic experiments. This difficulty can sometimes be avoided by assuming that the interfacial rate laws are simply related to the reaction stoichiometry or by trying to derive information on the interfacial reactions through fitting, to experimental data obtained in a mixed regime, analytical solutions of differential equations that take into account both diffusion and chemical reactions. Unfortunately, both of these procedures can lead to erroneous interpretations, as in only a few cases can the rate laws be correctly derived from stoichiometric considerations, and when too many variables (e.g., rate constants, rate laws, and diffusional parameters) are simultaneously adjusted to fit experimental data, many alternative models can usually satisfy the same equations. Therefore, it is important that the boundary condition of the diffusion equations (i.e., the interfacial rate laws) be derived by separate, suitable experiments.

We now treat, in some detail, under the usual assumptions of steady-state diffusion and linear concentration gradients through the diffusion films, only one simple case of mixed regime. The case deals with a situation that is an extension of case 4 described in section 5.8. Here the additional complication of a slow interfacial reaction is introduced.

Case 6: *The interfacial partition reaction between the two phases of uncharged species is slow. The rate is controlled both by the slow partition reaction and by diffusion to and away from the interface of the partitioning species.*

Although the extraction process is still described by Eq. (5.65), the slow interfacial chemical reaction and the corresponding interfacial flux, J_i, are

$$A_i \underset{k_{-1}}{\overset{k_1}{\longleftrightarrow}} A_i(\text{org}) \tag{5.91}$$

$$J_i = k_1[A]_i - k_{-1}[\overline{A}]_i \tag{5.92}$$

Following the same derivation as for case 4 of section 5.8 and referring again to Fig. (5.8) (here, however, the concentrations $[A]_i$ and $[A]_i$ are at equilibrium only at the end of the extraction process), we obtain for the interfacial concentrations, equations identical with Eqs. (5.71) and (5.72). By substituting them into Eq. (5.92), keeping into account that at the steady state

$$J_i = J_w = J_o = J \tag{5.93}$$

and solving for J, we obtain

rate $= -d[A]/dt =$ rate (forward) $-$ rate (reverse) $=$

$$= S\, k_1[A](k_1\Delta_w + k_{-1}\,\Delta_o + 1)^{-1} - S\, k_{-1}[\overline{A}](k_1\Delta_w + k_{-1}\Delta_o + 1)^{-1} \qquad (5.94)$$

A comparison between Eq. (5.94) and Eq. (5.73) indicates that the presence of a slow interfacial chemical reaction shows up as an additional term in the denominator of the rate laws. Nevertheless, the same functional dependence on S, $[A]$, and $[\overline{A}]$ is exhibited by Eq. (5.73) and Eq. (5.94). This means that the rate law does not allow choice between partition kinetics controlled only by diffusion, or only by a slow partition reaction, or by a combination of the two.

When the rate constants are large numbers (fast reactions), both the numerators and the denominators of Eq. (5.94) can be divided by k_{-1}. Since $k_1/k_{-1} = K_{DA}$ and $(k_{-1})^{-1}$ can be neglected relative to the other terms in the sum ($K_{DA}\,\Delta_w + \Delta_o + k^{-1}$), an equation identical with Eq. (5.73) is obtained for which only diffusion controls the partition rate. At the other extreme, when interfacial film diffusion is very fast, and the reactions are very slow, we can set $1 \gg (k_1\Delta_w + k_{-1}\,\Delta_o)$ and the rate equation becomes equal to

$$\text{rate} = S\, k_1\,[A] - S\, k_{-1}\,[A] \qquad (5.95)$$

that is, the partition rate is controlled only by the slow chemical reaction.

5.10 EXPERIMENTAL TECHNIQUES

Several experimental techniques are available for measuring the kinetics of solvent extraction. They differ in the different efficiency of stirring which can be achieved in the two phases, and in the control of the hydrodynamic conditions in proximity to the interface and in the interfacial area. Most of these techniques can be grouped into five categories.

1. Highly stirred vessels: Those techniques in which the two phases are so highly stirred that droplets of one phase are dispersed in a continuum of the second phase (the dispersing phase).
2. Constant interfacial-area-stirred cells: Those techniques in which the two phases can be stirred to a variable extent, while a known and constant interfacial area is maintained.
3. Rotating diffusion cells: These techniques bring the two phases in contact at the surface of a rotating membrane filter having well characterized hydrodynamics.
4. Moving drops: These techniques measure the extraction that takes place when a single droplet of one phase travels along a vertical tube filled with the second phase.
5. Short-time phase contact methods: These techniques do not stir the two phases, and the contact time between them is very short.

5.10.1 Highly Stirred Vessels

Different versions of highly stirred vessels are used in kinetic studies. The common feature is that samples are removed from the extraction apparatus at different times to measure concentrations. The extraction occurs under conditions similar to those that are met in practical technological or analytical extractions.

Figure 5.10 shows an advanced version of this technique [16]. The apparatus incorporates a continuous monitoring of the rate and permits instantaneous data analyses that are accomplished by the use of a Teflon phase separator and an online microcomputer. The device allows measurement of extraction rates with half-lives as short as 10 s.

When the two phases are so highly stirred that one phase is dispersed into the other in the form of minute droplets, it is generally difficult to control the interfacial area through which the extraction takes place. This constitutes the major drawback to the use of this technique to derive unequivocal information

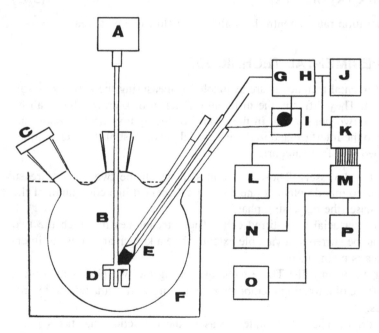

Fig. 5.10 Computer-assisted extraction kinetics-measuring apparatus for highly stirred phases: (A) high-speed stirrer; (B) stirrer shaft; (C) sample inlet; (D) Teflon stirring bar; (E) Teflon phase separator; (F) water bath; (G) flow-cell; (H) spectrophotometer; (I) peristaltic pump; (J) chart recorder; (K) A/D converter; (L) clock; (M) minicomputer; (N) dual-floppy disk drive; (O) printer, (P) plotter. (From Ref. 16.)

on the extraction regime. Typical drop diameters are 200 μm, that is, about 10–20 times larger than the claimed minimum thickness of the diffusion films usually reached in well-stirred biphasic systems.

5.10.2 Constant Interfacial-Area-Stirred Cells

In this technique, the two phases are stirred while the aqueous and the organic layers are positioned one on top of the other and the transfer occurs through the contact area [17–19]. The interfacial area, which coincides with the geometric horizontal section at which the contact between the two phases is established, is always kept constant, whereas the degree of stirring of the two phases is varied over the widest possible range. The major limitations of the technique is that the motion of the liquids in proximity of the interface and the interfacial turbulence are not well defined and vary with the equipment. Consequently, poorly defined diffusion films result. Nevertheless, experience has shown that by using well-designed equipment, reproducible data can be obtained, and film diffusion can be treated quantitatively by using the two-film theory. With these cells, ultimate film thicknesses of about 10^{-3} cm can be reached. Figure 5.11 shows a version of a constant interfacial-area-stirred cell.

Several different versions of extraction cells are used. The main difference between the various cells lies in the presence or absence of internal baffles that modify the internal forced convection, in the shape and dimensions of the stirrers that drive the forced convection, and in the presence of grids (or screens) in proximity to the interface that help to stabilize the interface when the two

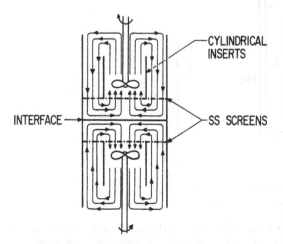

Fig. 5.11 Constant interfacial–area-stirred cell ("Lewis-type cell") for liquid–liquid extraction kinetic measurements.

phases are stirred. In presence of these baffles and grids, the range of variation of the stirring speed of the phases can be quite wide. The grids (or screens) transform most of the translation kinetic energy of the liquids into turbulent energy, preventing the rippling of the interface. The cells allow an independent change of the revolution number of the stirrers in the two phases. In this way, the different influence of the density and viscosity of the two phases can be taken into account, and the organic and aqueous phase can be stirred in such a way to have similar fluid dynamic conditions.

5.10.3 The Rotating Diffusion Cell

In this cell, the aqueous and the organic phase are brought in contact at the surface of a microporous membrane filter, the pores of which are filled with the organic phase [20]. The microporous filter is attached to a hollow cylinder filled with the same organic phase. The cylinder is dipped into the aqueous phase and is rotated by a motor. A scheme of the rotating diffusion cell is presented in Fig. 5.12. The cell has a filter with well-defined hydrodynamics on both

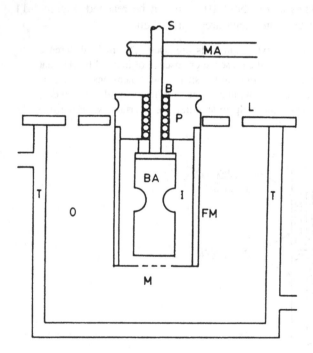

Fig. 5.12 Cross section of the rotating diffusion cell: B, bearing; BA, internal baffle; FM, filter mount; I, inner compartment; L, lid with holes; M, membrane; MA, mounting rod; O, outer compartment; P, pulley block; S, hollow rod; T, thermostated beaker. (From Ref. 20.)

sides. This allows description of the diffusional processes of the reactants and products of the extraction reactions on the aqueous and the organic side of the interface.

> The basic assumption is that the rotating filter creates a laminar flow field that can be completely described mathematically. The thickness of the diffusion boundary layer (δ) is calculated as a function of the rotational speed (ω), viscosity, density, and diffusion coefficient (D). The thickness is expressed by the Levich equation, originally derived for electrochemical reactions occurring at a rotating disk electrode:

$$\delta = 0.643 \, \omega^{-1/2} D^{1/3} v^{1/6} \tag{5.96}$$

> v is the kinematic viscosity. The organic phase inside the membrane is considered at rest, and the membrane thickness and porosity control the diffusion rate. The major drawbacks of this technique are difficulties in well characterizing the thickness and porosity of the membrane filter, in utilizing really thin filters (less than 10^{-3} cm), and that, although the hydrodynamics inside the rotating cylinder can be assimilated to those of a rotating disk, no studies have conclusively shown that the flows inside and outside the cylinder are the same.

A more recent version of this cell [21] makes use of a rotating capillary, containing a gelified aqueous phase. This eliminates the microporous membrane filter.

5.10.4 Moving Drops

In the moving drop technique (also described in Chapters 7 and 9), a drop of the organic or aqueous phase is produced at the end of a vertical column filled with the other phase. The drop travels along the tube, during which extraction occurs across the drop surface. By measuring the time of traveling, the drop size, and from the volume of collected drops, it is possible to evaluate the rate of extraction (see Chapter 9 for a detailed discussion of drop behavior and mass transfer).

> The technique can be applied to drops that move both upward and downward, the choice depending on the higher or lower density of the drops relative to the continuum liquid that fills the column. Although the technique allows one to control both drop size and interfacial area, there are some disadvantages that impose the use of great care both in the treatment of the experimental data and in the design of the apparatus. When the extraction is very rapid, much of the extraction can take place during drop formation. Further extraction can also take place in the stagnant pool where the drops are collected. On the other hand, when the extraction is very slow, very long columns are required to assure sufficient drop residence times. The hydrodynamic control of the single drops can also be very difficult because of lack of internal circulation and the presence of drop oscillations. When the drops are quite small and the

extractant adsorbs on the drop surface, the internal circulation can be absent or strongly reduced, and it is difficult to establish whether the degree of turbulence of the two phases in the moving drop method can ever be enough to ensure the measurement of a transfer rate in a kinetic regime or, on the opposite side, is so low that the extraction rate is always occurring in a diffusional regime.

5.10.5 Short-Time Phase-Contact Method

This method, which has been developed and extensively utilized in Russia [22,23], deals with nonstirred liquid phases and is based on measurements of the electrical conductivity of the aqueous phase and of its changes during extraction. In the experimental device, the two phases are initially kept separated inside a thermostated cell and then instantaneously contacted through a known contact area. At this moment, the concentration variation in one phase is recorded as a function of time, generally using conductometry.

> The aqueous phase is usually contained in a metal capillary. The capillary is pushed into the lower part of the cell, containing the organic phase, at time zero of the extraction. The method examines the reaction during very short time intervals (from 0.05–1 s) and is relatively simple, as the two phases are not stirred. The consumption of the reactants is very small (about 20 mm^3 per experiment) and permits one to make many determinations in a relatively short time (about 10 kinetic curves per hour). Unfortunately, the method proves inapplicable for concentrated aqueous solutions and when many species are involved in the extraction reaction.

This technique is based on a non–steady-state approach to diffusion. The non–steady-state diffusion is not treated in our simplified approach to extraction kinetics; the interested reader is referred to Yagodin and Tarasov [24] and Tarasov and Yagodin [25].

SYMBOLS

Parameter	Denoted by	First shown, in text order
Aqueous phase	No symbol, or index w	General
Organic phase	Symbols org, index o (not to be confused with zero 0) or a bar above the variable, constant, etc.	Eq. (5.27)
J	Diffusion flow, mass cm^{-2} s^{-1}	Eq. (5.1)
D	Diffusion coefficient, cm^{-2} s^{-1}	Eq. (5.1)
c	Concentration, mol cm^3	Eq. (5.1), Fig. 5.2
x	Diffusion coordinate, cm	Eq. (5.1)

Parameter	Denoted by	First shown, in text order
∂	Derivative	Eq. (5.1)
d	Also used for derivative	Eq. (5.3)
δ	Thickness of organic δ_o or aqueous film δ_w at interface	Eq. (5.3), Figs. 5.1 and 5.2
Δ	Diffusional parameter, $=\delta/D$, cm^{-1} s	Eq. (5.4)
R	Film resistance, transfer resistance, for indexed phase	Fig. 5.1, Eq. (5.64)
V	Volume of indexed reservoir	Eq. (5.5), Fig. 5.2
Q	Area of diffusion film, cm^2	Eq. (5.5)
γ	Interfacial tension, dynes cm^{-1}	Section 5.5
γ_2	Langmuir adsorption constant	Eq. (5.40)
Π	Inerfacial pressure, dynes cm^{-1}	Eq. (5.26)
k	Boltzmann constant	Eq. (5.26)
T	Temperature, kelvin	Eq. (5.26)
t	Time, seconds, s	Eq. (5.5)
	index 0 at beginning of experiment	Eq. (5.36)
	index eq at equilibrium	Eq. (5.28), (5.36)
X, Y, Z, W	Abbreviations for chemical compounds	Eq. (5.10)
k	Rate constant; k_{-1} rate constant of reverse reaction	Eq. (5.13), (5.29)
M^+	Metal cation (central atom in complex MB, or ML)	Eq. (5.22), (5.27)
L^-	Ligand anion (ligand in complex ML)	Eq. (5.22)
BH	Extractant	Eq. (5.27), (5.32)
B^-	(Organic) anion (ligand in complex MB)	Eq. (5.29)
K	Equilibrium constant (index D for distribution)	Eq. (5.18)
	index ass for association	Eq. (5.23)
	index ex for (overall) extraction	Eq. (5.28)
	index DB, distribution constant of extractant BH	Eq. (5.31)
	index a, acid dissociation constant of HB	Eq. (5.32)
	index DM distribution constant of complex MB	Eq. (5.33)
	index MB complex formation constant of MB	Chapter 3
ad	Index for species adsorbed at the interface	Eq. (5.37)
A	Any solute distributing between two phases	Eq. (5.65)

Parameter	Denoted by	First shown, in text order
α_2	Langmuir adsorption coefficient	Eq. (5.40)
S	Specific interfacial area, cm^{-1}	Eq. (5.38)

REFERENCES

1. Whitman, W. G. *Chem. Metal Eng.*, **1923**, *29*(4):146.
2. Whitman, W. G.; Davies, D. S. *Ind. Eng. Chem.*, **1924**, *16*:1233.
3. Bird, R. B.; Stewart, W. E.; Lightfoot, E. N. *Transport Phenomena*, John Wiley and Sons, New York, 1960.
4. Cussler, E. L. *Diffusion, Mass Transfer in Fluid Systems*, Cambridge University Press, Cambridge, 1984.
5. Astarita, G. *Mass Transfer with Chemical Reaction*, Elsevier, Amsterdam, 1967.
6. Frank-Kamenetskii, D. A. *Diffusion and Heat Transfer in Chemical Kinetics*, Plenum Press, New York, 1969.
7. Frost, A. A.; Pearson, R. G. *Kinetics and Mechnanism*, Wiley-Interscience, New York, 1956.
8. Moelwin-Hughes, E. A. *The Kinetics of Reactions in Solution*, Oxford University Press, London, 1950.
9. Eigen, M.; Wilkins, R. W. *Mechanisms of Inorganic Reactions*, Vol. 49, *Adv. Chem. Ser.*, American Chemical Society, Washington, DC, 1965.
10. Adamson, A. W. *Physical Chemistry of Surfaces*, John Wiley and Sons, New York, 1982.
11. Davies, J. T.; Rideal, E. K. *Interfacial Phenomena*, Academic Press, New York, 1961.
12. Danesi, P. R.; Chiarizia, R.; Vandergrift, G. F. *J. Phys. Chem.*, **1980**, *84*:3455.
13. Danesi, P. R.; Chiarizia, R. *The Kinetics of Metal Solvent Extraction*, Crit. Rev. Anal. Chem., **1980**, *10*:1.
14. Danesi, P. R. *Solv. Extr. Ion Exch.* **1984**, 2:29.
15. Danesi, P. R.; Vandegrift, G. E.; Horwitz, P.; Chiarizia, R. *J. Phys. Chem.*, **1980**, *84*:3582.
16. Watarai, H.; Cunningham, L.; Freiser, H. *Anal. Chem.*, **1982**, *54*:2390.
17. Lewis, J. B. *Chem. Eng. Sci.*, **1954**, *3*:218.
18. Nitsch, W.; Kahn, G. *Ger. Chem. Eng. Engl. Trans.*, **1980**, *3*:96.
19. Danesi, P. R.; Cianetti, C.; Horwitz, E. P.; Diamond, H. *Sep. Sci. Technol.* **1982**, *17*.961.
20. Albery, W. J.; Burke, J. E; Lefier, E. B.; Hadgraft, J. *J. Chem. Soc. Faraday Trans.*, **1976**, *72*:1618.
21. Simonin, J. P.; Musikas, C.; Soualhia, E.; Turq, P. *J. Chim. Phys.*, **1987**, *84*:525.
22. Tarasov, V. V.; Yagodin, G. A.; Kizim, N. F. *Dokl. Akad. Nauk. SSR*, **1971**, *45*: 2517.
23. Tarasov, V. V.; Yagodin, G. A. VINITI, *Khim. Teknol. Neorganich. Khim.*, **1974**, *4*.
24. Yagodin, G. A.; Tarasov, V. V. *Proc. Int. Solv. Extr. Conf.*, ISEC 71, Vol. 1,

Society of Chemical Industry, (Gregory, J. G.; Evans, B; Weston, P.C. Eds.), London, p. 888, 1971.

25. Tarasov, V. V.; Yagodin, G. A. *Interfacial Phenomena in Solvent E. Ytraction*. In *Ion Exchange Solvent Extraction*, *Vol.* 10, Ch. 4 (Marinsky, J. A. and Marcus, Y. Eds.), Marcel Dekker, New York, pp. 141–237, 1988.
26. Eigen, M. *Pure Appl. Chem.*, **1963**, 6:97.

6

Ionic Strength Corrections

INGMAR GRENTHE* Royal Institute of Technology,
Stockholm, Sweden

HANS WANNER Swiss Federal Nuclear Safety Inspectorate,
Villingen-HSK, Switzerland

Thermodynamic data always refer to a selected standard state. The definition given by IUPAC [1] is adopted in the NEA-TDB project as outlined in the corresponding guideline [2]. According to this definition, the standard state for a solute B in a solution is a hypothetical solution, at the standard state pressure, in which $m_B = m^o = 1 \, mol \cdot kg^{-1}$, and in which the activity coefficient γ_B is unity. However, for many reactions, measurements cannot be made accurately (or at all) in dilute solutions from which the necessary extrapolation to the standard state would be simple. This is invariably the case for reactions involving ions of high charge. Precise thermodynamic information for these systems can only be obtained in the presence of an inert electrolyte of sufficiently high concentration, ensuring that activity factors are reasonably constant throughout the measurements. The objectives of this chapter are to describe and illustrate various methods of extrapolation to zero ionic strength and to describe the methods preferred in the NEA review along with recommended values of the auxiliary parameters. By following these guidelines, the members of the NEA specialist teams performing the review will be assured of using the same theory consistently for the extrapolation to zero ionic strength.

*Retired.
By kind permission from the OECD NEA-TDB project, reprinted from Report TDB-2, *Guidelines for the extrapolation to zero ionic strength*, OECD Nuclear Energy Agency, Data Bank, F-91191 Gif-Sur Yvette, France. For latest revised version, see *Chemical Thermodynamics of Americium* Silva, R. J., Bidoglio, G., Rand, M. H., Robouch, P. B., Wanner, H., and Puigdomenech, I., Eds., Elsevier/North-Holland 1995.

The activity factors of all the species participating in reactions in high ionic strength media must be estimated in order to reduce the thermodynamic data obtained from the experiment to the standard state ($I = 0$). These estimates are based on the use of extended Debye-Hückel equations, either in the form of "specific ion interaction methods" or the Davies equation. However, the Davies equation (see section 6.2.4) should in general not be used at ionic strengths larger than $0.1 \, \text{mol} \cdot \text{kg}^{-1}$. The following forms of specific ion interaction methods have been elaborated in the past:

1. The Brønsted-Guggenheim-Scatchard approach (abbreviated "B-G-S equation" in this document), (see section 6.1).
2. The Pitzer and Brewer "B-method" (abbreviated "P-B" herein) (see section 6.2.1).
3. The Pitzer virial coefficient method, see section 6.2.2. Methods 1 and 2 are equivalent and differ only in the form of the denominator in the Debye-Hückel term. Method 3 requires more parameters for the description of the activity factors. These parameters are not available in many cases. This is generally the case for complex formation reactions.

The method preferred in the NEA Thermochemical Data Base review is the specific ion interaction model in the form of the Brønsted-Guggenheim-Scatchard approach.

One may sometimes have access to the parameters required for the Pitzer approaches, e.g., for some hydrolysis equilibria and for some solubility product data, cf. Baes and Mesmer [3] and Pitzer [4]. In this case, the reviewer should perform a calculation using both the B-G-S and the P-B equations and the full virial coefficient methods and compare the results.

6.1 THE SPECIFIC ION INTERACTION EQUATIONS

6.1.1 Background

The Debye-Hückel term, which is the dominant term in the expression for the activity coefficients in dilute solution, accounts for electrostatic, nonspecific long-range interactions. At higher concentrations, short-range, nonelectrostatic interactions have to be taken into account. This is usually done by adding ionic strength dependent terms to the Debye-Hückel expression. This method was first outlined by Brønsted [5,6], and elaborated by Scatchard [7] and Guggenheim [8]. The two basic assumptions in the specific ion interaction theory are:

Assumption 1: The activity coefficient γ_j of an ion j of charge z_j in the solution of ionic strength I_m may be described by Eq. (6.1).

$$\log_{10} \gamma_j = -z_j^2 D + \sum_k \varepsilon_{(j,k,I_m)} m_k \tag{6.1}$$

D is the Debye-Hückel term:

$$D = \frac{A\sqrt{I_m}}{1 + Ba_j\sqrt{I_m}} \tag{6.2}$$

A and B are constants that are temperature dependent, and a_j is the effective diameter of the hydrated ion j. The values of A and B as a function of temperature are listed in Table 6.1.

The term Ba_j in the denominator of the Debye-Hückel term has been assigned a value of $Ba_j = 1.5$, as proposed by Scatchard [9] and accepted by Ciavatta [10]. This value has been found to minimize, for several species, the ionic strength dependence of $\varepsilon_{(j,k,I_m)}$ between $I_m = 0.5\,m$ and $I_m = 3.5\,m$. It should be mentioned that some authors have proposed different values for

Table 6.1 Debye-Hückel Constants

$t(°C)$	A	$B(\times 10^{-8})$
0	0.4913	0.3247
5	0.4943	0.3254
10	0.4976	0.3261
15	0.5012	0.3268
20	0.5050	0.3276
25	0.5091	0.3283
30	0.5135	0.3291
35	0.5182	0.3299
40	0.5231	0.3307
45	0.5282	0.3316
50	0.5336	0.3325
55	0.5392	0.3334
60	0.5450	0.3343
65	0.5511	0.3352
70	0.5573	0.3362
75	0.5639	0.3371

Source: Ref. 13.

Ba_j, ranging from $Ba_j = 1.0$ [11] to $Ba_j = 1.6$ [12]. However, the parameter Ba_j is empirical and as such correlated to the value of $\varepsilon_{(j,k,I_m)}$. Hence, this variety of values for Ba_j does not represent an uncertainty range, but rather indicates that several different sets of Ba_j and $\varepsilon_{(j,k,I_m)}$ may describe equally well the experimental mean activity coefficients of a given electrolyte. The ion interaction coefficients listed later in the chapter in Tables 6.3–6.5 have to be used with $Ba_j = 1.5$.

The summation in Eq. (6.1) extends over all ions k present in solution. Their molality is denoted m_k. The concentrations of the ions of the ionic medium is often very much larger than those of the reacting species. Hence, the ionic medium ions will make the main contribution to the value of $\log_{10} \gamma_j$ for the reacting ions. This fact often makes it possible to simplify the summation $\sum_k \varepsilon_{(j,k,I_m)} m_k$ so that only ion interaction coefficients between the participating ionic species and the ionic medium ions are included, as shown in Eqs. (6.5)–(6.9).

Assumption 2: The ion interaction coefficients $\varepsilon_{(j,k,I_m)}$ are zero for ions of the same charge sign and for uncharged species. The rationale behind this is that ε, which describes specific short-range interactions, must be small for ions of the same charge, since they are usually far from one another due to electrostatic repulsion. This holds to a lesser extent also for uncharged species.

Equation (6.1) will allow fairly accurate estimates of the activity coefficients in mixtures of electrolytes if the ion interaction coefficients are known. Ion interaction coefficients for simple ions can be obtained from tabulated data of mean activity coefficients of strong electrolytes or from the corresponding osmotic coefficients. Ion interaction coefficients for complexes can either be estimated from the charge and size of the ion or determined experimentally from the variation of the equilibrium constant with the ionic strength. Ion interaction coefficients are not strictly constant but vary slightly with the ionic strength. The extent of this variation depends on the charge type and is small for 1:1, 1:2, and 2:1, electrolytes for molalities less than 3.5 m. The concentration dependence of the ion interaction coefficients can thus often be neglected. This point was emphasized by Guggenheim [8], who has presented a considerable amount of experimental material supporting this approach. The concentration dependence is larger for electrolytes of higher charge. In order to accurately reproduce their activity coefficient data, concentration-dependent ion interaction coefficients have to be used: (see Pitzer and Brewer [14]; Baes and Mesmer [3]; or Ciavatta [10]). By using a more elaborate virial expansion, Pitzer and coworkers [4, 15–21] have managed to describe measured activity coefficients of a large number of electrolytes with high precision over a large concentration range. Pitzer's model generally contains three parameters, or more, as compared to one in the specific ion interaction model. The use of the model requires the knowledge of all these parameters. The derivation of Pitzer coefficients for many metal complexes would require a very large amount of additional experimental work, since no data of this type are currently available.

The way in which the activity coefficient corrections are performed according to the specific ion interaction model is illustrated below for a general case of a complex formation reaction. Charges are omitted for brevity.

$$m\text{M} + q\text{L} + n\text{H}_2\text{O}(1) \rightleftharpoons \text{M}_m\text{L}_q(\text{OH})_n + n\text{H}^+ \tag{6.3}$$

The formation constant of $\text{M}_m\text{L}_q(\text{OH})_p$, $*\beta_{n,q,m}$, determined in an ionic medium (1:1 salt NX) of the ionic strength I_m, is related to the corresponding value at zero ionic strength, $*\beta^o_{n,q,m}$, by Eq. (6.4).

$$\log{}^*_{10}\beta_{n,q,m} = \log{}^*_{10}\beta^o_{n,q,m} + m\log_{10}\gamma_\text{M} + q\log_{10}\gamma_\text{L} + n\log_{10}\alpha_{\text{H}_2\text{O}}$$
$$- \log_{10}\gamma_{n,q,m} - n\log_{10}\gamma_\text{H} \tag{6.4}$$

The subscript (n,q,m) denotes the complex ion $\text{M}_m\text{L}_q(\text{OH})_n$. If the concentrations of N and X are much greater than the concentrations of M, L, $\text{M}_m\text{L}_q(\text{OH})_n$ and H, only the molalities m_N and m_X have to be taken into account for the calculation of the term $\sum_k \varepsilon_{(j,k,I_m)} m_k$ in Eq. (6.1). For example, for the activity coefficient of the metal cation M, γ_M, Eq. (6.5) is obtained.

$$\log_{10}\gamma_\text{M} = \frac{-z_\text{M}^2 0.5091\sqrt{I_m}}{1 + 1.5\sqrt{I_m}} + \varepsilon_{(\text{M,X},I_m)} m_\text{X} \tag{6.5}$$

Under these conditions, $I_m \approx m_\text{X} = m_\text{N}$. Substituting the $\log_{10}\gamma_j$ values in Eq. (6.4) with the corresponding forms of Eq. (6.5) and rearranging leads to

$$\log{}^*_{10}\beta_{n,q,m} - \Delta z^2 D - n\log_{10}a_{\text{H}_2\text{O}} = \log{}^*_{10}\beta^o_{n,q,m} - \Delta\varepsilon I_m \tag{6.6}$$

where

$$\Delta z^2 = (mz_\text{M} - qz_\text{L} - n)^2 + n - mz_\text{M}^2 - qz_\text{L}^2 \tag{6.7}$$

$$D = \frac{0.5091\sqrt{I_m}}{1 + 1.5\sqrt{I_m}} \tag{6.8}$$

$$\Delta\varepsilon = \varepsilon_{(n,q,m,\text{N or X})} + n\varepsilon_{(\text{H,X})} - q\varepsilon_{(\text{N,L})} - m\varepsilon_{(\text{M,X})} \tag{6.9}$$

Here $(mz_\text{M} - qz_\text{L} - n)$, z_M and z_L are the charges of the complex $\text{M}_m\text{L}_q(\text{OH})_n$, the metal ion M and the ligand L, respectively. Equilibria involving $\text{H}_2\text{O}(1)$

as a reactant or product require a correction for the activity of water, a_{H_2O}. The activity of water in an electrolyte mixture can be calculated as

$$\log_{10} a_{H_2O} = \frac{-\Phi \sum_k m_k}{\ln(10) \times 55.51} \tag{6.10}$$

where Φ is the osmotic coefficient of the mixture and the summation extends over all ions k with molality m_k present in the solution. In the presence of an ionic medium NX in dominant concentration, Eq. (6.10) can be simplified by neglecting the contributions of all minor species, i.e., the reacting ions. Hence, for a 1:1 electrolyte of ionic strength $I_m \approx m_{NX}$, Eq. (6.10) becomes

$$\log_{10} a_{H_2O} = \frac{-2m_{NX}\Phi}{\ln(10) \times 55.51} \tag{6.11}$$

Values of osmotic coefficients for single electrolytes have been compiled by various authors, e.g., Robinson and Stokes [22]. The activity of water can also be calculated from the known activity coefficients of the dissolved species.

In the presence of an ionic medium NX of a concentration much larger than those of the reacting ions, the osmotic coefficient can be calculated according to Eq. (6.12) [23].

$$1 - \Phi = \frac{A\ln(10)|z_+ z_-|}{m_{NX} \times (1.5)^3}\left[1 + 1.5\sqrt{m_{NX}} - 2\log_{10}(1 + 1.5\sqrt{m_{NX}})\right.$$
$$\left. - \frac{1}{1 + 1.5\sqrt{m_{NX}}}\right] + \frac{1}{\ln(10)}\varepsilon_{(N,X)}m_{NX} \tag{6.12}$$

The activity of water is obtained by inserting Eq. (6.12) into Eq. (6.11). It should be mentioned that in mixed electrolytes with several components at high concentrations, it is necessary to use Pitzer's equation to calculate the activity of water. On the other hand, a_{H_2O} is near constant (and $= 1$) in most experimental studies of equilibria in dilute aqueous solutions, where an ionic medium is used in large excess with respect to the reactants. The ionic medium electrolyte thus determines the osmotic coefficient of the solvent.

In natural waters the situation is similar; the ionic strength of most surface waters is so low that the activity of $H_2O(l)$ can be set equal to unity. A correction may be necessary in the case of seawater, where a sufficiently good approximation for the osmotic coefficient may be obtained by considering NaCl as the dominant electrolyte.

In more complex solutions of high ionic strengths with more than one electrolyte at significant concentrations, e.g., (Na^+, Mg^{2+}, Ca^{2+}) (Cl^-, SO_4^{2-}), Pitzer's equation may be used to estimate the osmotic coefficient; the necessary interaction coefficients are known for most systems of geochemical interest.

Note that in all ion interaction approaches, the equation for mean activity coefficients can be split up to give equations for conventional single ion activity coefficients in mixtures, e.g., Eq. (6.1). The latter are strictly valid only when used in combinations that yield electroneutrality. Thus, while estimating medium effects on standard potentials, a combination of redox equilibria with $H^+ + e^- \rightleftharpoons \frac{1}{2} H_2(g)$ is necessary (see Example 3).

6.1.2 Estimation of Ion Interaction Coefficients

6.1.2.1 Estimation from Mean Activity Coefficient Data

Example 1: The ion interaction coefficient $\varepsilon_{(H^+,Cl^-)}$ can be obtained from published values of $\gamma_{\pm,HCl}$ vs. m_{HCl}.

$$2 \log_{10} \gamma_{\pm,HCl} = \log_{10} \gamma_{+,H^+} + \log_{10} \gamma_{-,Cl^-}$$
$$= -D + \varepsilon_{(H^+,Cl^-)} m_{Cl^-} - D + \varepsilon_{(Cl^-,H^+)} m_{H^+}$$
$$\log_{10} \gamma_{\pm,HCl} = -D + \varepsilon_{(H^+,Cl^-)} m_{HCl}$$

By plotting $\log_{10} \gamma_{\pm,HCl} + D$ vs. m_{HCl}, a straight line with the slope $\varepsilon_{(H^+,Cl^-)}$ is obtained. The degree of linearity should in itself indicate the range of validity of the specific ion interaction approach. Osmotic coefficient data can be treated in an analogous way.

6.1.2.2 Estimations Based on Experimental Values of Equilibrium Constants at Different Ionic Strengths

Example 2: Equilibrium constants are given in Table 6.2 for the reaction

$$UO_2^{2+} + Cl^- \rightleftharpoons UO_2Cl^+ \tag{6.13}$$

The following formula is deducted from Eq. (6.6) for the extrapolation to $I = 0$:

$$\log_{10} \beta_1 + 4D = \log_{10} \beta_1^\circ - \Delta \varepsilon I_m \tag{6.14}$$

The linear regression is done as described in the NEA Guidelines for the Assignment of Uncertainties [24]. The following results are obtained:

$$\log_{10} \beta_1^\circ = 0.170 \pm 0.021$$
$$\Delta \varepsilon(13) = -0.248 \pm 0.022$$

Table 6.2 The Preparation of the Experimental
Equilibrium Constants for the Extrapolation to $I = 0$ with
the Specific Ion Interaction Method, According to
Eq. (6.13)

I_m	$\log_{10}\beta_1(\exp)^a$	$\log_{10}\beta_{1,m}{}^b$	$\log_{10}\beta_{1,m} + 4D$
0.1	-0.17 ± 0.10	-0.174	0.264 ± 0.100
0.2	-0.25 ± 0.10	-0.254	0.292 ± 0.100
0.26	-0.35 ± 0.04	-0.357	0.230 ± 0.040
0.31	-0.39 ± 0.04	-0.397	0.220 ± 0.040
0.41	-0.41 ± 0.04	-0.420	0.246 ± 0.040
0.51	-0.32 ± 0.10	-0.331	0.371 ± 0.100
0.57	-0.42 ± 0.04	-0.432	0.288 ± 0.040
0.67	-0.34 ± 0.04	-0.354	0.395 ± 0.040
0.89	-0.42 ± 0.04	-0.438	0.357 ± 0.400
1.05	-0.31 ± 0.10	-0.331	0.491 ± 0.100
1.05	-0.277 ± 0.260	-0.298	0.525 ± 0.260
1.61	-0.24 ± 0.10	-0.272	0.618 ± 0.100
2.21	-0.15 ± 0.10	-0.193	0.744 ± 0.100
2.21	-0.12 ± 0.10	-0.163	0.774 ± 0.100
2.82	-0.06 ± 0.10	-0.021	0.860 ± 0.100
3.5	0.04 ± 0.10	-0.021	0.974 ± 0.100

[a]Equilibrium constants for Eq. (6.13) with assigned uncertainties,
corrected to 25°C where necessary.
[b]Equilibrium constants corrected from molarity to molality units,
according to the procedure described in Ref. 2.
Note: The Linear Regression of this Set of Data is shown in
Fig. 6.1.

The experimental data are depicted in Fig. 6.1, where the area between the
dashed lines represents the uncertainty range that is obtained by using the
results in $\log_{10}\beta_1^\circ$ and $\Delta\varepsilon$ and correcting back to $I \neq 0$.

Example 3: When using the specific ion interaction theory, the relationship
between the normal potential of the redox couple UO_2^{2+}/U^{4+} in a medium
of ionic strength I_m and the corresponding quantity at $I = 0$ should be
calculated in the following way. The reaction in the galvanic cell

$$Pt, H_2|H^+\|UO_2^{2+}, U^{4+}, H^+|Pt \tag{6.15}$$

is

$$UO_2^{2+} + H_2(g) + 2H^+ \rightleftharpoons U^{4+} + 2H_2O \tag{6.16}$$

For this reaction

$$\log_{10} K^{\circ} = \log_{10} \left(\frac{a_{U^{4+}} \times a_{H_2O}^2}{a_{UO_2^{2+}} \times a_{H^+}^2 \times f_{H_2}} \right) \tag{6.17}$$

$$\log_{10} K^{\circ} = \log_{10} K + \log_{10} \gamma_{U^{4+}} - \log_{10} \gamma_{UO_2^{2+}} - 2\log_{10} \gamma_{H^+} \\ - \log_{10} \gamma_{f,H_2} + 2\log_{10} a_{H_2O}, \tag{6.18}$$

$f_{H_2} \approx p_{H_2}$ at reasonably low partial pressure of $H_2(g)$, $a_{H_2O} \approx 1$, and

$$\log_{10} \gamma_{U^{4+}} = -16D + \varepsilon_{(U^{4+}, ClO_4^-)} m_{ClO_4^-} \tag{6.19}$$

$$\log_{10} \gamma_{UO_2^{2+}} = -4D + \varepsilon_{(UO_2^{2+}, ClO_4^-)} m_{ClO_4^-} \tag{6.20}$$

$$\log_{10} \gamma_{H^+} = -D + \varepsilon_{(H^+, ClO_4^-)} m_{ClO_4^-} \tag{6.21}$$

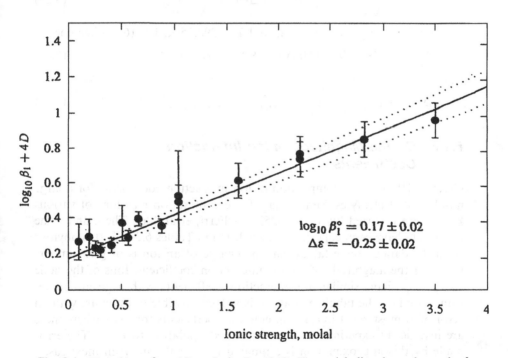

Fig. 6.1 Plot of $\log_{10} \beta_1 + 4D$ vs. I_m for Eq. (6.13). The straight line shows the result of the weighted linear regression, and the area between the dashed lines represents the uncertainty range of $\log_{10} \beta_1^{\circ}$ and $\Delta\varepsilon$.

Hence,

$$\log_{10}K^\circ = \log_{10}K - 10D + (\varepsilon_{(U^{4+},ClO_4^-)} - \varepsilon_{(UO_2^{2+},ClO_4^-)}$$
$$- 2\varepsilon_{(H^+,ClO_4^-)})m_{ClO_4^-} \tag{6.22}$$

The relationship between the equilibrium constant and the standard potential is

$$\ln K = \frac{nF}{RT}E \tag{6.23}$$

$$\ln K^\circ = \frac{nF}{RT}E^\circ \tag{6.24}$$

E is the standard potential in a medium of ionic strength I, E° is the corresponding quantity at $I = 0$, and n is the number of transferred electrons in the reaction considered. Combining Eqs. (6.22)–(6.24) and rearranging them leads to Eq. (6.25).

$$E - 10D\left(\frac{RT\ln(10)}{nF}\right) = E^\circ - \Delta\varepsilon m_{ClO_4^-}\left(\frac{RT\ln(10)}{nF}\right) \tag{6.25}$$

For $n = 2$ in the present example and $T = 298.15\,K$, Eq. (6.25) becomes

$$E[mV] - 295.8D = E^\circ[mV] - 29.58\Delta\varepsilon m_{ClO_4^-} \tag{6.26}$$

where

$$\Delta\varepsilon = (\varepsilon_{(U^{4+},ClO_4^-)} - \varepsilon_{(UO_2^{2+},ClO_4^-)} - 2\varepsilon_{(H^+,ClO_4^-)}) \tag{6.27}$$

6.1.3 On the Magnitude of Ion Interaction Coefficients

Ciavatta [10] made a compilation of ion interaction coefficients for a large number of electrolytes. Similar data for complexation reactions of various kinds were reported by Spahiu [25] and Ferri, et al. [26]. These and some other data have been collected and are listed in Tables 6.3–6.6. It is obvious from the data in these tables that the charge of an ion is of great importance for the magnitude of the ion interaction coefficient. Ions of the same charge type have similar ion interaction coefficients with a given counterion. Based on the tabulated data, it is judged possible to estimate, with an error of at most ± 0.1 in ε, ion interaction coefficients for cases where there are insufficient experimental data for an extrapolation to $I = 0$. The error made by this approximation is estimated to be ± 0.1 in $\Delta\varepsilon$ in most cases, based on comparison with $\Delta\varepsilon$ values of various reactions of the same charge type.

Table 6.3 Ion Interaction Coefficients $\varepsilon_{j,k}$ for Cations j with $k = Cl^-$, ClO_4^-, and NO_3^{-a}

$j \quad k\rightarrow$	Cl^-	ClO_4^-	NO_3^-
H^+	0.12 ± 0.01	0.14 ± 0.02	0.07 ± 0.01
NH_4^+	-0.01 ± 0.01	-0.08 ± 0.04^b	-0.06 ± 0.03^b
$ZnHCO_3^+$	0.2^c	—	—
$CdCl^+$	—	0.25 ± 0.02	—
CdI^+	—	0.27 ± 0.02	—
$CdSCN^+$	—	0.31 ± 0.02	—
$HgCl^+$	—	0.19 ± 0.02	—
Cu^+	—	0.11 ± 0.01	—
Ag^+	—	0.00 ± 0.01	-0.12 ± 0.05^b
UO_2^+	—	0.26 ± 0.03^d	—
UO_2OH^+	—	-0.06 ± 3.7^d	0.51 ± 1.4^d
$(UO_2)_3(OH)_5^+$	0.81 ± 0.17^d	0.45 ± 0.15^d	0.41 ± 0.22^d
UF_3^+	0.1 ± 0.1^e	0.1 ± 0.1^e	—
UO_2F^+	0.04 ± 0.07^f	0.29 ± 0.05^d	—
UO_2Cl^+	—	0.33 ± 0.04^d	—
$UO_2ClO_3^+$	—	0.33 ± 0.04^e	—
UO_2Br^+	—	0.24 ± 0.04^e	—
$UO_2NO_3^+$	—	0.33 ± 0.04^e	—
UO_2SCN^+	—	0.22 ± 0.04^e	—
NpO_2^+	—	0.25 ± 0.05^f	—
PuO_2^+	—	0.17 ± 0.05^f	—
$AlOH^{2+}$	0.09^g	0.31^g	—
Pb^{2+}	—	0.15 ± 0.02	-0.20 ± 0.12^b
Zn^{2+}	—	0.33 ± 0.03	0.16 ± 0.02
$ZnCO_3^{2+}$	0.35 ± 0.05^c	—	—
Cd^{2+}	—	—	0.09 ± 0.02
Hg^{2+}	—	0.34 ± 0.03	-0.1 ± 0.1^b
Hg_2^{2+}	—	0.09 ± 0.02	-0.2 ± 0.1^b
Cu^{2+}	0.08 ± 0.01	0.32 ± 0.02	0.11 ± 0.01
Ni^{2+}	0.17 ± 0.02	—	—
Co^{2+}	0.16 ± 0.02	0.34 ± 0.03	0.14 ± 0.01
$FeOH^{2+}$	—	0.38^h	—
$FeSCN^{2+}$	—	0.45^h	—
Mn^{2+}	0.13 ± 0.01	—	—
UO_2^{2+}	0.21 ± 0.02^i	0.46 ± 0.03	0.24 ± 0.03^i
UF_2^{2+}	—	0.3 ± 0.1^e	—
USO_4^{2+}	—	0.3 ± 0.1^e	—
$U(NO_3)_2^{2+}$	—	0.49 ± 0.14^j	—
Mg^{2+}	0.19 ± 0.02	0.33 ± 0.03	0.17 ± 0.01
Ca^{2+}	0.14 ± 0.01	0.27 ± 0.03	0.02 ± 0.01
Ba^{2+}	0.07 ± 0.01	0.15 ± 0.02	-0.28 ± 0.03

(*continued*)

Table 6.3   (*continued*)

$\frac{j \; k\rightarrow}{i}$	Cl^-	ClO_4^-	NO_3^-
Al^{3+}	0.33 ± 0.02	—	—
Fe^{3+}		0.56 ± 0.03	0.42 ± 0.08
Cr^{3+}	0.30 ± 0.03	—	0.27 ± 0.02
La^{3+}	0.22 ± 0.02	0.47 ± 0.03	—
$La^{3+}\rightarrow Lu^{3+}$	—	$0.47\rightarrow0.52^{h}$	—
UOH^{3+}	—	0.48 ± 0.08^{j}	—
UF^{3+}	—	0.48 ± 0.08^{e}	—
UCl^{3+}	—	0.59 ± 0.10^{j}	—
UBr^{3+}	—	0.52 ± 0.10^{e}	—
UNO_3^{3+}	—	0.62 ± 0.08^{j}	—
Be_2OH^{3+}	—	0.50 ± 0.05^{k}	—
Pu^{4+}	—	1.03 ± 0.05^{f}	—
Np^{4+}	—	0.82 ± 0.05	—
U^{4+}	—	0.76 ± 0.06^{e}	—
Th^{4+}	0.25 ± 0.03	—	0.11 ± 0.02

[a]*Source*: Refs. [10,24] unless specified. Uncertainties are 95% confidence level.

[b]Ion interaction coefficients can be described more accurately with an ionic strength-dependent function, listed in Table 6.5.

[c]Ferri et al. [26].

[d]Evaluated in NEA-TDB review on uranium thermodynamics [37].

[e]Estimated in NEA-TDB review on uranium thermodynamics [37].

[f]Riglet et al. [36], where the following assumptions were made: $\varepsilon_{(Np^{3+}.ClO_4^-)}\approx\varepsilon_{(Pu^{3+}.ClO_4^-)}=0.49$ as for other (M^{3+}, ClO_4^-) interactions, and $\varepsilon_{(NpO_2^{2+}.ClO_4^-)}\approx\varepsilon_{(PuO_2^{2+}.ClO_4^-)}\approx\varepsilon_{(UO_2^{2+}.ClO_4^-)}=0.46$.

[g]Hedlund [35].

[h]Taken from Spahiu [25].

[i]The coefficients $\varepsilon_{(M^{n+}.Cl^-)}$ and $\varepsilon_{(M^{n+}.NO_3^-)}$ reported by Ciavatta [10] were evaluated without taking chloride and nitrate complexation into account. See section 6.3.

[j]Evaluated in NEA-TDB review on uranium thermodynamics [37] using $\varepsilon_{(U^{4+}.ClO_4^-)}=(0.76\pm0.06)$.

[k]Taken from Bruno [38], where the following assumptions were made: $\varepsilon_{(Be^{3+}.ClO_4^-)}=0.30$ as for other $\varepsilon_{(M^{2+}.ClO_4^-)}$, $\varepsilon_{(Be^{2+}.Cl^-)}=0.17$ as for other $\varepsilon_{(M^{2+}.Cl^-)}$, and $\varepsilon_{(Be^{2+}.NO_3^-)}=0.17$ as for other $\varepsilon_{(M^{2+}.NO_3^-)}$.

6.2.   OTHER METHODS FOR IONIC STRENGTH CORRECTIONS

6.2.1   The Pitzer and Brewer Equation

The P-B equation is very similar to the B-G-S equation. The expression for the activity coefficient of an ion i of charge z_i takes the form

$$\log_{10}\gamma_i = \frac{-z_i^2 0.5107\sqrt{I_m}}{1+\sqrt{I_m}} + \sum_j B(i,j)m_j \qquad (6.28)$$

Table 6.4 Ion Interaction Coefficients $\varepsilon_{j,k}$ for Anions j with $k = Li^+$, Na^+, and K^{+a}

$j\ k\rightarrow$ $\downarrow$	Li^+	Na^+	K^+
OH^-	-0.02 ± 0.03^b	0.04 ± 0.01	0.09 ± 0.01
F^-	—	0.02 ± 0.02^c	0.03 ± 0.02
HF_2^-	—	-0.11 ± 0.06^c	—
Cl^-	0.10 ± 0.01	0.03 ± 0.01	0.00 ± 0.01
ClO_4^-	0.15 ± 0.01	0.01 ± 0.01	—
Br^-	0.13 ± 0.02	0.05 ± 0.01	0.01 ± 0.02
I^-	0.16 ± 0.01	0.08 ± 0.02	0.02 ± 0.01
HSO_4^-	—	—	—
		0.01 ± 0.02	
NO_3^-	0.08 ± 0.01	-0.04 ± 0.03^b	-0.11 ± 0.04^b
$H_2PO_4^-$	—	-0.08 ± 0.04^b	-0.14 ± 0.04^b
HCO_3^-	—	-0.03 ± 0.02	—
SCN^-	—	0.05 ± 0.01	-0.01 ± 0.01
$HCOO^-$	—	0.03 ± 0.01	—
CH_3COO^-	0.05 ± 0.01	0.08 ± 0.01	0.09 ± 0.01
$SiO(OH)_3^-$	—	-0.08 ± 0.03^c	—
$B(OH)_4^-$	—	-0.07 ± 0.05^b	—
$UO_2(OH)_3^-$	—	-0.09 ± 0.05^c	—
$UO_2F_3^-$	—	0.00 ± 0.05^c	—
$(UO_2)_2CO_3(OH)_3^-$	—	0.00 ± 0.05^c	—
SO_4^{2-}	-0.03 ± 0.04^b	-0.12 ± 0.06^b	-0.06 ± 0.02
HPO_4^{2-}		-0.15 ± 0.06^b	-0.10 ± 0.06^b
CO_3^{2-}		-0.05 ± 0.03	0.02 ± 0.01
CrO_4^{2-}		-0.06 ± 0.04^b	-0.08 ± 0.04^b
$UO_2(SO_4)_2^{2-}$		-0.12 ± 0.06^c	—
$UO_2(CO_3)_2^{2-}$		0.04 ± 0.09^c	—
PO_4^{3-}		-0.25 ± 0.03^b	-0.09 ± 0.02
$P_2O_7^{4-}$		-0.26 ± 0.05	-0.15 ± 0.05
$Fe(CN)_6^{4-}$		—	-0.17 ± 0.03
$U(CO_3)_4^{4-}$		0.09 ± 0.10^c	—
$UO_2(CO_3)_3^{4-}$		0.08 ± 0.11^c	—
$UO_2(CO_3)_3^{5-}$		-0.66 ± 0.14^c	—
$U(CO_3)_5^{6-}$		-0.27 ± 0.15^c	—

[a]*Source*: Refs. [10,34] unless specified. Uncertainties are 95% confidence level.
[b]Ion interaction coefficient can be described more accurately with an ionic strength-dependent function, listed in Table 6.5.
[c]Evaluated in NEA-TDB review on uranium thermodynamics [37].

Table 6.5 Ion Interaction Coefficients $\varepsilon_{(1,j,k)}$ and $\varepsilon_{(2,j,k)}$ for Cations j with $k=Cl^-$, ClO_4^-, and NO_3^- (First Part), and for Anions j with $k=Li^+$, Na^+, and K^+ (Second Part), According to the Relationship $\varepsilon=\varepsilon_1+\varepsilon_2\log_{10}I_m$ [a]

j $k\rightarrow$	Cl^-		ClO_4^-		NO_3^-	
	ε_1	ε_2	ε_1	ε_2	ε_1	ε_2
NH_4^+	-0.088 ± 0.002	0.095 ± 0.012			-0.075 ± 0.001	0.057 ± 0.004
Ag^+					-0.1432 ± 0.0002	0.0971 ± 0.0009
Tl^+			-0.18 ± 0.02	0.09 ± 0.02		
Hg_2^{2+}					-0.2300 ± 0.0004	0.194 ± 0.002
Hg^{2+}					-0.145 ± 0.001	0.194 ± 0.002
Pb^{2+}					-0.329 ± 0.007	0.288 ± 0.018

j $k\rightarrow$	Li^+		Na^+		K^+	
	ε_1	ε_2	ε_1	ε_2	ε_1	ε_2
OH^-	-0.039 ± 0.002	0.072 ± 0.006				
NO_2^-	0.02 ± 0.01	0.11 ± 0.01				
NO_3^-			-0.049 ± 0.001	0.044 ± 0.002	-0.131 ± 0.002	0.082 ± 0.006
$B(OH)_4^-$			-0.092 ± 0.002	0.103 ± 0.005		
$H_2PO_4^-$			-0.109 ± 0.001	0.095 ± 0.003	-0.1473 ± 0.0008	0.121 ± 0.004
SO_3^{2-}			-0.125 ± 0.008	0.106 ± 0.009		
SO_4^{2-}	-0.068 ± 0.003	0.093 ± 0.007	-0.184 ± 0.002	0.139 ± 0.006		
$S_2O_3^{2-}$			-0.125 ± 0.008	0.106 ± 0.009		
HPO_4^{2-}			-0.19 ± 0.01	0.11 ± 0.03	-0.152 ± 0.007	0.123 ± 0.016
CrO_4^{2-}			-0.090 ± 0.005	0.07 ± 0.01	-0.123 ± 0.003	0.106 ± 0.007
PO_4^{3-}			-0.29 ± 0.02	0.10 ± 0.01		

[a] *Source*: Refs. [10,34]. Uncertainties are 95% confidence level.

where the summation over j covers all anions for the case that i is a cation and vice versa. Tables of $B(I, j)$ are given by Pitzer and Brewer [14] and by Baes and Mesmer [3]. The Debye-Hückel term [see Eq. (6.2)] is different from that in the B-G-S equation. Apart from a slightly different value for A, the factor Ba_j has been chosen equal to 1.0 in the P-B equation compared to 1.5 in the B-G-S equation. The B-G-S equation is preferred to the P-B equation in the critical evaluations of the NEA-TDB Project for the reasons given in section 6.1.1.

6.2.2 The Pitzer Equations

The following text is only intended to provide the reader with a brief outline of the Pitzer method. This approach consists of the development of an explicit function relating the ion interaction coefficient to the ionic strength and the addition of a third virial coefficient to Eq. (6.1). For the solution of a single electrolyte MX, the activity coefficient may be expressed by Eq. (6.29) [15]:

$$\ln\gamma = |z_M z_X| f^\gamma + m \frac{(2\nu_M \nu_X)}{\nu} B^\gamma_{MX} + m^2 \left(\frac{2(\nu_M \nu_X)^{3/2}}{\nu} \right) C^\gamma_{MX} \qquad (6.29)$$

where ν_M and ν_X are the numbers of M and X ions in the formula unit and z_M and z_X their charges. m is the molality of the solution and $\nu = \nu_M + \nu_X$. In aqueous solutions at 25°C and 10^5 Pa, the following relations are given [16]:

$$f^\gamma = -0.392 \left(\frac{\sqrt{I_m}}{1 + 1.2\sqrt{I_m}} + 1.667 \ln(1 + 1.2\sqrt{I_m}) \right) \qquad (6.30)$$

$$B^\gamma_{MX} = 2\beta^{(0)}_{MX} + \frac{\beta^{(1)}_{MX}}{2I_m} (1 - (1 + 2\sqrt{I_m} - 2I_m)e^{-2\sqrt{I_m}}) \qquad (6.31)$$

$$C^\gamma_{MX} = \tfrac{3}{2} C^\phi_{MX} \qquad (6.32)$$

where f^γ is the Debye-Hückel term extended to include osmotic effects, the parameters $\beta^{(0)}_{MX}$ and $\beta^{(1)}_{MX}$ define the second virial coefficient (corresponding to ε in the B-G-S equation), and C^ϕ_{MX} defines the third virial coefficient. $I_m = \tfrac{1}{2}\sum m_i z_i^2$ is the ionic strength in molal units. In the case of 2:2 electrolytes, one could add to the second virial coefficient terms of the same form. Pitzer's equations have been extended to cover electrolyte mixtures [17], including terms allowing for the interactions of ions of the same charge sign. Equation (6.33) is an extension of Eq. (6.29) for a anions and c cations

[4]. Here a and a' cover all anions, c and c' cover all cations, and $(\sum mz) = \sum_c m_c z_c = \sum_a m_a |z_a|$.

$$\ln \gamma_{MX} = |z_M z_X| f^\gamma + \frac{2\nu_M}{\nu} \sum_a m_a \left(B_{Ma} + \left(\sum mz\right) C_{Ma} + \frac{\nu_X}{\nu_M} \theta_{Xa} \right)$$

$$+ \frac{2\nu_X}{\nu} \sum_c m_c \left(B_{cX} + \left(\sum mz\right) C_{cX} + \frac{\nu_M}{\nu_X} \theta_{Mc} \right)$$

$$+ \sum_c \sum_a m_c m_a \left(|z_M z_X| B'_{ca} + \frac{1}{\nu} (2\nu_M z_M C_{ca} + \nu_M \psi_{M_{ca}} + \nu_X \psi_{caX}) \right)$$

$$+ \frac{1}{2} \sum_c \sum_{c'} m_c m_{c'} \left(\frac{\nu_X}{\nu} \psi_{cc'X} + |z_M z_X| \theta'_{cc'} \right)$$

$$+ \frac{1}{2} \sum_a \sum_{a'} m_a m_{a'} \left(\frac{\nu_M}{\nu} \psi_{Maa'} + |z_M z_X| \theta'_{aa'} \right) \tag{6.33}$$

In Eq. (6.33) f^γ is the same as in Eq. (6.29); C_{MX}, B_{MX}, and its derivative with ionic strength, B'_{MX}, have the forms

$$B_{MX} = \beta^{(0)}_{MX} + \frac{\beta^{(1)}}{2I_m} (1 - (1 + 2\sqrt{I_m}) e^{-2\sqrt{I_m}}) \tag{6.34}$$

$$B'_{MX} = \frac{\beta^{(1)}_{MX}}{2I_m^2} (-1 + (1 + 2\sqrt{I_m} + 2I_m) e^{-2\sqrt{I_m}}) \tag{6.35}$$

$$C_{MX} = \frac{C^\phi_{MX}}{2\sqrt{|z_M z_X|}} \tag{6.36}$$

where $\beta^{(0)}$, $\beta^{(1)}$, C^ϕ are the same as for pure electrolytes. In Eq. (6.33) the θ terms summarize the interactions between ions of the same charge sign that are independent of the common ion in a ternary mixture, and the ψ terms account for the modifying influence of the common ion on these interactions. Pitzer [4] points out that higher order electrostatic terms (beyond the Debye-Hückel approximation) become important in cases of unsymmetrical mixing, especially if one of the ions has a charge of three or higher. Higher order electrostatic effects, on the other hand, were found to be unimportant for cases of symmetrical mixing and for pure unsymmetrical electrolytes.

Based on the cluster integral method [27], Pitzer [19] divided the difference terms θ_{MN} into two parts

$$\theta_{MN} = {}^s\theta_{MN} + {}^E\theta_{MN} \tag{6.37}$$

The $^E\theta_{MN}$ terms may be expressed as:

$$^E\theta_{MN} = \frac{z_M z_N}{4I_m}\left(J(x_{MN}) - \frac{1}{2}J(x_{MM}) - \frac{1}{2}J(x_{NN})\right) \tag{6.38}$$

$$^E\theta'_{MN} = -^E\frac{\theta}{I_m} + \frac{z_M z_N}{8I_m^2}\left(x_{MN}J'(x_{MN}) - \frac{1}{2}x_{MM}J'(x_{MM}) - \frac{1}{2}x_{NN}J'(x_{NN})\right) \tag{6.39}$$

where $x_{ij} = 6z_i z_j 0.392\sqrt{I_m}$ at 25°C and 10^5 Pa and $J(x)$ is an integral that can be approximated as

$$J(x) = -\frac{1}{6}x^2(\ln x)e^{-10x^2} + \left(\sum_{k=1}^{6}C_k x^{-k}\right)^{-1} \tag{6.40}$$

The first term is in fact important only at very low ionic strengths; thus for cases used in equilibrium analysis one has

$$J(x) = \left(\frac{4.118}{x} + \frac{7.247}{x^2} + \frac{4.408}{x^3} + \frac{1.837}{x^4} + \frac{0.251}{x^5} + \frac{0.0164}{x^6}\right)^{-1} \tag{6.41}$$

The $^s\theta_{MN}$ parameters can be evaluated from data on mixtures of electrolytes by calculating the differences between the experimental value of γ^{exp} and the value calculated with the appropriate values for all pure electrolyte terms and $^E\theta$ terms but zero values for $^s\theta$ and ψ terms. For the activity coefficient of MX in a MX-NX mixture one has Eq. (6.42) and equivalent expressions for other cases.

$$\frac{\nu}{2\nu_M m_N}\Delta\ln\gamma = \left(^s\theta_{MN} + \frac{1}{2}\left(m_X + \left|\frac{z_M}{z_X}\right|m_M\right)\psi_{MNX}\right) \tag{6.42}$$

This is the equation of a straight line with intercept $^s\theta_{MN}$ and slope ψ_{MNX}, when the left side of Eq. (6.42) is plotted against a function of the composition $\frac{1}{2}(m_X + \left|\frac{z_M}{z_X}\right|m_M)$.

The Pitzer equation for single electrolytes with the value of parameters collected in several publication may be used as a compact source of activity coefficient data. From the values $\gamma^0_{\pm jk}$ thus obtained, one may calculate, for example, $\varepsilon_{(j,k,I_m)}$ values using Eq. (6.1).

For the estimation of medium effects on solubility equilibria in mixtures of electrolytes involving ions of charge less than three, one may neglect θ and ψ terms in Eq. (6.33).

In equilibrium analysis studies carried out in the presence of an inert salt (medium salt NX) and small (trace) concentrations of the reactants, only the terms involving m_{NX} have to be considered in Eq. (6.33), while those involving m_i^{trace} can be neglected. Nevertheless, as the main difficulty there still remains the accurate estimation of the parameters of single electrolyte $\beta^{(0)}$, $\beta^{(1)}$, and C^ϕ for species such as metal ion complexes.

Equations for single ion activity coefficients [4], osmotic coefficients [17], and other thermodynamic quantities [28], as well as applications in different cases (e.g., H_2SO_4 and H_3PO_4 solutions) have been given by Pitzer and coworkers [4,20].

From Tables 6.3 and 6.4 it seems that the size and charge correlations can be extended to complex ions. This observation is very important because it indicates a possibility to estimate the ion interaction coefficients for complexes by using such correlations. It is, of course, always preferable to use experimental ion interaction coefficient data. However, the efforts needed to obtain these data for complexes will be so great that it is unlikely that they will be available for more than a few complex species. It is even less likely that one will have data for the Pitzer parameters for these species. Hence, the specific ion interaction approach may have a practical advantage over the inherently more precise Pitzer approach.

6.2.3 The Equations Used by Baes and Mesmer

Baes and Mesmer [3] use the function $F(I_m)$ proposed by Pitzer to express the ionic strength dependence of the ion interaction coefficient B_{MX} in Guggenheim's equations. For a single electrolyte

$$\log_{10} \gamma^o_{\pm MX} = -|z_M z_X| \frac{0.511\sqrt{I_m}}{1 + \sqrt{I_m}} + \frac{2\nu_M \nu_X}{\nu} B_{MX} m_{MX} \tag{6.43}$$

where

$$B_{MX} = B^\infty_{MX} + (B^o_{MX} - B^\infty_{MX})F(I_m) \tag{6.44}$$

$$F(I_m) = \frac{1 - (1 + 2\sqrt{I_m} - 2I_m)e^{-2\sqrt{I_m}}}{4I_m}, \qquad F(0) = 1, F(\infty) = 0. \tag{6.45}$$

The Pitzer function linearizes the dependence of the ion interaction coefficient on the ionic strength quite well, even in the cases of 4:1 and 5:1 electrolytes, where constant $\varepsilon_{(M,X)}$ values [see Eq. (6.1)] are not obtained at high ionic strengths. The parameters B^o_{MX} and B^∞_{MX} can be determined from a single electrolyte activity coefficient datum by calculating first B_{MX}. (Note that $B_{MX} \neq \varepsilon_{(j,k)}$, since the Debye-Hückel term in Eq. (6.43) does not have the factor 1.5 in the denominator.) By plotting $B_{MX}(I_m)$ values against $F(I_m)$, B^∞_{MX} is obtained as the intercept while B^o_{MX} is obtained from the slope of the straight line [see Eq. (6.44)]. The equation for a mixture is similar to

Eq. (6.43) and $B_{MX} = 0$ if M and X are of the same charge sign. In the case of equilibrium constant measurements, ΔB values are expressed by equations similar to Eq. (6.6).

The corresponding ΔB° and ΔB^∞ can be obtained together with the β° values from the system of equations

$$\log_{10}\beta(I_{m,n}) - \Delta z_i^2 \frac{0.511\sqrt{I_{m,n}}}{1 + \sqrt{I_{m,n}}} - q\log_{10}(a_{H_2O})_n$$
$$= \log_{10}\beta^\circ + \Delta B^\infty[1 - F(I_{m,n})]I_{m,n} + \Delta B^\circ F(I_{m,n})I_{m,n} \qquad (6.46)$$

where $\beta(I_{m,n})$ and $(a_{H_2O})_n$ refer to the values of β and a_{H_2O} at ionic strength $I_{m,n}$. From the values obtained for ΔB° and ΔB^∞ and equations similar to Eq. (6.9), one may estimate the unknown B°_{MX} and B^∞_{MX} values.

6.2.4 The Davies Equation

The Davies equation [29] has been used extensively to calculate activity coefficients of electrolytes at fairly low ionic strengths.

The Davies equation for the activity coefficient of an ion i of charge z_i is, at 25°C:

$$\log_{10}\gamma_i = -0.5102z_i^2 \left(\frac{\sqrt{I_m}}{1 + \sqrt{I_m}} - 0.3I_m \right) \qquad (6.47)$$

The equation has no theoretical foundation but is found to work fairly well up to ionic strengths of $0.1\ \text{mol} \cdot \text{kg}^{-1}$. It should not be used at higher ionic strengths. The Davies equation has a form similar to the B-G-S equation but with ion interaction coefficients equal to $0.153z_i^2$, i.e., 0.15, 0.61, and 1.38 for ions of charge 1, 2, and 3, respectively. These values do not agree very well with the tabulated ε values.

6.3 ION INTERACTION COEFFICIENTS AND EQUILIBRIUM CONSTANTS FOR ION PAIRS

Two alternative methods can be used to describe the ionic medium dependence of equilibrium constants:

- One method takes into account the individual characteristics of the ionic media by using a medium-dependent expression for the activity coefficients of the species involved in the equilibrium reactions. The medium dependence is described by virial or ion interaction coefficients as used in the Pitzer equations and in the specific ion interaction model.
- The other method uses an extended Debye-Hückel expression where the activity coefficients of reactants and products depend only on the

ionic charge and the ionic strength, but it accounts for the medium-specific properties by introducing ionic pairing between the medium ions and the species involved in the equilibrium reactions. Earlier, this approach has been used extensively in marine chemistry (See Refs. [30–32, 39]).

It can be shown that the virial type of activity coefficient equations and the ionic pairing model are equivalent, provided that the ionic pairing is weak. In these cases, it is in general difficult to distinguish between complex formation and activity coefficient variations unless independent experimental evidence for complex formation is available, e.g., from spectroscopic data, as is the case for the weak uranium(VI) chloride complexes. It should be noted that the ion interaction coefficients evaluated and tabulated by Ciavatta [10] were obtained from experimental mean activity coefficient data without taking into account complex formation. However, it is known that many of the metal ions listed by Ciavatta form weak complexes with chloride and nitrate ions. This fact is reflected by ion interaction coefficients that are smaller than those for the noncomplexing perchlorate ion (see Table 6.3). This review takes chloride and nitrate complex formation into account when these ions are part of the ionic medium and uses the value of the ion interaction coefficient $\varepsilon_{(M^{n+},ClO_4^-)}$ for $\varepsilon_{(M^{n+},Cl^-)}$ and $\varepsilon_{(M^{n+},NO_3^-)}$. In this way, the medium dependence of the activity coefficients is described with a combination of a specific ion interaction model and an ion pairing model. It is evident that the use of NEA-recommended data with ionic strength correction models that differ from those used in the evaluation procedure can lead to inconsistencies in the results of the speciation calculations.

It should be mentioned that complex formation may also occur between negatively charged complexes and the cation of the ionic medium. An example is the stabilization of the complex ion $UO_2(CO_3)_3^{5-}$ at high ionic strength.

6.4 TABLES OF ION INTERACTION COEFFICIENTS

Tables 6.3–6.5 contain the selected specific ion interaction coefficients used in this review, according to the specific ion interaction model described in section 6.1. Table 6.3 contains cation interaction coefficients with Cl^-, ClO_4^-, and NO_3^-; Table 6.4 anion interaction coefficients with Li^+, with Na^+ or NH_4^+, and with K^+. The coefficients have the units of $kg \cdot mol^{-1}$ and are valid for 298.15 K. The species are ordered by charge and appear, within each charge class, in standard order of arrangement [33].

In some cases, the ionic interaction can be better described by assuming ion interaction coefficients as functions of the ionic strength rather than as constants. Ciavatta [10] proposed the use of Eq. (6.48) for cases where the uncertainties in Tables 6.3 and 6.4 are ± 0.03 or greater.

$$\varepsilon = \varepsilon_1 + \varepsilon_2 \log_{10} I_m \tag{6.48}$$

For these cases, and when the uncertainty can be improved as compared to the use of a constant ε, the values ε_1 and ε_2 given in Table 6.5 should be used.

It should be noted that ion interaction coefficients tabulated in Tables 6.3–6.5 may also involve ion pairing effects, as described in section 6.3. In direct comparisons of ion interaction coefficients, or when estimates are made by analogy, this aspect must be taken into account.

6.5 CONCLUSION

The specific ion interaction approach is simple to use and gives a fairly good estimate of activity factors. By using size/charge correlations, it seems possible to estimate unknown ion interaction coefficients. The specific ion interaction model has therefore been adopted as a standard procedure in the NEA Thermochemical Data Base review for the extrapolation and correction of equilibrium data to the infinite dilution standard state. For more details on methods for calculating activity coefficients and the ionic medium/ionic strength dependence of equilibrium constants, the reader is referred to Ref. 40, Chapter IX.

REFERENCES

1. Lafitte, M. J. Chem. Thermodyn., **1982**, *14*, 805.
2. Wanner, H. Standards and conventions for TDB publications, TDB-5.1, Rev. 1, OECD Nuclear Energy Agency, Data Bank, Gif-sur-Yvette, France, 1989.
3. Baes, C. F., Jr.; Mesmer, R. F. *The Hydrolysis of Cations*; John Wiley and Sons, New York, 1976.
4. Pitzer, K. S. Theory: Ion interaction approach. In *Activity Coefficients in Electrolyte Solutions*, Pytkowicz, R. M., Ed., Vol. I, CRC Press, Boca Raton, Florida, **1979**; 157–208.
5. Brønsted, J. M. Studies of solubility: IV. The principle of specific interaction of ions. J. Am. Chem. Soc., **1922**, *44*, 877–898.
6. Brønsted, J. M. Calculation of the osmotic activity functions of univalent salts. J. Am. Chem. Soc., **1922**, *44*, 938–948.
7. Scatchard, G. Concentrated solutions of strong electrolytes, Chem. Rev., **1936**, *19*, 309–327.
8. Guggenheim, E. A. *Applications of Statistical Mechanics*; Oxford, Clarendon Press, 1966.

9. Scatchard, G. *Equilibrium in Solution Surface and Colloid Chemistry*; Harvard University Press, Cambridge, MA, 1976.
10. Ciavatta, L. The specific interaction theory in evaluating ionic equilibria. Ann. Chim. (Rome), **1980**, *70*, 551–567.
11. Guggenheim, E. A. Phil. Mag. **1935**, *19*, 588.
12. Vasilév, V. P. Influence of ionic strength on the instability constants of complexes. Russ. J. Inorg. Chem., **1962**, *7, 8*, 924–927.
13. Helgeson, H. C.; Kirkham D. H.; and Flowers, G. C. Theoretical prediction of the thermodynamic behavior of aqueous electrolytes at high pressure and temperatures: IV. Calculation of activity coefficients, osmotic coefficients and apparent molal properties to 600°C and 5 kb, Am. J. Sci., **1981**, *281*, 1249–1516.
14. Pitzer, K. S.; Brewer, L. Thermodynamics, 2nd Ed., McGraw-Hill, New York, 1961.
15. Pitzer, K. S. Thermodynamics of electrolytes: I. Theoretical basis and general equations. J. Phys. Chem., **1973**, *77*, 268–277.
16. Pitzer, K. S.; Mayorga, G. Thermodynamics of electrolytes: II. Activity and osmotic coefficients for strong electrolytes with one or both ions univalent. J. Phys. Chem., **1973**, *77*, 2300–2308.
17. Pitzer, K. S.; Kim, J. J. Thermodynamics of electrolytes: IV. Activity and osmotic coefficients for mixed electrolytes. J. Am. Chem. Soc., **1974**, *96*, 5701–5707.
18. Pitzer, K. S.; Mayorga, G. Thermodynamics of electrolytes: III. Activity and osmotic coefficients for 2–2 electrolytes. J. Sol. Chem., **1974**, *3*, 539–546.
19. Pitzer, K. S. Thermodynamics of electrolytes: V. Effects of higher-order electrostatic terms. J. Sol. Chem., **1975**, *4*, 249–265.
20. Pitzer, K. S.; Silvester, L. F. Thermodynamics of electrolytes: VI. Weak electrolytes including H_3PO_4. J. Sol. Chem., **1976**, *5*, 269–278.
21. Pitzer, K. S.; Peterson., J. R.; Silvester, L. F. Thermodynamics of electrolytes: IX. Rare earth chlorides, nitrates, and perchlorates. J. Sol. Chem., **1978**, *7*, 45–56.
22. Robinson, R. A.; Stokes, R. H. *Electrolyte Solutions*; Butterworth, London, 1959.
23. Stokes, R. H. Thermodynamics of solutions. In *Activity Coefficients in Electrolyte Solutions*, Pytkowicz, R. M., Ed., Vol. I, CRC Press, Boca Raton, Florida, 1979; 1–28.
24. Wanner, H. *Guidelines for the Assignment of Uncertainties*. TDB-3, OECD Nuclear Energy Agency, Data Bank, Gif-sur-Yvette, France, 1989.
25. Spahiu, K. Carbonate complex formation in lanthanoid and actinoid systems; Ph.D. thesis, Royal Institute of Technology, Stockholm; 1983.
26. Ferri, D.; Grenthe, I.; Salvatore, F. Studies on metal carbonate equilibria: VII. Reduction of the *tris* (carbonato) dioxourante(VI) ion, $UO_2(CO_3)_3^{4}$, in carbonate solution. Inorg. Chem., **1983**, *22, 21*, 3162–3165.
27. Friedman, B. L. *Ionic Solution Theory*, Interscience Publishers, New York, 1962.

28. Silvester, L. F.; Pitzer, K. S. Thermodynamics of electrolytes: VIII. High-temperature properties, including enthalpy and heat capacity, with application to sodium chloride. J. Phys. Chem., **1977**, *81*, 1822–1828.

29. Davies, C. W. *Ion Association*; Butterworths, Washington, 1962.

30. Johnson, K. S.; Pytkowicz, R. M. Ion association and activity coefficients in multicomponent solutions. In *Activity Coefficients in Electrolyte Solutions*; Pytkowicz, R. M., Ed., Vol. II, CRC Press, Boca Raton, Florida, 1979; 1–62.

31. Millero, F. J. Effects of pressure and temperature on activity coefficients. In: *Activity Coefficients in Electrolyte Solutions*; Pytkowicz, R. M., Ed., Vol. II, CRC Press, Boca Raton, Florida, 1979, pp. 63–151.

32. Whitfield, M. Activity coefficients in natural waters. In: *Activity Coefficients in Electrolyte Solutions*; Pytkowicz, R. M., Ed., Vol. II, CRC Press, Boca Raton, Florida, 1979; pp. 153–299.

33. Wagman, D. D.; Evans, W. H.; Parker, V. B.; Schumm, R. H.; Halow, I.; Bailey, S. M.; Churney, K. L.; Nuttall, R. L. The NBS tables of chemical thermodynamic properties; Selected values for inorganic and C1 and C2 organic substances in S1 units. J. Phys. Chem. Ref. **1982**, *11*, 1–392.

34. Ciavatta, L. Personal communication, Universita de Napoli, Naples, June 1988.

35. Hedlund, T. Studies of complexation and precipitation equilibria in some aqueous aluminium(III) systems; Ph.D. thesis, University of Umeå, Umeå, Sweden, 1988.

36. Riglet, Ch.; Robouch, P.; Vitorge, P. Standard potentials of the (MO_2^{2+}/MO_2^{+}) and (M^{4+}/M^{3+}) redox systems for neptunium and plutonium. Radiochim. Acta, **1989**, *46*, 85–94.

37. Grenthe, I.; Fuger, J.; Konings, R. J. M.; Lemire, R. J.; Muller, A. B.; Nguyen-Trung, C.; Wanner, H. *Chemical Thermodynamics of Uranium*, Wanner, H., Forest, I., Eds., Elsevier Science Publishers, Amsterdam, 1992.

38. Bruno, J. Stoichiometric and structural studies on the Be^{2+}-H_2O-$CO_2(g)$ system; Ph.D. thesis, Royal Institute of Technology, Stockholm, 1986.

39. Pytkowicz, R. M. Activity coefficients, ionic media, and equilibrium in solutions; in: *Activity Coefficients in Electrolyte Solutions*; R.M. Pytkowicz, Ed.; Vol. I; CRC Press, Boca Raton, Florida, 1979; 301–305.

40. Grenthe, I.; Puigdomenech, I. *Modelling in Aquatic Chemistry*; OECD/NEA, Paris, 1997.

7
Development of Industrial Solvent Extraction Processes

GORDON M. RITCEY Gordon M. Ritcey & Associates, Inc., Nepean, Ontario, Canada

7.1 INTRODUCTION

The development of a solvent extraction process from a given aqueous feed solution may be confined by several restraints. For example, the temperature, flow rate, acidity, and so forth may be (essentially) fixed, and the solvent extraction process developed must then be capable of accommodating these restrictions.

The initial bench-scale experimental investigations into solvent extraction processes are conducted with small apparatus, such as separating funnels. Following the successful completion of these tests, when the best reagent and other conditions for the system have been established, small-scale continuous operations are run, such as in a small mixer-settler unit. The data so obtained are used to determine scale-up factors for pilot plant or plant design and operation (see Chapters 7 and 8).

The need for pilot plant operations is considered by one school of thought to be an unnecessary expense in the development of a solvent process. The rationale here is that monies and time spent in piloting the process can be used to modify, if necessary, the full-scale plant. Thus, if few (or no) modifications are required, considerable savings can result. This approach is probably satisfactory if the process is very similar to an existing process and sufficient data are available on which to base the plant design and operation. The practical experience of the plant design engineers and operators is also of considerable importance in such cases. On the other hand, problems encountered in a plant operation that has been designed without piloting the process could result in the loss of considerable time and expense.

Of course, almost any process or plant can be made to work, even though the basic data and design are poor. The costs, however, may be astronomical! It should be pointed out, however, that the size of a pilot plant could vary from a few gallons per minute of total throughput to several hundred gallons per minute. Normally, one would expect that the size of a pilot plant would be in direct proportion to the size of the final plant.

This chapter comprises some selected excerpts taken from a comprehensive two-volume text on the subject coauthored by the author [1]. It specifically addresses the extraction of metals, but many of the (general) conclusions are equally valid for the extraction of organic compounds, with appropriate modifications.

7.2 INITIAL STUDIES

In the rapid determination of whether an extractant may be useful for the extraction of a particular substance or to determine the best extractant in a series of extractants, an approach known as "screening" may be employed. Much of the usefulness of this approach depends on the investigator's knowledge and experience of solvent extraction, since the tests carried out are minimal and are generally based on the relative properties of an extractant.

In screening tests, it is common practice to use approximately 0.1 mol dm^{-3} or 5 vol% solutions of extractant, and extraction tests are conducted at a phase ratio of 1, with a 5 min contact time. These conditions, of course, may be varied, but the object is to stay with as simple an approach as possible. Analysis of only the aqueous raffinate need be carried out, and distribution data calculated from this.

The initial requirement in the development of a solvent extraction process for the recovery or separation of metals from an aqueous solution is knowledge of the solution composition, pH, temperature, and flow rate. Both pH and temperature can be adjusted, within certain economic limits, before feeding to the solvent extraction circuit, but only in a few cases can the leaching or dissolution conditions be dictated by the extraction process. Consequently, no serious development work on the extraction process can be carried out before the leaching conditions or the type of feed solution are established.

Properties of the feed solution and the substance to be extracted will decrease the number of extractants that may be applicable. For example, in the extraction of metals, if no anionic metal species are present in the feed solution, there is little point in considering anionic (amine) extractants. On the other hand, if anionic metal species are present, then the amine type best suited to the extraction of the anionic species can be selected, knowing that primary amines generally extract anionic sulfate complexes best, whereas tertiary and quaternary amines extract anionic chloro-complexes best. In this way, a considerable saving in both time and money can be achieved.

A literature search to determine whether any existing processes are likely to be applicable should be undertaken. Here, the extent of the search will depend on the knowledge and experience of the investigator. After selecting the most suitable class of extractant, the next stage is to determine the various parameters of the extraction systems including the choice of diluent and modifier, which with the extractant make up the solvent. These should include specificity of the selected extractants for the metal (or compound) to be extracted; the requirements for the type of diluent and modifier; loading and stripping characteristics; pH dependency; kinetics of extraction, scrubbing, and stripping; concentration of extractant and phase ratio required; solubility of the solvent in the aqueous phase; and the physical attributes of the system. Many of these aspects are discussed in some detail in previous chapters.

7.3 SOLVENT PRETREATMENT OR CONDITIONING

Almost all solvent solutions used in solvent extraction processes require treatment, either before their initial use or after being stripped of the extracted species. After preparation of the solvent by mixing appropriate amounts of extractant, modifier, and diluent, the solution is treated by contacting with an appropriate aqueous solution before entering the extraction stage. Such pretreatment is generally referred to as equilibration of the solvent.

The object of equilibration is to provide a solvent that will effectively extract the required species, either because of the form of the active constituent of the solvent or by maintaining the necessary extraction pH. As an example of the former condition, consider the extraction of uranium using a tertiary aliphatic amine as extractant. Extraction of metal species by such amines is considered to occur by liquid ion exchange (see Chapters 3 and 4). For a tertiary amine to act in this manner, it must be first converted to an amine salt:

$$R_3N_{(org)} + H^+ + A^- \leftrightarrow R_3NH^+A^-_{(org)} \tag{7.1}$$

This is accomplished by contacting the solvent with an appropriate aqueous acid, for example, sulfuric acid:

$$2R_3N_{(org)} + H_2SO_4 \leftrightarrow (R_3NH)_2SO_{4(org)} \tag{7.2}$$

Having formed the amine salt, the solvent can be used for the extraction of uranium. This is shown as an ion exchange process in Eq. (7.3):

$$2[(R_3NH)_2SO_4]_{(org)} + UO_2(SO_4)_3^{4-}$$

$$\leftrightarrow [(R_3NH)_2UO_2(SO_4)_3]_{(org)} + 2SO_4^{2-} \tag{7.3}$$

in which the anionic sulfate moiety of the amine salt (organic phase) is exchanged for the anionic uranyl sulfate species (aqueous phase), to form a uranyl amine sulfate complex as the extracted species (anion exchange).

Stripping the uranium from the solvent can be accomplished by using either acid or alkaline solutions. If an alkali carbonate solution is used, the stripped solvent then requires equilibrating with sulfuric acid before recycling to the extraction stage. Sulfuric acid stripping obviates the need for such equilibration.

Acidic extractants that undergo a cationic exchange reaction with metal ions do not generally require equilibration before extraction, because the exchange involves protons that are already present in the extractant:

$$M^{n+} + nHX_{(org)} \leftrightarrow MX_{n(org)} + nH^+ \tag{7.4}$$

and, as before, if alkaline stripping is employed, equilibration with acid is required before recycling the solvent to the extraction stage. These equilibrium equations are discussed in detail in Chapters 3 and 4.

Problems tend to arise in cationic systems when a particular pH is to be maintained. The protons resulting from the exchange increase the acidity of the aqueous phase, which can alter (lower) considerably the extraction of a metal ion. For example, the separation of cobalt and nickel using di-2-ethylhexylphosphoric acid (DEHPA) as the extractant is carried out at pH 5–6. This system is discussed extensively in section 11.3, where it is pointed out that the use of a nonequilibrated solvent results in lowering the pH of the aqueous phase to such an extent that very poor cobalt extraction is achieved. It has been shown that if the ammonium salt of DEHPA is used, rather than the free acid, a pH ~5.5 can be easily maintained without having to add ammonia continuously throughout the extraction stage to maintain the optimum extraction pH (a messy operation). The exchange of ammonium ions for metal ions produces an ammonium salt (nitrate, sulfate, or other, depending on the anion present in the aqueous feed). These salts are almost neutral in aqueous solution, but they have a buffering effect. Thus, an aqueous feed pH as low as 3.5 can be used, since the equilibrium pH is maintained at between 5 and 6. The reactions involved are:

$$[(RO)_2PO(OH)]_{(org)} + NH_4OH \leftrightarrow [(RO)_2PO(ONH_4)]_{(org)} \tag{7.5}$$

$$2[(RO)_2PO(ONH_4)]_{(org)} + CoSO_4$$
$$\leftrightarrow [((RO)_2POO)_2Co]_{(org)} + (NH_4)_2SO_4 \qquad (7.6)$$

A similar situation occurs using the sodium salt of DEHP.

Stripping of cobalt from the loaded extract is accomplished by various acids, such as nitric acid:

$$[((RO)_2POO)_2Co]_{(org)} + 2HNO_3$$
$$\leftrightarrow 2[(RO)_2PO(OH)]_{(org)} + Co(NO_3)_2 \qquad (7.7)$$

and the solvent phase is then equilibrated with ammonia [Eq. (7.5)] for recycling to the extraction stage.

Commercially produced extractants are not pure and always contain some residual starting materials, by-products formed during manufacture, or degradation products. For example, DEHPA contains some 2-ethylhexanol (starting material) and some of the monoester acid. Tertiary amines invariably contain primary or secondary amines, and so on. If the impurities in the extractant used in the solvent are more soluble than the extractant itself, which is usually the case, they will be lost from the system to the aqueous phase after some recycling of the solvent. The concentration and solubility of such impurities will depend on the time required to remove them from the system. The problems associated with such impurities become particularly evident if their effects are not considered during bench-scale investigations. Thus, their presence in a solvent system can produce cruds, good or poor phase separation, and enhanced or poor loading characteristics. Such effects could result in abandoning a particular solvent because of poor chemical and physical characteristics; conversely, problems could arise when the process is tried on a continuous scale if the enhancing effects of the impurities result in poor extraction characteristics as a result of their loss to the raffinate. It is suggested, therefore, that all solvents be conditioned before use in bench-scale studies that will ultimately be developed into a continuous system.

Although there is no specific method of conditioning, a rule of thumb is to treat the freshly prepared solvent with a solution similar to that used in the test work. This may require several contacts with the aqueous solution, with stripping between contacts if a metal is extracted.

Inclusion of this practice in all solvent extraction studies will ensure that the solvent is not discarded as being unsuitable, and that problems in pilot plant or continuous operations are not the result of impurities that were present in the bench-scale test work.

It should also be pointed out that extractant quality might vary from sample to sample. This can be particularly frustrating in both development work and pilot plant operations. It is good practice to test every batch of extractant before use, to ensure that its characteristics are similar to the material previously used. Such testing will also provide the investigator with the knowledge that the reagent is as ordered, and is not another reagent entirely. It is not uncommon for the wrong material to be shipped.

Similar comments can be applied to diluents and modifiers. The investigator should be aware of the problems that can arise from impurities in the components that go to make up a solvent.

7.4 DISTRIBUTION DATA

7.4.1 Extraction

Distribution data can be obtained in two ways. The first employs variation of the phase ratio of the aqueous and organic phases; the second involves recontacting the organic phase with fresh aqueous phase until the saturation loading of the solvent is reached.

In the phase ratio variation method, the aqueous and organic phase are mixed mechanically until equilibrium is reached at phase ratios between about 1:10 and 10:1. After allowing the phases to disengage, they are separated and analyzed for the metal or metals of interest. It is necessary to remove any entrained solvent from the aqueous phase, and any aqueous phase from the organic phase, before analysis. This can be accomplished by centrifuging the phases or by filtering the aqueous phase through dry, coarse filter paper, and the organic phase through a silicone-treated phase-separating paper.

It must be emphasized that the equilibrium pH must be the same in all tests. One way of ensuring this is to check the pH of the aqueous phase after equilibration of the phases, adjust if necessary by the addition of acid or alkali, and then mix the phases again till equilibrium is reached.

A distribution isotherm is then constructed by plotting the metal concentration in the organic phase against the concentration in the aqueous phase, as a function of the phase ratio. An example of such an isotherm is shown in Fig. 7.1, for the extraction of nickel by DEHPA(Na) at pH 6, showing three different concentrations of extractant [1].

The second method of constructing an extraction isotherm initially requires the choice of a suitable phase ratio, and the organic and aqueous (feed) phases are contacted at this ratio until equilibrium is obtained. The phases are allowed to separate, and the aqueous phase is removed and analyzed. A small measured portion of the organic phase is also taken for analysis. Fresh aqueous feed is then added to the remainder of the organic phase, in an amount to give the same phase ratio as that originally used. The

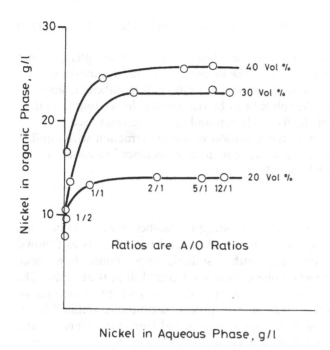

Fig. 7.1 Distribution isotherms for nickel at pH 6 for three different concentrations (vol%) of DEHPA(Na), in the kerosene-aqueous sulfate systems.

phases are again contacted until equilibrium is obtained, and the same procedure performed as just described. This process is repeated until saturation of the solvent with the metal is obtained. Here again, care must be taken to maintain the same pH throughout the procedure. One drawback to this method is that a large volume of the organic phase must be used initially to allow for sampling, especially if numerous contacts are required. Generally, the first method employing variation in the phase ratio is recommended.

The extraction isotherms constructed from these data can be used in the construction of a McCabe-Thiele diagram (see Chapter 8).

The loading capacity of the solvent can be obtained from either of these two methods for the particular concentration of extractant used. It cannot be assumed that the loading capacity is linear with increasing extractant concentration (see Fig. 7.1). If high concentrations (>10 vol%) are to be used, it is advisable to determine the loading capacities at these concentrations. This is readily done either by contacting a solvent several times with fresh aqueous solution, or by using a concentrated solution of the metal with a high (aqueous/organic) ratio, until maximum loading is achieved.

For most exploratory work, analysis of the organic phase is not necessary. If no volume change of the phases occurs and no third phase or crud is formed, the analysis of the aqueous raffinates is sufficient, since the metal concentration in the solvent can be readily calculated from the initial metal concentration of the feed solution and the phase ratio used.

The determination of the variation of metal extraction with equilibrium pH is carried out in much the same manner as described earlier, and is described in detail in Chapter 4.

7.4.2 Stripping

Data required for the construction of stripping isotherms are obtained in a manner similar to those for extraction. A general procedure is as follows: the loaded solvent is contacted with a suitable strip solution (e.g., acid, base, etc.) at an appropriate phase ratio until equilibrium is attained. The aqueous phase is then removed, fresh strip solution is added to the organic phase, and the procedure repeated. This process is continued until all (or as much as possible) of the metal has been stripped from the organic phase. Analysis of the aqueous strip liquors, if no volume changes occur, allows the calculation of the concentration of metal in the organic phase after each strip. However, if volume changes do occur, then the organic phases must also be analyzed. Then the stripping isotherm can be drawn and a McCabe-Thiele diagram constructed to determine the number of theoretical strip stages required at the phase ratio and strip solution concentration used.

7.4.3 Kinetics of Extraction and Stripping

The time required for equilibrium to be reached in both of these operations will need to be known to design a plant.

The time required for a system to reach equilibrium can be determined by shake-out tests, as described in earlier sections. Contact times are varied between about 0.5 and 15 min, at suitable intervals, and the extraction coefficient for each contact time plotted as a function of time. With this method, there is a lower practical limit on the contact time of about 0.25 min. These data will not be directly applicable to a continuous process because the rate of metal extraction is a function, in part, of the type and degree of agitation. However, a good idea of whether the extraction rate is sufficiently fast for the system to be suitable for use in a large contactor can be obtained. For example, if equilibrium is attained in less than 1 min, almost any type of contactor may be used.

A procedure has been reported using data on the kinetics of mass transfer to estimate the Murphree stage efficiencies in continuous counter-current mixer-settlers so that the number of real contacting stages can be calculated for a wide range of operating conditions [2]. Extensive test work can thus be minimized by using a limited set of statistically designed experiments to determine the effect on kinetics of mass transfer of the major operating variables. Figure 7.2 shows the noticeable effects on the increase in the rate coefficients as the stirring speed is increased in a Kelex-copper sulfate system. The dispersion of the solvent resulted in poorer kinetics than dispersion of the aqueous phase. There was little effect on the kinetics from change in phase ratio. Data were also presented for the effects on sedimentation and coalescence.

Stripping rates are also of importance in commercial operations, and these can be determined in a manner similar to those previously described.

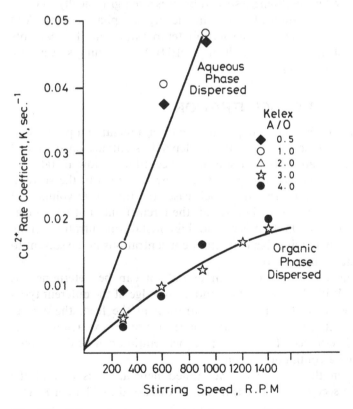

Fig. 7.2 Effect of agitation speed on the rate coefficient.

7.4.4 Scrubbing

The object of scrubbing a loaded solvent is to remove as much as possible of any unwanted coextracted metal. There are many possibilities of accomplishing this by varying the scrub solution. Normally, however, scrubbing is achieved by water, dilute acid, or base, or by an aqueous solution of a salt of the metal of primary interest in the solvent phase. Scrubbing tests are carried out in a manner similar to those described earlier with the loaded solvent being contacted with the scrub solution at the appropriate concentration, pH, phase ratio, contact time, and temperature. After separating the phases, the organic phase is usually analyzed to determine whether the unwanted metals, and how much of the metal of interest, have been removed. The best scrub solution and conditions are determined by varying the individual conditions.

The scrub raffinate may contain substantial amounts of valuable metals; thus, consideration should be given to where in the overall process this raffinate should be recycled.

Data obtained in scrubbing tests can be presented graphically, such as by plotting the concentration of metals in the organic phase against A/O ratio, salt concentration in the scrub solution, temperature, etc. Typical plots for the removal of nickel from cobalt in a DEHPA-containing solvent are shown in Figs. 7.3 and 7.4 [3].

7.5 EXTRACTANT CONCENTRATION

The concentration of extractant required to extract a metal at a phase ratio suitable for a given type of contactor will depend essentially on the metal concentration in the aqueous feed solution. Accordingly, low metal concentrations will not require high extractant concentrations in the solvent, since the large phase ratio needed in such cases (so that large volumes of unused solvent will not be cycling through the circuit) may not be compatible with the contactor. On the other hand, high metal concentrations in the feed would probably require high extractant concentrations in the solvent to maintain a suitable phase ratio.

The loading characteristics of an extractant can be determined as described above. With this information and knowledge of the different types of contactors available, the metal concentration in the feed, the extraction rate of the metal, the approximate concentration of extractant in the solvent can be determined. The optimum concentration can be determined only by operating a continuous circuit.

One problem that should be mentioned here again is that of the coextraction of a second (unwanted) metal, as discussed in Chapter 8. This

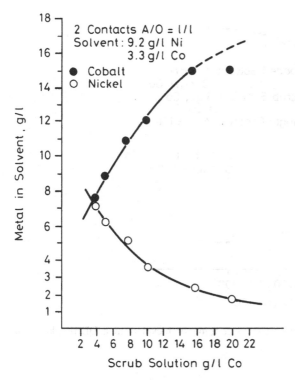

Fig. 7.3 Effect of scrub solution composition on the removal of nickel from cobalt in a DEHPA-kerosene solvent.

might occur if this second metal has extraction characteristics similar to those of the metal that is preferentially extracted. To inhibit coextraction, it is advisable to operate a circuit at as close to saturation loading of the solvent with the metal of primary interest as is practical; in this way, coextraction of other metals is depressed. As the extractant concentration in the solvent in a continuous operation is fixed, if fluctuations in the concentration of metal in the feed occur, any decrease in concentration will result in coextraction of the unwanted metal. Conversely, an increase in metal concentration would result in a loss of metal to the raffinate. This problem is usually minimized by using a large feed tank for the solvent extraction circuit to level out such fluctuations in metal concentration to the solvent extraction circuit.

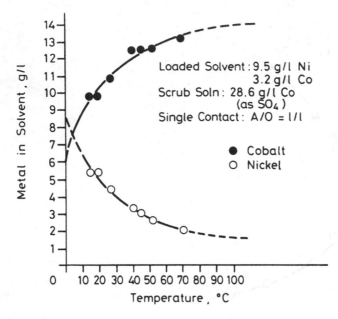

Fig. 7.4 Effect of temperature on the scrubbing of nickel from a DEHPA-kerosene solvent.

7.6 DILUENT AND MODIFIER REQUIREMENTS

Probably the first tests that should be carried out to determine the best diluent for the system are those concerned with coalescence rates of the phases. These can be done in several ways, but all comparisons should be made using the same experimental conditions.

One simple, but effective, method is to make up solvents with each of the diluents to be tested, with the same concentration of extractant (and modifier if necessary) in each diluent. Place measured volumes of the solvent and aqueous phases in a graduated beaker or suitable container, and mix for a given period at the same phase continuities. The degree of agitation should be the same for all tests and is best carried out with a mechanical shaker or stirrer [4]. As soon as the desired agitation time has elapsed, start a timer, and place the container in an upright position. Measure the length of time required for the primary and secondary phase breaks to occur. The rate of coalescence can also be determined by timing the rise of the coalescing band up the container.

The effects of different diluents on the extraction and stripping properties of a solvent can be determined by similar mixing tests as described above.

The formation of a third phase during the extraction of a metal by solvent (extractant and diluent) cannot be tolerated in liquid–liquid extraction processes for obvious reasons. Elimination of a third phase is usually accomplished by the addition of a modifier to the solvent, or by increasing the temperature of the system.

The amount of modifier required to prevent third phase formation can be determined in the following way. The aqueous and solvent phases are first contacted, and once the three phases have separated, the lower aqueous phase is drawn off and discarded. The modifier to be considered is then added from a burette in small increments to the two organic phases, and the mixture shaken after each addition. The amount of modifier required to produce a single organic phase is then used to calculate the amount required to be added to the solvent. Generally, about 2–5 vol% of modifier is needed, but more may be necessary if high concentrations of extractant are used in the solvent. Any effects of modifiers on the kinetics and equilibria of metal extraction and stripping can be determined by shakeout tests.

7.7 TEMPERATURE EFFECTS

Temperature can have a considerable effect on both the extraction and stripping properties of a solvent extraction system relative to equilibrium, kinetics, and metal separations [5] (see Chapters 3–5). Therefore, it is advisable to investigate these effects, especially when the solvent tends to be viscous or high loadings of metal are to be obtained in the solvent (Fig. 7.5).

7.8 STAGEWISE SEPARATIONS

Many of these experimental methods can be used in the development of systems more complex than the extraction of a single substance, such as metal separations. Metal separation processes may involve as few as two metals and as many as 15, as in the rare earths (see Chapter 11).

Consider the separation of two chemically similar metals having very similar extraction properties, such as cobalt and nickel. Some of the available commercial extractants show very poor discrimination between these metals. Obviously, a few stages of extraction will be insufficient to obtain a good separation. An approximate idea of how many stages are required can be obtained by contacting a solvent several times with fresh feed containing

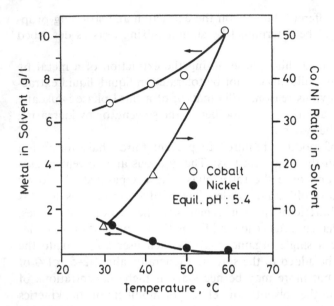

Fig. 7.5 The effect of temperature on the loading and separation of cobalt and nickel. Solvent: 15 vol% DEHPA; 5 vol% TBP; 80 vol% kerosene.

both these metals. Analysis of the solvent after each contact stage, and plotting these concentrations against the number of stages, will give a curve for each metal similar to those shown in Fig. 7.6. Without completing the curves, a good approximation of the number of stages required to provide a given Co/Ni ratio can be obtained by extrapolation.

In the example shown in Fig. 7.6, using an alkylphosphoric acid, about 60 stages are required to give a Co/Ni ratio of about 100. So many stages would be too many for mixer-settler operation, and other types of con-tactors would have to be considered. In this particular example, a sieve plate pulsed column has been shown to be very effective [3]. However, with the development of the alkylphosphonic and alkylphosphinic acids, the sep-aration of cobalt and nickel can be achieved in very few stages, owing to the high rejection of nickel (see Chapter 11).

It is a good plan to begin solvent extraction studies involving several metals with solutions containing only single metals to obtain the basic extraction data. It also provides time for any necessary analytical develop-ment to be carried out. However, synthetic solutions should not be worked on longer than necessary, as the actual plant liquor will almost certainly be much different. Another point to be made here is that in leaching, filtering, precipitation, and flotation, chemicals are added to aid these processes.

These chemicals can occasionally cause adverse effects, such as poor phase disengagement, during solvent extraction (see section 7.13).

7.9 SCALE-UP DECISIONS

As in any process development work, the data have to be examined critically to make certain that there is sufficient reason to carry the project on to the next phase. Thus, if a particular system has poor stripping characteristics, it is very doubtful whether any improvement can be made by going to a larger scale, in which case the investigation may have to be terminated. However, if the bench data have produced sufficient information to draw a conceptual flow sheet, then a decision can be made on whether to run a small-scale continuous operation or a pilot plant. Many other factors have to be considered in making this decision, such as the economics of the process, cost of

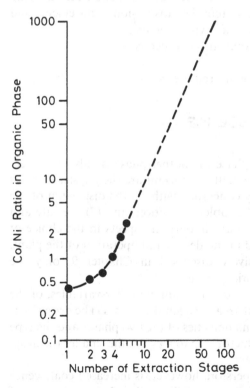

Fig. 7.6 Cobalt-nickel separation as a function of number of contact stages: concentration of cobalt and nickel $20 \, \text{g dm}^{-3}$ each in aqueous feed solution.

construction, materials and labor, sale of the anticipated product, possible by-products, environmental aspects, and so on.

In the development of solvent extraction systems, the initial considerations of the chemistry of the system will limit the choice of extractant, diluent, and modifier. Shake-out tests and a study of extraction and stripping characteristics will further limit the choice. Simple graphic construction of extraction and stripping isotherms will permit the design of a small-scale continuous operation in terms of the number of stages and flow rates.

In summary, the following parameters are required before considering scale-up to a larger system:

The solvent system: extractant, diluent, and, if necessary, modifier
Equilibrium data: extraction and stripping isotherms
Kinetic data, as these will govern, to a large extent, the type of contactor
    required
Approximate number of theoretical stages for extraction and stripping
The necessity or desirability of a scrub stage between extraction and stripping
The physical characteristics of the system: i.e., dispersion requirements and
    settling rates, emulsion tendencies, viscosity, etc.
Approximate solubility of the solvent in the aqueous phase
Temperature effects
Solvent pretreatment and conditioning requirements

7.10 DISPERSION, COALESCENCE, AND CONTACTORS

The transfer of a metal from one phase to another phase is only one aspect of the solvent extraction process; equally important are the physical aspects of the system. These are primarily concerned with (1) the dispersion of the phases on mixing, (2) the type of droplet formation, and (3) the rate and completeness of coalescence. These are important aspects in the choice of suitable contacting equipment and in the design and operation of the plant. Because these topics are extensively discussed in Chapter 9, only the essential points are summarized briefly here.

The rate of mass transfer is a function, among other variables, of the drop size distribution or interfacial area between the phases. The drop size is governed by the surface tension, and densities of the two phases and the type of agitation and design of the contactor. Up to a point, the smaller the drop, the greater the rate of mass transfer.

Coupled with increasing dispersion, however, is increased coalescence time and, therefore, the size requirements of the settler. Also, as the size of the droplets of the dispersed phase decreases, the droplets behave more and more like rigid spheres, and so the rate of mass transfer decreases. Thus, the kinetics, along with the proper dispersion and coalescence, will influence the choice of contacting equipment. In addition to drop size, other factors affecting coalescence include the temperature, viscosity, and density of the phases, and the presence of surface-active agents, or solids. The addition of wetting materials within the contactor or phase separation equipment will promote the coalescence rate.

Coalescence can be divided into primary and secondary break time. *Primary break time* is the time required for the two phases to meet at a sharply defined interface and provides a measure of the settler or phase separation requirement of the process. The *secondary break* concerns the fine haze of droplets and is the major cause of entrainment losses. The speed and completeness of phase disengagement have a marked effect on the capital and operating costs of the plant and are directly linked to the design and method of operating of the mixer. In general, but only to a point, higher extraction efficiencies are achieved as the speed of the mixer is increased and the resulting droplet size of the dispersion is reduced. However, the dispersion band thickness increases as the droplet size of the dispersion is reduced, and a larger settler area (with consequently larger inventories of solvent) are then required for a given plant throughput. Figure 7.7 shows the variation of dispersion band depth with the total flow/unit settler area under varying conditions of mixing, expressed by the term $n_R^3 d^2$ where n_R is the speed of the impeller (rev/s) and d its diameter (ft) [6].

Increased capital costs will be incurred if the mixer is designed to produce a high proportion of extremely small droplets, and this cost has to be balanced against the increased revenue obtained from the improved extraction efficiency. At upper levels of $N^3 D^2$, the problem of organic carry-over with the aqueous phase may become an important consideration. In addition to the directly calculable increase in settler costs when operating with very highly dispersed phases, the possibilities of emulsion formation and flooding caused by the accumulation of "crud" in the settler must be considered. Both of these become more likely as the settler duty becomes more severe. Many authors have cited design ideas on mixer settlers [7, 8].

In all practical mixing systems, a range of droplet sizes is produced by the mixer, and there will always be a proportion of droplets of the dispersed phase that will not settle out readily and, therefore, will remain entrained in the continuous phase leaving the settler. Impellers that consist of enshrouded radial vanes create very high shear rates within the mixer, particularly when they are run close to the bottom of the mixing tank to provide a pumping as well as a mixing action. These high-shear rates are the prime cause of the haze of fine droplets produced in many plants. Thus, the design

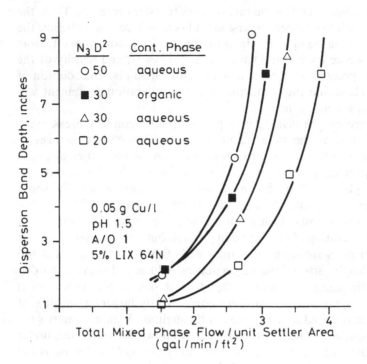

Fig. 7.7 Effect of variation of mixer N^3D^2 on settler dispersion band depth.

of the impeller blades is important in minimizing organic entrainment in the aqueous phase [7].

The major emphasis in the development and optimization of contactors has been to increase throughput while maintaining efficiency, and to increase efficiency without reduction in throughput [1,5]. Many types of contactors are available for achieving mass transfer, each with its own particular advantage so that there is no single contactor that is best for all solvent extraction processes, either technically or economically. A classification chart of various contactor types is given in Fig. 7.8 [1].

Mixer-settlers have been the more common type of equipment and, with the development of hydrometallurgy over the past 20 years, designs have improved considerably. To select the appropriate equipment, a clear understanding of the chemical and physical aspects of the process is required. Also the economics must be considered relative to the type of equipment to suit particular conditions of given throughput, solution and solvent type, kinetics and equilibrium, dispersion and coalescence, solvent losses, number of stages, available areas, and corrosion.

The design procedure for extracting equipment requires the evaluation of the following:

Number of stages of extraction, scrubbing, and stripping
Stage efficiency
Flow capacity
Droplet size

The operation of equipment is dependent on the dispersion and coalescence of the phases to achieve mass transfer. For example, systems with high mass transfer rates do not require the formation of the extremely fine droplets that systems of low rates of mass transfer usually require.

In stagewise equipment, the design and scale-up is simple and is often determined from bench data. The stage efficiency is usually high, and the capacity is determined by the settler design necessary to achieve coalescence of the dispersed phases. For differential contactors, such as columns, the flow capacity is determined by the droplet size and the type of internals (see

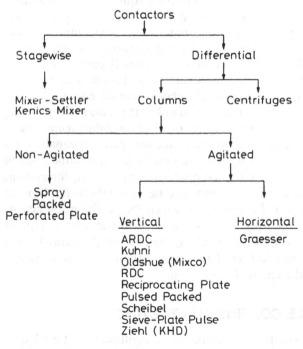

Fig. 7.8 Contactor classification.

Chapter 9). Differential contactors are more susceptible than mixer-settlers to axial and backmixing. If plug flow can be used, there would be little problem, but this rarely occurs. Backmixing can be caused by localized high velocities in the continuous phase, resulting in:

1. Entrained dispersed phase droplets, causing disperse-phase backmixing
2. Turbulent eddies in the continuous phase
3. Nonuniform velocity distribution normal to the direction of the flow

The net result of such deviations from countercurrent plug flow is to decrease mass transfer as well as the total throughput per cross-sectional area. If no allowance is made for backmixing, the resulting plant will be considerably underdesigned.

Some advantages and disadvantages of stagewise and differential contactors are shown in Table 7.1 [1].

The types of equipment used, which range from stirred tanks and mixer-settlers to centrifugal contactors and various types of columns, affect both capital and operating costs [9]. In the decision to build a plant, the choice of the most suitable contactor for the specific situation is most important. In some systems, because of the chemistry and mass transfer rates involved, several alternative designs of contacting equipment are available. In the selection of a contactor, one must consider the capacity and stage requirements; solvent type and residence time; phase flow ratio; physical properties; direction of mass transfer; phase dispersion and coalescence; holdup; kinetics; equilibrium presence of solids; overall performance; and maintenance as a function of contactor complexity. This may appear very complicated, but with some experience, the choice is relatively simple.

A comparison of contactors is not valid unless the same system is being compared under the same process conditions of phase ratio, temperature, solution composition, etc. Because not all contactors perform equally for a given process, comparative tests must be run under identical conditions to yield reliable physical and economic comparisons. In general, for systems requiring up to four theoretical stages, mixer-settlers probably offer the most economic solution; however, for systems in which the partition coefficients are small, therefore necessitating many stages of extraction, some type of differential contactor is usually preferred. An analysis of the many factors will generally result in the selection of the most economic contactor that is capable of performing the required separation.

7.11 SMALL-SCALE CONTINUOUS TESTING

Commercially available equipment for small-scale continuous test work on a solvent extraction process is limited. Generally, a series of small mixer-settlers

Table 7.1 Equipment Performance

	Mixer-settlers	Nonagitated differential	Agitated differential	Centrifugal
Advantages	Good contacting of phases Handle wide range flow ratio (with recycle) Low headroom High efficiency Many stages Reliable scale-up Low cost Low maintenance	Low initial cost Low operating cost Simplest construction	Good dispersion Reasonable cost Many stages possible Relatively easy scale-up	Handle low gravity difference Low holdup volume Short holdup time Small space requirement Small inventory of solvent
Disadvantages	Large holdup High power costs High solvent inventory Large floor space Interstage pumping may be necessary	Limited throughout with small gravity difference Cannot handle wide flow ratio High headroom Sometimes low efficiency Difficult scale-up	Limited throughout with small gravity difference Cannot handle emulsifying systems Cannot handle high flow ratio Will not always handle emulsifying systems, except perhaps pulse column	High initial cost High operating cost High maintenance Limited number of stages in single unit, although some units have 20 stages

having up to about 500 cm^3 capacity for each stage has been used. Much of the early work employed homemade equipment, and even now much of the equipment is constructed in the laboratory.

Several designs of small-scale mixer-settler units can be obtained commercially and can be assembled to provide several extraction, scrub, and strip stages. Regular turbine blades are used for pumping and mixing; thus the mixer-settlers are horizontal and the cascading principle is not necessary. The major problem with small-scale equipment made of Plexiglas is that it tends to craze quite rapidly and discolor. Consequently, it becomes difficult to observe accurately the interface level or the dispersion band in the settler. However, the ease with which mixer-settler units can be made out of Plexiglas means that they can be discarded and replaced by new ones when such conditions occur.

Mixer-settlers are the smallest continuous countercurrent equipment available. Other types of contactors can be purchased or constructed, but require larger volumes of solutions for operation. Hence, a Karr column is available having a 1 inch diameter column. Similarly, a sieve plate pulse column of similar dimensions can be readily constructed. However, scale-up from small diameter columns such as these is not too satisfactory [G. M. Ritcey, unpublished data]. The major problem appears to be that wall effects in small-diameter columns significantly influence both the physical and chemical characteristics of a solvent extraction system. Thus, a 2 inch diameter sieve plate column appears to be the minimum size for which wall effects have a minimum effect on the system. Other columns, with greater axial and backmixing, require a larger diameter column for scale-up. Whether a 2 inch diameter column can be considered small-scale or pilot plant size is debatable as the total flow capacity of such a column can be up to 0.5 US gal min^{-1} (1.8 dm^3 min^{-1}). The size of a pilot plant, as we have noted before, covers a wide range of throughput.

Other small-scale contacting equipment is available, e.g., centrifuges and the "raining bucket" or RTL contactor. The minimum flow rate of available centrifuges seems to be about 1 gal (US) min^{-1} (3.6 dm^{3}min^{-1}). A small-scale raining bucket contactor can be obtained that is suitable for bench-scale operations. The 4 inch (10 cm) diameter, 30 inch (0.76 m) long unit would have a maximum throughput of about 1 US gal hr^{-1} (3.6 dm^3 hr^{-1}).

In-line mixers manufactured by, for example, Kenics, Lightning, and Sulzer are also applicable for continuous small-scale testing of a solvent extraction process, and 1 inch diameter models are available. This mixer system can be used either horizontally or vertically. However, few data are available for this type of contactor, although they would appear to offer many possibilities, not only for liquid–liquid systems, but also for use in solvent-in-pulp processes (in the case of perhaps the Kenics and Lightning mixers) [G. M. Ritcey, unpublished data].

Because of the diversity of contacting equipment available, it is unlikely that all these contactors will be available in any one laboratory or pilot plant. Consequently, unless test work is carried out on similar contactors, the system may not be optimized. Since mixer-settlers are the easiest to construct, are simple to operate, and require little room and low-flow rates, these contactors are, in many cases, the only ones used to investigate a continuous solvent extraction process. This is by no means ideal and may result in abandonment of a process that, using another type of contactor, could be found to be entirely satisfactory.

The data obtained from small-scale continuous operations will be required to determine whether the process should be investigated on a larger scale, such as a pilot plant. These data should be sufficient to draw a conceptual flow sheet, which will include a number of stages for extraction, scrubbing [10], and stripping; the flow rates, size and type of equipment, and the various parameters considered earlier in this chapter. Another important aspect should also be considered in the continuous test work, and that is chemical analysis of both the aqueous and organic phases for their various components.

7.12 PILOT PLANT OPERATIONS

The pilot plant design should be small enough to provide the maximum amount of data at a minimum cost, while being large enough to provide sufficiently accurate design data to design the full-sized plant. The actual size and capacity of the pilot plant depends on many factors; thus, it is impossible to indicate optimum size-capacity parameters. For example, pilot plant operations at Duval Corporation, Arizona, used a $75 \, \text{gal} \, \text{min}^{-1}$ $(270 \, \text{dm}^3 \, \text{min}^{-1})$ flow rate, whereas those for the Bluebird operation (Arizona) used a total throughput of $0.2–0.6 \, \text{gal} \, \text{min}^{-1}$ aqueous at an O/A ratio of about 3 [total throughput approximately $0.8–2.4 \, \text{gal} \, \text{min}^{-1}$ $(3.0–9.0 \, \text{dm}^3 \cdot \text{min}^{-1})$].

Choice of contactors for pilot plant operations will probably have been made following bench-scale and small-scale test work, during which the kinetic and separation factors were determined. The design of the mixing equipment has to take into account the necessary volume throughput and the required contact time. Contact times can be extended in mixers by partial recycling. In mixer-settler operations, the settlers are normally designed with a length/width ratio of between 2 and 4, to minimize channeling effects and excessive velocity of the solutions in the settler. The area of the settler will depend on the coalescence rate of the dispersed phase and, therefore, will vary somewhat from one extraction system to another. In copper extraction circuits, settler area is usually based on a phase disengagement rate of $2\,\text{gal}\,\text{min}^{-1}\,\text{ft}^{-2}$ ($0.08\,\text{m}^3\,\text{m}^{-2}\text{min}^{-1}$), allowing for a 2 inch (5 cm) dispersion band. Solvent depth varies up to 1.5 ft (0.45 m).

Equipment chosen for pilot plant operations should closely approximate the commercially available equipment that will be used eventually. Thus, it would be of little use to operate a pilot plant using stagewise contactors if differential contactors were to be employed in the actual commercial operation.

Metering of flows is a most important consideration in a pilot plant, and these should be accurately controlled wherever possible. Thus, small upsets in flow conditions, resulting in a change in O/A ratio, can create many problems that do not become apparent until analyses are done on the various solutions.

Pilot plant operations, as we have noted previously, can vary between extremes of flow rates. It is necessary, therefore, that the feed and reagent volumes be large enough that the pilot plant may be operated for a sufficient length of time to obtain meaningful data. For example, if the aqueous feed to the solvent extraction (SX) circuit is being produced batchwise, variations between batches are bound to occur. Such variations should be controlled as much as possible. Batchwise production of the feed solution may be very different from feed to the actual plant, especially if the plant process involves continuous production of the feed to the SX circuit.

It must also be remembered that feed solutions that are allowed to stand for any length of time (such as when they are shipped from one place to another) may behave in quite different ways from a freshly prepared feed solution in the SX circuit. Furthermore, solutions shipped in cold climates may freeze during shipment, and this may result in problems in SX, such as emulsion formation, that would not occur if freshly prepared solutions were employed. The preparation and treatment of feed solutions for small-scale and pilot operations are important considerations in scale-up, and failure to recognize this may well result in considerable problems in the final plant.

Periodic sampling and analysis are essential to the success of the operation. With experience, pilot plant operators, in many cases, can quickly detect upsets without having to rely on analyses. For example, a change in color of the solvent or raffinate, in many cases, is an indication that a change in the operating conditions has occurred.

A pilot plant can be relatively inexpensive if construction and installation is done in-house, using readily available materials and equipment. Of course, the size of the pilot plant will determine whether in-house construction is feasible, as will the knowledge and experience of the personnel. On the other hand, it may prove necessary to use one of the several consulting companies who have experience in this field, or to have the pilot work performed by the manufacturers of the contactors chosen for the operation.

It will be instructive to consider briefly the methodology employed in the development of solvent extraction processes that have become operational. The development of the Bluebird Mine operation for the extraction of copper from dump leaching liquors by solvent extraction and the subsequent recovery of copper as either copper sulfate or cathode copper is used as an example. The initial investigations [11] included the following:

1. Bench-scale batch tests to determine the characteristics of the extraction of copper from sulfuric acid liquors by LIX 64
2. Continuous bench-scale operations to study the extraction of copper and other metals present in the leach liquor: effects of variables such as temperature, O/A ratio, solvent composition, extraction kinetics, settling characteristics, etc.
3. Determination of solvent losses from solubility, entrainment, and chemical degradation
4. Stripping characteristics and the production of copper sulfate crystal, saturated copper sulfate solution, powdered copper, or electrolytic copper from the strip liquors

Following the successful completion of this test work, pilot plant operations were initiated, having the following objectives:

1. To establish the economic feasibility of producing wire bar-grade copper by a combined SX–EW (EW = electrowinning) process
2. To develop the engineering data to enable estimates of operating and capital costs of a process capable of producing approximately 30,000 lb (13,600 kg) of copper per day
3. To develop the engineering data required for the design and operation of the commercial plant
4. To produce electrolytic copper for evaluation of product quality

Other objectives of the pilot plant operation include those given for bench-scale and small-scale continuous studies, together with the following:

1. Determining the effects of impurity build-up on recycle of the extraction raffinate to the heap leaching operation, and on recycle in the extraction and electrowinning circuits;
2. Obtaining data on the maximum capacity of settler units
3. Obtaining data on methods for controlling flow, pH, and metal concentration
4. Measuring the transfer of acid between the strip and extraction circuits and establishing methods for control

5. Establishing control methods between the solvent extraction and electrowinning circuits, for flow rates, acid concentrations, and effect of recycle liquors on copper purity
6. Establishing the optimum conditions for operation of the electrowinning circuit

This outline of the many factors that have to be investigated for the design and operation of an SX–EW plant is perhaps typical of the development of an SX process on which little or no prior work has been done. Today, with the knowledge and experience available, processes similar to those already in operation may be designed and scaled up more economically. The Bluebird Mines operation took about 4 years from initial bench-scale tests to plant operation.

7.13 SOLVENT LOSS, RECOVERY, AND ENVIRONMENTAL CONSIDERATIONS

Any economic advantage of a solvent extraction process over other separation processes for metals can be lost if the loss of solvent from the system becomes too high.

Solvent loss can occur in essentially five ways: by solubility in the aqueous phase; by entrainment; by volatilization; by degradation; and by loss in crud. In solvent-in-pulp (SIP) processes a further distinction can be made, that of sorption of the solvent on the solid particles in the system. Furthermore, there are the additional losses from sampling of the process and from the spillage that may be found in numerous operating plants. For most systems, loss of solvent by solubility in the aqueous phase is usually small, and little can be done about this type of loss. Entrainment losses can vary considerably, usually because of poor equipment design, instability in the system, poor diluent choice, etc. Loss of solvent by volatilization will depend on the temperature of the system and on how well the system is enclosed. Excessive agitation together with elevated temperatures will cause another type of loss, often associated with volatilization losses. This is loss by misting. Where volatilization losses mean the vaporization of the organic components at that temperature, misting will result in the loss of all components of the solvent as well as the aqueous phase. Sorption of solvent components onto solid particles seems to depend largely on the mineral composition of the solids, and extractant degradation is a function of the stability of the extractant, composition of the aqueous phase, and temperature. Crud losses are due to many causes, to be explained later, but primarily high shear, suspended solids, and composition of the aqueous phase.

The importance of solvent loss in a solvent extraction process cannot be overemphasized, because of the constraints imposed by economics and environmental considerations.

7.13.1 *Soluble Losses*

Soluble loss of a reagent (extractant, modifier, or diluent) from the solvent phase is an inherent part of the solvent extraction process, since all organic compounds are soluble, to some extent, in water. The conditions prevailing in the system can also promote solubility, which can be a particular problem if the composition and properties of the aqueous phase are inflexible. For example, the solubility of alkylphosphoric acid and carboxylic acid extractants is dependent on temperature, pH, and salt concentration in the aqueous phase.

Low-molecular-weight extractants can generally be expected to have uneconomic solubilities in most systems, but where high salt concentrations prevail, the solubility may be substantially lower and may be economic. This has been shown to be true for naphthenic and Versatic acids, which have high solubilities in water but appear to be economically useful when used in high ammonium sulfate liquors, such as those produced in the Sherritt-Gordon process for the extraction of cobalt and separation from nickel [12].

Three causes of extractant solubility in the aqueous phase may be distinguished: solubility of un-ionized and ionized extractant and metal-extractant species. For extractants such as acids, amines, and chelating reagents, their polar character will always result in some solubility in the aqueous phase over the pH range in which they are useful for metal extraction. Solubility depends on many factors including temperature, pH, and salt concentration in the aqueous phase, as discussed in Chapter 2.

Increasing the temperature of a system can be expected to increase the solubility of all the components of a solvent. Variation of pH is likely to affect only the polar components of the solvent. By increasing the salt concentration in the aqueous phase, most solvent components are expected to exhibit decreased solubility because of salting out. The solubility of a solvent component in the aqueous phase is also likely to increase with increase in the concentration of the component in the solvent.

The nature and composition of the aqueous phase can also provide conditions for solvent degradation or, more particularly, for extractant degradation. For example, the presence of oxidizing agents, such as nitric acid, could result in oxidation, especially if the extractant contains a readily oxidizable bond or functional group. Tertiary amines are usually quite susceptible to oxidation. Again, the presence of oxidizing agents such as

permanganate, vanadate, or chromate can give rise to extractant degradation. The most thoroughly investigated extractant for this is tributylphosphate (TBP), and considerable data have been reported on this reagent [13].
Generally, the degradation products are the dibutyl and monobutyl phosphates, which have substantial solubility in the aqueous phase. Such
degradation complicates the analytical determination of extractant losses.
Extractant degradation may not be apparent, and thus a decrease in metal
loading resulting from degradation may be wrongly considered to be due to
extractant loss to the raffinate.

7.13.2 Solubility of Extractants in Aqueous Solution

7.13.2.1 Carboxylic Acids

The solubility of carboxylic acid extractants in aqueous solution is a function of temperature, pH, and salt concentration. Most of the information on
the solubility of these reagents is limited to their solubility in water, and not
much practical information is available on actual leach liquors and similar
solutions.

Under conditions of low salt concentration in the aqueous phase, and
above a pH ~6, the solubility of these acids is economically prohibitive.
Even at lower pH values, the solubility could be considered high from the
point of view of pollution. For a C_7-C_9 fraction of aliphatic mono-
carboxylic acids, the solubility in water as a function of concentration in the
solvent phase is shown in Fig. 7.9 [14]. Here again, complete information is
unavailable, since no reference is made to pH or salt concentration. Presumably, the pH is in the range 4–6. However, it is interesting to note the
rapid increase in solubility in the aqueous phase as the extractant concentration is increased. This effect also applies to other extractants.

The solubility of Versatic 911 in water has been studied as a function
of both pH and salt concentration for a fixed ($0.5 \, mol \, dm^{-3}$) concentration
of reagent in the solvent [15] and is illustrated in Fig. 7.10. The use of high
salt concentrations ($4 \, mol \, dm^{-3}$ ammonium sulfate) has been shown to be
effective in keeping Versatic 911 losses at pH 7–8 to more economic levels
(<100 ppm) in the separation of cobalt from nickel [16].

7.13.2.2 Alkylphosphoric Acids

Available information on DEHPA indicates an effect of salt concentration
similar to that shown for the carboxylic acids [17]. Figure 7.11 indicates a
significant difference between sodium and ammonium salts and solutions
containing sodium and ammonium hydroxides. Moreover, below about
5 wt% salt concentration, the solubility becomes uneconomic. However, all
the data given in Fig. 7.11 are for alkaline systems. For acid systems

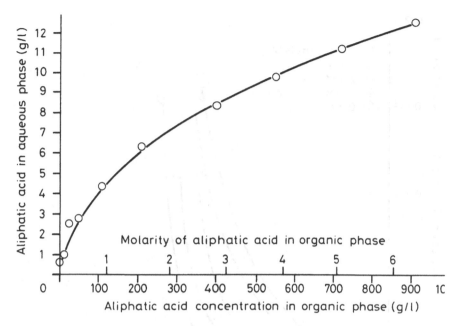

Fig. 7.9 Solubility of C_7–C_9 aliphatic carboxylic acids in water. (From Ref. 14.)

(H_2SO_4), the solubility of DEHPA from plant operating data is probably less than 30 ppm and varies somewhat with pH.

The solubility of the sodium salt of DEHPA in basic (NaOH) solution has been reported, together with the effect of temperature on the water solubility of this salt [18], (Figs. 7.12 and 7.13). It is evident that the presence of salts in the aqueous phase depresses the solubility of this extractant in water (Fig. 7.11). This has been confirmed in the extraction of cobalt with DEHPA(Na) at pH 5–6, for which a solubility of the extractant was found to be <50 ppm. Furthermore, the use of DEHPA in the extraction of cobalt from an ammoniacal (pH 11) system containing sodium sulfate showed no apparent loss of extractant after 10 contacts of a DEHPA–kerosene solvent with fresh aqueous solution [1]. Operation of pilot plants using DEHPA(NH_4) and DEHPA(Na) for the extraction of cobalt, at pH 5–6 and at 60°C, showed the loss of DEHPA to be less than 50 ppm [3]. Temperature also has a significant effect on the solubility of DEHPA(Na) (Fig. 7.13).

7.13.2.3 Amines

Aliphatic amines used in solvent extraction operations have low solubility in acidic aqueous solutions, generally below about 10 ppm. Solubility is

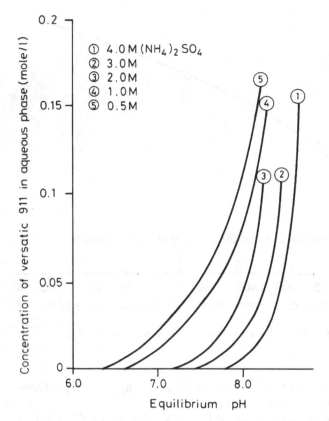

Fig. 7.10 Solubility of Versatic 911 ($0.50 \, mol \, dm^{-3}$) in ammonium sulfate solution. (From Ref. 15.)

a function, among other things, of the number of carbon atoms in the molecule. Thus, low-molecular-weight amines are substantially soluble in water, whereas those of molecular weight greater than about 200 are essentially insoluble. The degree of chain branching also affects solubility; the greater the branching, the lower the solubility. These effects are also shown by other extractant types.

Studies on the solubility of Amberlite LA-2, Adogen 368, and Alamine 336 [19] have shown that in acidic sulfate solutions (pH 1.4 and 1.8), the solubilities of these three amines tend to a limiting solubility of 10 ppm or less (Fig. 7.14). The initial high solubilities shown for each of these extractants are probably not due to solubility of the extractant, but to the presence of more soluble secondary and primary amines contained in the commercially prepared materials. Presumably, the method of analysis used for the

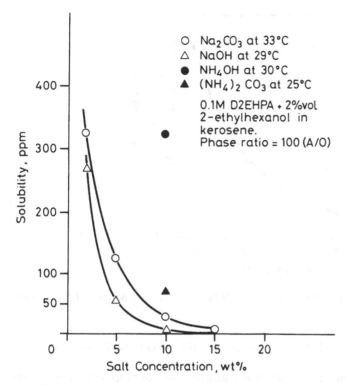

Fig. 7.11 Solubility of DEHPA in various aqueous solutions. (From Ref. 17.)

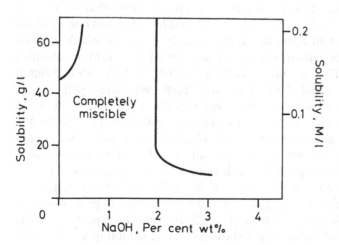

Fig. 7.12 Solubility of the sodium salt of DEHPA in NaOH (wt%) solution at 20°C.

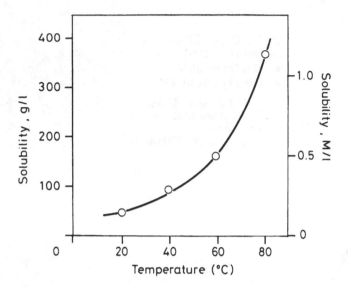

Fig. 7.13   Solubility of the sodium salt of DEHPA in water as a function of temperature.

determination of soluble amine loss provided total (primary, secondary, and tertiary) losses.

In most commercial operations, amine losses of more than 10 ppm are reported.

7.13.2.4   Chelating Extractants

Solubility data on the LIX and Kelex extractants indicate that these materials are poorly soluble in aqueous media. Accordingly, in plant operations at about pH 1.5, reported losses are approximately <15 ppm, which includes both soluble and entrainment losses as determined by inventory (detailed later in Table 7.6). The solubility of LIX 63, LIX 64, and LIX 64N in water at pH 4.8 has been reported at 5.8, 4.3, and 6.2 ppm, respectively [20].

The solubility of Kelex 100 has been reported to be 4.4 (pH 0.5), 1.4 (pH 1.0), and 1.6 (pH 1.5) ppm, in tests employing $0.5 \, mol \, dm^{-3}$ Kelex 100 in Solvesso 150 diluent for the extraction of copper from solutions containing about $30 \, g \, dm^{-3}$ of metals (Cu, Ni, and Co) [21]. Furthermore, over several months no change in the Kelex concentration was found in pilot plant operations with a $0.5 \, mol \, dm^{-3}$ Kelex solution [21]. At higher pH values (up to 9) Kelex solubility has been determined at <1 ppm in ammonium sulfate solutions up to $300 \, g \, dm^{-3}$ [22]. It would be expected that the amphoteric nature of Kelex would show increasing solubility at pH values above 7, and lower than about 1, similar to the variation in solubility of

8-hydroxyquinoline. Presumably, with high salt concentrations in the aqueous phase, solubility is suppressed. The one result at pH 0.5 (4.4 ppm) would suggest increasing solubility of this reagent as the pH of the system is decreased below 1.

7.13.2.5 Tributylphosphate (TBP)

Relative to other extractants, a substantial amount of information is available on the solubility of TBP in various aqueous media [23]. The solubility of this extractant in water is a function of temperature [24], and decreases with increase in temperature.

In aqueous acid solutions, solubility varies depending on the type of acid and on its concentration. Thus, in halo acids, the solubility increases markedly with acid concentration, a general order being HI > HBr > HCl. For nitric and phosphoric acids, solubility decreases up to about 8 mol dm^{-3} acid, and then begins to increase.

The effect of increasing TBP concentration in the organic phase results in an increase in its solubility in water. Furthermore, the diluent used also affects the TBP solubility.

7.13.3 Solubility of Modifiers

There are little or no data on the stability or solubility of the various reagents used as modifiers, except TBP. Neither is the relationship between solubility losses and concentration in the solvent known nor whether it is a function of only its solubility in the aqueous phase. Of those alcohols used as modifiers, the only solubility data appear to have been determined by

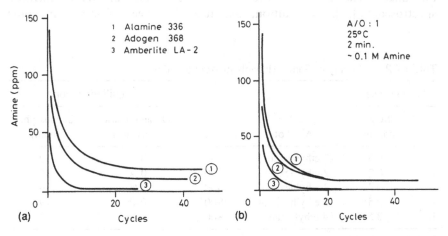

Fig. 7.14 Solubility of amines in acidic sulfate solution at (a) pH 1.4 and (b) pH 1.8.

Blake et al. [17], Table 7.2. Aqueous solubility decreases with increase in the number of carbon atoms and with a decrease in concentration in the organic phase. This observation is general and has been found to apply in other cases, for example, for C_8–C_9 carboxylic acids, for which the solubility is almost a linear function of the carboxylic acid concentration in the organic phase [25]. (see also Chapter 2).

7.13.4 Solubility of Diluents

Here again, few data are available on the solubility of the various diluents, in either its use or being studied for use in solvent extraction circuits. Although this is expected to be low (<2 ppm), what few data are available apply to water and not to solvent extraction raffinates (see, however Chapter 2).

7.13.5 Reagent Stability and Degradation

Degradation of solvent components can arise from various sources, such as oxidizing properties of the aqueous phase; too high a temperature of the system; aging and oxidation by atmospheric oxygen; and general instability of the component over prolonged periods of use. In this context, it is apparent that the LIX extractants are susceptible to dehydration resulting from the adjacent oximato and hydroxyl groups. Indeed, the manufacturer of the LIX reagents states in various publications that those containing LIX 63 should not be used above about 40°C. Undoubtedly, dehydration occurs and is readily observed. Thus, for example, if LIX 65N is heated, globules of water are formed. The ease of dehydration of the LIX reagents is also illustrated by the fact that, even under high vacuum, they cannot be distilled unchanged. The distillates always show the absence of OH by infrared spectroscopy [26]. In addition, oxidation has been observed in some

Table 7.2 Solubility of Some Alcohols in Acid Solution

Aqueous phase				Equilibrium conc	
pH	SO_4^{2-} (mol dm^{-3})	Alcohol	O/A ratio	Organic phase (wt/vol%)	Aqueous phase (ppm)
1	0.5	2-ethylhexanol	300	0.81	40
1	0.2	2-ethylhexanol	400	0.60	36
1.7	0.2	2-ethylhexanol	400	0.60	36
1	0.5	2-ethylhexanol	800	0.48	19
1	0.5	4-ethyloctanol	800	1.76	3

Source: Ref. 17.

extraction circuits [27]. Such degradation was shown to be catalyzed by the presence of transition metals (Co > Ni > Cu) and can be easily reversed by treatment with hydroxylamine.

Kelex 100, on the other hand, does not contain adjacent OH groups and can be distilled unchanged under vacuum. However, heating pure Kelex 100 at atmospheric pressure results in considerable darkening of the reagent, probably from decomposition at the double bond [28]. However, the reagent appears to withstand pressure reduction to produce metal powders [personal communication from A.W. Ashbrook, 1973].

The stability of these reagents may be compared, for example, to DEHPA, which has been used at 60°C for many months without decomposition [3], and is shortly to be used at 70–80°C in a solvent extraction plant [personal communication from A.W. Ashbrook, 1973]. However, the thiophosphinic acids are prone to oxidation; thus the commercial reagents CYANEX 302 and 301, respectively mono- and dithiophosphinic acids are both degraded. CYANEX 301 degrades in two stages, the first being the formation of the disulfide, which can be reversed by treatment with, for example, zinc powder or nascent hydrogen [29]. CYANEX 302 on the other hand is degraded in one step to elemental sulfur and CYANEX 272, particularly in the presence of iron and air.

Tertiary amines are susceptible to oxidation ($R_3N{:}O$), and some evidence has been found to suggest that oxidation does indeed occur after several months recycling in a uranium solvent extraction circuit [A. W. Ashbrook, unpublished data].

Diluent stability has apparently never been seriously considered, except for the solvent extraction processes involving radioactive materials. Generally, one would expect that straight-chain saturated aliphatic diluents would be very stable materials, and that with the introduction of other elements and bond types, such as oxygen and nitrogen and double bonds, the stability would be expected to decrease. Thus for a series of different hydrocarbons, the most stable or inert would be expected to be the *n*-paraffins, and the least stable those containing olefinic compounds. The acid stability of various hydrocarbons has been studied by contacting with $4 \, \text{mol dm}^{-3}$ nitric acid (25 vol%) for 7 days at 70°C. The order of stability of these conditions is shown in Table 7.3 [28]. However, in the petrochemical industry the degradation of hydrocarbons in the presence of cobalt has been observed [30]. This phenomenon has also been found in some cobalt solvent extraction circuits with the production of carboxylic acids. The presence of such potential metal extractants may have a deleterious effect on the overall process [31]. The problem can be minimized by the addition of an antioxidant, e.g., 2,6-di-*tert*-butyl-4-methylphenol (BHT) into the solvent.

Table 7.3 Stabilities of Pure Hydrocarbons

Type of hydrocarbon	Stability
n-Paraffins	Stable
Isoparaffins	Variable stability (low-boiling fraction, stable; high-boiling fraction, unstable)
Cycloparaffins	Moderately stable to unstable
Aromatics	Moderately stable to unstable
Mixed cycloparaffins and aromatics	Unstable
Olefins	Unstable

Source: Ref. 31.

7.13.6 Volatilization

Volatilization of solvent components can become a problem when the system is operated at elevated temperatures or in hot climates. The human toxicity of solvent components is a generally unknown factor and could be a problem in a system enclosed in a building.

Generally, one would expect that the most volatile component would evaporate first, and this would probably be the diluent. In several cases of operating simulated solvent extraction processes at temperatures up to 70°C, it has been noted that the diluent is rapidly volatilized [G. M. Ritcey and B. H. Lucas, unpublished data]. Problems of volatilization appear not to have occurred to any great extent in the past (perhaps the losses were not measured), but any trend to the use of elevated temperatures would require that this form of solvent loss be thoroughly investigated.

7.13.7 Entrainment

Entrainment of solvent in the aqueous raffinate, in the strip liquor, or in cruds can be the most serious cause of solvent loss. Here, it is expected that the ratio of solvent components in the entrained material will be the same as in the solvent. Entrainment losses can occur because of insufficient settling area or time allowed for phase disengagement; poorly designed or operated mixers; too much energy input into the mixing stage; lack of additives to suppress emulsion formation; poor diluent choice; high extractant concentration in the solvent; solids in the aqueous feeds causing crud formation, etc. Entrainment losses can be minimized by the incorporation of various devices in the circuit, such as skimmers, centrifuges, coalescers, after-settlers, flotation, foam fractionation, activated carbon, or others. Although flotation units are effective for the removal of entrained solvent and are not affected by the presence of solids, they are expensive to operate as regards reagents and power, whereas packed coalescence beds, although relatively inexpensive and effective, can be easily blocked by solids. Baskets containing polypipe pieces tend to be coated with a thin film in a reasonably short time, perhaps less than a month, and thus require expensive periodic maintenance. Centrifuges, although capable of high throughput, are expensive, have high power consumption, and may require frequent cleaning.

7.13.8 Crud

Common to all or most solvent extraction operations in the mining industry is the problem of stable formation of cruds. The crud can constitute a major solvent loss to a circuit and thereby adversely affect the operating costs. Because there can be many causes of crud formation, each plant may have a crud problem unique to that operation. Factors such as ore type, solution composition, solvent composition, presence of other organic constituents, design and type of agitation all can adversely affect the chemical and physical operation of the solvent extraction circuit and result in crud formation [32–34].

Crud is defined as the material resulting from the agitation of an organic phase, an aqueous phase, and fine solid particles that form a stable mixture. Crud usually collects at the interface between the organic and aqueous phases. Other names that have been used for the phenomenon are grungies, mung, gunk, sludge, and others.

The following section covers the general aspects of solvent losses by crud; its formation and characteristics and its treatment and prevention. Losses attributed to emulsion and crud formation, can in part be related to (1) nature of feed; (2) reagent choice; (3) equipment selection; and (4) method of operation, such as the droplet size, continuous phase, excessive turbulence, etc.

7.13.8.1 Possible Causes of Crud Formation

7.13.8.1.a Nature of Feed

The nature of the feed composition can be a major determining factor as to whether crud will form in the subsequent extractive operations [33,34]. The presence and concentration of certain cations, such as Fe, Si, Ca, Mg, or Al, with sufficient shear in the mixing process can produce stable cruds [32,34]. Solids must be absent from most solvent extraction circuits, and clarification is usually aimed at achieving about 10 ppm of solids. One of the major causes of crud is the lack of good clarification of the feed solution, with the result that solids get through to the solvent extraction circuit. The presence of colloids, such as silica [35], can produce stable emulsions and crud during the mixing of the phases to achieve mass transfer. Aged feeds can constitute a greater potential crud problem than fresh leach solutions [36]. In plants where bacteria have been prevalent because of favorable environmental conditions, crud has resulted, and expensive circuit modifications were subsequently required. The elimination of air to such circuits is often necessary to minimize bacterial and fungal growth [37]. Certain systems may have hydrolyzed compounds precipitating out of solution, and thus a crud results. In certain extraction systems, the anionic strength of the aqueous feed solution may be insufficient, so that stable emulsions occur when the two phases are mixed. If sufficient agitation is applied over a period, then crud can result. One other important cause of cruds in solvent extraction plants is dust from the air if drawn into the agitation in a mixer-settler circuit. Thus, vessels should be covered to prevent dust accumulation. Organic matter in the feed, such as lignin or humic acids, may also promote crud formation.

7.13.8.1.b Nature of Solvent

The choice of the extractant and solvent composition is an important aspect of the successful solvent extraction operation, but the possibility of crud due to the solvent composition must not be overlooked. Many systems require a modifier to improve phase disengagement, to assist in solubilizing the metal-organic species, and to reduce third-phase and emulsion tendency. If a solvent has the tendency to produce emulsions on mixing with the aqueous feed solution, which could cause cruds if colloids or suspended solids are present in the aqueous feed, then the cause may be due to several factors. Perhaps the system requires the addition of a modifier, a change to a different modifier, or a higher modifier concentration. Also, possibly the diluent type and composition may not be compatible with the system. An aromatic diluent or an aliphatic diluent with some aromatic content may be more desirable than a completely aliphatic diluent for that particular process. Frequently, in the solvent makeup, there are unreacted chemicals from the manufacturing process or, possibly, impurities from the containers used to transport the solvent components. The problems associated with such impurities become particularly evident if their effects are not considered during bench-scale investigations [1]. Their presence in a solvent system can produce cruds, good or poor phase separation, and enhanced- or poor-loading characteristics. Such effects could result in abandoning a particular solvent because of its poor chemical and physical characteristics.

Degradation of the solvent due to the presence of certain metals in the feed solution, use of oxidizing agents during stripping, high-temperature processing, and biodegradability all may result in decomposition products forming stable emulsions and cruds. Several uranium plants have reported the degradation of the isodecanol modifier to isodecanoic acid in the amine/ isodecanol system. Naphthenic acid extraction of copper from leach liquors also showed degradation to an insoluble crud composed of 23% of the extractant [38]. In the refining of uranium using tributylphosphate in contact with nitric acid, degradation products of mono- and dibutyl phosphates are produced. Amines are susceptible to degradation in the presence of oxidizing agents. Carboxylic acids and DEHPA have been reported to withstand low-pressure reductions in a process to produce metal powders [39]. The LIX reagents containing LIX 63 (e.g., LIX 64, LIX 64N, LIX 70, and LIX 73) are stable up to temperatures of 40°C, whereas those not containing LIX 63 (e.g., LIX 65N and LIX 71) can safely withstand higher temperatures.

7.13.8.1.c Equipment Selection

There is no universal contacting equipment suitable for all solvent extraction operations. Even within a plant, it may be completely wrong to select the same type of contactor for all stages of the extraction process. Therefore, for each plant the final selection is governed by the nature and composition of aqueous feed, the type of solvent and its composition, and how their respective physical characteristics affect the mixing process, flow patterns, and coalescence [9]. Naturally, the mass transfer efficiency must also be considered. Thus with an understanding of all the physicochemical variables that affect the extraction, as well as the minimization of emulsions and crud, the right equipment for that plant can then be selected [9,38]. That is, if the tendency to form emulsions and cruds is to be minimized, the type of equipment and the method of agitation used to achieve mass transfer is of concern [33]. Degradation of the solvent may also have to be considered in the equipment choice, depending on the chemical system. Thus in one plant, centrifugal contactors were chosen over mixer-settlers because of lower solvent degradation [32].

7.13.8.1.d Method of Operation

In the solvent extraction process, one of the major concerns should be the technique by which mass transfer is achieved. That is, the physical design and operation can contribute not only to high solvent losses, such as by entrainment, but also to the formation of stable emulsions and cruds. These depend on the nature of the dispersion, the type of droplet formation, and the rate and completeness of coalescence. These factors are important in the choice of suitable contacting equipment and in the design and operation of the plant. Different types of dispersion will be created for a particular system, depending on the type of contacting equipment selected and the energy input to the system. Differences will be indicated by the rates of mass transfer, drop size distribution, wetting of surfaces, sedimentation and coalescence rates, and entrainment. Subject to the physicochemical properties of the two dispersed phases such as viscosity, surface tension, and presence of solids or colloids; with increasing agitation and decrease in drop sizes a region of instability will be reached, followed by a stable emulsion. If solids or colloids are also present, then a crud will result. This is demonstrated in Fig. 7.15, showing the operating regions of pulsed columns [36]. The information is readily related to excessive turbulence in mixer-settlers, particularly of the pump-mix design, and in certain agitated columns where backmixing is severe. Flow patterns during mixing can influence emulsion tendency, which can be further influenced by the continuous phase. Thus, if solids are present in mixer-settler operations and excessive turbulence exists, it would be advisable to use a type of contactor more suited to the physicochemical characteristics of the system. Centrifugal contractors would also be an unwise choice if solids are present or crud formation is likely. Equipment such as the Graesser contactor (RTL contactor), pulse sieve-plate column, ARD contactor and possibly in-line mixers could be considered. Pulsed columns have been described in the extraction of uranium from ore leach slurries and in the presence of crud [36,40].

Choice of the continuous phase, organic, or aqueous, coupled with the contactor and flow patterns produced during the operation, may reduce the tendency of emulsification and crud formation if solids are present.

Although it is usually desirable to operate at saturation loading of the solvent, there are certain situations for which it is necessary to maintain

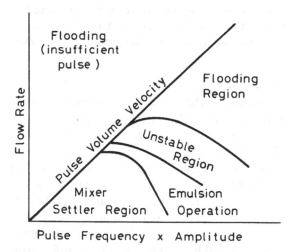

Fig. 7.15 Operating regions of a pulsed column.

less-than-saturation conditions. For example, in the extraction and separation of zirconium from hafnium in a nitric acid system, using TBP, the presence of the solvent and agitation results in a stable emulsion and crud [40]. Similar crud problems can occur in rare earth circuits using DEHPA, during which below-saturation loading has to be maintained to prevent crud formation [36,41].

7.13.8.2 Description of Some Cruds

The chemical and physical aspects of crud can differ for each separate operation and will vary in inorganic composition, organic content, color, and density. The composition of many cruds appears to have in common such constituents as Si, Al, Fe, P, SO_4, particles of gypsum, clay, and other fine particles together with the solvent. Often there is a direct relation between the feed liquor and the crud compositions, indicating possible aqueous carryover as well as inefficient clarification before solvent extraction [33]. Various researchers have reported on the formation of crud and its characterization in their circuits [42–45].

In some of the South African plants, a tarlike substance is generated and has been shown to contain aliphatic carboxylic acids. There is evidence of the presence of isodecanoic acid, from the oxidation of the isodecanol modifier. More than one plant in North America has experienced some crud problems when greater than 3% isodecanol was present. Possibly the vortex created on mixing is a contributing factor to the crud produced by the oxidation of the isodecanol in the presence of ferric iron, as a possible oxidant. In one early plant, the animal glue flocculent used for filtration of the leach pulp caused severe fungal growth in the solvent extraction circuit. The addition of a solvent-soluble fungicide (a derivative of benzothiazole) alleviated the crud caused by the fungus. Use of an aromatic diluent, instead of an aliphatic one, was also effective in minimizing the crud problem caused by bacteria.

In some scrub and strip circuits, the crud is mainly composed of silica, as well as inorganic sulfates. Also, if poor pH control is used in the uranium stripping circuits with ammonium sulfate, then uranium is a major constituent [33,46]. Such crud may be treated with dilute sulfuric acid, and recirculation through a pump results in the crud breaking down. There is evidence, in at least a few uranium circuits, that the presence of humic acids may be a possible cause of the crud problem [34,47]. Lignin appears to be another cause of crud formation [33,46]. Humic acids contained in the feed solution have also been implicated in the formation of waxy cruds in plants extracting uranium from phosphoric acid.

The presence of hydrolyzable compounds, which precipitate because of excessive agitation and high shear, can cause severe crud problems.

Zirconium presents a particular problem and is a major constituent of at least one crud in a uranium processing circuit. Also, as stated earlier, zirconium will tend to hydrolyze in an HNO_3–TBP circuit given the right conditions of shear and of energy input (proximity to saturation of the solvent and the type and construction of contactor, particularly in the coalescing zone). The use of Teflon plates in columns for such an extraction process definitely increased coalescence and decreased the tendency for hydrolysis and crud formation [40].

7.13.8.3 Treatment and Solvent Recovery

There is no universal treatment scheme that would be amenable to all cruds because of the difference in the causes of crud formation. Because crud originating in one plant is usually different from that obtained in another; or in another circuit in the same plant, so the treatment scheme adopted for solvent recovery will, of necessity, be different. Some cruds are broken down by settling and skimming, others by filtering [47], some by acidification or neutralization to an alkaline condition [34,42], and some are broken by extreme agitation such as cycling through a centrifugal pump. After such treatment procedures, if successful, the solvent that is released can be readily decanted for recovery. In some plants, a separate solvent treatment stage is necessary, with a wash such as Na_2CO_3 being used.

Thus, with cruds differing in their chemical and physicochemical history of formation, treatment to break down the crud and recover the solvent is accomplished only by the testing and evaluation of many possibilities. Some examples of crud descriptions have been reported [46].

7.13.8.4 Crud Prevention

Because crud is a difficult phenomenon to completely characterize, is often site-specific, and indeed can vary within an extraction, scrub, and strip-circuit operation, preventative measures are difficult to cite. The following are some of the methods that can be suggested for the prevention of crud. Again, it must be emphasized that because cruds have a different history of formation, one or more preventive measures may be necessary.

Solids in the feed were mentioned as one of the major causes of subsequent crud formation in the solvent extraction circuit. Good clarification, therefore, is necessary to minimize crud and thereby operating costs. Table 7.4 itemizes some benefits of good clarification [48]. In North America and South Africa, the objective is to obtain approximately 10 ppm suspended solids in the feed to extraction. This is usually obtained by the use of sand filters after a countercurrent decantation (CCD) circuit. At least one plant has reported that crud quickly developed when the sand filters were not in

Table 7.4 Benefits of Good Clarification

Better mixing efficiency
Less crud and, therefore, less solvent loss
Lower organic entrainment
Less difficulty with maintaining continuous phase
Decreased ion transfer to electrolyte (in copper processing)
  by suspended solids
Increased efficiency in the tank house
Decreased maintenance

operation [49]. New plants are attempting to achieve this objective by the use of Enviroclear thickeners, leaf clarifiers, and sand filters. Although sand filters are used successfully in some plants, other plants have not been able to achieve their objective. Insufficient frequency of backwashing the sand filters is a probable cause of poor clarification.

Because dust can cause crud if permitted in a mixer-settler circuit, particularly if the settlers are located in the open, adequate covers over the settlers should be provided.

Colloidal silica in some circuits such as $TBP–HNO_3$ in uranium reprocessing can be reduced by the addition of gelatin and heating to 80°C to coagulate, followed by centrifuge separation [33]. The addition of certain surfactants to lower the surface tension may also reduce cruds caused by colloidal silica [47], as well as break emulsions. Addition of sequestering agents was successful in eliminating the deposition of calcium–rare earth precipitate, in a rare earth–DEHPA circuit [50]. However, the addition of such reagents could also cause solvent degradation with continual cycling, so the approach must be carefully investigated before adoption. Surfactants and other compounds used to enhance the liquid–solid separation after leaching could also enhance emulsion and crud tendency.

Certain organic constituents, such as lignins and humic acids, may be solubilized in the aqueous feed and thus cause problems in the solvent extraction circuit [33]. Very little is yet known about this particular area of crud formation, but it has been found in phosphoric acid circuits that coagulation of humates with surfactants followed by filtration has been useful [51]. Lignin can be removed by passing the leach solution through a bed of activated carbon, and pressure treatment under oxygen at 200–250°C will destroy the lignin components.

Crud caused by bacterial and fungal growth may be minimized by elimination of as much air as possible to the system [52]. These growths, in association with any solids present in the feed liquor, lead to the formation of crud. Fungal growth in some circuits, arising from the isodecanol modifier in an amine-uranium system, was eliminated by using 35% Solvesso 150 aromatic diluent [53,54]. The aromatic diluent acts as a bactericide and fungicide. Commercial bactericides have also been used successfully.

Proper selection of the extractant as well as the other solvent components can minimize emulsions and crud formation. Freshly prepared solvents can often contain impurities that could subsequently cause operational problems. Therefore, all solvents should be conditioned before use. Having selected a suitable solvent system, cyclic tests should be performed to determine whether degradation of any of the solvent components is taking place, as any degradation products could cause crud formation in the circuit.

There are several items for consideration in the operation of the circuit to minimize crud formation. In certain systems, solvent saturation can result in the formation of gelatinous solids, as in the rare earth–DEHPA system [50,55] and in the zirconium–TBP circuit [56]. The phenomenon is partially due to the increase in viscosity as loading is reached. An increase in the O/A ratio thus results in a decrease in crud formation. As the viscosity increases, excessive agitation can produce stable emulsions.

Flow patterns can be altered by change of the continuous phase and, accordingly, the tendency for the formation of emulsions or cruds is altered. At one uranium refinery, the solvent is maintained in the continuous phase to produce flow patterns to reduce emulsion tendency [57].

Occasionally, the water used for solution makeup to the scrub or strip circuits, because of impurities, can cause subsequent emulsions and cruds. At one uranium plant in South Africa, deionized water was used to prepare the ammoniacal strip solution, rather than normal plant water, which tended to cause crud formation [53,54].

Equipment selection is important, as is also the proper operation of the contacting devices. It is generally recognized that high shear is the primary cause of droplet haze and subsequent emulsion and crud formation. Thus, the type and amount of agitation (shear) must be optimized for mass transfer while minimizing emulsion and crud formation. Table 7.5 indicates items of information that may be required in analysis of crud formation problems in a plant [32].

7.13.9 Extractant Losses

Table 7.6 shows some typical reported extractant losses occurring in solvent extraction processes. Most of these data were obtained from inventory calculation and, hence, represent total loss from the several ways, by which solvent components may be lost from the system [3,15,16,58–69].

Table 7.5 Possible Information Required for Consideration in Solving Plant Crud Problems

---

Ore: mineralogy and analysis
Leach: oxidant and quantity added:
    Possible effects on degradation, emulsion production,
      or crud stabilization
Liquid–solid separation:
  Type of separation (e.g., CCD)
  Type and quantity of surfactant added
  Use of sand filters or other types of clarifiers (e.g., anthracite) and
    frequency of regeneration
  Feed solution to solvent extraction:
    Suspended solids
    Dissolved solids
    Solution composition (especially silica, aluminum,
      molybdenum, zirconium)
    Presence of humic acids, lignin
Solvent extraction:
  Extraction circuit:
    Extractant
    Diluent
    Modifier and concentration
    Mixer design (e.g., baffling)
    Agitation (rpm and design, energy)
    Vortex in mixer
    Continuous phase
    Degradation of diluent, modifier and extractant
      (and surfactant from L/S separation)
    Settler design (entry to dispersion band,
      flow rate design, baffling)
    Velocity across settler
    Viscosity, surface tension of interfacial tension
    Presence of fungus or bacteria
    Presence of precipitates
  Stripping circuit:
    Mixer and settler operations as in "extraction" above
    Stripping agent
    Viscosity, surface tension, or interfacial tension
    Continuous phase
    Degradation of diluent, modifiers, and extractant
    Presence of fungus or bacteria
    Presence of precipitates

Table 7.6 Extractant Losses Reported for Some Processes

Metal extracted	Extractant	Extraction pH	Loss (ppm)	Ref.
Ni	Naphthenic acid	4.0	90	[58]
	Naphthenic acid	6.5	900	[59]
	Versatic 911	7.0	900	[15]
			300	Shell Chemicals
Co	Versatic 911	7.7	100	Shell Chemicals
Rare earths	DEHPA	2.0	7	[16]
Co	DEHPA	5.5–6.5	30	Personal comm.
U	Tert. amines	1.5–2.0	10–40	[3]
Cu	LIX 64	1.5–2.0	4–15	[60–63]
U	TBP	2.0	25–40	[64–67]
Cu	Kelex 100	1–2	1–10	Personal comm.
Hf	MIBK	$1.5 \, \text{mol} \, \text{dm}^{-3}$ HCl	20,000	[60–69]

7.13.10 Environmental Considerations

Perhaps the area of environmental pollution most pertinent to solvent extraction processing is that of water pollution and, consequently, the toxicity of solvent extraction reagents to aquatic life then become important. Thus, the primary consideration in the development of water quality criteria for these reagents is to determine their toxicity toward fish and their biodegradability [70].

Since different species of fish may differ widely in their tolerance to the same reagent, four fish types are usually used, such as the fathead minnow, bluegill, goldfish, and guppy. These four species have been used by Pickering and Henderson in the study of tolerance limits for a number of petrochemicals [71]. Median tolerance limits (TL_m) (LC50) denoting the concentration at which 50% mortality occurs were computed in different concentrations of solvent causing 50% mortality of the fish under the experimental conditions during a period of 96 hr. In general, the toxicity was greater when soft water was used in the tests. Similar static toxicity test results have been reported from a study performed in Canada on guppies and minnows in the presence of various extractants and diluents [G. M. Ritcey and B. H. Lucas, unpublished data; 72]. An ultrasonic probe was used to emulsify the 0.1% solution of the reagents in the test water to prepare the stock solution for the static tests. Guppies, each weighing 0.08–0.16 g, and fathead minnows weighing 0.73–1.89 g were evaluated in the program. The weight/volume ratio of fish/solution was $0.03 \, \text{g} \, \text{dm}^{-3}$ for guppies and $0.70 \, \text{g} \, \text{dm}^{-3}$ for minnows.

The results of these bioassays on extractants and modifiers are shown in Table 7.7a and for diluents in Table 7.7b. It is interesting to note the wide range of tolerance limits that are shown for the various extractants. Many show a toxic effect below their solubility level. Also, the toxicity of the Solvesso 150 and Isopar L diluents was acute compared with the other diluents examined.

Table 7.7a Acute Toxicity Static Bioassays on Extractants and Modifiers (TL_m 96 hr)

	TL_m (ppm)	
	Guppies	Minnows
Aliquat 336	0.18	
Primene JMT	0.70	
Kelex 100	1.10	
LIX 63	4.0	1.6
Alamine 336	10.0	4.0
Isodecanol	12.0	8.4
LIX 64N	15.0	2.7
TBP	18.0	9.6
LIX 70	32.0	15.0
Versatic 911	102.5	
DEHPA	173.0	

Table 7.7b Acute Toxicity Static Bioassays on Diluents (TL_m 96 hr)

Diluent	Effect	Concentration (ppm)
Oil-based livestock spray	nontoxic	1000
Napoleum 470	nontoxic	1000
Mentor 29	nontoxic	1000
DX 3641	nontoxic	1000
Isopar L	nontoxic	1000
Shell 140	nontoxic	1000
NS-144	total mortality	1000
Solvesso 150	minnows TM_m	10.7
	Guppies TM_m	14.7
Isopar E	TM_m	10.0

Acute toxicity tests are reported for the Cyanamid reagent Cyanex 272, showing TL_m values of $4.9\,\mathrm{g\,kg^{-1}}$ and $>2.0\,\mathrm{g\,kg^{-1}}$, respectively, for rats (oral) and rabbits (dermal) [73]. The TL_m 96 hr tests for bluegill sunfish and rainbow trout gave values of 45 and 22 ppm, respectively.

Tests by Mobil Oil Corporation [74] on dibutyl butylphosphonate showed that its toxic effects on rats were moderate at $3\,\mathrm{g\,kg^{-1}}$ (oral), and for rabbits it was nontoxic at $>5\,\mathrm{g\,kg^{-1}}$ (dermal) exposure. With DEHPA, acute toxicity was found for rats at concentrations of $1.4\,\mathrm{g\,kg^{-1}}$ (oral).

In most countries, the manufacturer now has to provide such toxicity data for approval before marketing.

7.14 ECONOMICS

Any successful solvent extraction process depends upon the selection of inexpensive extractants that can operate at the natural condition of the solution, with minimum loss of the organic phase to the aqueous solution. Also, an inexpensive means of recovery of the metal from the organic phase is necessary. In some cases, the cost of neutralization of the feed solution before solvent extraction processing or the cost of maintaining a buffered pH during extraction may prove excessive when coupled with solvent losses.

The economics of the solvent extraction process are very dependent upon "upstream" and "downstream" portions of the plant. Integration of the total processing step is essential to obtain maximum return. Variables, such as tonnage rates and changes in solution composition, can have a most significant result on the economics of solvent extraction. Generally, economic considerations may be divided into two major areas: (1) capital investments and (2) operating costs.

7.14.1 Capital Investments

Capital investment is primarily related to the size of equipment necessary for a given total throughput, taking into consideration the number of extraction, scrubbing, and stripping stages required. Usually, the building that may be required to house the process is an important cost factor in determining areas and height demanded by the solvent extraction equipment. The organic phase is a major cost inventory item and may help determine the choice of equipment. For example, in one system the solvent inventory was reduced from 75,000 to 3,000 gal by using centrifugal contactors instead of mixer-settlers [75]. In addition, because of the size of the centrifugal contactors, a smaller area was required, resulting in decreased building costs. Naturally, in some areas of the world, the process requires no building to house the equipment and, therefore, the capital costs of the process can be decreased.

Although the capital cost depends on such factors as flow rates, solvent inventory, equipment, and building, it should not be assumed that for a given chemical process involving solvent extraction, the capital costs will be the same in one area of the world as in another. In one area, something like three plants were installed for an identical process, the capital cost varying as much as 300% from the lowest to the highest [76]. For any given process, there are many possible designs for the plant, incorporating different types of equipment and instrumentation that result in variations in capital cost. Therefore, it must be emphasized that (1) careful selection of equipment in relation to adherent costs of solvent and building, and (2) critical evaluation of the engineering contractor's design are necessary to optimize the lowest possible capital cost of the solvent extraction operation.

In summary, capital costs can be influenced by the following:

1. The number of stages and, thus, the number of mixer-settlers or columns required
2. Kinetics and, therefore, the size of the mixing device
3. Coalescence and, therefore, the settling requirements, which dictate equipment and solvent inventory costs
4. Entrainment and, therefore, the cost of solvent recovery equipment
5. Flow rates and flow ratios
6. Building requirements
7. Ancillary equipment, such as pumps, piping, instrumentation, etc.
8. Equipment to meet environmental regulations
9. Engineering design

7.14.2 Operating Costs

Operating costs may be divided into eight general areas:

1. Preparation of the feed liquor may be required before solvent extraction. For example, precipitation of an unwanted metal and removal by filtration, or pH adjustment may be necessary.
2. Preequilibration of the solvent may be required. In some systems, this cost is minimal, but in others it may be high; for example, in uranium extraction from sulfate solutions using tertiary amines, the sulfuric acid preequilibration of the solvent before extraction is a few cents or less per pound of U_3O_8 produced. By comparison, in a TBP–HNO_3 system for the recovery of zirconium, the preequilibration costs, using nitric acid, amount to about 50 cents per pound of Zr produced.

3. In some extraction systems it may be necessary to remove undesirable coextracted metals by scrubbing. Whereas this process step might cost 5 cents per pound of metal produced in one system, in another, scrubbing may not be necessary.

4. Stripping costs may vary considerably in one process, compared with others. The choice of the stripping medium is often the major cost differential, and this choice will dictate the final product that is produced. The reagent stripping costs in the TBP–HNO$_3$ system are zero, since water is used, but by comparison, the costs for recovery of uranium from amines with sodium carbonate are 1–2 cents per pound of U$_3$O$_8$ produced. Acids lost in recycling tankhouse acid constitute another operating cost.

5. Solvent losses are a factor to be considered in every solvent extraction process. Each process will have a certain loss of reagent by solubility, because of the pH and salt content. In some cases, loss of solvent by degradation, such as by high acidity or alkalinity, may occur. Solvent loss by solubility or degradation can vary considerably from process to process. In addition to soluble and degradation losses, some plants will have extreme losses because of entrainment and crud formation. These losses can be minimized by proper design of the process equipment. The type of contactor used may actually be causing the high entrainment losses, as well as losses from crud formation. If the equipment is not redesigned, and this may be impossible, then expensive solvent recovery equipment may be necessary. Thus, it is important to carefully select both the extractant and the diluent, as well as the equipment, when optimizing the process to minimize solvent losses.

6. Labor and maintenance costs can usually be kept to the order of a few cents per pound of metal produced. However, where a circuit has been improperly designed and several manual operations are required, then labor and maintenance costs can be high. Excessive solvent entrainment losses, mentioned earlier, are added to labor costs. Any crud that forms and has to be removed can result in additional costs from labor requirements. For example, if incorrectly designed contacting equipment is chosen, which results in periodic shutdowns necessary for cleaning because of formation of cruds and precipitates, a major maintenance problem could arise.

7. Energy costs to the mixing circuits are influenced by the equipment type and flows.

8. Raffinates containing appreciable amounts of the metal one desires to recover can also account for increased operating costs. If subsequent treatment of the raffinate is required to meet environmental regulations, an additional cost is incurred.

7.15 OPTIMIZATION OF THE SOLVENT EXTRACTION PROCESS

Much of the optimization of the solvent extraction plant can be achieved in the pilot plant testing. As noted earlier on the subject of process design, one must investigate the dependence of the dispersion and coalescence characteristics and their effect on extraction and phase separation. Also, such variables as metal concentration, equilibrium pH (or free acidity or free basicity), salt concentration, solvent concentration (extractant, diluent, and modifier), and temperature have to be studied to determine their effect on mass transfer. Although many of the variables can be tested in the pilot plant, many circuits are not optimized until the full-scale plant is in operation.

In the optimization of the solvent extraction process for the recovery of copper using LIX 64N, Robinson [77] described the cost function in terms of the sum of the operating and capital costs. The operating costs were taken as resulting from losses of copper and solvent:

$$\text{Cost}_{(\text{oper})} = AC_c + S_l A \cdot (X_l C_x) + (1 - X_l \cdot C_{\text{sol}}) \tag{7.8}$$

where

A is the feed rate of the aqueous phase
C_c is the cost of copper
X_c is the concentration of copper in the raffinate
S_l is the solvent loss per litre
X_l is the volume concentration of LIX 64N
C_{xl} is the cost of LIX 64N
C_{sol} is the cost of diluent

The capital costs depend on the (1) number of stages of extraction and stripping, (2) concentration of LIX 64N, (3) size of stages, and (4) ratio of flow rates.

The size of settlers is determined by the rate of phase disengagement, which, here, is a function of the solvent concentration of LIX 64N and the aqueous phase feed rate. The *capital cost of extraction equipment* is a function of the number of tanks, the size of the tanks, and the solvent inventory [9]:

$$\text{Cost}_{(\text{equip})} = NK_l \frac{A}{S_{\text{ex}}} b + \frac{A}{NV} \cdot X_l C_{xl} + (1 - X_l) C_{\text{sol}} \tag{7.9}$$

where

N is the number of tanks
S_{ex} is the design settling rate

V is the volume of organic phase in each tank
k_l is a factor relating tank settling area to cost
b is another cost factor

The *costs function for stripping* is similar in form to that used for extraction. Thus *the total cost function*, $\text{Cost}_{(tot)}$ is given by:

$$\text{Cost}_{(tot)}t = \text{Cost}_{(oper)} + R\,\text{Cost}_{(cap)} \tag{7.10}$$

where R is the interest rate on capital, which includes equipment, buildings, and so forth.

Although it may seem desirable to use the maximum concentration of solvent mixture and operate at saturation loading, this may be impractical, or even impossible, in actual plant practice. For the solvent mixture, although increased loading will result in increased concentration of the extractant, often the increase will not be a linear relationship [78]. In addition, in some systems, increasing solvent concentration results in higher solvent losses in the aqueous phase because of viscosity effects, even at elevated temperatures. Indeed, as the viscosity increases, excessive agitation can produce stable emulsions, making the process inoperative. Thus, a compromise has to be developed between a reasonable solvent concentration and good operation with equipment that has been selected. Robinson [77] stated that an increase in concentration of the extractant could result in:

1. Higher loading in the organic phase, resulting normally in reduced stripping costs
2. Higher extraction efficiency for the same number of stages, or fewer stages for the same extraction efficiency
3. Increased operating costs, associated with solvent losses in the system
4. Increases or decreases in working capital associated with solvent inventory. The form is dependent on the interactions of Eqs. (7.8) and (7.9) and the concentration of extractant in the solvent

Some writers have stated that there is no point in maintaining operation at maximum solvent loading. This may well be true for some situations, but where a plant can run at saturation loading, the economics will favor this operation because of maximum utilization of solvent, which is an expensive inventory item. Generally, a slightly lower than saturation condition is run to allow changes in solution feed concentration and perhaps surges or change in flows. However, there are certain situations for which the solvent is maintained at considerably less than saturation. Two of these conditions are as follows:

1. In dump leaching of copper, it is not necessary to use extra stages of extraction to recover all the copper from the solution, since the raffinate will be recycled to the dump.
2. In the extraction and separation of zirconium from hafnium in a nitric acid system, using TBP, the system operates best if run at about 10% less than saturation [56]. As saturation of the solvent is approached, a zirconium compound precipitates in the presence of the solvent, causing cruds and emulsions. This problem is also encountered in rare earth circuits using DEHPA.

The optimum number of extraction stages is a function of the concentration of the active agent, the flow ratio, the value of unextracted metal, pH, and the interest on the capital investment.

Other variables that may affect extraction efficiency and, therefore, cost are the following [79]:

1. Viscosity of the organic phase as a function of the diluent (Fig. 7.16)
2. In a mixer-settler, the effect of agitation speed on the extraction or stripping
3. The effect of organic phase continuous versus aqueous phase continuous operation

The importance of optimization of as much of the solvent extraction circuit as early as possible in the flow sheet design has been noted earlier. Because of the nature of the equipment selected, optimization may well not be possible until the pilot plant or plant stage. However, if the choice of equipment favors mixer-settlers, optimization of the process may be possible early in the developmental stages. Such an optimization was shown for copper extraction in a joint investigation by the University of Bradford and CANMET [2]. Procedures for investigating mixer and settler performance using a batch-stirred tank have been described, with the objective of reducing experimental work to a minimum. Equations were developed from a set of statistically designed experiments to determine the effect of major operating variables on kinetics of mass transfer. These variables were impeller speed, stirring time, volume fraction of the phase dispersed, and the type of dispersion (organic or aqueous continuous). The effect of increase in the extractant concentration at constant dispersed phase holdup is shown in Fig. 7.17. Knowledge of densities, viscosities, interfacial tensions, and drop sizes was insufficient to explain the relationship of the measured velocities to each other. Additional factors must affect, therefore, the coalescence rate. The data obtained from the statistically designed set of experiments can be used to develop an empirical equation to predict the degree of copper extraction in a batch-stirred tank as a function of

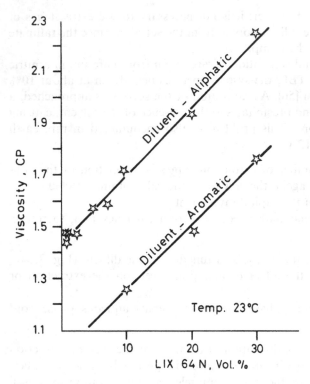

Fig. 7.16 Effect of diluent on viscosity of solvent.

stirring speed, stirring time, and phase ratio. From this data, the stage efficiency in a flow system may be estimated. Optimization of these parameters can result in providing sufficient data for the most economic design of large-scale mixer-settlers.

The design criteria necessary for optimization of the engineering and design of the solvent extraction process can probably be summarized as follows:

1. Clarification of feed liquors
2. pH adjustment and possible filtration
3. Heating requirements
4. Nominal feed flow rate
5. Solution grade expected
6. Production capacity demanded
7. Concentration of solvent
8. Selection of diluent and possible modifier

9. Stage requirements for extraction and, therefore, equipment choice and design based on total flow requirements
10. Scrubbing: solution type, stages, and equipment
11. Stripping: solution type, stages, and equipment
12. Possible recovery of metal as oxides or powders and expected purity
13. Solvent losses and possible recovery stages and equipment
14. Labor and maintenance
15. Complete utilization of the solvent
16. Dispersion: coalescence (equipment design)
17. Kinetics: choice and size of equipment

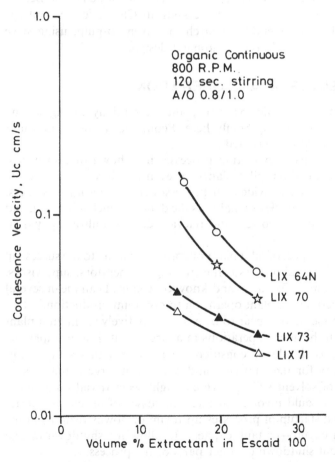

Fig. 7.17 Effect of extractant concentration on coalescence.

18. Equilibrium: number of stages
19. Possible by-products
20. Environmental considerations

7.16 SOLVENT IN PULP PROCESSING

If substantial reduction in processing costs is to be achieved in the mill, then the liquid–solids separation must be eliminated. This costly operation accounts for approximately 50% of the capital and operating costs of the mill. Such a possibility exists in the use of resin-in-pulp continuous ion exchange or solvent-in-pulp extraction from the leach slurry [1,80]. There currently is a renewed activity in several research laboratories to develop suitable equipment and technology to accomplish such a goal. The Beijing Research Institute of Uranium Ore Processing in China, for example, is presently engaged in considerable research in solvent-in-pulp, using sieve plate pulse columns [81] and other equipment design.

7.17 PLANT SCALE-UP AND OPERATION

Scale-up of a process from pilot plant to plant is essentially an engineering problem, and is discussed only briefly here. Economics is most important and that section should be consulted.

Experience in solvent extraction processes has shown that such processes can be scaled up from pilot plant—or even bench-scale—data quite reliably. This is particularly evident in processes employing mixer-settlers. However, scale-up will only be as reliable as the data on which it is based, and time spent in obtaining the correct and relevant data will always pay dividends.

One important aspect of pilot plant operations relating to the successful operation of a commercial process is the training of operators, supervisors, and analytical personnel. Expertise and knowledge gained can mean several months saved in getting the plant operation to optimum production.

Several processes that appear to be easy, or relatively so, in pilot plant testing can prove to be major operations in a large plant. For example, the following operations should be considered: shutdown and startup; additional tank capacity for the emptying and refilling of large contactors or settlers; recovery of solvent spillage, which might seem trivial on a small-scale operation but could involve considerable losses of expensive extractant; shutdown and startup of processes upstream or downstream from the solvent extraction circuit—or tank capacity to allow for shutdown of one unit process without shutdown of other parts of the process.

In general, the pilot plant should be operated as closely as possible to the conditions that might prevail in the commercial plant; conversely, the commercial plant should be based on data and conditions obtained in the pilot plant operations, and extrapolations to larger units, flows, equipment, and so on must be given serious consideration if the plant is to be a successful venture.

There are examples of successful operations that have not gone through the pilot plant stage but have been scaled up directly from bench-scale data. Thus, one process for the separation of (bulk) rare earths from an aqueous solution scaled up from only bench-scale information to contactors capable of flow rates of $500\,gal(UK)\,min^{-1}$ ($2.3\,m^3\,min^{-1}$). As it happened, these contactors were ideal for this particular process.

A change in the type of contactor between that used in the pilot plant to that used in the commercial operation can result in problems. This may occur because a particular large-scale contactor was not readily available. For example, in two plants separating chemically similar metals, although sieve plate pulse columns were found to be excellent contactors for the required separation, other columns were chosen for the commercial plant because pulse columns of sufficient size were not readily available. The columns chosen, at least in one case, resulted in very poor physical operations and process efficiency.

The answer here seems to be lead time; given sufficient lead time, most equipment manufacturers are capable of producing the required type of contactor to suit the process specifications.

Design, scale-up, and operating problems of some solvent extraction processes have been described in the literature, and make interesting reading for engineers and chemists involved in such operations [1,3,5,10,11,16,36, 40,59,61,82,83].

REFERENCES

1. Ritcey, G. M.; Ashbrook, A. W. *Solvent Extraction, Principles and Application to Process Metallurgy*, Parts 1 and 2; Elsevier Science Publishers, Amsterdam, (1979, 1984).
2. Slater, M. J.; Ritcey, G. M.; Pilgrim, R. F. Proc. ISEC'74, Lyon, Society of Chemical Industry, London, **1974**, *1*, 107–140.
3. Ritcey, G. M.; Ashbrook, A. W.; Lucas, B. H. Development of a solvent extraction process for the separation of cobalt from nickel, presented at the Annual AIME Meeting, San Francisco, 1972. CIM Bull., January 1975.
4. Murray, K. J.; Bouboulis, G. J. S/X carrier solvents, Proceedings of AICHE Symposium on Solvent Ion Exchange, Tucson, Arizona, May, 1973.
5. Lo, T. C.; Baird, M. H. I.; Hanson, C. Eds., *Handbook of Solvent Extraction;* John Wiley and Sons, New York, 1982.

6. Warwick, G. C. I.; Scuffham, J. B.; Lott, J. B. Proc. ISEC '71, The Hague, Society of Chemical Industry, London, 1971, 1373p.

7. Warwick, G. C. I.; Scuffham, J. B. Proceedings of International Symposium on Solvent Extraction on Metallurgical Processes, Antwerp, 1972, pp. 40–47.

8. Agers, D. W.; Dement, E. R. Proceedings of International Symposium on Solvent Extraction on Metallurgical Processes, Antwerp, 1972, pp. 31–39.

9. Ritcey, G. M. Proceedings of AIChE Symposium on Solvent Ion Exchange, Tucson, Arizona, May 1973.

10. Nelson, R. R.; Brown, R. L. *The Design of Metal Producing Processes*, Kibby, R. M., Ed., American Institute of Mining Metallurgy, New York, 1967, p. 324.

11. Miller, A. *The Design of Metal Producing Processes*, Kibby, R. M., Ed., American Institute of Mining and Metallurgy, New York, 1967, p. 337.

12. Kunda, W.; Warner, J. P.; Mackiw, V. N. Trans. CIM, **1962**, *65*, 21.

13. McKay, H. A. C.; Healy, T. V.; Jenkins, I. L.; Naylor, A. Eds., *Solvent Extraction Chemistry of Metals*, MacMillan, London, 1966.

14. Gindin, L. M.; Bobikov, P. I.; Rozen, A. M. Russ. J. Inorg. Chem., **1960**, *5*, 906, 1146.

15. Ashbrook, A. W. J. Inorg. Nud. Chem., **1972**, *34*, 1721.

16. Ritcey, G. M.; Lucas, B. H. Proc. ISEC'71, The Hague, Society of Chemical Industry, London, 1971, p. 463.

17. Blake, C. A.; Brown, K. B.; Coleman, C. F. USAEC Rep. ORNL-1903, 1955.

18. Slyusarev, D. F. Zh. Anal. Khim., 1972, *27*, 753.

19. Pich, H. C.; Santos, C. S.; Elias, J. T.; Alves, M. F.; Conceicao, E. H.; Viera, H. X. Proceedings Symposium on Recovery of Uranium, IAEA, Vienna, paper IAEA-SM-135/17, 1971.

20. Ashbrook, A. W. Anal. Chim. Acta, **1972**, *58*, 115.

21. Ritcey, G. M. Proceedings Second Annual Meeting, Canadian Hydrometallurgists of Metal. Soc., CIMM, Canadian Institute Minerals and Metallurgy, October 1972, p. 11.

22. Ritcey, G. M.; Lucas, B. H. CIM Bull., February 1974.

23. Shulz, W. W.; Navratil, J. D.; Kertes, A. S. *Science and Technology of Tributyl Phosphate*, CRC Press, Vols. 1–4, 1991.

24. Higgins, C. E.; Baldwin, W. H. Anal. Chem., **1960**, *32*, 233.

25. Rydberg, J.; Reinhardt, H. Chem. Ind., **1970**, p. 488.

26. Ashbrook, A. W. Commercial chelating solvent extraction reagents. 1. *o*-Hydroxyoximes; Purification and isomer separation/extraction. Metallurgy Division, Mines Branch, Dept. Energy, Mines and Resources, Canada, Report EMA 73-10, 1973.

27. Takahashi, M.; Ogata, T.; Okino, H.; Abe, Y. Paper abstracts ISEC'83, Denver, USA, pp. 359–360, 1983.

28. Ashbrook, A. W. Commercial chelating solvent extraction reagents, II. *n*-Alkenyl-8-hydroxyquinoline; Purification and properties, Metallurgy Division, Mines Branch, Dept. Energy, Mines and Resources, Canada Report EMA 73-34 1973.

29. Rickelton, W. A.; Mihaylov, I.; Love, B.; Louie, P. K.; Krause, E., U.S. Patent 5,759,512, June 2, 1998.
30. Rickelton, W. R.; Robertson, A. J.; Hillhouse, J. H., Solv. Extr. Ion Exch., **1991**, *9*(1):73–84.
31. Marston, A. L.; West, D. L.; Wilhite, R. N. *Solvent Extraction Chemistry of Metals*, McKay, H. A. C.; Healy, T. V.; Jenkins, I. L.; Naylor, A., Eds., MacMillan, London, 1966, p. 213.
32. Ritcey, G. M. *Hydrometallurgy*, 1980, *5*, 97.
33. Ritcey, G. M. Proceedings of 14th International Mineral Processing Congress, CIM, Toronto; Vol. 1, 1982.
34. Ritcey, G. M. International Solvent Extraction Conference, ISEC '83, Denver, AIChE, 1983, pp. 88–89.
35. Cao, S.; Sworschak, H.; Hall, A. Proc. ISEC '74, Lyon. Society of Chemical Industry, London, 1974, pp. 1453–1480.
36. Ritcey, G. M.; Slater, M. J.; Lucas, B. H. Proceedings of International Hydrometallurgy Symposium, AIME, Chicago, February 1973, AIME, New York, 1973, pp. 419–474.
37. Orth, D. A.; McKibben, J. M.; Scotten, W. C. Proc. ISEC'74, Lyon, Society of Chemical Industry, London, 1974, pp. 514–433.
38. Fletcher, A. W.; Hester, K. W. A new approach to copper-nickel ore processing. Paper presented at the Annual AIME Meeting, New York, February 1964.
39. Burkin, A. R. Proceedings of First Hydrometallurgy Meeting, CIM, Ottawa, October 1971.
40. Ritcey, G. M.; Joe, E. G.; Ashbrook, A. W. Trans. AIME, **1967**, *238*, 330–334.
41. Huppert, K. L.; Issel, W.; Knoch, W. Proc. ISEC'74, Lyons, 1974, The Hague, Society of Chemistry Industry, London, 1971, pp. 2063–2074.
42. Young, W. Crud in Gulf Minerals Rabbit Lake solvent extraction circuit, Paper at AIME Annual Meeting, New Orleans, February 1979.
43. Kowalik, T.; Cantwell, T. Crud problems in solvent extraction and strip circuits of Conquista uranium, Paper at AIME Annual Meeting, New Orleans, February 1979.
44. Bakshani, N.; Maurer, E. E.; Alien, M. P.; Degenhart, A. L. Crud at the Sohio uranium solvent extraction circuit. Paper at AIME Annual Meeting, New Orleans, February 1979.
45. Moyer, B.; McDowell, W. J. Proceedings of AIME International Hydrometallurgy Symposium, *Hydrometallurgy — Research, Development and Plant Practice*, Osseo-Asare, K.; Miller, J. D. Eds., Metallurgical Society of AIME, 1983, pp. 503–516.
46. Ritcey, G. M.; Wong, E. W. Proc.ISEC'88, Moscow, **1988**, 2:116.
47. Lucas, B. H.; Ritcey, G. M. CIM Bull. June 1975; Canadian Patent No.101759, September 1977; U.S. Patent 3,969,476, 1976.
48. Rossiter, G. Anamax Twin Buttes oxide plant operating experience-first year. Presented at the Arizona Section, AIME, Hydrometallurgical Division, Spring, 1976.

49. Abramo, J. A.; Lowings, S. W. H. Uranium processing at Exxon's Highland operation, Proceedings of AIChE Symposium on Solvent Ion Exchange, Tucson, Arizona, May 1973.
50. Lucas, B. H.; Ritcey, G. M. CIM Bull. January 1975.
51. Hurst, F. J. Recovery of uranium from wet-process phosphoric acid by solvent extraction, presented at Annual AIME Meeting, Las Vegas, February 1976.
52. Young, W. Davy Power-Gas, Paper at AIME Annal Meeting, New Orleans, February 1979, p. 59.
53. Meyburgh, R. G. J. S. Afr. Inst. Min. Metall., **1970**, October, 54–66.
54. Meyburgh, R. G. J. S. Afr. Inst. Min. Metall., **1971**, April, 190–197.
55. Gaudernack, B.; Braaten, O. Occurrence and extraction of rare earths in Norway, presented at the Ninth Rare Earth Research Conference, Virginia, October 10–14, 1971.
56. Ritcey, G. M.; Conn, K. Liquid-liquid separation of zirconium and hafnium, Eldorado Nuclear, R&D Division, Ottawa, Rep. T67-7, 1967.
57. Ryle, B. G. USAEC Rep. TID 5295, 1956.
58. Fletcher, A. W.; Flett, D. S. *Solvent Extraction Chemistry of Metals*, McKay, H. A. C.; Healy, T. V.; Jenkins, I. L.; Naylor, A. Eds., MacMillan, London, 1966, p. 359.
59. Fletcher, A. W.; Hester, K. D. Trans. AIME, **1964**, *229*, 282.
60. Tunley, T. H.; Faure, A. The Purlex process—a description of the pilot plant that led to the use of the process in South Africa. Paper presented at the Annual AIME Meeting, Washington, D.C., February, 1969.
61. Tremblay, R.; Bramwell, P. Trans. Can. Inst. Miner. Metall., **1959**, *62*, 44.
62. Bellingham, A. I. Proc. Australes. Inst. Miner. Metall., **1961**, *198*, 85.
63. Lloyd, P. J. J. S. Afr. Inst. Miner. Metall., **1961**, *62*, 465.
64. Dasher, J.; Power, K. L. Eng. Miner. J., **1971**, *772*(4), 111.
65. Rawling, K. R. World Miner., **1969**, *22*, 34.
66. McGarr, H. J. Eng. Miner., **1970**, *171*(10):79.
67. Agers, D. W.; House, J. E.; Swanson, R. R.; Drobnick, J. L. Trans. Soc. Miner. Eng., **1966**, *35*, 191.
68. Hartlage, J. A. Kelex 100—a new reagent for copper solvent extraction. Paper presented at Society of Mineral Engineers, AIME Fall Meeting, Salt Lake City, September 1969.
69. Flett, D. S. Miner. Sci. Eng., **1970**, *2*(3), 17.
70. Ritcey, G. M.; Lucas, B. H.; Ashbrook, A. W. Proc. ISEC'74, Lyon, Society of Chemical Industry, London, **1974**, *3*, 2873–2884.
71. Pickering, Q. H.; Henderson, C. J.WPCF, **1966**, *38*(9), 1419–1429.
72. Hawley, J. R. Use, characteristics, and toxicity of mine-mill reagents in Ontario, Ministry of Environment, Ontario, Toronto, 1974.
73. American Cyanamid Company, Technical data on cobalt-nickel separation using CYANEX 272 extractant, 1982.
74. Mobil Oil Corporation, Material safety data bulletin, 1981.
75. Chem. Eng. News, **1968**, *11*, 44–46.

76. Agers, D. W. Proc. First Hydrometallurgy meeting, CIMM Mines Branch, Ottawa, October 28–29, 1971.
77. Robinson, C. G. Proc. ISEC'71, Hague, Society of Chemical Industry, London, 1971, pp. 1416–1428.
78. Ritcey, G. M.; Ashbrook, A. W.; Lucas, B. H. CIM Bull., January 1975. U.S.Patent: 3,999,055, 1968.
79. Tunley, T. H.; Birch, C. P. The recovery of copper from sulphate leach liquors by liquid ion exchange with LIX64N, National Institute of Metallurgy, Johannesburg, South Africa, Report No. 140904.
80. Ritcey, G. M. Proc ISEC'88, Moscow, **1988**, *2*, 116.
81. Jintang, Wang; Renli, Zhang. Paper presented at Hydrometallurgy'81, Manchester, Society of Chemical Industry, June 1981.
82. Albinsson, Y.; Ohlsson, L. E.; Persson, H.; Rydberg, J. Appl. Radial. Isot. **1988**, *39*(2), 113.
83. Rydberg, J. Acta Chem. Scand., **1969**, *23*, 647.

8

Principles of Industrial Solvent Extraction

PHILIP J. D. LLOYD Energy Research Institute, University of Cape Town, Rondebosch, South Africa

8.1 INTRODUCTION

The theory of solvent extraction was considered in Chapter 1, and Chapter 7 covered the application of liquid–liquid extraction in industry. The principles underlying the design of industrial applications are addressed in this chapter.

At the very simplest level, an aqueous solution contains a valuable component to be recovered, and a number of other components from which the desired component should be separated. It is assumed that, as a result of laboratory studies such as those outlined in previous chapters:

1. A suitable extraction system has been identified that will extract the desired component selectively from the less desired components.
2. Suitable physical and chemical conditions for carrying out the extraction have been found.
3. The rates of extraction of the desired and possibly also some of the undesired components have been determined at least qualitatively.

To design a process to recover the valuable component, a number of questions which must be answered:

- What fraction of the desired component can be recovered?
- How much of the extractant must be added to achieve the desired recovery?
- How much of the undesired components are extracted with the desired component?

- Can anything be done to reduce the concentration of undesired components in the extract phase?
- Having extracted the desired component, how can it be recovered from the extract in a useable form?
- How, physically, can the extraction be carried out, and what type and size of apparatus is required?
- How much of the extract is lost to the raffinate, and can anything be done to recover it?

This chapter therefore outlines the methods for answering these types of questions, so that the solvent extraction process may be applied in practice and desired components may be recovered in an energy-efficient, environmentally safe, and economical way.

The nomenclature used in solvent extraction has been defined in Chapter 1 and is illustrated in Fig. 8.1. Not all of the steps shown in this figure will be found in every extraction process, but equally there may be occasions where it is necessary to add additional steps; for example, to recover the extractant from the scrub raffinate. So while Fig. 8.1 is not a completely general flow diagram it covers most of the processes likely to be found in practice. Variations of this flow sheet will become apparent during the remaining chapters.

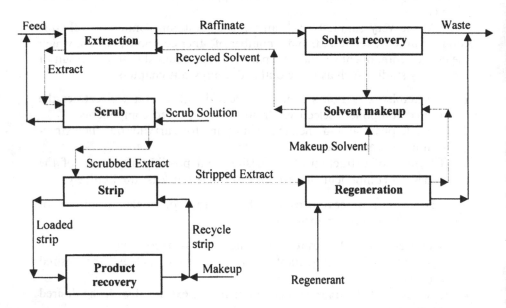

Fig. 8.1 Basic processes in solvent extraction.

Table 8.1 Effect of Phase Ratio in Single Stage Extraction

Phase ratio	Fraction extracted		Relative purity
Θ	E_A	E_X	E_A/E_X
0.1	0.500	0.010	50.50
1	0.909	0.091	10.00
10	0.990	0.500	1.98

Note: Assuming $D_A = 10$, $D_X = 0.1$.

8.2 EXTRACTION

8.2.1 *Single-Stage Extraction*

As described in Chapter 4, if a solution containing a desired component X is contacted with an immiscible solvent phase, then X distributes itself between the feed solution and the solvent according to:

$$D_X = [X_e]/[X_a] \tag{8.1}$$

where the subscript e represents the extract (solvent) phase and subscript a the aqueous phase.

If the phase volume ratio $\Theta = V_e/V_a$, then the fraction extracted, E_X, is given by:

$$E_X = D_x\Theta/(D_x \cdot \Theta + 1) \tag{8.2}$$

that is, *the fraction extracted in a single stage is a function of both the distribution ratio and the phase ratio.* This is an important finding. Much of the work in previous chapters has concentrated on the equilibrium distribution of species between the extract and raffinate phase. However, as soon as an answer to the question "What fraction of the desired component can be recovered?" is sought, the volumetric ratio between the two phases becomes almost as important as the equilibrium distribution. Consider now another simple case with a desired component A for which $D_A = 10$ and a contaminant X for which $D_X = 0.1$. Table 8.1 shows the effect of varying the phase ratio on (1) the fractions of A and X which are extracted, and (2) the ratio of the fractions extracted, which is a measure of the product purity.

This not only shows how the extent of extraction varies with phase ratio at a constant distribution ratio, but also how varying the phase ratio affects the relative purity of the product. Increasing the phase ratio by a factor of 100 nearly doubles the recovery, but drops the relative purity by a factor of over 25. Note that, in a single stage, it is not possible to achieve both high recovery and a high degree of separation simultaneously. Also, although the distribution ratios of the two species differ by a factor of 100, the relative purity is always less than 100.

A second phenomenon of great industrial importance is the effect of saturation of the solvent on the product purity. Implicit in the derivation of Eq. (8.2) is the assumption that D_X is constant. As discussed in Chapter 2, D_X is only a constant under ideal and constant conditions (usually at trace concentrations in both phases). It changes markedly as the concentrations vary in the two phases.

At higher concentrations, so much of an extractable component may be extracted that an appreciable fraction of the extractant is bonded to the extracted component, so that in turn the concentration of the free ligand in the extract phase is significantly reduced.

Industrial practice naturally requires the maximum use of the relatively expensive extractant, so that saturation of the extractant phase with reduction of the free ligand concentration to a minimum is the general rule. A model is thus needed to quantify the effect of a reduction in D_X as the concentration of the extracted species in the organic phase increases.

In many extraction systems, D_X is proportional to $[L]^n$, where: $[L]$ is the *free* ligand concentration, and the exponent n is determined by the number of ligand molecules per molecule of extracted complex. In these systems, therefore, we may write:

$$D_X = [X_e]/[X_a] = D^0\{[L_e]_{\text{tot}} - m \cdot [X_e]\}^n \tag{8.3}$$

in which:

D^0 is the distribution ratio of species X at trace concentrations, readily determined in the laboratory.

$[L_e]_{\text{tot}}$ is the total ligand concentration in the extract phase.

m approximates the ratio of ligand molecules to extracted molecules close to saturation.

n approximates the ratio of ligand molecules to extracted molecules close to infinite dilution.

The term in curved brackets on the right-hand side of the equation is the free ligand concentration.

It should be noted that m and n in Eq. (8.3) are sometimes equal, but often differ in value, which underlines the fact that this equation has no theoretical basis. It is merely a convenient way of representing much experimental data using three parameters that can readily be determined experimentally.

The equation can be fitted to a wide range of isotherms with quite small residual errors over the entire range of concentrations of interest. It is a liquid–liquid analogue of the Langmuir adsorption isotherm applicable to vapor–solid interactions, with the extractant concentration taking the place of the free surface area of the solid.

Where more than one component is extracted, then the free ligand concentration will be reduced by all components present in the organic phase. For example, consider the extraction of two components, A and B, that have similar chemistries of extraction. Assume for both components $m = n = 2$, that they have D^0 values of 10 and 1 respectively.

The equilibria for the distribution of A and B between the two phases are described by

$$D_A = [A_e]/[A_e] = 10 \cdot (1 - 2[A_e + B_e])^2 \tag{8.4a}$$

$$D_B = [B_e]/[B_a] = 1 \cdot (1 - 2[A_e + B_e])^2 \tag{8.4b}$$

while two mass balance equations

$$[A_F] = [A_a] + \Theta \cdot [A_e] \tag{8.4c}$$

$$[B_F] = [B_a] + \Theta \cdot [B_e] \tag{8.4d}$$

where subscript F indicates the feed, provide sufficient equations to solve simultaneously for the four unknowns $[A_a]$, $[A_e]$, $[B_a]$, and $[B_e]$ in terms of $[A_F]$, $[B_F]$, and Θ. Simple analytical solutions are available only for cases where $n = 1$. In the general case, it is easier to solve these sets of equations using spreadsheets and tools such as Solver in M-S Excel.

The results of one such set of calculations are shown in Fig. 8.2. The more extractable species A competes strongly for the ligand, so that the equilibrium curve of A in the presence of B is depressed only slightly below that for the extraction of A on its own. In contrast, the equilibrium curve for the extraction of B in the presence of A is depressed markedly relative to the extraction of free B. The effect of this is to improve the product purity significantly over what would have been possible at low concentrations of the extracted species in the extract. Thus, at high concentrations, saturation effects can improve product purity to a far greater extent than the equilibrium isotherms of the individual species would indicate superficially.

This is a key finding in understanding why solvent extraction is widely adopted in industrial practice. It is possible to achieve a higher purity of product than would be indicated by the separation between species when they are extracted individually. Clean separations between species with very similar extractabilities such as rare earth ions have proved practical by relying on saturation effects more than on the inherent separability of the species.

The two phenomena, namely the effect of phase ratio on the fractional recovery and purity and the effect of saturation on purity, have thus far been

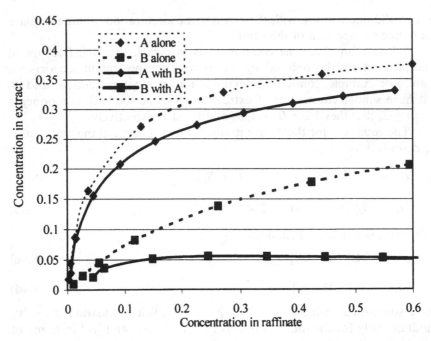

Fig. 8.2 Isotherms for the extraction of two species at a phase ratio of 1 and equal concentrations in the feed.

illustrated using only single-stage batch extraction. In both cases, if better separation was sought, then the recovery of the desired species was reduced. This explains why single-stage extraction is rarely adopted commercially. Multistage operation, outlined in Chapter 1.4.3, offers significant advantages, and in particular permits both high recovery and high purity of the product.

8.2.2 Repeated Extraction or "Cross-Flow" Systems

Returning to the example shown in Table 8.1, it is intuitively obvious that the low degree of extraction achieved when using a phase ratio $\Theta = 0.1$ could be improved if the raffinate could be reextracted with fresh solvent, and that if this reextraction were to be carried out at the same phase ratio, the second extract would show a similar purity. The first and second extracts could then be combined to give overall a better recovery than could be achieved in a single stage.

For a single, highly extractable component this argument is entirely valid. A mass balance over any one stage gives:

$$V_a[A_F] = V_e \cdot [A_E] + V_a \cdot [A_R] \tag{8.5}$$

which on rearranging becomes:

$$([A_F] - [A_R])/[A_E] = \Theta \tag{8.6}$$

where

$[A_F]$ is the concentration of A in the feed to any one stage,
$[A_R]$ is the concentration of A in the raffinate leaving any one stage, and
$[A_E]$ is the concentration of A in the extract that is in equilibrium with $[A_R]$

$[A_E]$ and $[A_R]$ are related via the isotherm, which is known, so the series of Eq. (8.6) governing each individual stage of extraction can be solved analytically. The properties of the solution can be readily understood graphically.

Figure 8.3 is a graphical construction that represents a series of equations such as Eq. (8.6), for the case $\Theta = 1$ and the isotherm of Eq. (8.4). From $[A_F]_1$, a line of slope $-1/\Theta = -1$ intercepts the equilibrium curve at $[A]_1$. A perpendicular from this intercept cuts the x axis at $[A_R]_1 = [A_F]_2$, from which a further line of slope -1 is drawn. Of course, Θ may be varied from

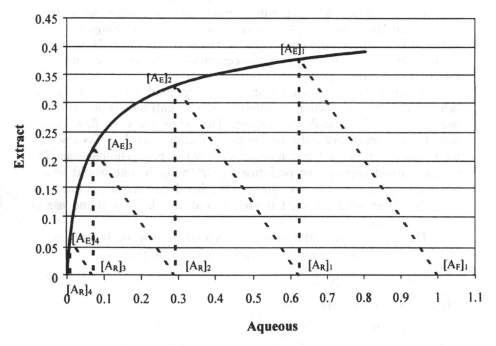

Fig. 8.3 Graphic construction for a four-stage cross-flow cascade.

Table 8.2 Extraction of a Single Component in a Cross-Flow Cascade at $\Theta = 1$

Stage no.	Aqueous		Extract			
	Feed	Raffinate	$[A_E]$	E_A	E_{TOT}	$[A_E]_{AVG}$
1	1.000	0.623	0.377	37.7%	37.7%	0.377
2	0.623	0.292	0.331	53.2%	70.8%	0.354
3	0.292	0.071	0.221	75.7%	92.9%	0.310
4	0.071	0.008	0.063	88.5%	99.2%	0.248
Single	1.000	0.079	0.230	92.1%	92.1%	0.230

stage to stage, but there is no particular advantage in doing so. Four stages are shown with a feed at a relative concentration of 1.

Table 8.2 gives the stage-by-stage performance of the four stages of repeated extraction, the cumulative extraction, E_{TOT}, and the average concentration of A_E, $[A_E]_{AVG}$, in the organic phases mixed together after each repeated extraction. In addition, Table 8.2 also shows the performance of a single stage in which the volume of extractant is the same as the total used in the four stages, i.e., a single extraction at a phase ratio of 4.

This illustrates that even with a moderate distribution ratio (10 in this case), useful recoveries (> 99% as shown by E_{TOT} for Stage 4) can be achieved in comparatively few stages by repeated extraction. However, as the number of stages increases, the concentration of the extracted species in the combined extract, $[A_E]_{AVG}$, drops significantly.

A single-stage extraction using the same total volume of solvent achieves only 92% extraction, and the extract concentration is only 0.23, vs. nearly 0.25 for the cross-flow extraction. The use of four cross-flow extraction stages is clearly preferable to a single extraction. Equally, of course, the use of more than four extraction stages, each with a proportionately smaller volume, would improve the performance. In the limit, one would seek a differential contacting process similar to the Soxhlet extractor employed for extraction from solid phases, but such a contactor has not found use in solvent extraction.

The mathematical treatment of a cross-flow cascade is straightforward. The mass, W_E, which is extracted in n stages is given by:

$$W_E = \sum V_{En} \cdot [C_E]_n \tag{8.7}$$

The cumulative recovery is given by:

$$E_{TOT} = W_E/([C_F] \cdot V) \tag{8.8}$$

while the average concentration in the combined extracts is given by:

$$[C_E]_{\text{AVG}} = W_E/(n \cdot V_E) \tag{8.9}$$

provided the phase ratio is kept the same in each repeated extraction; if not, $n \cdot V_E$ in Eq. (8.9) must be replaced by $\sum V_{En}$.

Now consider the effect of repeated extraction on the degree of separation that can be achieved from contaminants. Table 8.3 shows how the performance of the extraction deteriorates when a second component is present. Comparison with Table 8.2 shows that the percentage extraction of *A* in four stages drops to slightly over 95%, while the extraction of the impurity *B* rises to 37%. The purity of *A*, calculated as the ratio of *A* in the extract to the sum of *A* and *B* in the extract, becomes poorer as more stages of cross-flow extraction are added. *A* single stage, using the same total volume of extractant as used in the four stages of cross-flow extraction, performs nearly as well as the cross-flow cascade.

The reasons for this poor performance are clear. In stages 3 and 4 of the cascade, the desired component *A* is at low concentration in the extract phase, so there is a relatively high concentration of the free ligand available and the undesired component *B* is relatively highly extracted. Indeed, in the fourth stage, the concentration of *A* in the extract is lower than that of *B* even though *A* is 10 times more extractable than *B*.

Clearly, it would be preferable from the point of view of the purity of the product if there were high concentrations of *A* in both the extract and the raffinate of a stage. This can be achieved in countercurrent extraction, which allows both high recovery and the achievement of high product purity when properly designed.

8.2.3 Countercurrent Extraction

Consider the extract from the fourth stage of the cascade given in Table 8.3. The concentration of the desired species is 0.11, so it is clearly not saturated. Thus in principle it could form the feed to the third stage where the aqueous

Table 8.3 Extraction of Two Components in a Cross-Flow Cascade with $\Theta = 1$

Stage	$[A_R]$	$[A_E]$	$E_{\text{A.TOT}}(\%)$	$[B_R]$	$[B_E]$	$E_{\text{B.TOT}}(\%)$	Purity of $A(\%)$
Feed	1.000	—	—	1.000	—	—	—
1	0.662	0.338	33.8	0.951	0.049	4.9	87.4
2	0.371	0.291	62.9	0.882	0.069	11.8	84.2
3	0.158	0.212	84.2	0.778	0.104	22.2	79.1
4	0.047	0.111	95.3	0.630	0.147	37.0	72.0
Single	0.144	0.214	85.6	0.627	0.093	37.3	69.7

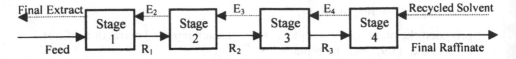

Fig. 8.4   Countercurrent extraction.

concentration is higher. Systems in which such a strategy is employed are
known as countercurrent extraction systems.

A typical flow arrangement is shown in Fig. 8.4, which clearly shows
how the name arises. A mass balance over any one stage 'i' gives:

$$V_E([A_E]_i - [A_E]_{i+1}) = V_A([A_R]_{i-1} - [A_R]_i) \qquad (8.10)$$

which on rearrangement yields:

$$\Theta = \frac{V_E}{V_A} = \frac{([A_R]_{i-1} - [A_R]_i)}{([A_E]_i - [A_E]_{i+1})} \qquad (8.11)$$

If the cascade is calculated by the same techniques as used for the cross-flow
cascade in the previous section, but with the layout shown in Fig. 8.5, the
organic concentration increasing from stage to stage and a phase ratio
$\Theta = 1$, the performance is shown graphically in Fig. 8.5. In this case, the
calculation has been carried out from the organic feed end, rather than the
aqueous feed end as in the case of Fig. 8.3, to allow for the increase in
organic-phase concentration. In every other respect the calculation is identical to that which led to Fig. 8.3. From $[A_R]_3$ a vertical line is drawn to
intersect the equilibrium curve, from which point a line of slope -1 is drawn
to reach the point $[A_R]_2$ on the axis. A vertical line from this point reaches the
equilibrium curve at point E, and again a line of slope -1 is drawn, but on
this occasion it is only carried to point G, because there is contact between
aqueous phase of concentration $[A_R]_1$ and organic phase not of zero concentration but of concentration $[A_E]_3$. The calculation then continues in the
same way for another stage, making three stages in all.

As the calculation continues, it is found impossible to proceed much
beyond the point marked A on the diagram, at which point the aqueous
concentration is about 0.34. The reason for this is that at a phase ratio
$\Theta = 1$ the cascade has too little organic phase for the duty it is being asked
to perform. Consider the rectangles $ABCD$ and $CEFG$ in Fig. 8.5. The
diagonals BD and EG have the same slope. The diagonals AC and CF thus
have the same slope, and AF is therefore a straight line. The point A has the
coordinates $[A_F]$; $[A_E]_1$ and the point F the coordinates $[A_R]_2$; $[A_E]_3$. Thus the
slope of the line

$$AF = \frac{([A_E]_1 - [A_E]_3)}{([A_F] - [A_R]_2)} = 1/\Theta \tag{8.12}$$

which is identical to Eq. (8.11) except it represents the mass balance over two stages.

Thus to achieve an extraction equivalent to that shown in Fig. 8.3 requires an overall mass balance for which, at the first stage, $[A_E]_1 = 0.377$ and $[A_F] = 1.0$; while at the last stage $[A_E]_3 = 0$ and $[A_R]_3 = 0.008$. Then by Eq. (8.11) over the whole cascade:

$$\Theta = \frac{(1.00 - 0.008)}{(0.377 - 0)} = 2.63$$

or the slope of the mass balance line = $1/2.63 = 0.38$.

Mass balance lines such as AF are important and are known as *operating lines* in countercurrent cascades. Clearly, an operating line cannot cross an equilibrium line. Wherever an operating line approaches an

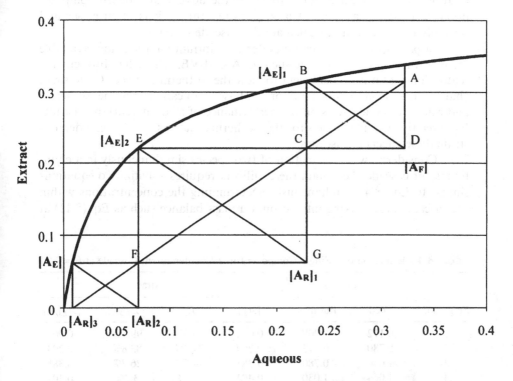

Fig. 8.5 Graphic construction for a countercurrent cascade.

equilibrium line, a *pinch point* is the result and the number of stages needed to achieve the desired degree of extraction approaches infinity.

In Table 8.4, the calculations behind Fig. 8.5 are repeated for a value of $\Theta = 2.63$. Comparison with the results of Table 8.2 indicates clearly that the countercurrent cascade offers a similar overall extraction in the same number of stages, together with an extract of significantly higher concentration.

The data of Table 8.4 are also shown in Fig. 8.6, which is shown as an equilibrium isotherm, an operating line, and a series of steps between the operating line and the isotherm. These steps are entirely equivalent to the lines establishing the mass balances for each stage in Figs. 8.3 and 8.5. For instance, the horizontal line AB represents $[A_F]_1-[A_R]_1$, while the vertical line BC represents $[A_E]_1-[A_E]_2$.

The graphical construction of an extraction isotherm, an operating line, and the stepwise evaluation of the number of stages in this manner is known as a *McCabe-Thiele diagram*. Historically, it found great application in a variety of mass transfer operations, from gas adsorption through distillation to solvent extraction. However, the advent of modern computational techniques has made it largely redundant, as it is often easier and certainly more accurate to calculate the cascade directly.

In part also this is because better equilibrium data are now available through the use of equipment such as AKUFVE, which has brought the realization that it is not sufficient to view the isotherm as fixed. Quite small changes in aqueous composition or in phase ratio can change the isotherms and cause dramatic effects on the performance of a countercurrent cascade. Even relatively crude models for the isotherms, such as Eq. 8.3, can demonstrate these effects in cascades.

Consider now the extraction of two species simultaneously in a countercurrent cascade. To obtain the equilibria requires solving two equations similar to Eq. (8.4) simultaneously. Determining the concentrations within the cascade means taking into account a mass balance such as Eq. (8.11) at

Table 8.4   Extraction of a Single Component in a Countercurrent Cascade at $\Theta = 2.63$

Stage no.	Aqueous		Extract			
	Feed	Raffinate	$[A_E]$	E_A	E_{TOT}	$[A_E]_{AVG}$
1	0.172	0.008	0.062	95.3%	99.2%	0.062
2	0.780	0.172	0.294	77.9%	83.8%	0.294
3	1.030	0.780	0.388	24.2%	26.7%	0.388
4	1.064	1.030	0.401	3.2%	3.2%	0.401

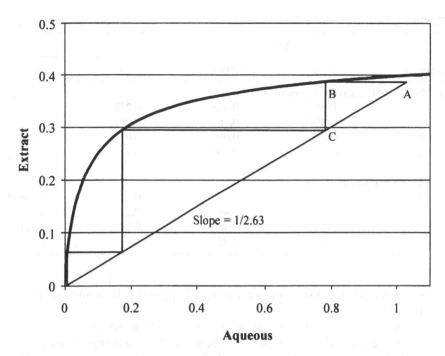

Fig. 8.6 McCabe-Thiele graphic construction for a countercurrent cascade.

each stage. Table 8.5 presents such calculations for the same two equilibria as used before, with A having $D = 10$ and B having $D = 1.0$. It should be noted that the calculation is not simple, and requires estimation of the equilibrium at each stage before the next stage is computed.

The behavior of the system is very counterintuitive. The more extractable component is relatively well behaved and a recovery of 99.4% is achieved in four stages. In contrast, the less extractable component behaves unusually. Its concentration *increases* in the aqueous phase between the first and the third stages. In the extract phase, it has a peak concentration in the extract from the fourth stage. Thus, the less extractable component circulates between the first and last stages, being "squeezed out" of the organic phase by the more extractable component, then being extracted back from the aqueous phase once the concentration of the more extractable component in the aqueous phase has fallen.

This is caused by the coupling of the two equilibria. The presence of high concentrations of the less extractable component so disturbs the equilibria that the more extractable component has a distribution ratio of only about 2.5 in the fourth stage of the cascade. If the less extractable component

Table 8.5 Calculation of a Countercurrent Cascade with Two Components, $\Theta = 3.0$

	Aqueous		Extract	
Stage no.	Feed	Raffinate	Feed	Extract
1. The more extractable component				
1	1.00	0.686	0.227	0.332
2	0.686	0.265	0.086	0.227
3	0.265	0.058	0.017	0.086
4	0.058	0.006	0.0	0.017
2. The less extractable component				
1	1.00	1.208	0.127	0.058
2	1.208	1.491	0.221	0.127
3	1.491	1.496	0.223	0.221
4	1.496	0.826	0.0	0.223

were not present, the distribution ratio would approach the infinite dilution value of 10.

Because of the high concentrations of the less extractable component within the cascade, the final separation between the two components is not particularly good. The final extract is only 85% pure. This could be improved by reducing the phase ratio, which would have the effect of loading the more extractable component sooner in the cascade, and thus squeezing out the less extractable component better. In practice, however, one cannot reduce the phase ratio too far without running the risk of the cascade becoming unstable, or requiring, for instance, temperature control. It is preferable to scrub the extract, as described in the next section.

It should be noted that this was a fairly severe test. Having a feed containing equimolar quantities of components whose extractability differs by a factor of only 10 is rare. Usually the differences in extractability are greater, or the less extractable component is present in low concentrations relative to the more extractable. In both these cases, good separations can be achieved in a countercurrent cascade.

The cascade can be drawn graphically in the McCabe-Thiele form, but the equilibria are so distorted that the exercise is not very valuable. This is a general finding in industrial practice. Graphical methods are adequate to give an indication of the number of stages likely to be needed for a particular duty. They fail when separations between similar species are sought.

Highly selective extractants are desirable, but often show poorer physical or chemical properties than less selective extractants. As the example above illustrates, a difference of a factor of only 10 in the distribution ratios gave a reasonably purified product in four countercurrent extraction stages, even when the contaminant was present at relatively high concentrations. One can improve the performance of the cascade if the desired product is present at a higher concentration than other extractable species. This is the basis for further purification by means of scrubbing of the final extract.

8.2.4 Extraction with Scrubbing

Where a high-purity product is sought, the same philosophy of saturating the aqueous phase with the unwanted component is used in scrubbing the extract. A recycle stream of the pure product is used to "scrub" the impure extract. In the process, the extract is saturated with the desired product, and the impurities are removed. The final raffinate from scrubbing is then recycled to the aqueous feed so that the now-contaminated desired product can be recovered. This is shown in Fig. 8.1.

In Table 8.6, the results of computations identical in principle to those described previously are given. The final extract from the countercurrent cascade described in Table 8.5 was contacted with a pure stream of the more extractable component at a high phase ratio. The high phase ratio is chosen to minimize the volume of aqueous phase that must be recycled to the feed.

The cascade is fed with an extract of the same composition as that resulting from the extraction shown in Table 8.5, with $E_A = 0.332$ and $E_B = 0.058$. In four stages, using a concentrated solution of A at $\Theta = 10$, $E_A = 0.350$ and $E_B = 0.00014$, the purity of the product is increased from 85–99.96%. Note that the composition of the scrub solution is the same as

Table 8.6 Calculation of a Countercurrent Cascade in Scrubbing, $\Theta = 10$

	Aqueous		Extract	
Stage no.	Feed	Raffinate	Feed	Extract
1. The more extractable component				
1	0.500	0.390	0.339	0.350
2	0.390	0.328	0.333	0.339
3	0.328	0.299	0.329	0.333
4	0.299	0.297	0.332	0.329
2. The less extractable component				
1	0.0002	0.0016	0.00014	0.00028
2	0.0016	0.0027	0.00028	0.00039
3	0.0027	0.0035	0.00039	0.00047
4	0.0035	0.0041	0.00047	0.0584

the composition of the final extract after scrubbing. In practice, additional purification can often take place during recovery of the components from solution after stripping, which allows the scrub solution to be even purer, and in turn allows the product purity to be higher than shown in this example.

Also of note in Table 8.6 is the composition of the final raffinate, with $A_A = 0.297$ and $A_B = 0.004$. This can clearly be recycled directly to the feed to the extraction cascade, where the fact that this stream is purer than the feed stream will further aid the achievement of product purity.

Techniques for achieving high-purity products by countercurrent extraction and scrubbing of the extract have proved essential for the production of nuclear-grade uranium. They have also found application in the separation of the rare earths and a number of other difficult separations. A feature of the operation of these systems is the need for close control of flow rates and even temperature in order to achieve a consistent product quality. The product quality is a very nonlinear function of the operating parameters. However, with modern control systems this disadvantage can be overcome.

8.2.5 Summary of Extraction

In this section, the first three questions posed in the introduction have been answered. It has been shown that:

- The fraction of the desired component that can be recovered is determined not only by its inherent extractability but also by the phase ratio and the presence of competing extractable components.
- The volume of solvent needed for a particular duty depends on the system adopted for the extraction; in general, a single-stage extraction requires more solvent phase than a cross-flow cascade, which in turn needs more than a countercurrent cascade.
- The purity of the product is determined not only by the inherent selectivity of the solvent system for the component sought, but also on the phase ratio, the concentration of the contaminants, and the performance of any scrubbing of the extract.

8.3 STRIPPING

The general principles established for extraction apply to stripping, although of course distribution ratios are sought that are significantly less than unity in order to accomplish the strip as efficiently as possible.

Stripping can equally be done in single-stage, cross-flow, or countercurrent systems. To illustrate how the overall concepts remain valid, the performance of a countercurrent cascade accepting as feed the scrubbed extract of Table 8.6 is calculated assuming that the equilibria involved are described by:

$$D_A = 0.1(1 - 2[A_e + B_e]) \tag{8.13a}$$

$$D_B = 0.01(1 - 2[A_e + B_e]) \tag{8.13b}$$

that is, as for Eq. (8.4) but with m not equal to n. Further, the cascade should strip over 95% of the desired component and yield a strip raffinate which is as concentrated as possible (assuming a solubility of A in the stripping solution of 2).

As before, the same mass balance limitations on phase ratio apply. Although this is a stripping cascade, it remains countercurrent in its structure. For an organic feed concentration of 0.350 (that is, the extract after scrubbing, as shown in Table 8.6), assuming >95% strip and a final aqueous of <2, the *maximum* phase ratio is $\Theta_{max} = 2/(0.95*0.350) = 6.02$. (Maximum in this case because stripping is in the reverse direction to extraction.) Choosing a phase ratio of 5.0 gives the results shown in Table 8.7.

Four stages are needed. The strip raffinate has a concentration of 1.71, and the stripped organic contains only 2% (0.0073/0.35) of the component A in the scrubbed extract, so the stripping efficiency is 98%. The less extractable (more readily stripped) component B was completely stripped in three stages.

Note that the stripped extract still contains traces of the desired component. This is a slight nuisance as the stripped extract will, in the normal course of events be recycled to the extraction stages where it will reduce the

Table 8.7 Calculation of a Countercurrent Cascade in Stripping, $\Theta = 5.0$

Stage no.	Aqueous		Extract	
	Feed	Raffinate	Feed	Extract
1. The more extractable component				
1	0.000	0.074	0.022	0.0073
2	0.074	0.232	0.054	0.022
3	0.232	0.601	0.128	0.054
4	0.601	1.713	0.350	0.128
2. The less extractable component				
1	0.000	0.0000	0.000	0.000
2	0.0000	0.0000	0.0000	0.0000
3	0.0000	0.00003	0.00001	0.0000
4	0.00003	0.00070	0.00014	0.00001

extraction efficiency in the first stage. This practice is fairly common in industrial practice, as it is more economic to lose a little of the desired component in the extraction stages than to have too dilute a strip raffinate by seeking a total strip.

The purpose of seeking a concentrated strip solution is to reduce the energy required to recover the product from the strip solution. In the case of metal salts, precipitation, electrolysis, direct reduction, and a host of other techniques may be used to generate the final product. In the case of the extraction of organic compounds, distillation, crystallization, or similar separation methods are used. In each case, the more concentrated the strip solution, the less energy is required to recover the desired components.

It sometimes happens that an undesired component is not stripped. It then builds up in the organic phase until it interferes seriously with the extraction. In this case, the solvent is given a more vigorous strip to regenerate its performance by removing the contaminant. A small sidestream may be bled continuously from the recycled organic phase and regenerated, or the contaminant may be allowed to build up in the extract for some time before the solvent is regenerated on a batch basis. Because of its vigorous nature, regeneration of the organic phase can be expensive, but sometimes the contaminant is very valuable—platinum, gold, and cobalt complexes have acted as contaminants on occasions, and their recovery has paid for the regeneration!

8.4 EQUIPMENT FOR CONTINUOUS CONTACT

8.4.1 Stagewise Extraction

Thus far in the discussion of industrial practice, we have referred to "stages" of extraction without defining a stage. Clearly these stages could be scaled-up versions of separatory funnels, but this is inefficient, because it implies batch rather than continuous operation. Industry prefers continuous operation because it is generally simpler to control automatically, and because it makes better use of labor.

The typical stage might be a mixer-settler. As its name suggests, this comprises some means for mixing the two phases and an adjoining means for separating them. Mixers can be stirred tanks, of a sufficient size to retain the mixed phases long enough to effect transfer of the desired species from one phase to the other. The mixer usually consists of some form of impeller or propeller in a tank that usually contains some means to prevent the mixed liquids from swirling and thus reducing the efficiency of mixing. In circular cross-section mixers, this usually takes the form of vertical baffles mounted on the wall of the mixer.

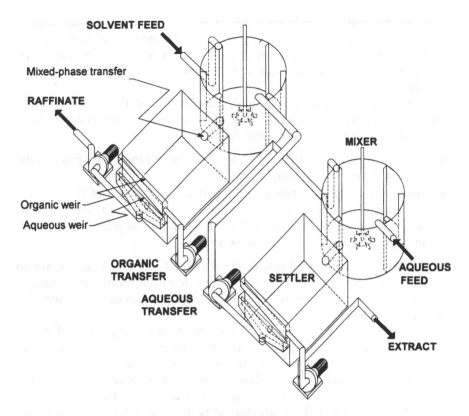

Fig. 8.7 Layout of two stages of a mixer-settler cascade.

Figure 8.7 illustrates two stages of a mixer-settler cascade. The aqueous feed enters the first stage, where it is mixed with partially loaded extract from the second stage. The mixed phases pass to the settler, where they are separated under gravity. The fully loaded extract overflows an upper weir and passes to scrubbing and/or stripping. The first-stage raffinate overflows a lower weir and is pumped to the second stage, where it is mixed with the solvent feed. After settling again, the final raffinate passes to treatment before discharge as waste.

There are many variants on this simple theme. For instance, many other methods for mixing have found use. In a design offered by Lurgi, the phases are mixed in what is essentially an axial flow pump, and then pass down a relatively long pipe where the turbulence of flow keeps the phases mixed while the extraction takes place. In another design, the individual phases are pumped and then join and pass through a static mixer. There are no particular physicochemical reasons for preferring one type of mixer to another. As long as they provide adequate interfacial area for the extraction to take place, without creating such small droplets that they will not settle efficiently, and provided there is sufficient residence time for the desired degree of extraction to take place, then one mixer will work as well as another.

There is, however, one physicochemical criterion that is important in industrial mixing, and that is ensuring that the correct phase is dispersed in the other. There are several reasons for this:

- It is sometimes found that mass transfer is more rapid if one phase is the dispersed phase rather than the other.
- Alternatively, the dispersed phase is chosen because, by definition, it will not contain droplets of the continuous phase. In this way the dispersed phase, after settling, will not entrain the continuous phase and entrainment losses from the settler will be reduced.

Whatever the reason for choosing the dispersed phase, it is important to ensure that the mixer will keep that phase dispersed during operation, as changes in the dispersed phase, i.e., phase inversion, can cause considerable operating problems.

Usually the continuous phase is the phase present in greater volume. It is possible to run for long periods with the greater volume phase dispersed, but phase inversion is always a risk in such circumstances. To overcome this risk, where it is desired that the lesser volume phase is continuous, then a portion of that phase may be recycled from the settler back to the same mixer to ensure that within each stage it is the greater volume, even if it is the lesser volume phase overall.

In large-scale operation, the volumetric flow of the phase to be dispersed is so large that it becomes necessary to disperse that phase into the mixed phases. Otherwise "blobs" of the dispersed phase will act locally as the continuous phase, and the intended continuous phase will be dispersed in the blobs before the shear forces in the mixer break them up. This can lead to excessive entrainment losses.

In some mixer-settler designs, the impeller is arranged both to mix the two phases and to provide the necessary energy to transfer the phases from one stage to the next, in which case it is known as a *pump-mix mixer*. The head required to move a phase from one stage to the next is small, so the impeller need not be efficient as a pump. Nevertheless, the design of impellers for the dual purpose of both mixing and pumping is more of an art than a science. Moreover, in full-scale operation it has been found difficult to start up cascades of pump-mix mixers and achieve equilibrium rapidly. Accordingly, this design has primarily found use in small-scale applications in the nuclear and pharmaceutical industries.

Settlers tend to be less varied in their design. They typically comprise a relatively large, shallow tank, rectangular in shape, with an inlet for the mixed phases at one end and two outlets for the separated phases at the other. Various devices are used to introduce the mixed phases gently into the settler, and to control the flow of the mixed phases while they separate, but these do not change the basic principle of separation under gravity.

The level of the heavy phase outlet within the settler controls the level of the interface (Fig. 8.8). At the interface, the static pressure due to light phase above the interface is determined by the density of the light phase, ρ_o, and the depth of the light phase, H. Similarly, the static pressure above the interface in the overflow leg is determined by the density of the heavy phase, ρ, and the height h of the weir above the interface. The pressure must be the same at the same elevation, so $h \cdot \rho = H \cdot \rho_o$, or

$$H = h \cdot (\rho / \rho_o) \tag{8.14}$$

Because the difference between the densities of the two phases is often small, and may vary from one stage to the next, H can vary strongly with h. Thus it is often necessary to make the height of the weir adjustable. Similarly, if there are significant differences in flow rate, then the depth of the liquid overflowing the weir has an effect on h and thus on the height of the interface.

The importance of these considerations is that the shallower the settler, the more difficult the interface control. Shallow settlers are desired because they reduce the inventory of the solvent. However, it is possible to make the settler so shallow that interface control can be lost. In one case, large, shallow settlers suffered from the effects of wind pressure, that caused such massive

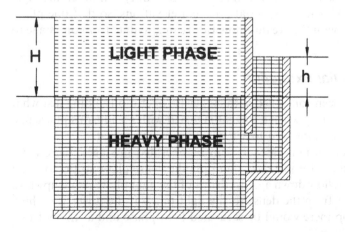

Fig. 8.8 The control of the level of the interface in a settler.

oscillations in the level of the interface that light phase often passed over the heavy phase weir.

The introduction of the mixed phases into the settler has been found to be important if clean separation is to be ensured. "Picket fences," vertical plates set at an angle to the flow from the mixer into the settler, have been used to calm the flow and ensure the spreading of the mixed stream across the width of the settler. Various packings have been employed to aid settling, but under industrial conditions they are liable to clog with adventitious material or "third phases."

Baffles placed across the settler, of progressively lower height from entrance to overflow, have been employed to hold back the mixed phases to permit them to separate. The mixed phases will spread rapidly right across a settler unless there is a baffle to hold them back. There have been many reports of "wedges" of mixed phase in small-scale settlers. These are never seen in industrial practice because considerations of pressure at a given point, as were used above to determine the height of the interface, show that a wedge is inherently unstable.

In one mixer-settler design, the mixed phases flow down a shallow trough placed over the settler, which gives them an opportunity to coalesce and separate before entering the settler. In this way, the capacity of the settler is markedly increased, with a concomitant reduction in the inventory of solvent required for a given duty.

The manner in which individual mixer-settler stages can be linked together to form countercurrent cascades is illustrated in Fig. 8.7. If each stage is on the same level, then some form of pump must be provided to move each phase from one stage to the next. As indicated earlier, it is sometimes convenient to use the mixer for this duty. Another arrangement has the individual stages set at different elevations, so that one phase (usually the phase with the greatest flow rate) can gravitate from one stage to the next.

8.4.2 Differential Extraction

Thus far we have been concerned with stagewise operations, doing just what is done in the laboratory when a solvent is mixed with an aqueous phase in a separatory funnel and allowed to settle before being separated.

While countercurrent cascades can be operated in stages, there is no need do so. Consider, for instance, pumping the dense phase to the top of a tower and letting it flow down against drops of the light phase rising upward. Assuming transfer from the dense to the light phase if the tower were high enough, at the top there would be saturated light phase and at the bottom, depleted heavy phase.

Clearly the concept of a stage has no meaning in such a tower. Instead, we deal with differential *transfer units*, which are a measure of the change in concentration per unit of difference in concentration (recall that the rate of extraction is determined largely by the difference between the actual and the equilibrium concentration of a solute, or "driving force").

At each point in the tower, a component A has an actual concentration $[A]$ and an equivalent equilibrium concentration $[A]_e$. Then the number of transfer units (NTU) required for the extraction is given a first approximation by:

$$\text{NTU} = \int_{[A]_r}^{[A]_f} d[A]/([A] - [A]_e) \tag{8.15}$$

where $[A]_f$ and $[A]_r$ are the feed and raffinate concentrations of A respectively. This is illustrated in Fig. 8.9. The driving force is $[A] - [A]_e$ and the inverse of the driving force is to be integrated between the feed and the raffinate concentrations (not shown in Fig. 8.9).

This integral clearly depends on the slope of the operating line and, as in the case of stagewise operations, if the operating line approaches the equilibrium curve too closely, then the driving force approaches zero and the inverse becomes very large. That is, when the operating line is close to

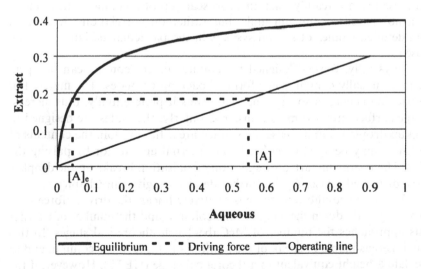

Fig. 8.9 Illustration of the determination of the driving force from equilibrium and operating lines.

the equilibrium curve and there is a pinch point, the number of transfer units becomes large.

Equation 8.15 refers to a *single phase*, and $[A]_e$ is related to $[A]$ at each point via the phase ratio (or operating line, which is the same thing). The NTU could be calculated for the extract phase instead of the aqueous phase, that is, either from the difference between $[A_R]$ and $[A_R]_e$ or from that between $[A_E]_e$ and $[A_E]$.

Where there is an analytical expression for the equilibrium such as Eq. (8.3), then Eq. (8.15) may be integrated directly. Otherwise it is necessary to perform the integration numerically or graphically.

Once the *number* of transfer units has been found, the height of the tower is determined from the product of the number and the *height* of each transfer unit (HTU). The HTU is determined by physical parameters such as the droplet size, the flow patterns in the tower, and the effect of any packing. These all affect the *rate* of mass transfer, which is addressed in Chapter 9. Very often the rate of mass transfer cannot be estimated from first principles, and it is necessary to estimate the height by determining the number of transfer units achieved and then dividing the actual height of the column employed by the number of transfer units, i.e.:

$$\text{HTU} = H/\text{NTU} \tag{8.16}$$

where H is the height of the column employed.

HTU is subjected to the effects of both radial and axial mixing, and these are not readily quantified, so scale-up of columns of this kind is often not based on fundamentals, but rather on correlations determined from detailed studies of several systems in the particular design of column chosen.

Physically, towers designed for countercurrent contact can be open, but more usually contain some form of packing or plates. The material of the packing is chosen so that one phase wets it preferentially, thus increasing the surface area for mass transfer. Similarly, the plates are designed to breakup droplets and increase the surface area. In addition, the contents of the tower may be agitated either by an internal agitator or by pulsing the fluids. The energy imparted by agitation or pulsation breaks up the droplets of the dispersed phase. Again, further details are given in Chapter 9.

When the equilibrium curve is relatively linear, the driving force does not vary greatly down the length of the column, and the number of transfer units approaches the number of McCabe-Thiele theoretical stages. In this case, it is reasonable to speak of the number of stages in the column, and to calculate a height equivalent to a theoretical stage (HETS). However, if the equilibrium curve and the operating line are far from parallel, the number of theoretical stages becomes a poor measure of the column's performance, and the number of transfer units should be used.

8.5 EXTRACTION EFFICIENCY

Solvent extraction is a kinetic process. The key variables in determining the rate of extraction are (1) the displacement of the system from equilibrium, also referred to as the driving force; (2) the area through which mass can be transferred, or the interfacial area; and (3) specific resistances in the interfacial region, particularly any slow interfacial reactions.

To a lesser extent, the rate is also affected by diffusion through the bulk liquids, but in general industrial practice there is sufficient turbulence to ensure that the bulk phases are well mixed. There is some control over interfacial area, though not too much flexibility is available because a very large interfacial area is associated with very fine droplets or very thin films. This may result in excessive loss of solvent by entrainment. These aspects are extensively discussed in Chapters 7 and 9.

There is little to be done about the displacement of the system from equilibrium (although it may be noted that the average displacement is maximal in a countercurrent cascade). As shown in section 8.4.2, the driving force can be increased by reducing the slope of the operating line, i.e., increasing the phase ratio θ, but this is generally not economical. Very little can be done about interfacial resistances once the extraction system has been chosen, although renewal of the surface during bubble coalescence and dispersion assists in overcoming some forms of this kind of resistance. Thus industry either has to make the time of contact long enough to ensure that equilibrium is essentially attained, or it has to accept the inherent inefficiency of a single stage, and employ more stages than would otherwise be needed. In practice, it employs the latter strategy.

8.5.1 The Efficiency of a Single Stage

Figure 8.5 gives a graphic construction for a series of equations of the type given in Eq. (8.11), for a single stage at equilibrium. The same basic equation governs a nonequilibrium stage, that is:

$$([A_R]_f - [A_R]_e)/([A_E]_e - [A_E]_f) = \theta \tag{8.17}$$

except that the product streams $[A_R]_e$ and $[A_E]_e$ are no longer at equilibrium, but are reduced by the inefficiency to $[A_R]_i$ and $[A_E]_i$. This is illustrated in Fig. 8.10.

8.5.2   The Efficiency of a Cascade

The concepts of section 8.5.1 are readily extended to determine the efficiency of a cascade. To illustrate this, the results of calculations identical to those of Tables 8.2 and 8.4, but with an 80% stage efficiency, are given in Table 8.8. The resultant McCabe-Thiele diagram is given in Fig. 8.11.

Comparison with the earlier data for equilibrium conditions shows that an extra stage is needed. In spite of this, the final raffinate is significantly higher than was achieved at 100% extraction efficiency. The influence of the inefficiency is clearly greater the more dilute the solutions.

Figure 8.11 shows how the equilibrium curve "shrinks" in the presence of inefficiencies. In multicomponent systems where there is mutual interference in extraction by several components, the effceny shrinkage comes on top of the other reductions in the equilibrium curve, and for this reason there is stress in such systems on achieving high efficiency.

8.6   SOLVENT LOSSES

Throughout, we have made the tacit assumption that the two phases, which for convenience, we have called the aqueous and the organic phases, are totally immiscible. For many systems this is a reasonable approximation, but

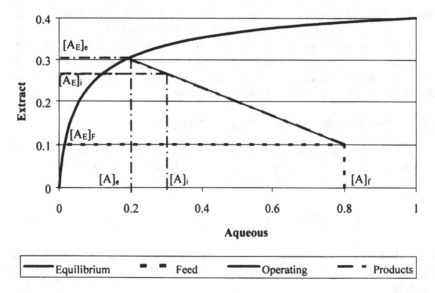

Fig. 8.10   Illustration of the effect of reduced efficiency of extraction.

Table 8.8 Extraction at 80% Efficiency, Phase Ratio 2.63

Stage no.	Aqueous			Extract	
	Feed	Equilibrium	Inefficient extract	Extract	Feed
1	0.105	0.004	0.024	0.037	0.006
2	0.402	0.031	0.105	0.150	0.037
3	0.875	0.284	0.402	0.330	0.150
4	1.038	0.834	0.875	0.392	0.330
5	1.064	1.031	1.038	0.402	0.392

for some systems mutual miscibility must be taken into account, particularly when the primary solute is organic.

The methods for doing so are described in Chapter 9. The basic principles remain unchanged — the primary difference is the choice of a consistent basis for calculation, such as a solvent-free basis. Graphic techniques based on triangular coordinates provide approximate answers, but modern computational techniques are to be preferred.

Some consideration should be given at this point to the need to prevent loss of the organic phase in the aqueous raffinate. This loss can arise by either solubility in the aqueous phase or by entrainment of droplets not fully settled. The solvent lost in this way can offer a finite environmental hazard and be an economic cost on the process.

Clearly the primary duty is good engineering practice, which is covered especially in Chapter 9. Often, however, additional security is provided in the following form:

- Additional settler capacity for final raffinate
- Extraction of residual organic phase using a third diluent, from which it is later separated, typically by distillation
- Coalescence on a solid wetted preferentially by the organic phase
- Flotation with air in the presence of surfactants

8.7 SUMMARY

In this chapter, we have seen the way in which laboratory studies of solvent extraction are adapted to industrial use. Starting from a batch extraction, it was shown how both recovery and product purity could be markedly influenced by the volumetric phase ratio, and how it was impossible to achieve both high recovery and high purity in a single stage. Cross-flow or repeated extraction was then evaluated, and it was shown how this could improve

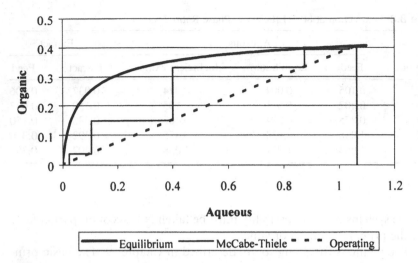

Fig. 8.11 McCabe-Thiele diagram showing the effect of reduced efficiency of extraction.

both recovery and purity, yet often resulted in mixed extract solutions that were too dilute to be processed without further upgrading.

The concept of countercurrent extraction was then introduced, and it was shown how the minimum phase ratio for a given degree of extraction was determined. Countercurrent extraction could yield both high recoveries and concentrated extracts, but studies on two-component extractions soon showed that product purity suffered.

This led to the discussion of scrubbing as an essential adjunct to countercurrent extraction where purity was important, and it was shown that washing the extract with a small amount of aqueous phase could improve purity markedly. Stripping was shown to follow the same underlying principles as extraction for achieving efficient removal of extracted species, and the need to choose phase ratios carefully to maximize the concentration of the desired species in the strip solution was stressed.

There followed a brief discussion of equipment for carrying out solvent extraction in industrial practice, both by stagewise and differential contact. Some of the first principles for the design of differential contactors were outlined and the part played by the efficiency of extraction in continuous equipment was discussed. Finally there was an outline of methods for the control of solvent loss which forms probably the most important environmental aspect of the application of solvent extraction.

9

Engineering Design and Calculation of Extractors for Liquid–Liquid Systems

ECKHART F. BLASS* Technische Universität Munich, Munich, Germany

9.1 INTRODUCTION

The previous chapters have demonstrated that liquid–liquid extraction is a mass transfer unit operation involving two liquid phases, the raffinate and the extract phase, which have very small mutual solubility. Let us assume that the raffinate phase is wastewater from a coke plant polluted with phenol. To separate the phenol from the water, there must be close contact with the extract phase, toluene in this case. Water and toluene are not mutually soluble, but toluene is a better solvent for phenol and can extract it from water. Thus, toluene and phenol together are the extract phase. If the solvent reacts with the extracted substance during the extraction, the whole process is called *reactive extraction*. The reaction is usually used to alter the properties of inorganic cations and anions so they can be extracted from an aqueous solution into the nonpolar organic phase. The mechanisms for these reactions involve ion pair formation, solvation of an ionic compound, or formation of covalent metal-extractant complexes (see Chapters 3 and 4). Often formation of these new species is a slow process and, in many cases, it is not possible to use columns for this type of extraction; mixer-settlers are used instead (Chapter 8).

This chapter explains in detail how this mass separation problem can be solved. We shall adhere to engineering symbols as listed at the end of this chapter (see section 9.10) to facilitate comparison with other chemical engineering texts. Reference to these symbols will be made in the text without any extensive explanations.

---

*Retired.

9.2   FUNDAMENTAL DESIGN OF EXTRACTORS

Returning to the example of removing phenol from coke wastewater by solvent extraction, in the first important step, the two liquids must come into contact. This can be achieved by dispersing one of the liquids into the other. When the resulting droplets have been in contact long enough with the raffinate phase to allow the extractable component to change phases, the two liquids have to be separated. The most favorable economic way of achieving this is by separating the drops in the gravity field. This requires that the liquids possess different densities. Figure 9.1 shows a simple design that implements the process steps for a continuous phase throughput (i.e., a spray column). Let us again use the system water–phenol–toluene as an example. The short table of physical properties in Fig. 9.1 shows that toluene is lighter than water. Thus, the flow direction of the two liquids in the gravity field is defined: the water flows from top to bottom, the toluene from bottom to top. Accordingly, the fluids must be transported to the column and removed by pumps. If the column is filled with water for the initial operation and if the water is always kept at the marked height by an overflow control during the following, stationary operation, the toluene breaks into drops at the feed point to become the dispersed phase, then rises in the continuous water phase. When the drops have reached the water interface, they gradually coalesce and form a continuous toluene phase that floats on the water phase. This water–toluene boundary is sometimes called the principal interface.

However, the column can also be operated in reverse by filling it with toluene and adjusting the principal interface at the bottom of the column. Then water is the dispersed phase and would break into drops at the feed point at the top of the column. These drops descend in the toluene phase and, at the bottom of the column, coalesce to a water homophase that is below the toluene phase. When needed, the principal interface can be adjusted somewhere between top and bottom of the column, whereby the heavier liquid is dispersed above and the lighter liquid below the principal interface. How is it decided which of the two liquids should be dispersed? Understanding the flow and mass transfer processes in the extractor, which are analysed in this chapter, provides the answer. At this point, only the important factors are listed; thus, the dispersed phase should be:

- The one with the higher viscosity, since a low viscosity of continuous phase makes a higher phase throughput possible because of the lower energy of dissipation
- The one with the higher flow rate to obtain the maximum mass transfer area
- The one with the lower interfacial tension between liquid and vapor for easier dispersion

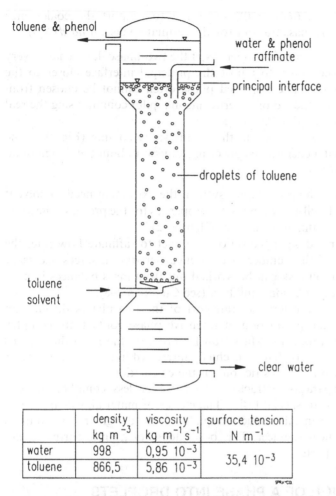

toluene & phenol

water & phenol raffinate

principal interface

droplets of toluene

toluene solvent

clear water

	density kg m^{-3}	viscosity kg m^{-1}s^{-1}	surface tension N m^{-1}
water	998	0,95 10^{-3}	35,4 10^{-3}
toluene	866,5	5,86 10^{-3}	

Fig. 9.1 Spray column for liquid–liquid extraction. The water phase flows from top to bottom, the toluene as the lighter phase from bottom to top. If the column is always filled with water at the marked height, the toluene breaks into drops at the feeding point. When the drops have reached the water interface, they coalesce and form a continuous toluene phase. The phenol transfers from the water phase to the toluene phase.

- The one that is more potentially explosive, more poisonous, and more expensive, since the holdup of the dispersed phase is only about 10–30%
- The extract phase, if the interfacial tension σ between the two liquid phases decreases with increasing concentration of the transitional

component x [i.e., $(d\sigma/dx < 0)$] so as not to support the coalescence between drops by mass transfer (or the raffinate phase, if $d\sigma/dx > 0$)

If the physical and volumetric properties of the two-phase flows change very much, it may be necessary to adjust the principal interface closer to the middle of the column. The dispersed phase often cannot be chosen from theory, but only with the aid of experiments in a pilot column using the real material system (see section 9.9).

Considering the processes in the simple spray column (Fig. 9.1), the main questions that a chemical engineer must answer about the design of an extractor can be listed.

1. Questions concerning the liquid system: the extraction needs a solvent of suitable and well-known physical properties for the process. The same is true for the raffinate phase (see Chapter 2).
2. Thermodynamic design: the ratio of extract to raffinate flow rate, the phase ratio, and the number of theoretical separation stages necessary for the separation task can be worked out from mass balances in connection with the solution equilibria (see Chapters 2–8).
3. Flow and mass transfer: the transport of the two phases through an extractor and the production of intensive phase contact are complex hydrodynamic problems. Mass transfer provides the main dimensions of the extractor. This chapter is chiefly interested in a suitable extractor design, but also in the restrictions of the calculation.
4. Design of apparatus: extractors are more or less complicated constructions fitted to special tasks. The choice of material, calculation of stability, and technical production quality of the design are special tasks of the mechanical engineer. This book does not go into further detail about these aspects.

9.3 DISPERSION OF A PHASE INTO DROPLETS

9.3.1 Gravitational Extractors Without Input of Mechanical Energy

To obtain a large transfer area between raffinate and extract phases, one of the two liquids must be dispersed into drops. Figure 9.2 demonstrates this process schematically at a single nozzle. Similar to a dripping water tap, individual drops periodically leave the nozzle when the volumetric flow rate of the dispersed phase is low. When the flow rate is higher, however, the liquid forms a continuous jet from the nozzle that breaks into droplets. Because of stochastic mechanisms, uniform droplets are not formed. If the polydispersed droplet swarm is characterized by a suitable mean drop

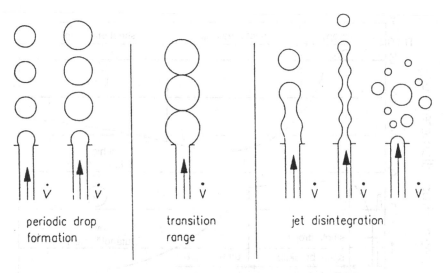

periodic drop
formation

transition
range

jet disintegration

Fig. 9.2 Mechanism of drop formation on a single hole is dependent on the throughput. Individual drops periodically leave the nozzle when the volumetric flow rate V is low. When the flow rate is higher, the liquid forms a continuous jet that breaks into droplets. (From Ref. 2.)

diameter, however, it is possible to describe and calculate the behavior of such a polydispersed droplet swarm by approximation. The Sauter diameter d_{32} can be used for flow and mass transfer calculations. It is defined by the measured drop size distribution as follows:

$$D_{32} = \frac{\sum_i n_{Pi} d_{Pi}^3}{\sum_i n_{Pi} d_{Pi}^2} \tag{9.1}$$

where n_{Pi} represents the number of drops of the class i of the diameter d_P. Figure 9.3a lists Sauter diameters calculated from drop size measurements at single nozzles with $d_N = 1.5\,mm$ for the system toluene (dispersed phase d)– water (continuous phase c) as a function of the toluene velocity v_N in the nozzle [1]. The results are valid for sieve trays (and perforated plates), if the distance between the holes is more than twice the Sauter diameter. Sieve trays are simple and frequently used dispersion devices. Experiments have shown that the flow rates through all the holes of a sieve tray are identical only if the flow rate is exactly high enough to form a jet. Since the sieve tray serves to distribute the toluene equally over the cross section of the column, only the curve for $v_N > 22\,cm\,s^{-1}$ is of interest (Fig. 9.3a). According to this curve, the liquid flow rate, and thus the velocity through the hole v_N should

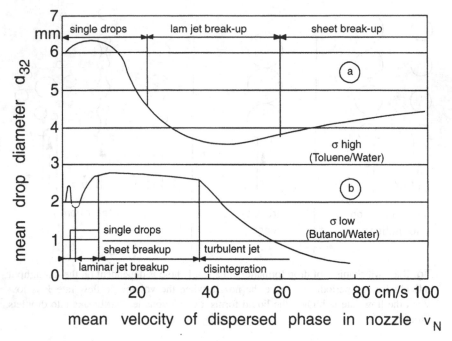

Fig. 9.3 Sauter mean diameter d_{32} calculated from drop size measurements at single nozzles of $d_N = 1.5\,\mathrm{mm}$ for the liquid systems (a) toluene (dispersed phase d) water (continuous phase c); and (b) butanol (d) water (c), is dependent on the mean velocity v_N of the dispersed phase in the nozzle. (From Ref. 5.)

be chosen such that the Sauter diameter reaches its lowest value. This gives the greatest possible drop surface d_s per unit volume and, thereby, the largest mass transfer area:

$$\left(\frac{A}{V}\right)_s \equiv a_s = \frac{6\varepsilon_d}{d_{32}} \tag{9.2}$$

when ε_d represents the holdup of the drops in the total volume.

From experiments, equations have been derived that enable calculation of the minimum velocity in the nozzle, the nozzle velocity, and the Sauter diameter at the drop size minimum. They provide the basis for the correct design of a sieve tray [3,4]. Figure 9.4a shows the geometric design of sieve trays and their arrangement in an extraction column. Let us again consider toluene–phenol–water as the liquid system. The water continuous phase flows across the tray and down to the lower tray through a downcomer. The toluene must coalesce into a continuous layer below each tray and reaches

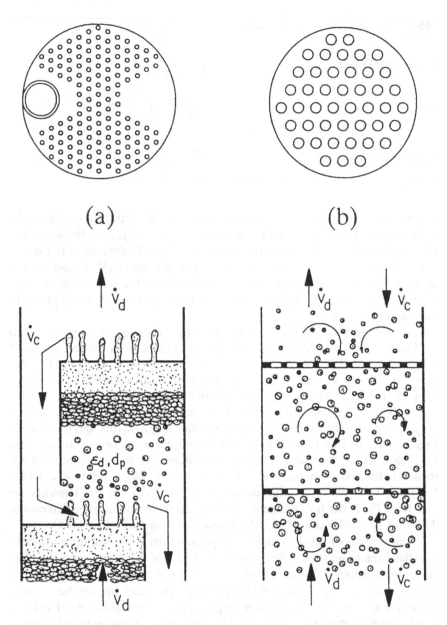

Fig. 9.4 Sieve trays with downcomers for liquid systems with (a) high interfacial tension and dual-flow trays for liquid systems; and (b) with low interfacial tension. In the case of downcomers, only the phase to be dispersed flows through the holes. The droplets are formed by jet disintegration. In dual-flow trays, both liquids flow through the same holes alternately. The larger drops split because of collision with the tray.

Table 9.1   Calculation of Drop Diameter for Dual-Flow Plates

Sauter mean diameter	$d_{32} = 1.2\sqrt{\dfrac{\sigma}{\Delta\rho g}\left(\dfrac{\nu_c \varepsilon_d}{\nu_d}\right)^{0.3}}$
Single-drop velocity	$\nu_e = 1.41\left(\dfrac{\Delta\rho g\sigma}{\rho_c}\right)^{0.25}$
Limiting criteria	$\dfrac{\Delta\rho g d_N^2}{\sigma} \leq \dfrac{1}{3}$

a height such that the static pressure caused by the buoyancy presses the liquid through the holes at the required velocity. Figure 9.3b shows the results of another liquid system (i.e., butanol–water). Compared with the system toluene–water, its interface tension σ is much lower [5]. Therefore, the drops are much smaller, and very low velocities at the hole cause the formation of a jet. Clearly, surface tension is a very important material property for extraction systems.

The extremely low liquid velocities at the hole require only very small stationary layers, which cannot be implemented in technical columns without difficulties. Therefore, sieve trays without downcomers are used (Fig. 9.4b). Here, both liquids must flow countercurrently through the same holes. Neither stationary layers nor liquid jets of the dispersed phase are formed. According to experimental observations, the bigger drops split up because they collide with the sieve trays. The drops flow stepwise through the holes in bigger collectives and alternate with the continuous phase flowing through the holes. Table 9.1 shows reliable, experimentally proved equations for medium drop sizes in such dual-flow trays and also a limiting criterion for sieve trays with or without downcomers according to Hirschmann and Blass [5]. If the hole diameter d_N obtained from the limiting criterion in Table 9.1 is smaller than 2 mm, dual-flow plates should be selected.

Sieve trays of this type define the possible size of drops, once the sieve tray has been designed for the intended flow rate of the dispersed phase. The same can be applied if tower packings are placed in a column to break up the dispersed phase by colliding with the packing. The drop size can be changed in an extractor during operation by supplying mechanical energy. The three most important operations for an additional energy input are mixing, pulsing, and centrifuging. Extractors with these systems are more flexible than spray and simple plate columns in that they can be adapted to changing liquid systems and operating conditions, although they are more expensive in initial costs and in maintenance.

9.3.2 Extractors with Liquid Pulsing

Pulsing means that either the whole liquid content of a sieve tray column is continually pushed up and down by a piston that moves to and fro, or the whole plate package is moved up and down [3]. Figure 9.5 illustrates the two extractor constructions schematically. They show about the same efficiency

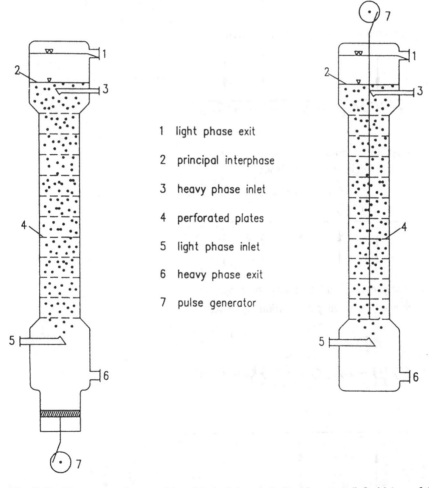

1   light phase exit

2   principal interphase

3   heavy phase inlet

4   perforated plates

5   light phase inlet

6   heavy phase exit

7   pulse generator

Fig. 9.5 Sieve tray columns with pulsing of the whole liquid content (left side) or of the whole plate package (right side, Karr column). The trays are meant to equalize the phase flows across the column cross section and to disperse the droplets as uniformly as possible.

if the plate geometry is identical. It also shows the principal interface in the upper part of the column so that the lighter phase is dispersed.

Figure 9.6 demonstrates what happens at the sieve trays when the whole liquid is pushed up and down. Three different situations can be observed with the aid of high-speed photographs. During the downward motion of the liquid pulse, larger drops do not follow the motion of the pulsing, but accumulate below the plate owing to their relatively high buoyancy. Thus, once the upward motion starts, the relatively large drops of this layer, together with the continuous liquid, are pushed through the holes.

1   assembly of drops during downstroke

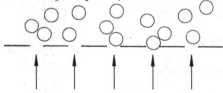

2   pushing of drops through the holes
      at the beginning of upstroke

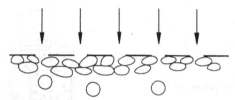

3   compressing and standstill of drops
      above the sieve plate during upstroke

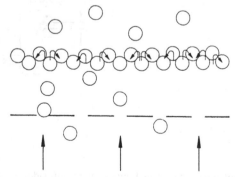

Fig. 9.6 Behavior of droplets near the sieve tray in a liquid-pulsed column during downstroke and upstroke of the liquid content.

The smaller drops follow. Both liquids accelerate in the holes, because the sum of the cross section of all the holes is less than half the column cross section. However, this motion is retarded within a short distance, whereby a zone of drop compaction results above the trays. These phenomena are modeled based on a balance of maximum and minimum kinetic energy and the cohesive energy of the droplets [1]. After that, the resulting equation for the maximum stable drop diameter in the field of pulsing is:

$$d_{P,\max} = \frac{4\sigma}{\rho_d(v_1^2 - v_2^2)} \tag{9.3}$$

where v_1 is the highest nozzle velocity, v_2 the velocity at a short distance above the sieve tray with an approximate value of zero.

New modeling was performed [6] taking into consideration first the influence of drop viscosity that produced a better agreement with the experimental results than Eq. (9.3). The authors presume that bigger drops have to be deformed in the sieve tray holes and therefore the circulation inside the drops is accelerated. Taking into account that the kinetic energy of drops is changing into surface energy and circulation energy, the authors derive Eq. (9.4) by balancing the energies for the stable drop size d_s at a sieve tray hole.

$$d_s = \frac{12\frac{d_N^2}{d_s^2}\sigma + 20.5\eta_d|v_{cN} + v_A|}{\rho_d(v_{cN} + v_A)^2} \tag{9.4}$$

where

d_s = stable drop diameter
d_N = hole diameter
v_{cN} = velocity of continuous phase in the hole
v_A = rising velocity of a drop, Eq. (9.15)
v_d, ρ_d = dynamic viscosity and mass density of dispersed phase

Figure 9.7a shows values of stable drop diameters that are calculated with Eqs. (9.3) and (9.4) for special conditions for the liquid system toluene–water as a function of the product of pulse amplitude, a, and frequency, f, the so-called pulse intensity. Figure 9.7b is a parity plot and shows that Eq. 9.4 describes the measured drop sizes for a wide range of velocities and parameters of the sieve tray geometry quite well.

The mean Sauter diameter, d_{32}, is clearly smaller than the maximum stable diameter. As experimentally proved by many authors, the ratio of these is between 0.4–0.6.

There are various further attempts to model the complex flow and drop-disintegration phenomena for drop size calculation. However, it has

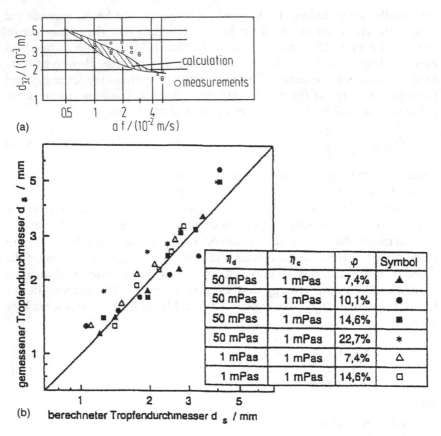

Fig. 9.7 (a) Diameters of stable drops calculated from Eqs. (9.3) and (9.4) for the liquid system toluene-water as a function of the pulse amplitude *a* and frequency *f*. (b) Parity plot of measured and calculated crop sizes. (From Ref. 6.)

not been possible so far to include mass transfer influences and the inter-action of dispersion and coalescence. Only empirical equations given in Ref. [1, Chapter 17] give the best fit to experimental data and have the broadest range of validity.

$$\frac{d_{32}}{\sqrt{\dfrac{\sigma}{\Delta\rho g}}} = C_1 e^{0.74} \left(\frac{h}{0.05\,meter}\right)^{0.10} \left[\exp\left(-3.00\frac{af\Delta\rho^{1/4}}{g^{1/4}\sigma^{1/4}}\right)\right.$$

$$\left. + \exp\left(-28.65\frac{af\Delta\rho^{1/4}}{g^{1/4}\sigma^{1/4}}\right)\right] \qquad (9.5)$$

The best values of the parameter C_I are 1.51, 1.36, and 2.01 for no mass transfer, $c \rightarrow d$ and $d \rightarrow c$ direction of transfer respectively. The product af is considered as the agitation variable in the equation, since the fit could not be improved if a and f were treated separately. The average absolute value of the relative deviation in the predicted values of d_{32} from the experimental points is 16.3%. Even in packed columns, the separation can be substantially improved by pulsing of the continuous phase resulting from greater shear forces that reduce the drop size and increase the interfacial area [1, Chapter 8].

9.3.3 Agitated Extractors

Extraction technique uses mixing tools mainly for adding mechanical energy. Combinations of stirred tanks and settlers (so-called mixer-settlers) have been used for many years, and with them the extraction efficiency of one theoretical stage of separation could be approximated. If higher separation efficiency is required, several of these stages are connected in series to form mixer-settler batteries. If the simple stage can be easily calculated, the evaluation of a battery provides no additional problems as described in Chapter 8, where the use of such batteries is extensively discussed. However, the initial costs, the cost for solvents, the need for energy and space, and the problem of interface control in each stage are considerably higher than for countercurrent columns [4]. Figure 9.8 shows sections of some frequently used stirred countercurrent columns [4]. All contain horizontal stator baffles that divide the column into successive sections, and each section has a mixer, fixed to a central shaft. Stirred extractors differ according to the types of mixer and stator baffles. The mixers are meant to disperse one phase into droplets and to mix the two phases in the section as efficiently as possible. The stator baffles must support the separation of the two liquids in each stage and provide the countercurrent flow of the two liquids from stage to stage and, at the same time, provide a free area as large as possible for the liquid flows passing through. Section 9.3 discusses this subject.

The problem of liquid dispersion by mixers is now discussed. In mixing, dispersion takes place in the turbulent shear field of the mixer. The intensity of the shear field is largely influenced by the geometry of the mixer. The power consumption P can be calculated by the equation:

$$Ne = \frac{P}{\rho_c n_R^3 d_R^5} \tag{9.6}$$

If the Newton number, Ne, in relation to the Reynolds number, Re_R, of the mixer is known from experiments and the rotor speed, n_R, the mixer diameter, d_R and the mass density ρ_c of the continuous liquid are given.

rotating disc contactor RDC

Oldshue—Rushton contactor

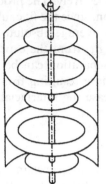

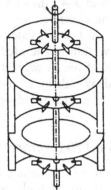

stirred cell extractor RZE

Kuehni—extractor

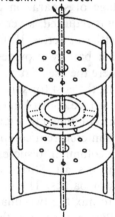

asymmetric disc contactor ARD

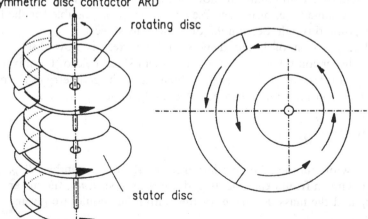

rotating disc

stator disc

$$Re_R = \frac{n_R d_R^2}{\nu_c} \tag{9.7}$$

Dispersion stirring devices always work in the turbulent flow regime; therefore, the statistic theory of turbulence [7] is important for understanding the processes during dispersion. To calculate the drop sizes in the turbulent shear field, an energy balance for a single drop is again used, as in the preceding section. According to Kolmogoroff [7], assuming local isotropic turbulence, the maximum drop diameter in the local isotropic shear field follows, which results in:

$$d_{P,max} \approx \left(\frac{\sigma}{\rho_c}\right)^{0.6} \Psi^{-0.4} \tag{9.8}$$

when Ψ is the energy per unit mass. The same proportionality applies for the Sauter diameter in stirred tanks. By using the energy per unit mass for mixers,

$$\Psi \approx n_R^3 d_R^3 \tag{9.9}$$

and the Weber number

$$We = \frac{\rho_c n_R^2 d_R^3}{\sigma} \tag{9.10}$$

it is possible to use:

$$\frac{d_{32}}{d_R} \approx We^{-0.6} \tag{9.11}$$

instead of Eq. 9.8.

The most important options for the validity of Kolmogoroff's theory are:

Local isotropic turbulence
Equal viscosities and equal densities of the two phases
Negligible influence of the drops on the surrounding phase
Negligible holdup of the drops

The first two options are fulfilled quite well in stirred extractors for a normal rotor speed and for many liquid systems. This is why empirical equations for the mean drop size in stirred extractors usually originate from

Fig. 9.8 Sections of several stirred countercurrent columns. Horizontal stator baffles divide the column into successive sections, each fitted with a mixer, all fixed to a central shaft. The mixers disperse one phase into droplets and mix the two phases in the section as efficiently as possible. The stator baffles must support the phase separation and the countercurrent flow of the liquids from state to stage.

Eq. 9.11 and differ in the proportionality constant. For details, the reader is referred to references 3 and 4. For drop size calculations, the empirical equations given in Ref. [1, chapter 17] that take into account all the influencing parameters are preferable.

Rotating Disc Contactor (RDC):

$$\frac{d_{32}}{D_R} = \left(\frac{C_1}{0.07 + \sqrt{Fr_R}}\right)\left(\frac{\eta_c}{\sqrt{\sigma\rho_c D_R}}\right)^{-0.12}\left(\frac{\rho_d}{\rho_c}\right)^{0.16}\left(\frac{D_R^2\rho_c g}{\sigma}\right)^{-0.59}$$

$$\times \left(\frac{h}{D_c}\right)^{0.25}\left(\frac{D_C}{D_R}\right)^{0.46} \tag{9.12}$$

The optimized values of C_1 are 0.63, 0.53, and 0.74 for no mass transfer, $c \rightarrow d$ and $d \rightarrow c$, respectively. The value of the holdup is ignored due to lack of data. Equation (9.12) predicts the drop size with an average absolute value of the relative deviation of 23%.

Kuehni columns:

$$\frac{d_{32}}{D_R} = C_1 e^{0.37} n_s^{-0.11}[0.14 + \exp(-18.73 Fr_R)]$$

$$\times \left(\frac{\eta_c}{\sqrt{\sigma\rho_c D_R}}\right)^{-0.20}\left(\frac{D_R^2\rho_c g}{\sigma}\right)^{-0.24} \tag{9.13}$$

The values of the constant C_1 are 9.81×10^{-2} for no mass transfer and $c \rightarrow d$ transfer, and 0.31 for $d \rightarrow c$ transfer. The stage number n_s, which varies from 2–17 in the present set of data, shows a rather weak effect on drop size. Equation (9.13) predicts the drop diameter with an average absolute value and relative deviation of 17.6%.

9.3.4 *Centrifugal Extractors*

A centrifugal field is a further method of providing mechanical energy for the generation and coalescence of droplets. Centrifugal extractors, which have cylindrical sieve trays arranged concentrically round a shaft, match best with the gravity extractors described earlier. They are sometimes called rotating columns. The most popular industrial version is the Podbielniak extractor, which was used first for penicillin extraction [8], that uses a horizontal shaft. Figure 9.9 shows a simple experimental centrifuge with a vertical shaft used to demonstrate the mode of operation. The rotor is covered with a thick glass disk [9], which makes it possible to record the processes in the rotor with a high-speed camera. In this design, the two phases must be fed into and taken out from the rotor with the aid of rotating sealing devices at the shaft. The heavy phase is fed at the rotor centre and conveyed to the outside by the centrifugal field force. The light phase is fed at the periphery of the rotor and

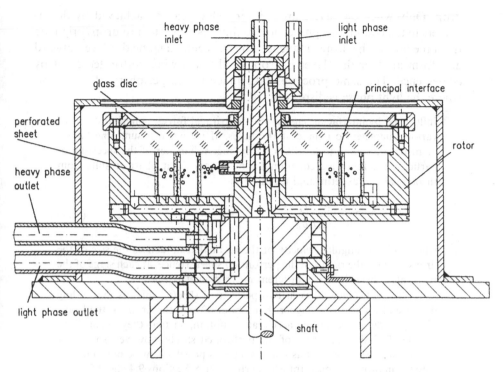

Fig. 9.9 Experimental centrifuge with cylindrical dual-flow perforated sheets arranged concentrically round a vertical shaft. The heavy phase is conveyed to the outside by centrifugal field force; the light phase is displaced to the inside by the heavy phase. With a control valve in the discharge pipe of the light phase, the degree of filling of the rotor with the two phases can be defined (i.e., the location of the principal interface). (From Ref. 9.)

displaced to the inside by the heavy phase. This causes a countercurrent flow of the two phases. With a control value in the discharge pipe of the light phase, the degree of filling the rotor with heavy and light phase can be defined (i.e., the location of the principal interface). This causes a high exit pressure on the light phase, which just compensates the higher centrifugal force of the heavy phase. The principal interface in Fig. 9.9 can be found at a medium radius of the rotor. The perforated sheets in Fig. 9.9 have no downcomers for the continuous liquid, which is common for technical centrifugal extractors. Then, both liquids must flow through the same holes, just as they do when dual-flow plates are used in the gravity field. However, other processes take place here. The heavy phase accumulates at the inside of the sheets as a very thin layer that then flows through a partial cross section of the holes countercurrently to the light phase and finally breaks from the hole as a single drop. Only when the layers are thicker, which can be achieved by down-comers in the continuous phase, do jets that disintegrate into drops originate from the holes. The drops, influenced by the centrifugal field, move outward and form another thin layer in front of the following perforated sheet by coalescence. The same processes take place at the periphery of the rotor, where the light phase is dispersed.

Influenced by interfacial tension and centrifugal forces, spherical drops of various diameters originate at the holes. If we again assume the Sauter diameter, according to Eq. (9.1), as the mean diameter of the spectrum of particles, the following equation for heavy and light phases results from theoretical and experimental results [10]:

$$d_{32} = 3.22\sqrt{\frac{\sigma}{r\omega^2\Delta\rho}\frac{\rho_c}{\rho_d}} \qquad (9.14)$$

where ω is the angular velocity of the rotor that influences the size of the drops, and r is the radius of the perforated sheet. Hence, at each stage of the perforated sheet there is a different mean diameter.

If the distance between two perforated sheets, which is the settling distance of the drops, is higher than about 40 mm, the drops of the heavy phase are deformed by the accelerated motion, so that they again disintegrate. Technical distances of the perforated sheets, however, should be lower than 40 mm so that as many stages as possible can be built into the rotor to improve the mass transfer performance. Sections 9.4 and 9.5.2 deal further with this topic. Drop swarms of the light phase, however, do not change their mean diameter during their retarded motion to the rotor shaft, no matter how far is the distance to the next sheet.

In addition to the rotating columns previously described, there are a number of other designs for centrifugal extractors, many originally developed for the separation of radioactive wastes in nuclear processes (see Chapter 12). They are both of the mixer-settler type, as discussed in section 9.3.3, and of the rotating column types.

In the simplest case, a liquid phase in a static mixer is dispersed into another phase and then fed into a centrifugal separator for phase separation (external mixer–centrifugal separator) [11,12]. The advantage of this design is that mixer and settler are separately optimized, and the centrifugal unit is relatively simple. At best, for each mixer-settler stage, a separation performance corresponding to a theoretical stage can be achieved; for a more efficient separation, several of this type of mixer-settler may be connected in series. Furthermore, the mixing can be integrated with phase separation in the centrifuge into an "internal centrifugal mixer-separator," as in the Robatel design. Figure 9.10a shows four such interconnected stages. These centrifuges are used, for example, by the French nuclear reprocessing

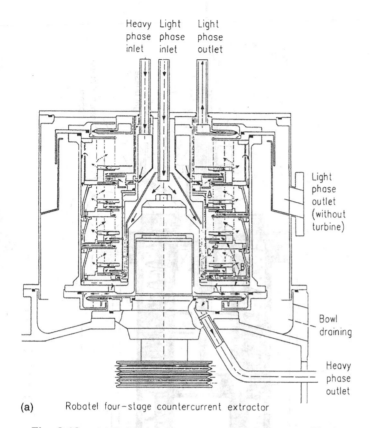

Heavy Light Light
phase phase phase
inlet inlet outlet

Light
phase
outlet
(without
turbine)

Bowl
draining

Heavy
phase
outlet

(a)   Robatel four-stage countercurrent extractor

Fig. 9.10 (a) Robatel four-stage centrifugal extractor. The heavy and the light liquids are introduced near the bottom and top, respectively. They flow countercurrently and leave the extractor at opposite ends. The extractor consists of a rotating bowl divided by baffles into horizontal compartments on top of one another. The stationary central shaft carries mixing disks (*A*) that run through each compartment and serve to mix and pump the two phases to the settling chamber of the stage. This chamber comprises a set of two overflow chutes (*B* for the heavy phase; *C* for the light phase), that stabilize the separating area independently from the outputs. Each compartment has connections to take the settled phases to the previous and following stages. The standard design has three to eight stages. (Courtesy of Robatel SLPI and Eries, Genas, France.) (b, see next page) Westfalia countercurrent centrifugal extraction decanter. The heavy liquid phase containing solids such as microorganisms and the solvent are fed into the rotating bowl through separate inlets at different parts of the bowl. Phase contact is effected in countercurrent flow in the feed area and in the spirals of the scroll. The separation zone is adjusted for variable liquid densities by means of the ring dam at the cylindrical bowl end. After mass transfer in the contact zone, the enriched solvent is discharged at the cylindrical bowl end under pressure by a centripetal pump. The raffinate phase discharges at the conical end of the bowl. All particles that sediment in the bowl are conveyed by the scroll to the conical end of the bowl and discharged together with the raffinate phase. (From Ref. 13.)

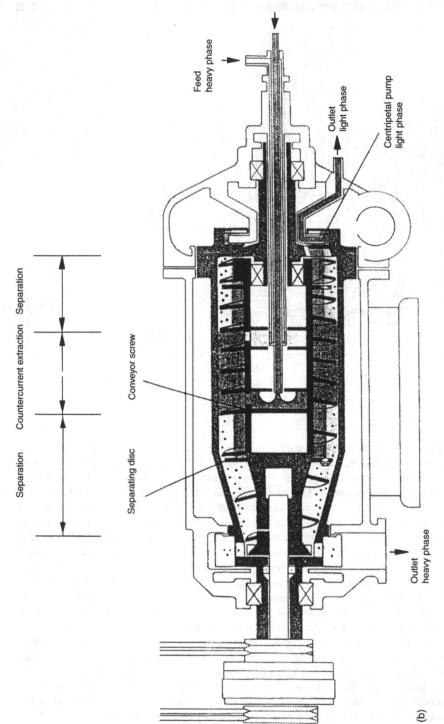

Fig. 9.10 (*continued*).

Feed heavy phase

Outlet light phase

Centripetal pump light phase

Separation

Countercurrent extraction

Separation

Conveyor screw

Separating disc

Outlet heavy phase

(b)

industry. In the United States, internal centrifugal mixer-separators are also used, although the design principle is different based on the Sharples type of centrifuges. Figure 9.10b shows the countercurrent centrifugal extraction decanter by Westfalia [13], used in the biochemical industry when the raffinate phase contains microorganisms.

9.4 FLOW OF PHASES AND LIMITS OF THROUGHPUT

In a nonpulsed, nonstirred, noncentrifuged flow, drops move through the continuous liquid driven by the density difference. Their stationary velocity of motion results from the interaction of buoyancy and drag forces. Figure 9.11 shows the behavior of individual drops in a static continuous liquid. The stationary relative velocity of the drops v_p is plotted as a function of their diameters d_p. Unless their interfaces are scrupulously clean, only very small drops behave like rigid spheres. When the drops are bigger, the shear forces of the continuous liquid acting on the drop surface cause a circulation inside the drop. Thus, the drops rotate in their surroundings and move faster than rigid drops. If the drops are even bigger, they are deformed elliptically, which makes their motion slower. Finally, when the drops reach a certain diameter, their shape is no longer stable, and the drop oscillates along and across its major axis, because periodic whirls form and separate from the drop. These oscillating drops move along coiled paths. Their axial velocity component, plotted in Fig. 9.11, remains nearly constant. The drop size can be increased only up to a maximum value $d_{p,\max}$. If the value is higher, the drop will break owing to drag forces.

The measured relation between individual drop velocity v_P, and drop diameter d_P is well represented by the following equation [2]:

1. For circulating drops:

$$Re_P = K_L^{0.15}(Ar^{0.523}K_L^{-0.1435} - 0.75)$$
$$\text{if } Ar > 1.83K_L^{0.275} \tag{9.15a}$$

2. For oscillating drops:

$$Re_P = K_L^{0.15}(4.18Ar^{0.281}K_L^{-0.00773} - 0.75) \tag{9.15b}$$

The transition point of the two equations is determined by:

$$Ar_{c-0} = 372.9K_L^{0.275} \tag{9.15c}$$

The basic terms v_P and d_P and the liquid properties are made non-dimensional in the equations according to the similarity theory:

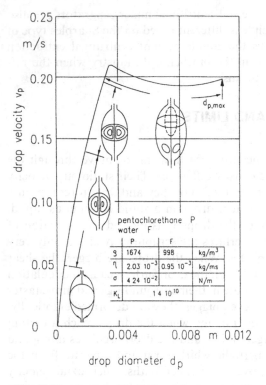

Fig. 9.11 Stationary relative velocity, v_p, of pentachloroethane drops in motionless water is dependent on drop diameter, d_p. Very small drops behave like rigid spheres, as shown. Larger drops have an internal circulation and are finally deformed elliptically. When they have reached a certain diameter, the drops in the end oscillate along and across their major axis. Their axial velocity is nearly independent of the diameter. Once the drop size is higher than a maximum value $d_{p,\text{max}}$, the drop will break, owing to the drag forces. (From Ref. 2.)

Particle Reynolds number: $Re_P = \dfrac{v_p d_P}{\nu_d}$ (9.16)

Archimedes number: $Ar = \dfrac{d_P^3 g \rho_c \Delta\rho}{\eta_c^2}$ (9.17)

Characteristic number of the continuous liquid: $K_L = \dfrac{\rho_c^2 \sigma^3}{\Delta\rho g \eta_c^4}$ (9.18)

Now the *two*-phase flow in a nonpulsed countercurrent column will be discussed. If the continuous liquid flows toward the drops, the drop velocity decreases, whereas the holdup of the drops in the flow system increases if the flow rate of the dispersed phase is maintained. Thus, the remaining flow area for the flow of the continuous liquid narrows. This interaction between flows of the two phases leads to a limit of operation when the buoyancy is not sufficient to overcome the flow resistance. This is a complex phenomenon on which the deformability of the drops is superimposed, so that modeling the real processes mathematically would not be particularly promising.

However, a simple model does work: the liquids should occupy a share of the cross section according to their holdup and should stream toward each other as layers. If their superficial velocity, that is, their volumetric flow rate, related to the complete column cross section, S, is:

$$\dot{\nu}_c = \frac{\dot{V}_c}{S}, \dot{\nu}_d = \frac{\dot{V}_d}{S} \tag{9.19}$$

they have an effective velocity of $\dot{\nu}_c(1 - \varepsilon_d)$ or $\dot{\nu}_d/\varepsilon_d$. Now the relative velocity v_r, between the two liquids is:

$$v_r = \frac{\dot{\nu}_d}{\varepsilon_d} + \frac{\dot{\nu}_c}{1 - \varepsilon_d} \tag{9.20}$$

Either v_r or ε_d can be calculated from Eq. (9.20), if there is one more relationship between the two terms. In Fig. 9.12, v_r related to the single-drop velocity v_P according to Eq. (9.15) is plotted as a function of the drop holdup for droplet swarms with the Archimedes number as a dimensionless term for the drop diameter for the measured values. It can be seen that the relative velocity constantly decreases, as the holdup of the drops, ε_d, increases and the size of the drops in the swarm decreases.

Many experiments have been published that attempt to predetermine the dependency of Fig. 9.12. To illustrate the main point, we use the simple equation:

$$\frac{v_r}{v_P} = (1 - \varepsilon_d)^k \tag{9.21}$$

The velocity of single drops v_P can be calculated by Eq. (9.15). The exponent k is defined by the empirical correlation:

$$k = 4.6 - 0.13 \ln Ar \tag{9.22}$$

which describes well-published experiments in the range $2.5 < k < 4.6$ [14].

The premises have now been provided for calculating the operation limits (i.e., the highest possible flow rates of a countercurrent column). This is reached as soon as no further increase of the flow rate can be performed by the density difference. The consequence is that the light phase leaves the column at the bottom instead of at the top and the opposite situation for the heavy phase (see Fig. 9.1). This load limit is also called the *flooding point*.

The mathematical formulation of this condition makes use of the fact that an increase of the flow rate cannot increase the drop holdup:

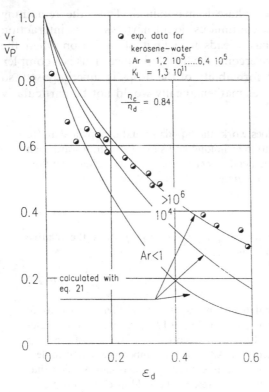

Fig. 9.12 Stationary relative velocity v_r, of the drop swarms related to the single drop velocity v_p is dependent on the drop hold-up, ε_d, and Archimedes number Ar. (From Ref. 2.)

$$\frac{d\varepsilon_d}{d\dot{v}_d} = \frac{d\varepsilon_d}{d\dot{v}_c} = \infty \qquad (9.23)$$

or

$$\frac{d\dot{v}_d}{d\varepsilon_d} = 0 \quad \text{when} \quad \dot{v}_c = \text{constant} \qquad (9.24a)$$

$$\frac{d\dot{v}_c}{d\varepsilon_d} = 0 \quad \text{when} \quad \dot{v}_d = \text{constant} \qquad (9.24b)$$

If Eq. (9.20) for the relative velocity v_r is used to calculate the superficial velocity, one obtains the equations for calculating the phase flow velocities at the flooding point:

$$\dot{v}_c = (1 - \varepsilon_d)^2 \left(v_r + \frac{dv_r}{d\varepsilon_d} \varepsilon_d \right) \qquad (9.25a)$$

$$\dot{\nu}_d = \varepsilon_d^2 \left(\nu_r - \frac{d\nu_r}{d\varepsilon_d}(1 - \varepsilon_d) \right) \qquad (9.25b)$$

or, using the dimensionless form [2]:

$$\dot{\nu}_c \left(\frac{\rho_c}{\Delta\rho\nu_c g} \right)^{1/3} = Ar^{-1/3}(1 - \varepsilon_d)^2 \left(Re_r + \frac{\varepsilon_d dRe_r}{d\varepsilon_d} \right) \qquad (9.26a)$$

$$\dot{\nu}_d \left(\frac{\rho_c}{\Delta\rho\nu_c g} \right)^{1/3} = Ar^{-1/3}\varepsilon_d^2 \left(Re_r - (1 - \varepsilon_d)\frac{dRe_r}{d\varepsilon_d} \right) \qquad (9.26b)$$

where the Reynolds number Re_r is calculated with the velocity ν_r.

Equations (9.26a) and (9.26b) together with Eq. (9.21) for ν_r and Eq. (9.15) for ν_I are the basis for flooding point diagrams shown in Fig. 9.13, where the ordinate and abscissa contain the phase flow rates in dimensionless form. The diagram shows the calculated flooding curve of circulating drops for six different Archimedes values for a liquid system with $K_L = 10^{11}$ and $\eta_c/\eta_d = 0.5$. In addition, it provides straight lines of constant holdup ε_d and of constant flow rate ratios $\dot{\nu}_c/\dot{\nu}_d$. If two of these five parameters are given, the other three can be taken from the diagram. The diagram can also be used for other values of K_L, since their influence on the flooding point is weak for circulating drops. The effect of other viscosity ratios is demonstrated on the flooding point curve for $Ar = 5 \times 10^5$ and can be transferred to other Ar values to estimate the results.

The diagram can also be used for oscillating drops, since according to Fig. 9.11, the rising velocities of oscillating drops hardly depend anymore on the diameter. The drop diameter attainable depends only on the liquid system, as stated in Eq. (9.15c). Figure 9.13 lists the Ar numbers according to Eq. (8.15c) for K_L, values of 10^8, 10^9, 10^{10}, and 10^{11}, and for these values the Ar lines are also plotted. These are the flooding curves for oscillating drops.

To use the flooding point diagram, first it is necessary to decide whether the drops produced in the extractor are circulating or oscillating. The mean diameter d_{32} (see Eq. 9.1) is used for the characteristic drop size. If the flow rate ratio is known from the thermodynamic design, the superficial velocities of both phases can be determined at the flooding point. The minimum column cross-sectional area and diameter necessarily follows directly from the superficial velocity at the flooding point with Eq. 9.19.

The method of calculation introduced in this chapter not only allows an exact determination of the column diameter for nonpulsed sieve tray columns, but also allows a good estimation of the diameters of pulsed and stirred extractors. For the latter, however, more exact specific equations exist for the flooding point, see for example [1,4].

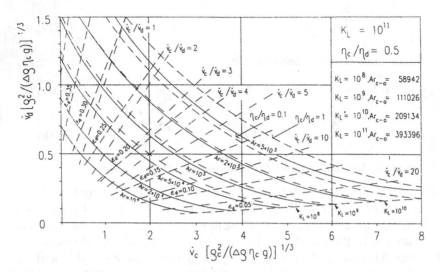

Fig. 9.13 Flooding point diagram of countercurrent extractors. The drawn lines for six different Ar values are valid for circulating drops. The dotted lines for three different K_L values are valid for oscillating drops. (From Ref. 2.)

At a maximum flow rate of the two phases, centrifugal extractors with cylindrical perforated plates also show a flooding point. In contrast to the processes in a gravity field, however, this is not caused by the countercurrent flow of the two phases between two perforated plates, since the holdup of the dispersed phase in the centrifugal field is very low owing to the high drop velocity. In this case, the two phases in the holes finally block each other so that the layer of the dispersed phase suddenly grows strongly and the dispersed phase also leaves at the continuous phase outlet.

The flooding diagram of Fig. 9.14 shows the flooding point caused by a flow rate of the two phases that is too high. There are two more flooding points that depend on the difference of the exit pressures of the two phases. How can they be explained? Section 9.3.4 mentions that the principal interface in the rotor can optionally be shifted by varying the exit pressure of the dispersed phase (i.e., the pressure difference of the two phases at the exit). If the two exit pressures are the same, the rotor must be filled with heavy phase. The heavy continuous phase then leaves at the exit of the dispersed light phase. On the other hand, at a specific excess pressure at the exit of the light phase, the rotor will be filled with light phase. Then, the light phase will leave at the exit of the heavy one. In both cases, the throughputs of both phases flooding points are reached. It is possible to predetermine them [15].

9.5 MIXING OF PHASES

The two liquid phases go different ways during their flow through the extractor and, therefore, have different residence times. Various reasons for this can be involved; for example, in a real droplet swarm, the various drop diameters lead to different rising velocities and motions, or the wake behind each drop drags the continuous phase in a direction opposite its original motion. Because of such phenomena, regions of different compositions are mixed, which reduces the possible mass transfer efficiency. These mixing processes can be observed only by a close analysis of the residence times of the phases in the extractor [16]. According to Fig. 9.15, this can be done by feeding a suitable sensor, for example, a colored tracer into the entering heavy feed. Then, at two cross sections, α and ω, the concentration of the color intensity, c_α or c_ω, respectively, is measured downstream as a function of time t. It can be noticed that the curve for the sensor widens downstream, as a result of the various mixing phenomena. The light phase can be traced in a similar way by measuring its residence time. It is essential that the tracer has the same flow behavior as the respective phase and that its amount is not altered during its residence time (e.g., no adsorption at solid surfaces, no extraction by the other phase and no reaction). Therefore, the functions $c_\alpha(t)$ or $c_\omega(t)$ are valid for both the tracer and the main phase.

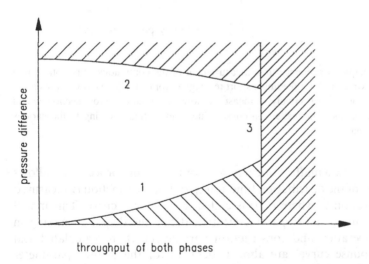

Fig. 9.14 Flooding point diagram of centrifugal extractors (see Fig. 9.9). Curve 1 marks the lower limit, when the rotor is filled with heavy phase; curve 2 marks an upper limit, when the rotor is filled with light phase; curve 3 marks the limit of the total throughput of both phases.

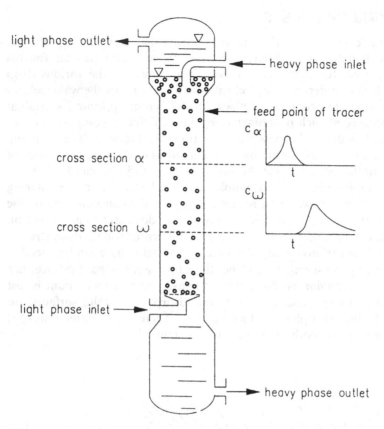

light phase outlet ←

← heavy phase inlet

← feed point of tracer

c_α

t

cross section α

c_ω

t

cross section ω

light phase inlet →

→ heavy phase outlet

Fig. 9.15 Principle of residence time measurement of the continuous heavy phase in a countercurrent extractor. After feeding a short test signal, for example, a colored tracer, the color intensity c_α or c_ω, respectively, is measured downstream at two cross sections α and ω as a function of time. The test signal curve widens downstream, owing to the various mixing phenomena.

Consider now a mathematical model for the mixing process that allows the calculation of the distribution function c_ω out of the function c_α measured at the cross section α, where $c_\omega(t)$ is called the response curve. This model must provide one or more parameters that define the degree of mixing in addition to operation and construction parameters. If the calculated and measured response curves are almost identical for the defined parameter values, not only is the model proved, but it will also have yielded a set of connected operation, construction, and mixing parameters. From a sufficient number of these measurements, one can work out a functional dependence for all of these parameters to predetermine the mixing processes in the various extractors.

9.5.1 Gravitational Extractors

The dispersion model is one of the frequently used models. It describes the "dispersion of the residence time" of the phases according to Fig. 9.16, for example, in one-dimensional flows by superimposing the plug profile of the basic flow with a stochastic dispersion process in axial direction, which is constructed by analogy to Fick's first law of molecular diffusion:

$$\dot{N}_D = -D_{ax} S \frac{dc}{dh} \tag{9.27}$$

Equation (9.27) defines the so-called axial dispersion coefficient D_{ax} as a model parameter of mixing. $\dot{N}_D$ is the dispersion flow rate, c the concentration of the tracer mentioned earlier, and S the cross-sectional area of the column. The complete mole flow rate of the tracer consists of an axial convection flow and the axial dispersion flow. The balance of the tracer amount at a cross section of the extractor leads to second-order partial differential equations for both phase flows at steady state. For example, for continuous liquids:

$$\frac{\partial c}{\partial t} = D_{ac,c} \frac{\partial^2 c}{\partial h^2} - \frac{\dot{v}_c}{1 - \varepsilon_d} \frac{\partial c}{\partial h} \tag{9.28}$$

The solution of this partial differential equation is known for special boundary conditions and measuring conditions at the cross sections α and ω

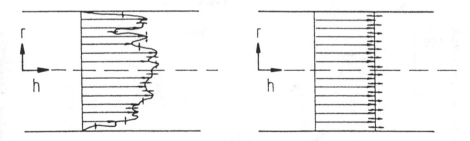

real flow conditions:
velocity profile with
axial and radial turbulence

model conception of material dispersion:
plug flow with superposition
of stochastic spreading in
axial direction

Fig. 9.16 Basic assumptions of the one-dimensional dispersion model. The dispersion of the residence time of the phases is modeled by superimposing the plug profile of the basic flow with a stochastic dispersion process in axial direction.

of the measuring zone [17]. The experimenter can then calculate the response curve in the cross section ω, $c_\omega(t)$, and can change the dispersion coefficient $D_{ax,c}$ as the fitting parameter so that the calculated response curve $c_\omega(t)$ agrees best with the measured one. Figure 9.17 for liquid pulsed sieve tray columns shows that the dispersion model gives a fine reflection of the process of mixing of the continuous phase. The measured and calculated response curves for a special value of $D_{ax,c}$ agree very well. However, there are large deviations for the dispersed phase, in spite of the best fit. This leads to the conclusion that the dispersion model provides an inadequate description of this process. Relevant literature offers correlations for dispersion coefficients

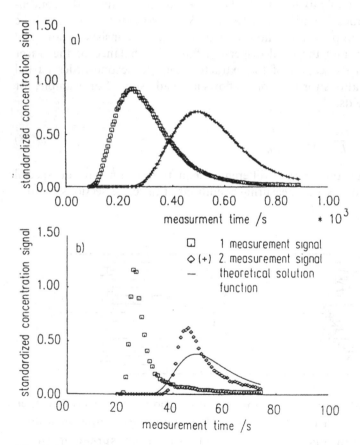

Fig. 9.17 Residence times of (a) continuous; and (b) drop phase: in a liquid pulsed sieve tray extractor according to measurements (squares and crosses) and according to calculations by the dispersion model (drawn lines). Measured and calculated response curves agree well for the continuous phase, but not for the dispersed phase. (From Ref. 14.)

in various extractors [1, Chapter 17]. For the continuous phase, they are usually between 1 and $5\,\mathrm{cm^2\,s^{-1}}$ (in some cases values as high as $20\,\mathrm{cm^2\,s^{-1}}$ are observed); for the dispersed phase they increase by a factor of 2–3.

The design engineer can use the dispersion coefficients determined in this way for the calculation of the real course of concentrations, c, of any component in the dispersed (d) and continuous (c) phases along the countercurrent column. If the mass transfer between the two phases, the actual task of an extractor, is included in the balance, the balance equations for an element of height dh of the extractor for stationary conditions is:

$$\dot{\nu}_c \frac{dc_c}{dh} + (1 - \varepsilon_d)D_{ax,c}\frac{d^2c_c}{dh^2} = -k_{od}a(c_d - c_d^*) \tag{9.29}$$

$$\dot{\nu}_d \frac{dc_d}{dh} + \varepsilon_d D_{ax,d}\frac{d^2c_d}{dh^2} = -k_{od}a(c_d - c_d^*) \tag{9.30}$$

The term on the right side of the equations quantitatively defines mass transfer from the dispersed to the continuous liquid, which is explained more fully in section 9.6. The coupled differential equation system can be analytically or numerically integrated for the appropriate practical boundary conditions.

Figure 9.18 shows schematically the influence of the flow dispersion on the concentration profile and the possible exit concentrations. Notice that dispersion changes the possible exit concentrations of the two phases negatively.

9.5.2 Centrifugal Extractors

For a specific operational condition, Fig. 9.19 shows a curve for the residence times of the dispersed heavy phase in a centrifugal extractor with three perforated sheets. These are plotted schematically one after the other from bottom to top on the right-hand side of the figure [9]. These sheets are fitted with downcomers and show the formation of jets of the dispersed phase in the centrifugal field. Extremely short residence times can be seen during jet formation and motion of the drops in the continuous phase. On the other hand, there is relatively long residence of the drops within the stationary layer of this phase in front of the following sheet, i.e., both before and after coalescence in the stationary layer. On the whole, the liquid molecules do not remain longer than about 40 s in a three-stage centrifuge, and usually not longer than about 2 min in a multistage industrial centrifugal extractor. This short residence time is one main attribute of centrifugal extractors.

Measurements of residence time show another special feature in that all layers of the two phases between the perforated plates are fully mixed, which means they behave like ideally mixed agitator tanks. This can be explained by

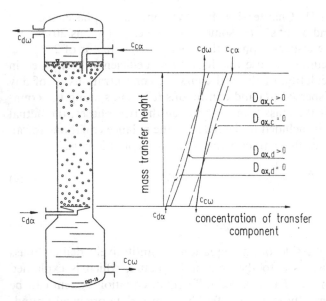

Fig. 9.18 The influence of flow dispersion on the concentration profiles of the two phases (mass transfer from the continuous phase to the dispersed phase).

the various peripheral flow velocities at different radii of the rotor and by the influence of the drops. If sufficient ideally mixed cells are connected in series, plug flow residence time behavior results. As mentioned at the beginning of section 9.5, there is no negative influence on the mass transfer by mixing and the motion of the droplet swarm approaches that of plug flow, since neither the drop sizes nor their velocity of motion are very different.

The dispersion model introduced in section 9.5.1 cannot be used for the demonstration of the mixing process in a multistage centrifugal extractor, because it can describe only the behavior of the extractor overall. There is no means of distinguishing between the various contributions to mixing from the single steps. Models, however, that consist of terms for plug flow patterns and ideal stirring tanks describe the real mixing behavior exactly [10].

9.6 MASS TRANSFER UNDER PHYSICAL CONDITIONS IN BINARY SYSTEMS

9.6.1 *Countercurrent Flow in a Gravitational Field*

We have now learned that the calculation of the flooding point provides the necessary diameter of the extractor, and that the calculation of the concentration profiles provides the possible output concentration for a given length of the extractor, or, conversely, the length of the extractor to obtain a desired output concentration of the phases. Equations (9.29) and (9.30) contain the mass transfer terms on the right-hand side of the equations.

The individual, or phase, mass transfer coefficients (i.e., k_d for the dispersed phase and k_c for the continuous phase) are defined as:

$$\dot{n}_c = k_c \Delta c_c \qquad\qquad (9.31a)$$

$$\dot{n}_d = k_d \Delta c_d \qquad\qquad (9.31b)$$

where $\dot{n}$ is the mole flux of the transitional component and Δc the concentration differences between the bulk phase and interface [18]. At steady state, the mole flux of the component from one phase to the interface must just be able to flow into the other phase. For this transitional state, an overall mass transfer coefficient k_o can be defined. To define k_o it must be assumed that the transitional resistance of the two phases is concentrated in one phase,

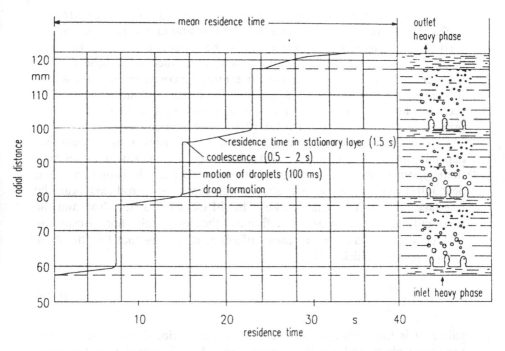

Fig. 9.19 The residence times of heavy droplets in a centrifugal extractor (see Fig. 9.9) with three concentric perforated sheets. Extremely short residence times are a main attribute of centrifugal extractors.

so that the bulk concentration of the resistanceless phase reaches as far as the interface, as in Fig. 9.20. An overall mass transfer coefficient can then be defined for the dispersed phase as well as the continuous phase:

$$\dot{n}_d = k_{od}(c_d - D_{eq}c_c) \tag{9.32a}$$

$$\dot{n}_c = k_{od}\left(\frac{c_d}{D_{eq}} - c_c\right) \tag{9.32b}$$

where $D_{eq} = (c_d/c_c)_{equilibrium}$ is the distribution ratio. Since both equations must provide the same transitional rate, it follows that, if interfacial resistances are neglected, the phase and overall mass transfer coefficients are connected by:

$$k_{od} = \frac{1}{(1/k_d) + (1/D_{eq}k_c)} \tag{9.33a}$$

$$k_{oc} = \frac{1}{(1/k_c) + (1/D_{eq}k_d)} \tag{9.33b}$$

It is generally agreed that mass transfer coefficients are only correlated for negligibly small convectional motion of the transitional component, which is vertical to the interface. However, when the mass transfer is mutual and equimolar, no such convections normal to the interface result; otherwise the transfer coefficient and the driving force must be corrected with the aid of theories of mass transfer [18]. The transitional rates and, accordingly, convectional flow rates normal to the interface are only low for the extraction process, so that the uncorrected Eq. (9.31) may be used.

The intensity of mass transfer shown by the mass transfer coefficient depends on the flow processes inside the drop or in its surroundings and, thereby, on the various life stages of the drops. During the drop formation, new interfaces and high concentration gradients are produced near the interface. The contact times between liquid elements of the drop and the surroundings that are near the surface are then extremely short. According to Fick's second law for unsteady diffusion, it follows that for the phase mass transfer coefficient [19]:

$$k = \sqrt{\frac{4D}{\pi t}} \tag{9.34}$$

where D is the diffusion coefficient of the transitional component in the respective phase and t is the contact time. The shorter the contact time, the higher the mass transfer coefficient. Thus, a repeated coalescence and disintegration of drops is an advantage. In all tray columns, pulsed or

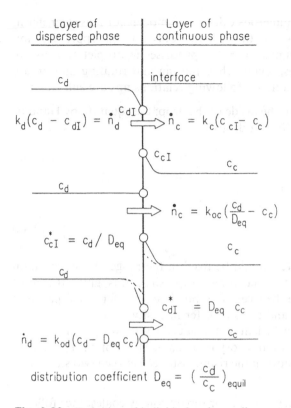

Fig. 9.20 Definition of individual and overall mass transfer coefficients.

nonpulsed sieve tray columns, and stirred columns, the disintegration zone is relatively small compared with the whole distance that the drops cover in the extractor. Thus, the influence of the mass transfer kinetics on the integral mass transfer during the drop formation can be neglected during the lifetime of the drops.

Next, we are interested in mass transfer during drop motion, which according to Fig. 9.11 is strongly dependent on drop size. When drops are very small, and therefore rigid, diffusion in the interior of the drops determines the mass transfer velocity. When the internal drops circulation occurs, however, mass transfer is considerably enhanced by the constant transport of liquid elements to the drop surface. The concentration gradient driving force is constantly kept at a high level. For oscillating drops, a further rise in the mass transfer velocity can be expected, since the interior of the drop is turbulently mixed.

The literature offers numerous calculation models for mass transfer in single liquid particles. However, they provide only a rough approximation to reality in industrial columns, since the processes in droplet swarms are much more complicated, especially when pulsing and rotating motion are superimposed. For estimation, the following relationships are sufficient:

1. For the mass transfer on the inside of the drop: the equation by Handlos and Baron [20] for oscillating drops:

$$k_d = C \frac{\nu_r}{1 + (\eta_d/\eta_c)} \tag{9.35}$$

with $c = 0.002$ [2], and

$$\nu_r = \sqrt{\frac{g d_P}{4} \frac{\Delta\rho}{\rho_c} + \frac{4\rho}{\rho_c d_P}} \tag{9.36}$$

The mean drop diameter should be fixed in the range of the transition region of circulating to oscillating drop, since this gives the most favorable compromise between the possible transfer area per unit volume of the column and mass transfer intensity.

2. For the mass transfer coefficient on the outside of the drop k_c, Eq. (9.34), according to the penetration theory by Highbie [19], obtains the contact time t as the quotient between the rising distance between two stages and the rising velocity.

After the motion of the drops, their separation by coalescence follows. During this process, the mass transfer is negligibly small compared with that where the drops are in motion, if the continuous liquid is not fed directly into the coalescence region. However, such a technique should be avoided so as not to disturb the settling process.

For simplicity, this section discusses only the mass transfer of one component in a liquid–liquid system with negligible miscibility of both liquids and with one transitional component. On the other hand, calculations must consider mass transfer rates of several components and more or less strong variation in the mass flows along the column, where both complicate the equation considerably [21–23]. Chemical reactions may cause further complications. Their kinetics can enhance the mass transfer coefficients and, therefore, the reaction equations have to be part of the mathematical model of the extractor [24,25].

9.6.2 Countercurrent Flow in a Centrifugal Field

The general definition of mass transfer coefficients according to Eqs. (9.31)–(9.33) is also valid for the mass transfer coefficient in a centrifugal field. The very short residence times in the various steps of flow through a stage indicate the use of the mass transfer coefficients according to Eq. (9.34) given by Highbie [19]. Otillinger and Blass [9] have proved that Eq. (9.34) can actually be used for the two mass transfer coefficients k_c and k_d in drop formation and motion, as well as in the coalescence phase, and for the mass transfer between liquid layers in a stage [1, Chapter 14]. Figure 9.21 shows this for a special situation, with four actual stages formed of concentric sieve trays. The number of theoretical stages is used as a measure for the performance of mass transfer plotted radially in the centrifugal extractor. The increase of the theoretical number of stages during drop formation and motion within the radial path of each stage can be clearly seen. The number of stages, depending on the coalescence and the contact of the liquid layers, can be observed at specific radii. The straight lines have been calculated using the mass transfer coefficient from to Eq. (8.34) using the residence times of the respective flow steps as contact times. The final points of the calculation

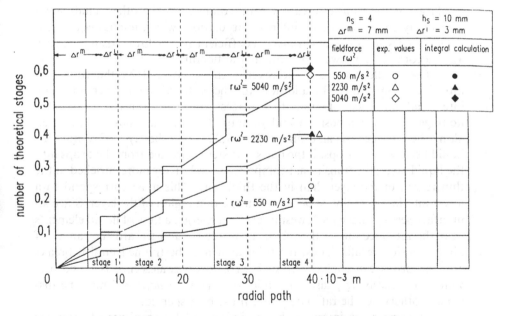

Fig. 9.21 The mass transfer performance of a centrifugal extractor (see Fig. 9.9) with four concentric sheets (stages). This is dependent on the radial path with the field force as parameter: n_s, number of stages; h_s, height of a stage; r_m, radial drop path in a stage; r_i, inactive radial distance (thickness of perforated sheet and of coalesced layer in a stage). (From Ref. 9.)

agree well with the measured results for the whole extractor (i.e., that for the whole radius of the rotor).

9.7 INFLUENCE OF INTERFACIAL ENERGIES ON FLOW AND MASS TRANSFER

Flow and mass transfer are influenced in many ways by the energetic conditions at the interface [26]. These conditions are described by the interfacial tension defined by the change of the free enthalpy with interfacial area. In general, interfacial tensions between liquids, and especially between liquid mixtures, cannot be calculated, but must be measured.

The presence of a transitional component reduces the interfacial tension for the majority of extraction liquid systems, so that $d\sigma/dx < 0$. Tensides, surface-active agents, drastically reduce the interfacial tension, even if their concentration is very small.

Interfacial tension is the parameter in equations influencing the drop size, as discussed in preceding sections. The smaller the value of σ, the smaller are the resulting drops, if all the other conditions are the same, and the larger is the transfer area per unit volume. On the other hand, small drops may show little or no internal circulation, which implies equivalent consequences for the mass transfer coefficient and a lower rising velocity and, accordingly, a lower flow rate at the flooding point.

These influences of σ, however, can be covered by the equations introduced previously. In addition to these, there are further energetic interfacial influences. During the process of mass transfer, local non-homogeneity of composition and perhaps temperature must be expected along the interface. As an example, Fig. 9.22 shows the layer of continuous liquid between two drops. If the mass transfer is directed from the drops into the liquid layer, the transitional component is relatively more enriched in the thin region of the layer than in the thick one. This must correspond to a lower interfacial tension at $d\sigma/dx < 0$ in the thin region of the layer. Since an interface seeks to have the lowest interfacial energy, a flow of liquid elements near the interface out of the region of lower tension toward the ranges of higher tension results. Thus, the thin layer may be thinned until it breaks and the drops can coalesce. If such drop–drop coalescence is to be prevented, the receiving phase for liquid systems with $d\sigma/dx < 0$ must be dispersed; otherwise, the raffinate phase must be dispersed.

Flow near the interface that is influenced by gradients of interface tension is called Marangoni convection. It may have further modes [27]. Thus, a low Marangoni convection in an interface, which results from small concentration differences, may be increased to a strong flow in the shape of rolling cells by mass transfer. These rolling cells transport liquid out of the

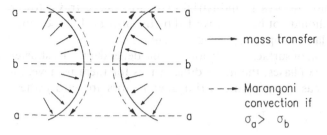

                                    ⟶ mass transfer

                                    - - -→ Marangoni
                                          convection if
                                          $\sigma_a > \sigma_b$

Fig. 9.22 Marangoni convection between two drops. If the mass transfer is directed out of the drops, the transfer component is relatively more enriched in the thin region of the layer. This corresponds to a lower interfacial tension for most of the extraction liquid systems ($d\sigma/dx < 0$). Thus, a flow of liquid near the interface out of the region of lower tension toward the ranges of higher tension results is called Marangoni convection. The coalesence of the two droplets is promoted.

deeper layers of the phase to the interface and thereby increase mass transfer. This is desired. In the past, criteria have been developed that allow the prediction of such Marangoni instabilities [26]; however, the number of correct and incorrect predictions is equal. Consequently, experiments should be carried out for a safe prediction of the interfacial instabilities of a particular liquid system under various conditions. The "kicking drop" method proves to be very useful. Drops are produced on a capillary in the continuous phase and, if Marangoni convection appears in the chosen direction of the mass transfer, a strong vibration of the drop can be observed. However, strong Marangoni convection can cause interfacial turbulence and the formation of very small drops. Emulsions like these cannot be separated in normal settlers and should be avoided a priori.

Finally, it must be noted that tensides that are adsorbed at the interface cause a stiffening of the interface. They hinder or even stop the inner circulation and oscillation of drops, and reduce the mass transfer intensity. Moreover, they form a barrier against the mass transfer, so that a further resistance term should be considered in the overall mass transfer process [28] in Eq. (9.33). Since the nature and concentration of tensides in industrial processes cannot be predicted, such phenomena cannot be taken into consideration during equipment calculations.

The wetting of solid surfaces by liquids is also due to interface energies. As a measure of the wettability, the wetting angle is used, which occurs at the contact line of the three phases between the solid and the wetting liquid. It can be measured with suitable optical systems by placing a drop on a plain, horizontal surface of the material used in the industrial apparatus. For total wetting, the wetting angle is 0°, rising to 180° if a wetting is not possible. All dispersion devices in industrial extractors, sieve trays, tower packings, or mixers should not be easily wetted by the dispersed phase, since this would result in large drops shaped like doughnuts.

These phenomena of surface energy all are time dependent. The shorter the contact time of the phases, the more difficult it is for them to develop. Therefore, flow and mass transfer in centrifugal extractors are hardly affected.

9.8 SEPARATION OF PHASES

Once mass transfer is completed, the drop phase must be separated from the continuous liquid. The basic event of the separation process is the coalescence of droplets producing a homophase. This can take place in a part of a countercurrent column especially provided for this purpose (see Figs. 9.1 and 9.5) or in a special settler (Fig. 9.23). If we wish to predetermine the separation process, the physical course of the droplet coalescence must be known. Figure 9.24 schematically illustrates the coalescence of a single drop

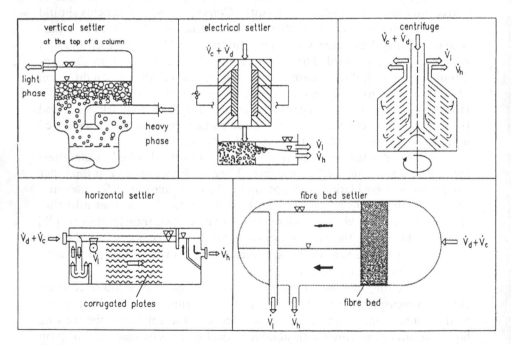

Fig. 9.23 Settlers for separation of liquid–liquid dispersions. V_h, volume flow of heavy phase; V_l, volume flow of light phase.

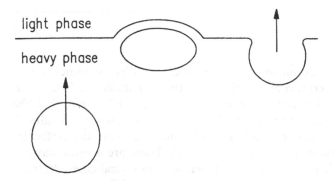

light phase

heavy phase

Fig. 9.24 Basic phenomena of drop coalescence at a horizontal interface. The drop has to reach the interface. A thin layer of the continuous phase remains between the drop and the interface. The thin layer has to drain until it breaks up. Then the drop can flow into its homophase. Mostly, the drainage process is the time-determining step of this process.

of lighter phase at a horizontal interface. The drop rises to the interface, where it is retarded, and deforms both itself and the interface. A thin layer of the continuous phase remains between the drop and the interface. This thin layer has to drain, forced by the contact pressure of the drop, until it reaches its critical thickness and breaks up. Then the drop can flow into its homophase. In most cases, this drainage process is the time-determining step of coalescence. If the drainage time can be calculated, the minimum value of the residence time τ of the drop phase in the settler is calculable from the volumetric flow rate of the dispersed phase, $\dot{V}_d$:

$$\tau = \frac{V}{\dot{V}_d} \qquad (9.37)$$

This is the reference value for the settler volume V. The literature contains many proposals for modeling the coalescence at horizontal interfaces and the settling time [see, e.g., Ref. 1, Chapter 13]. However, predetermination of the coalescence time proves to be very complicated, as different effects hinder the drainage of the thin layer. Especially, the influence of surfactants, which occur in every industrial extractor, cannot yet be evaluated using mathematical models. Surfactants produce surface tension gradients during the drainage process and, therefore, Marangoni convections back into the layer, which may lead to a drastic increase of the coalescence time [29,30].

For a practical settler design, the height H of the dense dispersion, which is below the principal interface and has to find space in the settler, is evaluated by a simple estimation formula:

$$H = C\left(\frac{\dot{V}_d}{S}\right)^{\nu} \tag{9.38}$$

The factor C and the exponent y must be determined by experiments with pilot settlers. These terms combine all the unknown parameters. The cross-sectional area, S, of the settler should be chosen in such a way that the superficial velocity, $\dot{v}_d$, of the dispersed phase does not exceed $5–9\,\mathrm{mm\,s}^{-1}$. Often aids for coalescing (e.g., packages of inclined plates or filter cartridges made of metal or plastic) (Fig. 9.23) are used. There are several ways of predetermining coalescence [31–33]. Furthermore, in special cases electrical fields and centrifuges are used to support coalescence [34].

9.9 DIMENSIONING OF LARGE-SCALE EXTRACTORS

9.9.1 The Problem of Scale-Up

The previous sections have pointed out that mathematical models of the processes must be proved by experiments, or adapted to experimental results with the aid of pilot extractors. For economic reasons, pilot extractors are usually much smaller than large-scale industrial apparatus. Pulsed pilot columns, for example, have a diameter between 30 and 250 mm, whereas industrial-size columns are up to 2500 mm and more in size. Thus, the question arises of whether or not the calculations or pilot experiments may be used for the dimensions of large-scale apparatus. This is a general problem for engineers.

Only a careful analysis of the differences that appear in construction, in the processes of flow and mass transfer and measuring and observing the behavior and the performance of large-scale extractors, can answer these questions. This provides formula of experience for the factors of scale-up that have been listed [4] for the countercurrent extractors described here.

Thus, the height of the column may need to be scaled up, but the column diameter may emerge from calculation of the flooding load. Large-scale circulations that are produced in large columns seem to be the reason for the decrease of the mass transfer performance as column diameter increases. This causes the axial mixing to be higher in large columns than in small ones. Such circulations are not prevented by inserts in the column. On the other hand, centrifugal extractors have flow and mass transfer behavior that agrees sufficiently between laboratory and large-scale equipment.

9.9.2 Problems of Industrial Liquid Systems

Clear test mixtures are used for laboratory experiments to be able to use fully defined, reproducible liquid systems. Such mixtures, according to agreement of the European Federation of Chemical Engineering [35], are listed in Table 9.2. They differ mainly in the interface tension. Compared with these, the variety of technical systems is very high [36]. To guarantee reliable dimensions for equipment, engineers must supplement their calculations by experiments that include the following:

1. The liquid system:

    Determination of the binodal solubility curve in a triangular diagram, the equilibria and their dependences on pH, temperature, salt content of the raffinate, etc.
    Determination of the ratio of solvent to raffinate flow rate
    Measurement of the separation time (e.g., in a shaking vessel) to evaluate the coalescence behavior
    Measurement of the physical properties of feed and product phases (e.g., mass densities, viscosities, surface tensions, and wettability) and their dependence on temperature on typical equipment materials
    Qualitative examination of the mass transfer intensity of a single droplet (kicking drop)

2. Pilot plants:

    Selection of the dispersed phase
    Preselection of a suitable extractor type on the basis of available information and testing at pilot scale
    Establishment of suitable operating conditions as, for example, the pulse intensity of pulsed extractors or the speed of rotation of centrifugal extractors
    Determination of the flooding point
    Observation of emulsion and dirty layers at the principal interface and their behavior
    Optimization of the operating conditions of the chosen apparatus
    Associated corrosion experiments in the pilot plant

3. Evaluation and storage of the results:

    Comparison of experimental and theoretical results
    Considerations on scale-up
    Final decision on extractor type
    Generalizing and storing of the experimental results for future application

Table 9.2 Important Properties of the Three EFCE Test Systems

			1		2		3
Phase I			Water				
Phase II			Toluene		n-Butanol		Butyl acetate
Formula			C_7H_8		$C_4H_{10}O$		$C_6H_{12}O_2$
Boiling point (1 bar)			110.4°C		117.5°C		126.09°C
Transferring component			Acetone		Succinic acid		Acetone
Formula			C_3H_6O		$C_4H_6O_4$		C_3H_6O
$\dfrac{\chi_t}{\text{mass-}\%}$	$\dfrac{\sigma}{\text{mN m}^{-1}}$		0 3.13 7.67	35.4 27.0 19.3	0 3.87 6.58	1.75 1.0 0.7	0 14.1 3.81 11.7 7.86 9.6
$\dfrac{\chi_t}{\text{mass-}\%}$	$\dfrac{\rho_I}{\text{kg m}^{-3}}$		0 3.13 7.67	997.8 993.7 987.8	0 3.56 6.25	985.6 995.6 1003.1	0 997.0 3.03 993.3 8.11 986.4
$\dfrac{\chi_t}{\text{mass-}\%}$	$\dfrac{\rho_{II}}{\text{kg m}^{-3}}$		0 3.13 7.67	866.5 864.5 862.6	0 3.56 6.25	846.0 866.5 881.4	0 882.1 3.03 879.4 8.11 873.6
$\dfrac{\chi_t}{\text{mass-}\%}$	$\dfrac{\eta_I}{c\,P}$		0 3.13 7.67	1.006 1.078 1.2	0 3.56 5.29	1.426 1.536 1.61	0 1.0237 3.03 1.11 9.11 1.28
$\dfrac{\chi_t}{\text{mass-}\%}$	$\dfrac{\eta_{II}}{c\,P}$		0 3.13 7.67	0.586 0.575 0.560	0 3.56 5.29	3.364 3.749 3.925	0.7345 03.03 0.72 8.11 0.6827
D_{eq} for $(D_{eq} = y_t/x_t)$			$0\% \leq \chi_t \leq 8\%$		$0\% \leq \chi_t \leq -6.5\%$		$0\% \leq \chi_t \leq 8\%$
			0.61–0.83		1.3–1.12		0.9–0.98
$\dfrac{\chi_t}{\text{mass-}\%}$	$\dfrac{D_I}{10^3 \text{ mm}^2\text{ s}^{-1}}$		0.59 3.45 5.96	1.14 1.07 1.01	0.69 3.66 5.21	0.57 0.52 0.47	0.03 1.093 31.86 0.598 74.15 1.68
$\dfrac{\chi_t}{\text{mass-}\%}$	$\dfrac{D_{II}}{10^3 \text{ mm}^2\text{ s}^{-1}}$		3.260.72 4.37	2.7 2.66 2.51	0.43 3.71 5.56	0.24 0.23 0.21	0.25 2.2 41.92 2.196 79.76 2.506

Source: Ref. 32.

9.10 NOMENCLATURE

Abbreviation	Unit	Physical quantity
A	m^2	Area, interfacial area
a	$m^2\,m^{-3}$	Interfacial area per unit volume
a	m	Pulsation amplitude
c	$kmol\,m^{-3}$	Concentration
D	$m^2\,s^{-1}$	Binary diffusion coefficient
D_{ax}	$m^2\,s^{-1}$	Axial dispersion coefficient
D_{eq}		Distribution ratio
d	m	Diameter
d_{32}	m	Sauter mean diameter
E	Ws	Energy
f	$1\,s^{-1}$	Frequency
g	$m\,s^{-2}$	Acceleration from gravity
H, h	m	Height
k	$m\,s^{-1}$	Mass transfer coefficient
$\dot{M}$	$kg\,s^{-1}$	Mass flow rate
$\dot{N}$	$kmol\,s{-}1$	Mole flow rate
$\dot{n}$	$kmol\,m^{-2}\,s^{-1}$	Mole flux
n_R	$1\,s^{-1}$	Rotor speed
n_P		Number of drops
n_S		Number of stages
P	kW	Power consumption
S	m^2	Cross-sectional area
t	s	Time, residence time
V	m^3	Volume
$\dot{V}$	$m^3\,s^{-1}$	Volumetric flow rate

9.10.1 Greek Letters

Δ		Difference
ε		Holdup
η	$kg\,m^{-1}\,s^{-1}$	Dynamic viscosity
ν	$m^2\,s^{-1}$	Kinematic viscosity
π	$3.1415\ldots$	
ρ	$kg\,M^{-3}$	Mass density
Σ		Sum
σ	$N\,m^{-1}$	Surface or interfacial tension
τ	s	Mean residence time
ψ	$kW\,kg^{-1}$	Specific energy dissipation

Nomenclature (*continued*)

Abbreviation	Unit	Physical quantity
ω	$1\,s^{-1}$	Angular velocity
$\dot{\nu}$	$m^3\,m^{-2}\,s^{-1}$	Superficial velocity
ν	$m\,s^{-1}$	Velocity
χ	$kmol\,kmol^{-1}$	Mole fraction

9.10.2 Dimensionless Numbers

All dimensionless numbers in this chapter are frequently used in chemical engineering, especially in heat and mass transfer and in multiphase flow [35]:

Archimedes number
$$Ar = \frac{d_p^3 g \rho_c \Delta\rho}{\eta_c^2}$$

$$Ar = \frac{Re^2}{Fr}\frac{\Delta\rho}{\rho} \quad \text{here encodes the drop}$$
diameter

Froude number
$$Fr = \frac{\nu^2}{dg}$$

Fr is the relation between inertia and gravity forces (not used in Chap. 9)

Liquid number
$$K_L = \frac{\rho_c^2 \sigma^3}{\Delta\rho g \eta_c^4}$$

$$K_L = \frac{Re^4 Fr}{We^3} \quad \text{combines important physical}$$
properties of the phases

Newton number
$$N_e = \frac{P}{\rho_c n_R^3 d_R^5}$$

N_e relates the power consumption with important parameters of mixing

Reynolds number of particle
$$R_e = \frac{\nu_P d_P}{\nu_d}$$

R_e is the relation of inertia and viscosity forces

Weber number of mixer
$$W_e = \frac{\rho_c n_R^2 d_R^3}{\sigma}$$

W_e is the relation of inertia and interfacial forces

9.10.3 Subscripts

c	Continuous phase
d	Dispersed phase
D	Flow dispersion
i	Phase i
i	Class of drop sizes
j	Jet formation
I	Interface
K	Column
L	Liquid phase
max	Maximum
min	Minimum
N	Nozzle hole
o	Overall
P	Particle
R	Stirrer
r	Relative
s	Swarm of droplets
t	Transfer component
z	Cell
α	Inlet
ω	Outlet
$*$	Equilibrium

REFERENCES

1. Godfrey, J. C.; Slater, M. J. *Liquid-Liquid Extraction Equipment;* John Wiley and Sons: Chichester, **1994**; 630.
2. Philhofer, T; Mewes, D. *Siebboden-Extraktionskolonnen;* Verlag Chemie: Weinheim, **1979**; 10.
3. Blass, E.; Goldmann, G.; Hirschmann, K.; Mihailowitsch, P.; Pietzseh, P. Ger. Chem. Eng. **1986**, 9, 222.
4. Lo., T. C.; Baird, M. H. I.; Hanson, C. *Handbook of Solvent Extraction;* John Wiley and Sons: New York, 1983.
5. Hirschmann, K.; Blass, E. Ger. Chem. Eng., **1984**, 7, 280.
6. Wagner, G.; Blass, E. Chem. Eng. Technol., **1999**, 21, 475.
7. Kolmogoroff, A. N. *Sammelband zur Theorie der statistischen Turbulenz;* Akademieverlag: Berlin, 1958.
8. Barson, N.; Beyer, G. H. Chem. Eng. Progr., **1953**, 49, 243.
9. Otillinger, E.; Blass, E. Chem. Eng. Technol., **1988**, 11, 312.
10. Otillinger, F. Fluiddynamik und Stoffübergang in Zentrifugalextraktoren [doctoral thesis]; TU Munchen, 1988.

11. Reinhardt, H. Chem. Ind., **1972**, 363.
12. Rydberg, J.; Persson, H.; Aronsson, P. O.; Selme, A.; Skarnemark, G. Hydrometallurgy, **1980**, *5*, 273.
13. Brunner, K. H.; Hemfort, H. Adv. Biotechnol. Processes, **1988**, *8*, 1.
14. Niebuhr, D. Untersuchungen zur Fluiddynamik in pulsierten Siebboden-Extraktionskolonnen [doctoral thesis], TU Clausthal, **1982**.
15. Schilp, R.; Blass, E. Chem. Eng. Commun., **1984**, *28*, 85.
16. Wen, C. Y.; Fan, L. T. *Models for Flow Systems and Chemical Reactors, Chemical Processes and Engineering;* Vol. 3, Marcel Dekker: New York, **1975**.
17. Krizan, P. Analyse des dynamischen Verhaltens dispenser Mehrphasensysteme bei ausgeprsgter Grof3raumstrbmung [doctoral thesis]; TU Munchen, **1987**.
18. Bird, R. B.; Stewart, W. E.; Lightfoot, E. N. Transport Phenomena; John Wiley and Sons: New York, **1960**; part III.
19. Highbie, R. Trans. Amer. Inst. Chem. Eng., **1935**, *31*, 365.
20. Handlos, A. E.; Baron, T. A.I.Ch.E.J., **1957**, *3*, 127.
21. Steiner, L. Rechnerische Erfassung der Arbeitsweisen von Flüssig-flüssig-Extraktionskolonnen, VDI-Verlag, Dusseldorf, Reihe 3 Nr. 154, 1988.
22. Pertler, M., Die Mehrkomponentendiffusion in nicht vollstaendig mischbaren Flussigkeiten [doctoral thesis]; TU Munchen, 1996.
23. Haeberl, M.; Blass, E. Chem. Eng. Res. & Des., **1998**, *77*, 647.
24. Moertens, M.; Bart, H.-J. J. Chem. Eng. Data, **2000**, *23*, 353.
25. Czapla, C.; Bart, H.-J. Chem. Eng. Technol. **2000**, *23*, 1058.
26. Sawistowski, H. Recent Advances in Liquid-Liquid Extraction, Hanson, C., Ed., Pergamon Press: Oxford, 1971.
27. Tourneau, M.; Wolf, S.; Stichlmair, J. Phase Boundary Correction in Liquid-Liquid Systems; DECHEMA monograph 136, **2000**.
28. Ollenik, R.; Nitsch, W. Ber. Bunsenges Phys. Chem., **1981**, *85*, 900.
29. Burrill, K. A.; Woods, D. R. J. Colloid Interface Sci., **1969**, *30*, 511.
30. Burrill, K. A.; Woods, D. R. J. Colloid Interface Sci., **1973**, *42*, 15.
31. Rommel, W.; Blass, E.; Meon, W. Chem. Eng. Sci., **1992**, *47*, 555.
32. Meon, W.; Rommel, W.; Blass, E. Chem. Eng. Sci., **1993**, *48*, 157.
33. Magiera, R.; Blass, E. Filtr. Sep. **1997**, *34*, 369.
34. Kriechbaumer, A. Entwicklung eines elektrischen Emulsionsspalters und Untersuchung zu den Koaleszenzvorgiingen und Stabilitdtsparametern im clektrischen Feld [doctoral thesis]; Graz, 1984.
35. Misek, T.; Berger, R.; Schroeter, J. Standard Test Systems for Liquid Extraction; Institution of Chemical Engineers: Rugby, 1985.
36. Hampe, M. J. Ger. Chem. Eng., **1986**, *9*, 251.

10

Extraction of Organic Compounds

RONALD WENNERSTEN   Royal Institute of Technology,
Stockholm, Sweden

10.1 INTRODUCTION

The separation of organic mixtures into groups of components of similar
chemical type was one of the earliest applications of solvent extraction. In
this chapter the term *solvent* is used to define the extractant phase that may
contain either an extractant in a diluent or an organic compound that can
itself act as an extractant. Using this technique, a solvent that preferentially
dissolves aromatic compounds can be used to remove aromatics from kerosene to produce a better quality fuel. In the same way, solvent extraction
can be used to produce high-purity aromatic extracts from catalytic refor-
mates, aromatics that are essentially raw materials in the production of
products such as polystyrene, nylon, and Terylene. These features have
made solvent extraction a standard technique in the oil-refining and petrochemical industries. The extraction of organic compounds, however, is
not confined to these industries. Other examples in this chapter include the
production of pharmaceuticals and environmental processes.

In these applications, solvent extraction constitutes an extraction stage
during which an organic phase is in contact with an aqueous phase or
another immiscible organic phase. The extract is then recovered by distillation or washing with an aqueous or organic phase.

Using new solvents and having a better understanding of the chemistry
involved in more specific interactions, solvent extraction has become a very
interesting separation technique for high-value organic chemicals (e.g.,
amino acids). Furthermore, liquid extraction using two phases with very high
water concentration has found applications for the separations of proteins.

Supercritical extraction with carbon dioxide as a solvent is also a most promising technique for a wide area of applications.

The development of better thermodynamic models has made it possible to simulate industrial processes from only a minimum of experimental data. This facilitates process integration and optimization, with a minimum of pilot plant tests.

10.2 DESIGN CRITERIA

10.2.1 General: Choosing an Organic Solvent

Although the solvent extraction process might seem simple from an engineering point of view and has relatively low equipment costs, the design of an economically attractive process requires a complex strategy that takes account of many dependent variables. There are numerous instances in which an improper comparison between solvent extraction and other separation techniques has been made because of a quite unsystematic investigation of the solvent selection step. The substance to be extracted, the solute, and the extracting solvent, made up of extractant and a diluent, interact in such a complicated way that the extraction step cannot be isolated from the recovery steps in the selection of the solvent. There are at least three integrated stages to be considered:

1. Liquid–liquid contact
2. Solvent recovery
3. Raffinate cleanup

In the extraction of organic components, as will be shown, these three steps can all be equally important for costs of operating and capital costs. Design calculations cannot be made based on distribution coefficients alone. Thus, several cases have shown that the optimum economic approach was neither obvious nor anticipated in preliminary studies.

The initial requirement in the development of a solvent extraction process for the recovery or separation of organic substances is knowledge of the feed composition and flow rate and of the restrictions on the product quality and raffinate composition. Starting with these conditions fixed, a preliminary solvent selection can be made from solvents representing different process configurations such as low-boiling solvents, high-boiling solvents, or solvents from which the solute is stripped back into an aqueous phase.

There is no "universal solvent," and solvent selection must be made individually for each separation problem. The specific choice must be based on a general knowledge of the different interactions in both phases. Among the desirable features for the extracting solvent are the following:

1. A high capacity for the species being extracted
2. Selectivity, dissolving to a large extent one of the key components without dissolving other components to any large extent
3. Low mutual solubility with water
4. Easy regeneration
5. Suitable physical properties, such as density, viscosity, and surface tension
6. Relatively inexpensive, nontoxic, and noncorrosive

Obviously, no solvent satisfies all these requirements, and the selection of a desirable solvent involves a compromise between these and other factors. When a preliminary selection has been made, a secondary screening can be performed based on simplified calculations of minimum energy requirement, since the capital costs for similar process configurations will not vary too much.

In the final selection, rigorous calculations must be carried out together with laboratory tests on phase separation properties, chemical stability, verification of equilibrium data used in the calculations, and so on. At this stage, the data obtained have to be examined critically to ensure that there is sufficient potential for the project to be carried to the next phase. If sufficient information has been collected for a conceptual flow sheet, a decision can then be made on whether or not to run a pilot plant. From pilot plant data, the process can be scaled up according to principles given in the literature. This outline of the principles of designing a solvent extraction process is general, but some of the individual steps require experience and effective aids, such as computer programs and experimental equipment. Some of the difficult steps are discussed in greater detail in the following examples.

10.2.2 Defining the Separation Problem

The industrial stream to be treated, the feed, will not be an analytical grade solute dissolved in water, but often contains several known and unknown substances, both organic and inorganic. To be able to make an initial selection of possible solvents, it is necessary to make a classification of the individual substances present in the feed and of the groups of substances with chemical similarities, for instance, paraffins, aromatics, salts, or others.

The following questions have to be answered:

What is the pH of the solution?
Which acids or bases are present?
Are there additives, such as surfactants, that will affect the phase separation properties?
Are solids present?

What is their nature?
Are there restrictions concerning product quality or raffinate compositions?
Are there any toxic components?
What is the stability of the substances?

When the foregoing criteria have been determined, one can proceed to the next step.

10.2.3 Primary Stage in Solvent Selection

It is advisable to begin with a literature search, preferably on a computer-based system or in a literature source book [1], to answer the question: what is the state of art concerning the extraction of the substance of interest or of similar materials? A preliminary selection of solvents must be based on an estimate of chemical interactions that could be important. These interactions are mainly ones of solute–solvent and solvent–solvent (see Chapter 2). Solute–solvent interactions are dependent on the natures of the solute and the solvent. The interaction is least when both are nonpolar; it becomes stronger as the polarity of the molecules increases; and it is strongest when there are interactions caused by formation of hydrogen bonds, charge transfer complexes, or coordination complexes. In reactive solvent extraction, it is desirable to find a solvent that can form strong, specific interactions with the solute to be extracted. The well-known rule of thumb that "like dissolves like" does not always apply. The solubilities of nonpolar substances are sometimes larger in polar than in nonpolar solvents.

The explanation for this could be that, although the solvent–solvent interactions may be extensive in a polar solvent, the gain in energy from the nonpolar solute interaction with a polar solvent can be still more so. In such a case, this larger gain outweighs the disadvantage of the elimination of the solvent–solvent interaction, and a higher solubility can be expected. The classification of the solvents according to their chemical structure and functional groups has been described in Chapters 2, 4, and 11 and is not treated here. If specific chemical interactions are present, as when an acid is extracted with a basic solvent, the complex formed may exhibit a low solubility in the solvent. Here, the solvent should comprise both an extractant (in this case a base) and a diluent that has a high solubility for the complex. The optimal diluent has the following characteristics:

It improves the distribution of the solute to be extracted.
It improves the selectivity.
It increases the density difference between the phases and lowers the viscosity of the solvent.

At this stage, it may not be possible to distinguish any promising process configuration for the recovery of the solvent. From the general criteria just given, a selection should be made of several solvents representing the following classes:

Solvents having a lower boiling point than the solute.
Solvents having a higher boiling point than the solute.
Solvents from which the solute could be stripped back into an aqueous phase, perhaps utilizing a temperature shift or a chemical reaction.

The recovery process must also be kept in mind, and fundamental vapor–liquid data, such as the formation of azeotropes, should be examined. Azeotropic data can be found in the literature [1], but are sometimes contradictory. Finally, solvents that are unstable, toxic, expensive, and high grade should be avoided, unless the product price is high and the feed flow rate is low.

10.2.4 Secondary Stage in Solvent Selection

As mentioned earlier, quantitative optimization of all solvent properties is a difficult and time-consuming procedure, since many process design alternatives must be considered for each solvent. Consequently, the solvent-screening strategy proposed here is to identify, as a primary step, a group of solvents that appears to have most promise as extractants for the solute. In the second step, computer simulations that use process simulation packages must be applied to quantify the effects of various process design alternatives. This involves collecting thermodynamic data through literature or experiments and fitting the data to a mathematical model.

10.3 PHASE EQUILIBRIA

Design of extraction processes and equipment is based on mass transfer and thermodynamic data. Among such thermodynamic data, phase equilibrium data for mixtures, that is, the distribution of components between different phases, are among the most important. Equations for the calculations of phase equilibria can be used in process simulation programs like PROCESS and ASPEN.

There are mainly four possibilities of acquiring the necessary phase equilibrium data for a certain system:

1. Data at the actual state (P,T-area of interest) are available from the literature.
2. Data are available at other states requiring different thermodynamic correlations for interpolation or extrapolation.

3. Data are available for similar systems at actual or other states—again requiring different thermodynamic correlations for interpolation or extrapolation.
4. Data are totally missing—experimental work is necessary to obtain data.

There are thousands of references and several source books on liquid–liquid equilibrium (LLE) data for organic systems in the literature [1]. However, the available data may be incomplete or unreliable, which makes experimental work necessary. Different thermodynamic models can then be used for interpolation and extrapolation.

10.3.1 Theory

Two main approaches can be taken when developing thermodynamic equations for the correlation and prediction of equilibrium data. Both are semiempirical in that they are based on simplifications of rigorous thermodynamic expressions, but include parameters that have to be fitted to experimental data.

The first method, which is the more flexible, is to use an activity coefficient model, which is common at moderate or low pressures where the liquid phase is incompressible. At high pressures or when any component is close to or above the critical point (above which the liquid and gas phases become indistinguishable), one can use an equation of state that takes into account the effect of pressure. Two phases, denoted α and β, are in equilibrium when the fugacity f (for an ideal gas the fungacity is equal to the pressure) is the same for each component i in both phases:

$$f_i^\alpha = f_i^\beta \quad \text{for } i = 1, 2, 3, \ldots n \tag{10.1}$$

(Reference states for both phases are chosen at the same temperature.)

From classic thermodynamics, the following relation can be derived [2]:

$$\ln \frac{f_i}{x_i P} = \ln \phi_i = \frac{1}{RT} \int_0^P \left[\left(\frac{\partial V}{\partial n_i} \right) T, P, n_j - \frac{RT}{P} \right] dP \tag{10.2}$$

$$\ln \frac{f_i}{x_i P} = \ln \phi_i = \frac{1}{RT} \int_V^\infty \left[\left(\frac{\partial P}{\partial n_i} \right) T, V, n_j - \frac{RT}{V} \right] dV - \ln \frac{Pv}{RT} \tag{10.3}$$

where x_i is the mole fraction of component i, and ϕ_i is the fugacity coefficient of component i in the mixture.

These equations are general and exact and express the fugacity of component i in a mixture as a function of the measurable quantities pressure (P), temperature (T), volume (V), and concentration (x).

The equilibrium condition is thus:

$$\phi_i^\alpha \cdot x_i^\alpha = \phi_i^\beta \cdot x_i^\beta \quad \text{for } i = 1, 2, \ldots n \tag{10.4}$$

To solve the integrals in Eqs. (10.2) and (10.3), it is necessary to have volumetric data; for example, an *equation of state* (EOS) of type:

$$V = f_1(T, P, n_1, n_2 \ldots) \tag{10.5}$$

or

$$P = f_2(T, P, n_1, n_2 \ldots) \tag{10.6}$$

Most of the equations of state are pressure explicit, and Eq. (10.6) can be used for equilibrium calculations. As the integration is made from V to ∞ the EOS has to be valid in the density range from zero to the actual density.

10.3.2 Equations of State

The EOSs are mainly used at higher pressures or when some of the components are near or above their critical point. The most commonly used *cubic equations of state* are the Peng-Robinson [3] and the Soave-Redlich-Kwong [4] equations.

The Peng-Robinson equation has the following appearance:

$$P = \frac{RT}{v - b} - \frac{a(T)}{v(v + b) + b(v - b)} \tag{10.7}$$

where v is the molar volume, a accounts for interactions between species in the mixture, and b accounts for differences in size between the species of the mixture.

For pure components:

$$b = 0.07780(RT_c/P_c) \tag{10.8}$$

$$a(T) = a(T_c)\alpha(T_R) \tag{10.9}$$

$$a(T_c) = 0.45724(R^2 T^2/P_c) \tag{10.10}$$

$$\alpha^{1/2}(T_R) = 1 + m(1 + T_R^{1/2}) \tag{10.11}$$

$$m = 0.37464 + 1.54226\omega - 0.26992\omega^2 \tag{10.12}$$

where T_c and P_c are the critical temperature and pressure, T_R is the reduced temperature ($T_R = T/T_c$), and ω is the acentric factor for component i.

For mixtures, one has to apply a mixing rule; for example, for a:

$$a = \sum_{i}^{N} \sum_{j}^{N} x_i x_j a_{ij} \tag{10.13}$$

and

$$a_{ij} = (1 - \delta_{ij})\sqrt{a_i a_j} \tag{10.14}$$

where δ_{ij} is an adjustable interaction parameter obtained from experimental data, and x_i and x_j are mole fractions of component i and j in the liquid phase.

For b the following mixing rule can be used:

$$b = \sum_{i}^{N} \sum_{j}^{N} x_i x_j b_{ij} \tag{10.15}$$

and

$$b_{ij} = (1 - \eta_{ij})\lfloor (b_i + b_j)/2 \rfloor \tag{10.16}$$

where η_{ij} is an adjustable parameter determined along with δ_{ij} by fitting the equation of state to equilibrium data. The cubic equations of state are further considered by Marr and Gamse [5].

10.3.3 Activity Coefficient Models

At low or moderate pressure, when the liquid phase is incompressible, an *activity coefficient model* (γ model) is more flexible to use than an equation of state. This method often works, even for strongly nonideal systems involving polar and associating components.

The fugacity of a component i in a liquid mixture is given by:

$$f_i^L = \gamma_i \cdot x_i \cdot f_i^0 \tag{10.17}$$

where f_i is the fugacity of component i at an arbitrary chosen reference state and γ_i is the activity coefficient that indicates the deviation from ideality.

The reference state is often chosen as pure component i as a liquid at temperature T and pressure P, which gives:

$$f_i^0 = P_i^0 \cdot \phi_i^0 \cdot \exp\left[\int_{P_i^0}^{P} (v_i/RT)dP \right] \tag{10.18}$$

where P_i is the vapor pressure for pure component i, ϕ_i is the fugacity coefficient for component i at saturation, and v_i is the molar volume for component i as a liquid.

For components at supercritical conditions, the reference fugacity, f_i, is usually taken as Henry's constant:

$$f_i^0 = H_i = \lim_{x_i \to 0} (f_i^L / x_i) \tag{10.19}$$

where H_i is a function of pressure and temperature. The activity coefficient γ_i is obtained from Gibbs "excess" energy as:

$$RT \ln \gamma_i = \left(\frac{\partial n_T G^E}{\partial n_i} \right) P, T, n_j \tag{10.20}$$

To calculate γ_i, an expression is required such as:

$$G^E = f_3(T, P, n_1, n_2 \ldots) \tag{10.21}$$

For most liquids at moderate pressures, the pressure dependence can be neglected and the temperature dependence is small. The most important parameter is the composition.

Many empirical and semiempirical equations for G^E have been published in the literature. The equations have to fulfill the boundary condition:

$$\gamma_i \to 1 \quad \text{as } x_i \to 1$$

for the reference state f_i° chosen as a pure component as a liquid and:

$$\gamma_i \to 1 \quad \text{as } x_i \to 0$$

for components in supercritical state, with reference state $f_i^\circ = H_i$.

The earliest equations for Gibbs excess energy, like Margules and van Laar, were largely empirical. More recent equations and NRTL and UNIQUAC are based on a semiempirical physical model, called the two-liquid theory, based on local composition. The molecules do not mix in a random way, but because of different bonding effects, the molecules prefer a certain surrounding. This results in a composition at the molecular level, the local composition, which differs from the macroscopic composition. The local mole fractions cannot be measured easily, but must be related to the overall composition. A simple correlation can be obtained from statistical thermodynamics for which the quotient of the local mole fractions equals the quotient of the total mole fractions times a Boltzmann's factor according to:

$$\frac{x_{21}}{x_{11}} = \frac{x_2}{x_1} \cdot \exp[(\mu_{21} - \mu_{11})/RT] \tag{10.22}$$

$$\frac{x_{12}}{x_{22}} = \frac{x_1}{x_2} \cdot \exp[(\mu_{12} - \mu_{22})/RT] \tag{10.23}$$

where $(\mu_{21}-\mu_{11})$ and $(\mu_{12}-\mu_{22})$ are two adjustable parameters that can be obtained by correlation to binary experimental data.

Activity coefficient methods work fairly well at temperatures well below the critical, at which the liquid phase is largely incompressible, and up to moderate pressures.

10.3.4 Liquid–Liquid Equilibria Using Activity Coefficient Models

Binary data can be represented with a $T-x$ diagram that shows the mutual solubility as function of temperature. Most of the binary systems belong to one of the classes in Fig. 10.1. For ternary systems, experimental data are usually obtained at constant temperature and given in ternary diagrams. There are many types of systems, but more than 95% belong to one of the two classes shown in Fig. 10.1.

The fugacity for two liquid phases in equilibrium can be calculated with Eq. (10.17). At equilibrium, the temperature in both phases is the same,

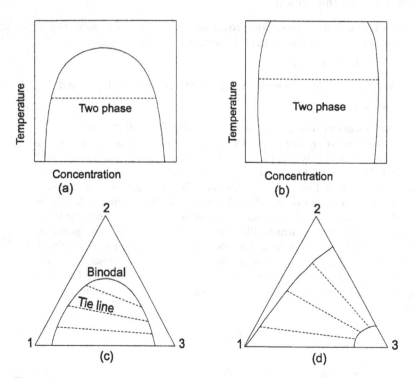

Fig. 10.1 Different types of liquid–liquid systems. (a), (b) Solubility as function of temperature for binary systems; (c), (d) ternary systems. (Dashed lines are examples of tie lines, which connect the two phases in equilibrium located at the binodal.)

which gives the same reference state for component i in both phases, and the equilibrium condition then reduces to

$$\gamma_i^\alpha \cdot x_i^\alpha = \gamma_i^\beta \cdot x_i^\beta \tag{10.24}$$

where α and β denote the two phases.

The activity coefficients can be calculated using any of the existing models if the binary parameters for all combinations of binary pairs are known. These parameters are obtained by fitting to experimental data. For ternary systems, one can either simultaneously fit all six parameters or first determine the parameters using binary data for those binary systems that have a phase separation and the rest of the parameters from ternary data.

In principle, as the γ models do not have an internal temperature dependence, the extrapolations in temperature must be carried out with great care. When extrapolating in temperature, the parameters in the γ models should be made temperature dependent by fitting to experimental data at different temperatures. Parameter estimation is obtained by minimizing an object function such as:

$$Q = \sum_{k=1}^{N} \sum_{i=1}^{n} [(X_{ik}^\alpha - X_{ik}^{*\alpha})^2 + (X_{ik}^\beta - X_{ik}^{*\beta})^2] \tag{10.25}$$

As an example, the system ethyl acetate–acetic acid–water can be used and equilibrium data using the UNIQUAC equation calculated [6]. Using the NRTL equation would have given similar results.

In principle, there are three situations:

1. Experimental liquid–liquid equilibrium (LLE) data are available.
2. Only experimental vapor–liquid equilibrium (VLE) data are available.
3. Experimental data are lacking.

In the first case, the six parameters in the UNIQUAC equation are obtained by fitting to experimental LLE data, shown in Fig. 10.2 as tie lines.
The binodal curve has been calculated and is shown in Fig. 10.2 as a solid line. Furthermore, the calculated distribution of acetic acid between both phases is shown in Fig. 10.3. From these figures, it can be seen that the fitting to experimental data is good.
In the second case, where LLE are lacking, VLE data are used to fit the parameters in the UNIQUAC model. These parameters, for the three binaries, were obtained from the literature in which VLE data were given at the following temperatures:

Ethyl acetate–acetic acid, 315 K
Ethyl acetate–water, $P = 20$ k_{Pa} (303–333 K)
Acetic acid–water, 313 K

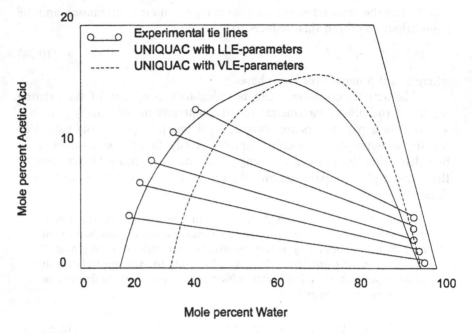

Fig. 10.2 Liquid–liquid equilibrium for the system ethyl acetate–acetic acid–water at 303 K.

As can be seen in Figs. 10.2 and 10.3, the agreement with experimental LLE is not very good.

Generally, one can say that parameters obtained from VLE should be used only for qualitative calculations. The result is considerably improved if the parameters for the binary ethyl acetate–water are calculated from solubility data.

When both VLE and LLE data are missing, it is not possible to assume ideal solutions. The separation into two liquid phases indicates a strong deviation from ideality. Here, the remaining possibility is to use some of the existing group contribution methods, such as UNIFAC, for which the parameters are based on LLE data.

10.4 INDUSTRIAL ORGANIC APPLICATIONS OF LIQUID EXTRACTION

10.4.1 Petroleum and Petrochemical Processing

Oil refineries and the petrochemical industry constitute one of the largest application areas for liquid extraction. The largest applications in this field are the treatment of lubricating oils and the separation of aromatic and aliphatic hydrocarbons. Other examples are the production of high-purity fiber-grade caprolactam, recovery of acrylic acid, and production of anhydrous acetic acid. The success of solvent extraction in the petroleum and petrochemical industry is due to careful process integration and also the development of large-scale column extractors.

The development of different processes and solvents for the *separation of aromatics from aliphatics* has reached a rather stable state. A number of different processes, some of them with capacities of several hundred thousand tons of aromatics per year, are in operation. The more important ones are listed in Table 10.1.

The Union Carbide Process

The process equipment consists mainly of two extraction columns with pulsating trays and four distillation columns according to Fig. 10.4 [7]. The feed, with a high content of aromatics, is pumped to the middle of the first extraction column where the aromatics are extracted with the solvent S1 (tetraethylene glycol). In the lower part, the extracted aromatics are washed with S2 (dodecane). The outgoing raffinate phase R1 (containing aliphatics, S2, and a small amount of S1) is distilled in the fourth distillation column. The distillate, D4 consists of pure aliphatics, whereas the bottom stream

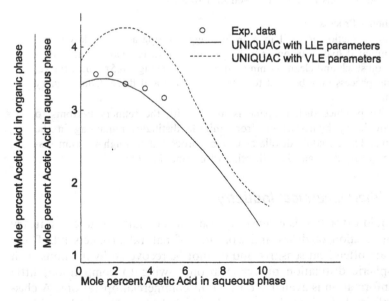

Fig. 10.3 Distribution coefficients for acetic acid.

Table 10.1 Different Processes for Separation of Aromatics from Aliphatics

Process	Extraction solvent	Extraction temp. (K)	Equipment
UDEX	Polyethylene glycols + water	423	Sieve plate column
Union Carbide	Tetraethylene glycol	373	Pulsed plate column
Shell	Sulfolane	393	RDC
DMSO	Dimethylsulfoxide	298	RDC
Arosolvan	N-Methylpyrrolidone	308	Vertical mixer-settler column
Formex	N-Formylmorpholine	313	Sieve plate column
Carmex	Methyl carbonate	298	RDC

S2 + ΔS1 is recirculated as reflux to the first extraction column. The leaving extract phase E1, containing aromatics and S1, is fed to the second extraction column where it is washed with more S2. The washed organic phase Sl is recirculated to the first column, while the extract phase E2, containing aromatics and S2 + ΔS1, goes to distillation.

The distillates from the distillation columns constitute the pure aromatics, while the bottom stream from the third column, containing S2 + ΔS1, is recirculated to the second extraction column. The operating conditions for this process are given in Table 10.2.

Re-refining Process

Solvent extraction is used in a technology for upgrading of low-value, contaminated hydrocarbons into high-value products such as base lubricating oil stock and clean-burning industrial fuels (Fig. 10.5). Thus, the re-refining process may be used to extract gasoline and diesel from refinery bottoms.

To produce fuel, propane is mixed with the refinery bottoms to separate heavy hydrocarbons from middle distillates remaining in the bottoms. These middle distillates can be processed through vacuum distillation to produce gasoline, diesel, and marine diesel [8].

10.4.2 Pharmaceutical Industry

Liquid–liquid extraction is extensively used in the pharmaceutical industry for the production of drugs and isolation of natural products [9]. These products are often heat sensitive and cannot be recovered by methods such as atmospheric distillation or evaporation. Owing to competition, little detailed information is available on current commercial operations. A classic example and the best documented, which has encountered problems general for pharmaceutical compounds, is the purification and concentration of penicillin [10].

Production of Penicillin

The fermentation broth containing penicillin is first filtered to remove mycelium and adjusted to pH 2–2.5 to convert it to the largely undissociated penicillinic acid before it is fed to the first extraction stage where it is contacted with, for example, butyl acetate (Fig. 10.6). The precise pH used is a compromise between the stability of the penicillin and the partition coefficient. The partition coefficient increases with pH, but unfortunately, it is accompanied by a rapid decrease in product stability [11]. To minimize the time the product is in the free-acid form, the feed is contacted with the extractant before acidification and the broth/extractant/acid efficiently contacted using an in-line mixer. It is also necessary to carry out the acidification and extraction at low temperature to minimize decomposition. Extraction takes place in countercurrent centrifugal contactors, e.g., a Podbeilniak extractor to ensure rapid throughput. Once in the organic phase the product is more stable, being removed from the aqueous acidic phase. However, because of entrained acid, it is still important to convert the penicillinic acid to a stable salt as soon as possible. This is achieved by mixing the extract phase with a 2% potassium phosphate solution at pH 6. The potassium penicillin is recovered by filtration and the solvent recirculated. Recovery of the organic phase from the aqueous raffinate is also very important to minimize costs and environmental impact. Butyl acetate, being a low-boiling solvent, can be recovered easily by distillation.

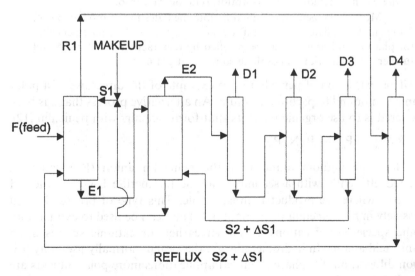

Fig. 10.4 The Union Carbide process.

Table 10.2 Some Operating Conditions for the Union Carbide Process

	First extraction column	Second extraction column
Equipment type	Pulsed plate column	Pulsed plate column
Number of ideal stages	10	7
Extraction temperature	373 K	373 K
Solvent	Tetraethylene glycol	Dodecane
Reflux	Dodecane	—
Sl/S2	9	1.43
Sl/F	7	—
Aromatics in feed	—	87 vol%
Recovery of benzene and toluene (*D1*, *D2*)	—	>99.5%
	—	—
Recovery of C8-aromatics (*D3*)	—	>98.5%
Purity of benzene	—	99.9%

Problems that can arise in the process include the formation of emulsions during extraction from the presence of surface-active impurities in the filtered broth. The effect of these can be minimized by introducing appropriate surfactants that can also reduce the accumulation of solids in the extraction equipment. In addition, other organic impurities are present that can be coextracted with the penicillin. It has been found that a number of these can be removed by adsorption onto active carbon.

Most of the penicillin is used as intermediates in the production of, for example, cephalosporin, but if it is necessary to produce pure penicillin for pharmaceutical use it can be purified by reextraction at pH 2–2.5 and further stripping with a phosphate solution at pH 6.

This extraction of penicillin is an example of the direct use of a polar organic compound to partition a solute. An alternative process that has been considered is to use organic reagents that form ionpairs with penicillin [12]:

$$R_4 N_{aq}^+ + P_{aq}^- \leftrightarrow R_4 N^+ P_{org}^-$$

Here the authors found that the penicillin anion (P^-) could be extracted efficiently with a secondary amine (Amberlite LA-2) in the pH range 5–7 where the product is most stable. This type of process is used extensively in hydrometallurgy (Chapter 11) and can be used to extract both anionic species using cations as shown earlier, or cationic species using organic acid anions. In hydrometallurgy, the system normally uses a hydrocarbon diluent, but for pharmaceutical applications more polar diluents are generally required.

Although the use of such chemically assisted extraction procedures is unlikely to displace the established extraction processes for commercial extraction of penicillin, there are a number of other systems in biotechnology where ion pair formation is used (section 10.5).

10.4.3 Separation of Isomers

Pure isomers are often used as starting products for fine chemicals (e.g., different drugs). The separation of isomers entails great difficulties because they often have boiling points differing by only a fraction of a degree and they have closely similar solubilities in many solvents. Solvent extraction processes for the separation of isomers, therefore, have to rely more on chemical reactions than on nonspecific physical interactions between the solute and the solvent.

Separation of *m*-Xylene from *o*- and *p*-Xylene
Xylenes react reversibly to form complexes with boron trifluoride (BF₃) in the presence of liquid hydrogen fluoride (HF). This fact can be utilized in the separation of *m*-xylene from its isomers, because the *m*-xylene complex is much more stable than the others and also more soluble in hydrogen

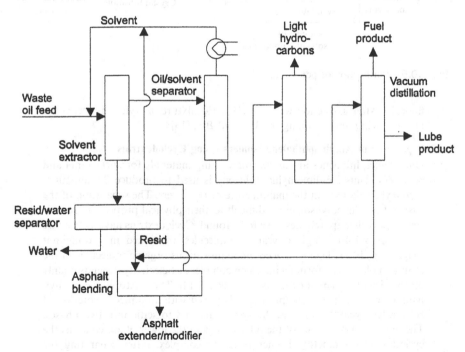

Fig. 10.5 The re-refining process.

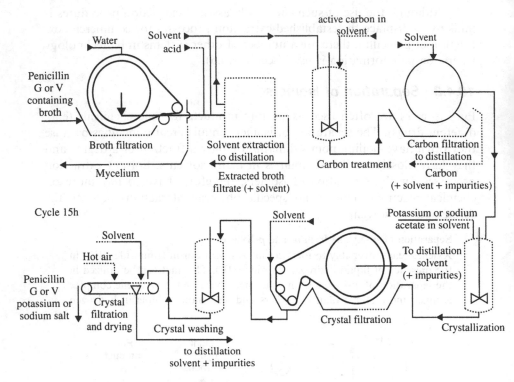

Fig. 10.6 Purification of penicillin.

fluoride. After extraction with the HF-BF₃ mixture, *m*-xylene is recovered by vaporizing and recycling the HF and BF₃ [13].

Separation of Alkylnaphthalene Isomers Using Cyclodextrins
Dimethylnaphthalenes are important starting materials for surfactants and polymers. Thus 2,6-dimethylnaphthalene is used to produce 2,6-dimethyl-carboxylic acid used in the manufacture of polyesters. The separation of the 2,6- and 2,7-dimethylisomers is difficult as their physical properties are very similar, so other approaches have to be found. Cyclodextrins have a structure where six α-1,4-D(+)-glucopyranose molecules are linked in a doughnut shape (Fig. 10.7). They can accommodate various organic molecules in the central apolar cavity forming inclusion complexes. Selectivity depends mainly on the size and shape of the guest molecule [14]. The mixture of dimethyl-naphthalenes in 1-methylnaphthalene is mixed with an aqueous solution of substituted cyclodextrins and the system allowed to settle into two phases. The upper phase consists of the oil sample and the lower phase contains the cyclodextrin/2,6-dimethyl isomer inclusion complex. After separating the phases, the aqueous phase is mixed with an organic solvent that dissociates

the inclusion complex and extracts the dimethylnaphthalene product. Analysis of the product showed 96% isomeric purity of the 2,6-dimethyl isomer from a 50:50 mixture of 2,6- and 2,7-isomers [15].

The use of cyclodextrins is still in the development stage and such systems have not yet been used commercially.

Dissociation Extraction

This is a separation technique for the separation of mixtures of organic acids or bases that exploits the differences in the dissociation constants of the components of the mixture. Application of dissociation extraction has been considered chiefly for the separation of organic acids and bases occurring in coal tar (e.g., *m*- and *p*-cresols) [16]. These isomers have very close boiling points, but they have appreciable differences in acid strength, as shown by their relative dissociation constants of 9.8×10^{-11} for *m*-cresol and 6.7×10^{-11} for *p*-cresol. If a mixture of these bases is partially neutralized by a strong base, there is a competition between the two acids for reaction with the base. *m*-Cresol, the stronger acid, reacts preferentially with the base to form an ionized salt, insoluble in organic solvents. The weaker acid, *p*-cresol, remains predominantly in its undissociated form and, accordingly, is more soluble in an organic solvent. When this process is applied with countercurrent flow of organic solvent and an aqueous base, the mixture of cresols may be separated. The *m*-cresol may be generated from its salt in the aqueous phase by treatment with a strong mineral acid.

10.4.4 Environmental Applications

Solvent extraction has many features that make this separation technique applicable for the removal of organic pollutants from wastewater. The organic solutes can be recovered, and there is thus a potential for economic credit to the operation. In comparison with biological treatment, it is not subject to toxicity instabilities. The disadvantage is that, even with solvents having low solubilities in water, solvent losses can be substantial owing to

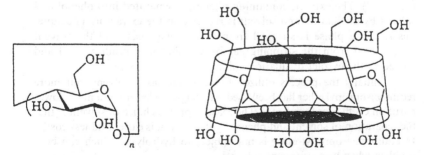

Fig. 10.7 Structure of cyclodextrin.

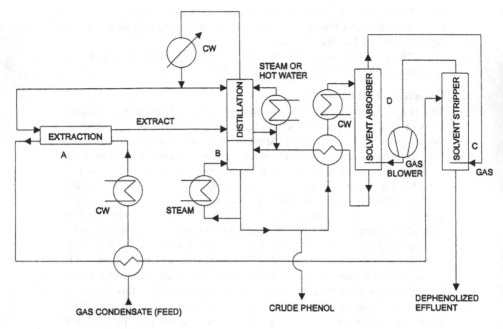

Fig. 10.8 The Phenosolvan process.

high flow rates of wastewaters. In practice, solvent extraction, therefore, has to be combined with other separation methods, such as distillation, to fulfill the environmental restrictions. Complex process considerations have then to be taken into account to evaluate the economics of different alternatives.

Phenosolvan Process

In this process developed by Lurgi [17], the phenolic effluent is contacted with the solvent in a multistage mixer-settler countercurrent extractor (Fig. 10.8). The extract, containing phenol, is separated into phenol and solvent by distillation and solvent is recycled to the extractor. The aqueous raffinate phase is stripped from solvent with gas, and the solvent is recovered from the stripping gas by washing with crude phenol and passed to the extract distillation column.

Initially the process utilized butyl acetate as a solvent, but more recently isopropyl ether has been used, although the latter has a much lower partition coefficient for phenol. The reason for this choice of solvent is that the separation of solvent and phenol by distillation is easier and less costly. In addition, isopropyl ether is not subject to hydrolysis, which can be a problem when using esters as solvents.

The Phenosolvan process is used in the treatment of wastewaters from plants involving phenol synthesis, coke ovens, coal gasification, low-pressure carbonization, and plastic manufacturing. The residual phenol content after dephenolization is usually in the range 5–20 mg dm^{-3} and plants treating 500 m^3 water per day are in operation.

10.5 APPLICATIONS OF LIQUID EXTRACTION IN BIOTECHNOLOGY

The recovery of products from biotechnological processes has traditionally been focused on bench-scale separation approaches, such as electrophoresis or column liquid chromatography. These methods are difficult to scale up to production levels and often become prohibitively expensive for medium- and low-value products.

Recently however, it has been recognized that liquid extraction is a potential method in the primary recovery of fermentation cell culture products, such as proteins and amino acids. The separation problem, however, is difficult because the product mixtures are often complex, including cell debris and enzymes. Proteins are not suitable for conventional solvent extraction because of incompatibility with organic solvents, but can be handled in aqueous two-phase systems or by extraction in reverse micellar systems (Chapter 15).

Although biotechnological processes often have been stated to be energy efficient in that the reaction temperature is low, it is important to realize that the product concentrations are low and that the product recovery step is often the most energy consuming.

10.5.1 Aqueous–Organic Systems

Solvent extraction has been proposed as an alternative for the separation of organic acids and amino acids from fermentation broth [18]. Thus, citric acid is currently produced in bulk by fermentation followed by recovery of the acid by precipitation as the calcium salt. Purification of the impure product is achieved by dissolving in sulphuric acid to precipitate calcium sulphate, leaving the citric acid in solution from which it is recovered by evaporation. An alternative process has been proposed using solvent extraction [18]. Citric acid can be extracted by both high molecular weight amines and organophosphate solvating reagents. Important criteria have been established for the processing of citric acid by solvent extraction: a good distribution coefficient, but <10 to allow for easy stripping with water, and a low tendency to form emulsions. Stripping with water is important because, although the citric acid could be recovered with a base, the formation of a citrate salt would require further processing as in the conventional flow sheet, thus removing the advantages of solvent extraction.

Wennersten found that the extractant system tri-*n*-butylphosphate (TBP) in Shellsol A diluent was temperature dependent with extraction decreasing with increasing temperature. Therefore, extraction from the broth at room temperature (22°C) and water stripping at 60°C provided an efficient process (Fig. 10.9). The extraction with long-chain amines was hampered by the formation of emulsions and thus poor separation. Emulsion formation depended on the diluent with hydrocarbons, e.g., Alamine 336/Shellsol A, given the best separation.

10.5.1.1 Carboxylic Acids

Acetic acid is an important intermediate organic tonnage chemical that may be produced by the petroleum industry and fermentation. The latter process requires the recovery of acetic acid from water solutions, and several techniques have been applied to this separation, including solvent extraction, azeotropic distillation, and extractive distillation. A comparison of economics between azeotropic distillation and solvent extraction combined with azeotropic distillation (Table 10.3) shows that the introduction of

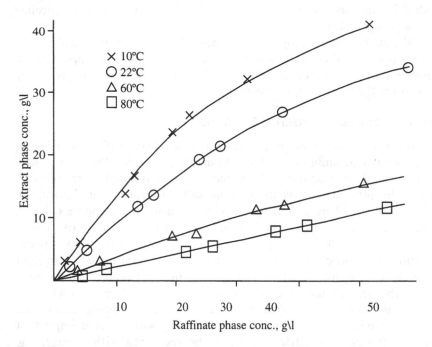

Fig. 10.9 Isotherms for citric acid extraction with TBP.

Table 10.3 Economics of Different Processes for Dehydration of Acetic Acid

	Plant capital (45,000 tons p.a. £m)	Steam usage (lb/lb acetic acid)	Operating cost (£/ton acetic acid)
Azeotropic distillation	0.5	6	4–5
Solvent extraction and azeotropic distillation	0.2–0.25	1.2–2.5	2–2.5

solvent extraction successfully introduces savings in plant investment and operating costs [19]. This process flow sheet is outlined in Fig. 10.10.

A number of studies have been carried out on the extraction of other important carboxylic acids from fermentation liquors [20,21].

10.5.1.2 Amino Acids

Amino acids, because they contain both carboxyl ($-COOH$) and amino ($-NH_2$) groups, can behave as cations at low pH and anions at high pH and have a zwitterion character at intermediate pH values. Although they have no net charge in this intermediate pH range, they are hydrophilic and so their solubility in nonpolar diluents is low. In addition, their extraction with carbon-bonded oxygen donor extractants is poor. Therefore, the most common method for extraction is to convert the amino acid into one of the ionic forms and use an appropriate ion-pair extractant. Thus extraction can be effectively carried out by a cationic extractant (HA), for instance, either a phosphoric acid extractant such as H(DEHPA), [22], or an organosulphonic acid such as H(DNNSA), where the extracted species is $(RNH_3COOH)^+ (A^-)$ [23]. The extent of extraction depends on the relative pK_a values of the two acids, with extraction favored when the extractant has a lower pK_a than the extractant. Thus glycine ($pK_a = 2.34$) is extracted by H(DNNSA) ($pK_a = 0.68$) over the pH range 1–5, but H(DEHPA) has a higher pK_a value (3.50), and little glycine extraction is found. However, some amino acids with two amino groups, e.g., histidine, form both singly and doubly charged cations and here the singly charged cation can be extracted by H(DEHPA) or even Versatic acid ($pK_a \sim 4.8$).

In high-pH media, the amino acid carries a net negative charge and so can be extracted by an anionic extractant, usually an alkylammonium salt, e.g., tri-*n*-octylamine hydrochloride (R_3NHCl). Here the extracted species is $(R_3NH)^+ (RNH_2COO)^-$ [24,25]. In both cases, stripping is easy using either dilute alkaline or acidic solutions respectively to regenerate the extracting species.

Of the two extracting systems, the use of an anionic extractant is more common to minimize coextraction of inorganic cations by the organic acids.

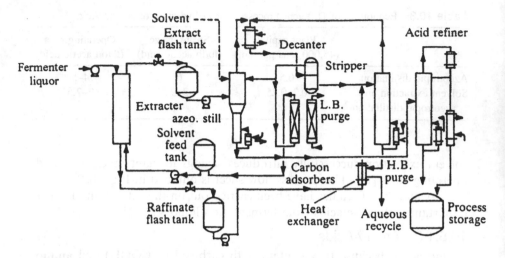

Fig. 10.10 Acetic acid flow sheet.

However, similar problems can also occur with the alkylammonium extractants, especially as the pH of the solution must be sufficiently high to produce an adequate concentration of the amino acid anions, i.e., more than two units above the pK of the amino acid. Thus, coextraction of hydroxyl ions competes with the amino acid anions and in addition lowers the pH of the aqueous phase and reduces the concentration of amino acid anions. To avoid such problems, a large buffer capacity is required in the aqueous phase with the buffer chosen to minimize coextraction of anions.

10.5.1.3 Ethanol

One of the problems associated with the production of ethanol by fermentation is the inhibition of the yeast activity with ethanol concentrations greater than about 12 wt%, thus defining an upper practical limit for ethanol production. Removing ethanol during fermentation by solvent extraction would allow a greater total product yield from a given batch of feed. The application of in-line extraction of ethanol from the fermentation broth has attracted interest. Problems do arise with direct extraction from broths with the effect of entrained organic solvents on microbial growth and the reduction in mass transfer of the ethanol by the presence of cell debris and other biological material. These problems can be minimized by suitable choice of the extracting solvent, e.g., long-chain aliphatic alcohols such as butanol, hexanol, etc., and removal of cell debris by filtration [26]. The process operates as a closed loop with the filtered broth circulating through a suitable column extractor before returning to the fermenter via a carbon column to remove residual organic solvent (Fig. 10.11).

10.5.2 Aqueous Two-Phase Systems

Under certain conditions, polymer incompatibility in aqueous solutions can lead to the formation of two phases with high water content. With such a system, it is possible to separate sensitive biological molecules, such as proteins, without denaturation, which would be the case for an ordinary aqueous–organic solvent system.

Intracellular enzymes and other microbial proteins are of increasing interest. Processes for their purification may be divided into five stages:

1. Disruption of the cells
2. Removal of cell debris
3. Concentration and enrichment
4. High-resolution purification
5. Concentration and finishing

An overall goal for these steps is to obtain a high yield while retaining the biological activity of the proteins. The required purity of the protein is determined by its end use. Enzymes that are to be used as technical catalysts require a lower purity than if they are to be used for analytic purposes or as pharmaceuticals.

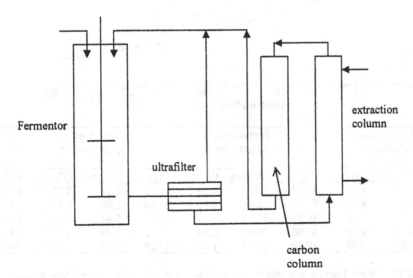

Fig. 10.11 Extraction of alcohol.

The phenomenon of aqueous two-phase polymer systems has been known since the 19th century. However, the utilization of aqueous two-phase systems as a separation method for biological materials was explored much later by Albertsson in the mid-1950s. Albertsson performed an extensive investigation on the ability of different water-soluble polymers to form aqueous two-phase systems and how biological macromolecules and cell particles partitioned in these phase systems [27].

The system polyethylene glycol (PEG)–dextran–water is still the most used and best-studied aqueous polymer two-phase system. A phase diagram for a typical two-phase system is shown in Fig. 10.12 for the PEG-dextran system. Both polymers are separately miscible with water in all proportions. As the polymer concentration increases, phase separation occurs, with the

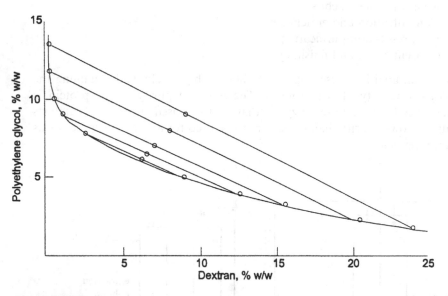

System	Total system polyethylene			Bottom phase polyethylene			Top phase polyethylene		
	Dextran	Glycol	H$_2$O	Dextran	Glycol	H$_2$O	Dextran	Glycol	H$_2$O
A	6.14	6.09	87.77	8.91	4.99	86.10	2.52	7.82	89.66
B	6.50	6.50	87.00	12.48	3.93	83.59	1.00	9.09	89.91
C	7.00	7.00	86.00	15.50	3.25	81.25	0.44	10.07	89.49
D	8.00	8.00	84.00	20.34	2.28	77.38	0.15	11.80	88.05
E	9.00	9.00	82.00	23.81	1.90	74.29	0.13	13.46	86.41

Fig. 10.12 Phase diagram and phase compositions of the dextran–polyethylene glycol system D 48-PEG 4000 at 293 K. All values are % w/w.

formation of an upper phase, rich in PEG, and a lower phase, rich in dextran, each phase consisting of more than 80% water. Within the two-phase region, any mixture of the three components split into two phases, with compositions dictated by the intersections of the tie line passing through the mixture point with the binodal.

The formation of two aqueous phases can be exploited in the recovery of proteins using liquid–liquid extraction techniques. Many factors contribute to the distribution of a protein between the two phases. Smaller solutes, such as amino acids, partition almost equally between the two phases, whereas larger proteins are more unevenly distributed. This effect becomes more pronounced as protein size increases. Increasing the polymer molecular weight in one phase decreases partitioning of the protein to that phase. The variation in surface properties between different proteins can be exploited to improve selectivity and yield. The use of more hydrophobic polymer systems, such as fatty acid esters of PEG added to the PEG phase, favors the distribution of more hydrophobic proteins to this phase. In Fig. 10.13, partition coefficients for several proteins in a dextran–PEG system are given [27].

Cells and cell particles generally partition almost entirely to one of the two phases or to the interface and one phase. Thus, it is possible to partition proteins away from the cell debris and, thereby, reduce the need for downstream processing.

The phase behavior of the polymers is also dependent on the type and concentration of salt present. Many times a sufficiently high concentration of salt in a single polymer solution can induce phase separation to form one salt-rich and one polymer-rich phase. Sodium and potassium phosphate are commonly used salts.

The efficiency of protein partitioning in aqueous two-phase systems can be considerably enhanced by the introduction of specific ligands into the phase system. The technique is called *affinity partitioning*, which is analogous to the well-known affinity chromatography concept widely used in analytical and preparative protein separations. In this technique, a small fraction of one of the phase-forming polymers is substituted by the same polymer with the specific ligand covalently attached to it. Since ligands interact strongly with specific proteins, the effect on the partition coefficient can be very dramatic, and there are examples when distribution coefficients have been changed by a factor of 10,000. The ligands that are used for affinity partition are mostly different dyes linked to a PEG molecule.

Recent developments in aqueous two-phase partitioning have been published by Raghavarao et al. [28].

Biotechnological processes, in general, are run at low concentrations of reactants and products. One reason for this is that most biocatalysts are

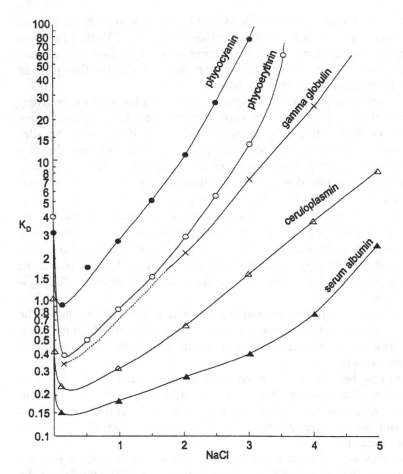

Fig. 10.13 Distribution coefficients of several proteins in the dextran–polyethylene glycol system in 0.0005 M KH_2PO_4, 0.0005 M K_2HPO_4, and with increasing concentration of NaCl; the latter is expressed as moles NaCl added per kilogram standard phase system.

sensitive to product inhibition. The reaction may stop when only a small fraction of the substrate has been converted into product. In such cases, there is a strong need for continuous extractive procedures during which the product is continuously removed. The integration of bioconversion and extraction is called *extractive bioconversion.*

The large differences in surface tension and dielectric constants between water and common organic solvents cause enzymes to unfold when they are exposed to the interface between the two solvents. This problem can be overcome through extractive bioconversions in aqueous two-phase systems.

A collection of cutting-edge methods intended to provide practical guidelines for those who are new to the area of separations in two-phase systems has been published by Hatti-Kaul [29].

10.5.3 Process Considerations

An important issue to be addressed in industrial applications of two-phase polymer systems for enzyme recovery is the economic or product quality requirement that the phase-forming polymers and salts be recycled.

In some rare cases, the polymer is accepted in the product (e.g., in a process for purification of enzymes for detergent purposes). This process utilizes the system PEG–Na_2SO_4–water, and more than 95% of the enzyme is extracted in a two-stage countercurrent process. No recycling of PEG is necessary, since the polymer is a part of the product. Usually, however, recycling of phase constituents is necessary, and several methods can be applied. As in the process described earlier, enzymes can be recovered from salt solutions by ultrafiltration, which also can be used to separate PEG from lower molecular weight products. If the product is charged, an ion-exchange chromatographic step could be sufficient for isolation. Several methods to precipitate the polymer are also available.

Because of the small density difference between the two aqueous phases, the phase separation properties of the system are poor. Solid-free systems can be separated in a few hours or overnight in simple cylindrical vessels. Centrifugal separators are usually necessary for phase separation when cell debris is present because of the relatively high viscosity of the bottom phase. The performance of centrifugal separators is greatly reduced by a high solid concentration in the top phase. Thus, when extracting proteins from disintegrated cells, conditions should be such that the protein is partitioned to the top phase and the cell material to the bottom phase.

10.5.4 Economic Considerations

In a study by Hustedt et al. [30], comparisons were made between the costs of liquid extraction recovery of formate dehydrogenase and conventional processing using centrifugation, precipitation, and column chromatography. The comparison was favorable for liquid extraction, and the major cost saving was the reduction in process time, compared with the conventional process. The dominating cost for the extraction process was that of phase polymers, in particular the dextran, which can contribute up to 90% of the materials cost in the two-phase system. This fact stresses the importance of developing methods for recovery of the phase constituents.

10.6 SUPERCRITICAL FLUID EXTRACTION

Solvent extraction processes usually run at ambient pressures and temperatures. If higher pressures are applied, it is mostly because a higher extraction temperature is required when equilibrium or mass transfer conditions are more favorable at an elevated temperature. Distillation, on the other hand, is usually carried out at higher temperatures and ambient pressures. To avoid thermal degradation, the pressure sometimes has to be lowered below ambient pressure. Distillation is based on the differences in vapor pressures of the components to be separated, whereas solvent extraction utilizes the differences in intermolecular interactions in the liquid phase.

Supercritical fluid extraction (SFE) is a separation technique that, to a certain extent, unites the principles of distillation and solvent extraction. SFE utilizes the very special properties of fluids at supercritical conditions, as is discussed below.

Figure 10.14 is a P-T diagram that illustrates the difference between conventional solvent extraction, supercritical extraction, and distillation. The choice of pressure and temperature area depends on the separation that is required.

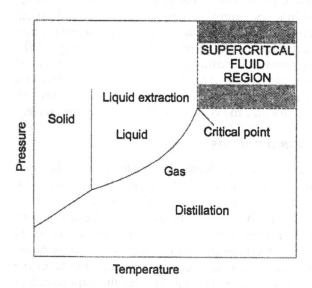

Fig. 10.14   Pressure-temperature (P-T) diagram for a pure component.

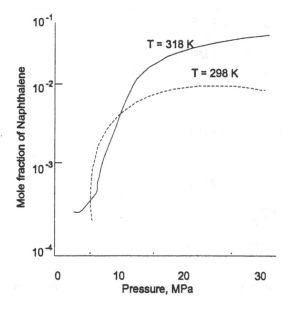

Fig. 10.15   Solubility of naphthalene in ethene at 298 K and 318 K.

10.6.1 *Physical Principles*

Figure 10.15 illustrates the solubility of naphthalene in ethene as a function of pressure at 298 and 318 K. The solubility of naphthalene increases dramatically when passing the critical point of ethene ($T_c = 282$ K, $P_c = 5$ MPa). The increasing solubility is due to the rapid increase of the density of ethene after passing the critical point.

Figure 10.16 illustrates the solubility of naphthalene in supercritical ethene as a function of temperature at different pressures. In Fig. 10.16 the temperature dependence of solubility is different in different pressure areas. At high pressure, an increase in temperature is followed by an increase in solubility, whereas at lower pressures the opposite effect occurs.

An increase in temperature at constant pressure, on one hand, leads to a decrease in solvent density, which would lower the solubility. On the other hand, an increase in temperature results in an increase in vapor pressure of naphthalene. At high pressures, the density dependence on temperature is small compared with the effect of vapor pressure, which results in an increased solubility. At lower pressures, the density effect dominates when increasing the temperatures, resulting in a decrease in solubility.

Figure 10.16 is characteristic for most mixtures in which the mutual solubility of the components is limited. From Fig. 10.16 it is

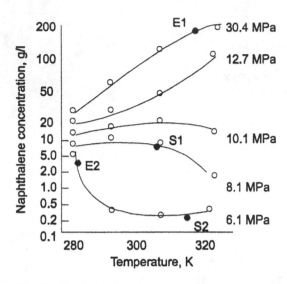

Fig. 10.16 Solubility of naphthalene in supercritical ethene as function of temperature at different pressures.

possible to derive simplified, general principles for extraction processes in the supercritical area. In Fig. 10.17, three different principles are shown schematically. Figure 10.17a illustrates the principle for extraction and solvent recovery through a change in pressure. In Fig. 10.16, the point E1 represents the state in the extractor (30 MPa). The extract phase is passed through an expansion valve at which the pressure is reduced to 8 MPa (point S1 in Fig. 10.16). Naphthalene now precipitates in the separator. The solvent is recompressed and returned to the extractor. In Fig. 10.17b, naphthalene is extracted at 6 MPa and 285 K (point E2 in Fig. 10.16). The extract phase is passed through a heat exchanger in which the temperature is raised to 315 K (point S2 in Fig. 10.16). Also, in this case, naphthalene precipitates in the separator. The solvent passes through another heat exchanger to lower the temperature and is then returned to the extractor. The extracted substances can also be recovered through adsorption (e.g., on activated carbon), which is shown in Fig. 10.17c.

The potential to vary the density of the solvent without passing any phase borders make the supercritical area very interesting from the point of view of separation. Through a gradual decrease in pressure it is possible, for example, to fractionate a mixture of substances.

10.6.2 Advantages and Disadvantages of Supercritical Extraction

The main disadvantage of SFE is the elevated pressure, requiring more expensive process equipment. The critical pressures, however, are below the pressures used in many high-pressure processes in the petrochemical industry today.

Table 10.4 gives critical data for the most common solvents used in high-pressure extraction. Table 10.5 illustrates the favorable mass transport properties that can be achieved in the supercritical area owing to a low viscosity and a high diffusivity, compared with the liquid phase.

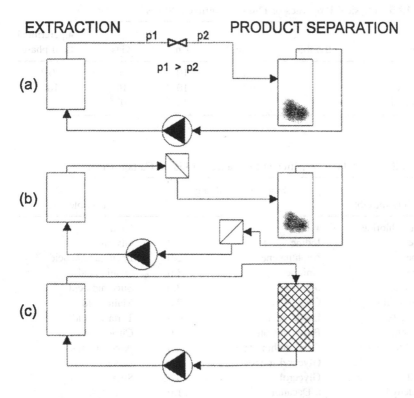

Fig. 10.17 (a) Extraction and separation through reducing the pressure; (b) extraction and separation through a change in temperature; and (c) extraction and separation through adsorption.

Table 10.4 Critical Properties for Some Solvents

Solvent	Critical Temp. (K)	Critical pressure (MPa)
Ethene	283.8	5.05
Trifluoromethane	299.8	4.69
Carbon dioxide	304.2	7.29
Ethane	305.4	4.82
Dinitrogen monoxide	309.7	7.17
Sulfur hexafluoride	318.8	3.77
Propene	365.8	4.54
Propane	370.0	4.24

Table 10.5 Physical Properties of Fluids in Different States

Property	Symbol	Liquid	Units	Gas	Supercritical fluid phase
Density	ρ	$g\,cm^{-3}$	10^{-3}	1	0.3
Diffusivity	D	$cm^2\,s^{-1}$	10^{-1}	$5 \cdot 10^{-6}$	10^{-3}
Dynamic viscosity	η	$g\,cm^{-1}\,s^{-1}$	10^{-4}	10^{-2}	10^{-4}

Table 10.6 Solubilities of Different Substances in Liquid Carbon Dioxide

Completely miscible	Fraction dissolving (% wt)		Insoluble
Stannous chloride	Water	0.1	Urea
Benzine	Iodine	0.2	Glycine
Benzene	Naphthalene	2.0	Phenylacetic acid
Pyridine	Aniline	3.0	Oxalic acid
Acetic acid	o-Nitroanisole	2.0	Succinic acid
Caprylic acid	Oleic acid	2.0	Malic acid
Ethyl lactate	Lactic acid	0.5	Tartaric acid
Amyl acetate	Butyl stearate	3.0	Citric acid
Glyceryl triacetate	Ethyl anthranilate	6.0	Ascorbic acid
Ethanol	Glyceryl monoacetate	1.0	Dextrose
Hexanol	Glycerol	0.05	Sucrose
Benzaldehyde	n-Decanol	1.0	—
Camphor	—	—	—
Limonene	—	—	—
Thiophene	—	—	—

The separation properties in SFE are dependent on the choice of solvents, as well as on the solutes. The most popular solvent, carbon dioxide, is a rather nonpolar solvent, which dissolves mainly nonpolar solutes. Solubilities of selected compounds in liquid carbon dioxide are given in Table 10.6. The solubility and selectivity can be altered by adding small amounts of polar solvents, called entrainers (e.g., water or ethanol).

10.6.3 Applications

Much of the interest in SFE has been focused on using carbon dioxide to extract different natural products from solid materials. Examples of large industrial processes in this area are decaffeinating coffee beans and hop extraction.

In the area of extracting solutes from aqueous solutions, many systems have been screened in feasibility tests that have used carbon dioxide as a solvent. A partial list of the solutes includes ethanol, acetic acid, dioxane, acetone, and ethylene glycol. The reason for these efforts has been potential low energy costs compared with distillation and the environmental advantages of using carbon dioxide.

Other systems that have been investigated are the separation of biocides from edible oils and fractionation of different components in vegetable oils. An example of the latter is given next. For an extensive literature survey of work done in the area of SFE, see references [31–34].

Separation of Mono-from Triglycerides
In Fig. 10.18, the quasi-ternary system carbon dioxide–acetone–glycerides at 13 MPa is shown [34]. As shown in Fig. 10.18, acetone increases the solubility of glycerides in the carbon dioxide phase. The entrainer has the function of increasing the solubility of the solute and enhancing the selectivity.

In Fig. 10.19, a simplified process scheme is given to illustrate the separation of a nonvolatile compound from a mixture, using an entrainer. The process consists of two columns. In the first column, the nonvolatile monoglycerides are separated from the mixture, and in the second column, they are separated from the solvent.

The mixture to be separated is fed, together with the entrainer, to the middle of the first column. Here, the solvent, carbon dioxide and acetone, is supercritical to provide high solubility of the monoglycerides. The supercritical phase leaves the top of column I and goes to the lower part of column II. In column II, the binary solvent entrainer is subcritical and in the bottom of this column, the monoglyceride leaves, together with the entrainer. Part of it is returned as reflux to column I, whereas the rest goes to distillation for the separation of acetone. With a bottom temperature of

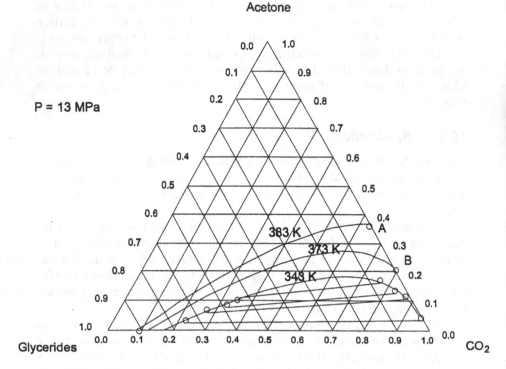

Fig. 10.18 Phase equilibrium for the quasi-ternary system CO$_2$–acetone–glycerides at 13 MPa and 343, 373, and 374 K.

353 K and a pressure of 13.5 MPa in column I and 383 K at the top of column II, 95% of the monoglycerides could be separated selectively from the mixture.

Supercritical fluid extractions with solid feed stocks are industrially carried out batchwise because of lack of equipment for feeding solid materials to a pressurized extractor.

With liquid feed solutions, however, it is possible to work in a manner analogous to traditional solvent extraction. Pressurized columns can be of the packed-bed type or agitated by magnetic stirrers. Because of the efforts of pilot plant tests, much of the scale-up work has to be carried out in laboratory extractors. From solubility measurements, it is possible to determine parameters in thermodynamic models (e.g., equations of state), which can be used for the simulation of large-scale applications.

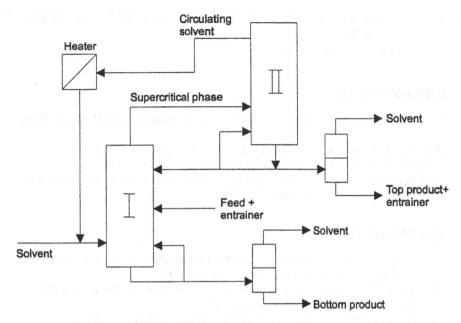

Fig. 10.19   Process for separation of a nonvolatile component from a mixture.

10.7   NOMENCLATURE

The symbols used follow standard textbooks for chemical engineering [see, Ref. 35].

f_i Fugacity of component i
x_i Mole fraction of component i in liquid phase
P Pressure
V Volume
v Molar volume
T Temperature
ϕ_i Fugacity coefficient of component i
n_i Number of moles of component i
a Adjustable parameter in equations of state that accounts for
            interaction between species in a mixture
b Adjustable parameter in equations of state that accounts for differences
            in size between species in a mixture
ω_i Accentric factor for component i
H_i Henry's constant for component i
G^E Gibbs excess energy
γ_i Activity coefficient for component i

α_I Parameter in the Peng-Robinson EOS Equation 10.11 or phase [e.g., Eq. (10.1)]
β Phase [e.g., Eq. (10.1)]

BIBLIOGRAPHY

Kennedy, J. F.; Cabral, J. M. S. *Recovery Processes for Biological Molecules*; Wiley: 1993.
Hatti-Kauk, R. Ed., *Aqueous Two-Phase Systems*; Humana Press, 2000.
Schugerl, K. *Solvent Extraction in Biotechnology*; Springer-Verlag: 1994.
Thornton, J. D. Ed., *Science and Practice of Liquid-Liquid Extraction*; Oxford Science Publications: Vol. 2, chapters 3–5, 1992.

REFERENCES

1.   Dortmunder Datenbank DDB (DETHERM database) [online]. Available from STN International, DECHEMA e.V. server, 2001.
2.   Smith, J. M.; Van Ness, H. C. *Introduction to Chemical Engineering Thermodynamics*; McGraw-Hill: New York, 1975.
3.   Peng, D. Y.; Robinson, D. B. Ind. Eng. Chem. Fundam., 1976, *15*, 59.
4.   Soave, G. Chem. Eng. Sci., 1972, *27*, 1197.
5.   Marr, R.; Gamse, T. Chem. Eng. Process., 2000, *39*, 19.
6.   Abrams, D. S.; Prausnitz, J. M. A.I.Ch.E., 1975, *21*(1), 116.
7.   Somekh, G. S.; Proc. ISEC 71, Lyon; Society of Chemical Industry: London 1971, *1*, 323.
8.   Kane, L.; Romanow-Garcia, S.; Nakamura, D. Hydrocarb. Process., 1995, *74*, 36.
9.   Verrall, M.S. in *Science and Practice in Solvent Extraction*, J. D. Thornton, Ed., Oxford Science Publications, Vol. 2, Chapter 3, 1992.
10. *Kirk-Othmer Encyclopedia of Chemical Technology*, 4th Edition 1993; Vol. 10, pp. 125ff.
11. Reuben, B. G.; Sjoberg, K. Chemtechnol., 1981, *11*, 315.
12. Reschke, M.; Schurgel, K. Chem. Eng. J. 1994, *28*, B1-29.
13. Bailes, P. J. Chem. Ind., 1977, 69.
14. Bender, M. L.; Komiyama, M. *Cyclodextrin Chemistry*; Springer-Verlag: Berlin, 1978.
15. Uemasu, I. Value Adding Through Solvent Extraction (Proc ISEC'96) edited Shallcross, D. C., Paimin, R.; and Prvcic, L. M. Ed., Univ. Melbourne Press, 1966; Vol. 2, 1635p.
16. Pratt, M. W. T. *Handbook of Solvent Extraction*. Lo, T. C., Baird, M. H. I., and Hanson, C. John Wiley and Sons: New York, 1983, p. 605.
17. Wohler, E. *Removal and Recovery of Phenol and Ammonia from Gas Liquor;* Lurgi Kohle Mineraloltechnik GmbH, 1978.
18. Wennersten, R. J. Chem. Tech. Biotechnol., **1983**, *33*, 85.

19. Lloyd-Jones, E. Chem Ind., 1967, 1590.
20. Kertes, A. S.; King, C. J. Biotechnol. Bioeng,. **1986**, *28*, 269–282.
21. Canari, R.; Hazan, B.; Bloch, R.; Eyal, A. M. Value Adding Through Solvent Extraction, (Proc ISEC'96) Shallcross, D. C., Paimin, R., and Prvcic, L. M., Univ. Melbourne Press, 1996, Vol. 2, 1517p.
22. Yagodin, G. A.; Yurtov, E.; Golubkov, A. S. Proc ISEC'86, Munich; DE-CHEMA: **1986**, *3*, 677.
23. Lukhezo, M.; Kelly, N. A.; Reuben, B. G.; Dunne, L. J.; Verrall, M. S. Value Adding Through Solvent Extraction, (Proc ISEC'96) Shallcross, D. C., Paimin, R., and Prvcic, L. M. Eds., Univ. Melbourne Press, *1*, 87 (1996).
24. Hano, T.; Matsumoto, M.; Ohtake, T.; Sasaki, K.; Hori, F.; Kawano, Y. J. Chem. Eng. Japan, 1990, *23*(6): 734.
25. Schugerl, K. *Solvent Extraction in Biotechnology*, Springer-Verlag, Berlin, 1994, p. 113.
26. Wang, H. Y.; Robinson, E. M.; Lee, S. S. Biotechnol. Bioeng. Symp., **1981**, *11*, 555–565.
27. Albertsson, P-A. Partition of Cell Particles and Macromolecules, Second Edition; John Wiley and Sons: New York, 1971.
28. Raghavarao, K. S. M. S.; Guinn, M. R.; Todd, P. Sep. Purif. Methods, 1988, *27*(1), 1.
29. Hatti-Kaul, R. *Aqueous Two-Phase Systems: Methods and Protocols*, Humana Press: Totowa, NJ 2000.
30. Hustedt, H.; Kroner, K.H.; Kula, M-R. Proc. Eur. Congr. Biotechnol., **1984**, *1*, 597.
31. Stahl, E.; Quirin, K.-W.; Gerard, D. *Dense Gases for Extraction and Refining*; Springer-Verlag: Berlin, 1987.
32. Bungert, B.; Sadowski, G.; Arlt, W. Ind. Eng. Chem. Res., 1998, *37*(8), 3208.
33. Teja, A. S.; Eckert, C. A. Ind. Eng. Chem. Res., **2000**, *39*(12), 4442.
34. Peter, S.; Brunner, G. Ger. Chem. Eng., **1978**, *1*, 26.
35. Reid, R. C.; Prausnitz, J. M.; Sherwood, T. K. The Properties of Gases and Liquids; McGraw-Hill: New York, 1977.

11

Solvent Extraction in Hydrometallurgy

MICHAEL COX* University of Hertfordshire, Hatfield, Hertfordshire, United Kingdom

11.1 INTRODUCTION

The use of solvent extraction as a unit operation in hydrometallurgy now extends to a wide range of metals from a variety of feed materials including low-grade ores, scrap and waste, and dilute aqueous solutions. The technology was pioneered in the 1940s for the extraction of uranium from its ores and, later, for the treatment of wastes from spent reactor fuel, still an important use of the technique today (see Chapter 12). The knowledge gained led to processes for the recovery of other high-value metals and the separation of elements such as the rare earths, zirconium–hafnium, and niobium–tantalum that, before the introduction of this technology, could be separated only by lengthy batch techniques with many recycle steps to obtain the desired purity. Gradually, solvent extraction was seen to have applications for the recovery of other less valuable, but important, metals such as cobalt and nickel. At the time, the process was confined to rather small operations. It was only after the development of selective chelating acidic reagents in the 1960s that liquid–liquid extraction was seen to be a commercially viable addition to the unit operations of hydrometallurgy and was able to compete with alternative processes like cementation.[†] Liquid–liquid

---

*Retired.

[†]*Cementation* is the process of recovery of metals from dilute aqueous solution by reductive precipitation using another metal with a more negative electrode potential, e.g., $Cu^{2+} + Fe^{o} \rightarrow Cu^{o} + Fe^{2+}$. The product, in this case "cement" copper, is relatively impure because of iron contamination. However, cementation can be used in conjunction with a solvent extraction flow sheet to remove small amounts of a metallic impurity, for example, removal of copper from a nickel solution by cementation with nickel powder. Here the dissolved nickel conveniently augments the nickel already in solution.

extraction is now an economic alternative to pyrometallurgy for metal extraction, especially when physical beneficiation of the ore to provide a suitable concentrate for smelting is difficult. The alternative hydrometallurgical route (Fig. 11.1) involves leaching of the ore to provide a leachate that, after any necessary solid–liquid separation, can be fed to the extraction circuit. Control of relative liquid flows allows the concentration of the desired metal and, thereby, economic processes for the recovery of metals from dilute aqueous wastes such as mine waters and wash solutions from the metal-plating industries.

Liquid–liquid extraction of metals has been used for many years as a concentration technique in analytical chemistry; consequently, it was logical to use the types of organic compounds employed in these systems for the development of industrial reagents for the same metals. However, because the overall requirements for these two end uses are different, modifications to the basic structure were necessary to take account of the particular requirements of hydrometallurgy, summarized by the following criteria:

1. The ability to transfer the desired metal selectively across the aqueous–organic interface in both directions
2. The ability of the reagent–diluent mixture to function efficiently with the proposed feed and strip solutions in terms of rates of operation and stability toward degradation
3. The ability of the reagent to perform with maximum safety to plant, personnel, and environment at minimum cost
4. The ability of the process to interface with other unit operations both upstream (leaching) and downstream (winning) in the overall extraction flow sheet

It can be seen that these criteria differ from those appropriate to analytical applications. Thus, in the latter, it is not essential to strip the metal from the organic phase, especially if a colorimetric method is being used. Also the chemistry of the feed solution is under the control of the analyst, who can change pH and add buffers and masking agents at will. Finally, because so little reagent is used, toxicity and cost are not prime considerations. Thus, although the general class of organic molecule will find application in both analytical and hydrometallurgical processes, the detailed structures of the compounds will be different. Examples of such modifications can be found in a number of texts [1–3], and factors associated with commercial reagent design have been discussed [3–6].

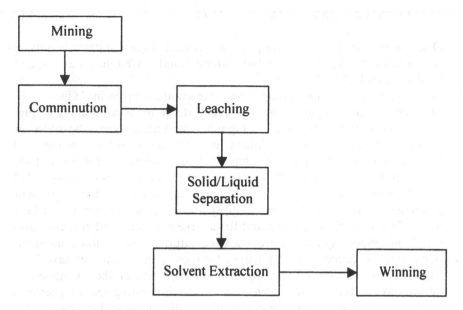

Fig. 11.1 Hydrometallurgical flow sheet using solvent extraction.

11.2 THEORY OF METAL EXTRACTION

Although a few metal compounds are sufficiently covalent to be extractable into an inert organic phase (e.g., halides of arsenic, antimony, gallium, and mercury), for a number of reasons, such as their ease of hydrolysis, this type of compound is unlikely to feature in a commercial extraction process. Normally, metal salts exist in aqueous phases as hydrated species and, as such, are incompatible with those nonpolar organic diluents normally used in hydrometallurgy. Therefore, to achieve extraction, species must be modified to make them more hydrophobic by such procedures as (1) reaction of the metal cation with suitable anions to produce a neutral complex, (2) formation of an ion pair, and (3) replacement of hydrated water molecules by an organic solvating reagent.

Reagents that are capable of such reactions are termed acidic, basic or ion pair, or solvating, and their modes of action are typified by the following equations:

$$M^{n+} + \overline{nHA} \longleftrightarrow \overline{MA_n} + nH^+ \tag{11.1}$$

$$MX_p^{(n-p)-} + \overline{(n-p)R_3NH^+X^-} \longleftrightarrow \overline{R_3NH_{(n-p)}MX_p} + (n-p)X^- \tag{11.2}$$

$$MY_q xH_2O + r\overline{B} \longleftrightarrow \overline{MY_q B_r} + xH_2O \qquad (11.3)$$

Discussion of the theory and application of such reagent systems is outside the scope of this chapter, but details can be found in Chapters 4 and 6, and in other texts [1,3,7,8].

A survey of current commercial extractants is given in Tables 11.1–11.4. The situation on availability of such extractants is continually changing, as modifications are always being made, and reagents may be added or deleted as the markets alter. Similarly, in the commercial world names and business interests of companies change. Thus, manufacturers are not included in this table. Commercial solvent extraction processes have expanded considerably since the 1960s, and there is an increasing use of solvent extraction technology in hydrometallurgy. Flow sheets have been devised for most of the transition elements and for a large number of other metals and metalloids, especially if these are available naturally only at low concentrations or are required at high purities. Of these elements, a few have been chosen for more detailed investigation on the basis of their commercial importance or because they demonstrate the interesting use of chemical properties to achieve a commercial process. Other texts and review articles are available that cover different elements, and these are included in the bibliography.

11.3 COBALT–NICKEL

Cobalt and nickel are both important elements for metallurgists, and their adjacent positions in the first-row transition series results in very similar chemical behavior, which causes problems in their separation. Any separation based on solvent extraction will be difficult unless some differences in chemical behavior can be exploited. Such differences include a higher rate of water exchange of the hexa-hydrated divalent cation for cobalt than for nickel, which might result in the development of a kinetic, rather than thermodynamic, separation. In addition, although the redox potentials of cobalt(II)–cobalt(III) and nickel(II)–nickel(III) couples are similar for the hexa-aquoion, on complexation, cobalt(II) is much more readily oxidized than nickel(II). Also, once formed, the cobalt(III) complexes are relatively inert to substitution. Finally, in the divalent state, cobalt has a strong tendency to form tetrahedral complexes in concentrated electrolytes (e.g., $CoCl_4^{2-}$), whereas in such aqueous solutions nickel(II) retains hexa-coordination (e.g., $Ni(H_2O)_6Cl_2$).

These general differences in chemical behavior have been exploited to provide the solvent extraction processes currently used or proposed for cobalt–nickel separation. All these processes remove cobalt from nickel

Table 11.1 Commercial Acidic Extraction Reagents

Reagent class	Structure	Extractants
Carboxylic acids		Naphthenic acids
		Versatic acids $R_1 = R_2 = C_5 =$ Versatic 10 $R_1 = R_2 = C_4–C_5$ Versatic 911
Phosphorus acids		
Phosphoric acids	$R_1 = OH; R_2 = C_4H_9CH(C_2H_5)CH_2O-$	mono-2-ethylhexylphos- phoric acid
	$R_1 = R_2 = C_4H_9CH(C_2H_5)CH_2O-$	di-2-ethylhexylphosphoric acid (DEHPA)
	$R_1 = R_2 = p-CH_3(CH_2)_6CH_2C_6H_5O-$	di-p-octylphenylphosphoric acid (OPPA)
Phosphonic acids	$R_1 = C_4H_9CH(C_2H_5)CH_2-$; $R_2 = C_4H_9CH(C_2H_5)CH_2O-$	2-ethylhexylphosphonic acid mono-2-ethylhexyl ester (PC88A), (P-507)
Phosphinic acids	$R_1 = R_2 = C_4H_9CH(C_2H_5)CH_2-$	di-2-ethylhexylphosphinic acid (P-229)
	$R_1 = R_2 = CH_3(CH_2)_3CH_2CH(CH_3)CH_2-$	di-2,4,4-trimethylpentyl- phosphinic acid (CYANEX 272)
Thiophosphorus acids	$R_1R_2P(S)OH$	
	$R_1 = R_2 = CH_3(CH_2)_3CH_2CH(CH_3)CH_2-$	di-2,4,4-trimethylpentyl- monothiophosphinic acid (CYANEX 302)
	$R_1R_2P(S)SH$ $R_1 = R_2 = CH_3(CH_2)_3CH_2CH(CH_3)CH_2-$	di-2,4,4-trimethylpentyl- dithiophosphinic acid (CYANEX 301)
Sulfonic acid		5,8-dinonylnaphthanenesul- phonic acid DNNSA

Table 11.2 Commercial Chalating Acidic Extractants

Reagent class	Structure	Extractants
α-Hydroxyoximes	$R^1C(NOH)CH(OH)R^2$	5,8-diethyl-7-hydroxydodecan-6-oxime (LIX63)
β-Hydroxyaryloximes		

R^1 = phenyl	$R^2 = C_9H_{19}$	LIX65N
	LIX65N + LIX63 (1%)	LIX64N
$R^1 = CH_3$	$R^2 = C_9H_{19}$	LIX84
$R^1 = H$	$R^2 = C_9H_{19}$	P1
	P1 + nonyl phenol	P5000 series
	P1 + tridecanol	PT5050
	P1 + ester modifier	M5640
	$R^2 = C_{12}H_{25}$	LIX860
	LIX84 + LIX860	LIX984
	LIX980 + tridecanol	LIX622
8-Quinolinol		

R = 5,5,7,7-tetramethyl-1-octenyl- (pre-1976)		
= 4-ethyl-1-methyloctyl- (post 1976)		Kelex 100
R = unknown side chain		Kelex 108
R = unknown saturated alkyl		LIX26
β-Diketones	$R^1COCH_2COR^2$	
	$R^1 = R\text{-}C_6H_5$ $R^2 = CH_3(CH_2)_5$	LIX54

solution, as no commercial flow sheet has yet been devised to remove nickel selectively from cobalt. Two different solvent extraction systems are used commercially involving (1) anion exchangers and (2) acidic chelating extractants, and these will be considered in turn.

11.3.1 Anion Exchangers

As noted earlier, cobalt(II) in strong electrolyte solutions will readily form tetrahedral anionic complexes that can be extracted into an immiscible

Table 11.3 Commercial Basic Extractants

Reagent class	Structure	Extractants
Primary amines	RNH_2	
	$R = (CH_3)_3C(CH_2C(CH_3)_2)_4$	Primene JMT
Secondary amines	R^1R^2NH	
	$R^1 = C_9H_{19}CH = CHCH_2$	
	$R^2 = CH_3C(CH_3)_2(CH_2C(CH_3)_2)_2$	Amberlite LA-1
	$R^1 = CH_3(CH_2)_{11}$	
	$R^2 = CH_3C(CH_3)_2(CH_2C(CH_3)_2)_2$	Amberlite LA-2
	$R^1,R^2 = CH_3(CH_2)_{12}$	Adogen 283
	$R^1,R^2 = CH_3CH(CH_3)CH_2(CH_2)_6$	HOE F2562
Tertiary amines	$R^1R^2R^3N$	
	$R^1,R^2,R^3 = CH_3(CH_2)_7$	Trioctylamine
		Alamine 300
	$R^1,R^2,R^3 = C_8-C_{10}$ mixture	Alamine 336
		Adogen 364
		Hostarex A 327
	$R^1,R^2,R^3 = (CH_3)_2CH(CH_2)_5$	Tri-isooctylamine
		Adogen 381
		Alamine 308
		Hostarex A 324
	$R^1,R^2,R^3 = (CH_3)_2CH(CH_2)_7$	Tri-isodecylamine
		Alamine 310
	$R^1,R^2,R^3 = CH_3(CH_2)_{11}$	Adogen 363
		Alamine 304
	$R^1,R^2,R^3 = CH_3(CH_2)_{12}$	Adogen 383
	$R^1,R^2,R^3 = C_{28}H_{57}$	Amberlite XE204
	$R^1 = CH_3(CH_2)_7, R^2 = CH_3(CH_2)_9$	
	$R^3 = CH_3(CH_2)_{11}$	Adogen 368
Quaternary amines	$R^1R^2R^2N(CH_3)^+ Cl^-$	
	$R^1,R^2,R^3 = C_8-C_{10}$ mixture	Aliquat 336
		Adogen 464
		HOE S 2706

Table 11.4 Commercial Solvating Extractants

Reagent class	Structure	Extractants
Carbon-oxygen extractants		
Amides	$R^1CONR_2^2$	
	$R^1 = CH_3$,	N503
	$R^2 = CH_3(CH_2)_5CH(CH_3)$	
	$R^1 = R^2$ unknown	A101
Ethers	$CH_3(CH_2)_3OCH_2CH_2OCH_2$- $CH_2O(CH_2)_3CH_3$	Dibutylcarbitol (Butex)
Ketones	$(CH_3)_2CHCH_2COCH_3$	Methyl isobutylketone (MIBK)
Phosphorus-oxygen extractants		
Phosphorus esters	$R^1R^2R^3PO$	
	$R^1,R^2,R^3 = CH_3(CH_2)_2CH_2O$	Tri-n-butylphosphate (TBP)
	$R^1,R^2 = CH_3(CH_2)_2CH_2O$, $R^3 = CH_3(CH_2)_2CH_2$	Dibutylbutylphosphonate (DBBP)
	$R^1,R^2 = CH_3O$, $R^3 = CH_3$	Dimethylmethylphosphonate
Phosphine oxides	$R^1R^2R^3PO$	
	$R^1,R^2,R^3 = CH_3(CH_2)_6CH_2$	Tri-n-octylphosphine oxide (TOPO) CYANEX 921
Phosphorus-sulfur extractants		
Phosphine sulfides	$R^1R^2R^3PS$	
	$R^1,R^2,R^3 = CH_3(CH_2)_6CH_2$	CYANEX 471X
Carbon-sulfur extractants R^1R^2S		
	$R^1,R^2 = C_6H_{13}$	Dihexylsulfide
	$R^1,R^2 = C_7H_{15}$	Diheptylsulfide

organic phase by amines or quaternary ammonium compounds. The most commercially important aqueous-phase ligand is chloride, although thiocyanate has also been studied by INCO Canada. Nickel/cobalt ratios of 2000:1 in the nickel raffinate and a cobalt product with a 1000:1 ratio over nickel are readily achievable. In chloride medium, the degree of formation of $CoCl_4^{2-}$ and, hence, the level of extraction, depend on the chloride concentration. This is shown in Fig. 11.2, which also shows the level of separation possible from nickel ($>10^4$) at a hydrochloric acid concentration of 8–10 mol dm^{-3}. This extraction process can be represented by Eq. (11.4), and the reaction can be easily reversed by water washing of the loaded organic phase.

$$CoCl_4^{2-} + \overline{2R_3NH^+Cl^-} \longleftrightarrow \overline{(R_3NH)_2CoCl_4} + 2Cl^- \qquad (11.4)$$

Figure 11.3 illustrates the effect of hydrochloric acid concentration on cobalt extraction, with a maximum cobalt extraction observed at approximately $9 \, mol \, dm^{-3}$. Above this concentration, the HCl_2^- ion competes with $CoCl_4^{2-}$ for the extractant; replacement of hydrochloric acid by a metal chloride solution, such as LiCl, eliminates this maximum. Fig. 11.3 also shows that extraction depends on the concentration and type of extractant; thus, extraction is greater for tertiary > secondary > primary amines. This anion-exchange process has been used commercially in Norway (Falconbridge Nikkelwerk) and Japan (Sumitomo) for cobalt–nickel separation.

11.3.2 Acid Extractants

For simple acidic extractants, such as carboxylic or sulfonic acids, the similarity in formation constants does not produce cobalt–nickel separation factors ($S_{Ni}^{Co} \sim 2$) sufficiently large for commercial operation (Fig. 11.4). Data for pH versus extraction for some chelating acid extractants does seem to offer the possibility of separation [e.g., for the hydroxyoxime Acorga P50, the pH_{50} for nickel(II) is 3.5 and for cobalt(II) 5.0]. Normally, this pH difference would be suitable for a separation process, but this particular system has hidden complications. The rate of nickel extraction is very slow compared with cobalt and, in addition, although cobalt is initially extracted

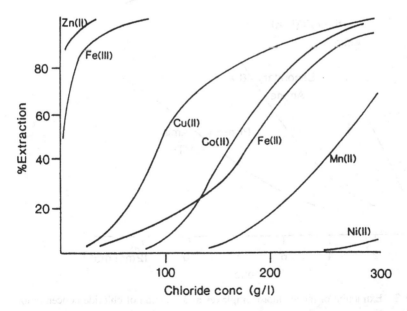

Fig. 11.2 Extraction of metal chlorocomplexes by amines. (From Ref. 2.)

as the divalent species, atmospheric oxidation to cobalt(III) readily occurs, which prevents normal stripping processes. A similar oxidation of the loaded organic phase has been observed with substituted quinolines and β-diketones. The complete exclusion of air from the circuit enables one to obtain high levels of cobalt stripping, but this complicates the operation. It has also been shown that, if cobalt(III) already exists in the aqueous phase, then hydroxyoximes and some β-diketones do not extract cobalt at all, although it has been reported that Kelex 100 will extract cobalt(III). This behavior suggests a possible cobalt–nickel separation route by the formation of stable cobalt(III) species, such as cobaltammines, in the aqueous phase and extraction of nickel by, for example, hydroxyoximes.

The oxidation of cobalt(II) in the organic phase can be minimized by the addition of donor molecules, such as the solvating extractants,

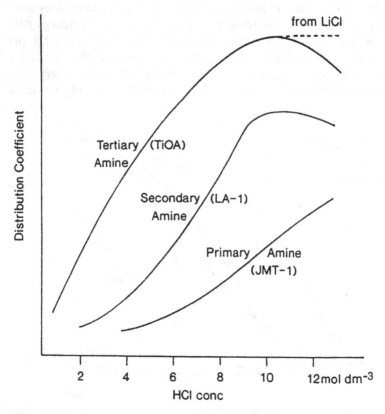

Fig. 11.3 Extraction of metal chlorocomplexes as a function of chloride concentration. (From Ref. 2.)

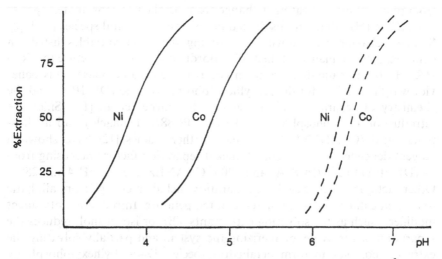

Fig. 11.4 Extraction of cobalt and nickel with naphthenic acid ($0.5 \, mol \, dm^{-3}$), dashed lines, and a mixture of naphthenic acid ($0.5 \, mol \, dm^{-3}$) and isotridecanal oxime ($1.0 \, mol \cdot dm^{-3}$), both systems in toluene. (From Ref. 10.)

tri-*n*-butylphosphate (TBP), tri-*n*-octylphosphine oxide (TOPO), and carboxylic acids. In addition to stabilizing cobalt(II), various degrees of synergism can be obtained, enhancing the separation of nickel from cobalt. Although separation factors of up to 150 have been observed with the system LIX* 63–carboxylic acid, again, there are problems with the slow rate of nickel extraction. By using a nonchelating oxime in place of LIX 63, separation factors of about 200 have been obtained, without any reduction in the rate of nickel extraction or oxidation of cobalt (see Fig. 11.4). The structure of the extracted complex for both metals is $MR_2(oxime)_2$, and relative ligand field stabilizing energies for nickel and cobalt have been proposed as the mechanism for synergism [9,10]. One problem with the use of this system in continuous operation is that the acid strip solution tends to hydrolyze the oxime.

Earlier it was noted that nickel and cobalt could be extracted by carboxylic and sulfonic acids, with nickel being extracted at the lower pH. However, with alkylphosphorus acids, a selectivity reversal is observed, with cobalt being favored under acid conditions. The cobalt–nickel separation factor has been shown to depend upon metal concentration, reagent structure, diluent, temperature, and the presence of a diluent modifier. Thus, with increasing cobalt concentration the color of the extractant phase changes

*LIX is a trademark of Cognis, previously Henkel.

from pink to blue, indicating a change from octahedral coordinated species (e.g., $[CoR_2(HR)_2]$) to an unsolvated polymeric tetrahedral species $[CoR_2]_n$. Similar behavior was shown by changing the other variables listed. In contrast, nickel remains octahedrally coordinated, with a structure $[Ni(R_2)-HR_x(H_2O)_{4-x}]$, maintaining a green color in the organic phase. This behavior was first noticed for di-2-ethylhexylphosphoric acid (DEHPA), and the chemistry of separation was elucidated by Barnes et al. [11]. Since the introduction of a phosphonic acid ester (PC88A, Daihachi) and a phosphinic acid (CYANEX* 272, Cytec) further studies [12] have shown a reagent dependence on the cobalt–nickel separation factor, increasing from 14 (DEHPA) to 280 (PC88A) and 7000 (CYANEX 272), at pH 4 and 25°C. Other factors that increased the separation include a change from aliphatic to aromatic diluents and an increase in temperature. Introduction of diluent modifiers, such as the solvating extractants TBP or isodecanol, reduces the separation factor by depolymerizing the system and partially solvating the extracted complex to form octahedral species. Di-2-ethylhexylphosphoric acid has been proposed commercially to separate cobalt–nickel with either column contractors or mixer-settlers [13], and PC88A has been incorporated into a flow sheet [14]. A problem with these circuits arises from oxidative degradation of the diluent by cobalt, which depends on the aromatic content of the diluent, on the temperature, and on cobalt loading [15]. This degradation can be eliminated by introduction of antioxidants such as nonyl phenol. It is surprising that such oxidative degradation of diluents has only recently been observed, when it is well known that cobalt compounds can act as oxygen carriers and oxidation catalysts. In addition to the diluents, oxidation of hydroxyoximes and 8-quinolinols has also been reported.

11.3.2.1 Nickel Laterite Processing

This discussion has centered on the use of solvent extraction to separate cobalt from nickel, as until recently the use of solvent extraction in nickel flow sheets has been minimal. However, with the increasing interest over the past decade in the development of hydrometallurgical processes for nickel lateritic ores, the use of solvent extraction to produce high-purity nickel products has become important. The flow sheets that have been developed are interesting as they indicate a number of different ways of using solvent extraction. Lateritic ores are formed by the chemical weathering of surface rocks and two main types are found, limonitic ores containing large amounts of iron (40%) and typically 1.4% nickel and 0.15% cobalt, and saprolitic ores that contain less iron (~15%) and cobalt (0.05%) but more nickel (2.4%) and silica. The saprolitic ores are amenable to pyrometallurgical

---

*CYANEX is a trademark of Cytec Inc., previously Cyanamid.

treatment usually to produce ferronickel, but the lower content of nickel and higher levels of cobalt in the limonitic ores makes them more suitable to hydrometallurgical processing. As the metal values are widely distributed through the limonitic ore, preconcentration is not feasible so that large amounts of the ore have to be leached, giving large volumes of a dilute leach liquor. Two reagents are used for leaching, ammonia in the Caron process, which dissolves the metal as ammoniates e.g., $[Ni(NH_3)_6]^{2+}$, and acid, which gives hydrated cations. Details of the complete hydrometallurgical flow sheets are outside the scope of this chapter and can be found elsewhere [13]. Here discussion will concentrate on the solvent extraction flow sheets to illustrate how the technology can separate the desired values from a range of other metals present in the leach liquor.

11.3.2.1a   Ammonia Leachate

The ammonia leach leaves the iron as an insoluble residue that is removed by filtration. The filtrate is contacted with a ketoxime reagent (e.g., LIX84I, Cognis) to extract the nickel [14].

$$Ni(NH_3)_x^{2+} + \overline{2H(oxime)} \longrightarrow \overline{Ni(oxime)_2} + 2NH_4^+ + (x-2)NH_3$$

$$(11.5)$$

However, this will also extract cobalt(II) under the same conditions and once extracted the cobalt is oxidized to cobalt(III) and cannot be stripped. Therefore, to avoid this problem the cobalt is oxidized prior to extraction. In this way, nickel can be selectively extracted from a feed solution ($9\,g\,dm^{-3}$ Ni and $0.3\,g\,dm^{-3}$ Co) leaving about $0.01\,g\,dm^{-3}$ Ni in the raffinate. Nickel is subsequently stripped from the loaded organic phase with strong ammonia. One of the main disadvantages of this flow sheet is that the cobalt(III) causes some oxidation of the reagent and a process of reoximation is required [14]. Also, ammonia transfers to the organic phase, which again represents an operating loss. However the introduction of solvent extraction raises the nickel content of the products to >99.5% nickel as compared to 85–90% nickel with the earlier flow sheet that did not use solvent extraction [13].

An alternative route to cobalt/nickel separation following ammonia leaching in the Caron process is to produce a mixed Co/Ni carbonate product. This can then be dissolved in sulfate media and processed using one of the acidic extractants described later [13].

11.3.2.1b   Acid Leaching

Acid leaching is usually carried out using sulfuric acid under pressure to dissolve the majority of the iron minerals and to release the cobalt and nickel. If this is carried out at 150–250°C, then the iron(III) is precipitated as haematite or jarosite, reducing the amount of iron in the leachate. The leachate produced from acid leaching contains higher concentrations of other elements that can cause problems in the solvent extraction circuit, especially calcium, magnesium, and silica. Silica is a particular problem because of the tendency to promote crud formation (see Chapter 8). Various flow sheets involving different reagent have been proposed to treat the leachate. The three main flow sheets will be considered in the following sections.

11.3.2.1c Bulong Process

The Bulong operation is situated near Kalgoorlie in Western Australia and in the year 2000 produced 9000 t/a nickel and 729 t/a cobalt. Following the pressure leach, the leachate is neutralized with lime to precipitate residual iron, chromium, and aluminum. The clarified filtrate is contacted with CYANEX 272, containing TBP to inhibit third phase formation, at pH 5.5–6.0 and 50°C to extract the cobalt, zinc, and manganese and maximize the rejection of nickel (Fig. 11.5). Nickel is then extracted from the raffinate with a carboxylic acid (e.g., Versatic 10) at pH 6.2–7.0. Coextracted calcium and magnesium are scrubbed using the loaded strip solution and report to the raffinate. The nickel is finally stripped from the organic phase with recycled electrolyte from the nickel electrowinning* circuit. Coextracted impurities are removed from the cobalt circuit initially by sulfide precipitation of cobalt, copper, and zinc from the loaded strip liquor. The sulfide cake is then redissolved in acid and the zinc extracted with DEHPA, the copper removed by resin ion exchange and the cobalt electrowon.

Problems associated with this flow sheet include cross contamination of the organic phases and aqueous solubility losses of the Versatic acid.

11.3.2.1d Cawse Process

Purification of the pressure acid leachate from treatment of the Cawse limonitic ore involves the precipitation of cobalt and nickel hydroxides with magnesium oxide following prior neutralization with limestone to remove iron and other impurities (Fig. 11.6). The hydroxide cake is selectively leached with ammonia, rejecting any iron, manganese, or magnesium but redissolving any copper and zinc in the precipitate. The cobalt is then oxidized to cobalt(III) prior to extraction, to prevent coextraction with nickel with the hydroxyoxime LIX 84I (Cognis). The nickel is stripped with dilute sulfuric acid and electrowon. Copper and some cobalt build up in the organic phase and these levels are controlled by including a bleed stream,

---

*Electrowinning is an electrolytic process where cathodic reduction is used to recover metal from an electrolyte using an inert anode. The process differs from electro-refining where the anodic reaction is dissolution of the metal, that is then reversed at the cathode.

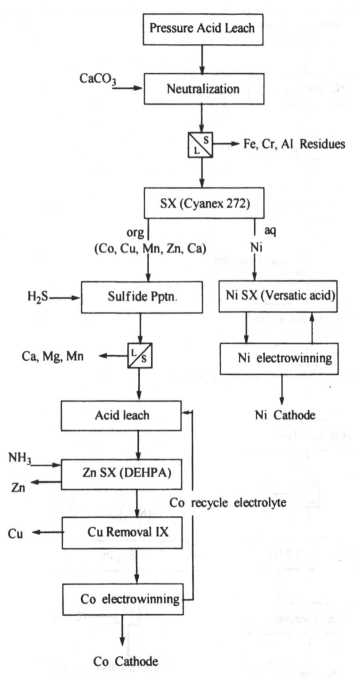

Fig. 11.5 Bulong process flow sheet (simplified).

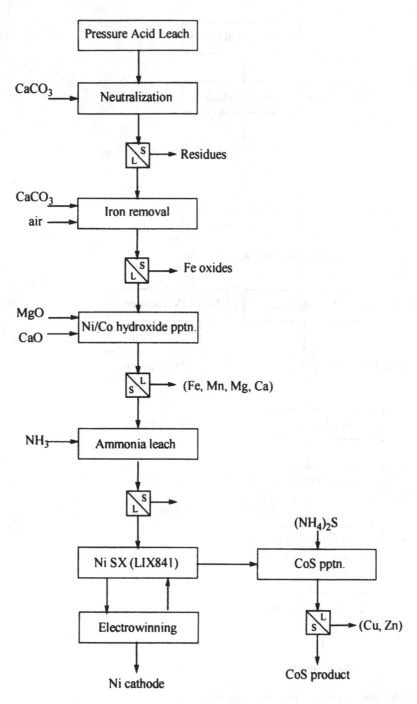

Fig. 11.6 Cawse process flow sheet (simplified).

from which the copper is removed with a more concentrated acid strip and cobalt by stripping in the presence of a reducing agent. One other major problem is the oxidative degradation of the oxime extractant by the cobalt(III) requiring the addition of a reoximation step to maintain extractant capacity. The cobalt in the raffinate is recovered by sulfide precipitation. The Cawse mine produced 8500 t/a nickel and 1900 t/a cobalt in 2000 at a cost in 1999 of $0.11 kg^{-1} after cobalt credits, the lowest cost production in the world.

11.3.2.1e Murrin-Murrin Process

The Murrin-Murrin mine is also near Kalgoorlie in Western Australia. This is the largest of the three projects, with a first-phase production of 45,000 t/a nickel and 3000 t/a cobalt. Future expansion has been suggested to 115,000 t/a nickel and 9000 t/a cobalt. The pressure acid leachate, following removal of the insoluble residue, is treated with hydrogen sulfide under pressure to precipitate nickel and cobalt and reject impurity elements. The sulfides are then dissolved in a pressure oxygen leach to produce a sulfate liquor that is treated with CYANEX 272. Initially zinc is removed at low pH, followed by a pH adjustment and extraction of cobalt (Fig. 11.7). Any coextracted nickel is scrubbed from the loaded organic phase with a dilute cobalt sulfate solution and the cobalt stripped with dilute sulfuric acid. Both the cobalt and nickel are recovered from solution as powders by hydrogen reduction that also produces ammonium sulfate as a by-product. The main problem in this flow sheet has been the crystallization of nickel ammonium sulfate because of the high nickel concentration (100 g dm^{-3}) in the feed to extraction. This severely reduced the production from the plant and the problem was solved by preloading the extractant with nickel prior to extraction by contact with the nickel sulfate raffinate.

11.3.3 Summary

Cobalt and nickel can be separated by several solvent extraction processes, the choice depending on the total flow sheet requirements, on the ratio of the metals in the feed solution, and on the required purity of the products.

The various systems can be summarized as follows:

1. Extraction of cobalt from chloride media by amines
2. Extraction of cobalt by organophosphorus acids, especially CYANEX 272, from acidic solutions
3. Extraction of nickel after prior oxidation to cobalt(III) using hydroxyoximes
4. Extraction of nickel by mixtures of simple oximes and carboxylic acids

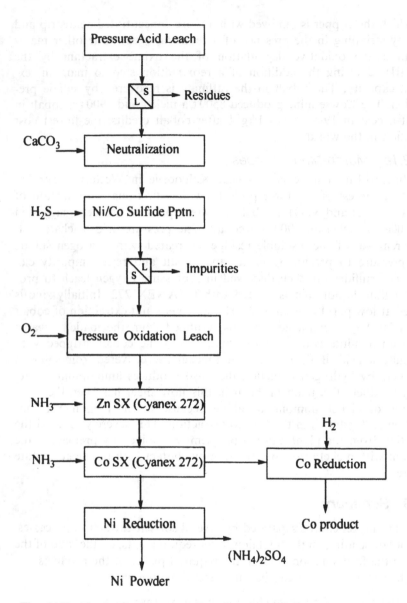

Fig. 11.7 Murrin-Murrin process flow sheet (simplified).

11.4 COPPER

There has been a long history of the extraction of copper by a hydro-metallurgical route, with the Rio Tinto process being operated in Spain from the mid-18th century. In the 19th century, the use of acid leaching of low-grade oxidized ores was practiced in Russia and the United States, with the resulting copper solution being reduced to metal by scrap iron. This process of leaching–cementation requires an adequate supply of acid and iron, and it produces only a low-grade copper powder that requires further treatment by smelting before marketing. The introduction of electrowinning of copper from acid solution provided a number of advantages over cementation [2] in that the product, cathode copper, had a much higher value than the cement copper and also, during electrowinning some of the sulfuric acid was regenerated for leaching. However, the demands placed on the feed liquor to electrowinning were severe enough to limit the intro-duction of the leach-electrowin route. Since high concentrations of other metals such as iron and magnesium have effects on the current efficiency and nature of the copper deposit, concentrations of these and several other elements should be as low as possible. This is difficult for iron and mag-nesium, in particular, because of their wide occurrence in mineral deposits. Solvent extraction provides a possible process whereby a pure feed to electrowinning could be obtained if a suitable extractant for copper was available.

Until recently, the favored leaching acid had been sulfuric acid, in which the metals exist as cationic species in solution. Examination of selectivity series for the common extractants available in the 1950s (Fig. 11.8) indicates that separation of copper from magnesium, for example, would be easily achieved by carboxylic or phosphoric acids, but with both these extractants, iron(III) would also be extracted with the copper. Although it would be possible to precipitate iron as the hydroxide by addition of alkali, this would complicate the process. However, as copper extraction does not occur below pH 3, the addition of alkali is necessary to achieve optimum extraction. This increases the overall costs of the extraction circuit and prevents the direct return of the extraction raffinate to leaching. Thus, it was obvious that to take advantage of the benefits of the process, new extractants that could selectively extract copper from iron at the pH of the leach liquor would be required. The search for such reagents was based on extractants used in analyti-cal chemistry, such as benzoin oximes, 8-quinolinol, and β-diketones (see Table 11.3). The use of these reagents in commercial operations is described individually.

11.4.1 Hydroxyoxime Extractants

The first commercial reagents designed specifically for copper extraction were an aliphatic α-hydroxyoxime (LIX63) and a β-hydroxyoxime based on benzophenone (LIX64). LIX63, with pH_{50} values of 3–4, was unable to extract copper from commercial leach liquors without alkali addition. Also, this reagent extracted iron(III) at lower pH values than copper. These disadvantages were largely overcome by LIX64, which was able to extract copper selectively from iron in the pH range 1.5–2.5, but with the disadvantage of a slow rate of extraction. This was solved by the addition of a small amount of LIX63. The mixture, LIX64N, for many years was the extractant of choice for commercial copper extraction from acid solution. The extracted complex had a 2:1 oxime/metal stoichiometry with the two oxime molecules linked by hydrogen bonds to give an overall planar configuration (Fig. 11.9). The copper could be stripped with sulfuric acid (150–200 g dm^{-3}), thus interfacing with spent electrolyte from the electrowinning

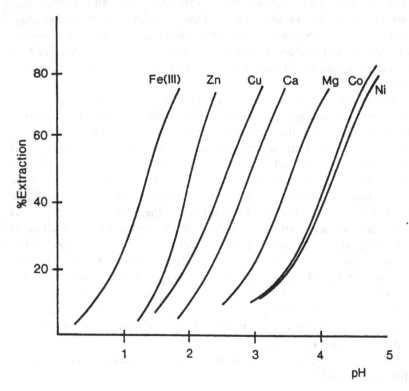

Fig. 11.8 Extraction of metals from sulfate media by DEHPA. (From Ref. 4.)

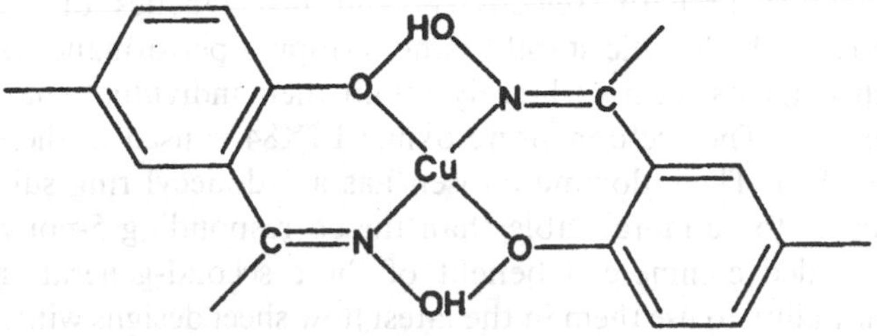

Fig. 11.9 Structure of *bis*(5-methylsalicyldoximato)copper(II).

circuit. Because of the commercial success of these reagents, other manu-facturers produced modifications of the basic β-hydroxyoxime structure with slightly different characteristics. Accordingly, chlorination of the benzene ring gave a reagent (LIX70) capable of extracting copper at pH 1, but re-quired concentrated acid ($> 300\,\mathrm{g\,dm}^{-3}$) for stripping. Other developments centered on modification of the oximic side chain from phenyl (LIX64, LIX65N) to methyl (SME529 Shell) and hydrogen (P50 ICI/Acorga), which varied both equilibrium and kinetic performance. The importance of these reagents in the development of large-scale commercial operations is described in several texts [1–3]. These references also contain detailed descriptions of some of the commercial operations.

More recently, a series of second-generation commercial reagents have become available that have been designed to produce a solvent phase to match the requirement of particular feedstocks or equipment. It is now generally accepted that the aldoximes, based on salicylaldehyde oxime, have a number of advantages in terms of fast extraction kinetics and high extractant strength, allowing the extraction of copper from highly acidic leachate with good selectivity over iron(III). However, increasing the ex-tractant strength increases the acid strength required for stripping, which reduces the overall copper transfer when interfaced with a normal tank-house spent electrolyte. Two approaches have been proposed to overcome these difficulties. Avecia with their Acorga* reagents have showed that addition to the hydroxyoxime of nonyl phenol (P5000 series) or tridecanol (PT5050) improves the stripping performance, with little effect on extrac-tion. More recently, other reagent modifiers, such as long-chain alkyl esters, have been added to P50 to produce the M series (Acorga), which are claimed to show improved commercial performance by reduction of entrainment and crud formation [16,17]. An alternative to this approach developed by Cognis involves a 1:1 mixture of an aldoxime and ketoxime, thereby

---

*Acorga is a trademark of Avecia, previously ICI, now owned by Cytec (2003).

combining the extractive strength, selectivity, and fast kinetics of the aldoxime with the proved hydrolytic stability and stripping performance of the ketoximes. Both reagents seem to largely retain their individual characteristics in the mixture. The acetophenone oxime LIX84 is used in these mixtures (see Table 11.3). The aldoxime chosen has a 5-dodecyl ring substituent that is reported to be more stable than the corresponding 5-nonyl derivative [18]. An added commercial benefit of these second-generation reagents has been the ability to use them in the latest flow sheet designs with a general change of circuit configuration from the three extraction–two strip of the 1970s, to either a two extraction–two strip for copper feed concentrations $>3 \, \mathrm{g \, dm^{-3}}$ or two extraction–one strip for feeds 0.5–$3 \, \mathrm{g \, dm^{-3}}$. This change has gradually occurred as operators have gained experience in running large extraction circuits and have become more concerned over capital expenditure than running costs. Table 11.5 shows the 2001 worldwide production capacity of copper solvent extraction plants, and approximately 30% of copper production now uses the leach–solvent extraction–electrowinning process. It is interesting to note the number of plants with a capacity $>100,000$ tons that have been constructed in Chile since 1990. The total production capacity in Chile of 1,430,000 tons represents 56% of the world's capacity for leach–solvent extraction–electrowon copper.

An alternative circuit for copper extraction involves ammoniacal leaching, and hydroxyoximes have been studied for this application [2]. However, because of their tendency to extract ammonia, these reagents have been superceded by β-diketones.

11.4.2 Quinoline-Based Extractants

A 7-alkyl-substituted derivative of 8-quinolinol (Kelex 100, Schering; see Table 11.3) was also designed for the selective extraction of copper from iron(III) in acid media. Here, the separation was based on the slow extraction of iron(III) relative to copper, presumably related to steric hindrance induced by substitution in the 7-position. With the short contact times and low pH values found in copper extraction circuits, separation factors >1000 were obtained. If the extraction was run to equilibrium, then iron(III) was preferred to copper. The reagent had an advantage over hydroxyoximes in its ability to extract copper from high-concentration, high-acid feeds, and it did not require high acid concentrations for stripping. However, unfortunately the quinolinic nitrogen atom is readily protonated and thus takes up acid on stripping, which either would be transferred to the extraction circuit or would have to be removed by a water wash, thereby complicating the circuit. Therefore, it is unlikely that Kelex 100 would be used for commercial copper extraction.

Table 11.5 Production Capacity of Acid Leach Copper SX-EW Plants (2001)[a]

Name / location	Leaching method	Capacity (tons / year)	Startup
USA			
Nord Copper Johnson Camp	Heap (oxide)	3,000	1990
Arimetco, NV[b]	Heap (oxide)	2,500	1989
Asarco Ray, AZ	Dump/Heap (oxide/sulf)	41,000	1979
Asarco Silver Bell	Heap/Dump (oxide/sulf)	20,000	1997
BHP, Miami, AZ	In situ/Agit. (sulf/tails)	12,500	1976
BHP, Pinto Valley, AZ	Dump (oxide/sulf)	8,750	1981
BHP, San Manuel	Heap/In situ (oxide)	18,000	1986
Burro Chief, Tyrone, NM	Heap/Dump (oxide/sulf)	71,500	1984
Cyprus Bagdad. AZ	Dump (oxide/sulf)	14,000	1970
Cyprus Miami, AZ	Heap (oxide/sulf)	76,000	1979
Cyprus Sierrita, AZ	Dump/Heap (oxide/sulf)	26,000	1987
Cyprus Tohono, AZ[b]	In situ (oxide)	21,900	1981
Phelps Dodge, Chino, NM	Heap (oxide/sulf)	59,700	1988
Phelps Dodge, Morenci, AZ	Heap/Dump (oxide/sulf)	355,000	1987
Equitorial Mineral Park, AZ	Dump (oxide/sulf)	2,500	1992
Equatorial Tonopah[b]	Heap (sulfide)	26,000	2000
Peru			
Cyprus Cerro Verde	Heap (oxide/sulf)	73,000	1977
Southern Peru (Toquepala)	Dump/Heap (oxide/sulf)	56,000	1995
Centromin	Mine Water (oxide/sulf)	5,000	1980
Mexico			
Cananea I and II, Mexico	Dump (sulfide)	40,000	1980
La Caridad, Mexico	Dump (oxide/sulf)	24,000	1994
Canada			
Gibraltar, Canada[b]	Dump (oxide/sulf)	5,475	1986
Africa			
ZCCM, Zambia	Agitation (oxide/tails)	90,000	1974
Sanyati Project (Reunion)	Heap (oxide)	3,500	1995
Bwana Mkubwa	Agitation (tails)	10,000	1998
Australia			
BHAS	Agitation (sulfide)	4,380	1984
Girilambone	Heap (sulfide)	20,000	1993
Gunpowder[c]	Heap (sulfide)		1990
Mt. Gordon	Pressure (sulfide)	45,000	1998
Olympic Dam	Agitation (oxide)	21,900	1989
Nifty	Heap	16,500	1993
Cloncurry[b]	Heap (oxide/sulf)		1996
Murchison United (Mt.Cuthbert)	Heap (oxide/sulf)	5,500	1996

(continued)

Table 11.5   (*continued*)

Name / location	Leaching method	Capacity (tons / year)	Startup
Chile			
Andacollo, Chile	Heap (oxide/sulf)	20,000	1996
Punta Del Cobre (Biocobre)	Heap (oxide/sulf)	8,000	1993
Cerro Colorado	Heap (oxide/sulf)	100,000	1994
Cerro Dominador	Heap (oxide)	1,000	1999
Chuquicamata Ripios Plant	Dump/Vat (oxide)	171,000	1988
Chuquicamata LGS	Dump (sulfide)	20,000	1993
Collahuasi	Heap (oxide)	50,000	1998
Dos Amigos, Chile	Heap (sulfide)	10,200	1997
El Abra, Chile	Heap (oxide)	220,000	1996
El Soldado	Heap (oxide)	7,500	1992
El Salvador I and II	Heap (oxide/sulf)	25,000	1994
El Teniente	In situ (sulfide)	5,000	1986
Escondida Oxide	Heap (oxide)	125,000	1998
Ivan Zar	Heap (oxide/sulf)	10,000	1995
Lomas Bayas	Heap (oxide)	60,000	1998
Los Bronces	Heap (oxide)	5,000	1998
Mantoverde	Heap (oxide)	42,000	1995
Michilla Lince I and II	Heap (oxide)	50,000	1992
Quebrada Blanca	Heap (sulfide)	75,000	1994
Radomiro Tomic	Heap (oxide)	260,000	1997
Santa Barbara, Chile	Vat (oxide)	42,000	1995
Tocopilla	Heap (oxide)	3,500	1984
Zaldivar	Heap (oxide/sulf)	120,000	1995

[a]Data kindly supplied by Dr. G. A. Kordosky (Cognis).
[b]Shutdown.
[c]Partially incorporated into Mt. Gordon.

Another commercial reagent based on quinoline is LIX34, which is a derivative of 8-alkylarylsulfonamidoquinoline. It has good extractant and stripping performance and, unlike Kelex 100, does not protonate easily. However, the cost of production outweighed these advantages.

11.4.3   β-Diketone Extractants

Although β-diketones cannot compete with hydroxyoximes for the extraction of copper from acid media, they do have advantages for extraction from ammoniacal leach solutions because, unlike hydroxyoximes, they do not transfer ammonia. The extraction follows Eq. (11.6), and the extent of extraction depends on the concentration of ammonia and ammonium salts.

$$Cu(NH_3)_4^{2+} + \overline{2HA} \longleftrightarrow \overline{CuA_2} + 2NH_4^+ + 2NH_3 \qquad (11.6)$$

Kinetics of extraction are fast and stripping is possible with solutions of low acidity and high copper concentration, as found with spent tankhouse liquors. A process using the reagent LIX54 for the recovery of copper from printed circuit board etch liquors has been very successful commercially (see Chapter 14). With these reagents, a higher copper transfer is achieved than with the hydroxyoximes, as illustrated by a tenfold reduction in solvent extraction plant size when changing from LIX64 to LIX54 for a leachate from a lead dross process. Some problems have been found with the long-term use of LIX54 in ammoniacal circuits due to the formation of a ketimine [Eq. (11.7)] that reduces the capacity of the LIX54 and could not be adequately removed by a normal clay cleanup treatment. Only one compound was identified, suggesting that the reaction could be inhibited by steric hindrance about the carbonyl group. Synthesis of compounds where the α-hydrogen atoms on the C_7H_{15} group were substituted by alkyl groups removed this problem.

$$C_6H_5COCH_2COC_7H_{15} + NH_3 \longrightarrow C_6H_5COCH_2C(NH)C_7H_{15} \quad (11.7)$$

11.4.4 Nitrogen-Based Extractants

Recent interest in chloride hydrometallurgy in the processing of complex sulfide ores has resulted in the development of a reagent to remove copper from such leachates. Here, there is a requirement for an extractant to operate without pH control and with a high selectivity over a wide range of metals and metalloids. Ion-pair extractants, the normal choice for chloride media [see Eq. (11.2)], are not suitable for this application, as copper has a lower tendency to form chloroanions than other elements in the leachate, such as iron(III), zinc, and lead. A reagent based on a pyridine carboxylic ester derivative operates as a solvating reagent for $CuCl_2$, according to Eq. (11.8).

$$CuCl_2 + \overline{2B} \longleftrightarrow \overline{CuCl_2B_2} \qquad (11.8)$$

Unlike the amine extractants, this compound does not protonate easily and provides excellent selectivity over those metals and metalloids that exist as chloroanions in this medium. Extraction of the latter species becomes significant only above 8 mol dm^{-3} chloride, as protonation becomes significant. This extractant [Acorga CLX 50], which contains 50% of the active ingredient, has been proposed for an integrated leach, extraction, electrowinning circuit (Cuprex process) [19].

11.5 PRECIOUS METAL EXTRACTION

The traditional route for the extraction and separation of precious metals consists of a series of selective dissolution–conditioning and precipitation steps that are generally inefficient in terms of the degree of separation that can be achieved. Even when conditions allow the complete precipitation of the desired element from solution, the precipitate will still be contaminated by entrained mother liquor, thereby requiring extensive washing, which generates wash streams containing small amounts of these valuable elements. Consequently, the traditional route required multiple precipitation and redissolution steps to provide the desired product purity, which, with the consequent wash streams, involved increased recycling and retreatment. This is not only very labor-intensive, but it also locks up considerable amounts of these elements, which, in turn, places an economic constraint on the whole operation. In some circumstances, a year can elapse between dissolving a sample of scrap at the start of the flow sheet and the emergence of the pure elements for marketing. During this time, the value of the elements is locked up in the plant and, hence, becomes a financial liability on the operation.

To overcome some of these problems and to produce a flow sheet that could accept a wide range of feed materials, the major refining companies embarked on research programs that could replace the traditional routes. These have led to the production of three overall flow sheets that reflect the chemical complexity of these elements and the difference in the raw material feed to the different refineries. To understand these schemes, it is necessary to outline some of the fundamental chemistry of the precious metals.

These elements are noble metals and, as such, can be dissolved only with great difficulty. The usual leaching agent is hydrochloric acid, with the addition of chlorine to increase the solution oxidation potential. This strong chloride medium results in the almost exclusive formation of aqueous chloroanions, with, under certain circumstances, the presence of some neutral species. Very seldom are cationic species formed in a chloride medium. However, these elements do possess a range of easily accessible oxidation states and, with the possibility of a number of different anionic complexes that are dependent on the total chloride concentration, this provides a very complicated chemistry. A summary of the most important chloro complexes found in these leach solutions is given in Table 11.6, from which the mixed aquochloro and polynuclear species have been omitted. The latter are found especially with the heavier elements.

The chloro species also differ in their rates of substitution, which, in general terms, are slower than those of the base metals and follow the order Au(I), Ag(I)$\gg$Pd(II) > Au(III) > Pt(II) > Ru(III)$\gg$Rh(III) > Ir(III) > Os(III)

Table 11.6 Common Chloro Species of the Precious Metals

Metal	Oxidation state	Major chloro species	Comments
Silver (Ag)	I	AgCl	Insoluble
		$AgCl_2^-$	High HCl concentration
Gold (Au)	I	$AuCl_2^-$	Very stable
	III	$AuCl_4^-$	
Platinum (Pt)	II	$PtCl_4^{2-}$	Conversion IV to II slow
	IV	$PtCl_6^{2-}$	Most common species, kinetically inert
Palladium (Pd)	II	$PdCl_4^{2-}$	Most common
	IV	$PdCl_6^{2-}$	Conversion II to IV difficult
Iridium (Ir)	III	$IrCl_6^{3-}$	Both species stable and
	IV	$IrCl_6^{2-}$	conversion IV to III easy
Rhodium (Rh)	III	$RhCl_6^{3-}$	
Ruthenium (Ru)	III	$RuCl_6^{3-}$	Complex equilibria between III and IV
	IV	$RuCl_6^{2-}$	depends on redox potential and chloride concentration
Osmium (Os)	III	$OsCl_6^{3-}$	
	IV	$OsCl_6^{2-}$	Os(IV) more stable than Ru(IV)

$\gg$Ir(IV), Pt(IV), with the states from Rh(III) being termed *inert*. Thus, kinetic factors tend to be more important, and reactions that should be possible from thermodynamic considerations are less successful as a result. On the other hand, the presence of small amounts of a kinetically labile complex in the solution can completely alter the situation. This is made even more confusing in that the basic chemistry of some of the elements has not been fully investigated under the conditions in the leach solutions. Consequently, a solvent extraction process to separate the precious metals must cope with a wide range of complexes in different oxidation states, which vary, often in a poorly known fashion, both in kinetic and thermodynamic stability. Therefore, different approaches have been tried and different flow sheets produced.

It can be seen that the three commercial flow sheets (Figs. 11.10–11.12) do have common features. Hence, because gold and silver can be easily separated from the remaining precious metals, they tend to be removed from the flow sheet at an early stage. In addition, significant quantities of gold and silver exist, in either primary ores or waste materials that are not associated with the platinum group metals. Silver often occurs as a residue from the chloride leaching operation and so is removed first, followed by gold, which can be extracted as the tetrachloroauric(III) acid, $HAuCl_4$, by a solvating reagent.

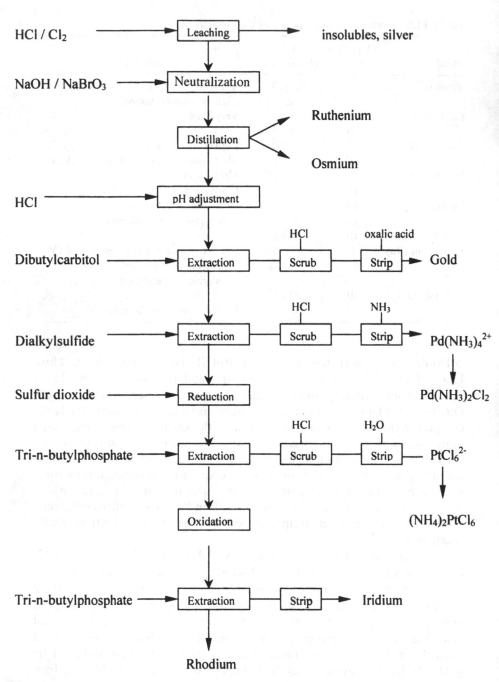

Fig. 11.10 INCO flow sheet for extraction of precious metals.

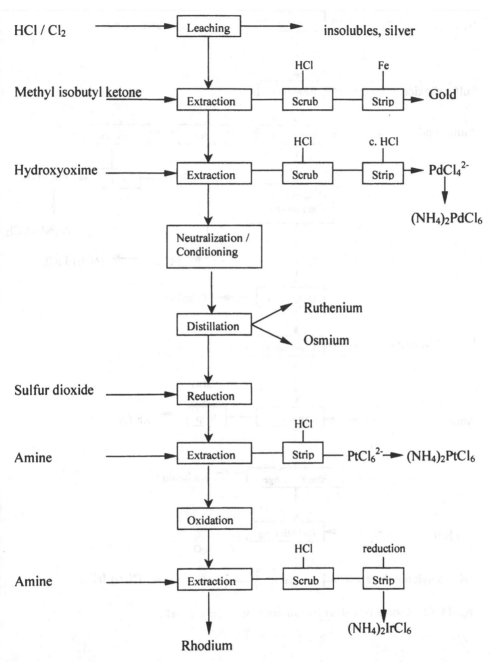

Fig. 11.11 Matthey Rustenberg flow sheet for extraction of precious metals.

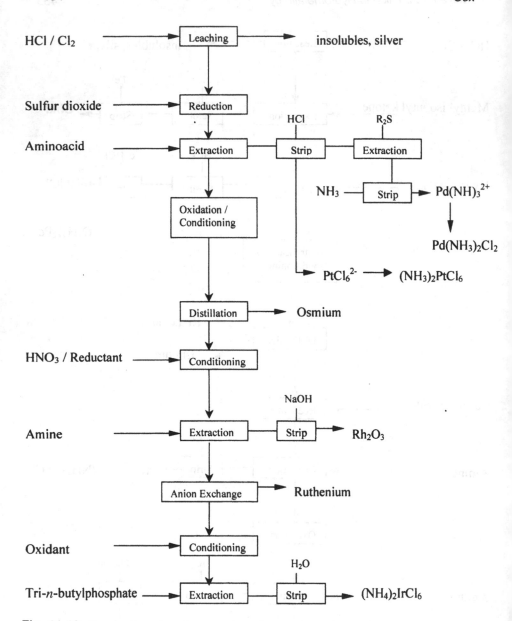

Fig. 11.12 Lonrho flow sheet for extraction of precious metals.

Notice that none of the flow sheets uses solvent extraction exclusively. Because the aqueous chemistry of osmium and ruthenium is very complex, most operators remove these elements by distillation of the tetraoxides, MO_4. Also, it has been advantageous to use ion exchange to separate and concentrate rhodium. The various extraction routes for individual elements are discussed in the following sections.

11.5.1 Gold

Gold exists in the leach solution as $AuCl_4^-$, and all the processes rely on the extraction of this ion or the parent acid by ion-pair or solvating extractants, respectively. The early studies by Morris and Khan [20] led to the adoption of dibutylcarbitol (diethyleneglycol dibutylether) by INCO in a gold refinery several years before the complete solvent extraction refining flow sheet was produced. The extractant has a high selectivity for gold (Fig. 11.13) and coextracted elements can be easily scrubbed with dilute hydrochloric acid [22]. Because of the high distribution coefficient, recovery of gold is best achieved by chemical reduction. Thus, oxalic acid at 90°C produces easily filtered gold grain. However, the slow kinetics necessitate a batch process. An alternative solvating extractant, 4-methyl-2-pentanone (methyl iso-butylketone; MIBK) is used by Matthey Rustenberg Refiners (MRR). This has a lower selectivity for gold (Fig. 11.14), so a simple acid scrub is not sufficient to remove impurities [22]. However, there is an advantage to the overall flow sheet in that some of the base elements, such as tellurium, arsenic, and iron, are removed at an early stage. Again, gold is recovered by chemical reduction with, for example, iron to give an impure product for subsequent refining. A third type of solvating extractant, long-chain alcohols have also been proposed, the most effective being 2-ethylhexanol, which can be easily stripped at ambient temperature with water [24]. Gold, like silver, being classified as a soft acid, should prefer interaction with soft donor atoms like sulfur. Therefore, it is no surprise to find that extraction with the trialkylphosphine sulfide, CYANEX 471X, is very efficient and provides excellent separation from elements such as copper, lead, bismuth, and other metalloids found in anode slimes [22]. Stripping can be achieved using sodium thiosulfate, which allows the separation of recovery and the reduction to the metal.

Amine salts can also be used to extract anionic species, but, in general, these are less selective for gold. However, Baroncelli and coworkers [25] have used trioctylamine (TOA) to extract gold from chloroanions of, for example, copper, iron, tin, and zinc, in aqua regia solution. The selectivity is attributed to the presence of $TOA \cdot HNO_3$ in the organic phase. Coextracted metals are scrubbed with dilute nitric acid, and gold is recovered by

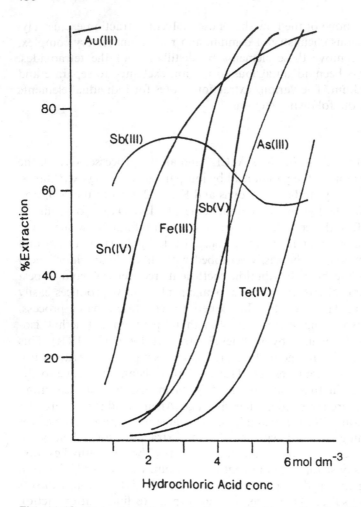

Fig. 11.13 Extraction of elements from a chloride medium by dibutylcarbitol. (From Ref. 21.)

a hydrochloric acid-thiourea solution. The same authors also used N,N'-di-n-butyloctanamide as a solvating extractant with aqua regia solutions. The protonated forms of Kelex 100 and LIX26, both acting as ion pair extractants, have been proposed by Demopoulos et al. [26] for gold extraction, with recovery as gold grain by hydrolytic stripping.

Alkaline cyanide solutions are another common lixiviant for gold, giving complex cyano species in solution. These anionic complexes may be extracted from aqueous solution by amines as ion pairs [27,28]. Recently, another extractant has been developed specifically to extract gold and silver from these media. The reagent, LIX79, is a trialkylated guanidine (Fig. 11.15) that can extract both metals rapidly from solutions within the approximate pH range 9–11.5 (gold) and 9–10 (silver). At higher pH values the metals are stripped from the organic phase as the guanidine becomes deprotonated [29,30].

Environmental concern over the use of cyanide leaching has stimulated the development of several alternative leaching processes. Of those tested, thiosulfate is the most promising, although the cost of this reagent is a disadvantage. Once again the gold is present in the leachate as an anionic

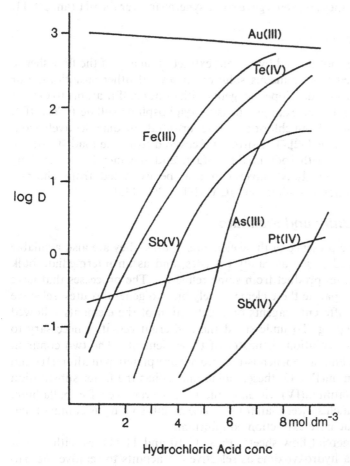

Fig. 11.14 Extraction of elements from chloride medium by MIBK. (From Ref. 23.)

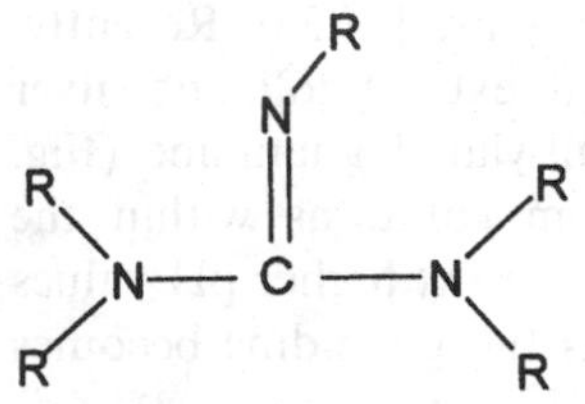

Fig. 11.15 Structure of LIX79.

complex, $[Au(S_2O_3)_2]^{3-}$, so extraction is possible with either amines, at pH < 8.0, or solvating reagents, e.g., TBP (pH > 10) [31]. It was also found that a mixture of these reagents gave some synergism over the pH range 9–11.

11.5.2 Silver

Although silver is not treated by solvent extraction in any of the flow sheets, silver is recovered from aqueous solution in several other situations. For these processes, Cytec developed reagents with donor sulfur atoms to extract this "soft" element. For example, tri-isobutylphosphine sulfide (CYANEX 471X) extracts silver from chloride, nitrate, or sulfate media selectively from copper, lead, and zinc [32]. The silver is recovered from the loaded organic phase by stripping with sodium thiosulfate, and the metal recovered by cementation or electrolysis. Silver can also be extracted from chloride solution by a dithiophosphinic acid (CYANEX 301) [33].

11.5.3 Palladium and Platinum

In addition to the presence of these elements in ores, they are also available from recycled feeds, such as catalyst wastes, and as an intermediate bulk palladium–platinum product from some refineries. The processes that have been devised to separate these elements rely on two general routes: selective extraction with different reagents or coextraction of the elements followed by selective stripping. To understand these alternatives, it is necessary to consider the basic solution chemistry of these elements. The two common oxidation states and stereochemistries are square planar palladium(II) and octahedral platinum(IV). Of these, palladium(II) has the faster substitution kinetics, with platinum(IV) virtually inert. However even for palladium, substitution is much slower than for the base metals so long as contact times are required to achieve extraction equilibrium.

Two commercial flow sheets (Figs. 11.10 and 11.11) use either di-*n*-octyl sulfide or a hydroxyoxime as selective extractants to remove the palladium, followed by tri-*n*-butylphosphate (TBP; INCO) and an amine

(MRR) to recover platinum. The extraction of palladium by the sulfide is slow with 2–3 hr to achieve equilibrium, which inhibits flow sheet design. However, it has been shown that tertiary amines can act as a kinetic accelerator, reducing equilibrium times to minutes, but no information has been published on the effect of these amines on palladium–platinum separation. High loading of the sulfides is possible, and stripping is easy with aqueous ammonia. The overall extraction and stripping equations are as follows:

$$PdCl_4^{2-} + \overline{2R_2S} \longleftrightarrow \overline{(R_2S)_2PdCl_2} + 2Cl^- \tag{11.9}$$

$$\overline{(R_2S)_2PdCl_2} + 4NH_3 \longleftrightarrow Pd(NH_3)_4^{2+} + \overline{2R_2S} + 2Cl^- \tag{11.10}$$

Neutralization of the strip solution with hydrochloric acid gives $Pd(NH_3)_2$-Cl_2 as product. One of the problems that has emerged is the formation of di-*n*-hexylsulfoxide [34] by oxidation of the sulfide. This may cause several problems including extraction of iron(III) that is strongly dependent on the HCl concentration. The iron can easily be stripped by water. There have also been indications of a buildup of rhodium in the extract phase that again can be explained by the extraction of anionic rhodium species by the sulfoxide. One benefit from the presence of the sulfoxide is that the rate of palladium extraction is increased by the presence of the protonated sulfoxide at high acidities; however, this kinetic enhancement is less that found with TOA · HCl, which remains protonated even at low acidities.

In this flow sheet, the aqueous raffinate from extraction is acidified to 5–6 mol dm^{-3} with hydrochloric acid to optimize platinum extraction by the solvating extractant TBP. The coextraction of iridium is prevented by reduction with sulfur dioxide, which converts the iridium(IV) to the (III) species, which is not extractable. Once again, kinetics are a factor in this reduction step because, although the redox potentials are quite similar, [Ir(IV)/(III) −0.87 V; Pt(IV)/(II) −0.77 V], iridium(IV) has a relatively labile d^5 configuration, whereas platinum(IV) has the inert d^6 arrangement. The species H_2PtCl_6 is extracted by TBP, from which platinum can be stripped by water and recovered by precipitation as $(NH_3)_2PtCl_2$.

The second selective extractant route (that of MRR) uses a hydroxyoxime to remove palladium. The actual compound used has not been specified, but publications refer to both an aliphatic α-hydroxyoxime and an aromatic β-hydroxyoxime. The α-hydroxyoxime LIX63 has the faster extraction kinetics, but suffers from problems with stripping. For the β-hydroxyoxime, a kinetic accelerator in the form of an amine (Primene JMT*) has been proposed. The precise mode of operation of this accelerator is unknown, but it may be a similar process to that proposed for the sulfide

---

*Trademark of Rohm and Hass.

system [i.e., a form of proton-transfer catalysis where the protonated amine extracts $PdCl_4^{2-}$, which then reacts with the hydroxyoxime in the organic phase: Eq. (11.11)].

$$PdCl_4^{2-} + \overline{R_3NHCl} \longleftrightarrow \overline{(R_3NH)_2PdCl_4} + \overline{2HA}$$
$$\longleftrightarrow \overline{PdA_2} + \overline{2R_3NHCl} + 2Cl^- \qquad (11.11)$$

Good loading of palladium and palladium–platinum separation factors of 103–104 are found. Stripping of the palladium is easy with hydrochloric acid, and $(NH_3)_2PdCl_4$ is obtained by precipitation. The raffinate from palladium extraction (see Fig. 11.10) is treated to remove osmium and ruthenium and, again, sparged with sulfur dioxide to reduce iridium(IV), following $PtCl_6^{2-}$ species removed by an amine. Stripping of the platinum is difficult because of the high-distribution coefficient, but techniques that have been successful include the use of strong acids, such as nitric or perchloric, to break the ion pair, or alkali to deprotonate the amine salt. The choice depends very much on the amine chosen; for example, if a quaternary amine is used, the deprotonation reaction is not available, and a strong acid has to be used. On the other hand, with other amines, the possibility of forming inner sphere complexes such as $Pt(RNH_2)_xCl_{(6-x)}^{+4-(6-x)-}$ is an important consideration, as these are nonextractible and cause a lockup of platinum in the organic phase. The formation of such complexes is enhanced by a deprotonation step that releases the free amine into the organic phase. As the ease of formation of platinum–amine complexes decreases in the order primary > secondary > tertiary, the reason for the choice of the latter for platinum extraction can be understood.

The phosphine sulfide CYANEX 471X has been proposed as an alternative to the foregoing reagents for palladium–platinum separation. Here, the faster extraction of palladium affords separation factors of 10:1, with a 10 min contact time. The chemistry of the process is similar to that of the alkyl sulfides. Various other separations have been published; for example, TBP will extract platinum(IV) chloroanions preferentially to palladium(II).

An alternative general process involves coextraction of palladium and platinum, followed by selective stripping (see Fig. 11.12). A novel amino acid extractant, made by the reaction between chloracetic acid and Amberlite* LA-2, a secondary amine, is used to extract the two elements from the leach liquor. However, in the given flow sheet, no selective stripping is employed, both elements being stripped by hydrochloric acid. The resulting chloroanions are then separated by using di-n-hexylsulfide to extract $PdCl_4^{2-}$.

---

*Trademark of Rohm and Hass.

Other coextraction-selective strip routes that have been proposed include the following:

1.   Extraction with a secondary amine, followed by stripping of the palladium by hydrazine, and bicarbonate to strip platinum by deprotonation of the amine ion pair:

$$\overline{(R_3NH_2)_2PdCl_4} + 2N_2H_4 \longleftrightarrow Pd(N_2H_4)_2Cl_2 + \overline{2R_3NH_2Cl}$$

$$(11.12)$$

$$\overline{(R_3NH_2)_2PtCl_6} + 2HCO_3^- \longleftrightarrow PtCl_6^{2-} + \overline{2R_3NH} + 2H_2O + 2CO_2$$

$$(11.13)$$

2.   Extraction with a tertiary amine, Alamine* 336, followed by palladium stripping with thiourea (Tu), and platinum with thiocyanate:

$$\overline{(R_3NH_2)_2PdCl_4} + 4Tu \longleftrightarrow Pd(Tu_2)Cl_2 + \overline{2R_3NHCl} \qquad (11.14)$$

$$\overline{(R_3NH)_2PtCl_6} + 2NCS^- \longleftrightarrow PtCl_6^{2-} + \overline{2R_3NHNCS} \qquad (11.15)$$

However, both of these above systems suffer from the possibility of lockup of metal in the organic phase, with the formation of unstrippable complexes, problems with stripping and metal recovery from the strip solutions.

3.   Extraction with 8-quinolinol derivatives, which can act as either a chelating acid or, after protonation, as an anion extractant. It was observed [35] that palladium extraction was favored at low acidity, but platinum was extracted best at high acidity (Fig. 11.16). At about $2\,mol\,dm^{-3}$ acid, the two metals were extracted equally. This suggested the following extraction equations:

$$PdCl_4^{2-} + \overline{2H_2ACl} \longleftrightarrow \overline{PdA_2} + 6Cl^- + 4H^+ \qquad (11.16)$$

$$PtCl_6^{2-} + \overline{2H_2ACl} \longleftrightarrow \overline{(H_2A)_2PtCl_6} + 2Cl^- \qquad (11.17)$$

The reaction to form the palladium complex is similar to that reported for amine salts, although here, because a bidentate chelating ligand is used, no chlorine atoms are retained in the complex, and the system is easy to strip. Also, as both reactions involve initial ion pair extraction, fast kinetics are observed with 3–5 min contact time to reach equilibrium at ambient temperature. The extraction conditions can be easily adjusted in terms of acidity to suit any relative metal concentrations and, because the reagent is used in the protonated form, good selectivity over base metals, such as iron and copper,

---

*Trademark of Cognis.

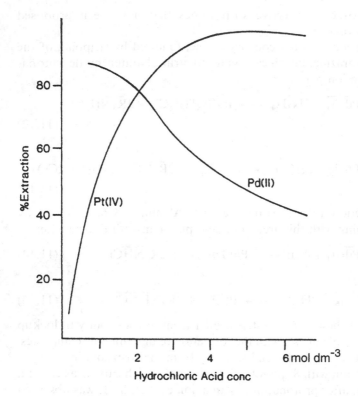

Fig. 11.16 Extraction of platinum(IV) and palladium(II) by LIX26 (5%), isodecanol (5%) in Solvesso (90%). (From Ref. 35.)

which must be extracted by the chelate mechanism, is found. Differential stripping is possible from the loaded organic phase by variation of the acidity of the strip solution; that is, dilute solutions to release platinum and more concentrated for palladium, and separation factors >200 can be obtained.

4. Finally, another development is the use of N-alkylamides that selectively extract platinum from strong acid media, presumably by ion pair formation between the protonated amide and the chloroanions:

$$\overline{RCONHR'} + H^+ \longleftrightarrow \overline{RCONH_2R'^+} \qquad (11.18)$$

$$\overline{2RCONH_2R'^+} + PtCl_6^{2-} \longleftrightarrow \overline{(RCONH_2R')_2PtCl_6} \qquad (11.19)$$

Stripping is easy with dilute acid or water, which deprotonates the amide and releases the metal anion. The process favors metals in high oxidation states [e.g., platinum(IV) or iridium(IV)], so that reduction of any iridium must precede the extraction of platinum and palladium, as with other systems. The platinum–palladium separation factors depend on the amide structure, but are rather small, about 25–55, so that scrubbing of the loaded organic phase is required to give a good separation [36].

11.5.4 Indium and Rhodium

In the commercial flow sheets, these elements are left in the aqueous raffinate after platinum and palladium extraction. Indium can be extracted in the +IV oxidation state by amines (see Fig. 11.11), or TBP (see Figs. 11.10 and 11.12). However, although the separation from rhodium is easy, the recovery of iridium may not be quantitative because of the presence of nonextractable iridium halocomplexes in the feed solution. Dhara [37] has proposed coextraction of iridium, platinum, and palladium by a tertiary amine and the selective recovery of the iridium by reduction to Ir(III). Iridium can also be separated from rhodium by substituted amides [S(Ir/Rh): 5×10^5).

Rhodium is recovered in all the examples by precipitation from the raffinate after ion exchange, but solvent extraction with an amine followed by thiourea stripping has been suggested [37]. The problem in these systems is that extraction is favored by low acidity and low chloride concentrations so under the conditions found in the commercial flow sheets rhodium extraction would be low [38,39]. In addition, stripping from the organic phase is difficult with the possibility that the rhodium remains in the organic phase as the dimer, $Rh_2Cl_9^{3-}$. Extraction with 8-hydroxyquinoline extractants has also been studied. Here extraction increased with acidity and chloride concentration to a maximum of 40%. This is explained by the formation of the $RhCl_6^{3-}$ species that is then extracted as an ion pair by the protonated 8-hydroxyquinoline in a similar way to palladium discussed earlier [40]. However, once again stripping was a problem. A critical review explaining these problems in more detail has been published [41].

11.6 LANTHANIDE EXTRACTION

The lanthanide group of elements (Table 11.7) is very difficult to separate by traditional methods because of their similar chemical properties. The techniques originally used, like the precious metals, included laborious multiple fractional recrystallizations and fractional precipitation, both of which required many recycle streams to achieve reasonably pure products. Such techniques were unable to cope with the demands for significant quantities of certain pure compounds required by the electronics industry; hence, other separation methods were developed. Resin ion exchange was the first of these

Table 11.7 Properties of Lanthanide Elements

Element	Atomic number	Atomic mass	Ionic radius (Ln^{3+}) (Å)	
Lanthanum (La)	57	138.9	1.061	
Cerium (Ce)	58	140.1	1.034	
Praseodymium (Pr)	59	140.9	1.013	
Neodymium (Nd)	60	144.2	0.995	Light
(Promethium (Pm)	61	—	0.979	earths
Samarium (Sm)	62	150.4	0.964	
Europium (Eu)	63	152.0	0.950	
Gadolinium (Gd)	64	157.3	0.938	Middle
Terbium (Tb)	65	158.9	0.923	earths
Dysprosium (Dy)	66	162.5	0.908	
Holmium (Ho)	67	164.9	0.894	
Erbium (Er)	68	167.3	0.881	Heavy
Thulium (Tm)	69	168.9	0.869	earths
Ytterbium (Yb)	70	173.0	0.858	
Lutetium (Lu)	71	175.0	0.848	
(Yttrium (Y)	39	88.9	0.88	

used commercially, but it suffered from the restrictions of low resin capacity that required large columns, and also the system had to use a batch operation. This process has now been largely superceded by solvent extraction. Both ion exchange and solvent extraction separations are based on the steady decrease in size across the lanthanide elements, which result in an increase in acidity or decrease in basicity with increasing atomic number. This causes a variation in formation coefficient of any metal–extractant complex, allowing preferential binding to an ion exchange resin, or extraction of the complex into an organic phase (see Chapters 3 and 4). The variations in formation coefficients and hence, separation factors, between adjacent elements, are small; so many equilibria and thus a large number of extraction units are required. Even then, separation is possible only by recycling both the exit streams. This configuration, rarely used outside lanthanide production, is essential to the preparation of pure products or the separation of adjacent elements. The operation is shown in Fig. 11.17. The bank of mixer-settlers is fed close to the midpoint, with a mixture of components, and by returning the exit streams to the process, 100% reflux is obtained. The components are force-fed to build up their concentration, and the reflux is continued until the desired degree of separation is achieved. Then, the operation is stopped and the individual mixer-settlers emptied to give the required products.

Figure 11.18 gives a typical separation of lanthanum, neodymium, and praseodymium from nitrate media, using TBP as extractant. The number of extraction units can be varied according to the products required; for example, lanthanum can be concentrated in the aqueous raffinate and neodymium in the organic extract in holding tanks external to the mixer-settlers, thereby decreasing the number of units required. It should be noted that yttrium, although not a member of the lanthanide series, often accompanies them in ores. The charge/size ratio of yttrium places it between dysprosium and holmium, although, depending on the extraction system, it can behave as either a light or heavy element.

The lanthanides occur in a number of ores, but as the relative concentration of individual elements differ according to the source, several extraction schemes have been developed to take advantage of this natural variability. To avoid confusion caused by such variations, this section will consider only the more general extraction schemes. The first operation in the flow sheets involves leaching of the ore and, in most cases, the leaching agent is one of the common mineral acids. The choice of extractant depends on the acid used; for example, the formation of nitrate complexes allows the use of solvating extractants (e.g., TBP), whereas the carboxylic or organophosphorus acids may be used with either hydrochloric, nitric, or sulfuric acid media. Occasionally, mixtures of two extractants will improve the separation of individual elements.

Lanthanides are also found as minor components in other ores, particularly in association with uranium or in phosphate rock. These are often coextracted with the major product and can be economically recovered from the waste streams resulting from the uranium or phosphoric acid extraction.

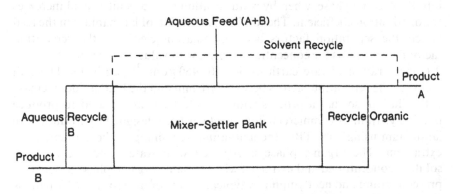

Fig. 11.17 Mixer-settler bank under reflux configuration.

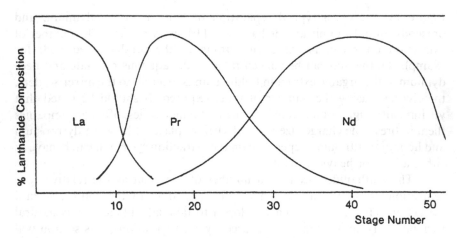

Fig. 11.18 Concentration profiles for mixer-settler bank under total reflux: system TBP (50%) from nitrate medium. (From Ref. 42.)

11.6.1 Solvating Extractants

The extraction of the lanthanides from nitrate media with TBP [Eq. (11.20)] depends on several factors including pH, nitric acid concentration, ionic strength, and concentration of the elements in the aqueous phase.

$$Ln(NO_3)_3 \cdot xH_2O + \overline{3TBP} \longleftrightarrow \overline{Ln(NO_3)_3 \cdot 3TBP} + xH_2O \qquad (11.20)$$

The situation is complicated by the ability of TBP to extract nitric acid as $HNO_3 \cdot xTBP$, which then competes with the lanthanides, so the process is optimized in terms of high pH, but low nitric acid concentration. Increased ionic strength has the advantage of salting out the aqueous phase complexes into the organic phase, whereby neutralization of excess nitric acid increases the distribution coefficient. The total concentration of lanthanides in the feed affects the separation factors between adjacent elements; the separation factor decreasing as the concentration rises, with values in the range $S = 0.7–2$ (concentration of rare earth oxides $60–460 \, g \, dm^{-3}$) (see Table 11.8). In contrast, the separation factors between yttrium and the lanthanides increase with dilution, so that a process with a dilute feed can be used to produce purified yttrium. A commercial flow sheet has been designed to produce pure lanthanum using 50% TBP, the lanthanum remaining in the raffinate after extraction. The organic phase is stripped with water, and the aqueous solution concentrated and extracted with the same organic phase to separate praseodymium and neodymium. A typical profile of the mixer-settler bank is shown in Fig. 11.18. Other solvating organophosphorus extractants have been studied to a small extent and these show some processing advantages with, for example, an organophosphonate showing higher separation factors than TBP.

11.6.2 Amine Extractants

Amines provide an alternative method of extraction from nitric acid media, extracting the nitrato-complexes of the lanthanides as ion pairs. Because of competition from nitric acid, the process has to operate above pH 2, usually with metal nitrates as salting out agents. The extraction also depends on the amine structure, with asymmetric compounds giving the highest distribution coefficients. Again, separation from yttrium is favored, with separation factors in the range of 5–20. An interesting process to produce pure yttrium from a mixed lanthanide feed depends on this element behaving as a heavy lanthanide in nitrate media and as a light lanthanide in thiocyanate solution. Thus, using a nitrate solution and an amine extractant, the light elements can be extracted from yttrium and the heavy lanthanides; then, using thiocyanate media, yttrium can be extracted from the remaining elements in the raffinate (Fig. 11.19).

11.6.3 Acidic Extractants

Both carboxylic (e.g., Versatic acid) and organophosphorus acids have been used commercially to extract the lanthanides. The extraction follows the formation of the metal–extractant complex [Eq. (11.21)] and so depends on the pH of the feed.

$$Ln^{3+} + \overline{3H_2A_2} \longleftrightarrow \overline{LnA_3(HA_3)_3} + 3H^+ \qquad (11.21)$$

Table 11.8 Separation Factors for Adjacent Lanthanides Using TBP from Nitrate Medium

Aqueous phase conc. (Ln_2O_3 g dm^{-3})	Separation factors					
	Sm/Nd	Gd/Sm	Dy/Gd	Ho/Dy	Er/Ho	Yb/Er
460	2.26	1.01	1.45	0.92	0.96	0.81
430	—	—	1.20	0.96	0.65	—
310	2.04	1.07	1.17	0.94	0.82	—
220	1.55	0.99	1.08	0.89	0.78	—
125	1.58	0.82	0.92	0.83	0.72	—
60	1.40	0.78	0.89	0.77	0.70	0.63

Source: Ref. 42.

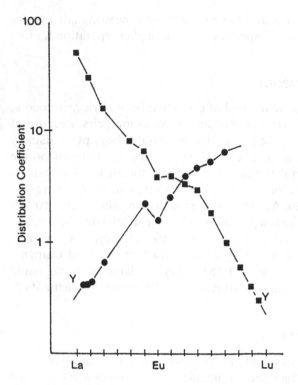

Fig. 11.19 Distribution coefficients for lanthanide extraction with Aliquat 336 from nitrate (solid squares) and thiocyanate (solid circles) media. (From Ref. 1.)

Organophosphorus acids, of which DEHPA has received the most attention, are more powerful extractants than Versatic acid and extract the heavy lanthanides even at low pH values. This does give problems with stripping, and the heaviest elements, ytterbium and lutetium, require 20% hydrofluoric or 50% sulfuric acid as strip solutions. Variation of pH thus allows the separation of heavy, including yttrium, from light elements. Separation factors between adjacent elements follow the order phosphinic ≫ phosphonic > phosphoric > carboxylic acids, although few data are available for the first two extractants (Table 11.9). In addition, with organophosphorus acids, higher separation factors are found from chloride than from sulfate or nitrate media. Thus, parameters that can be used to design a flow sheet include type and concentration of extractant, pH of the feed solution, nature of anion (organophosphorus acids), and nature and concentration of the strip solution.

Table 11.9 Separation Factors for Adjacent Lanthanides Using Acidic Extractants

	Ce/La	Pr/Ce	Nd/Pr	Sm/Nd	Eu/Sm	Gd/Eu	Tb/Gd	Dy/Tb	Ho/Dy	Er/Ho	Tm/Er	Yb/Tm	Lu/Yb
Versatic 911[a]	3.00	1.60	1.32	2.20	—	1.96	—	—	1.17	1.28	—	—	—
DEHPA[a]	—	—	1.40	4.00	2.30	1.50	5.70	2.00	1.80	2.40	3.40	—	—
DEHPA[b]	5.24	1.86	1.32	9.77	2.14	1.99	—	—	—	—	—	—	—
Phosphonic acid[b]	13.8	3.47	1.51	13.80	2.45	1.86	6.92	3.23	2.17	2.82	2.63	3.71	1.74
Phosphinic acid[b]	16.98	1.86	1.32	13.80	2.14	1.32	6.02	2.82	1.74	2.63	3.47	2.63	1.51

[a]Source: Ref. 42.
[b]Source: Ref. 43. (All organophosphorus extractants used in this work had 2-ethylhexyl substituents.)

11.6.3.1   Versatic Acid

Extraction takes place by a cation-exchange mechanism with maximum extraction occurring close to the pH of hydrolysis (i.e., about pH 6). No preference is shown for a particular anion in the feed solution, and separation factors are less dependent on the concentration of elements in the feed than for TBP. The high cerium–lanthanum separation factor indicates that Versatic acid would be a better extractant than TBP to produce pure lanthanum. The heavy elements are strongly extracted with yttrium lying between dysprosium and gadolinium. A process has been developed to produce pure yttrium by first using Versatic acid to remove the light elements in the raffinate. The extract is then stripped and the concentration of the strip phase adjusted so that, on reextraction with TBP, yttrium behaves as a light element, and separation from the remaining heavy elements is achieved. However, the small separation factors still require both extractions to be performed under total reflux in a large (56-stage) mixer-settler train.

11.6.3.2   Organophosphorus Acids

Di-2-ethylhexylphosphoric acid has received most attention, but both the organophosphonic and organophosphinic acids potentially offer some processing advantages. Organophosphorus acids are more powerful extractants than Versatic acid and, thus, extract at lower pH values. Consequently, a $0.1 \, mol \, dm^{-3}$ DEHPA solution completely extracts all the elements above europium at pH 2 and, as noted earlier, this makes the heavy elements difficult to strip. Although higher separation factors are found from chloride than from nitrate or sulfate media, increasing the acidity of the chloride solution decreases these values. A similar, but smaller, dependency of separation factors on acidity has also been found with sulfate systems.

Several different flow sheets have been developed around DEHPA, which, as discussed earlier, vary according to the precise composition of the feed material. A typical process using a nitric acid leach of a xenotime ore follows as an example.

The leach liquor is first treated with a DEHPA solution to extract the heavy lanthanides, leaving the light elements in the raffinate. The loaded reagent is then stripped first with $1.5 \, mol \, dm^{-3}$ nitric acid to remove the elements from neodymium to terbium, followed by $6 \, mol \, dm^{-3}$ acid to separate yttrium and remaining heavy elements. Ytterbium and lutetium are only partially removed; hence, a final strip with stronger acid, as mentioned earlier, or with 10% alkali is required before organic phase recycle. The main product from this flow sheet was yttrium, and the yttrium nitrate product was further extracted with a quaternary amine to produce a 99.999% product.

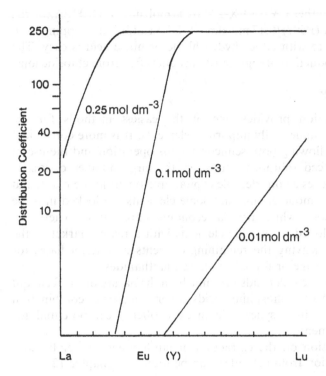

Fig. 11.20 Effect of DEHPA concentration on extraction of lanthanides. (From Ref. 2.)

Changing the concentration of extractant will give a different split of the lanthanides (Fig. 11.20); accordingly, the flow sheet can be modified to take advantage of variations in feed composition or to produce a particular element as product.

The rate of extraction of the lanthanides by the 2-ethylhexyl ester of 2-ethylhexylphosphonic acid (e.g., PC88A, P507) is slower than for DEHPA but the higher distribution coefficients allow extraction from more dilute solutions. Also, as acidic stripping is easier than with DEHPA, the reagent has been used commercially to separate lutetium from terbium and ytterbium.

Recently emphasis has shifted to the use of organophosphinic acids for the extraction of lanthanides [44] with CYANEX 272, providing significant advantages over other organophosphorus acid extractants for the heavy lanthanides with greater average separation factors and easier stripping. Furthermore, a 1:1 mixture of CYANEX 272 and a carboxylic acid gives significant synergism for the heavy lanthanides with higher loadings

($0.2\,\text{mol}\,\text{dm}^{-3}$) than either CYANEX 272 ($0.08\,\text{mol}\,\text{dm}^{-3}$), DEHPA, or the phosphonic acid ester (P507) ($0.17\,\text{mol}\,\text{dm}^{-3}$) [45]. In addition, stripping from the synergistic mixture with either hydrochloric or nitric acid is easy. The latter enables the production of high-purity products free from chloride ions.

11.6.4 Summary

Liquid–liquid extraction provides one of the easiest methods for the separation of these commercially important elements. It is more convenient than ion exchange, allowing both semicontinuous operation and the use of more concentrated feed solutions. However, the large number of mixer-settlers required imposes a considerable capital investment in the plant, and their use in the reflux mode entails that some elements are locked into the process for long periods, which also has economic implications. Therefore, flow sheets are developed to provide elements with a ready market at the earliest opportunity, leaving the remaining elements as intermediates for separation at a later stage or for sale as mixed lanthanides.

The choice of reagent depends very much on the nature of the feed, but the introduction of organophosphinic acids either alone or in combination with carboxylic acids in a synergistic mixture offer great potential for commercial development.

Further discussion on the extraction of lanthanides, especially concerning their separation from actinides, can be found in Chapter 12.

11.7 CONCLUSION

Solvent extraction is now a proven technology for the commercial extraction, separation, and concentration of a wide range of metals both from primary and secondary sources (see Chapter 14). In recent years, there has been a reduction in the development, production, and marketing of new commercial extractants as the overall costs of such activities increases. However, the use of established reagents in new hydrometallurgical applications continues to expand.

REFERENCES

1. Lo, T. C.; Baird, M. H. I.; Hanson, C. *Handbook of Solvent Extraction*; Wiley-Interscience: New York, 1983.
2. Ritcey, G.; Ashbrook, A. W. *Solvent Extraction*, Part 2; Elsevier; Amsterdam, 1979.
3. Ritcey, G.; Ashbrook, A. W. *Solvent Extraction*, Part 1; Elsevier; Amsterdam, 1984.
4. Cox, M.; Flett, D. S. Chem Ind., **1987**, p. 188.

5. Dalton, R. F.; Price, R. *Proceedings of Oslo Symposium 1982*; Society of Chemical Industry: London, paper 11–23, 1982.
6. Hudson, M. Hydrometallurgy, **1982**, *9*, 149.
7. Marcus, Y.; Kertes, A. S. *Ion Exchange and Solvent Extraction of Metal Complexes*; Wiley-Interscience: New York, 1969.
8. Sekine, T.; Hasegawa, Y. *Solvent Extraction Chemistry*; Marcel Dekker: New York, 1977.
9. Preston, J. S. Hydrometallurgy, **1983**, *10*, 187.
10. Preston, J. S. Hydrometallurgy, **1983**, *11*, 105.
11. Barnes, J. E.; Stetchfield, J. H.; Williams, G. O. R. J. Inorg. Nucl. Chem., **1976**, *38*, 1065.
12. Nishimura, S. *Extraction Metallurgy '81*; Institute of Mining and Metallurgy: London, p. 404, 1981.
13. Sole, K. C.; Cole, P. M. *Ion Exchange and Solvent Extraction*; Marcus, Y. and Sengupta, A. K. Eds.; Marcel Dekker: New York, 2002; Vol. 15, 143.
14. Mackenzie, J. M. W.; Virnig, M. J.; Boley, B. D.; Wolfe, G. A. *ALTA 1998 Nickel Cobalt Pressure Leaching and Hydrometallurgy Forum*, Melbourne; ALTA Metallurgical Services, 1988.
15. Flett, D. S.; J. Chem. Tech. Biotechnol. **1999**, *74*, 99.
16. Dalton, R. F.; Seward, G. W. *Proceedings of Reagents in the Minerals Industry*, Jones, M. J. and Oblatt, R., Eds., Institute of Mining and Metallurgy: London, 1984, 106.
17. Dalton, R. F.; Seward, G. W. *Proceedings ISEC '86*; Munich; DECHEMA: Frankfurt-am-Main, *2*: 11, 1986.
18. Kordosky, G. A.; Olafson, S. M.; Lewis, R. G.; Deffner, V. L.; House, J. E. Sep. Sci. Technol., **1987**, *22*, 215.
19. Dalton, R. F.; Price, R.; Henmana, E.; Hoffmann, B. *Separation Process in Hydrometallurgy*, Davies, G. A. Ed.; Ellis-Horwood: Chichester; 1987, 466.
20. Morris, D. F. C.; Khan, M. A. Talanta, **1968**, *75*, 1301.
21. Rimmer, B. Chem. Ind. **1974**, 63.
22. Soldenhoff, K.; Wilkins, D. *Solvent Extraction for the 21st Century, Proc. ISEC'99*; Society of Chemical Industry: London, 2001, *1*, 777.
23. Cleare, M. J.; Charlesworth, P.; Bryson, D. J. J. Chem. Tech. Biotechnol., **1979**, *29*, 210.
24. Grant, R. A.; Drake, V. A. *Proc ISEC2002*, Cape Town; 2002, 940.
25. Baroncelli, F.; Carlini, D.; Gasparini, G. M.; Simonetti, E. *Proceedings ISEC '88*, Moscow; Vernadsky Inst.; Academy of Sciences USSR: Moscow, 1988; Vol. 4, 236.
26. Demopoulos, G. P.; Pouskouleli, G.; Ritcey, G. M. *Proceedings ISEC '86*, Munich; DECHEMA: Frankfurt-am-Main, 1986; Vol. 2, 581 pp.
27. Villaescusa, I.; Miralles, N.; de Pablo, J.; Salvado, V.; Sastre, A. M.; Solv. Extn Ion Exch. **1993**, *11*, 613.
28. Argiropoulos, G.; Cattrall, R. W.; Hamilton, I. C.; Paimin, R. *Value Adding Through Solvent Extraction, Proc. ISEC'96*; University of Melbourne, 1996; Vol. 1, 123 pp.

29. Kordosky, G. A.; Sierokoski, J. M.; Virnig, M. J.; Mattison, P. L. Hydrometallurgy; **1992**, *30*, 291.
30. Virnig, M. J.; Kordosky, G. A. *Value Adding Through Solvent Extraction, Proc. ISEC'96*; University of Melbourne, 1996; Vol. 1, 311 pp.
31. Zhao, Jin; Wu, Zhichun; Chen, Jiayong. *Value Adding Through Solvent Extraction, Proc. ISEC'96*; University of Melbourne, 1996; Vol. 1, 629 pp.
32. Capela, R. S.; Paiva, A. P. *Proc. ISEC2002*; Cape Town, 2002; 335 pp.
33. Facon, S.; Rodriguez, M. Avila; Cote, G.; Bauer, D. *Solvent Extraction in the Process Industries, Proc ISEC93*; Elsevier Applied Science, 1993, Vol. 1, 557 pp.
34. Du Preez, A.; Preston, J. S. *Proc. ISEC2002*; Cape Town, 2002, 896 pp.
35. Pouskouleli, G.; Kelebek, S.; Demopoulos, G. P. *Separation Processes in Hydrometallurgy*; Davies, G. A. Ed.; Ellis-Horwood: Chichester, 1987, 175 pp.
36. Grant, R. A.; Murrer, B. A.; European Patent No. 0210004, 1986.
37. Dhara, S. C. *Proceedings of Precious Metals Mining, Extraction, and Processing*; Kudryk, V. Corrigan, D. A. Laing, W. W. Eds.; Metallurgical Society AIME, 1984, 199 pp.
38. Dolgikh, V. I.; Borikov, P. I.; Borbat, V. F.; Kouba, E. F.; Gindin, L. M. Tsvet Metal. **1967**, *40*, 30.
39. Fedorenko, N. V.; Ivanova, T. I.; Russ. J. Inorg. Chem., **1965**, *10*, 387.
40. Benguerel, E.; Demopoulos, G. P.; Cote, G.; Bauer, D. Solv. Extn. Ion Exch. **1994**, *12*, 497.
41. Benguerel, E.; Demopoulos, G. P.; Harris, G. B. Hydrometallurgy, **1996**, *40*, 135.
42. Brown, C. G.; Sherrington, L. G. J. Chem. Technol. Biotechnol., **1979**, *29*, 193.
43. Preston, J. S.; du Preez, A. C. Mintek Report M378; Council for Mineral Technology: Randburg, South Africa, 1988.
44. Li, D. *Solvent Extraction for the 21st Century, Proc ISEC'99*; 2000; Vol. 2, 1081 pp.
45. Yuan, C.; Ma, H.; Pan, C.; Rickelton, W. A.; *Value Adding Through Solvent Extraction, Proc ISEC'96*; 1996, Vol. 1, 733 pp.

BIBLIOGRAPHY

In addition to the list at the end of Chapter 1, texts concerned with applications of solvent extraction in hydrometallurgy include the following:

Burkin, R. (ed.). Critical Reviews in Applied Chemistry, Vol. 17; *Extractive Metallurgy of Nickel*, John Wiley and Sons: Chichester, 1987.
Chapman, T.W. Extraction—metals processing. In *Handbook of Separation Process Technology*, Chap. 8 Rousseau, R.W. Ed.; John Wiley and Sons: New York, 1987, 467.
Murthy, T. K. S.; Koppiker, K. S.; Gupta, C. K. Solvent extraction in extractive metallurgy. In *Recent Developments in Separation Science*;

Li, N.N. and Navratil, J.D. Eds.; CRC Press: Boca Raton, Florida, 1986, Vol. 8, 1 p.

Osseo-Asare, K.; Miller, J. D. Eds. *Hydrometallurgy; Research, Development and Plant Practice*; Metallurgical Society AIME; New York, 1982.

Proceedings of Extraction Metallurgy Conferences, published by Institution of Mining and Metallurgy, London, 1981, 1985, 1989.

Schulz, W. W.; Navratil, J. D. Eds. *Science and Technology of Tributyl Phosphate*, Vols. 1 and 2; CRC Press: Boca Raton, Florida, 1987.

Szymanowski, J. *Hydroximes and Copper Hydrometallurgy*; CRC Press: Boca Raton, Florida, 1993.

Yamada, H.; Tanaka, M. Solvent extraction of metal carboxylates. Adv. Inorg. Chem. Radiochem. **1985**, *29*, 143.

12

Solvent Extraction in Nuclear Science and Technology

CLAUDE MUSIKAS* Commissariat à l'Energie Atomique, Paris, France

WALLACE W. SCHULZ Consultant, Albuquerque, New Mexico, U.S.A.

JAN-OLOV LILJENZIN* Chalmers University of Technology, Göteborg, Sweden

12.1 INTRODUCTION

In the year 2000, 15% of the world's electric power was produced by 433 nuclear power reactors: 169 located in Europe, 120 in the United States, and 90 in the Far East. These reactors consumed 6,400 tons of fresh enriched uranium that was obtained through the production of 35,000 tons of pure natural uranium in 23 different nations; the main purification step was solvent extraction. In the reactors, the nuclear transmutation process yielded fission products and actinides (about 1000 tons of Pu) equivalent to the amount of uranium consumed, and heat that powered steam-driven turbines to produce 2,400 TWh of electricity in 2000.

Different countries have adapted various policies for handling the highly radioactive spent fuel elements:

1. in the *once through cycle*, the fuel elements are stored in temporary facilities, usually with the intention to finally deposit them in underground vaults; and
2. in the *recycling concept*, the spent fuel pins are dissolved in acid and "reprocessed" to recover energy values (i.e., unused uranium and other fissionable isotopes, notably plutonium), while the waste is solidified and stored in underground vaults.

In some countries, the main purpose of reprocessing is to recover plutonium for weapons use. The main separation process in all known reprocessing plants is solvent extraction.

---

*Retired.

As all uranium-burning reactors produce plutonium, which may be used for nuclear weapons, nuclear power for electricity production has become a political issue. Thus, some countries have decided to abstain from using nuclear energy, while others efficiently pursue it. This decision is partly affected by fear of the effects of nuclear radiation. That and connected problems are widely discussed among the public and decision makers. However, in this chapter we only provide a technical/scientific overview of current industrial-scale applications of solvent extraction chemistry and technology in the nuclear industry. A principal focus is on solvent extraction processes developed to partition and concentrate long-lived actinides and certain fission products from nuclear waste solutions, with the purpose of reducing handling difficulties and risks from the nuclear waste.

12.2 BACKGROUND

12.2.1 Historical Aspects

Nuclear science and technology have played key roles in development of liquid–liquid solvent extraction for the industrial-scale separation of metals. The high degree of purity of materials needed for nuclear applications, together with the high value of these materials, favors solvent extraction as the separation method of choice when compared with other methods. The need for a continuous, multistage separation system is unique to reprocessing of spent nuclear reactor fuels because of the requirement for nearly quantitative recovery and separation of fissile uranium and plutonium from some 40 or more fission products belonging to all groups of the periodic table. Only a brief account of the historical development and application of solvent extraction processes and technology in the nuclear field is included here. Numerous other authors have reviewed various aspects of this history; readers interested in historical details should consult references [1–8].

In 1942, the Mallinckrodt Chemical Company adapted a diethylether extraction process to purify tons of uranium for the U.S. Manhattan Project [2]; later, after an explosion, the process was switched to less volatile extractants. For simultaneous large-scale recovery of the plutonium in the spent fuel elements from the production reactors at Hanford, United States, methyl isobutyl ketone (MIBK) was originally chosen as extractant/solvent in the so-called Redox solvent extraction process. In the British Windscale plant, now Sellafield, another extractant/solvent, dibutylcarbitol (DBC or Butex), was preferred for reprocessing spent nuclear reactor fuels. These early extractants have now been replaced by tributylphosphate [TBP], diluted in an aliphatic hydrocarbon or mixture of such hydrocarbons, following the discovery of Warf [9] in 1945 that TBP separates tetravalent cerium from trivalent rare earths. Tributylphosphate has several major advantages over ethers and ketones, especially in its stability toward nitric acid, the preferred acid for nuclear fuel reprocessing [10]. Tributylphosphate allows easy and quantitative recovery of U(VI) and Pu(IV) from nitric acid solutions because of the high distribution ratios of these elements. Salting-out agents, such as $Al(NO_3)_3$, used in early extraction processes are no longer necessary. Furthermore, TBP has a low toxicity.

The solvent extraction process that uses TBP solutions to recover plutonium and uranium from irradiated nuclear fuels is called Purex (plutonium uranium extraction). The Purex process provides recovery of more than 99% of both uranium and plutonium with excellent decontamination of both elements from fission products. The Purex process is used worldwide to reprocess spent reactor fuel. During the last several decades, many variations of the Purex process have been developed and demonstrated on a plant scale.

In 1948, high-molecular-weight alkylamines were introduced as extracting agents by Smith and Page [11]. Industrial-scale processes for recovery of uranium from ore leach liquors were developed at the U.S. Oak Ridge National Laboratory [12]. Plutonium purification processes that were based on extraction by trilaurylamine [TLA] were developed in France [13] and were used for a time in the French UP2 Purex plant at the La Hague site.

Solvent extraction research has been a major program in many national institutions involved in nuclear technology. Research in U.S. laboratories during the 1960s led to the synthesis of new organophosphorus extractants and to detailed investigations of their extraction chemistry. Monofunctional extractants such as di-(2-ethylhexyl)phosphoric, -phosphonic, or -phosphinic acids were first investigated [14] at the Argonne National Laboratory (ANL, Chicago). The limited affinity of these monofunctional extractants for actinide ions in acidic media was an incentive to investigate the behavior of bifunctional extractants for the separation of the tri-, tetra-, and hexavalent actinides. Efficient new organophosphorus molecules were later synthetized in the U.S. [15] and in the former Soviet Union [16]. However, because poor back extraction of some metal ions [e.g., U(VI), Pu(IV), and Am(III)] by these polyfunctional organophosphorus molecules made them unsuitable for process applications, dihexyldiethylcarbamoylmethylenephosphonate (DHDECMP) and octylphenyldiethylcarbamoylphosphine oxide (CMPO), which have one phosphoryl and one amide functional group, showed promise for use and are still under study for actinide cation separations. As the best process is always a compromise between several parameters, changes in the preference of the engineers occurred over time due to the technical progress made [17].

In the European Community (EC) countries, two additional criteria for extractant selection are thought to be important. The first is related to the amount of incombustible waste produced in the process, which can be reduced if the organic solvents are completely incinerable: i.e., they must only contain C, H, O, and N atoms. The second is related to the degradation products of the extractant. To avoid full solvent regeneration before each solvent recycle, the degradation products must not interfere strongly with the extraction and back extraction of the major actinides: dialkylphosphoric acids are formed as degradation products of trialkylphosphates, -phosphonates, or -phosphine oxides, and they prevent the back extraction of metals at the low acidities needed in some parts of the process. Thus, the degradation products must be carefully eliminated before each solvent recycle, leading to generation of additional waste volumes.

12.2.2 Chemical Aspects

A primary goal of chemical separation processes in the nuclear industry is to recover actinide isotopes contained in mixtures of fission products. To separate the actinide cations, advantage can be taken of their general chemical properties [18]. The different oxidation states of the actinide ions lead to ions of charges from +1 (e.g., NpO_2^+) to +4 (e.g., Pu^{4+}) (see Fig. 12.1), which allows the design of processes based on oxidation reduction reactions. In the Purex process, for example, uranium is separated from plutonium by reducing extractable Pu(IV) to nonextractable Pu(III). Under these conditions, U(VI) (as UO_2^{2+}) and also U(IV) (as U^{4+}), if present, remain in the TBP phase.

In waste management programs, it might be desirable to isolate the minor actinides (Np, Am, Cm) and recycle them in nuclear reactors (see later flow sheets). These actinides appear together with the trivalent lanthanides, which have undesirable nuclear properties (e.g., high neutron reaction cross sections), and thus it becomes necessary to remove the trivalent lanthanides. Unfortunately, the trivalent lanthanides behave chemically like the trivalent actinides, because their similar ionic radii decrease with the atomic number in each series and overlap, resulting in complexes of very similar strength, leading to very small separation factors of the 4f (lanthanide) and 5f (actinide) M^{3+}–X^- complexes.

The actinide elements are "hard acid cations" and their bonds are principally ionic; thus, ionic interactions are the main driving force for bond formation. Consequently, "type A ligands" such as oxygen donors form complexes with the actinide ions much better than the "type B ligands" (see Chapter 3). To produce different stabilities in the complexes of the ions of the 4f and 5f series, the bonds must have some covalent character. As the degree of covalency increases with the decrease of the Pauling electronegativity of the donor atoms in the ligand, ligands with donor atoms less electronegative

Oxidation state (ion)	Ac	Th	Pa	U	Np	Pu	Am	Cm	Bk	Cf	Es	Fm	Md	No
I (M$^+$)													+	
II (M^{2+})						+				+	+	+	++	•
III (M^{3+})	•	++	+	++	++	++	•	•	•	•	•	•	•	++
IV (M^{4+})		•	++	++	++	•	++	++	++	+				
V (MO$_2^+$)			•^{1}	++	•	++	++							
VI (MO$_2^{2+}$)				•	++	++	++							
VII (MO$_4$(OH)$_2^{3-}$)*					++	++	+							

Fig. 12.1 Oxidation states of the actinide elements: • most stable ions in aqueous solutions; ++ oxidation states observed in aqueous solutions; +, unstable ions observed only as transient species. *In solids precipitated from alkaline solutions.

than oxygen have more covalency. In addition, the trivalent actinides have slightly more covalency than the lanthanides [19]. This has led to the search for new extractants containing more covalent donor atoms, such as sulfur or nitrogen. Such extractants have been shown to provide better separation of the trivalent ions of the 4f and 5f transition series (see section 12.9.1).

After the separation of the actinides from the high-level waste, it is desirable to remove certain other fission products from the nuclear wastes. Some Cs^+ and Sr^{2+} are low-charged cations that react well with macrocyclic ligands (e.g., crown ethers, calixarenes). Research to synthesize and investigate the properties of macrocyclic ligands for application in nuclear waste treatment has been an active effort internationally. Some of the results obtained are discussed in section 12.7.

Extractants that lead to the formation, in the organic phase, of complexes in which the metal is completely surrounded by oxygen atoms are often quite selective for the actinides. This is true with TBP, which selectively extracts actinides from nitrate solutions, forming complexes such as $Th(NO_3)_4(TBP)_2$ and $UO_2(NO_3)_2(TBP)_2$ (see Fig. 12.2). Nitric acid is preferred over other strong inorganic acids because it is the least corrosive acid for the stainless steel equipment used in large-scale reprocessing of spent nuclear fuels. Its low complexing affinity for the d transition ions and its ability to fill two coordination sites favors binding with large actinide ions.

12.3 URANIUM PRODUCTION

The process used for recovery of uranium from its ores depends on the nature of the ore. All the processes include a leaching step that solubilizes the metal. Solvent extraction is used most frequently for the recovery and

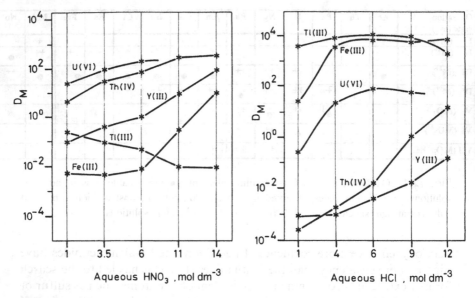

Fig. 12.2 Distribution ratios of actinide ions and some selected metals between undiluted TBP and HNO_3 or HCl solutions.

purification of uranium from the leaching liquors [20–22]. The production processes and their relative importance are listed in Table 12.1. The extraction processes are subject to variation in different mining areas, so the data in Table 12.1 should be considered only as illustrative of uranium solvent extraction processes that have been used. Today, the demand for mining new uranium for use in nuclear reactors is not as high as before. Orders for new nuclear reactors are few and, in some countries, uranium is available from dismantled nuclear weapons. Presently the price of uranium is low and the production from mines is constant, or in many countries decreasing.

12.3.1 Recovery from Sulfuric Acid Leach Liquors

Two extractants are used to recover uranium from sulfuric acid leach liquors: di-(2-ethylhexyl)phosphoric acid (HDEHP) in the Dapex process and commercial mixtures of trioctyl to tridecylamines (e.g., Adogen 364 or Alamine 336) in the Amex process. The chemical principles on which these processes are based are given in Table 12.2. Details, such as the nature of the diluent, extractant concentration, and washing and stripping conditions, vary from plant to plant. The advantages and disadvantages of the Dapex and Amex processes are compared in Table 12.3. The Amex process is more

Table 12.1 Commercial Processes for Recovery of Uranium from Ores

Type of ores/main locations	Leachant	Purification process	Relative importance in the production of U
$CaCO_3$, basic, U present mainly as U(VI) Grants (New Mexico), Rio Algon (Utah)	Na_2CO_3	Precipitation $Na_2U_2O_7$ (no solvent extraction steps)	11
Weakly basic to acidic ores Canada, Australia, U.S.A. (contain U(IV))	H_2SO_4 + oxidant	Dialkylphosphoric acids or trialkylamine extraction of U(VI) from H_2SO_4 liquors	80
High-grade pitchblende ores Zaire, Canada, Czechoslovakia, South Africa	HNO_3	TBP extraction of U(VI)	Few%
Phosphates Florida	H_2SO_4 leads to a phosphoric acid solution that contains 50–200 ppm of U	Synergistic extraction of U(VI) by HDEHP-TOPO or extraction of U(IV) by dialkylphosphoric acids mixtures	Few % will vary with U price. The phosphate rocks are the second largest potential source of U after sea water

Table 12.2 Chemical Basis for the Amex and the Dapex Processes for U(VI) Recovery from H_2SO_4 Leach Liquors

	Amex	Dapex
Extraction	$UO_2^{2+} + SO_4^{2-} + 2(R_3NH)_2SO_4 \leftrightarrows (R_3NH)_4UO_2(SO_4)_3$ $R_3N = $ Alamine 336 or Adogen 364 (trialkylamines with alkylgroups with 8 to 10 carbons)	$UO_2^{2+} + 2(HDEHP)_2 \leftrightarrows UO_2(HDEHPDEHP)_2 + 2H^+$
Stripping	*Acidic stripping* $(R_3NH)_4UO_2(SO_4)_3 + 4HX \leftrightarrows$ $4R_3NHX + UO_2^{2+} + 3HSO_4^- + H^+$ $HX = HCl$ or HNO_3 *Neutral stripping* $(R_3NH_4)_4UO_2(SO_4)_3 + (NH_4)_2SO_4 + 4NH_3 \leftrightarrows$ $4R_3N + UO_2(SO_4)_3^{4-} + 6NH_4^+ + SO_4^{2-}$ *Alkaline stripping* $(R_3NH)_4UO_2(SO_4)_3 + 7Na_2CO_3 \leftrightarrows$ $4R_3N + UO_2(CO_3)_3^{4-} + 4HCO_3^- + 3SO_4^{2-} + 14Na^+$	*Alkaline stripping* $UO_2($HDEHP DEHP$)_2 + 4Na_2CO_3 \leftrightarrows UO_2(CO_3)_3^{4-} +$ $4NaDEHP + H_2O + CO_2\uparrow + 4Na^+$ (TBP must be added to avoid the third-phase formation owing to low organic phase solubility of NaDEHP.) *Acidic stripping* $UO_2($HDEHP$)_2 ($DEHP$)_2 + 2H^+ \leftrightarrows 2($HDEHP$)_2 + UO_2^{2+}$

Table 12.3 Comparison of the Amex and Dapex Processes

Point of comparison	Amex process	Dapex process
Coextracted	Mo(VI), Zr(IV)	Fe(III), Th(IV), V(IV), Ti(IV), Mo(VI), Rare Earths
Stripping	Simple, many possibilities	With Na_2CO_3 which necessitates addition of TBP to avoid the separation of NaDEHP in a third phase
Extractant stability	Rather low stability	More stable than amines
Diluent	Tendency to form third phases with aliphatic hydrocarbons. Long-chain alcohols as phase modifiers or aromatic diluents are necessary	Many diluents can be used
Extraction kinetics	Rapid	Slow
Phase disengagement	Slow (sensitive to suspended solids)	Rapid

widely used than the Dapex process because of the greater selectivity of trialkylamines than HDEHP for uranium in H_2SO_4 solution.

12.3.2 Recovery from Nitric Acid Leach Liquors

High-grade pitchblende ores are leached with nitric acid to recover uranium. Extraction of uranium from nitrate solutions is usually performed with TBP. TBP-based solvents are used in several areas of the nuclear industry, especially for reprocessing of spent nuclear fuels and for refining the uranium product of the Amex and Dapex processes. Extraction of uranium by TBP solvents is described in sections 12.3.4 and 12.5.

12.3.3 Recovery from Phosphoric Acid Solutions

The concentration of uranium contained in phosphate rocks (50–200 ppm) is higher than that in seawater (see section 12.3.5). Even though economic recovery of uranium from phosphate rock is difficult, several phosphoric acid plants include operation of uranium recovery facilities.

Various processes have been used for uranium extraction from phosphoric acid solution; their main features are listed in Table 12.4. The HDEHP-TOPO process is increasingly preferred over others because of the stability of the extractant and the well-understood chemistry of the process.

Table 12.4 Main Industrial Processes for the Recovery of U from Wet Phosphoric Acid

Extractant	Mechanism of extraction	Stripping	Remark
Octylpyrophosphoric acid (H_2OPPA)	$U^{4+} + 2H_2OPPA \leftrightarrows U(OPPA)_2 + 4H^+$ (high distribution ratios)	In 25% $H_2SO_4 + 10$–20% HF: $U(OPPA)_4 + 4HF \leftrightarrows 2H_2OPPA + UF_4$	Cheap but unstable extractant. Difficulties in stripping.
Octylphenyl orthophosphoric acids (mono plus di) H_2MOPA + HDOPA	$U^{4+} + (HDOPA)_2 + H2MOPA \leftrightarrows (DOPA)_2 \, MOPA + 4H^+$ (high distribution ratios)	Oxidation to U(VI): $U(DOPA)_2 \, MOPA + \frac{1}{2}O_2 + 3H_2O \leftrightarrows UO_2^{2+} + H_2MOPA + (HDOPPA)_2 + 2OH^-$	Cheap extractant. Difficulty of stripping. Phase disengagement problems.
HDEHP–TOPO	$UO_2^{2+} + 2(HDEHP)_2 + TOPO \leftrightarrows UO_2 \, (DEHPHDEHP)_2 + 2H^+$ (rather low distribution ratios)	Reduction to U(IV) in H_3PO_4 or use of $(NH_4)_2CO_3$: $UO_2(HDEHP \, DEHP)_2TOPO + 3(NH_4)_2CO_3 + 2NH_3 \leftrightarrows 4NH_4DEHP + TOPO + UO_2(CO_3)_3^{4-} + 4NH_4^+$	Rather expensive extractant. Must oxidize U(IV) to U(VI) before extraction.

Today, phosphoric acid for use in the pharmaceutical or food industry must be free from uranium and other harmful metals. In the future, the processing of phosphate rock for production of fertilizers may be an important source of uranium because uranium removal may become compulsory to avoid its dissemination into the environment from the use of phosphate-based fertilizers.

12.3.4 Recovery as a By-Product from Production of Base Metals

Large-scale winning of copper by acidic leaching of copper ores sometimes results in waste solutions containing appreciable amounts of uranium. The uranium bearing aqueous raffinate from copper extraction is usually a dilute sulfuric acid solution. Uranium can be recovered using the same technique as described in section 12.3.1. A typical example is uranium production at the Olympic Dam mine in Australia, where the copper ore bodies are estimated to contain a total of over a million metric tons of uranium.

12.3.5 Recovery from Seawater

The chemical properties of U, such as the stability of the U(VI) ions as UO_2^{2+} and its water soluble anionic carbonato complexes, $UO_2(CO_3)_x^{(2x-2)-}$, where $x = 2$ or 3, make sea water the largest single source of U on earth. Its low concentration, $3.3 \, mg \, per \, m^3$, is the main problem for its recovery. Estimates of the price of U from seawater have shown that the cost is several orders of magnitude higher than for recovering U from high-grade ores. However, recent research on calixarenes [23] has showed that this type of macrocycle can achieve very high separation factors between U(VI) and other cations if the number of functional groups on the calixarenes corresponds to the number of ligands needed to saturate the coordination sphere of the UO_2^{2+} ion in its equatorial plan. This is achieved with calixarene(5) or (6). High separation factors are needed for the recovery of U as the seawater contains many other elements. Figure 12.3 shows the structure of two lipophilic p-*tert*-butylcalixarenes(6), bearing carboxylate or hydroxamate groups on the lower rim, which have been considered as extractants; some separation factors are reported in Table 12.5. It can be seen that the best functional group for the calixarene(6) is the hydroxamic group, CONHOH, probably because, at the pH of seawater, neutral U(VI) species, $UO_2(calix(6))H_4$, are formed in the organic phase; with carboxylate groups the lower pK_a of $R-COOH$ favors anionic species. It appears that with the carboxylate extractant Ni^{2+}, Zn^{2+}, and Fe^{3+} have an important negative effect upon the U(VI) extraction, but with the hydroxamate only Fe^{3+} in excess affects the U(VI) extraction.

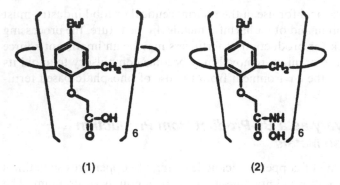

(1)   **(2)**

Fig. 12.3 Developed formula for calixarene(6) selective extractants for U(VI) extraction from sea water. (From Ref. 23.) (1) Carboxylate functions; (2) hydroxamate functions.

12.4 RECOVERY OF THORIUM FROM ITS ORES

Thorium, which consists mainly of the isotope ^{232}Th, has often been proposed as a secondary source of nuclear energy [24,25]. Irradiation of ^{232}Th with neutrons produces ^{233}Th ($t_{1/2}$ = 22.4 min), a beta-emitter that decays to fissionable ^{233}U. The world reserves of ^{232}Th are about three times larger than those of U. However, because the U nuclear fuel cycle was developed first and is easier to use for nuclear energy production, the thorium cycle has not received as much attention. However, thorium must be considered as a future energy source. Fission of thorium produces less hazardous waste, because the amount of long-lived transuranium isotopes is about 100 times less than in the uranium cycle.

Accelerator-driven reactors projects based on the Th cycle have been proposed [26].

Table 12.6 contains, in a simplified way, the composition, location, and treatment of main thorium ores. The purification of thorium by TBP extraction is illustrated in Fig. 12.4. This purification takes place after the dissolution of Th in nitric acid, generally from a hydroxide cake. When the Th is dissolved in sulfuric acid, purification is achieved by extraction with long-chain alkylamines.

12.5 REPROCESSING OF IRRADIATED NUCLEAR FUELS

12.5.1 Irradiated Uranium

The initial objective for reprocessing irradiated nuclear fuels [27,28] was to recover and purify ^{239}Pu for military applications. However, with the introduction of commercial nuclear power reactors, reprocessing of the irradiated fuels was considered as the best use of the available uranium resources. Reprocessing also allowed recovery of valuable fissionable nuclides that could be recycled in fast breeder reactors.

12.5.1.1 Purex Process Flow Sheet

Irradiated UO_2 is dissolved in nitric acid, resulting in a *dissolver solution* with the approximate composition listed in Table 12.7. This is treated by the Purex process. The main steps in the conventional Purex process are shown schematically in Fig. 12.5. All existing plants listed in Table 12.8 use some variation of the Purex process. Typically, the extractant composition (percentage TBP, diluent) and the extraction equipment (i.e., pulse columns, mixer-settlers, etc.), vary from plant to plant. However, the upper concentration limit is 30% TBP to prevent a phase reversal due to the increased density of the fully loaded solvent phase.

Figure 12.5 illustrates the basic components of the Purex process; three purification cycles for both uranium and plutonium are shown. High levels of beta and gamma radioactivity are present only in the first cycle, in which 99.9% of the fission products are separated. The other two cycles, based upon the same chemical reactions as the first cycle, obtain additional decontamination and overall purity of the uranium and plutonium products.

Distribution ratios of the actinides in largest concentration and valency state, and some important fission products, which form the basis for the design of the Purex process flow sheet, are shown in Table 12.9.

A typical flow sheet for the first cycle of the Purex process is shown in Fig. 12.6. The main steps are as follows:

Table 12.5 Comparison of the Extraction Selectivity of Two p-*tert*-butylcalix(6)arene, Uranophiles, Bearing 6 Carboxylates 1(6) or 6 Hydroxamates 2(6) Groups on the Lower Rim

Metal ion	M^{x+}/UO_2^{2+}	% UO_2^{2+} extr. by 1(6)	% UO_2^{2+} extr. by 2(6)
None	0	100	100
Mg^{2+}	1000	100	100
Ni^{2+}	10	77	98
Zn^{2+}	12	51	96
Fe^{2+}	10	0	66
Fe^{3+}	1	0	100

Note: The Organic Phase is $10^{-4}\,mol\,dm^{-3}$. Calixarene in $CHCl_3$. Aqueous Phase is $2 \times 10^{-5}\,mol\,dm^{-3}$ $UO_2(CH_3COO)_2$ Buffered to pH 5.9 by $10^{-2}\,mol\,dm^{-3}$ acetate. Phase Volume Ratio Org/Aq = 0.2.

Table 12.6 Commercial Processes for the Recovery of Th from Its Main Ores

Types of ores (main locations)	Ore composition	Leaching	Purification
Monazites (Brazil, India, Australia, South Africa, United States)	(La, Ce, Th) PO_4	Hot concentrated NaOH or H_2SO_4	Dissolution of oxides into HNO_3 and then TBP; alkylamines for Th in sulfate solutions
Thorianite Uranothorianite	ThO_2 (U,Th)O_2	HNO_3 or H_2SO_4	TBP for nitrates Alkylamines for sulfates
Thorite	$ThSiO_4$	Hot H_2SO_4	Alkylamines
Uranothorite (United States, Canada)	(U,Th)SiO_4	Hot H_2SO_4	Alkylamines

1. *Coextraction* of 99.9% of Pu and U and some 90% of ^{237}Np from 3–4 mol dm^{-3} HNO_3 solution into 30 vol% TBP, diluted with a mixture of aliphatic hydrocarbons or, occasionally, with pure dodecane. Fission products and trivalent actinides (Am, Cm) remain in the aqueous raffinate.

2. *U-Pu partitioning:* U(VI) remains in the TBP phase when Pu(IV) is reduced to Pu(III) and stripped (*back extracted*). Several reducing agents [e.g., U(IV), Fe(II), and hydroxylamine nitrate (NH$_2$OH · HNO$_3$)] can be used; all have their advantages and disadvantages. Reduction of Pu(IV) to Pu(III) in the U−Pu partition step is a process that has to be carefully controlled, because inadequate reduction can result in accumulation of plutonium in the partitioning equipment and increase hazards from nuclear criticality:

    (a) U(IV) solutions have to be prepared at the plant site and usually have to be stabilized against oxidants by addition of a nitrite scavenger such as hydrazine (N_2H_4), which can lead to the formation of highly explosive hydrazoic acid (HN_3). Although the use of U(IV) as reducing agent has the advantage of adding no new salts to the high-level waste, it does increase the solvent load and can lead to third phase formation. As it follows the uranium stream in the Purex process, it is not suitable in the second and third cycles, where Pu is purified from traces of U.

    (b) Instead other reducing agents, such as ferrous sulfamate or hydroxylamine, are preferred [29]. Ferrous sulfamate, Fe(SO$_3$NH$_2$)$_2$, rapidly reduces Pu(IV) to Pu(III) and the sulfamate ion reacts with HNO_2 to prevent the HNO_2-catalyzed oxidation of Pu(III) by

NO_3^-. However, ferrous sulfamate adds some undesirable inorganic salts to the aqueous high-level waste.

(c)   Hydroxylamine is a kinetically slow reducing agent and has to be used in conjunction with other reagents, such as U(IV) and hydrazine.

(d)   Electrolytic reduction of Pu(IV) to Pu(III) appears to be a very useful method of reduction. However, there is no large-scale plant experience, and a nitrate scavenger, such as hydrazine, is often necessary.

3.   *Uranium stripping:* Dilute HNO_3 solutions at 45–50°C are used to remove uranium from the TBP phase. Traces of the fission products ruthenium and zirconium are eliminated in the second and third cycles of the Purex process. Also, in the second and third cycles, neptunium and the last traces of plutonium are removed from the uranium product.

12.5.1.2   Solvent Degradation and Regeneration

Routine regeneration of the TBP solvent is an important step in the Purex process. The TBP, as well as the diluents, are degraded by hydrolysis and

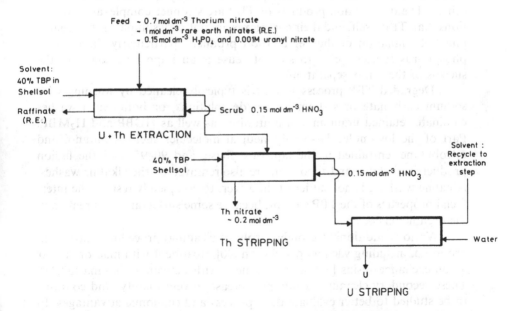

Fig. 12.4   Simplified flow sheet used in the recovery of thorium from its ores.

Table 12.7 Composition of Dissolver Solution from Several Reactor Fuel Types

Fuel type characteristics	Graphite gaseous reactor	Pressurized water reactor	Fast breeder reactor
Irradiation MW day per metric ton	4,000	33,000	80,000
Activity β,γ after 150 days of cooling (Ci per kg of fuel)	1,400	4,500	18,000
U(VI), $mol\,dm^{-3}$	1.05	1.05	0.85
Pu(IV), $mol\,dm^{-3}$	0.00275	0.01	0.22
HNO_3 $mol\,dm^{-3}$	3	3	3
Zr, $mol\,dm^{-3}$	0.0013	0.0095	0.0184
Ru, $mol\,dm^{-3}$	0.0005	0.0058	0.02
Fission products (kg per metric ton)	4.16	35	87
Transplutonium elements, $mol\,dm^{-3}$		0.0002	0.0028

Note: 1 Ci (Curie) corresponds to 3.7×10^{11} Bq (Becquerel, or dis./s).
Source: Ref. 82.

radiolysis. Degradation of TBP results in products such as dibutylphosphoric acid (HDBP), monobutylphosphoric acid (H_2MBP), and phosphoric acid. Depending upon the type of diluent, hydrolytic and radiolytic attack produces nitrato esters, nitro compounds, carboxylic acids, ketones, and others. The degradation products of TBP are stronger complexants for cations than TBP itself, and their complexes are not very soluble in the organic phase. Elimination of the degradation products, particularly the organophosphorus species, prior to solvent reuse is an important factor in the success of the Purex separations.

Degraded TBP process solvent is typically cleaned by washing with sodium carbonate or sodium hydroxide solutions, or both. Such washes eliminate retained uranium and plutonium as well as HDBP and H_2MBP. Part of the low-molecular-weight neutral molecules such as butanol and nitrobutane, entrained in the aqueous phase, and 90–95% of the fission products ruthenium and zirconium are also removed by the alkaline washes. Alkaline washing is not sufficient, however, to completely restore the interfacial properties of the TBP solvent, because some surfactants still remain in the organic phase.

Various additional (secondary solvent cleaning) procedures have been proposed, including vacuum distillation [30], treatment with macroreticular anion-exchange resins [31], and treatment with activated alumina [32,33]. These secondary cleanup operations increase solvent quality, and continue to be studied to better evaluate their process and economic advantages. In the French UP3 Purex plant, vacuum distillation is used to regulate TBP concentration and solvent quality with beneficial effects. The latter include

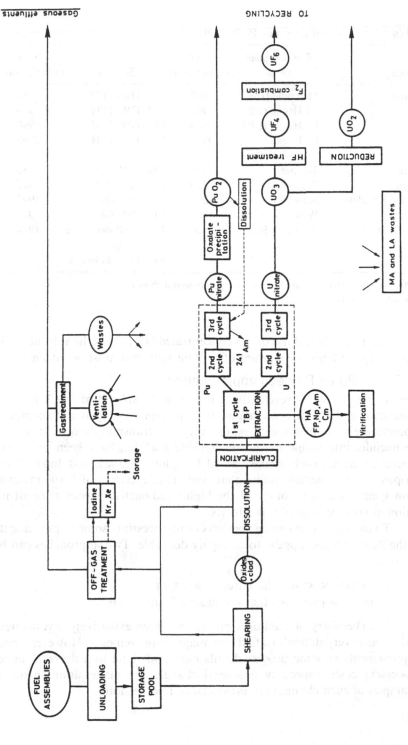

Fig. 12.5 Schematic representation of the reprocessing of light-water reactor spent fuels. The solvent extraction steps are shown in the dotted frames (HA, MA, LA, for high, medium, and low activity wastes, respectively).

Table 12.8 Principal Operating PUREX Plants

Country	Site and plant name	Capacity (ton per year)	Solvent	Year commissioned
France	Marcoule (UPI)	400	30% TBP–TPH[a]	1958
	La Hague (UP2)	400	30% TBP–TPH	1976
	La Hague (UP3)	800	30% TBP–TPH	1990
	La Hague (UP2–800)	800	30% TBP–TPH	1992
India	Tarapur	100	30% TBP–Shellsol	1982
Japan	Tokia Mura	210	30% TBP–dodecane	1977
United Kingdom	Sellafield	1200	30% TBP–OK[b]	1992
	Windscale	1000	20% TBP–OK	1971
United States	Savannah River	1800	7.5% TBP–dodecane or 30% TBP–dodecane	1954

[a]TPH, acronym for hydrogenated tetrapropylene, a branched dodecane.
[b]OK, odorless kerosene.

retention of metals and use of less concentrated Na_2CO_3 in the solvent wash, which helps to limit the amount of sodium salts that must be vitrified.

12.5.1.3 Purex Process Improvements

Reprocessing of spent reactor fuels has been the subject of much international debate [34]. Strategies ranging from no reprocessing at all to complete reprocessing with, in some cases, recovery and transmutation of long half-life nuclides into stable or shorter half-life nuclides, have been proposed. Several countries, such as the United Kingdom, France, and Japan, have adopted an intermediate policy that consists of recovery of fissile material from spent fuels and storage of the highly radioactive wastes after vitrification in deep geological repositories.

To increase the economic benefits of reprocessing, some improvements in the Purex process appear to be highly desirable. Two approaches can be followed:

1. make changes within the Purex process, or
2. develop new processes that are more advantageous.

The chemistry of the Purex process has been extensively investigated, and it seems very difficult to find new major improvements. However, some improvements in some process details may contribute to a decrease in reprocessing costs, especially if several changes are made simultaneously. Examples of such changes are presented in next sections.

12.5.1.3a Decreased Number of Cycles

A decrease in the number of uranium and plutonium purification cycles from three to two, or even one, would be highly advantageous. First-cycle decontamination factors of uranium from neptunium and from the fission products ruthenium and zirconium must be significantly improved to realize such a decrease.

12.5.1.3b Improved and Simplified Separation of Neptunium-237

Two routes are available to decontaminate uranium and plutonium from neptunium [35]:

1. Route all neptunium as Np(V) to the high-level waste in the first extraction cycle
2. Allow most of the neptunium to coextract with U(VI) as Np(IV) or Np(VI) and separate it in downstream uranium purification cycles

The latter alternative is currently followed in Purex plants in France and the United Kingdom and was used for years in the Purex plant at the U.S. Hanford site. Drake [36] has discussed laboratory and plant-scale experience in coextracting neptunium and uranium and in separating neptunium from uranium. Simplified Purex process operations would be achieved by diverting neptunium to the high-level waste in the first extraction contactor. Results of bench-scale experiments to find effective reagents for establishing and maintaining nonextractable Np(V) in Purex process extraction feed have been discussed by Drake. Much more work is necessary to

Table 12.9 Distribution of Various Metallic Species Between 30% TBP–Dodecane, 80% Saturated with U(VI), and HNO_3 Solutions at 25°C

Metal[a]	$3 \, mol \, dm^{-3}$ HNO_3	$1 \, mol \, dm^{-3}$ HNO_3	$0.01 \, mol \, dm^{-3}$ HNO_3
U(IV)	3.3	1.5	0.8
Np(VI)	0.8	0.4	—
Np(V)	Negligible	Negligible	Negligible
Np(IV)	0.12	0.04	Hydrolyzes
Pu(IV)	0.7	0.18	Hydrolyzes
Zr(IV)	0.01	0.0015	Hydrolyzes
Ru	0.0013	0.006	—
Ce(III)	0.0004	0.004	
Sr(II)	Negligible	Negligible	Negligible
Cs(I)	Negligible	Negligible	Negligible

[a]Tracer level concentrations except for uranium.

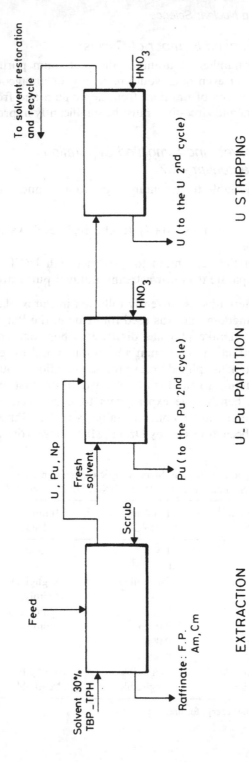

Fig. 12.6 Flow sheet for the first purification cycle in the Purex process (TPH is hydrogenated tetrapropylene, a commercial branched dodecane).

develop reductants that can be used successfully and routinely in operating Purex plants. However, removal of all Np from the high-level raffinate before vitrification may be required in the future. In that case, it would be desirable to extract Np as fully as possible in the first extraction cycle.

12.5.1.3c Partitioning of Technetium-99

Long-lived ($t_{1/2} = 2.1 \times 10^5$ years) ^{99}Tc, present as TcO_4^- in Purex process HNO_3 feed solutions, is partially coextracted with uranium and plutonium in the first cycle. Unless separated in the Purex process, ^{99}Tc contaminates the uranium product; subsequent processing of the $UO_2(NO_3)_2$ solution to UO_2 can release some of the technetium to the environment. The presence of technetium in the purification steps as well as in the uranium product causes several other complications. Thus it is desirable to route all Tc into the high-level waste. Efforts in this direction have been described in some recent flow sheets [37].

12.5.1.3d Improved Waste Management

Techniques for improving Purex process waste management could have a substantial financial impact on the costs of fuel reprocessing. Figure 12.7

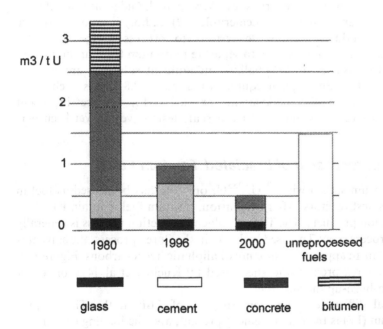

Fig. 12.7 Volumes of wastes containing long half life isotopes generated by the re-processing of spent nuclear fuels as a function of time. (From Ref. 38.) Comparison with the nonreprocessing option.

shows the volumes of deep underground wastes generated for 1 ton of U reprocessed in the UP3 plant from its start in 1980 [38]. The volumes of each kind of waste are given and the volume necessary to store unreprocessed spent fuel is shown for comparison. Several improvements have been made since 1980, but more volume reduction is desirable, particularly in the amount of sodium in spent solvent washes and in solutions used to decontaminate process equipment. Further, it is desirable from an environmental point of view to remove all alpha emitters and long-lived ^{137}Cs and ^{90}Sr from Purex process acid waste solutions, so-called waste partitioning; discussed in sections 12.6 and 12.7, respectively.

12.5.1.3e   Replacement of Tributylphosphate with an Alternative Extractant

From the time the Purex process was first developed, various groups in many countries have suggested candidate extractants to replace TBP in the Purex process. For reasons discussed by McKay [10], all of these attempts to replace TBP failed, in spite of advantages claimed for the replacement reagents. French [39] and Italian [40] workers have investigated the properties of several N,N-alkylamides with a view to their use in a Purex-type process. The following properties of N,N-dialkylamides are worthy of note: (1) they are completely incinerable; (2) radiolytic and hydrolytic degradation products are not deleterious to process performance; (3) reducing agents are not necessary to separate plutonium from uranium; and (4) physical properties of N,N-dialkylamide-diluent solutions are suitable for use in existing Purex plant equipment. Figure 12.8 shows a chemical flow sheet using an N,N-dialkylamide extractant for reprocessing spent light-water reactor fuels, but pilot-plant scale tests have not yet been performed.

12.5.2   Reprocessing of Irradiated Thorium Fuels

Natural thorium spiked with ^{233}U, ^{235}U, or ^{239}Pu has been used as fuel in several large test reactors. After irradiation, thorium fuels contain Pa, ^{233}U, Np, and fission products. The Thorex solvent extraction process is generally used to reprocess spent Th-based fuels. As in the Purex process, the solvent is TBP diluted in an appropriate mixture of aliphatic hydrocarbons. Figure 12.9 shows the Thorex process flow sheet used by Küchler et al. [41] for reprocessing high-burn-up thorium fuel.

Several difficulties arise from the use of TBP in the Thorex process. Thorium (IV) is the major element present, and the loading capacity of a 30% TBP–dodecane solution is considerably lower for the tetravalent actinides than it is for U(VI); thus, a larger organic/aqueous phase ratio

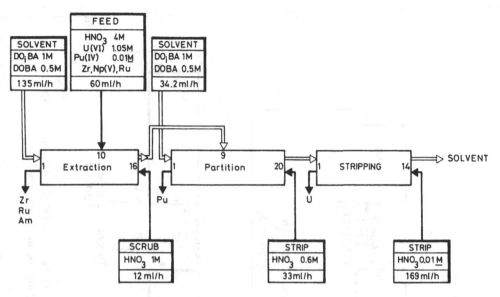

Fig. 12.8 Flow sheet of the first purification cycle of U and Pu in a process using an *N,N*-dialkylamide as extractant. (DOBA is $C_3H_7CON(CH_2CHC_2H_5C_4H_9)_2$; DOiBA, $(CH_3)_2CHCON(CH_2CHC_2H_5C_4H_9)_2$; diluent TPH (branched dodecane).

must be used to achieve suitable recovery of the thorium. Distribution ratios of Th(IV), Pa(V), and Zr(IV) are such that a good separation of Th(IV) is possible only at low aqueous feed acidity. Two extraction-strip cycles are necessary with fuels containing high quantities of hydrolyzable fission products. In the first cycle, carried out with acidic aqueous feed, thorium and uranium are separated from the bulk of the fission products. In the second cycle, performed with an acid-deficient feed solution, the final purification of uranium and thorium is achieved. The performance of this two-stage process is shown in Table 12.10.

If reprocessing of spent Th fuels is needed in the future, many drawbacks of the Thorex process can be avoided by using *N,N*-dialkylamides instead of TBP. It is known from studies of Pu(IV)–U(VI) separation with amides that U(VI)–Th(IV) separation in acidic medium is easier than with TBP. The advantages of amides over phosphates could probably be considered more freely for Thorex than for Purex plants because no Thorex plant has been built yet. The extraction behavior of Pa(V) has not been yet investigated but in view of the general chemical properties of the two extractants, smaller distribution ratios are expected with the *N,N*-dialkylamides.

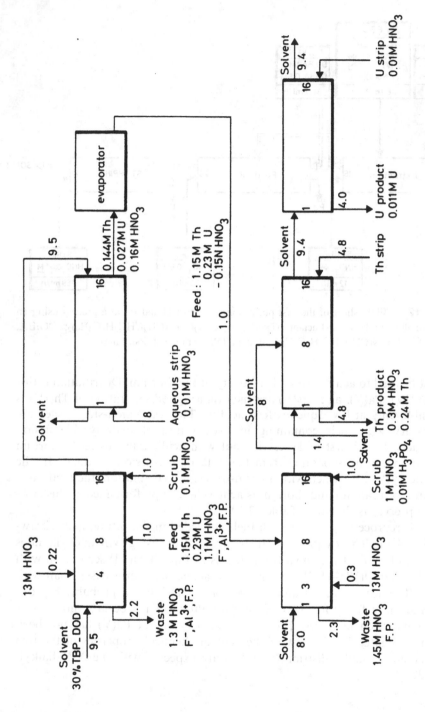

Fig. 12.9 Two-stage acid Thorex process for highly irradiated fuels. Numbers in the frames indicate stage number, whereas numbers on the lines indicate flow volumes relative to the feed volume (DOD and FP are dodecane and fission products).

12.6 SOLVENT EXTRACTION OF ACTINIDES FROM WASTE SOLUTIONS

Several kinds of radioactive wastes are produced by the nuclear industry, including those classified as high-level (HLW), low-level (LLW), and transuranic (TRU) wastes. Typically, the latter are those solid and liquid wastes that contain $3.7 \ 10^6$ Bq kg^{-1} (100 nCi g^{-1}) or more of transuranic nuclides, with a half-life ($t_{1/2}$) greater than 20 years. In several countries, TRU wastes must be eventually disposed of in deep geological repositories. Conversely, low-level wastes, which contain $<3.7 \ 10^6$ Bq kg^{-1} (<100 nCi g^{-1}) of transuranium elements can qualify for disposal in less expensive near-surface facilities. For economic reasons, therefore, there continues to be great interest in several countries to develop efficient and practicable solvent extraction processes for nearly quantitative extraction of all actinides, including, in particular, trivalent americium and curium, from any and all HNO$_3$ waste solutions. In contrast with tetravalent and hexavalent actinides, which are well extracted by the TBP and other monofunctional organophosphorus reagents, Am^{3+} and Cm^{3+} are poorly extracted from strongly acid and unsalted solutions. It is desirable to find reagents that will effectively extract the trivalent actinides, as well as any residual IV and VI actinides that may be present.

Several classes of reagents that have been or are currently being investigated for their ability to extract actinides from both HNO$_3$ and HCl waste solutions are discussed in the next sections.

Table 12.10 Performance of Two-Stage Thorex Process and a Uranium Purification Cycle

| Contaminant | Decontamination factor | | Third uranium cycle |
| | Thorex two-stage | | |
	Thorium	Uranium	
Total gamma	6×10^4	10^5	170
Total beta	3×10^5	2×10^6	3×10^3
^{144}Ce	6×10^6	3×10^7	2×10^5
^{106}Ru	5×10^4	5×10^4	170
^{95}Zr	6×10^5	6×10^6	2×10^4
^{233}Pa	103	2×10^4	

Source: Ref. 32.

12.6.1 Bifunctional Organophosphorus Reagents

Bifunctional organophosphorus compounds of the generic formula

$$(R'O_p)(RO_m)PO\text{-}(CH_2)_n\text{-}PO(RO_m)(R'O_p)$$

are known to form complexes with metal ions [42]. These compounds include diphosphonates ($m = 1, p = 1$), diphosphinates ($m + p = 1$), or diphosphine oxides ($m = p = 0$). Higher distribution ratios for $+3$ actinides from strong HNO_3 solutions are obtained when $n = 1$, which corresponds to the ideal size of a molecule for chelation of metal ions.

Bifunctional organophosphorus reagents in which one of the $P{=}O$ groups is replaced by an amide ($N{-}C{=}O$) group have the generic formula:

$$(R'O_p)(RO_m)PO\text{-}(CH_2)_n\text{-}CONR_1R_2,$$

with m and $p = 0$ or 1 and $0 < n < 5$ (see Appendix D, 13 and 14). These carbamoyl-organophosphorus compounds are much more effective than diphosphonates, diphosphinates, and diphosphine oxides in extracting Am^{3+} and Cm^{3+} as well as (IV) and (VI) actinides from strong HNO_3 media. Replacement of one $P{=}O$ group with an amide group also allows much greater flexibility in selecting hydrocarbon diluents and reagents for stripping of actinides.

Two carbamoyl-organophosphorus extractants have been intensively tested for their usefulness in plant-scale removal of any and all actinides from HNO_3 waste solutions: dihexyl-N,N-diethylcarbamoylmethylphosphonate [$(C_6H_{13}O)_2POCH_2CON(C_2H_5)_2$] and octyl(phenyl)-$N,N$-di-isobutyl-carbamoylmethylphosphine oxide [CMPO; $C_8H_7(C_6H_5)POCH_2CON(CH_2 \cdot CH(CH_3)_2)_2$]. The latter extractant is used in the Truex process [43]. Figure 12.10 illustrates a reference Truex process flow sheet for use with an HNO_3 waste solution. In the Truex process, CMPO is diluted with TBP in either an aliphatic hydrocarbon or a halogenated hydrocarbon (e.g., tetrachloroethylene) diluent. The TBP serves as a modifier, which greatly increases the organic phase solubility of CMPO–metal complexes.

Nash et al. [44] have identified the principal hydrolytic and radiolytic degradation product of CMPO to be $C_8H_{17}(C_6H_5)POOH$. Small concentrations of this acidic degradation product substantially increase the difficulty of stripping actinides from the Truex process solvent. The washing of degraded Truex process solvents with dilute Na_2CO_3 solutions restores the quality of the solvent to near-pristine condition.

Of the many radioactive and inert constituents in waste solutions, only ^{99}Tc and lanthanides coextract with actinides into the Truex process solvent. Trivalent lanthanides follow Am^{3+} and Cm^{3+} when the latter are stripped with dilute HNO_3; depending on the disposition of the americium and

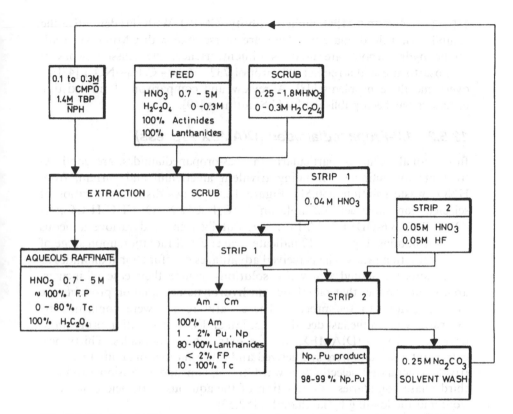

Fig. 12.10 Generic Truex process flow sheet for use with nitric acid wastes (NPH is a mixture of aliphatic hydrocarbons).

curium, separation of the associated lanthanides may be necessary. The ^{99}Tc follows uranium in the Truex process and co-strips with U(VI) into dilute Na_2CO_3 solutions. Anion-exchange resin and solvent extraction procedures have been used for separating uranium and technetium.

The Truex process has been extensively and successfully tested in the United States with several different actual nitric and hydrochloric acid waste solutions, and overall decontamination factors (Truex process feed to raffinate) of better than 10^5 have been achieved. This is encouraging for the pursuit of investigations for HLW declassification to avoid the storage of large volumes of wastes in deep underground repositories. Several runs of the Truex process have also been made in Japan to study the partitioning of the actinides contained in HLW [45], but with less success (e.g., coextraction of Ru). Many bifunctional organophosphorus extractants have been tested in Russia [46], but for practical applications they have chosen an extractant close to CMPO, diphenyldibutyl-CMPO. With this derivative the solubility and third phase problems are worse than with CMPO when aliphatic hydrocarbons are used as diluent. Hence, the Russian workers propose to use a fluoroether "fluoropol-732" ($CF_3-C_6H_4-NO_2$) [47] to overcome these problems [48]. A review and comparison of the existing processes has been published by U.S. authors [49].

12.6.2 1,3-Propanediamides (DIAMEX Process)

Bifunctional diamides, particularly the 1,3-propanediamides, are good extractants for actinides, including trivalent americium and curium, from HNO_3 waste solutions [50,51]. Figure 12.11 shows distribution ratios of selected ions between $0.5 \, mole \, dm^{-3}$ ($(C_4H_9CH_3NCO)_2CHC_2H_5OC_6H_{13}$ [DMDB-(2-(3,6-OD),1,3-DA)P] in *tert*-butylbenzene and various aqueous HNO_3 solutions. Figure 12.12 indicates the effect of radiation upon some of the distribution ratios. The perceived advantages of bifunctional diamides as extractants for actinides in waste solutions include their ease of incinerability and the fact that radiolytic and hydrolytic degradation products are not deleterious to the process. These early results were confirmed by extensive work in the last decade [52,53] in an EU collaboration program to develop a process (DIAMEX) for the extraction of actinides. Thirty new extractant molecules were synthesized and rated in order of ability to extract trivalent actinides; selectivity toward the fission and corrosion products; third phase boundaries as a function of the aqueous nitric acid concentration; and the loading by lanthanides [52,53].

The development of the Diamex process started with N,N'-dimethyldibutyltetradecylmalonamide [$(CH_3C_4H_9NCO)_2CHC_{14}H_{25}$, DMDBTDMA], but an alternative molecule, N,N'-dimethyldioctylhexyloxyethylmalonamide [$(CH_3C_8H_{17}NCO)_2CHC_2H_4OC_6H_{13}$, DMDOHEMA] was found to present some advantages (and also disadvantages). Among the advantages is a better decontamination from Mo(VI) (99.7% removal compared to 95% for DMDBTDMA), Fe(III) (98.1% removal compared to 27%), and Ru (79% compared to 50%) in countercurrent bench tests. The disadvantages are mostly the formation of more degradation products, which were more soluble in the organic phase, due to the replacement of the $CH_3C_4H_9N$ group by $CH_3C_8H_{17}N$.

Several tests using countercurrent separation in mixer-settler or centrifugal extractors with simulated and genuine high-level waste showed that the recovery of An(III, IV, VI) is quantitative. The back extraction of An(III) was complete. These early flow sheets were not designed to strip U(VI) and Pu(IV). The distribution ratios of these ions at low acidities are lower than those measured with CMPO and one can guess that the stripping

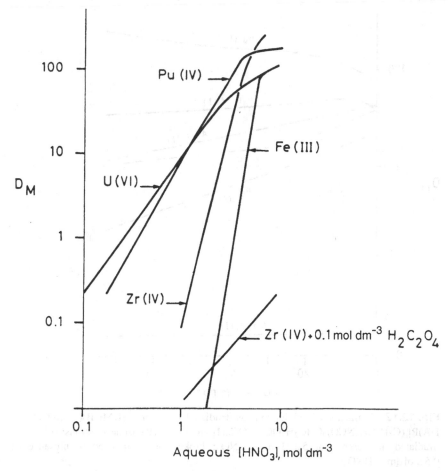

Fig. 12.11 Distribution ratios of various metallic species between $0.5 \, mol \, dm^{-3}$ $(C_4H_9CH_3NCO)_2CHC_2H_4OC_2H_4OC_6H_{13}$ [DMDB(2-3,6-OD,1,3-DA)P] *t*-butylbenzene solutions and HNO_3 solutions.

of Pu(IV) and U(VI) from Diamex will not be a difficult task. In addition, the degradation products of the solvent do not retain the actinides in the organic phase, as happens with the organophosphorus extractants.

Inactive metals in the wastes, such as Fe(III), Mo(VI), and Zr(IV), are retained in the aqueous feed by addition of suitable quantities of oxalic acid. As with CMPO, some Ru is extracted. Currently, the extraction of Pd, Tc, Np and their poor stripping remain a problem for which process modifications are necessary. The flow sheet in Fig. 12.13 and Table 12.11 has

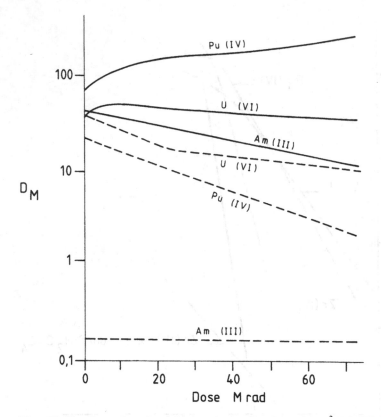

Fig. 12.12 Extraction of actinide ions by irradiated 1 mol dm^{-3} [DMDB-(2-3,6-OD,1,3-DA)P] (C$_4$H$_9$CH$_3$NCO)$_2$CHC$_2$H$_4$OC$_2$H$_4$OC$_6$H$_{13}$ into *t*-butylbenzene solutions: solid line, irradiated in presence of 5 mol dm^{-3} HNO$_3$; broken line, irradiated in presence of 0.5 mol dm^{-3} HNO$_3$.

been tested in a battery of 16 centrifugal extractors with a genuine raffinate from a small-scale Purex process at the European Institute for Transuranium Elements (ITU), Karlsruhe.

12.6.3 Monofunctional Organophosphorus Reagents

During development of the Truex process, various monofunctional organo-phosphorus reagents [e.g., TBP; DBBP (dibutylbutylphosphonate); HDEHP (di-(2-ethylhexyl)phosphoric acid)] were considered for removal of alpha-emitters from HNO$_3$ waste solutions. Two typical flow sheets for the use of HDEHP and TBP are shown in Fig. 12.14. Because of the poor performance of these two extractants, the processes were complicated. Distribution ratios of trivalent actinides (e.g., ^{241}Am) into TBP and DBBP are only high enough for practical applications from either highly salted or low-acid solutions, or very concentrated (i.e., 15–16 mol dm^{-3} HNO$_3$) solutions. Actual plant-scale experience with DBBP was very disappointing [56] because of the inability to adequately control the feed pH values in the range required for effective americium extraction.

The use of HDEHP for extraction of actinides from waste solutions also has several drawbacks, including extraction of trivalent ions at a pH at which tetravalent ions such as Zr(IV) and Pu(IV) are hydrolyzed, and difficulties in stripping tetravalent and hexavalent actinides. All in all, monofunctional organophosphorus reagents are vastly inferior to their bifunctional counterparts for extracting Am(III) and other actinides from strong HNO$_3$ media.

Many of these drawbacks were circumvented in a process developed at Chalmers University, which was successfully demonstrated in continuous operation using the old high-level raffinate concentrate from Purex processing of low burn-up fuel. Although operation was very easy and stable, the process flow sheet was complicated using bromoacetic acid, HDEHP, TBP, and lactic acid [57–59].

12.7 EXTRACTION OF Sr-90 AND Cs-134, -135, -137 FROM ACIDIC WASTES

To facilitate near-surface disposal of nuclear wastes, efficient removal of the $t_{1/2} = 30$ year fission product ^{90}Sr is desirable. The SREX process [60]

Table 12.11 Decontamination Factors for Extraction (Mass in the Feed to Mass in Raffinate) and for the Stripping [(Mass in the Feed Minus Mass in the Raffinate)/Mass in the Organic Out] in DIAMEX Hot Test with Flow Sheet in Fig. 12.13

Element	Extraction	Stripping	Element	Extraction	Stripping
Zr(IV)	1	—[a]	Mo(VI)	1	—[a]
Tc(VII)	120	30[b]	Ru(II)	1.3	61
Pd(II)	157	96	La(III)	167	>46000
Ce(III)	211	>22500	Pr(III)	216	>44000
Nd(III)	157	>50000	Sm(III)	110	—[a]
Eu(III)	113	>300000	Gd(III)	49	>300000
Np(V, VI)	29	29[b]	Am(III)	258	>25100
Cm(III)	301	3150			

[a]Below ICP-MS determination limit.
[b]Accumulation in the back extraction section.

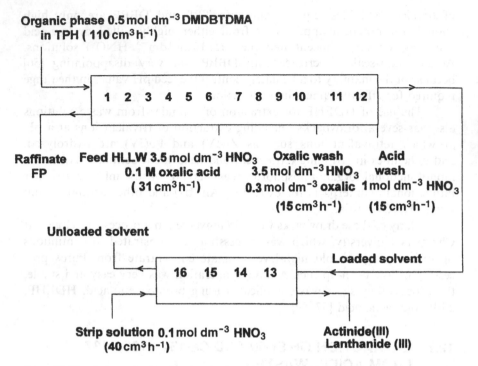

Fig. 12.13 Schematic drawing for the hot test of the DIAMEX process in a 16-stage centrifugal extractor battery. (From Ref. 53.)

provides such removal from highly acidic solutions. The process employs an *n*-octanol solution of the commercially available macrocyclic ether (cf. Appendix D, Fig. 21), di-*tert*-butyldicyclohexyl-18-crown-6 ($DtBuCH_{18}C_6$) as the extractant for ^{90}Sr. Batch contacts with simulated waste solutions show that the SREX process solvent is highly selective for strontium; of the many contaminants only barium and technetium coextract with strontium to any extent. Irradiation tests confirm that the SREX process solvent is satisfactorily resistant to radiolysis and hydrolytic attack is also minimal. The SREX process has been successfully tested in both batch contacts and continuous countercurrent tests with simulated waste solutions. An overall ^{90}Sr decontamination factor (SREX process feed to SREX process raffinate) of 4600 was achieved from Truex wastes.

    In the past, the extraction of ^{90}Sr and ^{137}Cs was investigated for some practical applications (heat and gamma ray sources) but the main interest today is for decreasing the thermal power and the potential hazard of nuclear waste in underground repositories. Results of extractions with some

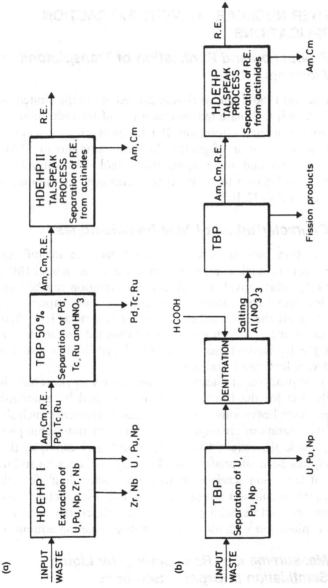

Fig. 12.14 Schemes for partitioning Purex high-level waste. (a, data from Ref. 54; b, data from Ref. 55.)

other macrocyclic extractants are collected in Table 12.12, and the formulae of some of the extractants are shown in Fig. 12.15.

12.8   OTHER NUCLEAR SOLVENT EXTRACTION APPLICATIONS

12.8.1   Production and Purification of Transplutonium Elements

During the period 1960–1980, work was carried out in the United States [64] and in France [65] to isolate macro quantities of transplutonium elements obtained from irradiated plutonium. Particular emphasis was placed upon separation and purification of gram to kilogram amounts of ^{241}Am for use in smoke detectors and other applications [66]. The solvent extraction processes used varied over time; two typical chemical separation flow sheets are illustrated in Fig. 12.16.

12.8.2   Characterization of New Radioisotopes

Solvent extraction techniques are frequently used to identify and characterize products of nuclear reactions, because the elemental (M) distribution ratios (D_M) that provide the scientific information can be measured at extremely low metal concentrations, as described in Chapter 4, providing information about the chemical properties of the metal ion. Figure 12.17 illustrates an example of such studies. Sophisticated radiometric analytical techniques greatly facilitate measurement of distribution ratios of radionuclides at very low concentrations.

Often the products of nuclear reactions have very short half-lives. This is especially true for the heaviest elements obtained by bombardment of heavy targets with heavy ions. To identify and characterize such short-lived nuclides, fast separations are required; solvent extraction techniques are well suited to provide the required fast separations. For example, the SISAK method [68] has been successfully used in conjunction with in-line gas jet separators at heavy ion accelerators to identify short half-life actinide isotopes produced by collision of heavy atoms. The Sisak method involves use of centrifugal contactors, with phase residence times as low as tenths of a second, in conjunction with in-line radiometric detection equipment.

12.8.3   Measurement of Radioactivity by Liquid Scintillation in Organic Solutions

Organic media are suitable for detecting and counting radioactivity by scintillation methods, because the quenching of scintillation in organic

Table 12.12 Distribution Ratios of Cs^+ and Sr^{2+} Between Aqueous Nitric Acid Solutions and Various Organic Solvents

Solvent and extraction reaction	Aqueous phase	D_M	Refs.
0.01 mol dm^{-3} dicarbollide (CoDC) in nitrobenzene	1 mol dm^{-3} HNO$_3$ + 0.03 mol dm^{-3} polyethylene glycol	Sr^{2+} 15.3	[61]
M^{n+} + nHCoDC $\leftrightarrows$ M(CoDC)$_n$ + nH$^+$	0.1 mol dm^{-3} HNO$_3$ + 0.03 mol dm^{-3} polyethylene glycol	Sr^{2+} 569	[61]
0.01 mol dm^{-3} dicarbollide (CoDC) in nitrobenzene	0.5 mol dm^{-3} HNO$_3$ 0.1 mol dm^{-3} HNO$_3$	Cs^+ 22 Cs^+ 102.6	[61]
1 mol dm^{-3} TBP + 0.1 mol dm^{-3} HDNNS + 0.05M crown 1(C1) into kerosene: M^{n+} + m(HNO$_3$)C1 + nHDNNS $\leftrightarrows$ M(C1)$_m$(DNNS)$_n$ + nH$^+$ + mHNO$_3$	3 mol dm^{-3} HNO$_3$ 0.5 mol dm^{-3} HNO$_3$	Sr^{2+} 0.06 Cs^+ 1.59 Cs^+ 1	[62]
1 mol dm^{-3} TBP + 0.1 mol dm^{-3} HDNNS + 0.05 mol dm^{-3} crown 2(C2) in kerosene: M^{n+} + m(HNO$_3$)C2 + nHDNNS $\leftrightarrows$ M(C2)$_m$(DNNS)$_n$ + nH$^+$ + mHNO$_3$	3 mol dm^{-3} HNO$_3$ 0.5 mol dm^{-3} HNO$_3$	Sr^{2+} 1.98 Cs^+ ~0 Sr^{2+} 2.57	[62]
10^{-2} mol dm^{-3} di n-octyloxy calix[4]arene crown 6 (CA1) in orthonitrophenyl hexyl ether: Cs^+ + NO$_3^-$ + CA1 $\leftrightarrows$ CsNO$_3$CA1	1 mol dm^{-3} HNO$_3$	Cs^+ 33	[63]
10^{-2} mol dm^{-3} di n-octyloxycalix[4]arene crown 6 (CA1) in orthonitrophenyl hexyl ether: Cs^+ + NO$_3^-$ + CA1 $\rightleftharpoons$ CsNO$_3$CA1	1 mol dm^{-3} HNO$_{3+}$ 4 mol dm^{-3} NaNO$_3$	Cs^+ 25	[63]
10^{-2} mol dm^{-3} benzyloxycalix[8]arene N-diethylamide (CA2) into orthonitrophenyl hexyl ether: Sr^{2+} + 2NO$_3^-$ + CA2 $\leftrightarrows$ Sr(NO$_3$)$_2$CA2	1 mol dm^{-3} HNO$_3$	Sr^{2+} 20 Cs^+ <.001	[63]
10^{-2} mol dm^{-3} benzyloxycalix[8]arene N-diethylamide (CA2) in orthonitrophenyl hexyl ether: Sr^{2+} + 2NO$_3^-$ + CA2 $\leftrightarrows$ Sr(NO$_3$)$_2$CA2	1 mol dm^{-3} HNO$_{3+}$ 4 mol dm^{-3} NaNO$_3$	Sr^{2+} 5 Cs^+ 0.1	[63]

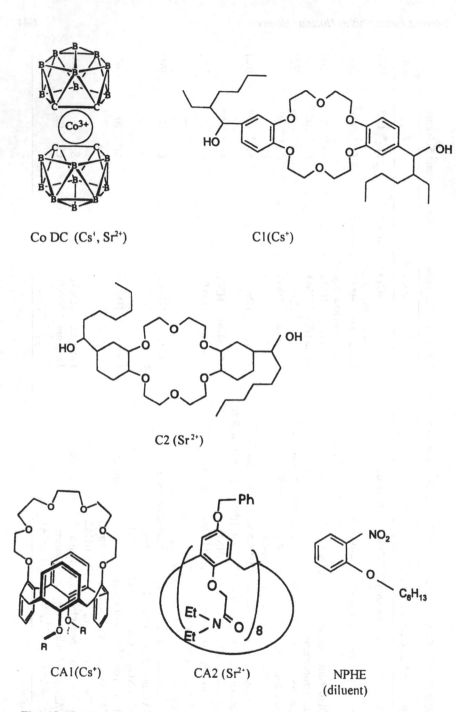

Co DC (Cs⁺, Sr²⁺)

Cl(Cs⁺)

C2 (Sr²⁺)

CA1(Cs⁺)

CA2 (Sr²⁺)

NPHE
(diluent)

Fig. 12.15 Formulae of the ligands used for Cs(I) or Sr(II) extraction from acidic media and cited in Table 12.12.

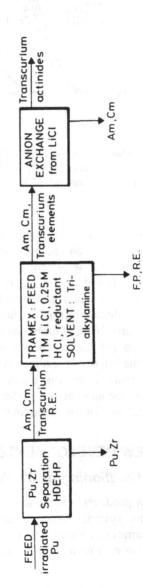

Fig. 12.16 Flow sheets used for separation of transplutonium elements from irradiated Pu: (a) France (Fontenay-aux-Roses); (b) United States (Oak Ridge National Laboratory). (Data from (a) Ref. 64; (b) Ref. 65.)

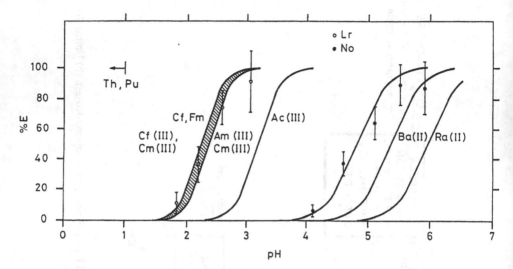

Fig. 12.17 Determination of the oxidation state of submicro traces of ^{256}Lr (open circles) and ^{255}No (solid circles). Solvent: $0.2\,\mathrm{mol\,dm^{-3}}$ TTA–MIBK; aqueous phases: buffers at various pHs. (From Ref. 67.)

environments is much less than in water. This is used in the PERALS method [69], in which radionuclides are transferred by solvent extraction into an organic phase of aromatic molecules. In the Perals system an electronic device based on the delay of counting after detection allows the discrimination of alpha and beta particles. This extends the range for the detection of alpha-emitting nuclides in solution. The good yield of scintillation measured by the number of photons detected by energy absorbed also allows the consideration of low-resolution alpha spectrometry, and this also improves the interest in radioactivity analysis by scintillation. The absorption of neutrinos in organic media followed by the counting by scintillation of the radionuclides formed is the basis of several projects in experimental physics dealing with the fundamental properties of neutrinos [70].

12.9 NEW DEVELOPMENTS

12.9.1 Partitioning of Wastes

The fission products contain some 50 elements located in the middle part of the periodic system. Some of the fission products are produced in considerable amounts, though their radioactivity is rather low [71], and they were therefore thought to have a potential value on the mineral market.

Some other fission products were considered as suitable for radiation sources (heat sources, radiography, etc.). Thus, processes were designed to isolate these from the rest of the fission products. Today, there is little interest in such applications.

However, there is renewed interest in separating some key nuclides (e.g., ^{90}Sr, ^{99}Tc, ^{129}I, ^{134,135,137}Cs), some of which have important medical applications. However, the prime interest comes from the desire to generate wastes that could be more easily and safely stored in deep geological repositories. In the most favorable situation, they could be safely stored in relatively inexpensive near-surface engineered facilities. The separation of the remaining actinides (traces of Pu, Np, Am, Cm, etc.) and of the most hazardous long-lived fission products is referred to as *partitioning*. The partitioned radionuclides would either be transmuted (e.g., actinides, ^{99}Tc, ^{129}I) in nuclear reactors or accelerators to form stable or short-lived nuclides, or stored (^{90}Sr and ^{137}Cs) in hardened engineered near-surface facilities for 300–500 years to allow them to decay to negligible levels. Solvent extraction is thought to be a key technique for partitioning and such partitioning could improve the economics of final waste handling.

The current conventional approach (Fig. 12.18a) to partition radionuclides from Purex process acidic high-level waste involves a sequential series of separation processes to remove selected nuclides or groups of radionuclides. An example of this approach is the United States "clean use of reactor energy" (CURE) concept [72] that utilizes the Truex process to remove actinides, ^{99}Tc, and lanthanides; the Srex process to remove ^{90}Sr from the raffinate from the Truex process; and precipitation of cesium phosphotungstate from the Srex process raffinate by addition of phosphotungstic acid. Results of batch contacts to test the Truex and Srex process portions of the Cure concept were discussed previously. The cesium phosphotungstate precipitation process has been used extensively on a plant scale at the U.S. Hanford site [73].

The Cure concept and the Japanese OMEGA concept [74] are quite similar, although they will likely employ different reagents. For example, for removal of actinides from Purex process high-level waste, Japanese scientists are investigating the properties of di-isodecylphosphoric acid. The Japanese Omega concept also involves removal of valuable rhodium and palladium from the high-level waste.

Even though the Cure and Omega concepts are year-2000-state-of-the-art, they have a number of potential disadvantages including the need to handle the highly radioactive waste in three or more extraction cycles. Improved partitioning processes that will accomplish the separations in fewer steps are highly desirable. For example, a solvent extraction process that provides for quantitative recovery of both ^{90}Sr and ^{137}Cs in a single

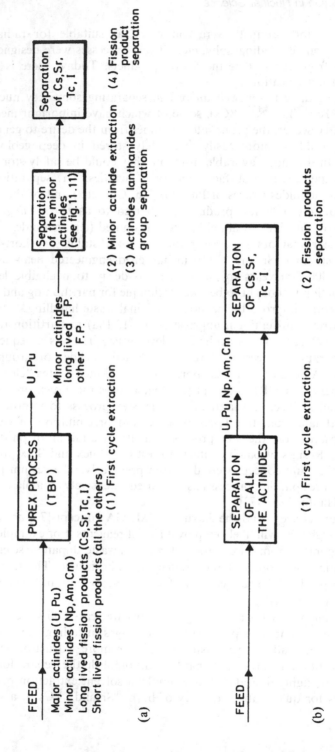

Fig. 12.18 Concepts for partitioning radionuclides from Purex process high-level waste: (a) today's technology; (b) tomorrow's technology. Note: Numbers in parentheses indicate separations that must be performed remotely.

extraction cycle is very much needed. In addition, a desirable goal is to find, if possible, an extractant that will extract (III), (IV), and (VI) actinides, plus TcO_4^-, and reject the lanthanides to the aqueous raffinate. One such program is the SPIN (solvent partition–incineration) program [75] for separation of the trivalent actinides from the trivalent lanthanides. This separation is necessary because nuclear incineration of the actinides requires that the neutron-consuming lanthanides be removed.

Recent investigations [76,77] indicate that certain soft donor ligands containing nitrogen and sulfur atoms extract trivalent actinides better than trivalent lanthanides from nitric acid solutions. However, the particular extractants studied then were not considered suitable for plant-scale use because of poor chemical stability [e.g., di(2-ethylhexyl)dithiophosphoric acid] against radiation or relatively high aqueous phase solubility (*tris*-pyridyl-triazine from dinonylnaphthalene sulfonate TPTZ mixtures into CCl_4). These results, however, encourage further investigations of soft donor ligands for use in preferential extraction of (III) actinides, as well as (IV) and (VI) actinides, from HNO_3 solutions. The search for good extractants remains a challenge for the future, because ligands bearing soft donor atoms will bind better with the *d* transition elements contained in the fission and corrosion products.

12.9.2 Trivalent Actinide-Lanthanide Group Separations

Considerable research has been done within the EU through the SPIN program [51,52,62,75] to improve methods for the separation of the minor actinides and some of the fission products from HLW. Some of the results have been already discussed. The goal was to optimize the formula of the soft donors (N and S) extractants in order to use them in large-scale separations of trivalent actinides from trivalent lanthanides. Both types of donors were studied. For the nitrogen donors the research started from *tris*-pyridyl-triazine. Substituting three *tert*-butyl groups on the pyridine rings [tri-(*tert*-butylpyridine)triazine] increased the lipophilicity of the extractants and allowed two drawbacks to be overcome, i.e., the solubility of the nitrogen donor in the aqueous phase as an acidic species and the rather high pH for extraction of trivalent ions. The solubility of the nitrogen ligand in aqueous solution became negligible and the influence of the branched alkyl substituents on the Am(III), Eu(III) distribution ratios can be seen in Fig. 12.19. The results shown indicate that the design of a flow sheet can be considered if an alternative to α-bromocapric acid can be found. This acid is undesirable because its bromine atom can lead to the production of corrosive HBr by radiolysis or hydrolysis. α-cyanocarboxylic acids have been

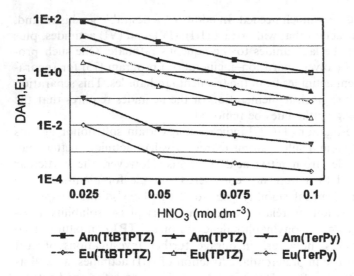

Fig. 12.19 Distribution ratios of Am(III) and Eu(III) between 1 mol dm^{-3} α-bromocapric acid + 0.02 mol dm^{-3} nitrogen ligand into TPH as a function of aqueous HNO$_3$ concentration. (From Ref. 52.)

tested, but their low solubility in TPH is another drawback [53]. Altogether 78 nitrogen donor molecules were synthetized. Among them are the tridentate nitrogen donors from the family of the 2,6-*bis*-(5,6-dialkyl-1,2,4-triazine-3-yl)-pyridine prepared by Kolarik et al. [78].

Sulfur donors belonging to the dialkyl-thiophosphorous acids were also investigated, and nine molecules have been considered. Separation factors between Am(III) and Eu(III) as high as 17,000 have been found and confirm the ability of sulfur donors to promote intergroup 5f-4f separations of trivalent ions. As indicated by the Pauling electronegativities of N and S, the S donors should bind better to the 5f trivalent ions than the N donors. With the individual sulfur donors or synergistic mixtures investigated, the extraction reactions were:

$$M^{3+} + 3R_2PSSH \leftrightharpoons M(R_2PSS)_3 + 3H^+$$

$$M^{3+} + 3R_2PSSH + nR'_3POM(R_2PSS)_3(R'_3PO)_n + 3H^+$$

The many laboratories involved in this work allowed a large experimental program to be set up to test the behavior of these soft donor extractants in countercurrent separations. The trivalent actinide-lanthanide separation process was named SANEX and four chemical systems have been explored (Fig. 12.20):

SANEX 1: Aqueous feed Acidity 0.01 mol dm^{-3} HNO$_3$

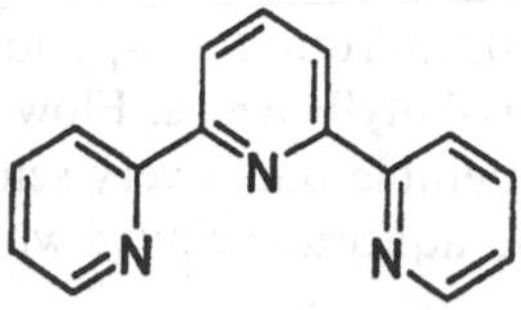

0.02 mol dm^{-3} Terpy + 1 mol dm^{-3} C$_8$H$_{17}$-CHBr-COOH into t-butyl benzene

SANEX 2: Aqueous feed Acidity 0.2 mol dm^{-3} HNO$_3$

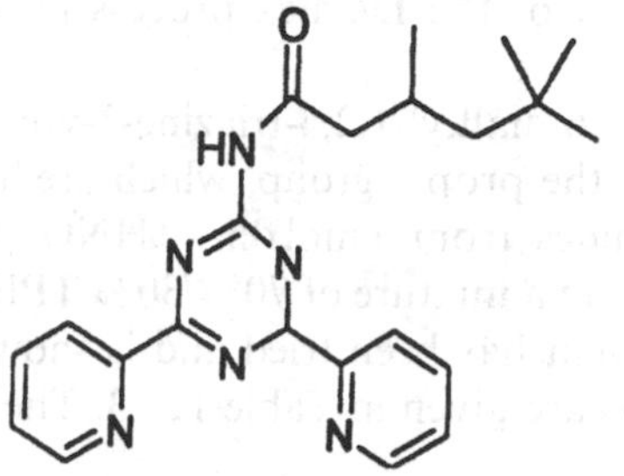

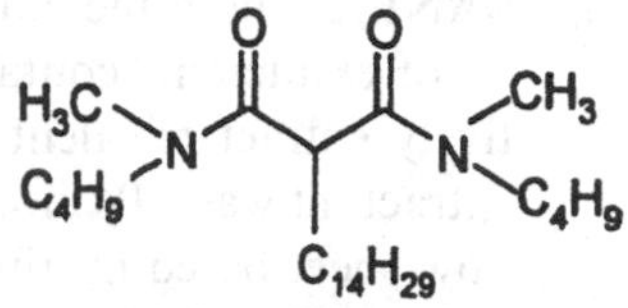

0.1 mol dm^{-3} BADTZ   +   0.5 mol dm^{-3} DMDBTDMA
+ 1 mol dm^{-3} C$_8$H$_{17}$-CHCN-COOH into t-butyl benzene

SANEX 3: Aqueous feed Acidity 1 mol dm^{-3} HNO$_3$

0.04 mol dm^{-3} n-PrBTP

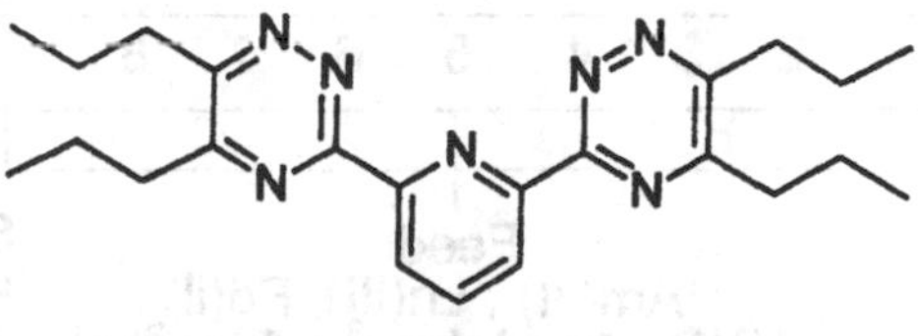

into TPH -octanol (70/30% vol.)

SANEX 4: Aqueous feed Acidity 0.5 mol dm^{-3} HNO$_3$

0.5 mol dm^{-3} BCDTP   + 0.25 mol dm^{-3} TOPO
                              into t-butyl benzene

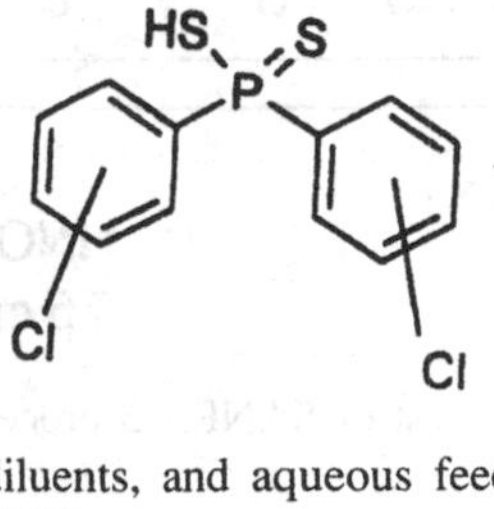

Fig. 12.20 Extractants, diluents, and aqueous feeds for the trivalent 5f–4f ion group separations used in the SANEX processes.

1. SANEX 1 was based on the preferential extraction of trivalent actinides over lanthanides using as solvent $0.02\,mol\,dm^{-3}$ *tris*-pyridine (Tpy) + $1\,mol\,dm^{-3}$ α-bromocapric acid in *tert*-butylbenzene. Flow sheet calculations showed that this process is operable but is very sensitive to the flow ratio (organic/aqueous) and the aqueous acidity, which must be maintained around $0.01\,mol\,dm^{-3}$.

2. SANEX 2 used a solvent composed of $0.1\,mol\,dm^{-3}$ 4,6-di-(pyridin-2-yl)-2-(3,5,5-trimethylhexanoylamino)-1,3,5-triazine (BADPTZ) + $1\,mol\,dm^{-3}$ α-cyanodecanoic acid + $0.5\,mol\,dm^{-3}$ DMDBTDMA with TPH as diluent. The aqueous feed was a mixture of actinides and lanthanides, mimicking the genuine output of the Diamex process in $0.2\,mol\,dm^{-3}$ HNO_3.

3. SANEX 3 used the 2,6-*bis*-(5,6-dialkyl-1,2,4-triazin-3-yl)pyridine family of extractants containing the propyl group, which are able to selectively extract trivalent actinides from $1\,mol\,dm^{-3}$ HNO_3 [53,78]. The extractant was $0.04\,mol\,dm^{-3}$ in a mixture of 70%/30% TPH/octanol. A flow sheet based on this solvent has been tried and is shown schematically in Fig. 12.21; the results are given in Table 12.13. The feed was an

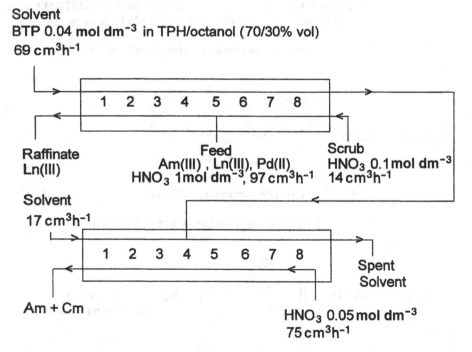

Fig. 12.21 Flow sheet for the hot test of SANEX 3 process using n-Pr-BTP.

Table 12.13 Decontamination Factors, *DF*, for the Extraction (Mass in the Feed/Mass in the Actinide Back-Extraction) in the SANEX 3 Hot Test Depicted in Fig. 12.21

Element	% in raffinate	% in actinide strip	% in solvent	*DF* feed/strip
Am(III)	0.11	98.34	1.71	1.017
Cm(III)	0.15	93.89	8.07	1.065
Y(III)	114[a]	1.8	<0.06	54
La(III)	119[a]	<0.05	<0.05	>2000
Ce(III)	118[a]	<0.03	<0.03	>3800
Pr(III)	120[a]	<0.05	<0.05	>1800
Nd(III)	119[a]	<0.02	<0.02	>6300
Sm(III)	114[a]	0.4	<0.08	246
Eu(III)	112[a]	2.3	<0.37	43
Gd(III)	108[a]	3.1	<0.28	32
RuNO(III)	109[a]	<1	0.66	>100
Pd(II)	0.09	0.97	75.5	103
Fe(II)	6.8	<2	73.4	>50

[a]Uncertainties of analysis are responsible for mass balances >100%.

active solution simulating the aqueous output of the Diamex process. It can be seen that Pd(II) and Fe(III) accumulate in the solvent, while Ru is not extracted. The decontamination factors for Eu and Gd are rather low [78], and kinetic or phase disengagement problems are probably responsible for the poor separation.

4. SANEX 4: The general properties of the sulfur donors, dithiophosphoric acids, published earlier [79] were confirmed [53]. More chemically stable molecules *bis*-(2,4,4-trimethylpentyl)dithiophosphinic acid, CYANEX 301 [80], or *bis*-(chlorophenyl)dithiophosphonic acid [81] were investigated. Flow sheets based on synergistic mixtures of $0.5 \, mol \, dm^{-3}$ *bis*-(chlorophenyl)-dithiophosphonic acid and $0.25 \, mol \, dm^{-3}$ TOPO, which permit rather acidic feeds ($0.5 \, mol \, dm^{-3}$ HNO_3) to be used, were also tested. The first experiments showed several problems such as formation of gels, decomposition of the extractant at the high acidities needed for stripping, and extraction and accumulation of Fe, Pd, Tc, and Np in the solvent. Low extraction and stripping kinetics prevent these experiments reaching the separation performance expected from the test-tube distribution ratios. However, very high Am(III)-Eu(III) separation factors, e.g., 17,500, were measured, indicating that the sulfur donors have a high potential for

trivalent actinide lanthanide group separation and must be considered further.

12.9.3 Uranium Mining–Milling Operations

As noted earlier (see section 12.3.1), the Amex process, which uses long-chain amines, is preferred over the Dapex process, which uses HDEHP, for solvent extraction of uranium from H_2SO_4 leach solutions. Because the surfactant properties of amine sulfates are conducive to formation of objectionable emulsions, the Amex process is very sensitive to the presence of solids in the H_2SO_4 leachate. For acceptable phase coalescence in the Amex process, feeds should contain no more than 20 ppm solids. The Dapex process can tolerate feeds containing as much as 100 ppm solids.

Thus there is considerable incentive to find extractants that could tolerate higher quantities of solids in H_2SO_4 leach liquors. Stripping of uranium from the Amex process extractant and subsequent regeneration of the amine solvent also consume considerable quantities of acid and base. Recovery of uranium from H_2SO_4 solutions would be simplified if a convenient neutral extractant could be found. An extractant with better selectivity for vanadium and molybdenum than HDEHP and long-chain amines is also desirable.

Radium, thorium, and other radionuclides accumulate in uranium mill tailings. The potential environmental effects of these radionuclides has become of increasing concern to the public. In the future, it may be necessary to modify existing uranium recovery processes to accommodate removal of radium and perhaps other radioactive decay products of uranium.

Thorium is produced in smaller amounts than uranium, but if its production increases in the future the tailings problem will be very similar. The rare earth industry also produces comparable radioactive effluents because many minerals that contain rare earths (e.g., monazites) also include Th and U.

12.10 SUMMARY

Solvent extraction technology has an unchallenged role in the nuclear fuel cycle and such technology has been aggressively developed and implemented over the last five decades to a fully mature and complete status. Incentives, either technical or economic, to make any short-term changes or perceived improvements to present advanced engineering-scale fuel cycle practice are not apparent and likely do not exist. Long-term changes in fuel processing strategies, e.g., implementation and transmutation of actinides and selected fission products, have been suggested by some. But even in this case, workable and efficient solvent extraction systems employing either organophosphorus or organonitrogen-based reagents for removing and separating actinides from high-level waste solutions are known and await only pilot-scale testing and demonstration in countercurrent equipment with fully radioactive feed solutions, should a decision be made to implement a transmutation approach.

Of course, even without engineering-scale economic and technical drivers, studies of new nuclear solvent extraction technology can and will proceed at academic and national laboratory institutions. Areas for which new technology could be beneficial include, among others, development of extractants that can be readily incinerated; detailed information concerning the kinetics of extraction of various solutes; and perhaps, development of contactors with very short residence times. Extraction kinetics must be more carefully investigated in the future to be able to take advantage of kinetic differences, especially between the actinides and the *d*-transition elements.

REFERENCES

1. Schulz, W.; Burger, L. L.; Navratil, J. D. Eds. *Science and Technology of Tributyl Phosphate, Vol. 3, Applications of Tributyl Phosphate in Nuclear Fuel Processing*; CRC Press: Boca Raton, Florida, 1990.
2. Hewlett, G.; Anderson, O. E. The New World 1939/1946; Pennsylvania State University Press: State College, Pennsylvania, 1962.
3. Long, T. *Engineering for Nuclear Fuel Reprocessing*; American Nuclear Society: La Grange Park, Illinois, 1978.
4. Wymer, R. G.; Vondra, B. L. Eds. *Light Water Nuclear Reactor Fuel Cycle*; CRC Press: Boca Raton, Florida, 1981.
5. Stoller, S. M.; Richards, R. B. Eds. *Reactor Handbook*, 2nd Ed.; Interscience Publishers: New York, 1961.
6. Lo, T. C.; Baird, M. H. I.; Hanson, C. Eds. *Handbook of Solvent Extraction*; John Wiley and Sons: New York, 1983.
7. McKay, H. A. C.; Miles, J. H.; Swanson, J. L. *Science and Technology of Tributyl Phosphate*; Schulz, W. W., Burger, L. L., Navratil, J. D., Eds.; CRC Press: Boca Raton, Florida, Vol. 3, Chap. 1, 1990.
8. Danesi, P. R. *Development in Solvent Extraction* Alegret, S. Ed.; Ellis Horwood: Chichester, U.K., 1988; Chap. 12.
9. Warf, J. C. J. Am. Chem. Soc., **1949**, *71*, 3257.
10. McKay, H. A. C. *Science and Technology of Tributyl Phosphate*; Schulz, W. W. Navratil, J. D. Eds.; CRC Press: Boca Raton, Florida, 1990; Vol. 1, Chap. 1.
11. Smith, L. E.; Page, J. E. J. Chem. Soc., **1948**, 67.
12. Coleman, C. F.; Brown, K. B.; Moore, J. G.; Crouse, D. J. Ind. Eng. Chem., **1958**, *50*, 1756.
13. Chesne, A.; Koehly, G.; Bathellier, A. Nucl. Sci. Eng., **1963**, *17*, 557.

14. Peppard, D. F. Liquid-liquid extraction of metal ions. In *Advances in Inorganic Chemistry and Radiochemistry*; Academic Press: New York and London, 1966; Vol. 9, p. 1.
15. Schulz, W. W.; Navratil, J. D. Solvent extraction with neutral bidentate organic phosphorus reagents. In *Recent Developments in Separation Science*, Li, N. N. Ed.; CRC Press: Boca Raton, Florida, 1982; Vol. VII, pp. 31–72.
16. Myaesoedov, B. F.; Chmutova, M. K. New methods of transplutonium elements isolation, purification and separation. In *Separation of f Elements*, Nash, K. L. Choppin, G. R. Eds.; Plenum Press: New York, 1995; p. 11.
17. Saran, M. S. Solv. Ext. Ion Exch., **1989**, 7, 783.
18. Arhland, S.; Liljenzin, J. O.; Rydberg, J. *Comprehensive Inorganic Chemistry*; Trotman-Dickenson A. E. Ed.; Pergamon Press: New York, Vol. 5, 1973; p. 465.
19. Musikas, C.; Le Marois, G.; Fitoussi, R.; Cuillerdier, C. *Actinide Separations*; Navratil, J. D.; Schultz, W. W. Eds.; ACS Symposium Series 117, 1980; p.131.
20. Lloyd, P. J. D. *Handbook of Solvent Extraction*; Lo, T. C. Baird, M. H. I. Hanson, C. Eds.; John Wiley and Sons: New York, 1983; p. 763.
21. Ritcey, G. M.; Ashbrook, A. W. *Solvent Extraction, Principles and Applications to Process Metallurgy*; Part 2; Elsevier Scientific: New York, 1979; Chap. 4.18.
22. Benedict, M.; Pigford, T. H.; Levi, H. W. *Nuclear Chemical Engineering*; 2nd Ed.; McGraw-Hill: New York, 1981; Chap. 5.8.
23. Nagasaki, T.; Shinkai, S. J. Chem. Soc. Perkin Trans. **1991**, 2, 1063.
24. Sherrington, L. *Handbook of Solvent Extraction*; Lo, T. C. Baird, M. H. I. Hanson, C. Eds.; John Wiley and Sons: New York, 1983; Chap. 25.6.
25. Benedict, M.; Pigford, T. H.; Levi, H. W. *Nuclear Chemical Engineering*; 2nd ed.; McGraw-Hill: New York, 1981; Chap. 6.8.
26. Carminati, F.; Klapisch, R.; Revol, J. P.; Roche, Ch.; Rubio, J. A.; Rubbia, C. Report CERN/AT/93-47(ET) November 1993.
27. Naylor, A.; Wilson, P. *Handbook of Solvent Extraction*; Lo, T. C. Baird, M. H. I. Hanson, C. Eds.; John Wiley and Sons: New York, 1982; p. 784.
28. Benedict, M.; Pigford, T. H.; Levi, H. W. *Nuclear Chemical Engineering*; McGraw-Hill: New York, 1981; p. 466.
29. Miles, J. H. *Science and Technology of Tributyl Phosphate*; Schulz, W. W. Burger, L. L. Navratil, J. D. Eds.; CRC Press: Boca Raton, Florida, 1990; Vol. 3, Chap. 1, Part 2.
30. Guillaume, B.; Germain, M.; Puyou, M.; Rouger, H. *Proceedings of International Conference on Nuclear Fuel Reprocessing, RECORD 87*; Soc. Francaise d'Energie Nucleaire: Paris, 1987; Vol. 1, p. 433.
31. Schulz, W. W. Macroreticular Ion Exchange Resin Cleanup of Purex Process Solvent; U.S. Atomic Energy Commission Report ARH-SA-58, Atlantic Richfield Hanford Co.; Richland, Washington 1970.
32. Mailen, J. C. Nucl. Technol., **1998**, 83, 182.
33. Reif, D. J. Nucl. Technol., **1988**, 83, 190.
34. *Proceedings of the International Conference on Nuclear Reprocessing, RECORD 87*; Soc. Francaise d'Energie Nucleaire: Paris, Vol. 1, 1987.

35. Kolarik, Z.; Shuler, R. *Proceedings of Extraction'84*; Inst. Chem. Eng. Symposium Series No. 88, Pergamon Press: London, 1984; p. 83.
36. Drake, V. A. *Science and Technology of Tributyl Phosphate*; Schulz, W. W.; Burger, L. L.; Navratil, J. D. Eds.; CRC Press: Boca Raton, Florida, 1990; Vol. 3, Chap. 3.
37. Baron, P.; Boullis, B.; Germain, M.; Gué, J. P.; Miquel, P.; Poncelet, F. J.; Dormant J. M.; Dutertre, F. Extraction cycles design for La Hague plants, *Proceedings of the International Conference on Future Nuclear Systems Emerging Fuel Cycles and Waste Disposal Options, Global '93*; Sept.12–17, Seattle Washington, 1993.
38. Hugelmann, D.; Pradel, Ph. Revue Générale Nucléaire, **1995**, *1*, 30.
39. Musikas, C. Sep. Sci. Technol., **1998**, *23*, 1211.
40. Gasparini, G. M.; Grossi, G. Solv. Extr. Ion Exch., **1986**, *4*, 1233.
41. Küchler, L.; Schäfer, L.; Wojtech, B. Kerntechnik, **1970**, *12*, 327.
42. Schulz, W. W.; Navratil, J. D.; *Recent Developments in Separations Science*; Li, N. N. Ed.; CRC Press: Boca Raton, Florida, 1980; Vol. 7, p. 31.
43. Schulz, W. W.; Horwitz, E. P. Sep. Sci. Technol., **1988**, *23*, 1191.
44. Nash, K. L.; Rickert, P. G.; Horwitz, E. P. Solv. Extr. Ion Exch., **1989**, *7*, 655.
45. Ozawa, M.; Nemoto, S.; Toghashi, A.; Kawata, T.; Onishi, A. Solv. Extr. Ion Exch., **1992**, *10*, 829.
46. Prybylova, G. A.; Chmutova, M. K.; Nesterova, N. P.; Myaesoedov, B. F.; Kabachnik, M. I. Radiokhimiya **1991**, *33*, 70.
47. Myaesoedov, B. F.; Chmutova, M. K.; Smirnov, I. V.; Shadrin, A. U. *Global '93: Future Nuclear Systems: Emerging Fuels Cycles and Waste Disposal Options*, American Nuclear Society: La Grange Park, Illinois, 1993; p. 581.
48. Babain, V. V.; Shadrin, A. Yu. *New Extraction Technologies for Management of Radioactive Wastes in Chemical Separation Technologies and related Methods of Nuclear Waste Management*, Choppin, G. R.; Khankhasayev M. Kh. Eds.; NATO Science Series, 2- Environmental Security; Kluwer Academic Publishers 1999; Vol. 53.
49. Horwitz, E. P.; Shulz, W. W. *Metal Ion Separation and Preconcentration Progress and Opportunities* Bond, A. H. Dietz, M. L. Rogers, R. D. Eds., American Chemical Society, 1998.
50. Musikas, C.; Hubert, H. Solv. Extr. Ion Exch., **1987**, *5*, 877.
51. Nigond, L.; Musikas, C; Cuillerdier, C. Solv. Extr. Ion Exch., **1994**, *12*, 297.
52. Madic, C.; Hudson, M. J. *High Level Liquid Waste Partitioning by Means of Completely Incinerable Extractants*; European Commission, Nuclear Science and Technology, EUR18038 EN 1998; p. 208.
53. Madic, C.; Hudson, M. J.; Liljenzin, J. O.; Glatz, J. P.; Nannicini, R.; Facchini, A.; Kolarik, Z.; Odoj, R. *New Partitioning Techniques for Minor Actinides*; European Commission, Nuclear Science and Technology, EUR19149 EN 2000; p. 286.
54. Liljenzin, J. O.; Rydberg, J.; Skarnemark, G. Sep. Sci. Technol., **1980**, *15*, 799.

55. Cecille, L.; Dworschak, H.; Girardi, E.; Hunt, B. A.; Mannone, F.; Mousty, E. *Actinides Separations*; Navratil, J. D. Schulz, W. W. Eds.; American Chemical Society: Washington D.C., 1980; p. 427.

56. Schulz, W. W. *The Chemistry of Americium*; U.S. Department of Energy Technical Information Center; Oak Ridge, Tennessee, 1976; p. 203.

57. Liljenzin, J. O.; Persson, G.; Svantesson, I.; Wingefors, S. The CTH-process for HLLW treatment, Part I–General description and process design. Radiochim. Acta. **1984**, *35*, 155.

58. Persson, G.; Wingefors, S.; Liljenzin, J. O.; Svantesson, I. The CTH-Process for HLLW treatment, Part II—Hot test. Radiochim. Acta. **1984**, *35*, 163.

59. Persson, G.; Svantesson, I.; Wingefors, S.; Liljenzin, J. O. Hot test of a TAL-SPEAK procedure for separation of actinides and lanthanides using recirculating DTPA—Lactic acid solution. Solv. Extr. Ion Exch., **1984**, *2*(1), 89.

60. Horwitz, E. P.; Dietz, M. L.; Fisher, D. E. Solv. Extr. Ion Exch., **1990**, *8*, 557.

61. Rais, J.; Selucky, P.; Kyrs, M. J. Inorg. Nucl. Chem., **1976**, *38*, 1376, 1742.

62. Shuler, R. G.; Bowers, C. B. Jr.; Smith, J. E. Jr.; Van Brunt, V.; Davis, M. W. Jr. Solv. Extr. Ion Exch., **1985**, *3*, 567.

63. Dozol, J. F.; Scwing-Weill, M. J.; Arnaud-Neu, F.; Böhmer, V.; Ungaro, R.; van Veggel, F. C. J. M.; Wipff, G.; Costero, A.; Desreux, J. F.; de Mendoza, J. *Extraction and Selective Separation of Long-Lived Nuclides by Functionalized Macrocycles*; European Commission, Nuclear Science and Technology series, EUR19605 EN 2000; p. 198.

64. King, L. J.; Bigelow, J. E.; Collins, E. D. *Transplutonium Elements—Production and Recovery*; Navratil, J. D. Schulz, W. W. Eds.; ACS Symposium Series 161, American Chemical Society: Washington, D.C. 1981; p. 133.

65. Berger, R.; Koehly, G.; Musikas, C.; Pottier, R.; Sontag, R. Nucl. Appl. Technol., **1970**, *8*, 371.

66. Schulz, W. W. *The Chemistry of Americium*; U.S. Department of Energy Technical Information Center: Oak Ridge, Tennessee, 1976, p. 30.

67. Silva, R.; Sikkeland, T.; Nurmia, M.; Ghiorso, A. Inorg. Nucl. Chem. Lett., **1970**, *6*, 733.

68. Skarnemark, G.; Skälberg, M.; Alstadt, J.; and Björnstad, T. Nuclear Studies with the Fast In-Line Chemical Separation System SISAK, Physica Scripta, **1986**, *34*, 597.

69. McDowell, W. J. Radioactivity Radiochem., **1992**, *3*(2), 26.

70. Raghavan, R. S. Phys. Rev. Lett. **1997**, *78*, 3618.

71. Choppin, G.; Liljenzin, J. O.; Rydberg, J. *Radiochemistry and Nuclear Chemistry*; 3rd Ed.; Butterworth-Heinemann, 2002; p. 593.

72. *CURE: Clean Use of Reactor Energy*, U.S. Department of Energy Report WHC-EP-0268; Westinghouse Hanford Co.: Richland, Washington, 1990.

73. Schulz, W. W.; Bray, L. A. Sep. Sci. Technol., **1987**, *22*, 191.

74. Mukaiyama, T.; Kubota, M.; Takizuka, T.; Ogawa, T.; Mizumoto, M.; and Yoshida, Y. Partitioning and Transmutation Program OMEGA at JAERI, *Proceedings of the GLOBAL'95 conference*; Versailles, France, **1995**, *1*, 110.

75. Barré, J. Y.; Bouchard, J. French R&D strategy for the back-end of the fuel cycle. *Proceedings of the Global '93*; Seattle, Washington, September 1993; 12–17.
76. Musikas, C.; Vitart, X.; Pasquiou, J. Y.; Hoel, P. *Chemical Separations*; King C. J.; Navratil, J. D. Eds.; Litarvan Literature: Denver, 1986; Vol. 2, p. 359.
77. Musikas, C.; Vitorge, P.; Pattee, D. *Proceedings International Solvent Extraction Conference ISEC 83*; American Institute of Chemical Engineers: Denver, 1983; p. 6.
78. Kolarik, Z.; Müllich, U.; Gassner, F. Solv. Ext. Ion Exch., **1999**, *17*, 23, 1155.
79. Musikas, C. *Proceedings of the International Symposium on Actinides/Lanthanides Separations*; Honolulu, Hawaii, 1984; and *Actinide-Lanthanide Separation*; World Scientific: Singapore, 1985 p. 19.
80. Zhu, Y. Radiochim. Acta. **1995**, *68*, 95–98.
81. Modolo, G.; Odoj, R. J. Radioanal. Nucl. Chem. **1998**, *228*(1,2), 83.
82. Miquel, P. Bull. Inf. Sci. Tech., CEA **1973**, *57*, 184.

13
Analytical Applications of Solvent Extraction

MANUEL AGUILAR, JOSÉ LUIS CORTINA, and ANA MARÍA
SASTRE   Universitat Politècnica de Catalunya, Barcelona, Spain

13.1 INTRODUCTION

The role of solvent extraction in analytical chemistry has steadily increased since the mid-1950s as a powerful separation technique applicable both to trace and macro levels of materials. Work in this field has provided the basis for a rich store of analytical methodology characterized by high sensitivity and selectivity as is described in Chapters 2–4. Developments of new extractants and their application to separation of a growing variety of compounds are recognized as an important area of analytical chemistry. Advances in this field over the last 50 years have been reported in a large number of publications, among them several monographs and reviews [1–5]. Because of the great range of concentrations (from weightless trace levels of carrier-free radioisotopes to macro levels of several weight percent of metal ions) for which quantitative separations by solvent extraction are applicable, this technique is equally useful in analytical, preparative, and process chemistry. Solvent extraction has been also used as a separation step in many analytical techniques and methods in response to the new problems posed in many other fields such as medicine, biology, ecology, engineering, etc. Among these, automatic methods of analysis have gained a notable momentum and have motivated the development of a large number of commercial instruments.

Solvent extraction has also played a major role in sample pre- or posttreatment to improve selectivity and sensitivity. Initially simple schemes of liquid–liquid extraction were used as separation methods for the clean-up and preconcentration of samples, mainly because of its simplicity, reproducibility, and versatility. Later, the distribution between two liquid phases was used as an efficient tool in chromatographic separation processes. More recently, the same principles have been implemented using supercritical fluids in preconcentration, cleanup, and column separation schemes.

In the past decades, the principles and the properties of solvent extraction reagents were used to develop extraction chromatography techniques, to design different types of membrane electrodes, and to prepare and develop impregnated materials. In such materials where the extractant is placed in a solid matrix, commonly an organic polymer or an inorganic adsorbent, advantage is taken of their improved properties when high volumes of aqueous samples are to be treated.

13.2 ROLES OF SOLVENT EXTRACTION IN ANALYTICAL CHEMISTRY

The analytical process can be defined as a set of operations separating the untreated, unmeasured sample from the results expressed as required in accordance with the "analytical black box" concept. However, following modern schemes of analysis, the total analytical process can be defined by a set of operations as shown in Fig. 13.1 [6]. According to these ideas, the so-called preliminary operations comprise a series of steps such as sampling, sample preservation and treatment (e.g., dissolution, disaggregation), separation techniques, development of analytical reactions, and transfer of an adequate portion of the treated sample to the detector. The second stage of the analytical process requires the use of one or several instruments to generate pertinent information. The resulting analytical signal (optical, electrochemical, thermal, etc.) should be unequivocally related to the presence, amount, or chemical structure of one or several analytes. Solvent extraction has been extensively used in two of these steps in the analytical process: step 4, sample preparation (pretreatment, separation) and step 5, which is more concerned with measurement. Details of procedures and methodologies are described in the following paragraphs.

13.3 SOLVENT EXTRACTION IN SAMPLE PREPARATION AND PRETREATMENT STEPS

In sample preparation or sample pretreatment steps there are a number of important operations that may include: dissolution of the sample, transformation of the elements into specific inorganic forms, conversion of the

1) General statement of the problem

2) Specific analytical statement of problem and definition of objectives

3) Selection of procedure

4) Sampling and preservation

5) Sample preparation (pretreatment, separation of interferences)

6) Measure (signal measurement and transduction)

7) Evaluation of data (data handling and treatment)

8) Conclusions

9) Report

Fig. 13.1 General scheme of the analytical process.

analytes into alternative chemical species, separation of the analyte from other chemical species present, and preconcentration. The significance of these preliminary operations is generally very critical because they could be the source of major errors that may hinder analyte preconcentration and elimination of matrix effects. On the other hand, these procedures can be rather complex and time consuming and thus require ample dedication. Then, the first stage of the analytical process decisively influences the precision, sensitivity, selectivity, rapidity, and cost. Sample pre- and post-treatments in analytical schemes are intended to:

1. Improve selectivity, by removal of interfering species from the sample matrix
2. Improve sensitivity by means of preconcentration
3. Prevent the deterioration of the analytical system by a sample cleanup step

13.3.1 Operation Modes

Liquid–liquid extraction is by far the most popular separation method for the cleanup and preconcentration of samples because it is simple, reproducible, and versatile. There are several ways to achieve these objectives, from the original discontinuous ("batch") and nonautomatic techniques to continuous separation techniques incorporated with automated methods of analysis. The methodologies can be classified into two general types:

13.3.1.1 Discontinuous Methods (Batch)

Discontinuous methods are performed in conventional separating funnels in one or more steps. In ultratrace analyses, tapered or specially profiled quartz tubes are recommended because of their easier cleaning (more compact size), the introduction of less contaminating material, and easier centrifugation in the case of difficulties with phase separation. Shaking must be continued until equilibrium is reached, which may last seconds, minutes, or (rarely) hours, depending on the physicochemical properties of the system; more than 2–5 min requires a mechanical shaker. Microscale extraction carried out in autosampler tubes, followed by direct automatic introduction of the organic phase into the atomizer, is recommended.

13.3.1.2 Continuous Methods

Continuous methods can be performed by different procedures, such as solvent recirculation, extraction chromatographic techniques, or countercurrent chromatography. The introduction of extractors in continuous methods such as segmented flow analysis (SFA), flow injection analysis (FIA), and completely continuous flow analysis (CCFA) that are used as pre- or postcolumn devices in chromatographic separation techniques (liquid chromatography) have been used widely in the last decade. Continuous-flow extraction involves segmenting of an aqueous stream with an organic solvent and separation of the phases using a membrane separator [7]. This option is seldom used in practice because the preconcentration factors are small and solvent consumption is large. An interesting possibility is offered by countercurrent extraction, in which gravitational forces retain the organic phase while the aqueous phase is pumped through it. The basic components of a continuous liquid–liquid extractor are shown in Figure 13.2.

As can be seen in this figure, a continuous liquid–liquid extraction system consists of three main parts and performs the following functions:

Fig. 13.2 General scheme of a continuous solvent extraction contactor.

1. It receives the two streams of immiscible phases and combines them into a single flow with alternate and regular zones of the two phases (solvent segmenter).
2. It facilitates the transfer of material through the interfaces of segmented flow in the extraction coil, the length of which, together with the flow rate, determines the duration of the actual liquid–liquid extraction.
3. It splits, in a continuous manner, the segmented flow from the extraction coil into two separated phases (phase separator).

The functions of these three parts are based on the same fundamental principle, i.e., the selective wetting of internal component surfaces by both organic and aqueous phases. In general, it is found that organic solvents wet Teflon surfaces whereas aqueous phases wet glass surfaces.

13.3.2 Solvent Extraction for Separation Steps

Solvent extraction can facilitate the isolation of analyte(s) from the major component (matrix) and/or the separation of the particular analyte from concomitant trace or minor elements. Extraction is usually a fast and simple process, that demands only very simple equipment but having as a disadvantage its rather low preconcentration coefficient. In practice, the separation/preconcentration step consists either of selectively removing the matrix without affecting the analyte(s) or of isolating the analyte(s) without affecting the matrix. There is a variety of separation/preconcentration methods available [8]. The most popular include solvent extraction, precipitation, sorption and chromatographic techniques, volatilization, and electrodeposition. The choice is dictated by the sample to be analyzed, the analytes and concentration levels to be determined, and the characteristics of the determination technique. In any case, the incorporation of a separation/preconcentration step increases the analysis time, may result in losses of the analyte(s), and at the same time raises demands both on the purity of reagents used and the analytical expertise required.

The selectivity of the extraction is expressed by the separation factor S, which is derived from the individual distribution coefficients (D_1, D_2) for two species (1, 2):

$$S = D_1/D_2 \tag{13.1}$$

This coefficient gives quantitative information for the separation of both species. Working with mixtures of extractants, additional experimental information is needed and the synergistic coefficient (SC) defined by [9]:

$$SC = \log D_{12}/(D_1 + D_2) \tag{13.2}$$

where D_{12} denotes the distribution coefficient of the species for the mixture of extractants. In analytical chemistry, for an extraction system based on a pH-dependent extraction reaction, as in the case of acidic extractants, the value of SC has been estimated by the following equation:

$$SC = n \, \Delta pH_{50} \tag{13.3}$$

where n is the charge of the metal ion and ΔpH_{50} is the difference of pH corresponding to 50% extraction when the total concentration of the extraction system is the same for the single system and for the mixtures.

13.3.3 Preconcentration for Trace Element Determination

Preconcentration is an operation in which the relative ratio of trace components vs. the macro component is increased; it is aimed, typically; for overcoming limited detection characteristics of the instrumental technique. The efficiency of this operation, the preconcentration factor (PF), is defined in terms of recovery as:

$$PF = [A_T]/[A_T^o] \tag{13.4}$$

where $[A_T]$ and $[A_T^o]$ are the concentrations of the microcomponent in the concentrate and in the sample respectively. The need for preconcentration of trace compounds results from the fact that instrumental analytical methods often do not have the required selectivity and/or sensitivity. Thus, the combination of instrumental techniques with concentration techniques significantly extends the range of application of the instrument. The chemical techniques used in preconcentration can provide, in many cases, analyte isolation, as well as high enrichment factors. The concept of sample preconcentration prior to determination could apply to many situations other than just concentration enrichment and minimization of matrix effects. Selection of a preconcentration scheme based on any liquid–liquid extraction step will depend upon on the type of analyte and/or sample matrix.

Concentration factors reasonably achieved by solvent extraction have practical limits. However, considering a reasonable practical limit of extracting a $100 \, cm^3$ sample with $1–2 \, cm^3$ of organic phase, a concentration factor of 100 to 50 can be achieved. One of the limitations of the preconcentration factor in batch solvent extraction is the difficulty of obtaining good separation of the small volume of organic phase. Significant portions of the organic solvent may adhere to the walls of the vessel, requiring repeated washings, which decreases the concentration factor.

Thus, to conduct successful analyses for many organic and inorganic compounds at trace concentrations, it is necessary to extract these compounds and use a concentration step prior to analysis. Many of the techniques developed for preconcentration are described in specialized books [10]. Proper choice of the extracting solvent can often be the critical step in the procedure.

13.3.3.1 Preconcentration of Inorganic Compounds

There has been a growing demand for highly sensitive analytical methods of trace components in complicated matrices. Recent advances of new ligands, chelating reagents, and instrumentation have improved detection and specificity. However, in the presence of an interfering matrix, or when the concentration of the component of interest is far less than the instrumental detection limit, it is necessary to concentrate the trace component with high selectivity beforehand. Although many types of reagents can be used to extract a wide variety of metal ions from different media, chelating agents are typically widely used. It is interesting to note that a considerable number of methods for specific extractions with reagents such as 8-hydroxyquinoline have been developed in the past, basically because the instrumentation at that time was not sufficiently specific and so prior separation was required. For example, interference in atomic absorption determinations are common, and so separations of interfering species are frequently required. Group preconcentration of trace elements has the objective of isolating the maximum number of elements in a single step using the minimum number of reagents, with the excess reagents being retained in the aqueous phase. On the other hand, by suitable choice of the appropriate pH and sometimes using appropriate masking agents, these reagents may become highly selective and even specific for a particular species. In some cases exchange techniques in which a less stable metal chelate is the source of the chelating agent may be very useful. Extraction reagents available for this purpose fall into several basic categories (solvating, chelating, ion exchangers, etc.) as described in Chapter 4.

A series of preconcentration techniques based on the use of a solid phase carrying a liquid metal extractant has been developed [11,12]. These solid phases substitute the solvent (liquid phase) carried of the liquid–liquid extractant with a solid phase. Such impregnated materials have been prepared using different types of solid supports: paper, resin beds, gels, polyurethane foams, flat and hollow fiber membranes, clays, silica, etc. The major advantage of most of these solid adsorbents is that the functional group is immobilized on a solid substrate, therefore providing the possibility of either batch extraction of the analytes from solution or using the solid phase in a column.

13.3.3.2   Preconcentration and Separation of Organic Compounds

A complete scheme for trace analysis of organic compounds generally consists of sampling, extraction, prefractionation, and analysis by gas chromatography (GC) or gas chromatography/mass chromatography (GC-MS). Determination of organics at parts per trillion (ppt) levels can be performed by combining sensitive and selective detection with sample preconcentration. Increasing the degree of preconcentration can make the limits of detection extremely small. Many standard methods of analysis [13] often include preconcentration as an integral part of the sampling and extraction procedure. It is especially important for environmental samples where many toxic and carcinogenic compounds are distributed from a wide variety of sources. Because of their serious damaging effects at low levels, to assess their environmental impact it is necessary to achieve the greatest possible analytical sensitivities. In some cases, detection limits as low as a few ppt are required. In addition to the polynuclear aromatic hydrocarbons (PAHs), much attention has been focused on many halogenated pollutants such as pesticides, polychlorinated biphenyls (PCBs), trihalomethanes (THMs), polychlorinated dibenzofurans (PCDFs), and dibenzo-*p*-dioxines. Such substances can be found in air, water, and solid and biological samples. Most preconcentration techniques fall into two classes: solvent extraction followed by solvent reduction or sorbent trapping with subsequent solvent elution or thermal desorption. There are many variations of these methods, and they are frequently used in combination.

13.3.3.2a   Solvent Extraction Reduction

Solvent extraction reduction is most frequently performed mainly in connection with the extraction of solid and biological samples by liquid partition. Extractions are typically accomplished using a Soxhlet apparatus that provides the benefits of multiple extractions. By repeated distillation and condensation of the solvent, the apparatus allows multiple extractions using the same (small) volume of solvent. Soxhlet extraction has been a standard method for many decades, and it is often the method against which other extraction methods are measured and verified [14].

Preconcentration will only give precise results for quantitative work if the initial extraction technique gives high, or at least known and reproducible, recoveries of the desired compounds from the initial sample. As typically, several cycles are needed, and the solvent containing the extracted compounds must be concentrated to a small volume. This is normally carried out by evaporation under reduced pressure. Volatile compounds may be lost in this procedure. However, for many applications the compounds of interest are nonvolatile compared to the extracting solvent. In such applications, the use of a rotary evaporation technique is rapid and straightforward, although severe losses of even nonvolatiles can be experienced without careful sample handling. The major problem involves the physical handling of the sample, as even the very small volume losses to the glass walls of the recovery flask or the disposable pipettes commonly used for sample transfer may result in significant and nonreproducible component loss.

13.3.3.2b Accelerated Solvent Extraction (ASE^{TM})

Accelerated solvent extraction is a relatively new extraction technique using equipment that holds the sample in a sealed high-pressure environment to allow conventional solvents to be used at higher temperatures [15]. Using a higher temperature without boiling, ASE allows smaller volumes of solvent to be used in a single-stage extraction. Extraction kinetics are also faster, so the entire process is much faster than Soxhlet extraction. After heating, the cell is allowed to cool to below the normal boiling point of the solvent and pressure is applied to the cell to force the solvent and extracted materials through a filter.

13.4 SOLVENT EXTRACTION AS A MEANS OF ANALYTICAL DETERMINATION

In the previous section, the role of solvent extraction was limited to preparing the analyte for subsequent analysis. A large majority of procedures that use solvent extraction in chemical analysis are used in this fashion. However, the extraction itself, or rather the distribution ratio characterizing it, may provide an appropriate measured signal for analysis. Examples of this use of solvent extraction are found in spectroscopy, isotope dilution radiometry, and ion-selective electrodes using liquid membranes. In the latter case, electrochemical determinations are possible by controlling the local concentration of specific ions in a solution by extraction.

13.4.1 Analytical Methods Based on Spectrophotometric Detection

The oldest application of solvent extraction in spectrophotometric determinations uses extraction from the original aqueous solution and subsequent back-extraction into a second aqueous phase. Here the extractant provides only separation or concentration, as in the case of Np(IV) and Pu(IV) determination [16]. However, as only a few element species (e.g., MnO_4^-, CrO_4^{2-}) are capable of absorbing light in the UV-VIS range, usually all spectrophotometric methods are based on reactions of analytes with color-forming reagents. The role of the most important organic reagents has been discussed [17] and a comprehensive dictionary is available [18].

Chelating reagents are the most popular. Although most of these reactions were initially developed for aqueous phase measurement, in many cases taking into account the low aqueous solubility of the analyte-chromogenic reagents, extraction into an organic phase was used for direct spectroscopic measurements. Among the most important reagents used are the following [19].

Dithizone (diphenylthiocarbazone) is a weak acid insoluble in water at pH <7 but readily soluble in CCl_4 and $CHCl_3$. It reacts with many metal ions to form chelates that can be extracted into the organic phase from which the excess of the green reagent is stripped with dilute NH_3. The most stable dithizonates (Pt, Pd, Au, Ag, Hg, and Cu) are extracted from strongly acid solutions. Other metals (Bi, Ga, In, Zn) are extracted from weakly acid media and some (Co, Ni, Pb, Tl, Cd) from neutral or alkaline media. Some compounds are extracted rapidly (Ag, Hg, Pb, Cd) and others more slowly (Pd, Cu, Zn), while a few (Rh, Ir, and Ru) require prolonged heating to be formed.

Azo dyes contain an azo link between two aromatic rings possessing a *ortho*hydroxy group. The most important reagents include PAN, PAR, and Arsenazo III and generally they offer high sensitivity for the majority of transition metals.

Chelating dyes include triphenylmethane reagents (e.g., Pyrocatechol Violet, Eriochrome Cyanine R, Chrome Azurol S, Xylenol Orange) and xanthene reagents (fluorones, e.g., Gallein, Pyrogallol Red, phenylfluorone, and salicylfluorone). They form chelates with most metals. Ionic surfactants make it easier to dissociate the protons of chelating triphenylmethane reagents and facilitate reactions with easily hydrolyzable metals (Be, Al, Fe, Se, Ti, Zr), leading to very sensitive but poorly selective methods.

Nonchelating dyes include basic triphenylmethane dyes (e.g., Brilliant Green, Malachite Green, Crystal Violet), xanthene dyes (e.g., Rhodamine B, Rhodamine 6G), azine dyes (e.g., Methylene Blue), and acid dyes (e.g., Eosin, Erythrosin). These are intensely colored and when paired with an oppositely charged analyte ion lead to high sensitivities.

Crown ethers are not chromogenic unless they contain a pendant chromogen able to dissociate a proton in a basic medium. The resulting anion interacts strongly with the crown-complexed cation compensating the electric charge. The formation of a zwitterion leads to a hydrophobic extractable species with a considerably shifted absorption maximum compared with the protonated species. This allows the same spectrophotometric determination to be used for a large number of metal ions, provided the appropriate crown compound is used in each case. Another method involves using specific macrocyclic reagents that incorporate a chromophoric group, even if the latter is not involved in the complexation of the metal ion. This latter use of crown compounds is a development of an earlier known method, used when no reagent could be found, which combines selectivity for a certain metal and spectrophotometric sensitivity. In such cases, the analyte is first extracted by a highly selective extractant, after which the reagent providing the photometric determination is added to the extractant phase [20].

13.4.2 X-Ray Fluorescence

Although this method is generally used on solid samples, it may be beneficial in some cases to combine the procedure with solvent extraction. In addition to the usual advantages of separation and preconcentration (before the solid sample is prepared), it may be easier to prepare the required solid sample by evaporation or crystallization from an organic extract phase rather than from an aqueous solution [21]. Of special interest is the technique that uses an extraction system that is solid at room temperature, but has a relatively low melting point. Liquid–liquid extraction is performed at a temperature above the melting point and after separation of the phases, the organic melt is allowed to solidify, to give a purified concentrate of the analyte [22].

13.4.3 Atomic Absorption Spectroscopy and Flame Emission Spectroscopy

It may seem that the high selectivity of these techniques would make a separation process superfluous. However, preconcentration, especially the removal of the solution medium, is often essential in atomic absorption determinations. Extraction into an organic phase decreases the detection limit by increasing the rate of introduction of the solution into the flame, since most organic solvents have a lower viscosity and surface tension than aqueous solutions. In addition, the sensitivity of the determination may be influenced favorably by changes in flame temperature and composition [23]. Using the combination of solvent extraction-flame atomic absorption, attention must be paid to the problems of vapor toxicity or noncombustibility of the extractant or its diluent. For example, benzene or chlorine containing solvents should always be avoided. Favored extraction systems are solutions of a chelating agent such as oxine (8-hydroxyquinoline), a ketone (methylisobutylketone), or butyl acetate [24].

The advantages of solvent extraction in combination with atomic absorption apply equally well for flame emission spectroscopy. In addition, the latter analytical method often requires separation of the analyte from a large excess of other components. This may be achieved either by extracting the elements to be determined or by carrying out the spectrometric analysis of an aqueous solution from which the interfering components have been removed by extraction [25].

13.4.4 Polarography (Voltammetry)

Solvent extraction has also been used to enhance the selectivity of polarographic determinations. Such measurements are normally carried out in aqueous solutions, and extraction followed by back-extraction has been widely used. However, it may be unnecessary to perform a back-extraction if the organic extractant phase has a sufficiently high dielectric constant to dissolve sufficient background electrolyte for a voltammetric determination or if the organic phase can be diluted with suitable polar solvents, such as methanol or acetonitrile [26].

13.4.5 Activation Analysis

Activation analysis is the main radiometric method used in analytical chemistry. The sample is irradiated in a neutron flux, and the resulting specific radioactivity of the analyte determined. Although high-resolution gamma spectrometers would seem to eliminate the need for highly selective separation of individual gamma-emitting radioactive elements, group separation still may be necessary. Solvent extraction can be used in combination with activation analysis in two ways: either by separating the analyte from interfering components and preconcentrating before irradiation, or by performing the neutron activation and separating the analyte before the radiometric determination. The latter method has the advantage of eliminating the need for a blank correction. However, when the interfering components become highly radioactive and the resulting radioisotopes are long-lived, it is preferable to remove them before irradiation.

13.4.6 Isotope Dilution Radiometry

In this type of radiometric analysis, a tracer quantity of a radioactive isotope is added to the analyte, which is then partly extracted using a specific extractant. Since the extractant may be considered as reacting totally with the analyte, the ratio of radioactivity in both phases provides the concentration of analyte in the sample. This method, first developed by Stary [2], has proved to be useful in several systems.

13.4.7 Liquid Scintillation Counting

Carbon-14 and tritium are radioisotopes with β-emissions of very low energy that are extremely difficult to detect with any form of window counter, due to self-absorption of the β-particles and absorption within the counter window. To reduce the self-absorption losses, it is desirable to mix the active sample homogeneously with the detecting material. This can be done by counting the sample in the gaseous phase. The gaseous activity can then be intimately mixed with the filling gas of any type of gas ionization detector, thus minimizing the effect of β-absorption and resulting in high counting efficiencies. In this, the radioactive sample and a scintillator material are both dissolved in a suitable solvent, and the resulting scintillations are detected and counted [27]. The method is called liquid scintillation counting. If a compound containing an α- or β-emitting isotope is dissolved in a solvent such as toluene, the radioactive emissions result in the formation of electronically excited solvent molecules. If the solution also contains a small amount of a suitable scintillator, the excited solvent molecules rapidly transfer their excitation energy to the scintillator, forming electronically excited scintillator molecules, which then relax by the emission of photons. All scintillator solutions contain:

1. a solvent;
2. a primary solute, the scintillator material; and may contain
3. a secondary solute.

The functions of the solvent are to keep the scintillator or solute in solution, and to absorb the decay energy of the radioisotope for subsequent transfer to the solute. Solvents fall broadly into three categories:

1. Effective solvents, e.g., aromatic hydrocarbons such as toluene and xylene.
2. Moderate solvents, e.g., many nonaromatic hydrocarbons.
3. Poor solvents; unfortunately this includes virtually everything else including the most common laboratory solvents such as alcohols, ketones, esters, and chlorinated hydrocarbons.

13.4.8 Other Methods

Several other analytical procedures exist in which solvent extraction may be applied. Thus extraction has been used in a limited number of analyses with procedures such as: (1) luminescence (fluorimetry), where, for example, the detection limit of rhodamine complexes of gallium or indium can be increased by extraction [28]; (2) electron spin resonance using a spin-labelled extractant [29]; and (3) mass spectrometry, where an organic extract of the analyte is evaporated onto pure Al_2O_3 before analysis [30].

Several thousand articles have been published on analytical procedures for chemical elements in which solvent extraction is involved. The scope of this chapter limits us to only one example for each element (Table 13.1).

Table 13.1 Elements Determined by an Analytical Procedure Involving Solvent
Extraction

Element	Extractant	Analytical technique	Ref.
Alkali metals	crown ether compounds	Spectrophotometry (AAS)	30
Lithium	dipivaloylmethane	Spectrophotometry (VIS)	31
Potassium	crown ether compounds	Spectrophotometry (VIS)	5
Rubidium/Caesium	nitromethane	Spectrophotometry (VIS)	32
Copper	dithizone	Isotope dilution	33
	diethyldithiocarbamate	Spectrophotometry (VIS)	17
Silver	diethyldithiocarbamate	Isotope dilution	34
	dithizone	Spectrophotometry (VIS)	17
Gold	diphenyldipyridylmethane	Activation analysis	35
	dithizone	Spectrophotometry (VIS)	17
Beryllium	β-diketones	Spectrophotometry (AAS)	36
Magnesium	8-hydroxyquinoline	Spectrophotometry (VIS)	37
Calcium	tributylphosphate/CCl_4	Spectrophotometry (VIS)	38
Strontium	polyethylene glycol	Activation analysis	39
Barium	crown ether compounds	Spectrophotometry (AAS)	40
Zinc	dithizone	Spectrophotometry (VIS)	41
	diethyldithiocarbamate	Spectrophotometry (VIS)	17
	thioxine	Molecular fluorescence	17
Cadmium	diethyldithiocarbamate	Polarography	42
	thioxine	Molecular fluorescence	17
Mercury	dithizone	Spectrophotometry (VIS)	43
Aluminium	8-hydroxyquinoline	Spectrophotometry (VIS)	44
Gallium	diisopropyl ether	Activation analysis	45
	thioxine	Spectrophotometry (VIS)	17
	8-hydroxyquinoline	Spectrophotometry (VIS)	17
Indium	β-mercaptoquinoline	Polarography	46
	thioxine	Molecular fluorescence	17
	8-hydroxyquinoline	Spectrophotometry (VIS)	17
Thallium	diethyldithiocarbamate	Spectrophotometry (VIS)	47
	8-hydroxyquinoline	Spectrophotometry (VIS)	17
Scandium	mesityl oxide	Spectrophotometry (VIS)	48
Yttrium	diantipyrylmethane	Complexometry	49
Lanthanides	diethylhexylphosphoric acid	Activation analysis	50
Thorium	trioctylphosphine oxide	Spectrophotometry (VIS)	51
	8-hydroxyquinoline	Spectrophotometry (VIS)	17
Uranium	quaternary amine	Spectrophotometry (VIS)	52
	diethyldithiocarbamate	Spectrophotometry (VIS)	17
	8-hydroxyquinoline	Spectrophotometry (VIS)	17
	thenoyltrifluoroacetone	Spectrophotometry (VIS)	17
Germanium	dibutyl ether	Spectrophotometry (AAS)	53
Tin	diisopropyl ether	Activation analysis	54

Table 13.1 (*Continued*)

Element	Extractant	Analytical technique	Ref.
Lead	diethyldithiocarbamate	Polarography	42
	thioxine	Spectrophotometry (VIS)	17
	dithizone	Spectrophotometry (VIS)	17
Titanium	monolaurylphosphoric acid	Isotope dilution	55
Zirconium	trioctylphosphine oxide	Spectrophotometry (VIS)	56
Hafnium	thenoyltrifluoroacetone	Spectrophotometry (VIS)	57
Phosphorous	butanol	Spectrophotometry (VIS)	58
Arsenic	carbon tetrachloride	Spectrophotometry (AAS)	59
	thioxine	Spectrophotometry (AAS)	17
Antimony	dithiocarbamate	Spectrophotometry (AAS)	60
	thioxine	Spectrophotometry (AAS)	17
Bismuth	dithiocarbamate	Spectrophotometry (VIS)	61
	thioxine	Spectrophotometry (VIS)	17
	dithizone	Spectrophotometry (VIS)	17
Vanadium	Cupferron	Spectrophotometry (AAS)	62
	8-hydroxyquinoline	Spectrophotometry (VIS)	17
	thenoyltrifluoroacetone		17
Niobium	tetraphenylarsonium salt	Spectrophotometry (VIS)	63
	8-hydroxyquinoline	Spectrophotometry (VIS)	17
Tantalum	methylisobutyl ketone	Spectrophotometry (AAS)	64
Selenium	diethyldithiocarbamate	Spectrophotometry (AAS)	65
Tellurium	methylisobutyl ketone	Spectrophotometry (AAS)	66
Chromium	methylisobutyl ketone	Activation analysis	67
	8-hydroxyquinoline	Spectrophotometry (VIS)	17
Molybdenum	N-benzoyl-N-phenyl hydroxylamine	Polarography	68
	8-hydroxyquinoline	Spectrophotometry (VIS)	17
Tungsten	methylisobutyl ketone	Spectrophotometry (AAS)	69
Manganese	tetramethylene-dithiocarbamate	Spectrophotometry (AAS)	70
	8-hydroxyquinoline	Spectrophotometry (VIS)	17
Technetium	tetraphenylarsonium chloride	Spectrophotometry (VIS)	71
Rhenium	ALIQUAT 336	Activation analysis	72
Iron	acetylacetone	Polarography	33
	diethyldithiocarbamate	Spectrophotometry (VIS)	17
Cobalt	1-nitroso-2-naphthol	Isotope dilution	73
	diethyldithiocarbamate	Spectrophotometry (VIS)	17
Nickel	dimethylglyoxime	Polarography	74
	diethyldithiocarbamate	Spectrophotometry (VIS)	17
	thenoyltrifluoroacetone	Spectrophotometry (VIS)	17
Platinum	diphenyldithourea	Spectrophotometry (VIS)	74
	diethyldithiocarbamate	Spectrophotometry (VIS)	17
	8-hydroxyquinoline	Spectrophotometry (VIS)	17

(*Continued*)

Table 13.1 (*Continued*)

Element	Extractant	Analytical technique	Ref.
Palladium	diphenyldithourea	Spectrophotometry (VIS)	74
	diethyldithiocarbamate	Spectrophotometry (VIS)	17
	8-hydroxyquinoline	Spectrophotometry (VIS)	17
Rhodium	diphenyldithourea	Spectrophotometry (VIS)	74
	8-hydroxyquinoline	Spectrophotometry (VIS)	17
	thenoyltrifluoroacetone	Spectrophotometry (VIS)	
Ruthenium	8-hydroxyquinoline	Spectrophotometry (VIS)	17

13.5 SOLVENT EXTRACTION–BASED MATERIALS FOR ANALYTICAL APPLICATIONS

13.5.1 Solvent-Impregnated Resins

In the last decade, the development of new impregnated materials with chelating and complexing properties has acquired great importance. These materials are prepared principally by the simple immobilization of complexing organic reagents by adsorption onto conventional macroporous polymeric supports (polar and nonpolar). They provide two important advantages over conventional ion exchange resins: the possibility of selecting the functional group to be immobilized and the facility for continuous column operation.

Such macromolecular resins containing an extractant within the lattice of the polymer were developed to bridge the gap between the two techniques of solvent extraction and ion exchange and are roughly classified into extractant-impregnated sorbents and Levextrel resins [12]. The former are prepared by soaking a polymeric resin in a diluent containing the extractant, then evaporating the diluent (dry impregnation method) or retaining the diluent (wet method). Levextrel resins are prepared by adding an extractant to the mixture of styrene monomers during bead polymerization with divinylbenzene. This method has increased in importance mainly because of its adaptability to preconcentration, separation, and/or determination of analytes. Additionally, they could be used in continuous-flow systems and in solid phase spectrophotometry by using chemical reactions that occur at interfaces (e.g., solid/liquid or gas/solid). Furthermore, most of the analytes could be analyzed directly on the solid matrix using X-ray spectrometry, neutron activation analysis, molecular absorption, fluorescence spectroscopy, or isotopic dilution methods. Alternatively, analytes may be eluted from the column and the analysis completed on the solution. In this context, atomic absorption (AAS) or inductively coupled plasma spectroscopy (ICP) are mainly used.

The use of impregnated resins in the preconcentration and separation of trace metal ions provides the following advantages:

- The active part of the resin (complexing ligand) can be selected to be compatible with the nature of the metal ion and the matrix of the sample, and with the analytical procedure to be applied.
- The impregnation procedures of the complexing molecules are simple.
- Their structure and composition could be compatible with integrated detection systems, when used in solid phase spectroscopic measurements.

13.5.1.1 Solid-Phase Spectrophotometry Applications

Solid-phase spectrophotometry (SPS) is a technique based on the preconcentration of the species of interest onto a solid, aided by complexants or other reagents, and subsequent measurement of the spectrophotometric properties of the species in the solid phase [75]. Depending on the spectrophotometric responses of the analytes or analyte complexes, several procedures based on absorbance and fluorescence measurements have been developed. In the first case, the absorbance of a resin containing the analyte fixed as a colored chromogenic species is measured directly. In the second case, solid phase fluorimetry (SPF), the diffuse reflected fluorescence, is measured. Most procedures using color or fluorescence measurement are based on the addition to the sample solution of a resin impregnated with a highly specific chromogenic agent for the analyte.

Ion exchanger colorimetry has been used as a sensitive and rapid method for vanadium analysis by immobilization of 2[2-(3-5-dibromopyridyl)azo]-5-dimethylaminobenzoic acid onto an ion exchanger resin AG1X2 [75]. Solid phase fluorimetry can be useful for the analysis of very dilute solutions in water analysis or trace metal determination; thus a chelating 8-(benzene-sulfonamido)quinoline, immobilized on Amberlite XAD2 support, has been used for the spectrofluorimetric determination of Zn(II) and Cd(II) [76].

13.5.1.2 Applications for Fiber Optic Chemical Sensors

The development of chemical sensors based on optical measurements has grown steadily in importance during the last decade. While a large variety of devices are possible, they share a common feature in multiple applications, i.e., an immobilized reagent phase that changes its optical properties upon interaction with an analyte on either a continuous or reusable basis. Systems in which chelating liquid extractants are immobilized onto solid polymeric supports have been used for chemical sensing. Particular attention has been given to fiber-optic devices for measuring and controlling metal ions and organic compounds in aqueous media. The applications of such devices have covered areas such as environmental applications, industrial process control, and biomedical and clinical applications. Small sensors based on immobilization of acid-base indicators such as Bromothymol Blue, Bromophenol Blue, Bromocresol Purple, Phenolphtalein, Phenol Red, Chlorophenol Red Alizarin, on nonionic macroporous supports, Amberlite XAD-2 and Amberlite XAD-4, have been developed and used for pH measurements [77]. These fiber-optic probes provide advantages over conventional electrodes in safety, reliability, applicability, and cost. A sensor based on perylene dibutyrate adsorbed onto Amberlite XAD4 has been characterized in some depth [78]. Fiber optic sensors, particularly fluorescent sensors, have become the object of considerable interest among researchers in recent years. Their performance is based on the change in fluorescent properties of organic reagents immobilized on a solid matrix upon contact with solutions of metals in a continuous system.

13.5.2 Liquid Membranes in Analytical Chemistry

The use of liquid membranes in analytical applications has increased in the last 20 years. As is described extensively elsewhere (Chapter 15), a liquid membrane consists of a water-immiscible organic solvent that includes a solvent extraction extractant, often with a diluent and phase modifier, impregnated in a microporous hydrophobic polymeric support and placed between two aqueous phases. One of these aqueous phases (donor phase) contains the analyte to be transported through the membrane to the second (acceptor) phase. The possibility of incorporating different specific reagents in the liquid membranes allows the separation of the analyte from the matrix to be improved and thus to achieve higher selectivity.

Solvents used in liquid membranes should have special characteristics such as low aqueous solubility, as a thin film of solvent is in contact with large volumes of aqueous solutions, and low viscosity to provide large diffusion coefficients in the liquid membrane. Furthermore, the analyte should have large partition coefficients between the donor and the membrane phase to give good extraction recovery and, at the same time, interfering substances in the sample should have low partition coefficients for efficient cleanup.

Two configurations of liquid membranes are mainly used in analytical applications: flat sheet liquid membranes that give acceptable extraction efficiencies and enriched sample volumes down to 10–15 µL, and hollow fiber liquid membranes that allow smaller enriched sample volumes. Flat sheet liquid membrane devices consist of two identical blocks, rectangular or circular in shape, made of chemically inert and mechanically rigid material (PTFE, PVDF, titanium) in which channels are machined so that when assembled the channels face each other. The channels are about 0.10–0.25 mm deep, 1.5 mm wide, and differ in length from 15 cm to 250 cm to provide volumes from 12 µL to 1000 µL. In the smallest devices the channels can be U-shaped grooves, while in other separators they are arranged in spirals with the feed inlet on the periphery and the outlet in the center (Fig. 13.3). The impregnated membrane, which is prepared by soaking the

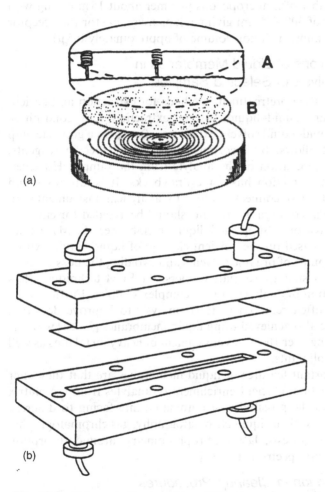

Fig. 13.3 Schematic diagram of two different membrane units. (a) Membrane separator unit composed of two machined blocks of PTFE or Ti (A), and PTFE membrane (B), impregnated with stationary liquid. (b) Membrane unit. The PTFE membrane is placed between the two blocks made of titanium. The two channels (donar and acceptor) that are formed have a nominal volume of 12 µL.

support in the organic phase, is clamped tightly between the planar surfaces of the two blocks. Such devices can be coupled to chromatographic columns, but are too large for on-line connection to packed capillary liquid chromatography or capillary electrophoresis. To decrease the volume while preserving the enrichment efficiency requires shallower channels with the possibility of clogging problems. In such cases, porous hollow fiber modules as supports for the liquid membrane can be used. The hollow fiber module may have only one fiber of a microporous polymer about 15 mm long with an internal diameter of 300–500 μm giving a lumen volume for the acceptor phase of 1.9 μL and annular donor volume of approximately 0.5 μL.

13.5.2.1 Applications of Liquid Membranes in the Analysis of Selected Samples

The most frequently used pretreatment methods for extraction and enrichment of analytes are liquid–liquid solvent extraction and solid-phase extraction (SPE). Liquid–liquid solvent extraction often gives a good cleanup from the matrix. It allows, by incorporating different specific reagents, improvement of the separation of the analyte from the sample. However, liquid–liquid solvent extraction has some drawbacks. It is laborious and difficult to automate and connect on-line to analytical instruments. In addition, large amounts of organic solvents should be avoided for environmental and health reasons. Supported liquid membranes (SLM) are an attractive alternative based on the efficient cleanup of liquid–liquid extraction, using small amounts of organic solvents and avoiding the possibility of emulsion formation. Using appropriate carriers, the SLM technique offers very selective extraction of analytes in very complex samples. Furthermore, preconcentration is often required in trace analysis to improve detection limits and this can be also achieved using liquid membranes [79–81] with the possibility of obtaining over 100 times enrichment of heavy metal ions as well as various organic pollutants.

The most important features of liquid membranes are that they offer highly selective extraction, efficient enrichment of analytes from the matrix in only one step, and the possibility of automated interfacing to different analytical instruments such as liquid chromatography, gas chromatography, capillary zone electrophoresis, UV spectrophotometry, atomic absorption spectrometry, and mass spectrometry [82].

13.5.2.1a Application in Cleanup Procedures

Many analyses of organic compounds in liquid samples require selective cleanup and concentration. Direct on-line coupling of sample preparation to the analytical instrumentation minimizes sample handling and thereby the risk for contamination or loss of analyte. Also, on-line coupling makes automation of the process possible, resulting in more reproducible and economical analysis. Membrane extraction techniques are then well suited to automated interfacing with various separation techniques (chromatography, capillary electrophoresis) as well as with various detection techniques such as UV-VIS and flow-through sensors.

   1. *Liquid membrane sample cleanup coupled to gas–liquid chromatography.* Several types of analysis including environmental [83–85] and biological samples [86–88] that need both cleanup and concentration prior to chromatographic determination can be advantageously pretreated by liquid membranes. Thus on-line combination of supported liquid membrane extraction in a hollow fiber configuration and column liquid chromatography with a phenol oxidase-based biosensor as a selective detector has been used for the determination of phenols in human plasma. The phenols are selectively extracted into a porous PTFE membrane impregnated with a water-immiscible organic solvent [89]. Also a liquid membrane in a miniaturized hollow fiber module coupled on-line to packed capillary liquid chromatography enables the use of very small volumes and has been used to determine amines in plasma using 6-undecanone as extractant.

   2. *Liquid membrane sample cleanup coupled to capillary zone electrophoresis.* Capillary zone electrophoresis (CZE) is a technique that can be used for determination of a great variety of analytes and it is especially suitable when only small sample volumes are available and high separation power is needed. Biological samples are difficult to analyze by CZE due to their complexity and the low concentration of analytes. Thus, before CZE determination, a cleanup procedure is necessary to eliminate the interfering matrix. On the other hand, to improve the detection limit of CZE, a preconcentration step is needed [90]. The use of a liquid membrane of 6-undecanone in a PTFE support has been used successfully as a prior step before CZE determination of bambuterol in human plasma to obtain a high degree of cleanup from the plasma and thus avoid adsorption problems in the CZE capillary [91]. The samples were concentrated more than 1000 times, allowing the detection limit to be lowered about 400 times. A hollow fiber miniaturized supported liquid membrane can be used as sample pretreatment for on-line connection to CZE for determining basic drugs in human plasma [92]. The analyte is extracted from the outside of the hollow fiber (feed phase) through the liquid membrane containing the organic solvent into the strip phase, which flows through the fiber lumen and can then be injected into the CZE capillary.

13.5.2.1b Application to Enrichment Procedures

1. *Liquid membrane enrichment coupled to mass spectrometry.* Membrane introduction mass spectrometry (MIMS) is an established method of sample

analysis that couples rapid introduction via a semipermeable membrane with the sensitivity and specificity of a mass spectrometer. Membranes are chosen to enrich the analyte concentration in the sample stream entering the mass spectrometer while rejecting the bulk of the mobile phase.

The membranes used are typically composed of cross-linked silicones and are suitable for on-line monitoring of volatile organic and inorganic compounds [93–94]. An alternative material is microporous PTFE, which has more rapid responses as well as lower selectivities and higher fluxes of the mobile phase compared to nonporous silicone membranes. More recently, developments in membrane introduction systems include the use of liquid membranes composed, for example, of a polyphenyl ether diffusion pump fluid [95–96]. This membrane has the advantage that it can take any desirable analyte and the selectivity can be modified using appropriate reagents.

2. *Liquid membrane enrichment coupled on-line with ion chromatography.* Low molecular mass carboxylic acids in low concentrations in air or soil samples can be determined by ion chromatography coupled on-line to a selective enrichment system consisting of a supported liquid membrane, impregnated with tri-*n*-octylphosphine oxide in di-*n*-hexyl ether [97–98]. The system allows the determination of these carboxylic acids at micromolar levels in the presence of interfering ions such as nitrite, chloride, sulfate, iron, and aluminum.

3. *Liquid membrane enrichment coupled on-line with atomic absorption spectrophotometry.* Metal ions can be readily determined by atomic absorption spectrometry. Nevertheless, preconcentration is sometimes required and supported liquid membranes can be used as an attractive alternative to other pretreatment methods such as liquid–liquid extraction [99–101]. The transport of the metal species across the liquid membrane is generally performed by carriers in the membrane such as alkylphosphorous acids, long-chain alkylamines, or chelating reagents. These carriers react with metal ions to give neutral species that are soluble in nonpolar solvents and diffuse across the membrane. In the case of cationic metal species, cationic exchanger extractants or chelating extractants are used frequently as carriers. In such systems, the metal ions are transported from the aqueous donor solution to the acceptor phase and protons are countertransported from the acceptor to the donor phase. The driving force for the process is the pH gradient between the donor and acceptor phases and this allows the metal ions to be transported against their concentration gradient with the result that the metal ions concentrate in the acceptor phase.

In the case of metal ions present as anionic complexes in the donor phase, solvating or ion-pairing extractants can be used as carriers. Here the metal ions and counterions are cotransported across the membrane from the donor to the acceptor phase. By using a complexing or reducing agent in the acceptor phase, it is possible to back-extract the metal ions into the acceptor phase.

Metal ions such Cu^{2+}, Cd^{2+}, and Pb^{2+} can be preconcentrated from water samples using liquid membranes containing 40% w/w of di-2-ethylhexylphosphoric acid in kerosene diluent in a PTFE support. The liquid membrane can be coupled on-line to an atomic absorption spectrometer and has been shown to be stable for at least 200 h with extraction efficiencies over 80%, and enrichment factors of 15 can be obtained. A liquid membrane has also been used for sample cleanup and enrichment of lead in urine samples prior to determination by atomic absorption spectrometry [100]. The experimental setup for metal enrichment is shown in Fig. 13.4. Lead was enriched 200 times from urine [80] and several metals were enriched 200 times from natural waters [88]. Using hollow fiber

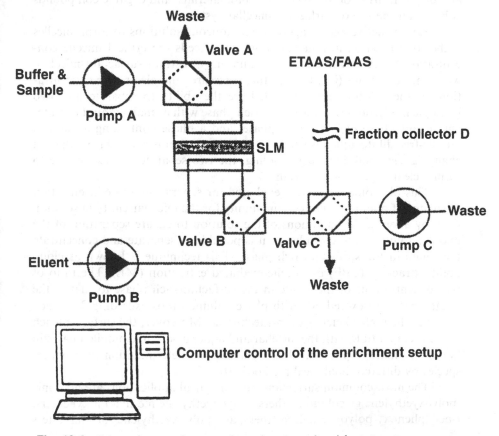

Fig. 13.4 Schematic setup for a membrane-based metal enrichment system.

geometry and crown ethers in the membrane, enrichment factors up to 3000 have been obtained for lead [102].

Other applications of supported liquid membranes have been related to metal speciation. For example, recently a system for chromium speciation has been developed based on the selective extraction and enrichment of anionic Cr(VI) and cationic Cr(III) species in two SLM units connected in series. Aliquat 336 and DEHPA were used respectively as carriers for the two species and graphite furnace atomic absorption spectrometry used for final metal determination. With this process, it was possible to determine chromium in its different oxidation states [103].

13.5.3 Micelles in Analytical Chemistry

Another of the new techniques for extractive preconcentration, separation, and/or purification of metal chelates, biomaterials, and organic compounds is based on the use of surfactant micellar systems.

Surface-active agents aggregate in aqueous solutions to form micelles if the concentration in aqueous solutions exceeds the critical micelle concentration (CMC). Dilute aqueous solutions of certain surfactant micelles, when the conditions (i.e., temperature, pressure, and electrolyte concentration) of the solution are changed, have the ability to separate into two isotropic liquid phases: a surfactant-rich phase with a small amount of water (surfactant phase or coacervate phase) and a phase containing an almost micelle-free dilute aqueous solution. This separation is reversible so that on changing the conditions, e.g., cooling, the two separated phases merge to form a clear solution once again.

This phenomenon can be exploited for separation and concentration of solutes. If one solute has certain affinity for the micellar entity in solution then, by altering the conditions of the solution to ensure separation of the micellar solution into two phases, it is possible to separate and concentrate the solute in the surfactant-rich phase. This technique is known as cloud point extraction (CPE) or micelle-mediated extraction (ME). The ratio of the concentrations of the solute in the surfactant-rich phase to that in the dilute phase can exceed 500 with phase volume ratios exceeding 20, which indicates the high efficiency of this technique. Moreover, the surfactant-rich phase is compatible with the micellar and aqueous-organic mobile phases in liquid chromatography and thus facilitates the determination of chemical species by different analytical methods [104].

The most common surfactants for analytical applications are nonionic (polyoxyethylene glycol monoethers, polyoxyethylene methyl-n-alkyl ethers, t-octylphenoxy polyoxyethylene ethers, and polyoxyethylene sorbitan esters of fatty acids) that demonstrate cloud point behavior with increasing solution temperature, and zwitterionic surfactants (ammonioethylsulfates, ammoniopropylsulfates, ammoniopropanesulfonates, phosphobetaine, dimethylalkylphosphine oxides) that show cloud point behavior on decreasing solution temperature. Thus, cloud point temperature depends on the structure of the surfactant and its concentration and range from −0.5C to 120C [105,106]. In a homologous series of polyoxyethylated surfactants, the cloud point temperature increases with the hydrocarbon chain length and increasing length of the oxyethylene chain. At a constant oxyethylene content in the surfactant molecule, the cloud point temperature is lowered by decreasing the molecular mass of the surfactant and by branching of the hydrophobic group.

The most important advantage of cloud point extraction is that only small amounts of nonionic or zwitterionic surfactants are required and consequently the procedure is less costly and more environmentally benign than other conventional extraction techniques such as liquid–liquid extraction and solid–liquid extraction [107,108]. Moreover, CPE offers the possibility of combining extraction and preconcentration in one step.

13.5.3.1 Experimental Protocols

Operation procedures are performed by adding a small volume of a concentrated nonionic or zwitterionic surfactant solution to an aqueous sample ($50-100 \, cm^3$) containing the analyte, taking into account that the final surfactant concentration must be greater than its CMC value so that micelles are present in solution. The analyte, depending on its affinity to the micelles, is incorporated into the micellar aggregates in solution. The solution is heated (in the case of nonionic surfactants) or cooled (with zwitterionic surfactants) until the cloud point temperature is reached and the solution is allowed to settle (settling temperature) in a thermostatted bath set at a temperature above or below that of the cloud point (for nonionic and zwitterionic surfactants respectively) until the phases separate. As the density of both phases in some cases is quite similar, centrifugation is often recommended to facilitate the physical separation. The analyte is concentrated in the surfactant-rich phase in a small volume ($50-400 \, \mu L$) [105,106]. Figure 13.5 shows the steps involved in CPE prior to analysis.

The surfactant selected for CPE technique should not have too high a cloud point temperature. In practice, it is possible to obtain almost any desired temperature by choosing an appropriate mixture of surfactants, as cloud point temperatures of mixtures of surfactants are intermediate between those of the two pure surfactants, or by the choice of an appropriate additive (i.e., salts, alcohols, organic compounds) [105].

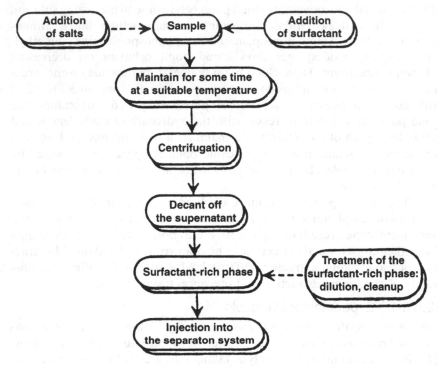

Fig. 13.5 Steps involved in cloud point extraction (CPE) prior to HPLC, GC, and CE analysis.

13.5.3.2 Applications of Cloud Point Extraction to the Analysis of Selected Samples

Cloud point extraction has been applied to the separation and preconcentration of analytes including metal ions, pesticides, fungicides, and proteins from different matrices prior to the determination of the analyte by techniques such as atomic absorption, gas chromatography, high performance liquid chromatography, capillary zone electrophoresis, etc.

1. *Cloud point extraction of metal ions.* The use of cloud point extraction as a separation technique was first introduced by Watanabe for the extraction of metal ions forming sparingly water soluble complexes [109]. Since then, the technique has been applied successfully to the extraction of metal chelates for spectrophotometric, atomic absorption, or flow injection analysis of trace metals in a variety of samples [105–107,110]. Other metal complexes such as $AuCl_4^-$ or thiocyanato-metal complexes can be extracted directly using nonionic surfactants such as polyoxyethylene

nonyl phenyl ether (PONPE) or Triton X prior to their determination by ICP or visible spectrometry [111,112]. The main advantages of cloud point extraction of metal ions include the simplicity of the extraction procedure and the possibility of obtaining high preconcentration factors ranging from 10 to 100 allowing the development of analytical methods for determining metal ions at very low concentrations. For example, Pd(II) can be determined in a surfactant phase of Triton X-100 with coproporphyrin III as a complexing agent in a procedure based on phosphorescence at room temperature with detection limits of $20\,nmol\,dm^{-3}$. Some parameters for the cloud point extraction of metals with different surfactant micelles are given in Table 13.2 [105,107].

2. *Cloud point extraction from biological and clinical samples.* The most frequent use of CPE is for the separation and purification of biological analytes, principally proteins. In this way, the cloud point technique has been used as an effective tool to isolate and purify proteins when combined with chromatographic separations. Most of the applications deal with the separation of hydrophobic from hydrophilic proteins, with the hydrophobic proteins having more affinity for the surfactant-rich phase, and the hydrophilic proteins remaining in the dilute aqueous phase. The separation of biomaterials and clinical analytes by CPE has been described [105,106,113].

3. *Cloud point extraction of environmental samples.* More recently, CPE has been used for sample preparation and preconcentration of organic analytes such as pesticides [114,115], herbicides, polycyclic aromatic hydrocarbons (PAH) [116], polychlorinated biphenyls (PCBs) [117], and phenols [118] in environmental samples prior to their determination. The compatibility of the surfactant-rich phase from CPE with micellar or conventional hydroorganic mobile phases allows subsequent determination of the analyte by thin layer chromatography, HPLC [119,120], micellar electrokinetic capillary chromatography, or CZE [121]. Cloud point extraction is a rapid, simple, sensitive, and efficient sample pretreatment for trace environmental analysis and offers some advantages over conventional liquid–liquid extraction technology in terms of enhanced detection limits from the large preconcentration factors, elimination of analyte losses during evaporation of solvents used in liquid–liquid extraction, and elimination of toxic solvents [107].

13.5.4 Ion-Selective Electrodes

An important advance in ion-selective electrodes (ISEs) and related systems was based on the concept of polymeric liquid membranes developed by Eisenman [122]. The principle of this approach was to incorporate an organic compound as the ionophore into a polyvinyl chloride membrane

Table 13.2 Summary of Cloud Point Extractions of Metals Chelates Using Nonionic Surfactants Micelles

Metal ion	Ligand	Nonionic surfactant	Experimental conditions
			pH; CF^a, $\%E^b$
Ni(II)	TAN^c	Triton X-100	pH 7.0 (phosphate); $C_F = 30$
	PAN	Triton X-100	pH 5–6
Zn(II)	PAN	PONPE-7.5	pH 10 (carbonate); $C_F = 40$
	QADI	PONPE-7.5	pH 9; NH_3; $C_F = 40$
Ni(II), Zn(II), Cd(II)	PAMP	PONPE-7.5	
Au(III)	HCl	PONPE-7.5	HCl; $\%E > 95$
Ni(II), Cd(II), Cu(II)	PAMP	PONPE-7.5	pH 5.6
Ni(II)	PAN	OP^d	pH 6.0; $C_F = 15–25$
Transition metal ions	TAN	PONPE-7.5	
Fe(III), Ni(II)	TAC	Triton X-100	
U(VI)	PAN	Triton X-114	pH 9.2; $\%E = 98$
U(VI), Zr(IV)	Arsenazo	Tween 40^e	pH 3; $\%E > 96$
Cu(II), Zn(II), Fe(III)	Thiocyanate	PONPE-7.5	$\%E = 72.5–96.8$
Er(III)	CMAP	PONPE-7.5	$C_F = 20$
Gd(III)	CMAP	PONPE-7.5	$C_F = 3.3$; $\%E = 99.88$
Cd(II)	PAN	Triton X-114	$C_F = 60$
Ni(II), Zn(II)	PAN	Triton X-114	
Ru(III)	Thiocyanate	Triton X-110	$C_F = 5–10$
Au(III)	HCl	$PONPE-10^f$	$\%E > 90$
Ga(III)	HCl	$PONPE-7.5^f$	$\%E \approx 90$
Ag(I), Au(III)	DDTP	Triton X-114^f	$C_F = 9–130$
Cu(II)	LIX54	Igepal CO-630^f	

[a]Concentration factor.
[b]Percent extracted.
[c]Abbreviations for ligands: CMAP: 2-(3,5-dichloro-2-pyridylazo)-5-dimethylaminophenol; DDTP: o,o-diethyldithiophosphoric acid; LIX54: dodecylbenzoylacetone; PAN: 1-(2-pyridylazo)-2-naphthol; QADI: 2-(8-quinolazo)-4,5-dipheylbenzimidazole; PAMP: 2-(2-pyridylazo)-5-methylphenol; PAP: 2-(2-pyridylazo)-phenol; TAC: 2-(2-thiazoylazo)4-methylphenol; TAN: 1-(2-thiazoylazo)-2-naphthol.
[d]OP surfactants refer to (polyethyleneglycol octylphenyl ethers).
[e]In this extraction, the concentrated surfactant-rich phase was a solid rather than a liquid.
[f]Abbreviations for surfactants: Igepal CO-630: nonylphenoxypoly(ethylenoxy)ethanol; PONPE-7.5: polyoxyethylene(7.5)nonylphenyl ether; PONPE-10: polyoxyethylene(10)nonylphenyl ether; TRITON X: t-octylphenoxypolyoxyethylene ether.

together with an appropriate plasticizer and additive to provide the membrane with the properties of a liquid phase. Ionophores are lipophilic, electron-rich complexing agents capable of reversibly binding ions and transporting them across organic membranes, and the ultimate performance of such devices depends strongly on the choice of ionophore. The mode of action of these electrodes is therefore a reversible ion exchange process accompanied by diffusion and migration effects.

Potentiometric sensors, such as ion-selective electrodes, ion-selective microelectrodes, and ion-selective field effect transistors (ISFET) are frequently used in analytical systems. Moreover, the recently introduced ion-selective optodes based on absorbance or fluorescence measurements provide additional sensors for a broad range of monitoring applications [123,124]. In these systems, the active constituent of the liquid membrane is an ion-pairing or complexing agent that is selective for a limited number of specific ions. This reagent is dissolved in a low-polarity organic solvent, which must also have a low vapor pressure to minimize losses through evaporation, high viscosity to prevent rapid loss during flow across the membranes, and low solubility in the aqueous sample solution with which it comes into contact. The complexing reagents are subdivided into two groups:

1. Electrically charged ligands, often called liquid ion exchangers, which are ionizable organic molecules of high molecular weight
2. Electrically neutral complexing agents, ionophores, that are capable of enveloping metal ions in a "pocket" of oxygen or nitrogen ligands and can serve as selective extractants for cations

The construction of a liquid membrane electrode is rather similar to that of the glass electrode, in that it requires an internal reference electrode. In addition, a solid support is required into which the active liquid reagents are incorporated. This support may be a porous polyvinyl chloride (PVC) plug, a porous polymeric film, or a commercial membrane. Many different designs have been suggested to solve the problem of interposing the liquid membrane between the aqueous solutions. The support prevents leakage of the organic solvent into the aqueous solution and the same time maintains efficient contact with the analyte solution. Two of these designs are illustrated in Fig. 13.6.

The electromotive force (emf) of liquid membrane electrodes depends on the activity of the ions in solution and their performance is similar in principle to that of the glass electrode. To characterize the behavior of liquid membrane electrodes, the linearity of the emf measurements vs. concentration of a certain ion in solution is checked. Additional performance data are the Nernstian slope of the linear range and the pH range over which the potential of the electrode is constant.

Eisenman [122] pioneered the development of this type of electrode and created the theory on which the measurements are based, and investigated the selectivity of certain membranes. Equally significant were the contributions of Buek, Durst, Worf [126], Koryta [127], and Simon, which were compiled into an important monograph [128].

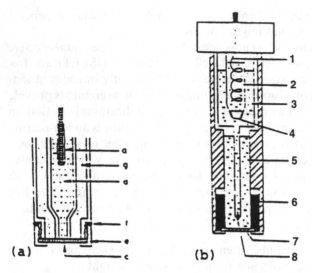

Fig. 13.6 Schematic diagrams of two liquid membrane electrodes: (a) a commercial Onion electrode; (b) an improved version developed by Szczepaniac and Oleksy [125].

The ion selectivity of a membrane can be established by measuring the potential difference between two identical reference electrodes. One electrode is immersed in the specimen solution, the other in a reference solution, and the membrane is interposed between them. The composition of solution 2 is constant, and, if both solutions contain monovalent ions, the ion-exchange process can be described as follows:

$$J^+(\text{aq}) + I^+(\text{membrane}) \Longleftrightarrow J^+(\text{membrane}) + I^+(\text{aq}) \tag{13.5}$$

and the potential difference is given by:

$$E = E_0 + 0.059 \log \{a_i + K_{ij}a_j\} \tag{13.6}$$

where E_0 is a constant related to the reference electrodes and the reference solution, a_i and a_j are the activities of ions i and j in the specimen solution, and K_{ij} is the selectivity constant of the membrane.

Inspection of Eq. (13.6) shows that the selectivity behavior of a liquid membrane is specified completely by the membrane selectivity constant, K_{ij}, which in turn is dependent on the equilibrium constant of Eq. (13.5) and on the mobility of ions i and j within the membrane. For the case in which the membrane consists of a neutral carrier [129], the exchange reaction can be presented as:

$$J^+(\text{aq}) + IS^+(\text{membrane}) \Longleftrightarrow JS^+(\text{membrane}) + I^+(\text{aq}) \tag{13.7}$$

If the ligand (neutral carrier) is capable of completely enveloping the cations, the terms K_{ij} and E_0 of the membrane depend only on the ratio of the complex stability constants K_{js}^*/K_{is}^* of the ions with the ionophore.

Liquid membrane electrodes are not capable of being specific for only one ion in solution. There is always some interference from other ions in solution with the given analyte. The selectivity coefficient provides an indication of the ability of an electrode to measure a particular ion in the presence of another ion. The response of an electrode to an interfering ion can be included in the Nernstian equation:

$$E = E'_0 + 0.059/z' \log \{a_A + K_{AB}a_B^{z'/z}\} \tag{13.8}$$

where:

E'_0 = a constant emf, including that of the reference electrode and the standard potential of the ion being measured

a_A = activity of species A with charge Z'

a_B = activity of interfering ion, B, with charge Z

$K_{A/B}$ = selectivity constant of the electrode for A over B

13.5.4.1 Liquid Ion Exchangers

Liquid ion exchange carriers must again be subdivided into cation- and anion-specific ion exchangers. For cations, extensive use is made of high molecular mass organic anions, such as sulfonates, carboxylates, thio-carboxylates, and phosphonates. An important example of this membrane type is the calcium-selective electrode, in which the liquid membrane consists of long-chain alkylphosphates [130–131]. These compounds are dissolved in di-*n*-octylphenylphosphate, which enhances the selectivity for calcium relative to magnesium and the other alkaline earth metals. A mercury(II) ion-selective electrode uses a thiophosphonyl derivative of thiobenzamide as the complexing agent [125]. For anions, stable ion pairs are formed with large cationic organic molecules, such as substituted organophenanthroline nickel or iron compounds dissolved in decanol.

Of special interest is the nitrate electrode, which has found many applications in quantitative analysis of nitrates in biological fluids and agricultural products. A nitrate-selective electrode [132] based on a tributyl-octadecylphosphonium ligand is also selective for perrhenate and perchlorate ions.

13.5.4.2 Neutral Carrier-Based Cation-Selective Electrodes

A number of polyether molecules have been described that form membranes selective for four monovalent and for six divalent cations [129,133,134]. Inspection of Table 13.3 shows that two classes of ligands can be distinguished:

Table 13.3 Some Commercial Ion Exchange Electrodes

Ion	Ligand	Area of application	Refs.
Li^+	Trioctylphosphine oxide + neutral carrier	Lithium in blood serum and pharmaceuticals	136
K^+	Valinomycin dissolved in diphenyl ether	Potassium in feldspar, urine, blood serum, seawater, vegetables	137
NH_4^+	Nonactin or dinactin	Activity of nitrate reductase, ammonium in mineral water, fruit juice, beer, urine, sewage water	138
Ca^{2+}	Phosphates and phosphonates	Protein binding studies, calcium in sewage water, mineral water, blood serum, biological fluid	131
NO_3^-	Tributyl octadecyl-phosphonium nitrate	Nitrate in agricultural products, soils, foods	132

the first is the macrotetrolides (antibiotics), such as nonactin and dinactin, which have eight oxygen atoms in a flexible ring and are specific for ammonium ions. Another antibiotic that has assumed considerable importance is valinomycin, which has a high selectivity for potassium across the membrane of living cells. The second class of ligands is composed of synthetic crown ethers, some acrylic polyethers, and cryptands. Figure 13.7 demonstrates how different geometric arrangements of these polyethers provide compounds that systematically bind ions, from Li^+ to Pb^{2+}.

Five liquid membrane electrodes (Table 13.3) are now commercially available and have found wide application in the testing of electrolytes in biological and technological systems. All five electrodes perform well in the concentration range over which the Nernstian slope is maintained, i.e., from $10^{-1}–10^{-5} \, mol \, dm^{-3}$. These electrodes to a certain extent have replaced in both chemical and clinical laboratories the more traditional instrumental methods of analysis, such as flame photometry and atomic absorption spectrometry. There are, of course, many more liquid membrane electrodes, but the availability of satisfactory solid electrodes has greatly restricted their development and practical application.

13.6 SOLVENT EXTRACTION–BASED TECHNIQUES

The term *chromatography* now embraces a variety of processes that are based on the differential distribution of the components in a chemical mixture between two phases. The difference between extraction processes involving a single equilibrium of two bulk phases and chromatography is that, in the latter technique, species are separated, not by discrete extraction steps, but by continuous equilibration between two phases, one that is stationary and the other mobile. The same fundamental laws of phase equilibrium apply to both, extraction and chromatography. In principle, the latter is a multiple-extraction process in which the mobile phase moves continuously over the fixed phase in a chromatographic column.

13.6.1 Liquid–Liquid Partition Chromatography

The liquid–liquid partition chromatography (LLPC) method involves a stationary liquid phase that is more or less immobilized on a solid support, and a mobile liquid phase. The analyte is therefore distributed between the two liquid phases. In conventional LLPC systems, the stationary liquid phase is usually a polar solvent and the mobile liquid phase is an essentially water-immiscible organic solvent. On the other hand, in reversed-phase chromatography (RPC), the stationary liquid is usually a hydrophobic

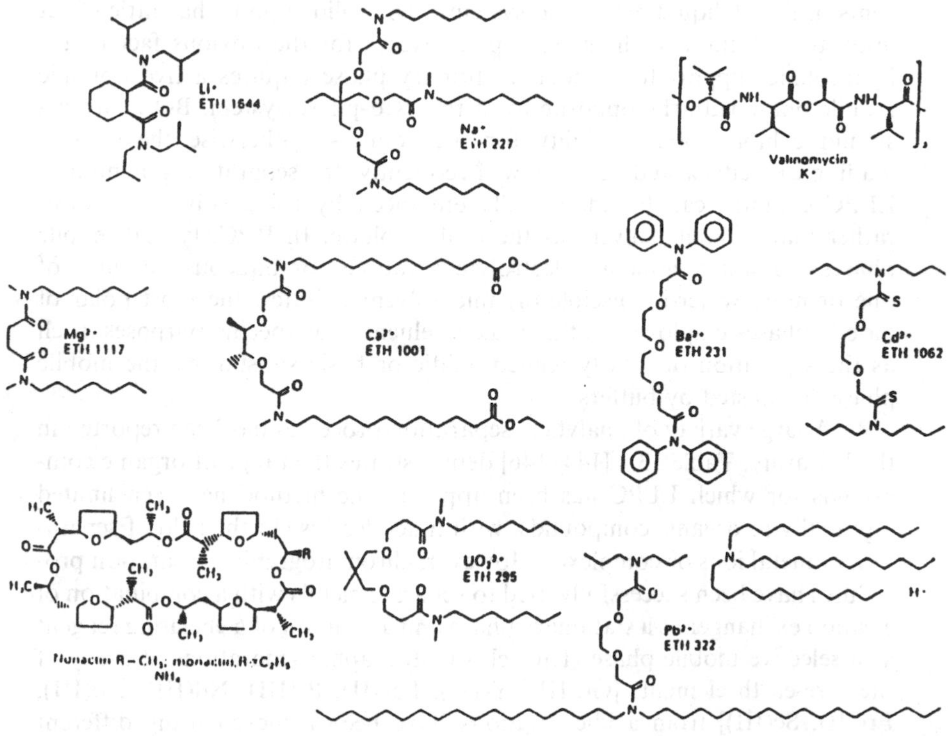

Fig. 13.7 Structures of cation-selective carriers. (From Ref. 133.)

solvent, whereas the mobile liquid is the polar solvent, most frequently water or an aqueous mixture with polar organic solvents [139–143].

For conventional or "normal," in contrast with reversed-phase, LLPC, many materials have been used as the solid support for the stationary liquid. In addition to silica gel, which was the first and is still the most popular material, a variety of other adsorbents that adsorb the polar solvent such as cellulose powder, starch, alumina, and silicic acid have been used. The more recent practice of HPLC has greatly simplified the technique in providing column stability for repeated use and for treatment of large volumes.

Reversed-phase chromatography is the predominant technique in HPLC, and chemically bonded silica gel supports are made specifically for the nonpolar stationary phase. In the last decade, as many as 60% of the published LLPC techniques refer to RPC. The reasons for this involve the significantly lower cost of the mobile liquid phase and a favorable elution order that is easily predictable based on the hydrophobicity of the eluate.

The selection of solvents for LLPC is similar to the selection of solvents in liquid–liquid extraction systems. The solid support has little effect upon the selection of the solvent pair, except for the obvious fact that a hydrophilic support for a polar stationary phase requires a hydrophobic mobile phase with the opposite for a reversed-phase system. Both solvents should exhibit good solubility for the solute(s); otherwise the column loading capacity would be too low. Frequently, the separation potential of LLPC columns can be additionally enhanced by using solvent mixtures rather than a single solvent as the mobile phase. In RPC, typical mobile phases are water, aqueous electrolyte solutions, or aqueous mixtures of one or more water-immiscible organic solvents. Water, the most polar of mobile phases employed, is the weakest eluent. For specific purposes, such as the separation of closely related acidic or basic substances, the mobile phase is adjusted by buffers.

A large variety of analytical separation processes has been reported in the literature. Table 13.4 [144–146] demonstrates the range of organic compounds for which LLPC has been applied. The method has been limited essentially to organic compounds, with much less use in the field of separation of metal ions or complexes. However, chromatographic separation procedures have been successfully used to separate metals with a combination of a cation exchanger as a stationary phase and a solution of a chelating reagent as a selective mobile phase. Thus chromatographic separations of many of the rare-earth elements [Gd(III), Y(III), La(III), Pr(III), Nd(III), Ho(III), Er(III), Sc(III)] from acidic solutions have been achieved using different acidic organophosphorous extractants (e.g., DEHPA, PC88A) as well as

Table 13.4 Some Applications of Liquid–Liquid Partition Chromatography (LLPC)

Eluate	Support	Stationary phase	Mobile phase
Aromatics	alumina	water	*n*-heptane
Aliphatic alcohols	silica acids	water	chloroform-CCl$_4$
Fatty acids	celite	aq. sulphuric acid	butanol-chloroform
Glycols	celite	water	butanol-chloroform
Phenols	silicic acid	water	*iso*octane
Amino acids	silica gel	water	dichloromethane
Steroids	silica acids	methanol/water	benzene-chloroform
Alkaloids	silica gel	water	acetonitrile-water
Urine	silica gel	aq. sulphuric acid	dichloromethane
Antibiotics	silica gel	water	dichloromethane
Pesticides	silica gel	water	*n*-heptane

neutral organophosphorous extractants (e.g., TOPO, TBP) immobilized onto polymer beads as stationary phases [150,151]. Similar approaches have been used in environmental control and determination of transuranium elements (TUE), especially Pu(IV) and Am(III), in various natural samples [152,153]. Among numerous sorbents for radioanalytical purposes, a new type of solid sorbent (TVEX) based on introducing an extractant into the polymer matrix during synthesis has been developed [141,142]. The behavior of uranium, plutonium, and transplutonium elements (TPE) using TVEXs with TBP, TOPO, DEHPA, and TOA, etc., has been studied. More recently, at the Argonne National Laboratory in the United States [156,157], a new family of impregnated resins based on the impregnation of Amberlite XAD7 with highly specific reagents for actinide analysis in environmental and biological samples has been intensively studied.

13.6.2 Supercritical Fluid Extraction

Supercritical fluids were soon found to be highly efficient extraction media, chiefly because of their high solvating power, their low viscosities (intermediate between a gas and a liquid), and their low surface tensions that enable their penetration deep into the extraction matrix. Supercritical fluid extraction (SFE) used in isolation is generally not selective enough to separate specific solutes from the matrix without further cleanup or resolution from coextracted species prior to qualitative and quantitative analysis. Consequently, for analytical applications, SFE is usually used in combination with chromatographic techniques to improve the overall selectivity in the isolation of specific solutes. The combined use of SFE with chromatographic techniques is quite recent.

The application of SFE for the preparation of samples in the analytical laboratory has received serious attention as a sample preparation step for extracting analytes of interest from a bulk matrix prior to their determination by other analytical methods including chromatographic, spectrometric, radiochemical, and gravimetric techniques.

Much of the current interest in using analytical-scale SFE systems comes from the need to replace conventional liquid solvent extraction methods with sample preparation methods that are faster, more efficient, have better potential for automation, and also reduce the need for large volumes of potentially hazardous liquid solvents. The need for alternative extraction methods is emphasized by current efforts to reduce the use of methylene chloride as an extraction fluid for environmental sample preparation [158]. The potential for applying SFE to a wide variety of environmental and biological samples for both qualitative and quantitative analyses is widely described in reviews [159–161] and the references therein. Analytical-scale SFE is most often applied to relatively small samples (e.g., several grams or less).

A good solvent for extraction should be selective so that it dissolves the desired analytes to a greater degree than other constituents in the sample matrix. It should be unreactive and stable, preferably nontoxic and from an economic point of view noncorrosive to equipment and inexpensive to buy. Many of these requirements are met by supercritical fluids such as carbon dioxide.

13.6.2.1   Off-Line and On-Line SFE

Analytical-scale SFE can be divided into off-line and on-line techniques. Off-line SFE refers to any method where the analytes are extracted using SFE and collected in a device independent of the chromatograph or other measurement instrument. On-line SF techniques use direct transfer of the extracted analytes to the analytical instrument, most frequently a chromatograph. While the development of such on-line SFE methods of analysis has great potential for eventual automation and for enhancing method sensitivities [159–161], the great majority of analytical SFE systems described use some form of off-line SFE followed by conventional chromatographic or spectroscopic analysis.

To perform off-line SFE, only the SFE step must be successful, and the extract can be analyzed at leisure by a variety of methods. The final product of an off-line SFE experiment typically consists of the extract dissolved in a few milliliters of liquid solvent, a form that is directly compatible with conventional chromatographic injectors. In contrast, the successful performance of on-line SFE requires that the SFE step, the coupling step (i.e., the transfer and collection of extracted analytes from the SFE to the chromatographic system), and the final chromatographic conditions be understood and controlled. With many on-line approaches the sample extract is committed to a single analysis, and the extract is not available for analysis by other methods. While off-line SFE has the advantages of simplicity and availability of commercial instrumentation, and provides extracts that can be analyzed by several instrumental methods, on-line techniques have greater potential for enhanced sensitivity and automation.

Off-line SFE is conceptually a simple experiment to perform and requires only relatively basic instrumentation. The instrumental components necessary include a source of fluid, most often CO_2 or CO_2 with an organic modifier, a means of pressurizing the fluid, an extraction cell, a method of controlling the extraction cell temperature, a device to depressurize the supercritical fluid (flow restrictor), and a device for collecting the extracted analytes.

13.6.2.2 On-Line Coupling SFE with Analytical Techniques

On-line supercritical fluid extraction/GC methods combine the ability of liquid solvent extraction to extract efficiently a broad range of analytes with the ability of gas-phase extraction methods to rapidly and efficiently transfer the extracted analytes to the gas chromatograph. The characteristics of supercritical fluids make them ideal for the development of on-line sample extraction/gas chromatographic (SFE-GC) techniques. SFE has the ability to extract many analytes from a variety of matrices with recoveries that rival liquid solvent extraction, but with much shorter extraction times. Additionally, since most supercritical fluids are converted to the gas phase upon depressurization to ambient conditions, SFE has the potential to introduce extracted analytes to the GC in the gas phase. As shown in Fig. 13.8, the required instrumentation to perform direct coupling SFE-GC includes suitable transfer lines and a conventional gas chromatograph [162,163].

Finally, supercritical fluid chromatography, in which a supercritical fluid is used as the mobile phase, was introduced by Klesper [164–166]. SFE directly coupled to SFC provides an extremely powerful analytical tool. The efficient, fast and selective extraction capabilities of supercritical fluids allows quantitative extraction and direct transfer of the selected solutes of interest to be accomplished to the column, often without the need for further sample treatment or cleanup. Extraction selectivity is usually achieved by adjusting the pressure of the supercritical fluid at constant temperature or, less often, by changing the temperature of the supercritical fluid at constant pressure. SFE coupled with packed column SFC has found

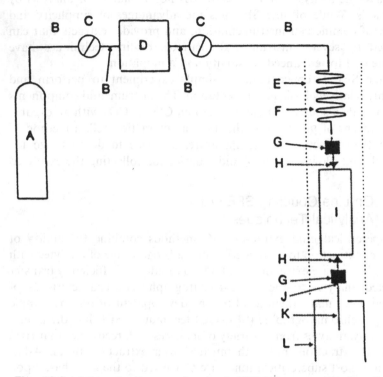

Fig. 13.8 Schematic diagram of a simple SFE-GC system showing all the required components. (Several manufacturers supply suitable components and specific suppliers are listed only for the reader's convenience.) Components are: (A) SFE grade extraction fluid source; (B) 1.5 mm o.d. stainless steel tubing (0.77 mm or smaller i.d.); (C) shut-off valves (SSI model 02–120 or equivalent, Supelco, Bellefonte, PA); (D) SFE pump; (E) SFE cell heater; (F) approx. 0.5 m long coil of 1.5 mm stainless steel tubing for fluid preheater; (G) 1.5 mm×1.5 mm tubing union (e.g., Parker or Swagelok brand); (H) finger-tight connectors (e.g. Slip-Free connectors from Keystone Scientific Bellefonte, PA, USA); (I) SFE cell; (J) restrictor connector ferrule (Supelco M2-A, Bellefonte, PA) which is used to replace the stainless steel ferrule in the outlet end of the tubing union 'G'; (K) 15–30 μm i.d. fused silica tubing restrictor (Polymicro Technologies, Phoenix, AZ); (L) GC injection port.

many applications in the last decade, particularly in connection with the determination of food compounds, drug residues, herbicide and pesticide residues, and polymer additives.

13.6.3 Solvent Extraction in Continuous Flow Injection Analysis

13.6.3.1 Fundamental of Flow Injection Analysis

Flow injection analysis is based on the injection of a liquid sample into a continuously flowing liquid carrier stream, where it is usually made to react to give reaction products that may be detected. FIA offers the possibility in an on-line manifold of sample handling including separation, pre-concentration, masking and color reaction, and even microwave dissolution, all of which can be readily automated. The most common advantages of FIA include reduced manpower cost of laboratory operations, increased sample throughput, improved precision of results, reduced sample volumes, and the elimination of many interferences. Fully automated flow injection analysers are based on spectrophotometric detection but are readily adapted as sample preparation units for atomic spectrometric techniques. Flow injection as a sample introduction technique has been discussed previously, whereas here its full potential is briefly surveyed. In addition to a few books on FIA [168,169], several critical reviews of FIA methods for FAAS, GF AAS, and ICP-AES methods have been published [170,171].

As noted earlier, the essence of flow injection is the controlled physical dispersion of an injected liquid sample into a continuous flowing unseg-mented liquid carrier stream that may contain a suitable reagent to produce a transient reproducible detector response proportional to the analyte con-centration. Thus, a typical FIA system consists of the following components:

1. A propulsion unit, usually a peristaltic pump, that produces a flow of one or several solutions (streams), either containing a dissolved reagent or merely acting as a carrier
2. An injection system whose function is the reproducible introduction of an accurately measured sample volume into the flow without stopping it
3. A mixing (reaction) unit
4. A flow cell for photometry detection

The simplest flow injection arrangement is the single-line configuration shown in Fig. 13.9. As can be seen, the carrier stream is propelled by a pump through a narrow conduit to an injector, then via the reaction manifold to the detector and finally to waste. Each injection yields a single peak at the detector, with the height, area, or width proportional to the analyte con-centration.

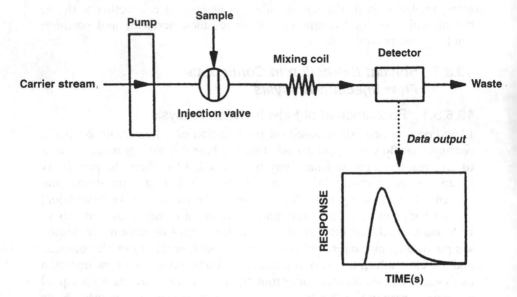

Fig. 13.9 Schematic diagram of a single-channel flow injection manifold, showing the transient nature of the signal output. (——) Liquid flow; (.....) data flow.

13.6.3.2 Liquid–liquid Extraction in Continuous Flow Analysis

A typical extraction manifold is shown in Figure 13.2. The sample is introduced by aspiration or injection into an aqueous carrier that is segmented with an organic solvent and is then transported into a mixing coil where extraction takes place. Phase separation occurs in a membrane phase separator where the organic phase permeates through the Teflon membrane. A portion of one of the phases is led through a flow cell and an on-line detector is used to monitor the analyte content. The back-extraction mode in which the analyte is returned to a suitable aqueous phase is also sometimes used. The fundamentals of liquid–liquid extraction for FIA [169,172] and applications of the technique [174–179] have been discussed. Preconcentration factors achieved in FIA (usually 2-5) are considerably smaller than in batch extraction, so FI extraction is used more commonly for the removal of matrix interferences.

Chemical reactions between the analyte and a reagent or combination of reagents can be implemented in continuous systems to obtain a reaction product that is capable of facilitating mass transfer and/or continuous detection that results in increased sensitivity or selectivity. These reactions could take place in different locations of the schemes defined previously, for example:

1. In a coil after the sample is merged with a reagent stream.
2. In the solvent segmenter when it is necessary to avoid precipitation of the extractable compound in the aqueous phase before it reaches the continuous extractor (SS).
3. In the extraction coil (EC), when the extraction reaction is based on the formation of metal chelates and the ligand is dissolved in the organic solvent. Here the reaction and the transfer of the chelate take place simultaneously.
4. In a coil located after the extractor using a reagent stream that is merged prior to the detector with the phase containing the analyte.

The derivatizing reactions used in these systems are not substantially different from those described in classical analytical determinations. The most common are ion-pair formation reactions, where the analyte forms an ion pair directly, e.g., determination of anionic surfactants by using ethyl violet or methylene blue, and the determination of codeine with picric acid and enalapril with bromothymol blue; or after reacting with a reagent to form a bulkier, charged species e.g., tetracyanatocobaltate or phosphomolybdate. Metal cations are typically transformed into extractable metal chelates by using such common ligands as described in section 13.4.1 (e.g., oxine, dithizone, dimethylglyoxime, etc.). Some of the most important applications of the technique are in the determination of surfactants in waters, quinizarin in hydrocarbons, amines in pharmaceuticals, and drugs in urine [177].

Typical FIA manifolds are shown in Fig. 13.10 with two general alternatives depending on whether injection takes place before or after the continuous extractor device. The most common situation is when prior injection of the sample take place. Figure 13.10(a) depicts a manifold for the determination of vitamin B_1 in pharmaceuticals [178], based on the oxidation of thiamine to thiochrome in a carrier of potassium ferricyanide in a basic medium (NaOH). The thiochrome is continuously extracted into a chloroform stream and the fluorescence of the organic phase is measured continuously.

Continuous multiextraction could be applied for the separation and determination of carcinogenic polynuclear aromatic compounds (Fig. 13.10b) in crude-oil-ash residues [179]. The ash sample dissolved in a cyclohexane stream merges with a DMSO stream in the first solvent segmenter. After passing through a glass extraction coil, the DMSO phase is mixed with a water stream of cyclohexane segmented with the aqueous DMSO phase. After passing through a Teflon extraction coil, the organic phase is carried through the flow cell of a fluorimeter.

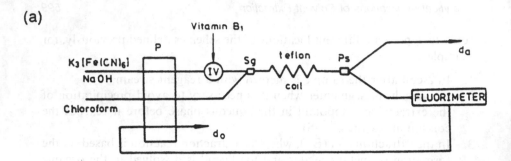

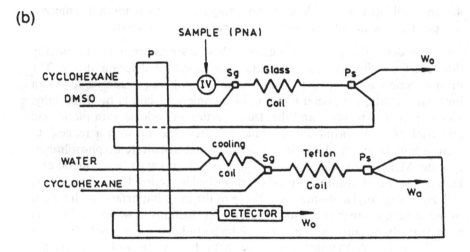

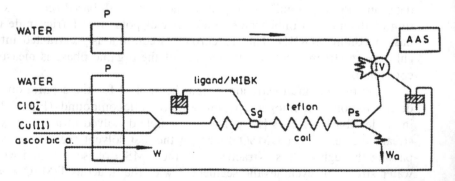

Fig. 13.10 Typical FIA extraction manifolds. Injection in front of the extractor components: (a) simple; (b) multiextraction; and (c) injection after the extraction has been carried out. (Based on systems from Refs. 178, 179.)

FIA can be used for the determination of traces of perchlorate in serum and urine by formation of an ion pair with cuproine-type chelates (CuL_2^+) and subsequent extraction with 6-methylpicolinealdehyde azine dissolved in MIBK (Fig. 13.10c). The sample is continuously added to the system and is mixed with another stream containing Cu(II) and ascorbic acid. This mixture merges with the organic solution in the solvent seg-menter. The organic phase stream containing the ion pair emerging from a T-type separator fills the loop of an injector valve. This volume is then injected into a water carrier in which copper is continuously determined and hence perchlorate (0.5 to 5 ppb) may be indirectly determined [179].

ACRONYM LIST

AAS	atomic absorption spectrophotometry
ASE	accelerated solvent extraction
CCFA	completely continuous flow analysis
CMC	critical micelle concentration
CPE	cloud point extraction
CZE	capillary zone electrophoresis
FAAS	furnace atomic absorption spectrophotometry
FIA	flow injection analysis
GC	gas chromatography
GC-MS	mass chromatography
GF-AAS	graphite furnace atomic absorption spectrophotometry
HPLC	high-performance liquid chromatography
ICP	inductively coupled plasma spectrophotometry
ICP-AES	inductively coupled plasma with atomic emission spectrophoto-metry
ISE	ion-selective electrode
ISFET	ion-selective field effect transistors
LLPC	liquid–liquid partition chromatography
ME	micelle-mediated extraction
MIMS	membrane introduction mass spectrometry
PAH	polynuclear aromatic hydrocarbon
PCDF	polychlorinated dibenzofuran
PCB	polychlorinated biphenyl
PF	preconcentration factor
PONPE	polyoxyethylene nonyl phenyl ether
PTFE	polytetrafluoroethene
PVC	porous polyvinyl chloride
REE	rare earth elements

RPC	reversed-phase chromatography
SC	synergistic coefficient
SFA	segmented flow analysis
SFE	supercritical fluid extraction
SLM	supported liquid membranes
SPE	solid-phase extraction
SPS	solid-phase spectroscopy
THM	trihalomethane
TUE	transuranium elements
UFA	unsegmented flow analysis
ALAMINE 336	mixture of trioctyl- and tridecylamine
ALIQUAT 336	methylated (quaternized) alamine 336
DEHPA	di-(2-ethylhexyl)phosphoric acid
MIBK	methylisobutylketone
DMSO	dimethylsulfoxide
PAN	1-(2-pyridylazo)-2-naphthol
PAR	4-(2-pyridylazo)resorcinol
PC88A	2-ethylhexylphosphonic acid 2-ethylhexyl ester
Oxine	8-hydroxyquinoline
SDVB	styrene divinylbenzene
TBP	tributylphosphate
TOA	tri-n-octylamine
TOPO	trioctylphosphine oxide
Thioxine	8-quinolinethiol

REFERENCES

1. Morrison, M. G.; Freiser, H. *Solvent Extraction in Analytical Chemistry*; John Wiley and Sons; New York, 1957.
2. Stary, J. *The Solvent Extraction of Metal Chelates*; Pergamon Press: Oxford, 1964.
3. De, A. K.; Khopkar, S. M.; Chalmers, R. A. *Solvent Extraction of Metals*; Van Nostrand Reinhold: London, 1970.
4. Alders, L. *Liquid-Liquid Extraction*; Elsevier: New York, 1995.
5. Zolotov, Y. A.; Bodnya, V. A.; Zagruzina, A. N. CRC Crit. Rev. Anal. Chem., **1982**, *14*, 2.
6. Keller, R.; Mermet, J. M.; Otto, M.; Widmer, H. M. Eds. *Analytical Chemistry*; Wiley-VCH, Verlag GmbH: Weinheim, Germany, 1998; p. 25.
7. Valcarcel, M. Liquid-Liquid Extraction in Continuous Flow Analysis. In *Developments in Solvent Extraction*; S. Alegret, Ed., Ellis Horwood Series in Analytical Chemistry: West Sussex, UK, 1988; p. 135.

8. Preconcentration of Trace Elements. In *Comprehensive Analytical Chemistry*; Lobinsky, L., Ed.; Vol. XXV, Elsevier: Amsterdam, 1992.

9. Aguilar, M. Graphical Treatment of Liquid-Liquid Equilibrium Data. In *Developments in Solvent Extraction*, Ed. Alegret, S. Ed., Ellis Horwood Series in Analytical Chemistry, West Sussex, UK, 1988; p. 87.

10. Tagushi, S.; Goto, K. *Pre-Concentration Techniques for Trace Elements*; Alfassi, Z. B.; Wai, C. M. Ed.; CRC Press: Boca Raton, Florida, 1989; p. 31.

11. Torre, M.; Marina, M. L. Crit. Rev. in Anal. Chem., **1994**, *24*, 327.

12. Cortina, J. L.; Warshawsky, A. *Ion Exchange and Solvent Extraction*, Marinsky, J. A.; Marcus, Y. Ed.; Marcel Dekker: New York, Vol. 13, 1996; 195.

13. Karasek, F. W.; Clement, R. E.; Sweetman, J. A. Anal. Chem., 1981, *53*(9), 1051A.

14. Webb, R. C. EPA Report EPA-660/475-003, June 1975.

15. Bowers, W. D.; Parsons, M. L.; Clement, R. E.; Kasarek, F. W. J. Chromatogr., **1981**, *207*, 203.

16. Lobinski, R.; Marczenko, Z. Eds., Spectrochemical Trace Analysis for Metals and Metalloids. In *Comprehensive Analytical Chemistry*, Volume XXX; Elsevier: Amsterdam, Volume XXX, 1996.

17. Cheng, K. L.; Ueno, K.; Imamura,T. *Handbook of Organic Analytical Reagents*; CRC Press: Boca Raton, Florida, 1982.

18. Townshend, A.; Burns, D. T.; Guibault, G. G.; Lobinski, R.; Marczenko, Z.; Newman, E. J.; Onishi, H. Eds. *Dictionary of Organic Analytical Reagents*; Chapman and Hall: London, 1993.

19. Lobinski, R.; Marczenko, Z. CRC Crit. Rev. Anal. Chem., **1992**, *23*, 55.

20. Kaiyanaraman, S.; Khopkar, S. M. Anal. Chim. Acta., **1975**, *78*, 231.

21. Murata, M.; Omatsu, M.; Mushimoto, S. X-ray Spectrom., **1984**, *13*, 83.

22. Maguar, B.; Lobanov, F. L. Talanta, **1973**, *20*, 55.

23. Cresser, M. *Solvent Extraction in Flame Spectroscopic Analysis*; Butterworths: London 1978.

24. Anders, U.; Mailer, D. Z. Anal. Chem., **1976**, *278*, 203.

25. Cordis, V.; Sigartau, G.; Pop, J. Z. Anal. Chem., **1976**, *279*, 355.

26. Fujinaga, T.; Nagaosa, Y. Chem. Lett., **1978**, *6*, 587.

27. Keller, C. *Radiochemistry*; Ellis Horwood Series in Physical Chemistry: Chichester, UK, 1988.

28. Zorov, N. B.; Gocci,T. F.; Winefordner, J. D. Talanta, **1976**, *23*, 455.

29. Zolotov, Y A.; Petrukhin, O. M.; Nagy, V. Y.; Volodarskii, L. B. Anal.Chim. Acta, **1980**, *115*, 1.

30. Zolotov, Y. A.; Kuzmin, N. M.; Petrukhin, O. M.; Spivakov, B. Y. Anal. Chim. Acta, **1986**, *180*, 137.

31. Guter, G. A.; Hammond, G. S. J. Amer. Chem. Soc., **1956**, *78*, 5166.

32. Fix, R. C.; Irvine, J. W. Jr., Lab. Nucl. Sci. Prog. Rep., May 1956.

33. Shigematsu, T.; Kudo, K. J. Radioanal. Chem., **1978**, *45*, 373.

34. Hoentsch, E.; Flachowsky, J. Isotopenpraxis, **1977**, *13*, 432.

35. Iqbal, M.; Ejaz, M.; Chaudhri, S. A.; Zamiruddin, A. J. Radioanal. Chem., **1978**, *42*, 335.

36. Jauniaux, M.; De Meyer, M.; Lejeune, W.; Levert, J. M. Bull. Soc. Chim. Belg., **1975**, *84*, 565.
37. Luke, C. L. Anal. Chem., **1956**, *28*, 1443.
38. Marczenko, Z.; Kalowska, H. Chem. Anal. (Pol.), **1976**, *21*, 183.
39. Shtefak, M.; Kirsh, M.; Rais, I. Zh. Anal. Khim., **1976**, *32*, 1364.
40. Takeda, Y.; Suzuki, S.; Ohyagi, Y.; Chem. Lett., **1978**, *12*, 1377.
41. Roberts, A. H. C.; Turner, M. A.; Syers, T. K. Analyst, **1976**, *101*, 574.
42. Karisson, R.; Gorton, L. Talanta, **1976**, *23*, 672.
43. Carry, J.; Menon, M. P. J. Radioanal. Chem., **1975**, *34*, 319.
44. Gentry, C. H. R.; Sherrington, L. G. Analyst, **1946**, *71*, 432.
45. Argollo, R. M.; Schilling, J. G. Anal. Chem. Acta., **1978**, *96*, 117.
46. Fujinaga, T.; Puri, B. K. Talanta, **1975**, *22*, 71.
47. Bode, H.; Z. Anal. Chem., **1955**, *144*, 165.
48. Kaiyanarman, S.; Khopkar, S. M. Anal. Chem., **1977**, *49*, 1192.
49. Charib, A.; Morris, D. F. C. Talanta, **1978**, *25*, 569.
50. Tomazic, B.; Branica, M. J. Radioanal. Chem., **1976**, *30*, 361.
51. Pollock, E. N. Anal. Chim. Acta., **1977**, *88*, 399.
52. Fantova, J.; Kliz, Z.; Cermakova, Z.; Suk, V. Chem. Listy, **1980**, *74*, 291.
53. Shimomura, S.; Sakurai, H.; Morita, I.; Mino, Y. Anal. Chem. Acta., **1978**, *96*, 69.
54. Steinnes, E. Radiochem. Radioanal. Lett., **1978**, *33*, 205.
55. Shamaev, Y. I.; Bogdanov, N. V. Zh. Anal. Khim., **1976**, *31*, 72.
56. Hasemann, K.; Bock, R. Z. Anal. Chem., **1985**, *274*, 185.
57. Huffman, E. H.; Beaufait, L. J. J. Amer. Chem. Soc., **1949**, *71*, 3179.
58. Wadelin, C.; Mellon, M. G. Anal. Chem., **1953**, *25*: 1668.
59. Fetchett, A. W.; Daughtrey, D.; Hunter, E. Anal. Chim. Acta., **1975**, *79*, 93.
60. Kamada, T.; Yamamoto, Y. Talanta, **1977**, *24*, 330.
61. Nakamura, K.; Ozawa, T. Anal. Chim. Acta., **1976**, *86*, 147.
62. Svehla, G.; Tolg, G. Talanta, **1976**, *23*, 755.
63. Lapenko, L. A.; Gibalo, I. M. Zh. Anal. Khim., **1976**, *31*, 481.
64. Spivakov, B. Y.; Orlova, Y. A.; Malyutina,T. M.; Kirilova, T. L. Zh. Anal. Khim., **1979**, *34*, 161.
65. Kamada, T. Shiroishi, T.; Yamamoto, Y. Talanta, **1978**, *25*, 15.
66. Chao, T. T.; Sanzolone, R. F.; Hubert, A. E. Anal. Chim. Acta., **1978**, *96*, 251.
67. Kudo, K.; Shigernatsu, T.; Kobayashi, K. J. Radioanal. Chem., **1977**, *36*, 65.
68. Bhowal, S. K.; Umiand, E. Anal Chem., **1976**, *282*, 197.
69. Musil, J.; Dolezal, J. Anal. Chim. Acta, **1977**, *92*, 301.
70. Roberts, R. F. Anal. Chem., **1977**, *49*, 1861.
71. Tribalat, S.; Beydon, J. Anal. Chim. Acta., **1953**, *8*, 22.
72. Fujinaga, T.; Lee, H. L. Talanta, **1977**, *24*, 395.
73. Fujinaga, T.; Puri, B. K. Indian J. Chem., **1976**, *14*, 72.
74. Zolotov, Y. A.; Petrukhin, O. M.; Shevehenko, U. N.; Dunina, U. Y.; Rukhadze, E. G. Anal. Chim. Acta., **1978**, *100*, 613.
75. Matsuoka, S.; Yoshimura, K.; Waki, H. Mem. Fac. Sci. Kyushu Univ., Ser. C., **1991**, *18*, 55–62.

76. Compaño, R.; Ferrer, R.; Guiteras, J.; Prat, M. D. Analyst, **1994**, *119*, 1225.
77. Kirkbright, G. F.; Narayanaswamy, R.; Welti, N. A. Analyst, **1984**, *109*, 15.
78. Narayanaswamy, R.; Sevilla, F. Anal. Chim. Acta., **1986**, *189*, 365.
79. Jönsson, J. Å.; Mathiasson, L. Trac-Trend. Anal. Chem., **1999**, *18*, 318.
80. Jönsson, J. Å.; Mathiasson, L. Trac-Trend. Anal. Chem., **1999**, *18*, 325.
81. Jönsson, J. Å.; Mathiasson, L. Trac-Trend. Anal. Chem., **1992**, *11*, 106.
82. Audunsson, G. Anal. Chem., **1986**, *58*, 2714.
83. Norberg, J.; Zander, Å.; Jönsson, J. Å. Chromatographia, **1997**, *46*, 483.
84. Nilvé, G.; Stebbins, R. Chromatographia, **1991**, *32*, 269.
85. Nilvé, G.; Audunsson, G.; Jönsson, J. J. Chromatogr., **1989**, *471*, 151.
86. Audunsson, G. *Anal. Chem.*, **1988**, *60*, 1340.
87. Lindegård, B.; Jönsson, J. Å.; Mathiasson. L. J. Chromatogr., **1992**, *573*, 191.
88. Thordarson, E.; Pálmarsdóttir, S.; Mathiasson, L.; Jönsson, J. Å. Anal. Chem., **1996**, *68*, 2559.
89. Norberg, J.; Emnéus, J.; Jönsson, J. Å.; Mathiasson, L.; Burestedt, E.; Knutsson, M.; Marko-Varga, G. J. Chromatogr. B, **1997**, *701*, 39.
90. Pálmarsdóttir, S.; Lindegård, B.; Deiningr, P.; Edholm, L. E.; Mathiasson, L.; Jönsson, J. A. J. Cap. Electrophor., **1995**, *2*, 185.
91. Pálmarsdóttir, S.; Mathiasson, L.; Jönsson, J. Å.; Edholm, L. E. J. Chromatogr. B, **1997**, *688*, 127.
92. Pálmarsdóttir, S.; Thordarson, E.; Edholm, L. E.; Jönson, J. Å.; Mathiasson, L. Anal. Chem., **1997**, *69*, 1732.
93. Bauer, S.; Solyom, D. Anal. Chem., **1994**, *66*, 4422.
94. Cisper, M. E.; Hembeergeer, P. H. Rapid Commun. Mass Sp., **1997**, *11*, 1454.
95. Johnson, R. C.; Koch, K.; Cooks, R. G. Ind. Eng. Chem. Res., **1999**, *38*, 343.
96. Johnson, R. C.; Koch, K.; Cooks, R. G. Anal. Chim. Acta., **1999**, *395*, 239.
97. Shen, Y.; Obuseng, V.; Grönberg, L.; Jönsson, J. Å. J. Chromatogr. A, **1996**, *725*, 189.
98. Grönberg, L.; Shen, Y.; Jönsson, J. Å. J. Chromatogr. A, **1993**, *655*, 207.
99. Djane, N. K.; Bergdahl, I. A.; Ndung'u, K.; Schütz, A.; Johansson, G.; Mathiasson, L. Analyst, **1997**, *122*, 1073.
100. Djane, K. N.; Ndung'u, F.; Malcus, G.; Johansson, G.; Mathiasson, L. Fresenius J. Anal. Chem., **1997**, *358*, 822.
101. Cox, J. A.; Bhatnagar, A. Talanta, **1990**, *37*, 1037.
102. Parthasarathy, N.; Pelletier, M.; Buffle, J. Anal. Chim. Acta., **1997**, *350*, 183.
103. Djane, N. K.; Ndung'u, K.; Johnson, K.; Sartz, H.; Törnström, T.; Mathiasson, L. Talanta, **1999**, *48*, 1121.
104. Scamerhorn, J. F.; Harwell, J. H. *Surfactant-Based Separations Processes*; Marcel Dekker: New York, 1989; p. 139.
105. Hinze, W. L.; Pramauro, E. Criti. Rev. Anal. Chem., **1993**, *24*, 133.
106. Carabias-Martínez, R.; Rodríguez-Gonzalo, E.; Moreno-Cordero, B.; Pérez-Pavón, J. L.; Garcia-Pinto, C.; Fernández Laespada, E. J. Chromatogr. A, **2000**, *902*, 195.
107. Quina, F. H.; Hinze, W. L. Ind. Eng. Chem. Res., **1999**, *38*, 4150.

108. Huddleston, J. G.; Willauer, H. D.; Griffin, S. T.; Rogers, R. D. Ind. Eng. Chem. Res., 1999, 38, 2523.
109. Watanabe, H.; Tanaka, H. Talanta, 1978, 25, 585.
110. García Pinto, C.; Pérez Pavón, J. L.; Moreno Cordero, B.; Romero Beato, E.; García Sánchez, S. J. Anal. Atomic Spectrometry, 1996, 11, 37.
111. Akita, S.; Rovira, M.; Sastre, A. M.; Takeuchi, H. Sep. Sci. Technol., 1998, 33, 2159.
112. Okada, T. Anal. Chem., 1992, 64, 2138.
113. Horvath, W. J.; Huie, C. W. Talanta, 1993, 9, 1385.
114. García Pinto, C.; Pérez Pavón, J. L.; Moreno Cordero, B. Anal. Chem., 1995, 67, 2606.
115. Stangl, G.; Niessner, R. Mikrochim. Acta., 1994, 113, 1.
116. Ferrer, R.; Beltran, J. L.; Guiteras, J. Anal. Chim. Acta., 1996, 330, 199.
117. Eiguren, A.; Sosa, Z.; Santana, J. J. Analyst 1999, 124, 487.
118. Calvo Seronero, L.; Fernández Laespada, M. E.; Pérez Pavón, J. L.; Moreno Cordero, B. J. Chromatogr. A, 2000, 897, 171.
119. Eiguren, A.; Sosa, Z.; Santana, J. J. Anal. Chim. Acta., 1998, 358, 145.
120. Moreno, B.; Pérez, J. L.; García, C.; Fernández, M. E. Talanta, 1993, 40, 1703.
121. Carabias-Martínez, R.; Rodríguez-Gonzalo, E.; Domínguez-Alvarez, J.; Hernández-Méndez, J. Anal. Chem., 1999, 771, 2468.
122. Eisenman, G. Ion Selective Electrodes; National Bureau of Standards (USA) Special Publications: Washington, D.C., 1969; p. 314.
123. Janata, J.; Josowitz, M.; de Vaney, D. M. Anal. Chem., 1994, 66, 207R.
124. Camman, K.; Lemke, U.; Rohen, A.; Sander, J.; Wilken, H.; Winter, B. Angew. Chem. 1991, 109, 519.
125. Szczepaniac, W.; Oleksy, J. Anal. Chim. Acta., 1986, 189, 237.
126. Worf, W. E. The Principles of Ion Selective Electrodes and of Membrane Transport; Akademiai Kiadó: Budapest, 1981; Elsevier: Amsterdam, 1981.
127. Koryta, J. Ion-Selective Electrodes; Cambridge University Press: London, 1975.
128. Freiser, H. Ion-Selective Electrodes in Analytical Chemistry; Plenum Press: New York, Vol. 1, 1978.
129. Pretsch, E.; Buchi, R.; Amman, D.; Simon, W. Essays on Analytical Chemistry; Pergamon Press: New York, 1977.; p. 321.
130. Svehia, G. Comprehensive Analytical Chemistry; Elsevier: New York, Vol. 11, Chap. 3, 1981; p. 346.
131. Hobby, P. C.; Moody, G. J.; Thomas, J. D. R. Analyst, 1983, 108, 581.
132. Geissler, M.; Kunze, R. Anal. Chim. Acta., 1986, 189, 245.
133. Simon, W.; Pretsch, E.; Morf, W. E.; Amman, D.; Oesch, U.; Dinten, O. Analyst, 1984, 109, 207.
134. Simon, W.; Morf, W. E.; Meier, P. C. Struct. Bonding, 1973, 16, 113.
135. Eisenman, G. Anal. Chem., 1968, 40, 311.
136. Xie, R. Y.; Christian, G. D. Analyst, 1987, 112, 61.
137. Pioda, L. A. R.; Stankova, Y.; Simon, W. Anal. Lett., 1969, 2, 665.

138. Scholer, R. P.; Simon, W. Chimia, **1970**, *24*, 372.
139. Mair, B. J. *Treatise on Analytical Chemistry, Part 1, Theory and Practice*; Kolthoff, I. M.; Elving, P. J.; Sandeli, E. B. Eds., Wiley-Interscience: New York, Vol. 3, 1961: p. 1469.
140. Pecsok, R. L.; Shields, L. D.; Cairns, T.; McWilliam, L. G.*Modern Methods of Chemical Analysis*, 2nd ed.; John Wiley and Sons: New York, 1968; pp. 41–70.
141. Huber, J. F. K. *Comprehensive Analytical Chemistry*, eds. Wilson, C. L.; Wilson, D. W.; Stroits, C. R. N. Elsevier: Amsterdam, Vol. 11B, 1968; p. 33.
142. Peters, D. G.; Hayes, J. M.; Hieftje, G. M. *Modern Chemical Analysis*; W.B. Saunders; Philadelphia, 1976.
143. Hamilton, R. J.; Sewei, P. A. *Introduction to High Performance Liquid Chromatography*; Chapman and Hall: London, 1977.
144. Snyder, L. R.; Kirkland, J. J. *Modern Liquid Chromatography*; John Wiley and Sons: New York, 1974.
145. Parris, N. H. *Instrumental Liquid Chromatography*; Elsevier: Amsterdam, 1976.
146. Lawrence, J. F.; Frei, R. W. *Chemical Derivitization in Liquid Chromatography*; Elsevier: Amsterdam, 1976.
147. Horvath, C. Ed. *High-Performance Liquid Chromatography*; Academic Press: New York, Vols. 1–3, 1980–1983.
148. Mulier, W. Kontakte (Merck), **1986**, *3*, 3.
149. Braun, R.; Ghersini, G. *Extraction Chromatography*; Elsevier: Amsterdam, 1975.
150. Wakui, Y.; Matsunaga, H.; Suzuki, T. M. Anal. Sci, **1988**, *4*, 325.
151. Muscatello, A. C.; Yarbro, S. L.; Marsh, S. F. *New Separation Chemistry Techniques for Radioactive Waste and other Specific Applications*; Cecille, L.; Casarci, M.; Pietrelli, L. Eds.; Elsevier Science Publishing: Amsterdam, 1991; p. 88.
152. Dietz, M. L.; Horwitz, E. P.; Nelson, D. M.; Walgren, M. Health Physics **1991**, *61*, 871.
153. Horwitz, E. P.; Dietz, M. L.; Nelson, D. M.; La Rosa, J. J.; Fairman, W. Anal. Chim. Acta., **1990**, *238*, 263.
154. Myasoedova, G. V.; Savvin, S. B. CRC Crit. Rev. Anal. Chem. **1988**, *17*.
155. Kremlyakova, N.; Barsukova, K. V.; Myasoedov, B. F. Radiochem. Radioanal. Lett., **1983**, *57*, 293.
156. Horwitz, E. P.; Dietz, M. L.; Fisher, D. E. Anal. Chem., **1991**, *63*, 522.
157. Horwitz, E. P.; Chiarizia, R.; Dietz, M. L. Solvent Extr. Ion Exch., **1992**, *10*, 313.
158. Westwood, S. A. Ed. *Supercritical Fluid Extraction and its Use in Chromatographic Sample Preparation*; Blackie Academic and Professional, Chapman and Hall: UK, 1993.
159. Veuthy, J. L.; Caude, M.; Rosset, R. Analysis, **1990**, *18*, 103.
160. Hawthorne, S. B.; Anal. Chem. **1990**, *62*, 633.
161. Vannoort, R. W.; Chervet, J. P.; Lingeman, H.; De Jong, G. J.; Brikman, A. U. Th. J. Chromatogr., **1990**, *505*, 45.

162. Hawthorne, S. B. Coupled (on-line) supercritical fluid extraction-gas chromatography. In *Supercritical Fluid Extraction and Its Use in Chromatographic Sample Preparation*; Westwood, S. A. Ed., Blackie Academic and Professional, Chapman and Hall; UK, 1993; p. 65.

163. Hawthorne, S. B.; Miller, D. J. J. Chromatogr. Sci, **1985**, *24*, 258.

164. Klesper, E.; Corwin, A. H.; Turner, D. A. J. Org. Chem., **1962**, *27*, 700.

165. Sugiyama, K.; Saito, M.; Hondo, T.; Seda, M. J. Chromatogr. **1985**, *332*, 107.

166. Anderson, I. G. M. Supercritical fluid extraction coupled to packed column supercritical fluid chromatography. In *Supercritical Fluid Extraction and Its Use in Chromatographic Sample Preparation;* Westwood, S. A. Ed., Blackie Academic and Professional, Chapman and Hall: UK, 1993; p. 112.

167. Foreman, J. K.; Stockweli, P. B. *Automatic Chemical Analysis*; Ellis Horwood: Chichester, UK, 1975.

168. Foreman, J. K.; Stockweli, P. B. *Topics in Automatic Chemical Analysis*; Ellis Horwood: Chichester, UK, Vol. 1, 1979.

169. Valcarecel, M.; Luque de Castro, M. D. *Flow-Injection Analysis*; Ellis Horwood: Chichester, UK, 1987.

170. Furman, W. B. *Continuous Flow Analysis: Theory and Practice*; Dekker: New York, 1976.

171. Coakley, W. A. *Handbook of Automatic Analysis. Continuous Flow Techniques;* Dekker: New York, 1982.

172. Ruzicka, J.; Hansen, E. H. *Flow-Injection Analysis*; Wiley: New York, 1981.

173. Valcarcel, M.; Gallego, M. Separation techniques. In *Flow Injection Atomic Spectroscopy*, Bruguera, J. L. Ed., Marcel Dekker: New York, 1989.

174. Luque de Castro, M. D.; Valcarcel, M. Trends in Analytical Chemistry, **1991**, *10*, 114.

175. MacLaurin, P.; Andrew, K. N.; Worsfold, P. J. Flow Injection Analysis. In *Process Analytical Chemistry*, McLennan, F.; Kowalski, B. R. Eds.; Blackie Academic and Professional: Glasgow, 1995; p. 159.

176. Valcarcel, M.; Luque de Castro, M. *Non chromatographic Continuous Separation Techniques*; Royal Society of Chemistry: Cambridge, UK, 1990.

177. Karlberg, B.; Thelander, S. Anal. Chim. Acta., **1978**, *98*, 1.

178. Shelly, D. C.; Rossi, T. M.; Warner, I. M. Anal. Chem., **1982**, *54*, 87.

179. Gallego, M.; Valcarcel, M. Anal. Chim. Acta., **1985**, *169*, 161.

14

The Use of Solvent Extraction in the Recovery of Waste

MICHAEL COX* University of Hertfordshire, Hatfield, Hertfordshire, United Kingdom

HANS REINHARDT MEAB Metallextraktion AB, Göteborg, Sweden

14.1 INTRODUCTION

Recently the protection of the environment has become increasingly important for industry with the requirement that the potential impact on the environment is considered for all aspects of industrial processes. Such considerations are supported by environmental legislation that controls all types of emissions as well as the treatment of wastes. Such legislation is based on global standards that have largely resulted from developments within the European Union, Japan, and the United States in collaboration with international conventions. Of these, the Basel Convention (1989) and the Earth Summit in Rio de Janeiro (1992) were significant in the control and prevention of wastes. In the case of liquid wastes that are most appropriate for treatment by liquid–liquid extraction, limits for discharge into the aqueous environment have been established by the three countries already mentioned. These limits depend on the particular country and sometimes on the industry. (See section 14.6.)

Environmental limits vary according to country, and in the future pressure will be applied to those countries where limits are considered too high. Indeed, there is a general trend for discharge limits to be reduced, with the concept of zero industrial discharge for certain metals being the ultimate aim for the future.

Metal pollution can arise from a number of different sources, for example:

---

*Retired.

- Acidic mine waters emerging naturally from underground mines
- Leachates from solid residues from the processing of primary minerals such as flotation and leach residues, and from landfill
- Metal treatment processing including coating (zinc), plating (chromium, nickel, copper), and pickling (hydrofluoric acid, iron)
- Chemical wastes from other industries: tanning (chromium), photographic (silver), electronic (copper, silver, gold, nickel, lead, cadmium)

Finally, there are many metal-containing solid wastes that may undergo leaching if disposed to land: spent catalysts (cobalt, nickel, vanadium); spent batteries (nickel, cadmium, lithium, lead); combustion ashes; etc.

Although this chapter concentrates on environmental problems associated with metals, the discharge of organic compounds is also subjected to stringent limits. The compounds that cause most concern are pesticides, herbicides, phenols, biphenyls, polychlorinated hydrocarbons (PCBs), and polyaromatic hydrocarbons (PAHs).

This chapter concentrates on the applications of solvent extraction. It should be remembered that other technologies could be applied to these wastes. These range from the well-established technologies of precipitation, adsorption, and electrolysis to those more recently developed, e.g., ion exchange and membrane processes: The flow sheet finally chosen will depend on a number of factors, including the precise composition of the waste, local expertise, disposal options, and of course the overall economics. In addition, this flow sheet may involve more than one technique. Thus one disadvantage of the use of solvent extraction, the possibility of entrained organic compounds in the raffinate, can be minimized or eliminated by the addition of a final adsorbent column before discharge.

14.2 SOLVENT EXTRACTION AND THE ENVIRONMENT

Before considering particular applications, the potential for the recovery of values from wastes is discussed. In other chapters the various applications of solvent extraction have been presented and the technology has been shown to be highly selective, easy to operate and control, and versatile in terms of scale. Therefore, it seems to be ideal for application in the treatment of wastes, both solid (after leaching) and liquid. As noted earlier, however, there are environmental limitations mainly caused by unintentional release of the solvent into the environment. These problems can be minimized by ensuring that the treated solutions are recycled within the plant and that any solutions discharged into the environment are treated appropriately to remove entrained organic compounds and residual toxic metals. Thus the main applications of solvent extraction in treatment of wastes are in the removal and recovery of values from waste streams to provide a process water for recycling. This reduces the effect on the environment by minimizing the amount of effluent for treatment and reduction in fresh water consumption.

The most important metalliferous liquid effluents where solvent extraction could be applied are from the various metal finishing operations: plating, pickling, etching, and the wash waters arising from the cleaning of work pieces. In the case of solids, in addition to scrap metal and alloy wastes from manufacturing operations, a number of other products use valuable and toxic metals and offer potential applications, e.g., spent automobile catalysts, Ni/Cd batteries, etc.

In the treatment of both types of waste, the objectives must be:

- To remove the metals effectively so as to produce a liquid waste stream capable of reuse or finally meeting environmental discharge limits
- To recover the metals for recycling within the plant, or at appropriate purity for sale
- To separate other impurities in a form for resale as by-products or safe environmental discharge

These principles are embodied in the concepts of *zero discharge* and *sustainable technology*.

14.3 RECOVERY OF METALS FROM WASTE: GENERAL ASPECTS

Successful use of solvent extraction in processes for the recovery of metals from waste depends extensively on successful pre- and posttreatments. Solvent extraction is a selective separation procedure for metals; however, it is limited to metals dissolved in aqueous media. It is obvious that solid wastes containing metals need a primary step of leaching to bring the metals in solution and it is very seldom that the resulting strip liquor from the solvent extraction process has a direct use. The technology covering these types of treatment is termed *hydrometallurgy*, defined as involving a sequence of chemical reactions of metals carried out in aqueous solutions.

To discuss the processes for the recovery of metals from waste, more than one hydrometallurgical procedure has to be considered, to establish general processing concepts for a broad range of feed materials, such as:

- Mine waters and leachates from residue dumps
- Plating, pickling, etching solutions, including rinse waters

- Neutralization sludge from the plating industries
- Flue dust from steel and brass mills
- Ash and scrap from industrial production

There are three main kinds of operations used within hydrometallurgy for metal recovery:

- Leaching (mixing and solid/liquid separation)—dissolution with acid, alkali, or bacteria, using redox, pressure (autoclave), etc.
- Purification and separation—solvent extraction, ion exchange, cementation, adsorption, liquid membranes, reverse osmosis, etc.
- Generation of products—evaporation, electrowinning, precipitation, crystallization

As can be seen from the list, there are many competing procedures for leaching, selective separation of the metals, and generation of products. Thus, the detailed design of a specific recovery process often involves a number of selected procedures, combined in unique ways. This chapter discusses processes containing a solvent extraction separation procedure.

14.4 WASTE RECOVERY PROCESSES INVOLVING SOLVENT EXTRACTION

To use solvent extraction for metal separation, it is first necessary to have the metals dissolved in an aqueous solution. Starting with a solid waste material, the initial treatment is therefore a leaching procedure to provide an aqueous solution from which the metals may be recovered [1–4]. The dissolved metal ions, together with all other ions in solution, interact with one another in various ways. The extent to which they do so and the resulting changes in behavior are of great importance in solution chemistry and are discussed in Chapter 2. In general, association occurs between ions of opposite sign, and the commonly used method of describing the reactions is the complex formation approach.

The interaction between the ions is regarded as a chemical equilibrium. The resulting product of such an association between two individual kinds of ions is called a complex species, and the extent to which these various complex species occurs are described in terms of the equilibrium constants.

It should be noted that in leaching the rate of reaction is of great importance, since the temperature used is relatively low and many factors combine to keep the rate slow. For practical reasons, leaching studies are performed on a homogenized sample, testing the temperature dependency of the rate of reaction (concentration and time) in various leaching mixtures. The most important variables in aqueous systems are pH and redox potential, together with the metal concentration. Oxidation is frequently used to make the solid to be leached unstable with respect to the solution and thus promote dissolution.

The importance of an optimal leaching procedure can be illustrated by considering a solvent extraction process for the production of 10,000 metric tons of copper. Commonly achieved leach solutions contain $2.5\,g\,dm^{-3}$ Cu in the feed solution. This value results in an aqueous feed flow from the leaching operation to the mixer-settler units of about $500\,m^3\,h^{-1}$. To this add $200\,m^3\,h^{-1}$ of organic solvent and allow organic recycle to achieve organic continuous conditions in the mixer, and the total liquid flow through the mixer-settlers will be over $1000\,m^3\,h^{-1}$. Now, if it is possible to design the leaching operation to provide a leach solution of $25\,g\,dm^{-3}$ Cu, then both the dimensions of the leaching and solvent extraction equipment can be drastically reduced.

14.4.1 Solvent Extraction from Acid Feed Solutions

As seen in Chapters 4 and 11, the extraction of many metals with an acid extractant (H^+R) is pH dependent, with, in general, the lower the pH, the lower the extraction of the metal. In other words, the D-value (or % extraction) decreases with decreasing pH (Fig. 11.8). In addition, the extraction of each metal ion (M^{n+}) liberates n hydrogen ions (H^+), from the extractant H^+R. This means that the pH decreases with extraction, resulting in decreased transfer of metal to the organic solvent. Thus, the more metal in solution and the higher acidity, the less is extracted, no matter how many stages are used.

For example, the extraction of nickel with di-(2-ethylhexyl)phosphoric acid (DEHPA) at a pH below 3 is <10%. If the extraction of nickel is carried out between pH 3–4, the extraction of about $0.3\,g\,dm^{-3}\,Ni^{2+}$ will decrease the pH to below 3 and the mass transfer of nickel will stop. For copper, the extraction is performed between pH 2–3 and thus, an amount of $3\,g\,dm^{-3}\,Cu^{2+}$ will decrease the pH from 3 to 2 and again extraction will stop.

To allow a higher extraction capacity (net transfer) and avoid the influence of liberated H^+ ions, simultaneous neutralization with NaOH directed into the mixer is used. Such an example is described later, where nickel is extracted from a cadmium electrolyte. Other means of controlling the pH includes the use of acid extractants in either alkaline (NH_4^+R) or neutral (Na^+R) form.

An increase of the net transfer of zinc was reported [9] for the extraction with DEHPA (50%) in kerosene of zinc from a feed solution containing $40 \, g \, dm^{-3}$ Zn, emanating from the leaching of steel furnace dust. The resulting raffinate, with a residual Zn concentration of $15–20 \, g \, dm^{-3}$, was recycled to leaching. Comparable net transfer conditions were obtained [5] in the modified Zincex process (MZP) for the processing of spent domestic batteries.

Similar high transfer of copper in a solvent extraction procedure has been recently reported [6]. After pressure leaching to liberate copper ($60 \, g \, dm^{-3}$) under conditions that minimize the free acid (pH 1.4) and iron in solution, copper was extracted to $10 \, g \, dm^{-3}$ in three stages with loading in the organic phase of $12.5 \, g \, dm^{-3}$.

The situation described is valid in sulfuric acid solutions, when no strong metal complexes are formed. However with hydrochloric acid the formation of metal chloride complexes makes the situation quite different. The influence of the chloride ion concentration on the extraction of some metal chloride complexes of general interest is shown in Fig. 11.2. For example, the separation of zinc and iron(III) can be achieved at a low chloride concentration, and later the separation of cobalt and nickel is described. The disadvantage of using the chloride system is concerned with stripping. For example, when a zinc chloride complex $ZnCl_4^{2-}$ is extracted with a tertiary amine, the stripped species corresponds to $ZnCl_2$, resulting in an increased chloride concentration and an interruption of the mass transfer. The consequence of this is that it is difficult to achieve concentrated zinc chloride strip liquors. Two examples are given next.

14.4.1.1 Extraction of Zinc from Weak Acid Effluents

The first example describes the extraction of zinc from weak acid solutions. In the manufacture of rayon, rinse waters and other zinc-containing liquid effluents are produced. The total liquid effluent in a rayon plant may amount to several m^3 per minute with a zinc concentration of $0.1–1 \, g \, dm^{-3}$ and pH normally 1.5–2. In addition to zinc, the effluent contains surface-active agents and dirt (organic fibers and inorganic sulfide solids). The use of both precipitation (OH^- and S^{2-}) and ion exchange has been reported to remove zinc from such effluents. In addition, solvent extraction has successfully been used to recycle the zinc back to the operation.

In the process [3,7], zinc is extracted to greater than 95% (pH > 2) with DEHPA (25%) in kerosene in two or three stages. Stripping is performed with a sulfuric acid solution (Figs. 14.1 and 14.2). By adjusting the net flow rate of the solution, the concentration of zinc in the strip solution may be increased to $50 \, g \, dm^{-3}$ or more. Thus, the resulting zinc sulfate solution can be reused directly in the spinning bath.

The extraction of zinc is pH dependent. The lower the pH the lower the extraction of zinc. In addition, the extraction of each zinc ion (Zn^{2+})

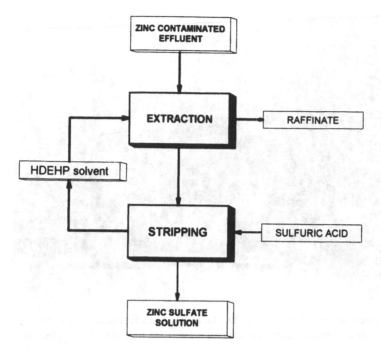

Fig. 14.1   Extraction of zinc from weak acid effluents.

liberates two hydrogen ions (H^+). Thus, the more zinc in solution, the less is extracted. This disadvantage means that using solvent extraction without some neutralization (pH > 2) results in a residual zinc concentration in the raffinate, which is too high to meet with environmental standards of today.

14.4.1.2   Extraction of Nickel from Cadmium Electrolytes

In the electrowinning of cadmium, nickel is an interfering element that has to be continuously removed from the electrolyte, a weak acidic cadmium sulfate solution. To handle the undesirable buildup of nickel contamination and at the same time obey environmental demands, the industry installed a solvent extraction "kidney" [8].

The nickel extraction can be performed with DEHPA under defined conditions. As described earlier for zinc, the extraction of nickel is pH dependent. The D-value for nickel decreases drastically with decreasing pH below 3.5. The extraction of each nickel ion (Ni^{2+}) liberates two hydrogen ions (H^+) from the extractant H^+ DEHPA, which means that only a very small amount of nickel can be transferred to the solvent before the extraction stops. However, by adding a neutralization reagent (NaOH) into the

Fig. 14.2   Zinc extraction plant at Enka AKZO, Netherlands. Full-scale plants have been built and operated at Svenska Rayon AB in Sweden and at Enka AKZO in The Netherlands.

mixer simultaneously with nickel extraction, this disadvantage can be avoid. Three large mixers in parallel are used to allow for the necessary reaction time (Fig. 14.3a and b).

The flow sheet is similar to the zinc block diagram with two extraction stages, containing three parallel mixers followed by a settler, and two stripping stages with normal mixer-settler arrangement. The pH is monitored and the acidity controlled by the addition of NaOH in the two first mixers. The extraction efficiency for nickel is better than 98%.

Another reported alternative to eliminate the extraction dependency on H^+ (pH) when extracting nickel with an acid reagent is to use the sodium form of the reagent (Na^+R) [9]. From a weak sulfuric acid rinse water containing $1-2 \, g \, dm^{-3}$ Ni, the metal could be effectively (> 99%) removed in two extraction stages. By loading the solvent with nickel, the transfer of sodium to the strip solution was minimized. Nickel was stripped from the loaded solvent with dilute sulfuric acid and recovery from the strip solution was best performed by electrowinning. The main problem with this procedure was high loss of extractant in the NaOH extraction reconditioning wash. A considerable reduction of this loss could be achieved by increasing the amount of sodium sulfate in the wash solution.

14.4.1.3 Extraction of Zinc and Chromium(III) from Electroplating Baths

The use of Zn-Cr(III) alloy plating has almost replaced the use of Cr(VI) in the electroplating industry due to its excellent corrosion resistance and its lower toxicity. Recently, a solvent extraction procedure for separating and selectively recovering the two metals, zinc and chromium, from electroplating wastewaters has been demonstrated [10].

Zinc is extracted and stripped with DEHPA in the conventional way as previously described. The major challenge lies in the extraction and stripping of Cr(III), which requires a higher pH than zinc. Chromium

(a)

Fig. 14.3 Nickel extraction plant with simultaneous neutralization.

(b)

Fig. 14.3 (*Continued*).

hydroxide precipitation places an upper limit on the pH that can be used. The extraction of Cr(III) with DEHPA is successful, but stripping of the organic solvent is difficult. Now, it was found that using NH_4^+ DEHPA (the ammoniated form), the extraction was similar to using the solvent in the acid form and all stripping problems were overcome.

The suggested flow sheet includes extraction of Zn with H^+ DEHPA ($0.1 \, mol \, dm^{-3}$) followed by the extraction of Cr(III) with NH_4^+ DEHPA ($0.1 \, mol \, dm^{-3}$). The stripping of both metals was performed with dilute sulfuric acid. By using the same reagent in both extraction circuits, contamination of the second solvent is avoided. A process for the recovery of Zn and Cr(III) with good separation from each other has been demonstrated. The treated wastewater contained the metals at levels meeting stringent environmental standards.

14.4.1.4 Extraction of Copper and Zinc from Brass Mill Flue Dust

Another reported example with two metals in solution is the recovery of copper and zinc from brass mill flue dust [11]. The material contains very little iron. The solid material is leached with sulfuric acid to produce a weak acid metal sulfate solution. The operation can be regarded as two separate inter-connected processes, one for copper recovery and one for zinc (Fig. 14.4).

In the *copper circuit*, copper-rich material is leached in a pH-controlled leach at 60°C to a final pH of about 2. Oxidizing conditions are maintained, e.g., by addition of MnO_2, to promote copper dissolution and to prevent cementation of metallic copper when the feed material contains metallic zinc (brass or iron). The leach solution, containing about $20\,g\,dm^{-3}$ zinc and $4\,g\,dm^{-3}$ copper, is filtered and fed to a solvent extraction circuit. Copper is extracted in four stages with a kerosene solution containing H^+R (10%). Stripping is performed in three stages with a sulfuric acid electrolyte. Copper metal is recovered from the electrolyte by electrowinning.

The *zinc circuit* consists of a similar pH-controlled leach at 60°C under oxidizing conditions. Zinc flue dust with low copper content is leached with the copper barren raffinate and with part of the zinc raffinate. The zinc leaching operation is maintained at about pH 2 for most of the leaching time and then slowly raised to a final pH of 4.5, reducing the iron level to below 10 ppm. The leach solution is filtered and cleaned from impurity metals such as Cu, Ni, and Cd by an ordinary cementation procedure, again filtered and finally fed to a solvent extraction circuit. Zinc is extracted in three stages with

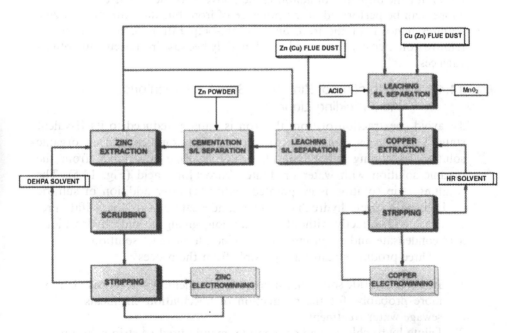

Fig. 14.4 Extraction of copper and zinc from weak sulfuric acid leach solutions.

a kerosene solution containing DEHPA (50%). The amount of zinc transferred in the extraction procedure can be maintained as high as 20–25 g dm^{-3}, because the remaining zinc in the raffinate is recycled to leaching. Stripping of zinc is performed in two stages with sulfuric acid electrolyte, and zinc metal is recovered from the electrolyte by electrowinning.

The process flow sheet was first tested for direct leaching of steel mill flue dust and production of zinc metal by electrowinning. The tests were performed in a continuously operating pilot plant, producing 10–20 kg/day zinc metal. The same pilot plant was then used for treating copper/zinc-rich brass mill flue dust in a closed loop operation, recycling all the zinc solvent extraction raffinate to the copper circuit leach section. In the zinc circuit leach section, only the amount of zinc rich dust necessary for neutralization of the copper solvent extraction raffinate was used. The results obtained from the pilot plant tests indicated contamination problems within the solvent extraction loops. The estimation of economic data showed a weak return on the assets compared with the alkali route, and sensitivity toward the raw material price.

A similar flow sheet has been suggested for the treatment of mine waters. However, the similarity is only apparent. The iron content in these waters (Fe 3–5 g dm^{-3}, Zn $\sim$ 1 g dm^{-3}, and Cu 0.5–1 g dm^{-3}, pH 2–2.5) is the reason for the difficulty of adapting the above flow sheet. The extraction of copper can be performed in the presence of iron, but the extraction of zinc requires removal of the iron, and the pre-preparation of the solution to remove iron is costly and complicated, mainly because iron occurs in both its valences.

14.4.1.5 Extraction of Zinc from Spent Hydrochloric Acid Pickling Liquors

To avoid coextraction of iron, the iron is initially reduced to its II-valent state. Then zinc is extracted as a zinc chloride complex into an organic solution containing tributylphosphate (TBP). Zinc is stripped from the organic solution with water or dilute hydrochloric acid (Fig. 14.5). The resulting strip solution is evaporated, either (1) after addition of sulfuric acid, giving a dilute hydrochloric acid condensate and a zinc sulfate precipitate, or (2) directly without any addition, giving a dilute hydrochloric acid condensate and a concentrate zinc chloride product solution.

Three product streams are possible from the process:

1. Iron (II) chloride solution, for possible treatment in a pyrolysis plant or, more probably, for the production of flocculation chemicals used in sewage water treatment
2. Dilute hydrochloric acid condensate, mainly used as strip solution

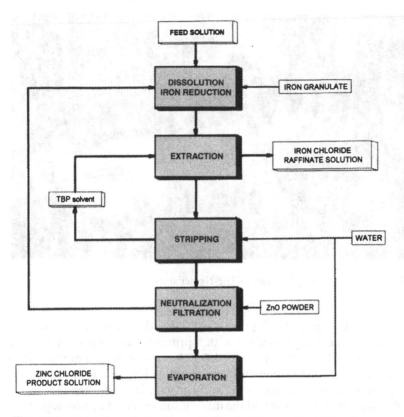

Fig. 14.5 Extraction of zinc from spent hydrochloric acid pickling liquors.

3. Zinc sulfate (solid precipitate) or a concentrated (about $250 \, g \, dm^{-3}$) zinc chloride solution

A pilot plant, containing two pulsed columns, one for extraction and one for stripping, and batchwise evaporation was in operation in Sweden during 1981. Pilot plant operations have also been performed in Holland (Fig. 14.6) and Germany. The experience from these tests shows that the process concept is technically practicable and well proven. The economics of the process, however, are strongly dependent on the cost for disposal of spent pickling liquors.

14.4.1.6 Extraction of Vanadium and Nickel from Soot and Fly Ash

In ash and soot residues from Orimulsion-fired power stations, the dominating metal constituents are vanadium, nickel, and iron, together with

Fig. 14.6   Zinc extraction pilot plant in The Netherlands.

high amounts of magnesium. A process [3,12] for the recovery of vanadium and nickel from these residues starts with a primary water wash to eliminate the large amount of magnesium (Fig. 14.7). The reason for this prewash is to recover and recycle magnesium back to combustion and to decrease the consumption of acid in the following sulfuric acid leaching step. The solubility of vanadium is very low and the dissolution of nickel can be neglected.

The residue after the water wash is leached at an elevated temperature by sulfuric acid, part of which is the recycled raffinate from the vanadium extraction. The leaching yield of vanadium (mainly IV-valent) is about 55% and of nickel about 95%. A final (post) leach with sodium hydroxide dissolves the remaining vanadium (mainly V-valent). The resulting leach solution, containing practically all the vanadium ($\sim$25 g dm^{-3}) and nickel ($\sim$12 g dm^{-3}) is fed to the solvent extraction circuit.

It may be mentioned that starting with ash and soot from crude oil-fired stations, the resulting metals are the same, but the main leaching residue is carbon. This residue is initially burned and the ash is leached again to increase the total yield of vanadium. In the same operation, the concentration of iron is reduced by precipitation of jarosite. During leaching, the redox potential is controlled by SO_2 addition to keep vanadium in its IV-valent state.

The extraction of vanadium is performed with a mixture of DEHPA and TBP in kerosene (Fig. 14.8). Half of the raffinate, with a somewhat increased acidity, is returned to leaching. The other half is further treated to extract nickel and to control the buildup of impurities. Owing to the high vanadium concentration in the feed, a large extraction capacity can be obtained and, consequently, the process flows will be less for a given vanadium throughput. The economical importance of this effect is decisive.

Vanadium is stripped from the organic solution with $1.5 \, mol \, dm^{-3}$ sulfuric acid to a concentration of about $50 \, g \, dm^{-3} \, V$. After an iron removal step with strong sulfuric acid, the organic solution is washed with water and recycled. From the vanadium strip liquor, ammonium polyvanadate (APV) is precipitated by oxidation and addition of ammonia. The APV slurry is thickened and pumped to a vacuum belt filter, where the APV cake is carefully washed with fresh water. The APV filter cake is dried and then calcined to vanadium pentoxide.

To the second half of the raffinate, ammonia is added to form a nickel ammonium complex, which is extracted with a hydroxyaryloxime (LIX84). Due to simultaneous formation of metal hydroxides, the solution is carefully

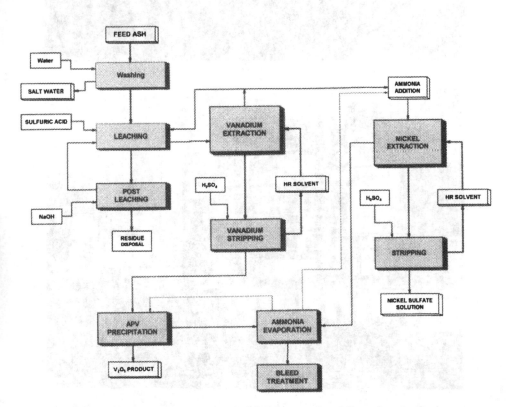

Fig. 14.7 Extraction of vanadium and nickel from soot and flue ash.

Fig. 14.8 Recovery of vanadium and nickel from soot, Stenungsund, Sweden.

filtered. After extraction of nickel, the resulting raffinate is further treated in a bleed treatment section. The organic solvent is scrubbed with weak sulfuric acid and then nickel is stripped by $1.5 \, mol \, dm^{-3}$ sulfuric acid. Flow rates of solutions and solvents are adjusted to give a very concentrated nickel sulfate strip liquor.

14.4.1.7 Recovery Cobalt and Nickel from Scrap Alloy

In a pilot plant [2,13], superalloy scrap containing Mo, W, Cr, Fe, Co, and Ni is pretreated in a furnace with carbon to transfer refractory metals (Mo, W, etc.) into carbides. The melt is granulated and the resulting material is charged into titanium baskets. Diaphragm-type electrolytic cells are used for anodic dissolution of the granulated material. Fe, Co, Ni, and small amounts of Cr are dissolved into a calcium chloride solution by the current. The metal carbides are not dissolved and remain as an anodic residue in the baskets.

The anolyte, containing about $75 \, g \, dm^{-3}$ Cl^- at pH 2, is fed to a solvent extraction circuit for the separation of Fe−Co and Fe−Ni with a tertiary alkyl amine, dissolved in kerosene. The separation is based on the tendency of the metals to form metal–chloride–amine complexes (Fig. 11.2). At low chloride ion concentration (about $75 \, g \, dm^{-3}$), Fe(III) is extracted to the organic solvent, while Co(II) and Ni(II) remain in the aqueous raffinate. If the chloride ion concentration is then increased to about $250 \, g \, dm^{-3}$, cobalt is extracted, leaving nickel behind in the raffinate.

Using an organic solution containing 25% Alamine 336 and 15% dodecanol in kerosene, it is possible to separate the metals one after the other by oxidizing iron to Fe(III) and altering the chloride ion concentration by evaporation. This was also the original object of the process. However, it was found that the complete oxidation of iron was complicated and, as the iron in ferro-cobalt and ferro-nickel did not ruin the market value of the products, the process finally used is shown in Fig. 14.9.

The anolyte is evaporated to increase the chloride concentration from $75–250 \, g \, dm^{-3}$ and cobalt, and most of the iron, is then extracted by the organic solvent. The nickel containing raffinate is diluted by condensate from the evaporation and returned to the Ni-cathode compartments in the diaphragm electrolytic cells for precipitation of nickel. Stripping of cobalt from the organic solvent is performed with the weakly acidic condensate. The strip liquor is fed to the Co-cathode compartment in the electrolytic cells.

The process described has been running in pilot plant scale operation for more than a year and the advantages of the process are summarized here:

- Using diaphragm-type electrolytic cells, electrical energy is used in an optimal manner for dissolution and deposition.

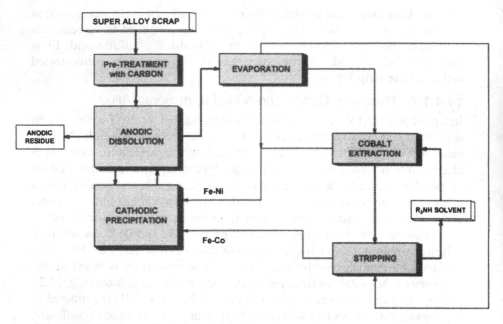

Fig. 14.9 Recovery of cobalt and nickel from scrap alloy.

- The amine-chloride system is well suited for the separation of the metals.
- Using an evaporator for raffinate concentration and production of strip solutions, a closed wet cycle is obtained and no aqueous waste pollutes the environment.
- The low consumption of chemicals and energy indicate an economic advantage over similar open cycle processes.

A limited amount of available waste and a price increase of the raw material resulted in the discontinuation of the project. This situation is not unique. Many projects suffer from the same problem, i.e., a waste material becomes a valuable raw material when it is in demand.

14.4.1.8 Extraction of Chromium(VI) from Surface Finishing Wastewater

Hexavalent chromate [Cr(VI)] is still used within the industry to meet critical high corrosion control and other metal surface finishing requirements. Cr(VI) is toxic and its control generates a hazardous, costly waste.

Recently, a solvent extraction process [14] for the recovery of hexavalent chromium from surface finishing process water, the A-LLX system, has been successfully demonstrated. The concentration level of Cr(VI) in the raffinate is low enough ($\leq 0.1\,\mathrm{mg\,dm^{-3}}$) for discharge to surface waters. Chromium recycling enables continued use of high performance chromium.

A tertiary amine (Alamine 336) is used as a selective extractant of Cr(VI). The extraction mechanism involves ion pairing between the amine and the dichromate anion, $HCrO_4^-$. After extraction, the Cr(VI)-loaded extractant is stripped with alkali to produce a strip liquor, containing 2% Cr(VI). This solution can be recycled.

14.4.1.9 Recovery of Zinc from Spent Batteries

Spent domestic batteries cause environmental repercussions for a number of reasons: the presence of soluble heavy metals; the increase in their use; the short life cycle, and the low efficiency of their recovery and recycling procedures. To improve the situation, a solvent extraction process plant has been built and operated in Spain [15]. It is based on the modified Zincex process (MZP) and has a capacity of processing 2800t/year of batteries in a continuous operation.

The process starts with pre-preparation of the feed material. The batteries are classified and pulverized and paper, plastics, and iron components are removed by physical procedures, leaving an impure zinc-manganese dust, which is treated by leaching in sulfuric acid to give a residue of mainly MnO_2 and graphite. Before the pregnant leach liquor, with a zinc content of $22\,\mathrm{g\,dm^{-3}}$, is purified by solvent extraction, metallic impurities like Hg, Ni, Co, Cd, etc., are removed by cementation.

Using a solvent extraction procedure, containing one depletion mixer-settler stage, three extraction, three washing, and two stripping stages, zinc is transferred to an acid sulfate strip liquor, containing about $140\,\mathrm{g\,dm^{-3}}$ zinc and suitable for electrowinning of high-quality zinc or evaporation to very pure zinc sulfate. The extractant reagent used is DEHPA, with a reported good selectivity for zinc over impurities in the feed liquor. In addition, the washing stages eliminate most of the entrained and coextracted Mn^{2+} and Cl^-. Only coextracted Fe^{3+} was left in the organic solvent after washing. Its concentration in the feed solution is <5ppm and this small quantity was eliminated by an HCl regeneration stage.

The economic evaluation of the plant indicates that for a feasible operation with the limited plant capacity, compensations for the treatment through direct consumers or through official organizations are required.

14.4.2 Extraction of Acids

Undissociated acids form adduct complexes with organic compounds, containing oxygen donor atoms. If the adducts formed are soluble in organic diluents like kerosene, the acids can be extracted from an aqueous solution by solvent extraction. This procedure has been described in more detail (See Chapter 10).

14.4.2.1 Recovery of Hydrofluoric and Nitric Acids from Stainless Steel Pickling Baths

A stainless steel pickling bath initially contains $2.2\,mol\,dm^{-3}$ HNO_3 and $1.6\,mol\,dm^{-3}$ HF. The bath is used until the iron concentration reaches $40-80\,g\,dm^{-3}$. At this stage, the bath contains about 50% unused acids in addition to the dissolved metals: iron, nickel, chromium, and molybdenum. In a used pickling bath, most of the acids are bound to the metals as complexes. By adding sulfuric acid to the pickling bath, sulfate will replace some of the nitrate and fluoride, leading to the formation of extractable, undissociated nitric and hydrofluoric acids.

The general procedure to remove monovalent acids from the aqueous solution by adding sulfuric acid and then extracting the released acids with an organic donor molecule, dissolved in an organic diluent, has been called acid exchange or the AX process [15,16]. The procedure was first presented for the treatment of spent stainless steel pickling liquids. The process consists of three main steps (Fig. 14.10):

1. Addition of H_2SO_4 and extraction of HNO_3 and HF with an organic solution, containing 75% TBP in kerosene, in a pulsed column operation
2. Precipitation of the metals (Cr, Fe, and Ni) left in the aqueous raffinate
3. Stripping HNO_3 and HF from the organic solution with water in a second pulsed column operation and recycling the acids back to the pickling bath

A plant for treating $600\,dm^3\,h^{-1}$ spent pickling liquid (corresponding to an annual production of 25,000 t stainless steel) was in operation for some years in Sweden (Fig. 14.11). The recovery operation was successful; however, the steel works are now closed.

Similar extraction procedures [17] using a mixture of trialkylphosphine oxides (CYANEX 923) for extraction of the mineral acids from electroplating and stainless steel pickling baths are reported as the ARSEP procedure.

The possibility of using nitric acid as an oxidizing agent in a leaching operation is often prohibitive because many reagents used in solvent extraction are not stable in an oxidizing solution. By the AX method, however, a nitric acid (nitrate) leach solution can be converted to a weak sulfuric acid (sulfate) solution, well suited for a solvent extraction treatment.

14.4.2.2 Purification of Green Phosphoric Acid (H₃PO₄)

By reacting apatite with sulfuric acid, phosphoric acid and gypsum will result. Most phosphoric acid is produced in this way and is normally used as raw material (green acid) to produce fertilizers. A small amount, however, is further refined to phosphoric acid of food grade quality.

The green phosphoric acid from a phosphoric acid plant (PAP) contains 25–30% H_3PO_4. The acid is heavily entrained with impurity cations, among which arsenic, cadmium, and uranium are the most toxic. In addition, anions, like chloride, fluoride, and sulfate must be considered. Selective extraction of phosphoric acid [18] as an adduct complex with TBP according to the block diagram in Fig. 14.12 has been used to produce acid of food grade.

The purification operation starts with addition of chemicals to precipitate arsenic and sulfate in a pretreatment step. The acid is then cooled and left for an appropriate aging time. A flocculent is added and the solid content is allowed to flock (precipitate) in a clarifier.

The clear phosphoric acid is fed to the solvent extraction operation, where about 60% of the acid is extracted countercurrently with an organic solvent containing TBP in kerosene. The resulting raffinate containing most

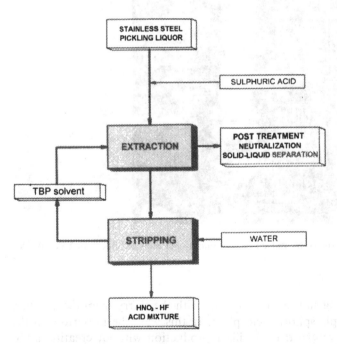

Fig. 14.10   Recovery of acids from stainless steel pickling baths.

Fig. 14.11 Acid recovery by pulsed plate column operation at Söderfors Steel Works, Sweden.

of the contaminating substances, together with the slurry from the clarifier, is returned to the phosphoric acid plant. In this way, the impurities in the green acid will be returned to fertilizer production without creating additional environmental problems.

After extraction, the organic solvent, loaded with phosphoric acid, is scrubbed with a phosphoric acid solution in equilibrium with the organic solvent. In this way, coextracted contamination is removed. The scrub liquor is recycled to extraction.

After adjusting the temperature to about 40°C, the organic solvent is treated with clean water to strip the phosphoric acid. This acid, containing about 30% H_3PO_4, is heated and water evaporated to produce concentrated (85%) phosphoric acid of food grade quality.

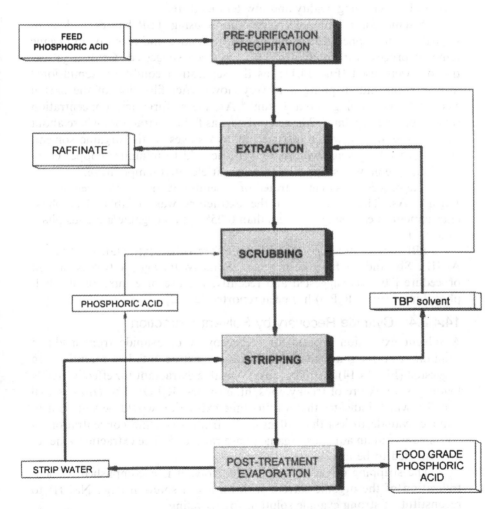

Fig. 14.12 Purification of phosphoric acid to food grade quality.

In addition, a process for the purification and production of citric acid has been reported, utilizing a similar flow sheet.

14.4.2.3 Removal of Arsenic Acid (H_3AsO_4) from Copper Electrolytes

The use of low-grade ores with large amounts of harmful impurities results in increased accumulation of these impurities in the electrolyte. One of the most harmful is arsenic, which seriously affects the quality of the copper cathodes. Arsenic occurs in the sulfuric acid electrolyte as arsenic acid, and as such can be extracted as an adduct complex with TBP. The extraction is favored by increasing acidity and low temperature.

Arsenic control by solvent extraction using TBP has been demonstrated [19] in copper electrorefining plant practice and recently an arsenic removal process with undiluted TBP has been suggested. Centrifugal extractors were used (Fig. 14.13), as the separation conditions (emulsions) in extraction and stripping were very slow. After filtration of the barren electrolyte, containing about $8\,g\,dm^{-3}$ As, the sulfuric acid concentration was raised to $180\,g\,dm^{-3}$. The electrolyte was fed to extraction, where about 50% of the arsenic was transferred in two stages to the organic solvent. After a scrubbing stage with dilute sulfuric acid to minimize copper coextraction, arsenic was stripped with water at elevated temperature.

The process was demonstrated on a semi-industrial scale—removal of $1\,kg\,h^{-1}$ As. The operation of the extractors was reliable. The phase entrainment in each stage was less than 0.05% at an organic/aqueous phase ratio of 1.

TBP only extracts As(V) and therefore other extractants active for As(III), Sb, and Bi have been tested. Some years ago, a two-extractant procedure [20] was suggested and recently, the use of a mixture of alkylphosphine oxides (R_3PO) has been reported [21,22].

14.4.2.4 Cyanide Recovery by Solvent Extraction

A solvent extraction process for the recovery of cyanide from acidified solutions using organophosphorus solvating extractants has recently been suggested (Fig. 14.14) [23]. The most favorable extractant for effective HCN loading is a mixture of trialkylphosphine oxides (R_3PO). The D_{HCN} is well over 10, which indicates that a four-stage extraction would be sufficient to remove cyanide to less than $10\,mg\,dm^{-3}$ from a starting concentration of $200\,mg\,dm^{-3}$ at an aqueous/organic phase ratio of 5. The extraction kinetics are reported to be fast.

The stripping of the organic loaded solvent is accomplished by neutralization of the organic solvent with an alkali solution (e.g., NaOH) to reconstitute a strong cyanide solution for recycling.

Fig. 14.13 Centrifugal extractor arrangement with five CENTREK EC-320 units.

14.4.3 Solvent Extraction from Ammoniacal Feed Solutions

Some years ago a hydrometallurgical route, called the AmMAR concept, was presented [24]. The concept (Fig. 14.15) represented a general processing route for a wide variety of feed materials. The detailed design of each specific process involved a specific number of known chemical operations, combined in unique ways. The main thread was the ammonia–ammonium chloride or carbonate leach solution and its extraordinary chemical flexibility.

The general outline of an ammoniacal process for treatment of a material containing two or more of the metals: iron, chromium, copper, nickel and zinc, includes the following process steps:

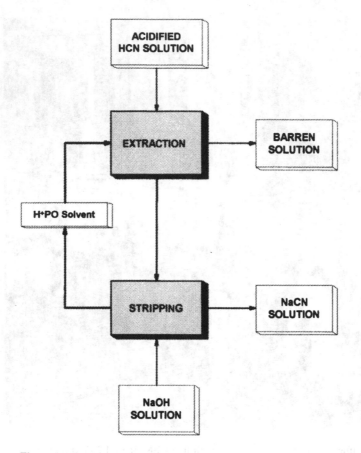

Fig. 14.14 Recovery of cyanide as NaCN.

1. Primary separation occurs in the leaching procedure, where the metals copper, zinc, and nickel are dissolved as metal ammonium chlorides or chloro complexes, while iron and chromium remain in the solid residue as hydroxides.
2. Copper and nickel are subsequently removed from the leach liquor by solvent extraction. After selective stripping, nickel sulfate is produced by crystallization and copper metal by electrowinning.
3. Excess ammonia is evaporated from the resulting extraction raffinate and zinc is precipitated as carbonate by the remaining ammonium carbonate in the solution or by addition of carbon dioxide. The evaporated ammonia is absorbed in the filtrate and recycled to the leaching step.

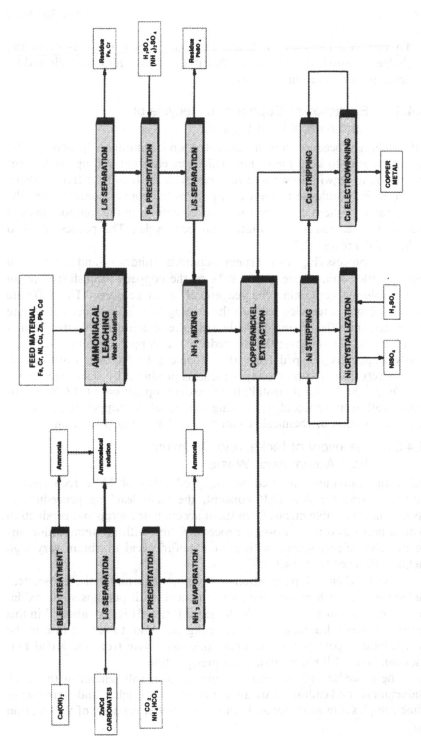

Fig. 14.15 General diagram of the AmMAR concept.

4. To maintain the water balance, especially when a water-containing sludge is treated, an additional effluent treatment with lime, followed by ammonia evaporation, is necessary.

14.4.3.1 Extraction of Copper and Recycle of Ammoniacal Etch Liquors

In the etching procedure, copper concentration is continuously increased by the etching and, to keep the etching efficiency constant and optimal, spent etchant has to be withdrawn and replaced with fresh etchant (replenisher). An alternative would be to remove a part of the copper content from the spent etchant, without changing the other conditions in the solution, in such a way that the barren etching solution can be recycled. This process is called the MECER process [2,25].

In the process (Fig. 14.16) spent etchant is withdrawn and fed through a mixer-settler unit, where about 30% of the copper is transferred to an organic solution containing the reagent H^+R in kerosene. The raffinate (barren etchant) is recycled back to the etching line. The copper containing rinse water, from the rinsing of the boards after etching, is also treated in a second extraction mixer-settler to reduce the copper content (<5 ppm). Finally, copper is stripped from the organic solution to a sulfuric acid copper electrolyte, and pure copper metal is produced by electrowinning.

Over 100 MECER installations are in operation at PCB manufacturers all over the world, recovering almost all the copper and reducing their consumption of chemicals to more than 95% (Fig. 14.17a and b).

14.4.3.2 Recovery of Nickel and Cadmium from Accumulator Waste

Nickel and cadmium are used in the production of NiFe rechargeable batteries. Using the AmMAR concept, the main leaching procedure to dissolve these valuable metals from spent accumulator scrap and production waste is performed in a two-step procedure, first with an ammonium carbonate solution and second with diluted sulfuric acid to obtain very high leaching efficiency (Fig. 14.18).

Due to chemical passivation in "old" waste materials, subsequent leaching of the leach residue with sulfuric acid, with perhaps some oxidation, is necessary to achieve very high yields. If the pH is kept about 3 in this second treatment, leaching of iron is negligible. Also, to reduce iron in the leach solution, spent potassium hydroxide electrolyte from discarded batteries can be used for iron hydroxide precipitation.

The procedure has several advantages: iron will remain in the leach residue; nickel and cadmium are almost completely leached and form metal-amine complexes in solution; and, finally, the buffer capacity of the solution

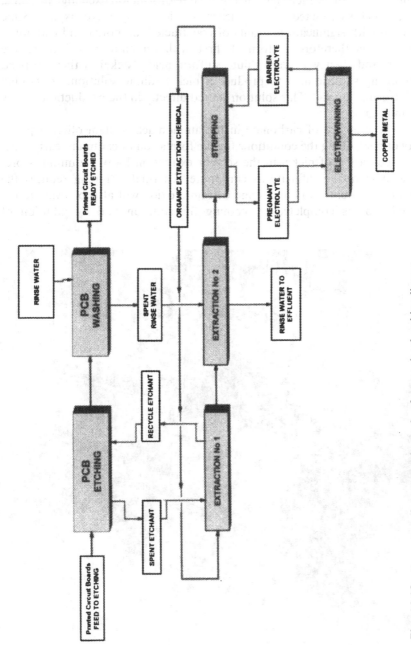

Fig. 14.16 Recovery of copper and recycle of ammoniacal etching liquors.

facilitates solvent extraction of metals with hydrogen ion exchange reagents. Thus, nickel is extracted from the leach solution using a hydroxyaryloxime (LIX 84) with significant amounts of coextracted ammonia and cadmium. Scrubbing is therefore performed, first with an ammoniacal carbonate solution and then with very dilute sulfuric acid. Nickel is then stripped with strong sulfuric acid to produce a nickel sulfate solution, containing $90\text{--}100\,\mathrm{g\,dm^{-3}}$ nickel. This solution is used directly in the production of new accumulators.

The presence of carbonate in the ammoniacal system offers a possibility of controlling the conditions for the formation of cadmium carbonate. After separation of nickel in the solvent extraction loop, cadmium is precipitated from the raffinate as carbonate. Thermal stripping reduces the ammonia concentration and some carbon dioxide will also be evaporated. Precipitation is completed by cooling the solution and by addition of

(a)

Fig. 14.17 Copper recovery units (MECER) integrated in the production of printed circuit boards. This photograph shows the solvent extraction unit.

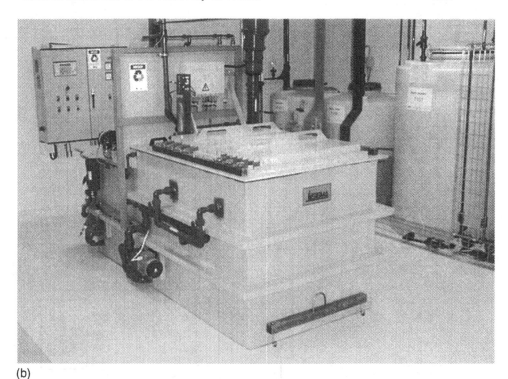

(b)

Fig. 14.17 (*Continued*). This photograph shows the electrowinning unit.

a carbon dioxide purge. After filtration of the cadmium carbonate, the filtrate is used for absorption of the stripped ammonia and the resulting solution is recycled to leaching. The cadmium carbonate precipitate is used directly in the main production.

It is necessary to bleed about 25% of the filtrate from the carbonate precipitation to maintain material balances. This large bleed is due to the sulfate added in the sulfuric acid leach and because the waste material contains considerable amounts of alkali (KOH) and water. The bleed needs special treatment to recover ammonia and to produce an environmentally acceptable effluent.

The process has been tested at pilot plant scale with good technical results. At the time, however, the amount of waste material available was not sufficient to justify the feasibility of a full-scale plant.

Alternatively, an acid route has also been reported [26], where the leaching is accomplished using a mixture of HNO_3 and HF, and separation and recovery of the nickel and cadmium is performed with solvent extraction. First, in the cadmium circuit the leaching solution is adjusted to pH 3 prior to

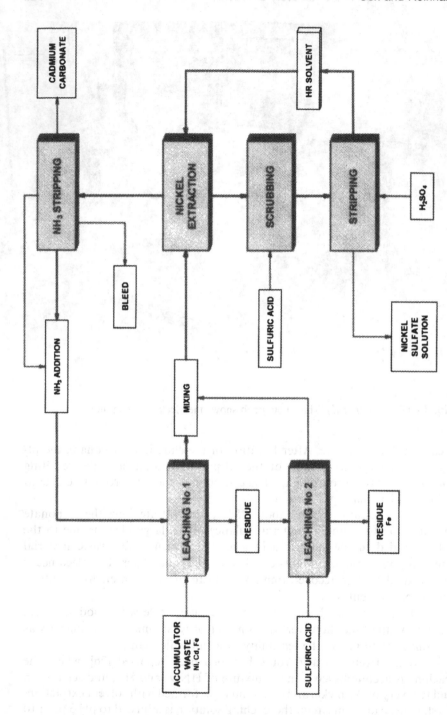

Fig. 14.18 Recovery of nickel and cadmium from accumulator scrap.

extraction with an ammonia preequilibrated NH_4^+ DEHPA. Following scrubbing with cadmium nitrate solution to remove coextracted Co, Ni, and NH_3 impurities, the scrubbed solvent is contacted with nitric acid to strip the cadmium and produce cadmium nitrate by crystallization. To extract nickel, the cadmium extraction raffinate is adjusted to about pH 5.5 and again contacted with NH_4^+ DEHPA. The loaded organic phase is scrubbed with nickel nitrate and stripping is performed with nitric acid solution at pH 1.0–1.7 from which nickel nitrate is recovered by crystallization.

14.4.3.3 Extraction of Ni, V, and Mo from Spent Desulfurizing Catalyst

In this process [27], the metals are extracted successively by the same extractant, employing suitable pH values for each metal. The spent catalyst is first roasted in an air stream at 300°C to remove oil and sulfur deposited on the surface of the catalyst. Then the metals are dissolved in an ammonium carbonate leach liquor at 80°C. Finally, the three metals are isolated and purified from one another using solvent extraction with an alkylmonothiophosphoric acid (MTPA) with a structure similar to DEHPA. The following flow sheet was suggested (Fig. 14.19):

14.4.4 Pyro-Hydrometallurgical Concept

In a centrally located recovery plant for the treatment of oxidic and hydroxidic metal containing waste (dust and sludge) from steel and metal works, including the plating industry, the main problem is the dominating amounts of iron. To avoid a secondary waste problem, special precautions have to be considered to create an iron product. One possibility is a pyrometallurgical treatment to produce pig iron or a master alloy.

Some years ago, the process, shown in Fig. 14.20, was suggested as a centrally located plant for treatment of metal-containing waste in Sweden. The operation is based on the use of a pyrometallurgical induction converter and hydrometallurgical AmMAR process technology.

In the pyrometallurgical treatment, dried and pulverized waste (flue dust and sludge) together with fluxes and carbon powder for the reduction, suspended in a small quantity of gas, are blown into the iron melt through an injection tuyere, situated below the metal surface in the converter. Due to the fine dispersion of the feed material and the intensive turbulence in the bath, a rapid reaction occurs between the injected metal oxides and the reducing carbon in the melt. Thus, the ferruginous powders are converted to pig iron or master alloys. In addition, the thermal energy in the flue gases from the furnace is used in the hydrometallurgical operation and for the drying of the input material.

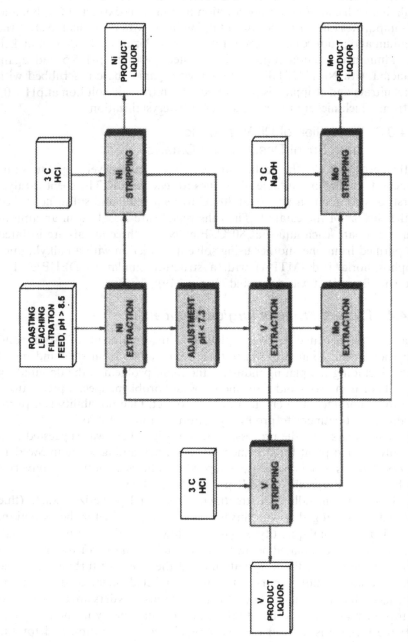

Fig. 14.19 Recovery of nickel, vanadium, and molybdenum from spent desulfurizing catalyst.

The first step in the hydrometallurgical treatment is leaching the feed material with an ammoniacal solution. Metals such as copper, nickel, and zinc form metal ammonium complexes and will therefore dissolve in the leach solution. Other metals like iron and chromium form insoluble hydroxides and will be found in the residue. This residue is dried and treated in the induction converter. Copper and nickel are separated from the filtered leach

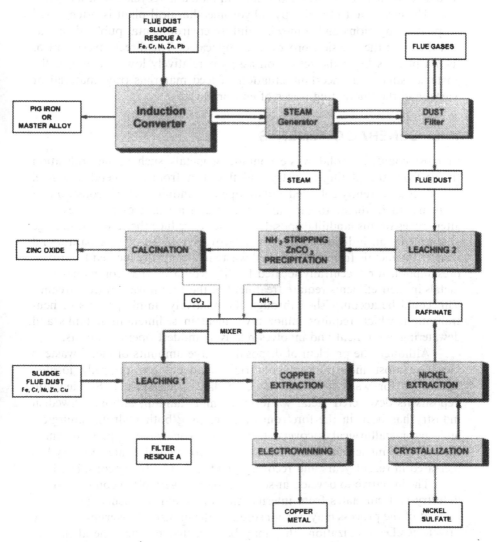

Fig. 14.20 Pyro-hydrometallurgical treatment.

solution by solvent extraction, either by sequential extraction of the metals, or by coextraction followed by selective stripping. Zinc coextraction can be neglected. Products are copper cathodes and nickel sulfate.

The secondary flue dust from the converter, containing mainly zinc and lead, is leached with the ammoniacal raffinate solution from the Cu/Ni extraction. Lead is left in the residue. Finally, zinc carbonate is precipitated by addition of CO_2 followed by thermal stripping of ammonia. The zinc carbonate is calcined to zinc oxide and ammonia is recycled to leaching.

The incentive to build a pyro-hydrometallurgical plant is often based on political opinions and somewhat diffuse environmental public demands and therefore the decision process is complicated. The direct investment of such a plant is high; the return on assets is relatively low, because of the unstable value and uncertain situation of feed materials (raw material or waste) and the fluctuating prices of end products.

14.5 GENERAL COMMENTS

In most countries, solid waste containing metals such as neutralization sludge from the plating industry and flue dust from the metal and steel industries is currently collected and dumped in landfill, where it constitutes a perpetual toxic threat to the environment and a waste of resources. The alternatives to this landfill disposal are either to reduce the rate of discharge at source by an individually designed recovery process or to separate and recover the metals from the collected waste in a centrally located facility. A presumption for a centrally located facility would be that companies with metals in their effluents require treatment of their total wastewater streams. This could be accomplished through the relatively simple process of neutralization, which requires minor investment in sedimentation tanks and dewatering equipment and involves relatively modest operation costs.

Although the problem of disposal of large amounts of metal waste is faced by most industrialized countries, relatively few centrally located operations for waste recovery have, to date, been started. Sweden still deposits its dewatered metal waste in a simple landfill, although Swedish industry has been in the forefront of developing both hydrometallurgical and pyrometallurgical recovery techniques. The same applies to most European countries; however, interest in environmentally safe recovery has increased in recent years and recovery plants are now being considered.

The initiative to develop at-source processes very often comes from an industry and emanates from internal environmental problems. The development of the process may be sponsored by the industry, government, or by private R&D organizations that may be attracted by the general market potential. The incentive of installing such a process in the industry is, besides the environmental advantages, the possibility of recycling expensive chemicals and of optimizing the main production.

The incentive to build a centrally located facility is often the result of an environmental problem that is tending to disturb the public at large. The initial money comes from governmental funds, although industry may actively participate. As industry usually has limited knowledge of such general recovery projects, R&D institutions or organizations predominantly perform the development work with generally close cooperation with the industry concerned.

The development conditions that are feasible for at-source processes are somewhat different from those of the centrally located facility regarding the freedom in feed materials and end products. This is especially characteristic for at-source processes, where the feed solution is dependent on optimal working conditions in the main process and therefore the composition is relatively constant and well defined. In addition, the product solution must have an accurate specification in order not to disturb the main production. So in the development work, a good knowledge of the main operation is necessary.

The strong link between at-source processes and the main operation is the cause of a large degree of fixed conditions. Absence of easily adjustable valves and an on/off mode of operation are recommended. The fact that the industry in question often has neither competent knowledge nor especially experienced personnel means that services and troubleshooting have to be arranged by an external organization. Certain independence in utility time of the auxiliary and main processes is arranged by built-in buffer capacities. On the contrary, the centrally located facility has to be rather flexible with respect to the physical conditions and to the composition of the feed material. This is very important, as this may be the decisive incentive for the investment. In addition, treatment of bleed streams is necessary and is often extensive.

It was earlier generally believed that the at-source approach with an individually designed recovery process entailed a considerable risk of investing in small, nonviable industries. Thus, it would appear that a centrally located facility was likely to be the only feasible way to solve the disposal needs of a large number of point sources. Such recovery facilities, based on hydrometallurgical treatment concepts, were already suggested 30 years ago. However, at that time, the feasiblity of these processes was doubtful. Pyrometallurgical treatment was marketed as an environmentally safe method for removing a number of different metals, such as copper, lead, and zinc from flue dust, wastewater sludges, or mining slimes and tailings. It has, however, been shown that such processes would probably not be economically justified in the case of the relatively small volumes of waste that were likely to be at hand.

In general, for the centrally located facility, direct investment is high and the decision time for the investment usually drastically increases with the size of operation. Big investment, a relatively low return, the unstable value and uncertain situation of feed materials (raw material or waste), and, finally, fluctuating prices of end products characterize a metal waste recovery process. These facts indicate the necessity of financial support from government to secure the realization of such a project.

The direct investment is relatively low for at-source processes, as they can be built in very small units. In addition, the operation costs are also comparably low because of the integration with the main operation. For example, administration costs can be almost neglected. The same is true for process supervision, especially at night, when it can be shared with the main operation. Usually, at-source solvent extraction processes have very low costs for chemicals and energy.

The economic outcome of the operation of a centrally located facility is comparable with all other hydrometallurgical operations and is highly dependent on price and quality of the metals produced. All products have to conform to commercial specifications. Adequate quality and quantity of cheap feed materials are also essential. It is very important to consider the fact that the transformation of waste and raw material usually involves a dramatic increase in the value of the feed material. The economic result cannot be isolated to an auxiliary process, but is the result of the combined effects on the total operation.

Considering the general conditions for recovery plants, a conclusion may be reached—it would be desirable to store all metal waste under controlled conditions and with no mixing or dilution. Future development will definitely result in new and more economic recovery procedures. As metal-containing wastes differ widely in nature and complexity, selective separation techniques such as solvent extraction will be of increasing importance.

14.6 APPENDIX: ENVIRONMENTAL LEGISLATION

Currently, legislation covering the discharge of metals into the environment is monitored, controlled, and devised by individual countries or regions. However, there is a move toward harmonizing of such legislation led by the United States, the European Union, and Japan.

The United States has the most extensive environmental regulatory programs that aim to minimize the release of metals into the environment. These regulations seek to control the emission of pollutants from specific industries and to define the emission limits according to the type of industry, the age and size of the plant, and the nature of the processes involved. In addition, different limits are set according to the following criteria: best practicable technology, best available technology, best conventional technology, new source performance standards, and pretreatment standards for existing and new sources. Details of the regulations and criteria, which are far too complex to be considered here, are given on the Internet at www.epa.gov/OST/guide.

The European Union is also a major contributor to environmental policy, with over 450 legislative measures. One of the early key directives (76/464/EEC) divides dangerous substances into two categories: List I or Black List and List II or Grey List. This was followed by directive 80/68/EEC controlling the discharge of these materials into groundwater and setting up authorization procedures. The drinking water directive (75/440/EEC) is concerned with the quality of surface water used for drinking and was followed by the bathing water directive (76/160/EEC). The directive on integrated pollution prevention and control (IPPC) (96/61/EEC) aims at preventing or minimizing environmental pollution. This directive not only covers emissions from a process but also considers energy efficiency, noise emissions, consumption of raw materials, and restoration of a site following closure. These factors are integrated into a pollution control strategy (best practical environmental option, BPEO) that covers the total impact of the process on the environment and includes the principles of sustainable development. The strategy does not emphasize the cost implications, and the emission limits would be based on best available techniques (BAT). Information concerning the various directives can be obtained from the Eur-Lex database, which can be accessed at the website //europa.eu.int/eur-lex/en/index.html and the various limits for pollutants from the Water Handbook at //europa.eu.int/comm./environment/enlarg/handbook/handbook.htm.

Environmental legislation in Japan followed a number of serious air and water pollution incidents, of which the Minimata incident of mercury pollution was one. It resulted in an environmental control law in 1967. This obliged the government and industries to control pollution, allocate the costs for pollution prevention, provide for relief of victims, and set procedures to solve pollution disputes. It also recognized the importance of economic development that seriously limited its effectiveness. This act was amended in 1970 and this clause concerning the link between the environment and economic development was removed. A number of laws followed controlling specific discharges to air and water, as well as waste management and soil pollution. In 1971, the Environmental Agency was created, resulting in a comprehensive system of pollution control in Japan. This agency is responsible for formulating basic legislation and coordinating the actions of related government departments in implementing environmental laws, such as the Water Pollution Control Law. It also has jurisdiction over industrial pollution control. Details of the standards imposed by the

Japanese environmental regulations can be found at www.env.go.jp/en/lar/ blaw/index.html for details of the basic Japanese environmental law and www.env.go.jp/en/lar/regulations/nes.html for the national effluent standards.

REFERENCES

1. Andersson, S. O. S.; Reinhardt, H. *Handbook of Solvent Extraction;* Lo, T.C. Baird, M. H. I. Hanson, C. Eds.; John Wiley and Sons, 1982; 751 pp.
2. Reinhardt, H. Chem. Ind., **1975**, 210.
3. Reinhardt, H. *Proc. Int. Waste Treatment and Utilization Conf.,* Waterloo, Canada, 1978; 83 pp.
4. Reinhardt, H. *Proc. Hydrometallurgy '81,* Soc. Chem. Ind.: London, Section F1, 1981.
5. Martin, D.; Diaz, G.; Garcia, M. A.; Sánchez, F. *Proc. ISEC 2002;* Cape Town, South Africa, 2002, 1045.
6. Sole, K. C. *Proc. ISEC 2002;* Cape Town, South Africa, 2002, 1033.
7. Reinhardt, H.; Ottertun, H.; Troeng, T. I. *Chemical Engineering Symposium Series 41,* I. Chem. E., Section W1, 1975.
8. MEAB Report. Extraction of nickel with simultaneous neutralization, 1989. www.meab-mx.se.
9. Flett, D. S.; Pearson, D. Chem. Ind., **1975**, 639.
10. Sze, Y. K. P.; Xue, L. Sep. Sci. Tech., **2003**, *38*, 405.
11. Andersson, S. O. S.; Reinhardt, H. *Proc. ISEC '77;* Canadian Inst. Min. Metal.: Toronto, 1979; Vol. 1, 798.
12. Ottertun, H.; Strandell, E. *Proc. ISEC '77;* Canadian Inst. Min. Metal.: Toronto, 1979; Vol. 1, 501.
13. Aue, A.; Skjutare, L.; Björling, G.; Reinhardt, H.; Rydberg, J. *Proc. ISEC '71;* The Hague, Soc. Chem. Ind. London, 1971; Vol. 1, 447.
14. Monzyk, B.; Conkle, H. N.; Rose, J. K.; Chauhan, S. P. *Proc. ISEC 2002;* Cape Town, South Africa, 2002; 755.
15. Martin, D.; Garcia, M. A.; Diaz, G.; Falgueras, J. *Proc. ISEC '99;* Barcelona, Spain, Soc. Chem. Ind. London, 2001; Vol. 1, 201.
16. Rydberg, J.; Reinhardt, H.; Lundén, B.; Haglund, P. *Proc. Int. Symp. Hydrometallurgy;* Chicago, 1973; 589.
17. Demster, J. H.; Björklund, P. *Proc. 4th Annual Meeting Hydromet. Sec. Metallurgical Soc.;* CIM; Toronto, 1974; 68.
18. Dias, G.; Martin, D.; Frias, C.; Pérez, O. *Proc. ISEC '99;* Barcelona, Spain, Soc. Chem. Ind. London, 2001; Vol. 2, 1449.
19. Kalujta, V. V.; Kuznetsov, G. I.; Kravchenko, A. N.; Lanin, V. P.; Miraevshy, G. P.; Pushkov, A. A.; Travkin, V. F.; Shklyar, L. I. *Proc. ISEC '88;* Moscow, 1988; Vol. 1, 252.
20. Dreisinger, D. B.; Leong, B. J. Y.; Balint, B. J.; Beyad, M. H. *Proc. ISEC '93;* York, UK: Soc. Chem. Ind. London, 1993; Vol. 3, 1271.

21. Wang, C.; Jiang, K.; Liu, D.; Wang, H. *Proc. ISEC 2002;* Cape Town, South Africa, 2002; 1039.

22. Cox, M.; Flett, D. S.; Velea, T.; Vasiliu, C. *Proc. ISEC 2002;* Cape Town, South Africa, 2002; 995.

23. Dreisinger, D.; Wassink, B.; Ship, K.; King, J.; Hames, M.; Hackl, R. *Proc. ISEC 2002;* Cape Town, South Africa, 2002; 798.

24. Reinhardt, H. *Proc. ISEC '93;* York, UK: Soc. Chem. Ind. London, 1993; Vol. 3, 1625.

25. Reinhardt, H.; Ottertun, H. Eur Patent Nr. 0 005 415, 1982. www.sigmamecer. com.

26. Ritcey, G. M. *Proc. ISEC '93;* York, UK: Soc. Chem. Ind. London, 1993; Vol. 1, 189.

27. Tsuboi, I.; Kunugita, E.; Komasawa, I. *Proc. ISEC '93;* York, UK: Soc. Chem. Ind. London, 1993; Vol. 3, 1319.

15

Recent Advances in Solvent Extraction Processes

SUSANA PÉREZ de ORTIZ and DAVID STUCKEY Imperial College, London, United Kingdom

15.1 INTRODUCTION

Recent developments in separation techniques are a response to the growing demands imposed by extraction processes that cannot be carried out economically using conventional technologies. Typical examples are the treatment of wastewaters containing low concentrations of pollutants, such as heavy metals, and the downstream separation of biological products. In the case of wastewaters, environmental regulations impose discharge concentrations of the order of a few parts per million, or even per billion, for certain pollutants. Therefore, treatment using conventional solvent extraction would require large volumes of solvent and several extraction stages to achieve the target concentrations, even in systems with large metal distribution coefficients (see Chapter 4).

The suitability of using solvent extraction for a given separation is determined by thermodynamic and kinetic considerations. The main thermodynamic parameter is the solute distribution ratio, D_M, between the organic and the aqueous phase. This is given by [Eq. (4.3), Chapter 4]:

$$D_M = [M]_{T,\text{org}}/[M]_{T,\text{aq}} \tag{15.1}$$

where $[M]_T$ is the sum of the concentrations of all M-species in a given phase, and the second subscript indicates the organic and the aqueous phase. The magnitude of D_M determines the feasibility of the separation as an industrial process; the higher D_M the better the solute separation.

Another consideration affecting the design of extraction processes is the extraction rate as it determines the residence time of the phases in the contactor, and consequently its size. The extraction rate in a two-phase system depends on the rate of interfacial transfer of species M, i.e., the interfacial flux J, and the interfacial area between the two liquid phases, Q. These are linked by the equation:

$$d[M_{t,aq}]/dt = JQ/V \tag{15.2}$$

where V is the total volume of the phases, and the subscript t indicates the contact time. Introducing the definition of specific interfacial area, a_s:

$$a_s = Q/V \tag{15.3}$$

Eq. (15.2) becomes:

$$d[M_{t,aq}]/dt = Ja_s \tag{15.4}$$

Taking as an example an extraction process with chemical reaction conducted in a conventional mixer-settler unit, the value of J will depend on the balance between the overall rate of chemical reaction and the mass transfer coefficients; i.e., J will in general depend on the concentrations of the reactants and on the degree of turbulence in the phases. For a given system, the interfacial area generated in the contactor, a_s, depends mainly on the degree of turbulence created by the power input into the mixer: the higher the power input, the smaller the drop sizes and the higher the value of a_s. Therefore, J and a_s are not independent, which makes the prediction of the rate of transfer in a contactor difficult.

Equations (15.1) and (15.4) provide the background for the study of the performance of solvent extraction processes.

15.2 NOVEL SOLVENT EXTRACTION PROCESSES

Conventional solvent extraction is a well-established technology for the separation of solutes from relatively concentrated feeds such as those found in the industrial production of chemicals and of metals by hydrometallurgy. Dilute streams, on the other hand, pose a challenge. Equation (15.1) indicates that treatment of these feeds using conventional liquid–liquid extraction requires a very large value of D_M, otherwise the organic phase volume would become unacceptably high from environmental and safety considerations. The novel solvent extraction technologies developed in the last few decades try to address these limitations. Their potential to improve the performance of conventional solvent extraction can be analyzed in a systematic way using Eqs. (15.1) and (15.4). Four novel technologies that have been developed in the last few decades will be discussed in the following sections:

- Liquid membranes
- Nondispersive solvent extraction

- Microemulsions
- Colloidal liquid aphrons

15.3 LIQUID MEMBRANES

One way of achieving both a large distribution coefficient and a reduction in the solvent duty was proposed and patented by Norman Li in 1968 [1]. The main feature of this new separation technique, called the liquid membrane process, was that it was a three-phase system consisting of two phases of a similar nature but different composition (aqueous–aqueous, organic–organic, gas–gas) separated by a third phase of a different nature and as insoluble as possible into the other two. The middle phase is the liquid membrane. Figure 15.1a shows the first configuration presented by Li, the single-drop liquid membrane, in which the membrane is formed by coating liquid drops or bubbles with a liquid film layer and by subsequently dispersing the resulting particles into a continuous liquid phase containing a solute. This configuration is now only of historical importance. There are two other configurations that have been more widely investigated due to their potential industrial application: (1) the emulsion liquid membrane, also called the surfactant liquid membrane; and (2) the supported liquid membrane.

In the emulsion liquid membrane configuration, the liquid membrane is formed by dispersing into the feed (phase 1) an emulsion of the stripping

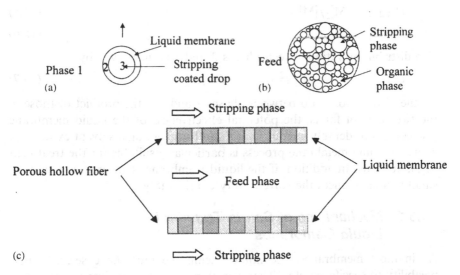

Fig. 15.1 Liquid membrane configurations: (a) Single-drop liquid membrane; (b) emulsion globule; (c) supported liquid membrane.

phase (phase 3) in an organic phase (phase 2) containing an emulsifying agent [2]. This configuration is shown in Fig. 15.1b. Here the liquid membrane is the continuous phase of the emulsion and the viability of the process depends primarily on the stability of the emulsion.

In the supported liquid membrane process, the liquid membrane phase impregnates a microporous solid support placed between the two bulk phases (Figure 15.1c). The liquid membrane is stabilized by capillary forces making unnecessary the addition of stabilizers to the membrane phase.. Two types of support configurations are used: hollow fiber or flat sheet membrane modules. These two types of liquid membrane configuration will be discussed in the following sections.

The formulation of the three phases must be such that the liquid membrane extracts the solute from one of the phases and the third phase strips it from the membrane. Thus extraction and stripping take place in the same contactor, and the stripping phase is where the solute is accumulated, instead of the organic phase as in the case of conventional solvent extraction. This allows for a middle phase of small volume that, being thin, behaves like a membrane.

The main advantage of this process becomes clear when applying Eq. (15.1) to the three-phase system. As the liquid membrane phase does not accumulate the solute, the distribution ratio that is relevant to the efficiency of this process is that between phases 1 and 3. At equilibrium, this equation applies to both pairs of phases: 2–1 and 3–2:

$$D_{M\,2-1} = [M]_2/[M]_1 \tag{15.5}$$

$$D_{M\,3-2} = [M]_3/[M]_2 \tag{15.6}$$

The distribution ratio between phases 3 and 1 is then given by:

$$D_{M\,3-1} = [M]_3/[M]_1 = D_{M\,2-1} \cdot D_{M3-2} \tag{15.7}$$

As the distribution ratio between phases 1 and 3 is the product of those in the two pairs of fluids, the potential effectiveness of the liquid membrane process is considerably greater than that of conventional solvent extraction. Thus the liquid membrane process is particularly suitable for the treatment of dilute feeds. In addition, if the liquid membrane is an organic phase, its small volume reduces the solvent duty considerably.

15.3.1 Mechanisms of Solute Transfer in Liquid Membranes

As in most membranes, the liquid membrane must have selective permeability to specific solutes. The overall mechanism of solute transfer consists of three steps: (1) extraction of the solute into the liquid membrane;

(2) diffusion of the extracted species through the membrane; and (3) stripping into the third phase. As in conventional solvent extraction, solute transfer into the membrane can be achieved by selective solubility or by chemical reaction with a component in the liquid membrane. Stripping at the interface of the liquid membrane with the third phase can be equally achieved by either physical or chemical mechanisms. The mechanism of solute separation requires analysis as it affects the conditions that control the overall rate of transfer between the first and the third phase.

Application of Eq. (15.1) to the liquid membrane process highlights one of the main advantages of the process, i.e., the high solute distribution coefficient that can be obtained between phases 3 and 1. However, another factor that must be considered when evaluating a separation process performance is the kinetics of transfer, which is given in a general form by Eq. (15.4). This equation indicates that the transfer rate in the contactor increases with both the interfacial flux and the specific interfacial area.

The two main mechanisms of solute separation by liquid membranes involve chemical reactions. In the case illustrated in Fig. 15.2a, the solute first dissolves in the liquid membrane, then diffuses toward phase 3 due to the buildup of a concentration gradient, and finally transfers to phase 3 at the

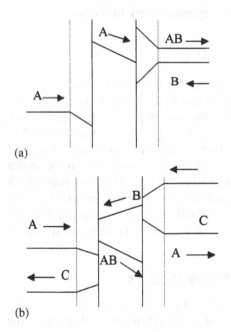

Fig. 15.2 Basic mechanisms of liquid membrane extraction: (a) type I facilitated transport $(A + B \rightarrow AB)$; (b) type II facilitated transport $(A + B \Leftrightarrow B + C)$.

second interface. Phase 3 contains a reagent that reacts irreversibly with the solute to form products that are insoluble in the membrane, and therefore incapable of diffusing back through the membrane, thus maintaining the solute gradient in the membrane. Ammonia, sulfidric acid, and phenol are among the various compounds that have been successfully removed using H_2SO_4 and NaOH as reactants in phase 3 [3,4]. A major disadvantage associated with this mechanism is the difficulty of achieving selective separations of solutes of similar size and chemical properties, as membranes are usually permeable to all of them.

The other type of chemical mechanism is more selective and is used when the solute is not soluble in the membrane phase, therefore requiring the addition of a selective reactant into the membrane to form a complex or an ion pair with the solute. The reaction product then diffuses across the membrane and at the second interface it reacts with a species added to phase 3 so that stripping also takes place by chemical reaction (Fig. 15.2b). This mechanism is called *carrier-mediated membrane transfer*. The reagent recovered from the reversed reaction then transfers back to the extraction interface. This is usually called the *reagent shuttle mechanism*.

A typical application of carrier-mediated transfer is the recovery of metal cations from aqueous phases. The overall reactions involved in the extraction and stripping stages can be represented by the following reversible reaction:

$$M^{n+}_{(aq)} + n.RH_{(org)} \leftrightarrow R_nM_{(org)} + n.H^+_{(aq)} \tag{15.8}$$

where M^{n+} is a metal cation of valence n, RH is an oil-soluble liquid ion-exchange reagent, and R_nM is the metal complex. In this example, the forward reaction takes place at the interface between phase 1 and the membrane, and the reverse reaction at the other membrane interface. Equation (15.8) provides the guidance for the formulation of both the liquid membrane and the stripping phase, as for a given concentration of metal ion in the feed a high concentration of reactant favors the forward reaction, whereas a high complex concentration and a low pH facilitate the reverse reaction. The latter indicates that one of the conditions required in order to improve the flux through the membrane is that the pH of phase 3 must be substantially lower than that of phase 1.

15.3.2 Emulsion Liquid Membrane Process

The emulsion liquid membrane (Fig. 15.1b) is a modification of the single drop membrane configuration presented by Li [2] in order to improve the stability of the membrane and to increase the interfacial area. The membrane phase contains surfactants or other additives that stabilize the emulsion.

Depending on the mechanism of extraction, the liquid membrane may also contain a carrier that reacts with the solute; the internal phase of the emulsion is the stripping phase and must be formulated accordingly. In this configuration, the liquid membrane is the continuous phase of the emulsion, and the extent of the interface between the feed and the liquid membrane depends on the size of the emulsion globules dispersed into the feed, which in turn depends on the physical properties of the phases and the mode and intensity of the mixing. The interfacial area on both sides of the liquid membrane is relevant to the rate of extraction, as indicated by Eq. (15.4). This equation shows that knowledge of the relevant interfacial area and of the interfacial flux J would allow the rate of extraction in the contactor to be calculated. However, both the size of the emulsion globules dispersed in a stirred contactor, which provides the extraction interfacial area, and the size of the emulsion droplets that leads to the calculation of the extent of the stripping interface are difficult to measure. Reported emulsion globule sizes measured in stirred tanks are of the order of 0.2–0.4 mm [5,6], whereas emulsion droplet sizes are in the range of 1–10 μm [3,7].

Figure 15.3 illustrates schematically the different stages of a continuous separation process using the emulsion liquid membrane. There are four main stages in the flow sheet: (1) emulsification of the stripping phase

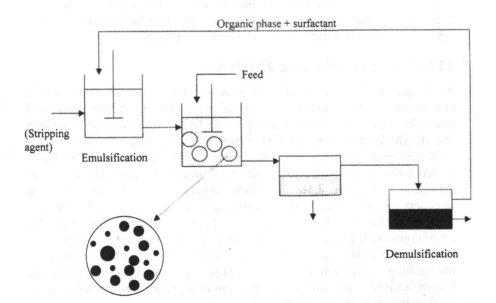

Fig. 15.3 Emulsion liquid membrane process.

with the liquid membrane phase; (2) dispersion of the emulsion into the feed; (3) separation of the emulsion from the raffinate phase; and (4) demulsification. This final stage separates the stripping solution that contains the species extracted from the feed, from the liquid membrane phase, which is recycled to the emulsification stage.

In common with the supported liquid membrane, the emulsion liquid membrane yields a solute partition coefficient of a higher order of magnitude than that obtained with the conventional solvent extraction process, thus allowing a high separation percentage from dilute feeds, and concentrated stripping solutions in just one contact stage. However, the process has its disadvantages; one is that the need to produce a stable emulsion requires the use of additives that slow down the rate of extraction and, even if their solubility is negligible, they may contaminate the raffinate. Another disadvantage is the problem posed by emulsion rupture in the contactor.

Emulsion rupture is usually due to emulsion swelling caused by the transport of the external phase into the emulsion. Three different mechanisms have been identified as causes of emulsion swelling: (1) occlusion due to entrainment of the external phase [8]; (2) secondary emulsification of the external phase caused by an excess of surfactant in the liquid membrane phase; and (3) external phase permeation through the liquid membrane [9,10]. The latter includes osmosis and, in the case of external aqueous feed, the transport of hydration water attached to complexes and water transport due to the presence of reverse micelles in the organic membrane. Swelling and emulsion rupture can be greatly decreased by including additives and especially designed components to the membrane phase.

15.3.3 Supported Liquid Membrane

As shown in Figure 15.1c, the supported liquid membrane consists of a microporous solid support impregnated with the membrane phase and placed between the two bulk phases [11–13]. In this case, the interfacial area and the thickness of the liquid membrane can be selected by choosing the solid membrane porosity, size, and thickness. The main advantages of supported liquid membranes over emulsion liquid membranes are their well-defined and easily measurable interfacial mass transfer area and membrane thickness, and the absence of surfactant additives that in general reduce the membrane flux and may contaminate the bulk phases. The main disadvantages are the difficulty in controlling the pressure on both sides of the membrane in order to avoid blowing the membrane out of the support, and the washing of the membrane from the support caused by shear forces. The former leads to contamination of the separated phases, and the latter to the need to reimpregnate the membrane.

15.4 NONDISPERSIVE SOLVENT EXTRACTION

Nondispersive solvent extraction is a novel configuration of the conventional solvent extraction process. The term *nondispersive solvent extraction* arises from the fact that instead of producing a drop dispersion of one phase in the other, the phases are contacted using porous membrane modules. The module membrane separates two of the immiscible phases, one of which impregnates the membrane, thus bringing the liquid–liquid interface to one side of the membrane. This process differs from the supported liquid membrane in that the liquid impregnating the membrane is also the bulk phase at one side of the porous membrane, thus reducing the number of liquid–liquid interfaces between the bulk phases to just one.

There are two different arrangements for the process. One uses two modules: one for extraction and the other for stripping, making it formally closer to conventional solvent extraction. The other configuration is closer to the liquid membrane process, as the three phases flow through the same module: the liquid membrane phase in the shell, and the feed and the stripping phase through the lumen of different fibers in the module. Therefore, this is a three-liquid phase system and although the liquid membrane may not be as thin as in the emulsion or supported liquid membrane configurations, extraction and stripping take place simultaneously in the same contactor, thus keeping the thermodynamic advantages of the three-phase system. A review of membrane-based nondispersive solvent extraction has been published by Pabby and Sastre [14].

The main benefits of nondispersive solvent extraction over the conventional process are: (1) it avoids the need of a settling stage for phase disengagement and the consequent risk of dispersed phase carryover; (2) the value of the interfacial area per unit volume can be much higher than in a liquid–liquid dispersion as there is no risk of phase inversion; and (3) the interfacial area is easily calculated and scale-up of the process is straightforward.

15.5 MICROEMULSIONS AND REVERSE MICELLES

The term *microemulsion* is applied in a wide sense to different types of liquid–liquid systems. In this chapter, it refers to a liquid–liquid dispersion of droplets in the size range of about 10–200 nm that is both thermodynamically stable and optically isotropic. Thus, despite being two phase systems, microemulsions look like single phases to the naked eye. There are two types of microemulsions: oil in water (O/W) and water in oil (W/O). The simplest system consists of oil, water, and an amphiphilic component that aggregates in either phase, or in both, entrapping the other phase to form the dispersion. The aggregates formed in the aqueous phase, called micelles, have their molecules orientated with their hydrophobic tails pointing to the interior of the aggregate and their hydrophilic head toward the continuous aqueous phase, whereas the aggregates in the organic phase, called reverse micelles, have the opposite orientation, as shown in Fig. 15.4. The microemulsion droplets are therefore the cores of either micelles or reverse micelles, stabilized by a surfactant layer. If more components besides the surfactant are present in the system, they may also be incorporated into the micelles or reverse micelles; these are then called mixed micelles or reverse micelles.

The ternary equilibrium diagram in Fig. 15.4 illustrates the effect of component concentration on the structure and the number of phases in

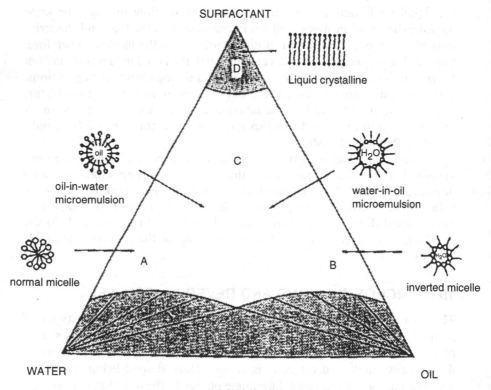

Fig. 15.4   Schematic ternary-phase diagram of an oil-water-surfactant microemulsion system consisting of various associated microstructures. A, normal micelles or O/W microemulsions; B, reverse micelles or W/O microemulsions; C, concentrated microemulsion domain; D, liquid-crystal or gel phase. Shaded areas represent multiphase regions.

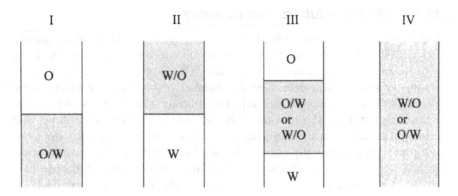

Fig. 15.5 Types of Winsor microemulsions.

a system consisting of an organic solvent, water, and a surfactant soluble in both. In the regions where the concentration of either water or organic solvent tends to zero, the system consists of one phase containing reverse micelles or micelles, respectively, with a negligible content of the third component in their cores. As the concentration of this component increases, more of its molecules transfer to the core of the micelles, forming either an O/W or a W/O microemulsion. In the multiphase region the system has two or more phases, depending on the components of the system and the characteristics of the amphiphilic molecule.

Winsor [15] classified the phase equilibria of microemulsions into four types, now called Winsor I–IV microemulsions, illustrated in Fig. 15.5. Types I and II are two-phase systems where a surfactant rich phase, the microemulsion, is in equilibrium with an excess organic or aqueous phase, respectively. Type III is a three-phase system in which a W/O or an O/W microemulsion is in equilibrium with an excess of both the aqueous and the organic phase. Finally, type IV is a single isotropic phase. In many cases, the properties of the system components require the presence of a surfactant and a cosurfactant in the organic phase in order to achieve the formation of reverse micelles; one example is the mixture of sodium dodecylsulfate and pentanol.

15.5.1 Critical Micelle Concentration

Critical Micelle Concentration (cmc) is the surfactant concentration below which the formation of reverse micelles does not occur, while the number of surfactant molecules per micelle is referred to as the aggregation number, n. The cmc is obtained through physical measurements, and varies from 0.1–1.0 mmol dm^{-3} in water or the nonpolar solvents.

15.5.2 Water Solubilization Capacity

The amount of water solubilized in a reverse micelle solution is commonly referred to as W_o, the molar ratio of water to surfactant, and this is also a good qualitative indicator of micelle size. This is an extremely important parameter since it will determine the number of surfactant molecules per micelle and is the main factor affecting micelle size. For an (AOT)/iso-octane/H_2O system, the maximum W_o is around 60 [16], and above this value the transparent reverse micelle solution becomes a turbid emulsion, and phase separation may occur. The effect of salt type and concentration on water solubilization is important. Cations with a smaller hydration size, but the same ionic charge, result in less solubilization than cations with a large hydration size [17,18]. Micelle size depends on the salt type and concentration, solvent, surfactant type and concentration, and also temperature.

15.5.3 Mechanisms of Solute Extraction
with Microemulsions

The Winsor II microemulsion is the configuration that has attracted most attention in solvent extraction from aqueous feeds, as it does not affect the structure of the aqueous phase; the organic extracting phase, on the other hand, is now a W/O microemulsion instead of a single phase. The main reason for the interest in W/O microemulsions is that the presence of the aqueous microphase in the extracting phase may enhance the extraction of hydrophilic solutes by solubilizing them in the reverse micellar cores. However, this is not always the case and it seems to vary with the characteristics of the system and the type of solute. Furthermore, in many instances the mechanism of extraction enhancement is not simply solubilization into the reverse micellar cores. Four solubilization sites are possible in a reverse micelle, as illustrated in Fig. 15.6 [19]. An important point is that the term *solubilization* does not apply only to solute transfer into the reverse micelle cores, but also to insertion into the micellar boundary region called the palisade. The problem faced by researchers is that the exact location of the solute in the microemulsion phase is difficult to determine with most of the available analytical tools, and thus it has to be inferred.

Some insights can be obtained from the mechanisms of extraction in two-phase systems. As in conventional solvent extraction, the mechanism of transfer of the solute is either physical or chemical. In conventional solvent extraction, physical transfer is used for species that prefer the organic phase, i.e., their distribution coefficient D allows the use of conventional solvent extraction. In some cases of low solubility in the organic phase, microemulsions have proved to enhance extraction. An important example in this

SURFACTANT

SOLUBILIZATE

1. SURFACE

2. PALLISADE

3. DEEP PALLISADE

4. CORE

Fig. 15.6 Possible solubilizate locations in a micelle. (From Ref. 19.)

category is the extraction of biological molecules, which is discussed in detail later.

15.5.3.1 Microemulsion Extraction with Chemical Reaction: Metal Ion Extraction

Conventional extraction with chemical reaction is used for solutes that are insoluble in the organic phase unless they react with a reagent present in that phase. An example of this is the extraction of metal ions described by Eq. (15.8). In this case, if the organic phase is replaced by a W/O microemulsion containing the reactant, there is usually extraction enhancement due to the solubilization of the metal complex in the microphase. There are two possible ways of forming a W/O microemulsion in the solvent phase:

1. The reactant forms reverse micelles in the organic phase leading to the formation of a microemulsion when this phase is contacted with the aqueous one (in which case, although perhaps unknown, the organic phase cannot be anything but a microemulsion).
2. The reactant does not form reverse micelles under the conditions of the process, in which case a surfactant, and sometimes also a cosurfactant, must be added to the organic phase in order to produce a reverse micellar phase. In this case the reverse micelles are usually mixed, i.e., they include in the micellar shell the reactant and the additives.

Only case (2) can provide a comparison between conventional and micellar extraction. The few comparisons reported in the literature on the metal extraction performance of microemulsions containing an extractant with that of the extractant on its own are, at first sight, contradictory. In some cases microemulsions produce both synergism and extraction rate enhancement with respect to the single surfactant, whereas in others they substantially reduce the metal distribution coefficient and the extraction rate, or leave them unchanged. An interesting example is found in the extraction with di(2-ethylhexyl)phosphoric acid (DEHPA). DEHPA does not form microemulsions in aliphatic solvents at pH 4 or below; however, it forms microemulsions on addition of a surfactant and a cosurfactant, e.g., sodium dodecylbenzene sulfonate and n-butanol. Bauer et al. [20] reported substantial improvements in the extraction of trivalent and quadrivalent metals with respect to the conventional system with a DEHPA microemulsion. However, Brejza and Pérez de Ortiz [21] obtained improvements in the extraction of Al(III) using the same microemulsion, in contrast with the extraction of Zn(II), which was reduced significantly with respect to the single DEHPA system. A similar different behavior of a microemulsion in the extraction of Bi(III) and Zn(II) was observed by Pepe and Otu [22]; they observed an increase in the extraction of the trivalent metal, but not for zinc. Brejza and Pérez de Ortiz [21] attempted to explain the contrasting effect of the microemulsion in the extraction of aluminum and zinc with DEHPA based on the different interfacial behavior of their complexes: the aluminum complex is more hydrophilic, therefore it may have a greater desorption energy than the zinc one, making the interface its preferred location, whereas the zinc complex is more soluble in the organic phase. Therefore, in the microemulsion system it would be preferentially solubilized in the palisade rather than in the micellar cores, leading to an enhancement of extraction due to the large increase in interfacial area produced by the microemulsion. On the other hand, the zinc complex is more soluble in the organic phase than in the aqueous phase or the palisade, and consequently the presence of the aqueous microphase does not improve its solubility in the extracting phase. This hypothesis remains to be confirmed.

One important point regarding microemulsion extraction is that the complexity of the system with its three phases, two interfaces, and usually unknown phase morphology makes the prediction of its performance quite difficult, particularly when dealing with solutes of diverse properties. The literature indicates that it does not always improve metal extraction, and in cases it may even hinder it due to the effect of the emulsifying additives.

15.5.3.2 Extraction of Biological Molecules

The mechanism of separation of biological molecules such as proteins and amino acids, and the parameters that affect the extraction distribution coefficient and the kinetics of extraction have been studied more extensively than the extraction of inorganic solutes. This is mainly due to the variety of size and structure of these molecules and, furthermore, to the fact that their characteristics may be adversely affected by their contact with solvents and surfactants.

15.5.3.2a Effect of System Parameters on Forward Transfer

The distribution of biomolecules between the aqueous and reverse micelle phases depends on system parameters such as pH, ionic strength, and salt type, all of which affect the physicochemical state of the protein and its interaction with surfactant head groups specifically, and the water pool in general. In addition to these factors, solvent structure and type, temperature, surfactant concentration, and cosurfactant play a significant role in determining the aggregation properties of a surfactant in the solvent such as size, and will influence protein partitioning behavior as much as they will affect the cooperative formation of the protein-micelle complex. Furthermore, protein size and hydrophobicity are also important in determining protein partitioning behavior in the reverse micelle phase. The following sections deal with the parameters known to influence protein extraction.

Effect of pH: The pH of a solution affects the solubilization characteristics of a protein primarily in the way in which it modifies the charge distribution over the protein surface. At pH values below its isoelectric point (pI), or point of zero net charge, a protein acquires a net positive charge, while above its pI the protein will be negatively charged. Thus, if electrostatic interactions are the dominant factor, solubilization should be possible only with anionic surfactants at pH values less than the pI of the protein; because at values above pI, electrostatic repulsion would inhibit solubilization. The opposite effect would be anticipated in the case of cationic surfactants.

Effect of salt type and concentration: The ionic strength of the aqueous solution in contact with a reverse micelle phase affects protein partitioning in a number of ways [18,23]. The first is through modification of electrostatic interactions between the protein surface and the surfactant head groups by modification of the electrical double layers adjacent to both the charged inner micelle wall and the protein surface. The second effect is to "salt out" the protein from the micelle phase because of the increased propensity of the ionic species to migrate to the micelle water pool, reduce the size of the reverse micelles, and thus displace the protein.

Effect of surfactant type and concentration: An increase in surfactant concentration results in an increase in the number of micelles rather than any substantial change in size, and this enhances the capacity of the reverse micelle phase to solubilize proteins. Woll and Hatton [24] observed increasing protein solubilization in the reverse micelle phase with increasing surfactant concentration. In contrast, Jarudilokkul et al. [25] found that at low "minimal" concentrations (6–20 mmol dm^{-3} AOT), reverse micelles could be highly selective in separating very similar proteins from fermentation broths, and the recovery in activity of up to 95% could be achieved from broths.

Effect of temperature: Luisi et al. [26] reported that the temperature markedly affected the transfer of α-chymotrypsin in a chloroform-trioctyl-methylammonium chloride (TOMAC) system. By increasing the temperature from 25–40°C, about 50% higher transfer yield was realized. No appreciable transfer of glucagone took place at room temperature, whereas transfer at 37°C was possible. These results contradict work by Dekker et al. [27], who studied the back stripping (desolubilization) of α-amylase from a TOMAC/isooctane/octanol/Rewopal HV5 system by increasing the temperature. This caused a decrease in W_o with increasing temperature and, as a result, the α-amylase was expelled from the reverse micelle phase.

Effect of affinity ligand: Woll et al. [28] reported that the solubilization of concanavalin A in AOT/isooctane increased by introducing an affinity surfactant, octyl-β-D-glucopyranoside as a cosurfactant. Further studies by Kelley et al. [29] using an affinity cosurfactant such as octyl glucoside for concanavalin A, lecithin for mycelin basic protein, and alkyl boronic acids for chymotrypsin were carried out, and the results show an increase in the amount of protein extracted at the same operating pH and salt concentration. Because of the large number of system parameters that can influence separation efficiency, a method is needed to optimize the parameters to maximize removal and the recovery of activity (not necessarily the same thing). Jarudilokkul et al. [25] used response surface methodology (RSM) to optimize the separation of three proteins (cytochrome c, ribonuclease A, and lysozyme) from a fermentation broth; the parameters found to be the most influential were AOT concentration, pH, and temperature.

15.5.3.2b Effect of System Parameters on Backward Transfer

There are two classes of parameters that influence the efficiency of back extraction: first, the parameters that govern the forward extraction such as pH, salt type and concentration, surfactant type and concentration, and protein type and concentration; and second, the pH, salt type and concentration of stripping solutions, and extraction temperature.

Effect of pH: It is obvious that in order to recover the protein from reverse micelles, the pH of the stripping solution needs to change toward the pI, which will result in a reduction of the protein interaction with the oppositely charged head groups. The extent of protein recovery from reverse micelles increases with increasing pH for anionic surfactants; however, for cationic surfactants the opposite is true.

Effect of surfactant concentration: Increases in the surfactant concentration will lead to an enhancement of protein extraction due to an increase in the number and size of the reverse micelles. The work of Hentsch et al. [30] showed that the back transfer of α-chymotrypsin decreased with decreasing AOT concentration. They suggested that this decrease was due either to α-chymotrypsin being trapped in the emulsion, or to denaturation.

Salt type and concentration: For back-extraction, increases in pH are not enough to strip the protein out from reverse micelles; this is also due to the size exclusion effect resulting from a decrease in the reverse micelle size [31,32]. This means that high salt concentration and salts that form small reverse micelles favor back transfer. Most of the work reported in the literature used KCl solution, normally $1.0\,mol\,dm^{-3}$ KCl coupled with a pH around 7.5. Marcozzi et al. [23] also showed that the back transfer efficiency of α-chymotrypsin depends on the salt type and concentration used in the forward transfer.

Counterion extraction: Due to the relative slowness of back extraction based on the methods above, the back-extraction of proteins encapsulated in AOT reverse micelles was evaluated by adding a counterionic surfactant, either TOMAC or DTAB, to the reverse micelles [33]. This novel backward transfer method gave higher backward extraction yields compared to the conventional method. The back-extraction process with TOMAC was found to be 100 times faster than back-extraction with the conventional method, and as much as three times faster than forward extraction. The 1:1 complexes of AOT and TOMAC in the solvent phase could be efficiently removed using adsorption onto montmorillonite so that the organic solvent could be reused.

15.5.4 Extraction Kinetics with Micellar Systems

For the scale-up of reverse micelle extractions, it is important to know which factors determine the mass transfer rate to or from the reverse micelle phase. So far most work has concentrated on the kinetics of solubilization of water molecules [34,35], protons [36], metal ions [20,35,37,38–40], amino acids [41], and proteins [8,35,42,43]. There are two separate processes: forward transfer, which is transfer of solute from the aqueous to the reverse micelle phase, and back transfer, which is the antithesis of the first one.

The most commercially important mechanism of all is the kinetics of solute transfer from an aqueous to a reverse micelle phase. The kinetics of extraction of metal ions have not received the same research attention as the extraction capacity of W/O microemulsions. As the mechanism of extraction of metal ions is chemical, the effect of creating a microemulsion in an organic phase that contains the reactant can be measured experimentally. Results indicate that, as in the case of extraction equilibrium, the rate of extraction may increase substantially by the presence of the microemulsion as compared with the conventional system [20,38,44] or decrease it to negligible levels [44]. Brejza found that the effect of the microemulsion on the kinetics of extraction depends on the characteristics of the metal ion and its complex, e.g., valence and degree of complex hydration, and on the solubility of the complex in the organic phase [44]. Thus there are no general rules as to the advantage of forming a microemulsion in a system that contains a reagent.

In the case of protein transfer, early work by Bausch et al. [42] found that transport was controlled by convective processes in the aqueous phase, and the mass transfer coefficient increased with increasing surfactant but depended strongly on stirring speed. Poppenborg et al. [45] evaluated the kinetic separation of lysozyme and cytochrome in a Graesser contactor and Lewis cell, and with a low rotor speed (2–3 rpm), low temperature (4°C), and a pH close to the pI of both proteins (pH 9–10), about 83% of the lysozyme and only 11% of the cytochrome c was extracted into the reverse micellar phase after 30 minutes. This optimum separation was based on the effect pH changes have on the extraction kinetics, and the rate of cytochrome c extraction was reduced much more than for lysozyme when the pH approached the pI of cytochrome c, and differed markedly from the effect the pH had on phase distribution. The extraction rate measured in the Graesser contactor differed from that measured in the Lewis cell, i.e., lysozyme was extracted faster than cytochrome c and this observation indicates that different steps of the reverse micellar transfer mechanism are controlling the transfer, depending on the way the phases are contacted.

The kinetics of back extraction are equally important to obtain a better understanding of the mechanism of solute transfer, and to determine the rate-limiting step for the process. Such information is crucial for the rational design of an extraction apparatus and, as discussed above, Jarudilokkul et al. [33] showed that counterion extraction resulted in remarkable increases in the back-extraction of proteins.

15.5.5 Micellar Extraction Potential Applications

15.5.5.1 Extraction from Synthetic Mixtures

The ease that certain protein mixtures can be separated using reverse micelle extraction was clearly demonstrated by Goklen and Hatton [46], Goklen [31], and Jarudilokkul et al. [25], who investigated a series of binary and ternary protein mixtures. In two cases, they were able to quantitatively extract cytochrome c and lysozyme from a ternary mixture of these proteins with ribonuclease A. Woll and Hatton [24] investigated the separation of a mixture of ribonuclease A and concanavalin A, and showed that the system behaved ideally and that there was no interaction between the proteins.

15.5.5.2 Extraction of Extracellular Enzymes

Rahaman et al. [47] demonstrated the use of reverse micelles (AOT/ isooctane) for the recovery of an extracellular alkaline protease (M_W, 33 k Dalton; pI, 10) from a whole fermentation broth. Purification factors as high as 6 and yields of 56% were achieved in a three-stage cascade. The combination of a cascade with a higher aqueous/organic ratio, and the use of true cross-flow designs, shows promise for purification without dilution. Krei and Hustedt [48] also demonstrated the application of the reverse micelle technique by extracting an α-amylase broth of *Bacillus licheniformis* using a CTAB/isooctane/5% octanol reverse micelle system. In a two-step extraction, they were able to reduce the protein concentration by a factor of 10 with a purification factor of 8.9, and a maximum yield of 89% α-amylase activity. Jarudilokkul et al. [49] extracted lysozyme from egg white, and while extractions as high as 98% were achievable, a variety of demulsifiers added to the mixture could actually enhance yields substantially.

15.5.5.3 Extraction of Intracellular Enzymes

Reverse micelles of CTAB in octane with hexanol as cosurfactant were reported to be able to lyse whole cells quickly and accommodate the liberated enzyme rapidly into the water pool of surfactant aggregates [50,51]. In another case a periplasmic enzyme, cytochrome c553, was extracted from the periplasmic fraction using reverse micelles [52]. The purity achieved in one separation step was very close to that achieved with extensive column chromatography. These results show that reverse micelles can be used for the extraction of intracellular proteins.

15.5.6 Process Consideration and Scale-Up

The liquid–liquid extractors developed for conventional liquid–liquid extraction are, in principle, also suitable for this application. At present, process development has only centered on the use of mixer-settlers, centrifuges, spray columns, and membrane extractors, and Lye [53] demonstrated that it was possible to carry out protein extraction with reverse micelles in a spray tower. He used a small tower to carry out batch extractions and showed that existing correlations predicting mass transfer were an order of magnitude too high due to the rigid interface of the reverse micelle droplet. However, Jarudilokkul et al. [54,55] have evaluated the use of a Graesser contactor ("raining bucket") to extract lysozyme from egg white. This type of contactor allows for the countercurrent flow of the reverse micelle phase and the pregnant aqueous mother liquor under very low shear conditions, thereby minimizing emulsion formation. These workers characterized the mass transfer performance of the contactor and developed an integrated system of separation and back-extraction [56].

Nevertheless, despite many advances in understanding the basic processes controlling the separation of organic solutes using reverse micelles, and in the design and operation of contactors, the use of reverse micelles has still not been scaled up to an industrial-sized unit.

15.6 COLLOIDAL LIQUID APHRONS

Colloidal liquid aphrons (CLAs), obtained by diluting a polyaphron phase, are postulated to consist of a solvent droplet encapsulated in a thin aqueous film ("soapy-shell"), a structure that is stabilized by the presence of a mixture of nonionic and ionic surfactants [57]. Since Sebba's original reports on biliquid foams [58] and subsequently "minute oil droplets encapsulated in a water film" [59], these structures have been investigated for use in predispersed solvent extraction (PDSE) processes. Because of a favorable partition coefficient for nonpolar solutes between the oil core of the CLA and a dilute aqueous solution, aphrons have been successfully applied to the extraction of antibiotics [60] and organic pollutants such as dichlorobenzene [61] and 3,4-dichloroaniline [62].

15.6.1 Preparation, Structure, and Stability of CLAs

Polyaphron phases are prepared by gradually introducing the organic solvent/nonionic surfactant solution from a burette into a stirred $2 \, cm^3$ reservoir of aqueous/surfactant solution. Due to the influence of preparation conditions on the stability of the polyaphron phase, a consistent preparation procedure needs to be used; the typical organic addition rate is $\sim 0.5 \, cm^3 \, min^{-1}$, with a stirring speed of 400 rpm using an 18 mm magnetic stirrer bar. The typical container in which polyaphron preparation takes place is a $30 \, cm^3$ glass jar with an internal diameter of 35 mm, and a temperature of $\sim 22 \pm 2°C$. The resulting polyaphron phases are highly viscous and cream colored, with a phase volume ratio (PVR = V_{org}/V_{aq}) of around 4 ($\phi = 0.8$), and are very stable, with no phase separation evident over a period of months.

Although potential applications for CLAs have been investigated, little work has been carried out to either confirm or refute Sebba's proposed structures for polyaphron or CLA phases (Fig. 15.7). Note the terminology that is used in this chapter.

Upon manufacture, the initial creamy-white phase consists of an aggregate of individual aphrons having a structure resembling that of a biliquid foam termed a *polyaphron* [57,58], and data on the structure and

Outer surface of shell

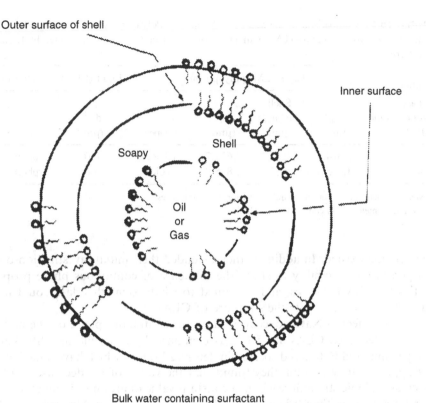

Inner surface

Shell

Soapy

Oil
or
Gas

Bulk water containing surfactant

Fig. 15.7 Proposed structure of CLA. (From Ref. 57.)

stability of this phase is obtained when the polyaphron is not dispersed in an aqueous phase. Upon dispersion of a polyaphron in a continuous aqueous phase, however, the individual aphrons become separated to form spherical droplets [58]; these are termed CLAs. Obviously the properties of the poly-aphron will depend on the aphron formulation and method of manufacture [58,63], while the properties of the CLAs will, in addition, be influenced by the nature of the continuous phase in which they are dispersed. Light scattering has previously been used to determine the size of dispersed CLAs and colloidal gas aphrons (CGAs) [63,64] though, due to their opaque nature, it is not applicable to the study of polyaphron phases.

Lye and Stuckey [65] used the techniques of cryoultramicrotome TEM and DSC to analyze polyaphrons to test Sebba's proposed structure for these phases (Table 15.1). Results from these methods were also compared with those obtained by light scattering of dispersed CLAs to see if any

Table 15.1 Comparison of the Sizes and Shapes of Polyaphrons as Determined by Electron Microscopy (Cryo-TEM) with Those of Dispersed CLAs as Determined by Light Scattering

Initial	Cryo-TEM (polyaphron)				Light (CLA)	scattering
Phase volume ratio (PVR)	$d_{oil}{}^a$ (μm)	Shell thickness (μm)	$d_{ov}{}^b$ (μm)	Shape	$d_{ov}{}^b$ (μm)	Shape
10	3.5–4.6	0.03	3.6–4.7	polyhedral	4.0	spherical
4	1.5–5.5	0.15	1.8–5.8	ovoid	7.0	spherical

[a]Mean diameter of oil core measured for major and minor axes.
[b]Overall diameter of oil core and soapy shell.

correlation existed. In addition, they extended their initial investigation on dispersed CLA stability to study the influence of continuous phase properties on CLA half-lives, and examined to what extent this data could be used to further elucidate the structure of CLAs.

The effect of NaCl concentration in the continuous phase on the half-lives of dispersed CLAs at 25°C was examined by Scarpello and Stuckey [66]. Compared to CLAs dispersed in deionized water, which have a half-life of approximately 60 min, they found that the values of $t_{1/2}$ decreased with increasing ionic strength reaching a constant value of around 15 min at ionic strengths greater than 0.3 mol dm^{-3}. The data presented by Matsushita and co-workers [64] for CGAs dispersed in 3.4 mmol dm^{-3} NaCl also indicates a decrease in $t_{1/2}$ for CGAs formulated from cationic or anionic surfactants, but no significant change for those formulated from a nonionic surfactant. Their values of $t_{1/2}$ are estimated to be around 4 min, which indicates that dispersed CGAs are considerably less stable than the CLAs investigated here. Addition of salts to CGA or CLA polyaphrons, especially those of polyvalent ions, has previously been shown to reduce the stability of non-dispersed polyaphron phases, and even break them [67].

Since the stability of CLAs displays a strong dependence on ionic strength, due to electrostatic interactions associated with the surfactant head groups, pH should also have an influence on CLA half-lives. The effect of continuous phase pH on the stability of CLAs dispersed in deionized water at 25°C was investigated by Lye and Stuckey [65]. Above a pH of 6–7, $t_{1/2}$ values were essentially constant, but began to decline as the continuous phase became more acidic. At low pH values, the excess hydrogen ion concentration led to protonation of the sulfonate head groups of the SDS molecules located at the outer soapy-shell interface. This would have two effects on aphron stability. First, it would reduce the surface charge on the aphrons and hence the energy barrier to droplet coalescence. It had previously been found that the zeta potential of a CLA suspension falls from $-45\,mV$ at pH 8.4 to $-36\,mV$ at pH 4 [68]. Secondly, protonation of the head groups reduces the polarity of the surfactant monomers, making it energetically less favorable for the hydrophobic tails of the surfactant molecules to remain in an aqueous environment. Experimentally, it is easy to show that the solubility of SDS in 20 mmol dm^{-3} buffer solutions rapidly decreases below pH 7, which may also be responsible for decreasing CLA stability at low pH. If the collision-coalescence mechanism proposed is correct, then the stability of CLAs should depend upon the temperature at which this process occurs. This was indeed found experimentally with $t_{1/2}$ values falling from 96 to 4 min for a corresponding increase in temperature of the continuous phase from 10°C to 60°C [65]. An Arrhenius plot resulted in a linear relationship between Ln k and 1/T. Performing linear regression yields a single value of the activation energy, E_a, for the collision process of 50 kJ mol^{-1}. A value for E_a of this magnitude would suggest that, over the temperature range investigated, the collision-fusion process of the CLAs is controlled by both diffusion and chemical reaction.

Concerning the structure of dispersed CLAs, the model originally proposed by Sebba [57] of a spherical oil-core droplet surrounded by a thin aqueous film stabilized by the presence of three surfactant layers is, in our opinion, essentially correct. However, there is still little direct evidence for the microstructure of the surfactant interfaces. From an engineering point of view, however, there is now quantitative data on the stability of CLAs which, together with solute mass transfer kinetics, should enable the successful design and operation of a CLA extraction process.

15.6.2 Formation of CLAs

On a more practical level, to use CLAs and CGAs in PDSE it is important to understand the influence of key parameters such as solvent type and polarity, and surfactant type (hydrophilic/lipophilic balance, HLB) and concentration, on the formulation and stability of CLAs and CGAs. These are discussed next.

A variety of nonpolar to moderately polar solvents has been evaluated for their ability to form stable CLAs (Table 15.2) [68]. As can be seen, as the solvent becomes more polar the aphron size increases and it becomes more unstable. Hence, the influence of the surfactant HLB number was evaluated using a moderately polar solvent, *n*-pentanol [68]. This demonstrated that if the HLB number of the surfactant is high enough, then it is possible to formulate stable CLAs.

Table 15.2 Preparation of CLAs from Different Solvents

Solvent	Solubility in water (wt %)	CLA stability[a]	CLA size (diameter, μm)	PVR obtained
n-Decane	52 ppb	Very stable	14.0	20
n-Octane	6.6×10^{-7}	Very stable	10.8	20
Isooctane	—	Very stable	9.6	20
n-Hexane	0.00123	Very stable	9.6	20
Decalin	<0.02	Very stable	13.1	20
Kerosene	—	Very stable	11.1	20
p-Xylene	0.0156	Stable	29.8	20
Toluene	0.0515	Stable	30.8	20
Benzene	0.1791	Stable	7.2	20
Decan-1-ol	Insoluble	Stable	25.1	10
Octan-1-ol	0.0538	Stable	27.9	5
n-Pentanol	2.19	Unstable	—	—

[a]CLAs made with 0.4% (w/v) sodium dodecyl benzene sulphonate (SDBS) in the aqueous phase, and 1% (w/v) Softanol-30 in the organic phase.

15.6.3 Kinetics of Solute Extraction and CLA Separation

Measuring the kinetics of solute extraction using CLAs is difficult for two reasons: first, due to their complex interfacial structure, i.e., the soapy shell, they are impossible to mimic in an apparatus with a defined interfacial area such as a Lewis cell; second, due to their large interfacial area, when they are dispersed with a solute with a high partition coefficient such as erythromycin, the partition equilibrium is virtually instantaneous. For example, at a pH of 10, erythromycin ($pK_a = 8.6$) has a partition coefficient of 170, and with $0.5 \, g \, dm^{-3}$ of erythromycin at this pH and a phase ratio of 100 partitioning is extremely rapid, and accurate rates of extraction are impossible to measure. Within the errors of measurement, as expected, mixing does not seem to have any effect on mass transfer or partitioning since the system is likely to be interfacially controlled. Using "typical" interfacial mass transfer rates for this type of system, together with a surface area obtained through particle size analysis, partition equilibrium should be achieved within 10^{-2} seconds. Hence, due to the large surface area, contact times for complete extraction of nonpolar solutes could be reduced to seconds in an in-line pipe contactor using CLAs. Even at a phase volume ratio of 50:1, 82% extraction was found, and a two-stage contactor would result in 96% extraction with concentration factors of around 100.

The kinetics of back-extraction of erythromycin at pH 6 are an order of magnitude slower and measurable, with incomplete recovery of the solute [69]. The reason for this is not entirely clear, but it is likely that the erythromycin forms a complex with the surfactant, which results in a dissociation reaction, slowing the back-extraction and resulting in less than 100% recovery. In addition, since the viscosity of the organic phase is 13 times greater than water, the diffusivity of the antibiotic in the organic phase is low, resulting in a longer extraction time from the stagnant solvent droplet.

After extraction, the solute-laden CLAs need to be separated from the mother liquor so that they can be back stripped. Hence attempts were made to filter the solute-rich CLAs from the aqueous phase using cross-flow microfiltration [70]. The filtration characteristics of the CLAs as indicated by the flux, CLA size, and concentration showed that they are completely retained by the membrane and do not foul the membrane surface. Using this system, the CLAs could easily be concentrated up to 30% w/v at low pressures, and the permeate stream remained totally clear. The CLAs appear to maintain their structural integrity because only $3-4 \, \text{mg dm}^{-3}$ of SDS was measured in the permeate.

15.6.4 Potential Applications of CLAs and CGAs

As previously discussed, CLAs can be used to extract any nonpolar solute from antibiotics to polluting chlorinated organics; however, polar solutes that do not partition well cannot be extracted. Because of this, "liquid ion exchangers" such as the alkylamines (Aliquat and Alamine 336) have been incorporated into the solvent phase to extract these polar solutes. Initial studies have shown that it is possible to incorporate as much as 50% of these reagents into a CLA without drastically influencing its stability [66]. In addition, the intriguing phenomenon of increased water inclusion in the CLA during formulation has been observed. Some evidence seems to suggest that this may be due to the formation of reverse micelles. If this was the case, then there is a possibility that quite polar solutes could be extracted using these "mini-liquid emulsion membranes," but more work on this phenomenon is required.

Another potential use for CLAs is in two-phase reactors where the substrate is very nonpolar, and the product is also poorly water soluble and perhaps unstable. Dispersing CLAs containing high concentrations of substrate into a fermenter will enable the fermentation to proceed rapidly without substrate limitation, while the product is removed rapidly back into the CLA. Problems with cells accumulating at the interface of the CLAs should not occur due to the strong negative charge on the CLA, and the generally negative charge on most bacterial cells. This situation is commonly faced in new biotransformation reactions, and even when the product is polar the use of CLAs will enhance the transfer of substrate.

The final intriguing use of CLAs is in the immobilization of enzymes in the soapy shell in order to carry out an enzymatic reaction. Thus the hydrolysis of *p*-nitrophenyl acetate to *p*-nitrophenol has been demonstrated by immobilizing a lipase into the shell of a CLA. The CLAs were then pumped through a cross-flow membrane, where they were separated and recycled, with the product appearing in the permeate [70].

15.7 CONCLUDING REMARKS

All the novel separation techniques discussed in this chapter offer some advantages over conventional solvent extraction for particular types of feed, such as dilute solutions and the separation of biomolecules. Some of them, such as the emulsion liquid membrane and nondispersive solvent extraction, have been investigated at pilot plant scale and have shown good potential for industrial application. However, despite their advantages, many industries are slow to take up novel approaches to solvent extraction unless substantial economic advantages can be gained. Nevertheless, in the future it is probable that some of these techniques will be taken up at full scale in industry.

REFERENCES

1. Li, N. N. U.S. Patent 3,310,794, November 12, 1968.
2. Li, N. N. Ind. Eng. Chem. Process Des. Dev., **1971**, *10*(2), 215–221.
3. Cahn, R. P.; Franfeld, J. W.; Li, N. N.; Naden, D.; Subramanian, K. N. *Recent Developments in Separation Science*; Li, N. N. Ed., CRC Press: Boca Raton, Florida, Vol. VI, 1981; p. 51.
4. Cahn, R. P.; Li, N. N. Sep. Sci., **1974**, *9*(6), 505–519.
5. Sharma, A.; Goswami, A. N.; Rawat, B. S. J. Membr. Sci., **1991**, *60*, 261.
6. Gallego Lizon, T.; Perez de Ortiz, E. S. Ind. Chem. Eng. Res. Dev., **2000**, *39*, 5020–5028.
7. Marr, R.; Kopp, A. Int. Chem. Eng., **1982**, *22*, 44–60.
8. Kinugasa, T.; Tanahashi, S. I.; Takeuchi, H. Ind. Eng. Chem. Res.,**1991**, *30*, 2470.
9. Colinart, P.; Delepine, S.; Trouve, G.; Renon, H. J. Membr. Sci.,**1984**, *20*, 167–187.
10. Ding, X. C.; Xie, F. Q. J. Membr. Sci.,**1991**, *59*, 183–188.
11. Bloch, R. *Hydrometallurgical Separations by Solvent Membranes*; Flynn, J. E. Ed., Membrane Science and Technology, Plenum Press; New York, 1970.
12. Danesi, P. R. Sep. Sci. Technol., **1984–1985**, *19*(11–12), 857–894.

13. Danesi, P. R. *Proc. ISEC '86*, Munich: DECHEMA, 1986; 527–535 pp.
14. Pabby, A. K.; Sastre, A-M. *Ion Exchange and Solvent Extraction*; Marcus, Y. and Sengupta, A. K. Eds., Marcel Dekker: New York, 2002; vol. 15, 331–469.
15. Winsor, P. A. *Solvent Properties of Amphiphilic Compounds*; Butterworth Scientific: London, 1954.
16. Zulauf, M.; Eicke, H. F. J. Phys. Chem., **1979**, *83*(4), 480.
17. Leodidis, E. B.; Hatton, T. A. Langmuir, **1989**, *5*, 741.
18. Leodidis, E. B.; Hatton, T. A. *The Structure and Reactivity in Reverse Micelles*; Pileni, M. P. Ed., Elsevier; Amsterdam, 270, 1989, 270p.
19. Osseo-Asare, K.; Kenney, M. E. *Proc ISEC '80*, Liege, 1980, 1, paper no. 80, 121.
20. Bauer, D.; Cote, G.; Komornicki, J.; Mallet-Faux, S. Can. Soc. Chem. Eng., **1989**, *2*, 425.
21. Brejza, E. V.; Perez de Ortiz, E. S. J. Colloid Interface Sci., **2000**, *227*, 244–246.
22. Pepe, E. M.; Otu, E. O. Solv. Extr. Ion Exch., **1996**, *14*(2), 247.
23. Marcozzi, G.; Correa, N.; Luisi, P. L.; Caselli, M. Biotechnol. Bioeng., **1991**, *38*, 1239.
24. Woll, J. M.; Hatton, T. A. Bioprocess Eng., **1989**, *4*, 193.
25. Jarudilokkul, S.; Poppenborg, L. H.; Stuckey, D. C. Sep. Sci. Tech., **2000**, *35*, 503–517.
26. Luisi, P. L.; Bonner, F. J.; Pellergrini, A.; Wiget, P.; Wolf, R. Helv. Chim. Acta, **1979**, *62*, 740.
27. Dekker, M.; van't Riet, K.; Van Der Pol, J. J.; Baltussen, J. W. A.; Hilhorst, R.; Bijsterbosch, B. H. J. Chem. Eng., **1991**, *46*, B69.
28. Woll, J. M.; Hatton, T. A.; Yarmush, M. L.; Biotechnol. Prog., **1989**, *5*(2), 57.
29. Kelley, B. D.; Wang, D. I. C.; Hatton, T. A. Biotechnol. Bioeng., **1993**, *42*(10), 1199.
30. Hentsch, M.; Menoud, P.; Steiner, L.; Flaschel, E.; and Renken, A. Bio/Technol., **1992**, *6*(4), 359.
31. Goklen, K. E. Ph.D. Thesis, Massachussetts Institute of Technology, 1986.
32. Dekker, M.; van't Riet, K.; Baltussen, J. W. A.; Bijsterbosch, B. H.; Hilhorst, R. Laane, C. in *Proc. 4th European Conf. on Biotechnology*, Amsterdam, 1987, 2, 507.
33. Jarudilokkul, S.; Poppenborg, L. H.; Stuckey, D. C. Biotech. Bioeng., **1999**, *62*(5), 593–601.
34. Battistel, E.; Luisi, P. L. J. Colloid Interface Sci., **1989**, *128*(1), 7.
35. Nitsch, W. Plucinski, P. J. Colloid Interface Sci., **1990**, *136*(2), 338.
36. Albery, W. J.; Choudhery, R. A.; Atay, N. Z.; Robinson, B. H. J. Chem. Soc. Faraday Trans. 1, **1987**, *83*(8), 2407.
37. Savastano, C. A.; Osseo-Asare, K.; Pérez de Ortiz, E. S. *Separation Processes in Hydrometallurgy;* Davis, G. A. Ed., SCI; London, 1987; p. 89.
38. Kim, H. S.; Tondre, C. Sep. Sci. Technol., **1989**, *24*, 485.
39. Plucinski, P.; Nitsch, W. Ber. Bunsenges. Phys. Chem., **1989**, *93*, 994.
40. Plucinski, P.; Nitsch, W. J. Phys. Chem., **1992**, *97*, 8983.
41. Plucinski, P.; Nitsch, W. Langmuir, **1994**, *10*, 371.

42. Bausch, T. E.; Plucinski, P. K.; Nitsch, W. J. Colloid Interface Sci., 1992, 150(1), 226.
43. Dekker, M.; van't Riet, K.; Bijsterbosch, B. H.; Fijneman, P.; Hilhorst, R. Chem. Eng. Sci., 1990, 45(9), 2949.
44. Brejza, E. V. Doctoral Thesis, University of London, 1994.
45. Poppenborg, L. H.; Brillis, A.; Stuckey, D. C. Sep. Sci. Tech., 2000, 35, 843–858.
46. Goklen, K. E.; Hatton, T. A. Proc. ISEC '86, Munchen, 1986, 3, 587.
47. Rahaman, R. S.; Lee, J. Y.; Cabral, J. M. S.; Hatton, T. A. Biotechnol. Prog., 1988, 4(4), 218.
48. Krei, G. A.; Hustedt, H. Chem. Eng. Sci., 1992, 47(1), 99.
49. Jarudilokkul, S.; Paulsen, E.; Stuckey, D. C. Biosep., 2000, 9, 81–91.
50. Laane, C.; Dekker, M. Proc. 6th Int. Symp. on Surfactants in Solution; Mittal, E. L., Ed., Plenum Press: New York, 9, 1989; 9p.
51. Giovenco, S.; Verheggen, F.; Laane, C. Enzyme Microb. Technol., 1987, 9, 470.
52. Jarudilokkul, S.; Poppenborg, L. H.; Valetti, F.; Gilardi, G.; Stuckey, D. C. Biotech. Techniques, 1999, 13, 159–163.
53. Lye, G. Y. Ph. D Thesis, University of Reading, 1993.
54. Jarudilokkul, S.; Paulsen, E.; Stuckey, D. C. Biotech. Bioeng. 2000, 69, 618–626.
55. Jarudilokkul, S.; Paulsen, E.; Stuckey, D. C. Biotech Progress, 2000, 16, 1071–1078.
56. Jarudilokkul, S.; Stuckey, D. C. Sep. Sci. Technol., 2001, 36, 657–670.
57. Sebba, F. Foams and Biliquid Foams—Aphrons; John Wiley: New York, 1987.
58. Sebba, F. J. Colloid Interface Sci., 1972, 40, 468.
59. Sebba, F. Colloid Polymer Sci., 1979, 257, 392.
60. Lye, G. J.; Stuckey, D. C. Separations for Biotechnology III; Pyle, D. L. Ed., SCI: London, 1994; 280–286.
61. Wallis, D. A.; Michelsen, D. L.; Sebba, F.; Carpenter, J. K.; Houle, D. Biotech. Bioeng. Symp. 1985, 15, 399.
62. Lye, G. J.; Poutiainen, L. V.; Stuckey, D. C. Biotechnology '94, 2nd Int. Symposium on Environmental Biotechnology; I. Chem. E., II, 1994; 25–27.
63. Matsushita, K.; Mollah, A. H.; Stuckey, D. C.; del Cerro, C.; Bailey, A. I. Colloids Surfaces, 1992, 69, 65–72.
64. Chaphalkar, P. G.; Valsaraj, K. T.; Roy, D. Sep. Sci. Technol., 1993, 28, 1287.
65. Lye, G. J.; Stuckey, D. C. Colloids Surfaces, A: Physiochem. Eng. Aspects, 1998, 131(1–3), 113–130.
66. Scarpello, J. T.; Stuckey, D. C. J. Chem. Tech. Biotech., 1999, 74, 409–416.
67. Save, S. V.; Pangarkar, V. G. Chem. Eng. Comm., 1994, 127, 35.
68. Stuckey, D. C.; Matsushita, K.; Mollah, A. H.; Bailey, A. I. Third Int. Conf. on Effective Membrane Processes: New Perspectives; University of Bath, U.K., 1993.
69. Lye, G. J.; Stuckey, D. C. Chem. Eng. Sci., 2001, 56, 97–108.
70. Lye, G. J.; Pavlou, O. P.; Rosjidi, M.; Stuckey, D. C. Biotech Bioeng., 1996, 51, 69–78.

16

Computational Chemistry in Modeling Solvent Extraction of Metal Ions

JERZY NARBUTT Institute of Nuclear Chemistry and Technology, Warsaw, Poland

MARIAN CZERWIŃSKI Zawiercie University of Administration and Management, Zawiercie, Poland

16.1 INTRODUCTION

Great progress in computational techniques and methods in the last decade of the 20th century has made it possible to perform quantum mechanical (QM) calculations on large molecular systems composed of several dozens or even hundreds of atoms. General ab initio schemes based on Hartree-Fock-Roothaan (HFR) methods [1], formerly used for small molecules composed of light elements, now are becoming applicable for metal complexes, including even heavy metals. Calculations based on the density-functional theory [2] have appeared particularly fruitful in this respect. These methods have also been applied in solvent extraction chemistry. Reviews have been presented, e.g., [3,4], on the computational methods—from purely empirical molecular mechanics (MM) approaches to advanced ab initio calculations rendered accessible in the form of commercial software packages: Cerius^2 [5], Gaussian 98 [6], ADF [7], Spartan [8], HyperChem [9], etc. These methods and programs have been widely used for structural optimization of metal complexes and calculation of their energy of formation, problems of primary importance in solvent extraction.

Numerical methods have been used in solvent extraction chemistry for treating experimental data for many years. As shown in Chapter 4 (see section 4.14.3), modeling of extraction processes in terms of a set of assumed equilibria with adjustable (best-fit) coefficients was widely used to identify and characterize the major species formed in these processes. The improvement of this approach, directed toward studying systems of greater diversity and complexity, including corrections for thermodynamic activity effects, is still an object of advanced research [10,11]; this problem is, however, outside the scope of this chapter. Another direction has been focused on data mining methods coupled with theoretical combinatorial chemistry, i.e., the analysis of vast collections of experimental data in order to establish a relationship between the structure of an ionophore and its extraction properties, as well as to design the most selective separation systems for the ions. Attempts to create an information system on solvent extraction based on large experimental databases and on structure-property relationships are continuing making possible empirical evaluation of the distribution ratios of metal ions in given extraction systems [12]. However, the main goal of computational methods in this field is nowadays the analysis of the extraction process both on the molecular level and from the thermodynamic point of view. The knowledge of the species involved, charge distribution, bond character, energies, extraction mechanism, etc. allows the interpretation and understanding of the phenomena studied, prediction of properties of the compounds, and the optimization of solvent extraction systems, e.g., by designing new extractants (ligands) with improved selectivity. Selective separations of metal ions are due to differences in solvation energy of metal-containing species in both liquid phases. Therefore, solvation phenomena are crucial when analyzing solvent extraction equilibria. Equally important is the formation of the metal-containing species, which can undergo different solvation (hydration) in both phases. These two broad fields—thermodynamics of solutions and coordination chemistry—are the main subjects of this chapter.

16.2 COMPUTATIONAL STUDIES ON STRUCTURES OF METAL COMPLEXES

To characterize the composition and structure of metal complexes formed in extraction processes (either in the aqueous phase or at the interface), various experimental methods are used. Theoretical methods become helpful in complementing the results if the spectroscopic data are not sufficient to fully describe the structure, if crystals suitable for diffraction studies are not available, etc. Moreover, the calculations can result in reliable structures of the complexes or ligands in solution, which are often different from those observed in the solid state.

16.2.1 Molecular Mechanics Modeling

Significant successes in studying structures of metal complexes have been achieved by using MM modeling—simple enough to carry out calculations using computers of low computing power, but not very accurate. The MM methods, based upon classical physics, use simple expressions and parameters describing interatomic forces in the molecule and also between molecules. To find the optimized structure of a molecule, the total strain energy accompanying the formation of the molecule must be minimized. This total strain energy includes the sum of the energies of deformation of all bond lengths and bond angles; the torsional energies related to barriers to rotation of the molecule fragments about bonds; and the energies related to non-bonded interactions between the molecule fragments, i.e., terms describing van der Waals (Lennard-Jones potential) and long-range electrostatic (Coulomb potential) interactions [Eq. (16.1)]:

$$E = \sum_{\text{bonds}} K_r(r - r_{\text{eq}})^2 + \sum_{\text{angles}} K_\theta(\theta - \theta_{\text{eq}})^2 + \sum_{\text{dihedrals}} V_n[1 + \cos(n\phi)]$$

$$+ \sum_{\text{nonbonded}} \left[\frac{A_{ij}}{R_{ij}^{12}} - \frac{B_{ij}}{R_{ij}^6} + \frac{q_i q_j}{\varepsilon R_{ij}} \right] \qquad (16.1)$$

The initial set of parameters for a selected force field, including the ideal bond lengths and bond angles, the related force constants, the effective barriers to rotation, point charges and van der Waals parameters of the atoms, and effective dielectric constant of the medium, are obtained from experimental data and/or QM calculations, and their performance is tested by comparing the calculated and experimental structures and energies. If necessary, the procedure is repeated to minimize the errors in the calculated output [13]. In some MM models a separate term is added, describing hydrogen-bonding potentials; however, the hydrogen bond energies calculated in such a way are usually approximate.

Optimizing the geometry of metal complexes in the gas phase neglects environmental effects (solvent, concentration), important in the analysis of both structure and stability of the complexes because of their interactions with neighboring molecules. To study complex formation in solution, some MM models use the parameterization of the entire force field, based on experimental data. In this case, the environment is implicitly included in the model and the environmental effects are isotropic [14]. Another way to account for the environment is to calculate through-space attractive and repulsive intermolecular interactions, by the use of the last right-hand-side term of Eq. (16.1), parameterized for the whole system, and, if related, also the term describing hydrogen bonding.

MM methods, originally developed for organic compounds, have been modified to describe metal complexes [13–17] where polarization interactions must be considered and where a variety of geometrical configuration is observed. This structural diversity reflecting the existence of multiple local energy minima is due to ligand-dependent effects observed in complexes of certain transition metal (TM) ions: Jahn-Teller effect, spin crossover [18,19], and stereochemical activity of valence s-electron pairs in some p-block heavy metal ions in lower oxidation states [20]. Various types of potential energy equations, and methods developing the parameters, are used to determine force fields for coordination compounds. It is generally assumed and confirmed for a variety of coordinated ligands that the parameters for metal complexes are transferable from the organic force fields [13]. Modification and extension of MM by the effect of ligand field stabilization energy on the strain energy in complexes of certain TM ions (the LFMM model) makes it possible automatically to take into account Jahn-Teller distortions, and to reproduce with a reasonable accuracy both the structures and the associated d-d transition energies [18,19]. The not very high but often sufficient accuracy of MM calculations is recompensed by lesser requirements for computational resources, shorter calculation time, and a possibility to deal with larger systems.

16.2.2 Quantum Mechanical Calculations

Accurate QM calculations (based on ab initio methods) on molecular systems as large as metal complexes in condensed phases and/or including conformational analysis are still limited by the resources of even the most powerful supercomputers. Solution of HFR equations [1], essential in the ab initio method, requires the calculation of n^4 integrals, where n denotes the number of basis orbitals. This very large number (in the case of large systems) causes serious problems and requires simplification of the method. Thus a significant reduction of the number of electrons (and the basis orbitals) in a molecule is possible by assigning all its electrons to two subsets: (1) electrons in the inner shells (core electrons), and (2) valence electrons in the atoms. The calculations are then carried out only for the much lower number of valence electrons; the effect of core electrons being included in the effective core potential (ECP), which represents the interaction of the core electrons with the valence electrons.

Semiempirical methods are widely used, based on zero differential overlap (ZDO) approximations which assume that the products of two different basis functions for the same electron, related to different atoms, are equal to zero [21]. The use of semiempirical methods, like MNDO, ZINDO, etc., reduces the calculations to about n^2 integrals. This approach, however, causes certain errors that should be compensated by assigning empirical parameters to the integrals. The limited sets of parameters available, in particular for transition metals, make the semiempirical methods of limited use. Moreover, for TM systems the self-consistent field (SCF) procedures are hardly convergent because atoms with partly filled d shells have many energetically close electronic states, in contrast to systems consisting of *s*- and *p*-block atoms only.

Because of these difficulties, great interest arose in the last decade in methods free of such limitations, based on the density functional theory (DFT). The DFT equations contain terms that evaluate—already at the SCF level—a significant amount (ca. 70%) of the correlation energy. On the other hand, very accurate DFT methods require calculation of much fewer integrals (n^3) than ab initio, which is why they have been widely used in theoretical studies of large systems. The DFT [2] is based upon Hohenberg-Kohn (HK) theorems, which legitimize the use of electron density as a basis variable [22].

In 1976 Warshel and Levitt introduced the idea of a hybrid QM/MM method [23] that treated a small part of a system (e.g., the solute) using a quantum mechanical representation, while the rest of the system, which did not need such a detailed description (e.g., the solvent) was represented by an empirical force field. These hybrid methods, in particular the empirical valence bond approach, were then used to study a wide variety of reactions in solution. The combined QM/MM methods use the MM method with the potential calculated ab initio [24].

All these methods have found applications in theoretical considerations of numerous problems more or less directly related to solvent extraction. The MM calculated structures and strain energies of cobalt(III) amino acid complexes have been related to the experimental distribution of isomers, their thermodynamic stability, and some kinetic data connected with transition state energies [15]. The influence of steric strain upon chelate stability, the preference of metal ions for ligands forming five- and six-membered chelate rings, the conformational isomerism of macrocyclic ligands, and the size-match selectivity were analyzed [16] as well as the relation between ligand structures, coordination stereochemistry, and the thermodynamic properties of TM complexes [17].

16.2.3 Structures of Ligands and Complexes, Ligand Design

Extensive MM calculations on metal complexes, supplementing extraction experiments, helped to explain how the structure of various crown ether ligands affects their binding and extraction ability toward alkali and alkaline earth metal ions, contributing to the design of new ligands [25,26] and synergistic systems [27] of improved selectivity. Complementarity of bi- and polydentate ligands to the metal ion, crucial for ligand design, is controlled by numerous factors including appropriate donor atoms (HSAB principle and the nature of metal-ligand bond), the number of donor atoms that satisfy the coordination and geometric requirements of the ion, the geometry of ligands (e.g., bite size), and orientation of the donor atoms [28].

More advanced semiempirical molecular orbital methods have also been used in this respect in modeling, e.g., the structure of a diphosphonium extractant in the gas phase, and then the percentage extraction of zinc ion-pair complexes was correlated with the calculated energy of association of the ion pairs [29]. Semiempirical SCF calculations, used to study structure, conformational changes and hydration of hydroxyoximes as extractants of copper, appeared helpful in interpreting their interfacial activity and the rate of extraction [30]. Similar (PM3, ZINDO) methods were also used to model the structure of some commercial extractants (pyridine dicarboxylates, pyridyloctanoates, β-diketones, hydroxyoximes), as well as the effects of their hydration and association with modifiers (alcohols, β-diketones) on their thermodynamic and interfacial activity [31–33]. In addition, the structure of copper complexes with these extractants was calculated [32].

Semiempirical and ab initio calculations were used to interpret experimental data and to find structure-complexation relationships in the extraction of transition metals and lanthanides [34–38]. The calculations contribute to understanding the factors that affect the strength of metal–ligand bonding, and to designing new ligands. For example, the extraction ability and selectivity of some β-diketonate ligands toward lanthanides is shown to be a function of ligand structure (bite size) and the energy of intramolecular hydrogen bonds [37,38]. Semiempirical PM3-tm and DFT B3LYP calculations in the gas phase made it possible to determine the experimentally unavailable molecular structure of some heterometallic TM complexes extracted by trioctylphosphine oxide into hexane [39].

Semiempirical QM calculations of the geometry, charge distribution, bond orders, dipole moments and HOMO energy for a series of monoamide extractants in the gas phase, carried out at the Hartree-Fock level, were supplemented by MM calculations on uranyl complexes with these ligands [40,41]. The observed qualitative relationships between the calculated values and the experimental distribution ratios of the metal have emphasized the importance of electron density on the coordinating atoms or groups, and steric effects in the ligands. Ligand basicity is a factor that affects the complex stability. Semiempirical calculations appeared insufficiently accurate to calculate basicities of several malonamides, the extractants of trivalent lanthanides and actinides; therefore advanced ab initio methods were used, which brought about a good agreement between the calculated (relative) values and the experimental data [42]. Conformational ab initio analysis of the protonated terpyridine extractant used for lanthanide/actinide separation (ion-pair extraction from strongly acidic solution) has confirmed stabilization of the *trans,trans* form of the terpyridine cation in aqueous solution by the formation of intermolecular hydrogen bonds to solvent and/or to accompanying anions [43].

Gas phase QM calculations at the HF level, tested by DFT B3LYP studies and completed by molecular dynamic simulations (see section 16.3), made it possible to estimate the stability of lanthanide complexes with various neutral mono- and bidentate ligands bearing amide and phosphoryl groups, which were considered potential extractants of lanthanide and actinide cations from nuclear waste solutions [44]. It has been demonstrated that the stability of the complexes depends on the basicity of the ligand binding sites, on the ligand structure (i.e., on the chelate effect and the size of the chelate ring), on the hardness of the lanthanide cations, and on the properties of the counterions. Ab initio calculations on various carbonyl and phosphoryl compounds have allowed the determination of factors influencing the selectivity of $Am^{3+}-Eu^{3+}$ separations in extraction by dialkyldithiophosphinic acid (CYANEX 301) with the abovementioned species as coextractants [45].

16.2.4 Complexes of Heavy Metals

QM studies on the compounds of heavy metals, actinides in particular, require the incorporation of relativistic effects that disturb the periodicity of physical and chemical properties observed for lighter elements [46]. Accurate relativistic calculations on compounds containing heavy atoms can be performed only for very small molecules; the calculations on larger species require a certain approximation. This can be achieved not only by the approximate solution of the Dirac-Schrödinger equation adjusted to a many-electron system, but also in a simpler way through the use of quasi-relativistic or fully relativistic effective core potentials (RECP). According to this approximation, the inner shells of the atom (core) are omitted and an additional potential energy term replaces the core electrons in the Hamiltonian (see section 16.2.2). The application of relativistic DFT methods, including new RECPs to molecular systems containing actinides, has been reviewed [47]. New methods for calculating RECPs are being tested [48]. This problem is not only theoretical but also has practical significance, because partitioning of trivalent actinides and lanthanides from liquid nuclear waste is a "hot" research task due to the necessity to isolate long-lived highly radiotoxic minor actinides (mainly Am and Cm) for further treatment [45].

The novel RECP integrals have been implemented inside the quantum chemistry DFT-based program MAGIC developed for studying large molecules containing heavy atoms, in order to design ligands with better selectivity for extractive separation of actinides from aqueous nuclear waste solutions [49,50]. The program also calculates the electrostatic solvent effects exerted on complex molecules placed in a cavity embedded in a dielectric continuum, which represents the bulk solvent (shown later). It has been shown that the calculated formation energies of fully optimized thenoyl-trifluoroacetonate complexes of plutonium and uranium in different oxidation states correspond to the experimental extraction data [51]. The DFT calculations on hydration of uranyl and plutonyl ions in the gas and aqueous phases have shown that the short-range interactions between the ions and their closest water molecules are very strong and involve an appreciable amount of charge transfer that could not be included in simple solvent cavity models. The $MO_2^{2+} \cdot 5H_2O$ hydrates appeared to be the most stable, their binding energy being, however, significantly greater in vacuo than in the aqueous solution [52]. This shows how important solvent effects are in the calculations of the formation energy of metal complex with lipophilic ligands, which are then extracted to the organic phase.

A fully relativistic, self-consistent field method based on the numerical solution of the Hartree-Fock equations with the approximation of local exchange instead of the HF nonlocal exchange potential, and with the self-consistent charge approximation, was used to calculate the electronic structure of anionic chloride and oxychloride complexes of Group 5 metals including the heaviest, dubnium (element 105). The calculation results explain the experimental extraction behavior of the metal ions in strongly acidic solutions. Based on the radii of the oxychloride complexes and using the Born model [Eq. (16.12)] to describe the extraction equilibria of the ion pairs, it was possible to predict the position of dubnium in the series of metals extracted to an organic phase [53]. Similar calculations for Group 6 metal ions lead to predicting the relative extraction behavior of seaborgium (element 106) within the group. It has been concluded that the difference between geometrical configurations and electron density distribution of Group 5 MO_2^{2+} ions and UO_2^{2+}, their actinide pseudoanalog, would result in different extraction behavior of Sg and U complexes [54]. Correct prediction of extraction properties is particularly important in studying chemistry of the heaviest elements, obtained on the scale of "one atom at a time," because solvent extraction is one of few experimental methods that can be used to study these "exotic" elements [55].

16.2.5 Solvent Effects

Solvent effects can be incorporated into two kinds of solvation models, either those that consider each solvent molecule as an individual molecular species (explicit models), or those that deal with the averaged effect of the solvent molecules through use of a coarse-grained description of solvent (e.g., dielectric models, implicit solvent models, etc.).

The explicit solvation models that consider solute molecules in a discrete environment of a number of individual solvent molecules can hardly be used by the available QM methods and hardware because such systems are too large. QM calculations involving solvent effects can be performed using implicit solvation models based on the continuum solvation theory, as reviewed, for example, by Cramer and Truhlar [56]. Instead of a discrete solvent medium consisting of a large number of solvent molecules, a continuous medium is considered. Its macroscopic properties, imposed by the dielectric constant of the solvent modeled, correspond to the properties of the bulk solvent. The medium is, however, homogenous, isotropic, and nonpolarizable, so it can well be replaced by an electric field. Into this field a solute molecule is placed, and its interactions with the field are studied. The substitution of this field for a large number of solvent molecules drastically decreases the number of orbitals in the system, i.e., the number of integrals to be calculated. The polarizing action of this field on the solute molecule modifies its charge distribution, dipole moment, energy of electrons, and to some extent its geometry. The modified solute affects in turn the field parameters, etc.; this is the principle of the self-consistent reaction field (SCRF) approach. Most current works on the implicit solvation models treat the solute either as quantum mechanically flexible, i.e., including all the previously mentioned effects, or as quantum mechanically rigid, including only the electronic relaxation.

The implicit solvation models do not take into account, however, interactions of the solute with solvent molecules that lead to formation of the first solvation shell around the solute. This shell may be defined either in terms of an integer number of solvent molecules, based on the solute–solvent distance or interaction energy, or in terms of distribution functions that result in an averaged noninteger number of solvent molecules. The interaction of bulk solvent with such a "supermolecule" (the solute with its first solvation shell) can be expressed by the terms due to cavity formation in the solvent, to electrostatic interactions, and, in certain models, also to dispersion and repulsion contributions. The presence of explicit solvent molecules in the first solvation shell requires the knowledge of their number and orientations after the supermolecule has been placed in the continuous medium. This can be done by the QM/MM methods based on Monte Carlo or molecular dynamics simulations, where the solute molecule, originally modeled by QM, is afterward treated as rigid and nonpolarizable. The introduction of the solute polarizability is possible but with a high computational cost [57]. Molecular modeling by molecular dynamics (MD) simulations are discussed in the following section. Solvent effects can be taken into account in MM calculations by parameterization of the force field, based on solution data [14,56].

QM calculations provide an accurate way to treat strong, long-range electrostatic forces that dominate many solvation phenomena. The errors due to the use of the approximate continuum solvation models can be small enough to allow the quantitative treatment of the solute behavior; therefore, this approach is widely used also for evaluating solvent extraction equilibria of organic molecules [56].

16.3 MOLECULAR SIMULATIONS

16.3.1 Monte Carlo and Molecular Dynamics Methods

Molecular dynamics simulations, which afford possibilities for microscopic insight into various chemical processes at the molecular level, play a particular role in solvent extraction, where they are used to model both complex formation in solution and its distribution between two liquid phases [58]. The advantage of MD consists in not being restricted to harmonic motion of a molecule (a set of atoms) about local minima. This allows the molecule under study to cross energy maxima and to adopt other stable configurations. Successive configurations of the system are generated by implementing Newton's laws of motion. The solution of these equations represents the evolution of positions of the molecule in time, i.e., its dynamic "trajectory," which can be used then to study time-dependent properties of the molecule, such as diffusion or reaction progress [59]. On the contrary, Monte Carlo (MC) techniques randomly move atoms (molecules) in such a way that alternative configurations are generated, which do not represent a time-dependent trajectory but correspond to a given energy [60].

Software packages such as AMBER [61] are widely used in MD simulations with the classical molecular force field where the potential energy is represented by Eq. (16.1). If the simulations relate to a solution, the intermolecular interactions are modeled using the last term, which describes nonbonded interactions (Coulomb and Lennard-Jones). In order to correctly treat long-range interactions that may become cut off to some extent at the boundaries of a finite size system, corrections are being introduced, e.g., by using the Ewald summation method for electrostatic interactions, or techniques allowing the user to avoid a discontinuity in potential energy and forces near the cutoff distance (shifted potentials, switching functions) [59,60]. The Ewald approach applying periodic boundary conditions makes it possible to obtain correct free energies of hydration from molecular calculations by extrapolation of relatively small simulation systems to infinite size [62].

An important advantage of the simulation methods consists in their explicit treatment of solvent effects as the effects of a set of individual molecules. To calculate the free energy of hydration of small hydrophobic solutes, the simulations have been successfully used, based on the energy of cavity formation in bulk water [62–65] (see section 16.4.2). The results correspond well with experiment in spite of simplifications of the model used. The use of a more sophisticated model that incorporates information on the locally tetrahedral structure of water improves the results [65,66]. Most applications, however, require calculations for large amphiphilic molecules, with both polar and nonpolar regions in close vicinity. In this respect, the water-structuring effects of carbohydrates were studied [67]. Such an approach may appear useful in solvent extraction studies for the analysis of interactions of metal chelates in the aqueous phase, where competition between hydrophobic hydration and outer-sphere hydration determines solvent extraction equilibria of coordinatively saturated metal complexes (see section 16.4.4).

16.3.2 Solvent Extraction Systems and Mechanisms of Extraction Processes

MD simulations are used to study solvent extraction processes of metal ions in two-phase systems modeled by two adjacent assemblies (boxes) consisting of hundreds of molecules of water (one) and of an immiscible organic solvent (the other). The contact surface of these two boxes represents the interface, which locally is molecularly sharp, but some mutual penetration of both kinds of solvents is observed [68,69]. MD simulations performed at the water/chloroform interface for a period of a few hundred picoseconds directly model, at the molecular level, the phenomena accompanying extraction of metal ions. It has been shown that large amphiphilic ionophore ligands—functionalized calix[4]arenes, phosphoryl-containing podands, cryptands—both free and complexed, adsorb at the interface (Fig. 16.1). This is due to the decrease in the free energy of the system when the hydrophilic parts of the solute interact with water, while the lipophilic parts interact with the organic solvent [70,71].

MD simulations coupled with free energy perturbation (FEP) calculations allow the calculation of the relative binding affinities of two different metal ions by the same ligand or of the same metal ion by two different ligands. The method consists in stepwise mutation of one solute to the other. Mutation in two solvents, e.g., water and chloroform, with the use of a thermodynamic cycle makes it possible to compute the changes in the free energy of solvation of the two solutes [58]. The relative free energies of transfer for two metal complexes can be calculated in this way (see section 16.4.2), thus enabling the determination of the extraction selectivity of certain extractants toward different metal ions. For example, such theoretical

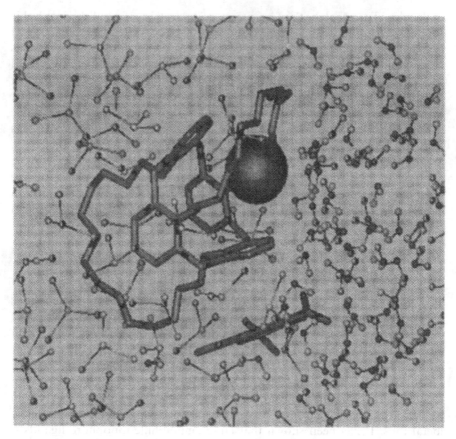

Fig. 16.1 Complex of Cs^+ Pic^- with calix-[4]-bis-crown-6 at the chloroform (left) and water (right) interface. A snapshot at 500 ps of MD. (A. Varnek and G. Wipff, unpublished).

calculations not only confirm the experimental fact that calix[4]-crown-6 ligands selectively extract Cs^+ over Na^+ ions from aqueous solutions, but also reliably explain this selectivity as the effect of the higher dehydration energy of the latter, required for complex formation [70]. This demonstrates again that solvent effects really play an important role in complexation selectivity. The role of the counterion (picrate, chloride) in the interfacial behavior of the crown [70], cryptand, and phosphoryl podand [71] complexes has also been pointed out. The structures of the complexes with alkaline earth and uranyl ions have been calculated [72].

MD simulations at the water/organic liquid interfaces give a powerful tool for studying mechanisms of the solvent extraction of metal ions, in particular described by the ion-pair model with complex formation at the interface. Snapshots of solutes taken at picosecond intervals allow us to observe the progress of the reaction, steady states and intermediate species (if any), and finally dynamic equilibrium. In many cases, this theoretical approach is irreplaceable as interfacial phenomena are very difficult to study experimentally [73] (see Chapter 5). MD simulations are also carried out to model extraction processes successfully without taking solvent effects into account. For example, gas phase MD studies on the mechanism of insertion of monovalent metal ions into calix[4]arene tubes have shown how the ions enter and migrate along the tube, and show the reason for the selectivity of the host molecule for the given ions [74].

It must be pointed out that in studies on the complexation of metal ions by extractants, ion solvation phenomena are of primary importance because upon complex formation the solvent molecules from the primary coordination sphere are replaced by binding sites of the ligands. MD simulations that poorly describe charge transfer and polarization effects are sometimes coupled with QM methods. For example, lanthanide solvation in acetonitrile solution was studied by both QM and MD simulations. The former, taking into account charge transfer and polarization, which are so important for this system, gives the structure, coordination numbers, and charge distribution on the Ln^{3+} solvates. The latter gives a remarkable dynamic feature of the primary solvation sphere of the ions, where the lifetime of acetonitrile molecules in the complex increases from Yb^{3+} to La^{3+}, inversely to the energy of cation–solvent interaction and unlike the aqueous phase behavior. The ion-binding selectivity by the pyridinedicarboxamide ligand, studied by MD-FEP simulations, has appeared to be determined to a great extent by the change in the free energy of cation solvation [75].

The limitations of MD simulations stem from the approximate character of MM methods; however, their relatively low computing requirements allow the study of large systems consisting of even 15,000–20,000 atoms. The size of such systems is, nevertheless, too small to take into account formation of aggregates (micelles, vesicles, etc.) that may play an important role in the mechanism of extraction in some special systems. This relatively small size of simulating boxes requires corrections to treat long-range interactions correctly (see section 16.3.1). Also, the force fields used in MD methods typically do not treat nonadditive effects correctly. The necessity of very short time simulation steps (in order not to lose important energy minima) makes the calculations long lasting.

16.4 EQUILIBRIA IN SOLVENT EXTRACTION OF METAL IONS

The results of QM calculations (except some semiempirical parameterized methods, e.g., AM1 and PM3) on the geometry of metal complexes (see section 16.2) are much more accurate and much better correlated with experiments than the calculated energy of intermolecular interactions, e.g., the energy of complex formation. That is because the latter is calculated as the difference between large total energies of products and substrates, and hence constitutes only a small fraction of the total energies. On the other hand, relative calculations for analogous complexes (formed by, e.g., different metals with the same ligand or different ligands with the same metal) are much more accurate [3]. To calculate Gibbs free energy, the entropy term must be calculated and deducted from the energy [Eq. (3.13)], which increases the error. Calculations on systems consisting of many molecules are even more difficult; therefore new methodological approaches are sought to overcome the limitations. Moreover, in relation to the solvent extraction of metal ions, the current QM calculations of the energy are usually fragmentary and do not correctly reflect the complicated process. This problem is discussed next using the example of a commonly accepted model of extraction.

16.4.1 Modeling of Solvent Extraction of Metal Ions

The process of transferring a metal ion M^{n+} from an aqueous solution to a water-immiscible organic solvent by an acidic chelating extractant HL that forms a lipophilic neutral complex ML_n with the ion can be represented as:

$$M^{n+} + nHL_{org} \leftrightarrow ML_{n.org} + nH^+ \qquad (16.2)$$

Both in this and in the subsequent equations the subscript org relates to a species in the organic phase (or to its concentration), while the lack of such a subscript denotes a species in the aqueous phase.

It has already been shown in Chapter 4 (section 4.2.1) that from the thermodynamic point of view the process described by Eq. (16.2) can be modeled by the sum of its partial processes (extraction steps), irrespective of whether they really proceed or not. That is because Gibbs free energy is the function of state and its total change does not depend on the reaction path. According to the complex formation–partition model [76], one can distinguish two main steps in extraction of metal ions:

1. Formation of a neutral complex by a given metal ion in the aqueous phase

$$M(H_2O)_k^{n+} + nL^- \leftrightarrow ML_n(H_2O)_{k-m} + mH_2O \qquad (16.3)$$

with the stability constant, β_n, of the complex, defined by Eq. (3.5). The concentration of "free" ligand in the anionic form, $[L^-]$, depends on the distribution of the extractant HL between the phases and its dissociation in the aqueous phase:

$$HL \leftrightarrow HL_{org} \tag{16.4}$$

$$HL \leftrightarrow H^+ + L^- \tag{16.5}$$

with the partition constant ($K_{D,HL}$) and the dissociation constant (K_a) of the extractant, defined by Eqs. (4.17) and (4.18), respectively.

2. Partition, i.e., transfer of the complex from the aqueous to the organic phase, accompanied [77] by at least partial dehydration

$$ML_n(H_2O)_{k-m} \leftrightarrow ML_{n,org} + (k-m)H_2O \tag{16.6}$$

with the equilibrium constant equal to the partition constant, K_{DM}, of the complex, also called the distribution constant, and defined by Eq. (4.1).

Taking into account all these relationships, one can express the extraction constant [the equilibrium constant in Eq. (16.2)] as:

$$K_{ex} = \beta_n K_{DM}(K_a)^n (K_{D,HL})^{-n} \tag{16.7}$$

When extraction of coordinatively unsaturated chelates proceeds in the presence of a coextractant (synergist), i.e., a lipophilic neutral electron donor, B, which additionally coordinates the metal ion, a third step must be added:

3. Adduct formation in the organic phase:

$$ML_{n,org} + pB_{org} \leftrightarrow ML_nB_{p,org} \tag{16.8}$$

with the equilibrium constant (the stability constant, $\beta_{n,p}$, of the adduct) given by:

$$\beta_{n,p} = \frac{[ML_nB_p]_{org}}{[ML_n]_{org} \cdot [B]^p_{org}} \tag{16.9}$$

The role of synergist is often played by undissociated HL molecules (self-adduct formation) when present in the system in a sufficient concentration.

16.4.2 Partition of Hydrophobic and Amphiphilic Solutes

The equilibrium distribution of a solute B between water and an organic solvent is described by the partition constant or by its function, the free energy of partition equal to the difference between the free energies of solvation and hydration:

$$\Delta G_p^0 = -RT \ln K_{D.B} = \Delta G_{solv}^0 - \Delta G_{hydr}^0 \tag{16.10}$$

The free energy of solvation (hydration) is defined as the free energy of transfer of the molecule from an ideal gas phase into a given solvent. Insertion of a solute molecule into a solvent requires the formation of a cavity of appropriate size and shape to accommodate the solute molecule. When using explicit solvation models we can relate the free energy of cavity formation in a solvent, e.g., water, to the excess free energy of equilibrium fluctuations of bulk water density, in particular, of excluding all the n_w bulk water molecules from a volume of a given size and shape which correspond to the size and shape of the solute molecule

$$\Delta G^{ex} = -RT \ln[p(n_w = 0)] \tag{16.11}$$

where $p(n_w = 0)$ denotes the probability of observing a solute-sized cavity in the bulk liquid [78]. The probability can be calculated either by Monte Carlo simulations (section 16.3.1) or from the information theory approach (section 16.5). Apart from cavity formation in water, there is a second component of the free energy of hydration, which originates in van der Waals interactions [typically approximated by the Lennard-Jones term in Eq. (16.1)] of the hydrophobic solute with surrounding water molecules. The free energy of hydrophobic hydration of numerous solutes—small neutral molecules—was calculated in this way [64,65]. For ionic solutes, the cumulate expansion method has been used to calculate free energies of charging [62,63]. A similar approach, based on the calculation of free energy of hydrophobic hydration in salt solutions, was used to study the effect of concentration of different salts (ion pairs) in aqueous solution on the chemical potential of a nonpolar solute. Molecular dynamic simulations of salt solutions resulted in reasonable values of Setchenow's salting-out and salting-in coefficients [79].

The hydrophobic hydration has been considered by molecular theories [80,81] in terms of water structuring. The challenge to the theories of hydrophobic effects is that the real solutes, in particular those dealt with in solvent extraction, are never entirely nonpolar. It has been recognized that hydrogen bonding to amphiphilic solutes in aqueous solutions disturbs the hydrophobic water structure around the solute. This is probably responsible for the observed nonadditivity of the effects of hydrophobic and hydrophilic fragments of numerous solutes, e.g., acetals and diethers, on the thermodynamic functions of their distribution between water and organic solvents, which strongly depend on the position of the hydrophilic fragments in the molecule [82].

The free energy of solvation of a solute can also be calculated using implicit solvation models (section 16.2.5), after parameterization. Carrying out such calculations for both water and an organic solvent allows the prediction of the partitioning behavior of a solute between these two liquid phases. Various continuous solvation models were used to evaluate partition constants of a large number of simple organic solutes in numerous two-phase systems [56]. The calculations were performed using various methods of calculation at different levels of theory, from MM [83] to HF-SCRF [84]. The calculations of the relative values of partition constants of two similar organic species in water/organic solvent systems were also made by MD simulations using statistical perturbation theory (section 16.3.2). An important advantage of this method consists in providing molecular-level details of the process and an insight into the structure of the species studied, and in making it possible to evaluate the mechanism of the process. However, given correct parameterization, the results of continuous solvation calculations seem to be more accurate than those obtained from MD-FEP simulations [85].

16.4.3 Partition of Ion Pairs

Extraction processes that proceed according to the model of ion-pair extraction are described by a formalism different from that presented in section 16.4.2, and are based on partition of single ions and their association in the organic phase [76] (see also section 2.6). The Born equation has been widely used to describe the transfer of an ion of the charge q and radius r from vacuum to the liquid (water) of the dielectric constant ε:

$$\Delta G = -\frac{q^2}{2r}\left(1 - \frac{1}{\varepsilon}\right) \qquad (16.12)$$

Numerous QM and MM calculations refer to this model. The free energy of hydration of the tetramethylammonium cation is equal to the sum of free energy of cavity formation in water [Eq. (16.11)], calculated by Monte Carlo simulations, and free energy of ion–water interactions [Coulomb and Lennard-Jones, Eq. (16.1)]. Small, in particular highly charged, ions strongly interact with water; therefore their free energy of hydration is large and negative [63]. Because the respective free energy of solvation in a given organic solvent (smaller ε) is much less negative, the free energy of transfer of a small single ion from water to the organic solvent is positive and resists the transfer. The driving force in this process is the negative free energy that accompanies breaking the hydrophobic hydration of large hydrocarbon fragments of, for example, tetraalkylammonium ions (tetrapropyl- and larger). Because of the principle of electroneutrality, the hydrophobic ions drag hydrophilic counterions into the organic phase in equivalent amounts; hence the experimental partition constant of an ion depends on the properties of the counterion. Concentration-dependent association of ions in an organic solvent of low dielectric constant shifts the equilibrium of the ion transfer toward the organic phase. Another important factor is interactions of ions with the components of any liquid phase. For example, the free energy of transfer of a complex polyanion from water to chloroform has been calculated (QM) based on the charge transfer theory [86].

The fact that ions of the same charge are never extracted alone but together with the equivalent amounts of oppositely charged ions, makes direct experimental determinations of single-ion partition constants impossible. In order to overcome this difficulty, the extrathermodynamic assumption was widely used that the partition constants of complex ions of the same structure but of opposite charge (e.g., the tetraphenylarsonium cation and the tetraphenylborate anion) are equal to each other [87]. Monte Carlo simulations of hydration of tetramethylammonium ion as a function of the ion charge have shown that the quadratic charge dependence is fulfilled as expected from the Born model, but that hydration is asymmetric in respect to the ion charge. Due to the structure of liquid water, well modeled in the calculations, negative ions are more favorably hydrated than positive ones [63]. This observation affords the possibility of modifying the extrathermodynamic assumption, although it is not clear how this computational result pertains to the very large tetraphenyl ions.

The combination of computational methods for calculations of the free energy of single-ion partition and the free energy of association [29] of ions in organic solvents can be helpful in further developing the partition–association model of extraction. A hypothetical thermodynamic cycle showing the role of ligand strain in ion-pair extraction has been presented for the example of lithium salts extraction by crown ethers [26]. The relative free energies of extraction of ion pairs in water/organic solvent systems can be calculated by using the MD-FEP simulations, as exemplified by several mono- and divalent cations with polydentate ionophores; it has been found that the calculated extraction selectivity corresponds to the experimental data [70].

16.4.4 Partition of Metal Chelates

In our opinion, the differences in partition constants of neutral complexes of metal ions of the same charge with the same number of the same ligands, extracted to inert organic solvents, are almost entirely due to the differences in hydration of these complexes in the aqueous phase or, more precisely, to the differences in the free energy of the dehydration accompanying the transfer of these complexes from the aqueous to the organic phase [77]. It must be pointed out that changes in hydration of metal complexes will be considered afterward not in terms of an integer number of water molecules attached or released, but rather in terms of free energy released upon transferring the complex from one liquid phase to the other. This can also be explained in terms of an averaged picture of dynamic equilibria between various differently hydrated forms of the same complex.

The hydrophobicity of metal complexes with organic ligands promotes their easy transfer from the aqueous to the organic phase. According to molecular theories, water structuring around nonpolar hydrocarbon regions of the solute (hydrophobic hydration) is accompanied by negative entropy and positive free energy changes, which depend on the volume or rather on the hydrophobic surface of the solute molecule [80,81]. In another approach, it is assumed that the free energy of cavity formation is more positive in water than in common organic solvents (see section 16.4.2). Both models predict a negative free energy change of removal of hydrophobic solutes from water, which means their spontaneous transfer to organic solvents. In contrast to hydrophobic hydration, specific hydration of metal complexes in the aqueous phase, either by water coordination to the central metal ion or by hydrogen bonding to the coordinated ligands, decreases the partition constant of the complex.

16.4.4.1 Inner-Sphere Hydration

The main contribution to hydrophilicity of coordinatively unsaturated metal complexes results from direct coordination of water molecules to the central metal ion, i.e., inner-sphere hydration of the complexes. The extent of the inner-sphere hydration depends on the properties of both the ion and ligand. For example, the smaller the ion within the Ln(III) series, the stronger its complex (e.g., *tris*-acetylacetonate) and the weaker its hydration in the inner sphere. This means that less energy is required to dehydrate the ion (and the complex) in the process of transferring from water to an organic phase. This conclusion may appear surprising in the light of increasing hydration energy across the series of Ln^{3+} ions [88] (see also Table 2.4). It may, however, be easily understood in terms of competition for the metal ion between ligands as, for example, between the strongly bonded acetylacetonate and rather weakly bonded water. The stronger the bonds with the chelating ligand [strengthened across the Ln(III) series], the weaker the bonds with the remaining water molecules. The effect of the strength of the metal–ligand bond on the inner-sphere hydration of the complex and, therefore, on its partition constant can also be illustrated by an example of zinc(II) extraction with chelating extractants belonging to the homologous series of β-diketones. Minute differences in the basicity of the ligands result in enormous changes in partition constants of the complexes. This has been explained in terms of decreasing coordinative unsaturation of the complexes with increasing ligand basicity, and the resulting decrease in their inner-sphere hydration [89].

In the calculations of the energy of hydration of metal complexes in the inner coordination sphere, one must consider hydrogen bond formation between the first-shell water molecules and those in bulk water, which leads to chains of hydrogen-bonded water molecules. Such hydrogen-bonded chains of ethanol molecules attached to the central metal ion have been found as a result of DFT B3LYP calculations on ethanol adducts to nickel acetylacetonate, where the calculated energy of hydrogen bonds correlated well with experimental data [90].

16.4.4.2 Outer-Sphere Hydration

Not only differences in the inner-sphere hydration of the complexes con-tribute to differences in partition constants. Even strongly lipophilic metal chelates with oxygen-donor organic ligands show some hydrophilic prop-erties, because not all electron pairs on the donor atoms are engaged in complex formation. Because of this, the solvent (water) molecules easily form hydrogen bonds with the donor atoms of coordinated ligands [91]. This outer-sphere hydration in the aqueous phase not only makes the complexes much more hydrophilic than hydrocarbons of similar molar volume, but also differentiates the complexes in respect to partition. This is illustrated by examples of the partition of neutral coordinatively saturated acetylacetonates of the $3d$ metal(III) ions (from Sc to Co) [92] and $4d$–$6f$ metal(IV) ions (from Th to Zr) [93]. In both series, the partition constants decrease with decreasing radius of the metal ion by over two orders of magnitude. A similar relationship has been observed in the series of 2,4-hexanedionates of $3d$ metal(III) ions [92]. According to our recent DFT calculations [94] based on the self-consistent isodensity polarized continuum model (SCI-PCM) [95], these differences are due to solvent effects exerted on the supermolecule (hydrate). The hydrate of a more covalently bonded chelate has a larger dipole moment; thus it more strongly interacts with bulk water.

It is worth mentioning that organic solvents with an acidic character also form hydrogen bonds with metal chelates. For example, chloroform and chlorophenols solvate acetylacetonates of trivalent metal ions with the same preference as water [96,97]. In the case of extraction of metal chelates into such solvents, the negative effect of their outer-sphere hydra-tion is cancelled to a large extent by the hydrogen bonding in the organic phase.

16.4.4.3 Adduct Formation

Coordinatively unsaturated metal chelates can be additionally coordinated in the inner sphere of the metal ion not only by water but also by other electron donor molecules. The presence in the extraction system of such lipophilic neutral species which form molecular adducts with the chelate in the organic phase [Eq. (16.8)] shifts the equilibrium of extraction [Eq. (16.2)] to the right. These coextractants, which alone poorly extract the metal ions, synergistically enhance the extraction by chelating ligands; therefore they are often called synergists. Synergistic extraction of metal ions by adduct (and self-adduct) formation to a significant degree compensates the effect of their inner-sphere hydration.

The authors' DFT calculations [98] have shown that the energy of adduct formation between metal(III) tropolonates and trioctylphosphine oxide (TOPO) depends not only on the size of the central metal but also on the electronic structure of the ion, in particular on the availability of empty d orbitals to form bonding molecular orbitals in the chelate. The bonding MOs in the adducts are easily formed with the participation of the unoccupied $(n-1)d$ orbitals of metal(III) ions of Group 3. The coordination bond between the TOPO molecule and the yttrium ion in the chelate appears to be much more complex (Fig. 16.2) than an expected simple σ bond. On the other hand, the energy of virtual nd orbitals of the Group 13 ions is too high to allow them to participate in bonding. The adducts (CN 7) of the p-block metal ions are therefore hypervalent compounds, of the bond order less than one [98]. This explains well the experimental observation that the TOPO adducts with tropolonates of Group 3 metals (Sc and Y) are much stronger than those of Group 13 metals (In and Tl), irrespective of the ionic radii. The contribution from the deformation energy into the energy of fully optimized structures of the adducts confirms the important role in the synergism of the ligand bite size, concluded in an earlier experimental work [99].

16.4.5 Selectivity of Metal Ion Separations

The selectivity of extractive separations of metal ions M^{n+} is commonly described by the separation factor (SF), defined as the ratio of the distribution ratios, D_M [Eq. (4.3)], of the ions in the same system. In the case of two metal ions A^{n+} and Z^{n+} extracted from the same aqueous solution by an acidic bidentate extractant HL, with no synergist in the system, one obtains:

$$SF_{A/Z} = \frac{\beta_{n,A} \cdot K_{DM,A} \cdot \sum_{i=0}^{n+1} \beta_{i,Z}[L]^i}{\beta_{n,Z} \cdot K_{DM,Z} \cdot \sum_{i=0}^{n+1} \beta_{i,A}[L]^i} \qquad (16.13)$$

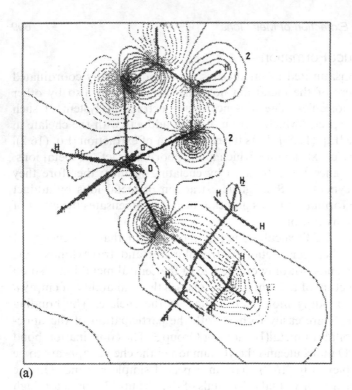

(a)

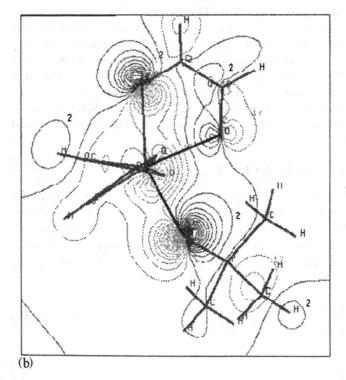

(b)

where β_i denotes the stability constant of ML_i, $M = A$ or Z; $K_{DM} = [ML_n]_{org}/[Ml_n]_{aq}$ denotes the partition constant of the neutral ML_n complex; and $[L]$ the equilibrium concentration of the anion L^- in the aqueous phase. For simplicity the charges of the species have been omitted.

The often used formulation of *SF* as the ratio of extraction constants of the metal ions [26,36–38] is justified only under special conditions, i.e., when the content of $[ML_i]^{n-i}$ complexes in the aqueous phase as compared to M^{n+} is insignificant (at low pH) and there is no adduct formation. The only (practically) form of the metal in the aqueous phase is then the uncomplexed ionic form, $[M(H_2O)_m]^{n+}$, while in the organic phase only the neutral complex $[ML_n]$ is present. Then Eq. (16.13) simplifies to the ratio of the respective β_n and K_{DM} values:

$$SF_{A/Z} = \frac{K_{ex,A}}{K_{ex,Z}} = \frac{\beta_{n,A} \cdot K_{DM,A}}{\beta_{n,Z} \cdot K_{DM,Z}} \qquad (16.14)$$

However, even this simplified formula does not justify the use of the ratio of stability constants of the extracted complexes as the only measure of selectivity of extractive separations. Such a widely used approach is obviously based on an implicit assumption that the partition constants of neutral complexes ML_n of similar metal ions are similar, so that their ratio should be close to unity. This is, however, an oversimplification because we have shown that the K_{DM} values significantly differ even in a series of coordinatively saturated complexes of similar metals [92,93]. Still stronger differences in the K_{DM} values have been observed in the series of lanthanide acetylacetonates, due to different inner-sphere hydration of the complexes (shown earlier), but in this case, self-adduct formation acts in the opposite direction [100,101] and partly compensates the effect of the differences in K_{DM} on *SF* (see also Fig. 4.15). Such compensation should also be observed in extraction systems containing coordinatively unsaturated complexes and a neutral lipophilic coextractant (synergist).

One can easily conclude that under certain conditions the differences in partition constants may exert a decisive effect on *SF*. The effect of differences in the stability constants of the complexes is significantly reduced if the aqueous phase concentration of the uncomplexed metal ions is not much higher than that of the $[ML_i]^{n-i}$ complexes. In the extreme case, when (at a moderate acidity) practically the only form of the metal in the aqueous

---

Fig. 16.2 3D contours of bonding molecular orbitals in the *tris*(2,3-butanedionato) yttrium(III)-TMPO adduct—a model of *tris*(tropolonato)yttrium(III)-TOPO: (a) MO of σ symmetry, $E_\sigma = -0.4006$ a.u.; (b) MO of π symmetry, $E_\pi = -0.2874$ a.u. (From Ref. 98; with permission from Wiley-VCH Verlag Weinheim, Germany.)

phase is the neutral chelate, ML_n, the β_n values influence neither the distribution ratios, nor the SF value, which then becomes dependent only on the K_{DM} ratio [77].

Of course, one has to recognize that for practical separation a high SF value is the prerequisite but not the sufficient condition. To separate selectively two metal ions, we must ensure their distribution ratios in the system to be respectively equally greater and less than unity. All these considerations show how many factors one must control to assure high selectivity of solvent extraction separations of metal ions. In the case of coordinatively unsaturated metal chelates, the increasing concentration of a synergist in the extraction system decreases, as a rule, the selectivity. For example, scandium and yttrium *tris*-tropolonates undergo partition with $K_{DM} = 0.381$ and 0.0020, respectively, in the system toluene/10^{-2} mol dm^{-3} tropolone/water (pH $= 5 \pm 1$) at 298 K [98]; therefore $SF_{Sc/Y} = 190$. Taking into account the stability constants [Eq. (16.9)] of their TOPO adducts [log$\beta_{3,1} = 4.27$ for Sc(trop)$_3$; log$\beta_{3,1} = 4.99$ and log$\beta_{3,2} = 7.39$ for Y(trop)$_3$] we arrive at the following calculated values for D_{Sc}, D_Y and $SF_{Sc/Y}$: 1.09, 0.022 and 50 at 10^{-4} mol dm^{-3} TOPO; 7.48, 0.246 and 30 at 10^{-3} mol dm^{-3} TOPO; 71.3, 6.87 and 10.4 at 10^{-2} mol dm^{-3} TOPO, respectively. In spite of the decrease in $SF_{Sc/Y}$ with increasing TOPO concentration, the optimum separation can be achieved at ca. 10^{-3} mol dm^{-3} TOPO. The effects become greater for ions that show greater differences in $\beta_{3,1}$, e.g., Sc^{3+} and In^{3+} (log$\beta_{3,1} = 4.27$ and 0.97 [98]). However, the unexpectedly increased selectivity in the synergistic Am^{3+}–Eu^{3+} system with CYANEX 301 and some O-bearing weak-base coextractants [45] shows that still other factors, for example different participation of the lanthanide and actinide f-orbitals in bonding, must be taken into account when considering the selectivity of separation.

There is, therefore, a good deal of work for theoretical chemists, dedicated to merging the calculations of the separate steps of extraction of metal ions into a whole. Only such a united approach will allow us to analyze correctly all the factors that affect the thermodynamics of extraction. The greatest challenge stems from the need to evaluate solvent effects by the use of more accurate, explicit solvation models. That will require quantum mechanical calculations on extremely large systems consisting of many hundred molecules, thousands of atoms.

16.4.6 Actual Calculations

16.4.6.1 MM and QM Methods

The reported QM and MM calculations on extraction of metal ions do not comprise as a rule the whole process described by the complex formation–partition model. The calculations rarely aim at the determination of the equilibrium of the process, because of the complexity of such a task. Most often, they are limited to the determination of the relative extractability of a given ion in a series of ions, or in a series of ligands. Therefore, in the majority of studies, the partition step becomes neglected and only the energy of complex formation is calculated as a qualitative or semiquantitative measure of metal extractability [25,32,36,51]. Sometimes the calculations become even more simplified and only the formation energy of a given conformation of the ligand is calculated as the measure of extractability of a given metal [26,37,38,45]. Another simplification consists in neglecting solvent effects on the stability of complexes and ligands, whereas both structure and stability of these species depend on solvent effects [26]. The calculations usually relate to complex formation in vacuo. Even those works that take into account dehydration of a given metal ion in the aqueous phase upon formation of the neutral complex [Eq. (16.3)] [36] still neglect bulk solvent effects, i.e., interactions of the complex with water molecules. These interactions that depend on the central metal ion affect to a different degree, not only the stability of the complexes but also their inner-sphere hydration (see section 16.4.4.2). Solvent effects exerted on simple molecules or ions were studied using both implicit and explicit solvation models (see section 16.2.5). The partition constants obtained (see section 16.4.2) remain in reasonable agreement with experiment [56,58]. The relative K_D values of simple organic solutes, calculated with the use of continuum solvation models, proved to be more accurate than those obtained from MD-FEP simulations, the average errors in $\log K_D$ for seven solute pairs being 0.50 and 1.01, respectively [85]. Unfortunately, to the authors' knowledge, similar calculations of partition constants have not yet been carried out on metal complexes.

We know of only one attempt to calculate all extraction equilibria for metal complexes, i.e., to analyze the whole metal ion extraction process including both complex formation and its two-phase distribution. In this attempt, Varnek and Wipff computed the relative free energy changes of extraction of several mono- and divalent cations with polydentate ionophores and counterions, using MD-FEP simulations [70]. For example, the calculated selectivity of extraction for the Ba^{2+}/Sr^{2+} pair from water to chloroform by a phosphoryl-containing podand ($\Delta\Delta G^0_{extr} = 1.5\,kcal\,mol^{-1}$) proved to be close to the experimental value ($\Delta\Delta G^0_{extr} = 1.2\,kcal\,mol^{-1}$) [70]. Despite the limitations of the method, this is the only methodologically correct approach used that makes it possible to deal with solvent extraction of metal ions. Other correlations based on the energy (or free energy) changes of complex formation as a measure of relative extractability of metal ions studied are oversimplified because they neglect the transfer of the complex from the aqueous to an organic phase. They are, however, often used because the QM methods applied for large systems that contain bulk solvent molecules still fail, whereas it is tempting to compare the calculation results with the experiment and to seek a correlation. We hope that the use of QM/MM methods and continuum solvation models may be helpful in overcoming such limitations. Also, statistical treatment of MD data for simple complexes that reversibly distribute between two liquid phases can contribute to the problem, provided the simulation systems (boxes) are large enough to model the properties of the bulk solvents.

To illustrate the importance of considering the whole extraction process, especially when coordinatively unsaturated complexes are extracted, let us consider extraction of trivalent lanthanides with β-diketones. The lanthanide contraction and the accompanying decrease in the coordination number (CN) make the complexes less and less coordinatively unsaturated across the lanthanide series, therefore less and less hydrated in the inner sphere of the metal ion [102]. This makes the partition constants of these complexes larger and larger, so that the K_{DM} values for the two extreme ions in the series (La–Lu) differ from each other by four orders of magnitude, approximately as much as the respective β_3 values [100,101,103]. (see also Fig. 4.15). It has been shown in a series of papers [37,38] that the ability of some β-diketones to extract lanthanides selectivity depend on the structure of the bidentate ligand, in particular on bite size, and on the energy of the intramolecular hydrogen bond in the enol form of the ligand. However, the experimental extraction data were correlated only with the calculated (semiempirical MNDO/H and ab initio) heats of formation of the anionic form of ligands, which affect only the thermodynamic stability of the lanthanide complexes. Because the respective differences in their K_{DM} values were disregarded, the observed agreement was of only qualitative character.

In another interesting work, the experimental extraction constants [Eq. (16.7)] of all the lanthanides, relative to that of lanthanum, have been correlated with the differences between the calculated (MM) total strain energies required for the formation of the respective metal(III) complexes with bis-alkylhydrogenphosphate ligands in the aqueous phase [36]. The calculations included the term responsible for dehydration of the metal ions, necessary to bind the ligands [Eq. (16.3)], but they are concerned with the process in vacuo, neglecting solvent effects. The partition step was completely disregarded. Therefore, the slope of the calculated energies vs. logarithm of the experimental extraction constants (1.26) has appeared much less than expected, which is $2.3RT = 5.71$ (the value reported in [36] is incorrect). This discrepancy shows that an important component of the energy has been omitted in the calculations. In fact, the inclusion of the term reflecting the differences in the partition constants of the complexes, most probably similar to those in the acetylacetonates (shown earlier), would significantly increase the slope and improve the correlation between the calculations and the experiment. It has been argued before that in the case of coordinatively unsaturated lanthanide complexes, the contribution from the partition step increases the differences in the K_{ex} values as well as the selectivity of lanthanide separation.

The necessity of considering the whole extraction process has been pointed out, however, calculations dealing with one particular extraction steps can also be important. For example, the authors' DFT calculations on adduct formation in extraction systems [98] well correlate with experimental data on the synergistic enhancement of partition constants of metal(III) tropolonates by trioctylphosphine oxide. In spite of simplifications (in contrast to the experimental conditions, the calculations were performed with models for tropolonate and TOPO, in the gas phase, and for 0 K), the computed order of the energies of adduct formation (Y > Sc > Tl > In; no Ga adduct) agreed with the experimental order of the adduct stability constants. The calculated energies of adduct formation for *tris*-tropolonates of scandium and yttrium are significantly less negative than the experimental free energy changes of adduct formation (which also include the entropy term), but the differences between the respective values, calculated $\Delta E_{add}(Y) - \Delta E_{add}(Sc) = -5.5\,kJ\,mol^{-1}$ and experimental $\Delta G_{add}(Y) - \Delta G_{add}(Sc) = -4.2\,kJ\cdot mol^{-1}$ [98] are comparable. Another work [94] points to a reasonable agreement between the relative values of standard enthalpy and free energy changes of transfer (heptane → water) of *tris*-acetylacetonates of scandium and cobalt(III); the experimental [92] and the calculated (DFT, continuous solvation model; see section 16.2.5) for model compounds.

16.4.6.2 Artificial Neural Networks

Another, very different calculation method, which belongs to the field of artificial intelligence, is till now only occasionally used in solvent extraction. It consists in the operation of artificial neural networks (NN), which are subject to appropriate procedures of construction and learning. NN computing techniques arose from attempts to model the functioning of the human brain and have evolved into a powerful tool for solving even the most complex, nonlinear problems, which are difficult to handle by standard modeling. This nonparametric method offers universal approximators able to represent any, even the most complex, implicit functional relationship. The calculations are based on the analysis of large sets of (empirical) data, part of which is used for training the system, the rest for testing. The widespread back propagation networks consist of large numbers of computational units, "neurons," connected by means of artificial synapses characterized by matrices of adjustable weights. These connections and the transfer functions of the neurons form distributed representations between the input and output data. The required representation is selected during the training procedure, when the weights are adjusted in a self-consistent process based on the minimization of errors generated by the discrepancy between the actual output of the network and the expected value used for training. An important advantage of NN is the universality of the applied methods, which are independent of detailed definition and of the data, within a given group of problems [104,105]. Numerous software packages related to NN operation, e.g., SNNS [106] and FlexTool [107], have been developed and used. The artificial neural networks are able to generalize the knowledge acquired and to self-improve the effectiveness of their performance by an evolutionary process.

In an excellent early review, Zupan and Gasteiger gave an essential background of the NN method and focused on its applications to various chemical problems, including process control and structure/activity relationships [108]. A few further works have shown the usefulness of the NN approach in various problems of solvent extraction, which are particularly difficult for modeling with the use of parametric empirical methods because of either complexity of the processes or lack of understanding of their nature. In fact, mass transfer processes involve a wide range of interactions between the chemical species in both liquid phases, and kinetics problems additionally entangle their course.

The ability of NN to recognize the structure/property relationships of crown ethers, widely used as extractants of alkali metal ions, was used for prediction of complexing properties of these ligands, which depend on their structure (ring size, number of benzo units) and on the metal ion. The reported average error of the training sets (10–11 ligands) was ± 0.27 $\log K$ units, and that of the testing sets (4–5 ligands) was ± 0.34 $\log K$ units [109]. The equilibrium distribution ratios of lanthanides in the HEH(EHP)/kerosene/HCl extraction system have been successfully modeled by NN as a complex function of the metal ion, its concentration, concentration of the extractant, phase ratio, and acidity of the aqueous phase [110,111]. A data set of 200 experimental points was used for training and testing. The system predicted the test data with an absolute average error of 11%, much better than most regression models based on the system parameters.

Such applications of NN as a predictive method make the artificial neural networks another technique of data treatment, comparable to parametric empirical modeling by, for example, numerical regression methods [e.g., 10,11] briefly mentioned in section 16.1. The main advantage of NN is that the network needs not be programmed because it learns from sets of experimental data, which results in the possibility of representing even the most complex implicit functions, and also in better modeling without prescribing a functional form of the actual relationship. Another field of

NN applications, perhaps more important, is process control. Processes that are poorly understood or ill defined can hardly be simulated by empirical methods. The problem of particular importance for this review is the use of NN in chemical engineering to model nonlinear steady-state solvent extraction processes in extraction columns [112] or in batteries of counter-current mixer-settlers [113]. It has been shown on the example of zirconium/hafnium separation that the knowledge acquired by the network in the learning process may be used for accurate prediction of the response of dependent process variables to a change of the independent variables in the extraction plant. If implemented in the real process, the NN would alert the operator to deviations from the nominal values and would predict the expected value if no corrective action was taken. As a processing time of a trained NN is short, less than a second, the NN can be used as a real-time sensor [113].

16.5 NEW INSIGHTS AND PERSPECTIVES

Various instruments of theoretical chemistry have been widely to describe separate steps of solvent extraction of metal ions. Because of the complexity of solvent extraction systems, there is still no unified theory and no successful approach aimed at merging the extraction steps. It has already been pointed out that the challenging problem for theoreticians dealing with solvent extraction of metals, in particular with thermodynamic calculations, is to evaluate correctly solvent effects by the use of the most accurate explicit solvation models and QM calculations. However, such calculations on extremely large sets consisting of hundreds or even thousands of molecules, necessary to model all aspects of the extraction systems, are still impossible due to both hardware and software limitations.

The methods applied to the theoretical description of complex formation, hydration, solvation, etc. discussed so far are characterized by different degrees of approximation made when solving the equations of Dirac-Schrödinger and Newton mechanics. But even the introduction of QM/MM-MD-MC methods does not allow one to make effective calculations. This is because it is common in this approach to neglect nonbonding interactions for pairs of atoms separated by distances greater than a selected cutoff value. The range of interactions in the solution is thus limited by the size of the system studied. This arbitrarily assumed division of the liquid phase into molecules that interact with one another and molecules which are neglected in the calculations and whose contribution is estimated by corrections makes the calculations possible. Nevertheless, in spite of this simplification such calculations are still very difficult, time-consuming, and expensive; moreover, the results obtained are often insufficiently accurate, inconsistent with the experiment, and therefore inconclusive.

The development of other methods with less computing requirements and free of such limitations seems to be necessary in order to understand better the processes that take place in the liquid phase, understand their molecular mechanism, predict the influence of various factors on their equilibria and kinetics, and evaluate correctly the thermodynamic functions for the whole process and its particular steps.

The recent theoretical approach based on the information theory (IT) in studying aqueous solutions and hydration phenomena [62–66] shows such a direction. IT is a part of the system based on a probabilistic way of thinking about communication, introduced in 1948 by Shannon and subsequently developed [114]. It consists in the quantitative description of the information by defining entropy as a function of probability

$$\eta = \sum_{i=1}^{\infty} p_i \ln\left(\frac{1}{p_i}\right) \tag{16.15}$$

The probability of cavity formation in bulk water, able to accommodate a solute molecule, by exclusion of a given number of solvent molecules, was inferred from easily available information about the solvent, such as the density of bulk water and the oxygen–oxygen radial distribution function [65,79].

MacKay's textbook [114] offers not only a comprehensive coverage of Shannon's theory of information but also probabilistic data modeling and the mathematical theory of neural networks. Artificial NN can be applied when problems appear with processing and analyzing the data, with their prediction and classification (data mining). The wide range of applications of NN also comprises optimization issues. The information-theoretic capabilities of some neural network algorithms are examined and neural networks are motivated as statistical models [114].

The applications of NN to solvent extraction, reported in section 16.4.6.2., suffer from an essential limitation in that they do not apply to processes of quantum nature; therefore they are not able to describe metal complexes in extraction systems on the microscopic level. In fact, the networks can describe only the pure state of simplest quantum systems, without superposition of states. Neural networks that indirectly take into account quantum effects have already been applied to chemical problems. For example, the combination of quantum mechanical molecular electrostatic potential surfaces with neural networks makes it possible to predict the bonding energy for bioactive molecules with enzyme targets. Computational NN were employed to identify the quantum mechanical features of the inhibitory molecules that contribute to bonding. This approach generates the relationships between the quantum mechanical structure of the inhibitory molecule and the strength of bonding. Feed-forward NN with back-propagation of error were trained to recognize the quantum mechanical molecular electrostatic potential at the entire van der Waals surface of a group of training molecules and to predict the strength of interaction between the enzyme and novel inhibitors [115]. On the other hand, the interactions on the microscopic level can be treated by a new kind of neural networks, quantum neural networks (QNN) [116]. Early publications concerned problems different from those discussed in this chapter. For example, QNN was applied to the recognition of handwritten numerals, and a reliability over 99% has been achieved [117]. One may expect that the QNN approach will also be helpful in describing intermolecular interactions accompanying the processes of extraction of metal ions.

The methods developed from either NN or IT undoubtedly reduce the requirements concerning the hardware that must be applied. In the NN approach it is sufficient to determine electrostatic potential, for which far less computer resources are required than for solving HFR equations supplemented with the energy of correlations, necessary for reliable calculation of interaction energy. Similarly, the calculation of information entropy (the IT case) based on the electron density is possible, and can be done with much shorter calculation time than in the case of HFR equations with the correlation energy.

Unfortunately, the application of these methods to the description of processes that take place in solutions, such as solvation, still requires the rejection of certain interactions between molecules that build the systems. This results from the tremendous number of interactions of the order of 2^N, where N denotes the number of molecules in the system. Optimization of such large systems consisting of mutually interacting elements requires special numerical procedures. The problem can be divided into classes according to the time required for their solution on digital computers. If there is an algorithm that solves a problem in a time which increases polynomially (or slower) with the increase in the problem, we call the problem a polynomial problem and classify it as belonging to the class P (polynominal). The class P is a subclass of the class NP (nondeterministic polynominal), which describes the processes taking place in solutions.

The computer time needed to solve an NP problem increases exponentially with the increase in N and it is impossible to solve it using classical computers. However, a solution has been theoretically possible since 1994, when Shor elaborated an algorithm that changed the increase of the computational time with the increase in the system size from exponential to polynomial [118]. The realization of the algorithm is possible basing on the construction of a "quantum computer" [119]. However, as long as the idea of construction of the quantum computer is far from being realized, it seems necessary to apply approximate algorithms that, in spite of some disadvantages, bring us closer to the understanding of the processes that take place in solutions. As far as the authors are aware, no QM approach to the explicit solvation models, in particular to solvation of metal complexes, has been used as yet.

ACKNOWLEDGMENTS

The authors would like to thank Alexandre Varnek for fruitful discussion, and also Mariusz Bogacki, Yizhak Marcus, Bruce Moyer, Claude Musikas, Jan Rydberg, and Slawomir Siekierski for their valuable comments on the draft of this chapter at various stages of writing. The work was supported in parts by the Institute of Nuclear Chemistry and Technology and by the Polish Committee for Scientific Research (KBN) under grant number 4 T09A 110 23.

REFERENCES

1. Roothaan, C. C. J. Rev. Mod. Phys. **1951**, *23*, 69.
2. Parr, R. G.; Yang, W. *Density-Functional Theory of Atoms and Molecules*; Oxford University Press: Oxford, 1989.
3. Deeth, R. J. Struct. Bonding **1995**, *82*, 1.
4. Rambusch, T.; Hollmann-Gloe, K.; Gloe, K. J. Prakt. Chem. **1999**, *341*, 202.
5. Cerius^2 Version 3.5; Molecular Simulations Inc., San Diego, CA.
6. GAUSSIAN 98; Revision A.5, Gaussian Inc., Pittsburgh, PA, 1998.
7. ADF Version 2.3; Scientific Computing and Modeling, Vrije Universiteit, Amsterdam.
8. Spartan Version 5.0.3; Wavefunction Inc., Irvine, CA.
9. HyperChem Version 5.02; Hypercube Inc., Gainesville, Florida.
10. Baes, C. F. Jr., Solv. Extr. Ion Exch. **2001**, *19*, 193.
11. Moyer, B. A.; Baes, C. F. Jr.; Case, F. I.; Driver, J. L. Solv. Extr. Ion Exch. **2001**, *19*, 757.
12. Varnek, A.; Wipff, G.; Solov'ev, V. P. Solv. Extr. Ion Exch. **2001**, *19*, 791.
13. Hay, B. P. Coord. Chem. Rev. **1993**, *126*, 177.
14. Boyens, J. C. A.; Comba, P. Coord. Chem. Rev. **2001**, *212*, 3.
15. Brubaker, G. R.; Johnson, D. W. Coord. Chem. Rev. **1984**, *53*, 1.
16. Hancock, R. D. Progr. Inorg. Chem. **1989**, *37*, 187.
17. Comba, P. Coord. Chem. Rev. **1993**, *123*, 1.
18. Deeth, R. J. Coord. Chem. Rev. **2001**, *212*, 11.
19. Deeth, R. J.; Foulis, D. L. Phys. Chem. Chem. Phys. **2002**, *4*, 4292.
20. Hancock, R. D.; Martell, A. E. Chem. Rev. **1989**, *89*, 1875.

21. Pople, J. A.; Santry, D. P.; Segal, G. A. J. Chem. Phys. **1965**, *43*, 129.
22. Hohenberg, P.; Kohn, W. Phys. Rev. B **1964**, *136*, 864.
23. Warshel, A.; Levitt, M. J. Mol. Biol. **1976**, *103*, 227.
24. Weiner, S. J.; Kollman, P. A.; Case, D. A.; Singh, U. C.; Ghio, C.; Alangona, G.; Profeta, S. Jr.; Weiner, P. J. Am. Chem. Soc. **1984**, *106*, 765.
25. Hay, B. P. *Metal Ion Separation and Preconcentration: Progress and Opportunities*; Bond, A. H. Dietz, M. L. Rogers, R. D. Eds.; ACS Symposium Series 716; American Chemical Society: Washington, D.C., 1999; p. 102.
26. Sachleben, R. A.; Moyer, B. A. *Metal Ion Separation and Preconcentration: Progress and Opportunities*; Bond, A. H. Dietz, M. L. Rogers, R. D. Eds.; ACS Symposium Series 716; American Chemical Society: Washington, D.C., 1999, p. 114.
27a. Sella, C.; Bauer, D. Solv. Extr. Ion Exch. **1992**, *10*, 579.
27b. Sella, C.; Bauer, D. Solv. Extr. Ion Exch. **1993**, *11*, 395.
28. Bond, A. H.; Chiarizia, R.; Huber, V. J.; Dietz, M. L.; Herlinger, A. W.; Hay, B. P. Anal. Chem. **1999**, *71*, 2757.
29. Hay, B. P.; Hancock, R. D. Coord. Chem. Rev. **2001**, *212*; 61.
30. Kopczyński, T.; Łozyński, M.; Prochaska, K.; Burdzy, A.; Cierpiszewski, R.; Szymanowski, J. Solv. Extr. Ion Exch. **1994**, *12*, 701.
31. Bogacki, M. B.; Jakubiak, A.; Prochaska, K.; Szymanowski, J. Colloids Surfaces A **1996**, *110*, 263.
32. Bogacki, M. B.; Jakubiak, A.; Cote, G.; Szymanowski, J. Ind. Eng. Chem. Res. **1997**, *36*, 838.
33. Kyuchoukov, G.; Bogacki, M. B.; Szymanowski, J. Ind. Eng. Chem. Res. **1998**, *37*, 4084.
34. Krueger, T.; Gloe, K.; Stephan, H.; Habermann, B.; Hollmann, K.; Weber, E. J. Mol. Model. **1996**, *2*, 386.
35. Stephan, H.; Gloe, K.; Krüger, T.; Chartroux, C.; Neumann, R.; Weber, E.; Möckel, A.; Woller, N.; Subklew, G.; Schwuger, M. J. Solv. Extr. Res. Develop. Japan. **1996**, *3*, 43.
36. Comba, P.; Gloe, K.; Inoue, K.; Krüger, T.; Stephan, H.; Yoshizuka, K. Inorg. Chem. **1998**, *37*, 3310.
37. Le, Q. T. H.; Umetani, S.; Suzuki, M.; Matsui, M. J. Chem. Soc. Dalton Trans. **1997**, 643.
38. Umetani, S.; Kawase, Y.; Le, Q. T. H.; Matsui, M. Inorg. Chim. Acta. **1998**, *267*, 201.
39. Torgov, V.; Erenburg, S.; Bausk, N.; Stoyanov, E.; Kalchenko, V.; Varnek, A.; Wipff, G. J. Mol. Struct., **2002**, *611*, 131.
40. Rabbe, C.; Madic, C.; Godard, A. Solv. Extr. Ion Exch. **1998**, *16*, 1091.
41. Rabbe, C.; Sella, C.; Madic, C.; Godard, A. Solv. Extr. Ion Exch. **1999**, *17*, 87.
42. Spjuth, L.; Liljenzin, J. O.; Hudson, M. J.; Drew, M. G. B.; Iveson, P. B.; Madic, C. Solv. Extr. Ion Exch. **2000**, *18*, 1.
43. Drew, M. G. B.; Hudson, M. J.; Iveson, P. B.; Russel, M. L.; Liljenzin, J. O.; Skalberg, M.; Spjuth, L.; Madic, C. J. Chem. Soc. Dalton Trans. **2002**, 2973.
44. Coupez, B.; Boehme, C.; Wipff, G. Phys. Chem. Chem. Phys. **2002**, *4*, 5716.

45. Ionova, G.; Ionov, S.; Rabbe, C.; Hill, C.; Madic, C.; Guillaumont, R.; Krupa, J. C. Solv. Extr. Ion Exch. **2001**, *19*, 391.
46. Pyykkö, P. Chem. Rev. **1988**, *88*, 563.
47. Schreckenbach, G.; Hay, P. J.; Martin, R. L. J. Comput. Chem. **1999**, *20*, 70.
48. Schautz, F.; Flad, H.-J.; Dolg, M. Theor. Chim. Acta **1999**, *99*, 231.
49. Skylaris, C.-K.; Gagliardi, L.; Handy, N. C.; Ioannou, A. G.; Spencer, S.; Willets, A.; Simper, A. M. Chem. Phys. Lett. **1998**, *296*, 445.
50. Willets, A.; Gagliardi, L.; Ioannou, A. G.; Skylaris, C.-K.; Simper, A. M.; Spencer, S.; Handy, N. C. Int. Rev. Phys. Chem. **2000**, *19*, 327.
51. Gagliardi, L.; Handy, N. C.; Skylaris, C.-K.; Willets, A. Chem. Phys. **2000**, *252*, 47.
52. Spencer, S.; Gagliardi, L.; Handy, N. C.; Ioannou, A. G.; Skylaris, C.-K.; Willets, A.; Simper, A. M. J. Phys. Chem. A **1999**, *103*, 1831.
53. Pershina, V.; Fricke, B.; Kratz, J. V.; Ionova, G. V. Radiochim. Acta **1994**, *64*, 37.
54. Pershina, V.; Fricke, B. J. Phys. Chem. **1996**, *100*, 8746.
55. Pershina, V. Chem. Rev. **1996**, *96*, 1977.
56. Cramer, C. J.; Truhlar, D. G. Chem. Rev. **1999**, *99*, 2161.
57. Gao, J. L. Acc. Chem. Res. **1996**, *29*, 289.
58. Kollman, P. Chem. Rev., **1993**, *93*, 2395.
59. Leach, A. R. *Molecular Modeling: Principles and Applications*, 2nd Ed.; Pearson Education Limited, 2001.
60. Allen, M. P.; Tildesley, D. J. *Computer Simulations of Liquids*; Clarendon: Oxford, 1987.
61. Case, D. A.; Pearlman, D. A.; Caldwell, J. C.; Cheatham, T. E. III; Ross, W. S.; Simmerling, C. L.; Daren, T. A.; Mertz, K. M.; Stanton, R. V.; Cheng, A. L.; Vincent, J. J.; Crowley, M.; Ferguson, D. M.; Radmer, R. J.; Seibel, G. L.; Singh, U. C.; Weiner, P. K.; Kollman, P. A. AMBER5; University of California: San Francisco, 1997.
62. Hummer, G.; Pratt, L. R.; Garcia, A. E. J. Phys. Chem. A **1998**, *102*, 7885.
63. Garde, S.; Hummer, G.; Paulaitis, M. E. J. Chem. Phys. **1998**, *108*, 1552.
64. Hummer, G.; Garde, S.; Garcia, A. E.; Paulaitis, M. E.; Pratt, L. R. J. Phys. Chem. B **1998**, *102*, 10469.
65. Hummer, G.; Garde, S.; Garcia, A. E.; Pratt, L. R. Chem. Phys. **2000**, *258*, 349.
66. Gomez, M. A.; Pratt, L. R.; Hummer, G.; Garde, S. J. Phys. Chem. B **1999**, *103*, 3520.
67. Behler, J.; Price, D. W.; Drew, M. G. B. Phys. Chem. **2001**, *3*, 588.
68a. Benjamin, I. Chem. Rev. **1996**, *96*, 1449.
68b. Benjamin, I. *Solvent Extraction for the 21st Century, Proc. ISEC'99*; Cox, M. Hildalgo, M. Valiente, M. Eds.; Soc. Chem. Ind.: London, 2001, *1*, 3.
69. Varnek, A.; Wipff, G. J. Mol. Struct. (Theochem) **1996**, *363*, 67.
70a. Varnek, A.; Wipff, G. J. Comput. Chem. **1996**, *17*, 1520.
70b. Varnek, A.; Wipff, G. Solv. Extr. Ion Exch. **1999**, *17*, 1493.
71. Varnek, A.; Troxler, L.; Wipff, G. Chem. Eur. J. **1997**, *3*, 552.

72. Nazarenko, A. Y.; Baulin, V. E.; Lamb, J. D.; Volkova, T. A.; Varnek, A. A.; Wipff, G. Solv. Extr. Ion Exch. **1999**, *17*, 495.
73. Szymanowski, J. Solv. Extr. Ion Exch. **2000**, *18*, 729.
74. Felix, V.; Matthews, S. E.; Beer, P. D.; Drew, M. G. B. Phys. Chem. Chem. Phys. **2002**, *4*, 3849.
75. Baaden, M.; Berny, F.; Madic, C.; Wipff, G. J. Phys. Chem. A **2000**, *104*, 7659.
76. Siekierski, S. J. Radioanal. Chem. **1976**, *31*, 335.
77. Narbutt, J. J. Radioanal. Nucl. Chem. **1992**, *163*, 59.
78. Reiss, H.; Frisch, H. L.; Lebowitz, J. L. J. Chem. Phys. **1959**, *31*, 369.
79. Kalra, A.; Tugcu, N.; Cramer, S. M.; Garde, S. J. Phys. Chem. B **2001**, *105*, 6380.
80. Nemethy, G.; Scheraga, H. A. J. Chem. Phys. **1962**, *36*, 3382.
81. Madan, B.; Sharp, K. J. Phys. Chem. B **1997**, *101*, 11237.
82. Gniazdowska, E.; Narbutt, J. J. Mol. Liquids **2000**, *84*, 273.
83. Sithoff, D.; Ben-Tal, N.; Honig, B. J. Phys. Chem. **1996**, *100*, 2744.
84. Li, J.; Hawkins, G. D.; Cramer, C. J.; Truhlar, D. G. Chem. Phys. Lett. **1998**, *288*, 293.
85. Best, S. A.; Merz, K. M. Jr.; Reynolds, C. H. J. Phys. Chem. B **1999**, *103*, 714.
86. Osakai, T.; Ebina, K. J. Electroanal. Chem. **1996**, *412*, 1.
87. Parker, A. J.; Alexander, R. J. Am. Chem. Soc. **1968**, *90*, 3313.
88. Marcus, Y. *Ion Properties*; Dekker: New York, 1997.
89. Narbutt, J. Polish J. Chem. **1993**, *67*, 293.
90. Polyakov, V. R.; Czerwiński, M. Inorg. Chem. **2001**, *40*, 4798.
91a. Narbutt, J. J. Inorg. Nucl. Chem. **1981**, *43*, 3343.
91b. Narbutt, J. J. Phys. Chem **1991**, *95*, 3432.
92. Narbutt, J.; Bartoś, B.; Siekierski, S. Solv. Extr. Ion Exch. **1994**, *12*, 1001.
93. Narbutt, J.; Fuks, L. Radiochim. Acta, **1997**, *78*, 27.
94. Czerwiński, M.; Narbutt, J. Phys. Chem. Chem. Phys. submitted.
95. Foresman, J. B.; Keith, T. A.; Wiberg, K. B.; Snoonian, J.; Frisch, M. J. J. Phys. Chem. **1996**, *100*, 16098.
96. Katsuta, S.; Imura, H.; Suzuki, N. Bull. Chem. Soc. Jpn **1991**, *64*, 2470.
97. Katsuta, S.; Yanagihara, H. Solv. Extr. Ion Exch. **1997**, *15*, 577.
98. Narbutt, J.; Czerwiński, M.; Krejzler, J. Eur. J. Inorg. Chem. **2001**, 3187.
99. Narbutt, J.; Krejzler, J. Inorg. Chim. Acta **1999**, *286*, 175.
100. Albinsson, Y.; Mahmood, A.; Majdan, M. Rydberg, J. Radiochim. Acta **1989**, *48*, 49.
101. Albinsson, Y. Acta Chem. Scand. **1989**, *43*, 919.
102. Habenschuss, A.; Spedding, F. H. J. Chem. Phys. **1980**, *73*, 442.
103. Suzuki, N.; Nakamura, S. Inorg. Chim. Acta **1985**, *110*, 243.
104. Kohonen, T. *Self-Organizing Maps*; Springer Verlag: Berlin, 1995.
105. Hayki, S. *Neural Networks: A Comprehensive Foundation;* Prentice Hall: Englewood Cliffs, NJ, 1994.
106. Zell, A.; Mamier, G.; Vogt, M.; Mache, N.; Hübner, R.; Döring, S.; Herrmann, K.-U.; Soyez, T.; Schmalzl, M.; Sommer, T.; Hatzigeorgiou, A.; Posselt, D.; Schreiner, T.; Kett, B.; Clemente, G. SNNS—*Stuttgart Neural*

Network Simulator User Manual, Version 4.0; University of Stuttgart, Institute for Parallel and Distributed High Performance Computing (IPVR), Department of Computer Science; Report June 1995. wwwra.informatik.uni-tuebingen.de/SNNS/.

107. FlexTool; Flexible Intelligence Group, LLC: Tuscalloosa, AL; 2003; www. flextool.com.

108. Zupan, J.; Gasteiger, J. Anal. Chim. Acta **1991**, *248*, 1.

109. Gakh, A. A.; Sumpter, B. G.; Noid, D. W.; Sachleben, R. A.; Moyer, B. A. J. Incl. Phenom. Mol. Recogn. Chem. **1997**, *27*, 201.

110. van Deventer, J. S. J.; Aldrich, C. *Value Adding Through Solvent Extraction, Proc. ISEC'96*; Shallcross, D.C. Paimin, R. Prvcic L.M. Eds. University of Melbourne, **1996**, *1*, 831.

111. Giles, A. E.; Aldrich, C.; van Deventer, J. S. J. Hydrometallurgy **1996**, *43*, 241.

112. Slater, M. J.; Aldrich, C. *Application of Neural Network and Other Learning Technologies in Process Engineering*; Mujtaba, I. M. Hussain, M. A. Eds.; Imperial College Press: London, 2001; 3.

113. Boger, Z.; Ben-Haim, M. *Solvent Extraction in the Process Industry, Proc. ISEC'93*; Logsdail, D. H. Slater, M. J. Eds.; Elsevier Applied Science, 1993, *2*, 1198.

114. MacKay, D. J. C. *Information theory, inference and learning algorithms*; Draft 3.1415, January 2003; www.inference.phy.cam.ac.uk/mackay/itprnn/book. html.

115. Braunheim, B. B.; Bagdassarian, C. K.; Schramm, V. L.; Schwartz, S.; D. Int. J. Quantum Chem. **2000**, *78*, 195.

116. Purushothaman, G.; Karayiannis, N. B. IEEE Trans. Neural Networks **1997**, *8*, 679.

117. Zhou, J.; Gan, Q.; Krzyżak, A.; Suen, C. Y. Int. J. Docum. Anal. Recogn. **1999**, *2*, 30.

118. Shor, P. W. Polynomial time algorithms for prime factorization and discrete logarithms on a quantum computer. *Proc. 35th Annual Symposium on the Foundations of Computer Science*; Goldwasser, S. Ed.; IEEE Computer Society Press: Los Alamos, CA, 1994, p. 124.

119. Bouwmeester, D.; Ekert, A.; Zeilinger, A. Eds. *The Physics of Quantum Information*; Springer Verlag: Berlin, 2001.

Appendix

A. FUNDAMENTAL CONSTANTS, BASIC AND DERIVED SI UNITS

A.1 Fundamental Constants

e   electron (proton) charge, $1.6022 \; 10^{-19}$ coulomb (C)

F   Faraday constant, $96,485 \; \text{C} \, \text{mol}^{-1}$

N_A Avogadro number, $6.022 \; 10^{23} \, \text{mol}^{-1}$

k_B Boltzmann constant, $1.3807 \; 10^{-23}$ joule/kelvin ($\text{J} \, \text{K}^{-1}$)

R   molar gas constant, $8.314 \, \text{J} \, \text{K}^{-1} \text{mol}^{-1}$ (0.08206 L atm $\text{mol}^{-1} \, \text{K}^{-1}$)

v_M molar volume of ideal gas, $0.02241 \, \text{m}^3 \, \text{mol}^{-1}$

A.2 Fundamental SI Units

Physical quantity	Symbol	Unit name	Symbol
Length	l	meter	m
Mass	m	kilogram	kg
Time	t	second	s
Electric current	I	ampere	A
Thermodynamic temperature	T	Kelvin	K

A.3 Derived SI Units as Used in Chemistry

Physical quantity	Symbol	Unit name, symbol, explanation
Physical mass (weight)	$m(w)$	gram, g; m_A is weight of pure substance A
Molar mass	M	molecular weight in atomic mass units (amu)
Amount of substance	n	moles, mol; $n_A = m_A M_A$
		1 mol = N_A molecules
Volume	V	liter, L, or dm^3 (SI-unit is m^3)
Concentration	c or [ ]	molarity, M, mol L^{-1}, or mol dm^{-3}
Density	ρ	$g/cm^3 = 10^3 kg/m^3$ (SI unit is kg/m^3)
Pressure	P	megapascal, MPa; 1 atm = 0.1013 MPa; 1 bar = 0.1 MPa
Temperature	$T\,^\circ C$	degrees Celsius; $0\,T^\circ C = 273.15\,K$
Energy	E	joule (watt seconds, Ws) J; 1 kilocalorie (Kcal) = 4.187 J; 1 electron volt (eV) = 1.602 2 10^{-19} J corresponds to 96.485 kJ mol^{-1}

B. SYMBOLS USED IN CHEMICAL REACTION FORMULAS

Attempts have been made to adhere to the following set of abbreviations for the reacting species to simplify their identification:

M   metal ion (central atom)
A   (preferably lipophilic) extractant anion (anion of HA)
L   (preferably hydrophilic) ligand in aqueous phase (monobasic)
X   any (likely noncomplexing) anion in aqueous phase
B   uncharged donor molecule (adduct former)
S   organic solvent molecules (solvating or nonsolvating)

Ionic charges are omitted, when not necessary for understanding. Other abbreviations commonly used:

A, B, C   for various solutes in general
A^{z-} anion in general, charge z^- (may be omitted)
C^{z+} cation in general, charge z^+ (may be omitted)
$C_{v+}A_{v-}$ electrolyte in general, with v^+ atoms of C and v^- atoms of A in a neutral salt (e.g., NaCl or $NaClO_4$).
e, and ne   for electrolyte and nonelectrolyte, respectively

Physical state:

Gaseous (g)
Solid (s)
Liquid (l), not recommended in this text
Aqueous [i.e., dissolved in water (aq)]
Organic [i.e., dissolved in organic solvent (org); not recommended (s)]
Water (w), to specifically refer to water molecules

B.1 Abbreviations for Organic Compounds

AA: acetylacetone; 2,4-pentanedione
BA: benzoylacetone
DCTA: 1,2-diaminocyclohexane tetraacetic acid
DBP: dibutyl phosphoric acid
DTPA: diethylenetriamine pentaacetic acid
DEHPA: di-2-ethylhexyl phosphoric acid
PC88A: (and Ionquest 801) 2-ethylhexyl phosphonic acid 2-ethylhexyl ester
EDTA: ethylene diaminetetraacetic acid
FAA: hexafluoroacetylacetone
HIBA: α-hydroxyisobutyric acid
LIX64: o-hydroxybenzophenone oxime
NTA: nitrilotriaminoacetic acid
TTA: thenoyltrifluoroacetone
TOA: trioctylamine; Aliquat-336 is mainly TOA
TDA: tridecylamine; Adogen 464 is C_8–C_{10} amine
TLA: trilaurylamine; C_{12}
TBP: tributyl phosphate
TOPO: trioctylphosphine oxide
RNH_3^+: organic primary ammonium ion
$R_2NH_2^+$: organic secondary ammonium ion
R_3NH^+: organic tertiary ammonium ion
R_4N^+: organic quaternary ammonium ion
R: organic radical in general

Protolytic compounds are written as monobasic acids (e.g., HA), but the proton may be excluded as an abbreviation (e.g., TTA or EDTA). For example, in the text HTTA and TTA may mean the same if not specified, as by the reaction HTTA $\leftrightarrows$ H^+ + TTA^-.

C. TERMINOLOGY FOR SOLVENT EXTRACTION EQUILIBRIA

Most chapters in this book use the following symbols, which follow IUPAC recommendations (see Refs. 1–4) where such recommendations are available.

C.1 Reference to Aqueous and Organic Phases

In the reaction below a metal ion M^{n+} reacts with n ligand anions L^- to form an uncharged complex ML_n. If the ions dissolve only in the aqueous phase, and the metal complex and undissociated acid HL dissolve only in the organic phase, IUPAC allows the reaction to be written in four different ways:

$$M^{n+}(aq) + nHL(org) \rightleftharpoons ML_n(org) + nH^+(aq) \; K_{ex} \qquad (C.1a)$$

$$M^{n+}_{aq} + nHL_{org} \rightleftharpoons ML_{n.org} + nH^+_{aq} \qquad K_{ex} \qquad (C.1b)$$

$$M^{n+} + nHL(org) \rightleftharpoons ML_n(org) + nH^+ \qquad K_{ex} \qquad (C.1c)$$

$$M^{n+} + n\overline{HL} \rightleftharpoons \overline{ML_n} + nH^+ \qquad K_{ex} \qquad (C.1d)$$

K_{ex} is the overall extraction constant [2]. Different authors favor different notations, all being used here (as in other literature). For example, the concentration of HL in the organic phase may be written $\overline{[HL]}$, $[HL]_{org}$, or $[HL]_o$, and in the aqueous phase $[HL]$ (with no index), $[HL]_{aq}$ or $[HL]_a$. Freiser and Nancollas [2] recommend against the combination $[HL]_o/[HL]_a$, which invites confusion.

Total concentration of species A in a phase is indicated either by $[A]_t$, $[A]_{tot}$, or c_A. "Original" value (e.g., total weight of substance, volume, concentration, or such, in the beginning of an experiment) may be indicated by an upper index 0 (e.g., $[A]^0$ or c_A^0. Total concentration and original concentration may mean the same.

C.2 IUPAC Terminology

IUPAC prefers "liquid–liquid distribution" or "liquid–liquid extraction" rather than "solvent extraction." We justify our preference for "solvent extraction" by its more common usage. IUPAC defines the following terms:

Distribution ratio is the "total analytical concentration of a substance in the organic phase to its total analytical concentration in the aqueous phase, usually measured at equilibrium." Symbol D. D shall be defined and, preferably, specified by an index; if the distribution of mercury is measured, the distribution ratio is written $D(Hg)$ or D_{Hg}. The term "partition ratio" is not used for the distribution ratio.

Distribution constant is symbolized by K_D. Thus for

$$A(aq) \rightleftharpoons A(org); K_D, K_D(A), K_{D,A}, K_{DA}, \text{ or } K_{DR} = [A]_{org}/[A]_{aq}$$

$$MA_z(aq) \rightleftharpoons MA_z(org); K_D, K_D(MA_z), K_{DMAz}, \text{ or } K_{DC} = \overline{[MA_z]}/[MA_z]$$

In this book we prefer a suffix to the index to identify if the distribution constant refers to the neutral reagent (or extractant) or to the uncharged metal complex (i.e., K_{DR} and K_{DC}, respectively).

Extraction equilibria (including *adduct formation*) must be defined by text, for example,

$$M^{n+} + nHA(org) + aS(org) \rightleftharpoons MA_nS_a(org) + nH^+ \ K_{ex}$$

Recovery factor is defined as the percentage ($R\%$), or fraction R, of the total quantity of a substance extracted under specified conditions. (In our text we also use percentage extraction, $E\%$.)

Solvent is "a liquid phase (usually an organic liquid) or the solution of an extractant in an organic diluent which is used to extract a substance from another liquid (usually aqueous)."

Diluent is "an inert (organic) solvent used to improve the physical properties (density, viscosity, etc.) or the extractive properties (e.g., selectivity) of the extractant. The diluent has negligible extractant properties for the substance to be extracted."

Extractant is "the reagent which forms a complex or other adduct in the solvent with the substance which partitions across the phase boundary of the extraction system. The extractant (extracting agent) may also partition."

Extract is "the separated phase (usually organic) containing the substance extracted from the other phase."

Stripping is "the process of back-extraction of the distribuend from the extract (usually into the aqueous phase)."

Distribuend is "the substance that is distributed between two immiscible liquids or liquid phases."

C.3 Complex Formation in Aqueous Solutions

The nomenclature used in this book is presented in section 3.1. It should be noted [2] that "For polynuclear complexes and complexes with several kinds of ligands, it may sometimes be practicable to use β with...multiple subscripts. Their general meaning must then be defined very clearly with a full reaction formula....Strict standardization in this field has not yet been achieved."

C.4 Other Equilibrium Constants Commonly Used in This Text

Q_M distribution ratio of M between nonaqueous and aqueous phase

D_M distribution ratio of M between organic solvent and aqueous phase
        (e.g., D_{La} for distribution of element La)

K_i stepwise formation constant for ith complex (e.g., ML_i)

K_{ex} overall extraction constant

K_a acid dissociation (concentration) constant; $pK_a = -\log K_a$

K_{al} first acid dissociation constant for multibasic acid

K_{adl} first adduct formation constant (organic phase)

K_{Dx} distribution constant for x (specified); for example, x = R, or = HA,
        for undissociated extractant (reagent), = C for neutral
        (e.g., metal-containing) complex

K_{sp} solubility product

K_w ionic product of water

β_n overall complex formation, or stability, constant for formation of nth
        complex (e.g., ML_n).

P thermodynamic distribution constant

s_{AC} solubility of AC in moles or grams per liter (mL^{-1} or gL^{-1})

K_{ass} association constant for ion pair formation

K_{BS} Bunsen coefficient for solubility of gas in a liquid

D. FORMULA STRUCTURES

Structural formulas of a number of commonly used extractants, and some typical chelate compounds.

1. Dinonylnaphtylsulfonic acid

2. Benzoic acid

3. Salicylic acid

4. Isopropyltropolone

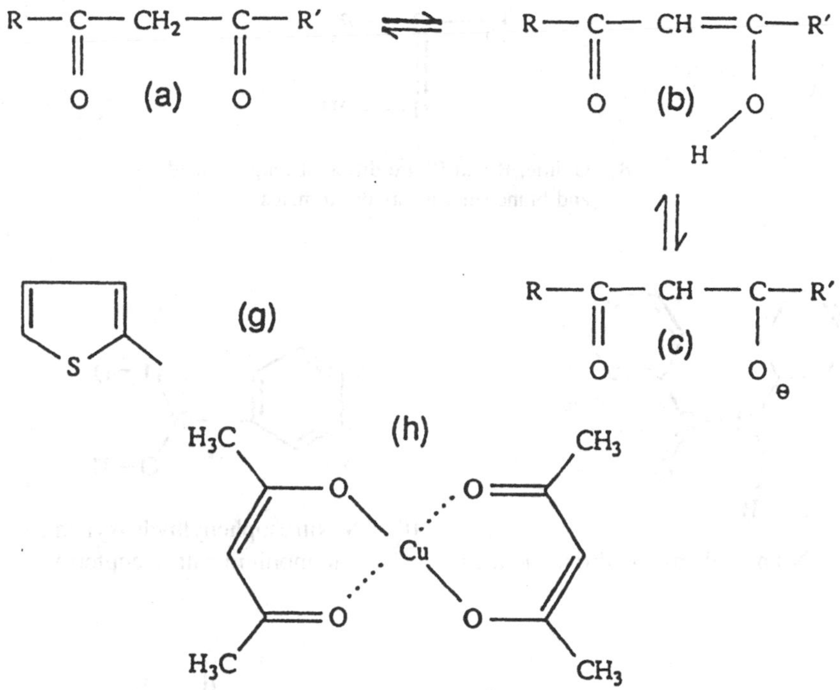

5. β-Diketones:
    (a) ketoform; (b) enol form; (c) enolate ion;
    (d) acetylacetone, 2,4-pentanedione, HAA: $R = R' = CH_3$;
    (e) trifluoroacetylacetone, HTFA: $R = CH_3$, $R' = CF_3$;
    (f) benzoylacetone, HBA: $R = CH_3$, $R' = C_6H_5$;
    (g) thenoyltrifluoroacetone, HTTA: R = formula (g), $R' = CF_3$;
    (h) bis-acetylacetonatocopper(II) complex; and
    (i) dibenzoylmethane: $R = R' = C_6H_5$.

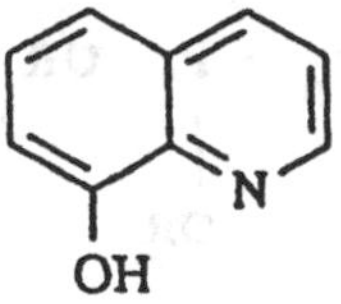

6. 8-Hydroxoquinoline, oxine

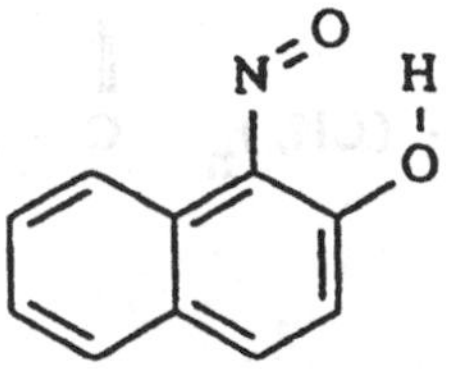

7. 1-Nitroso-2-naphthol

$$R_1 - \overset{\displaystyle \parallel}{\underset{\displaystyle N-OH}{C}} - R_2$$

8. Oxime: R and R' are different long-chained
    and branched aliphats or aromates

9. N-Phenylbenzo-hydroxamic acid

10. N-Nitrosophenylhydroxylamine
    (ammonium salt is cupferron)

$$R - \overset{\displaystyle \parallel}{\underset{\displaystyle O}{C}} - N\overset{\displaystyle R'}{\underset{\displaystyle R''}{}} \qquad R = -CH_2 - \overset{\displaystyle O}{\underset{\displaystyle R_2}{\overset{\parallel}{P}}} \overset{R_1}{}$$

(a)   (b)

11. (a) Amide R = R' = R'' = alkyl; and
    (b) carbamoylphosphinate: $R_1 = R_2$ = alkyl, R = phosphinate

$$RO - \overset{O}{\overset{\parallel}{\underset{\underset{RO}{\mid}}{P}}} - (CH_2)_n - \overset{O}{\overset{\parallel}{C}} - N\overset{R}{\underset{R}{}} \qquad RO - \overset{O}{\overset{\parallel}{\underset{\underset{RO}{\mid}}{P}}} - (CH_2)_n - \overset{O}{\overset{\parallel}{\underset{\underset{OR}{\mid}}{P}}} - OR$$

12. Carbamoylmethylene phosphonate   13. Dialkyl phosphonate

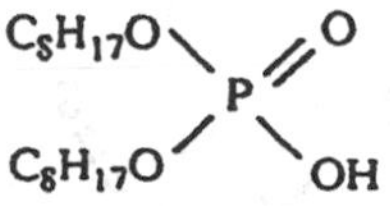

(a)

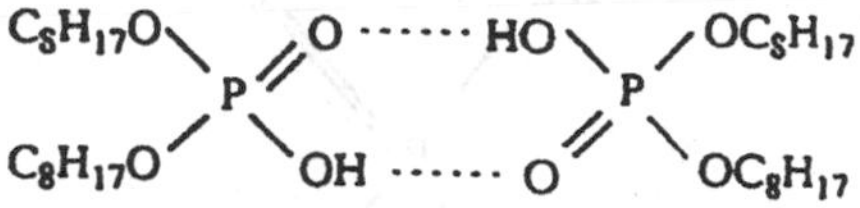

(c)

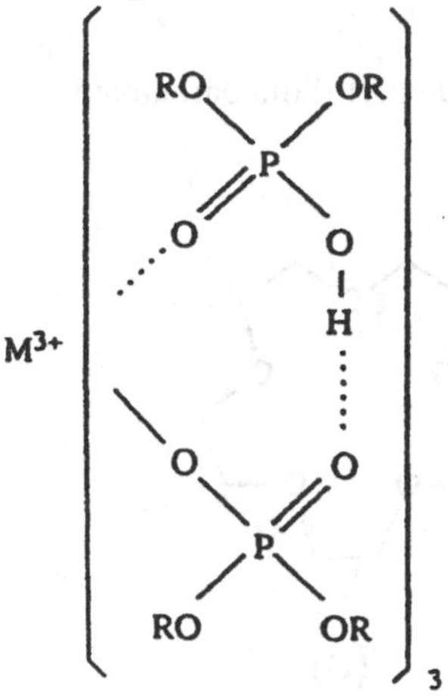

(b)

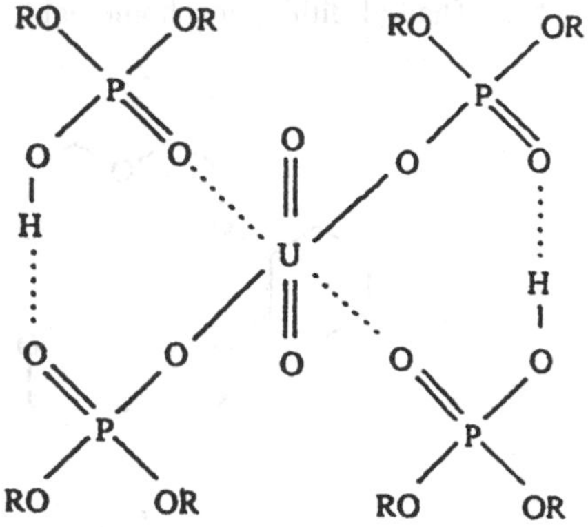

(d)

14. Di(2-ethylhexyl) phosphoric acid, HDEHP, D-2EHPA; R is C_8H_{17}:
    (a) monomeric form;
    (b) dimeric form;
    (c) $M(HA_2)_3$ complex formed with M(III); and
    (d) $UO_2(IIA_2)_2$ complex.

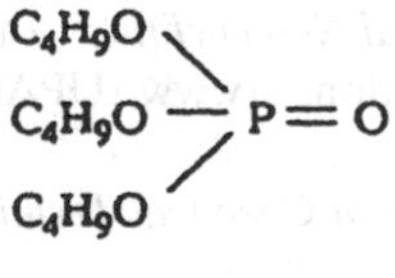

15. Tributylphosphate, TBP 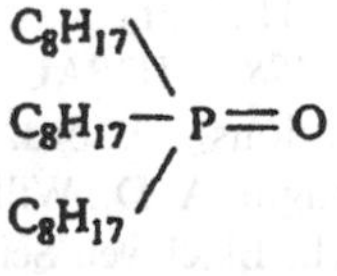   16. Trioctyl phosphine oxide, TOPO

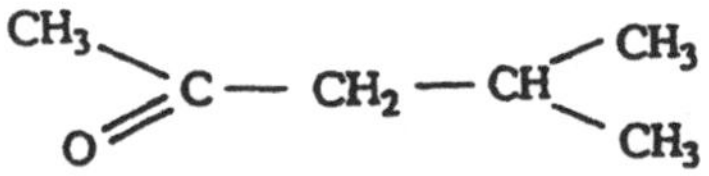

17. Methyl isobutyl ketone, MIBK, hexone 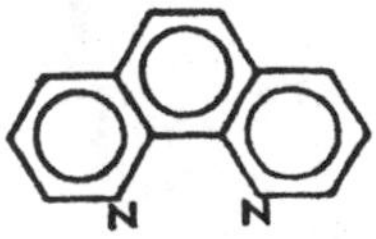   18. o-Phenanthroline

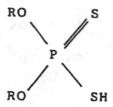

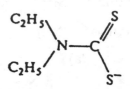

19. Dialkyl dithiophosphonic acid   20. Diethyl dithiocarbamate

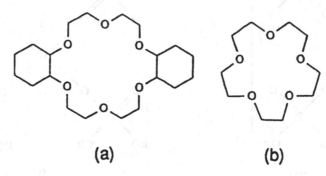

(a)   **(b)**

21. Crown ethers:
     (a) dicyclohexyl-18-crown-6, DC18-C6; and
     (b) 15-crown-5, 15-C5

REFERENCES

1. Högfeldt, E. *Stability Constants of Metal-Ion Complexes. Part A: Inorganic ligands*; IUPAC Chemical Data Series No 22; Pergamon Press: New York, 1982.
2. Freiser, H.; Nancollas, G. H. *Compendium of Analytical Nomenclature. Definitive Rules 1987*; IUPAC; Blackwell Scientific Publications (www.IUPAC.com/ publications): Oxford, 1987.
3. McNaught, A. D.; Wilkinson, A. *IUPAC Compendium of Chemical Terminology*, 2nd Ed.; Blackwell Science, 1997.
4. IUPAC. *Quantities, Units and Symbols in Physical Chemistry*, 2nd Ed.; Blackwell Science, 1993.

Index of Compounds

The key for the abbreviations contained within the index can be found on page 739.

ORGANIC COMPOUNDS

ABBREVIATIONS

Abbreviation	Meaning
AAS	atomic absorption spectrophotometry
act	activity
add	adduct
ads	adsorption/absorption
agg	aggregation
amm	ammonium
anal	analysis, analytical
appl	applications
aq	aqueous
calc	calculation(s)
centr	centrifugal, centrifuge, centrifugation
chem	chemical(stry)
chromat	chromatography
class	classification
coeff	coefficients
comp	compound(s)
compl	complexe(s)
compln	complexation
composn	composition
conc	concentration
const	constant(s)
contt	contactor
coord	coordination
corrl	correlations
def	definition
degrad	degradation
detn	determination
diag	diagrams
diff	diffusion(-al)
displ	displacement
diss	dissociation
dissol	dissolution
distrn	distribution
DN	donor number(s)
econ	economics
eff	effect
effic	efficiency
electr	electron, electrode, electrolyte
en	ethylenediamin
eql	equilibrium
eqn	equation(s)
equip	equipment
exch	exchange
exp	experiments
extrn	extraction(s)
fact	factor
form	formation

fr	from	purif	purification
funct	function	reactn	reaction
gravit	gravitational	rel	relations
hydrn	hydration	reproc	reprocessing
interact	interaction	SCF	supercritical fluid
irrad	irradiated	sel	selective, -ity
ISE	ion selective electrodes	selfadd	selfadduct
kin	kinetic	sepn	separation(s)
LLPC	liquid partition	SLM	supported liquid
	chromatography		membranes
mech	mechanics, mechanisms	soln	solution
met	metal, metallic,	solub	solubility
	metallurgical	solv	solvent, solvation
meth	method	spectr	spectroscopy, -metry
misc	miscibility	stab	stabilit(y)ies
mixn	mixing	struct	structure(s),
mixt	mixture(s)		stereochemistry
nucl	nuclear	syst	system/systematics
org	organic(s)	techn	technique,
oxidn	oxidation		technological
param	parameters	temp	temperature
pot	potential	thermod	thermodynamic
prep	preparation	tox	toxicity
proc	process(es)	transf	transfer
proj	project	vap	vaporization
prop	properties	w	with

Subject Index

The Key for the abbreviations contained within the index can be found on page 739.

Printed in the United States
by Baker & Taylor Publisher Services

Printed in the United States
by Baker & Taylor Publisher Services